LYSOSOMES

Biology, Diseases, and Therapeutics

Edited by

FREDERICK R. MAXFIELD
JAMES M. WILLARD
SHUYAN LU

WILEY

Published by John Wiley & Sons, Inc., Hoboken, New Jersey
Published simultaneously in Canada

Library of Congress Cataloging-in-Publication Data:

Names: Maxfield, Frederick R., editor. | Willard, James M., 1955- , editor. | Lu, Shuyan, 1971- , editor.
Title: Lysosomes : biology, diseases, and therapeutics / edited by Frederick R. Maxfield, James M. Willard, Shuyan Lu.
Other titles: Lysosomes (Maxfield)
Description: Hoboken, New Jersey : John Wiley & Sons, Inc., [2016] | Includes bibliographical references and index.
Identifiers: LCCN 2016006834 (print) | LCCN 2016008790 (ebook) | ISBN 9781118645154 | ISBN 9781118978306 (pdf) | ISBN 9781118978313 (epub)
Subjects: | MESH: Lysosomes
Classification: LCC QH603.L9 (print) | LCC QH603.L9 (ebook) | NLM QU 350 | DDC 571.6/55–dc23
LC record available at https://lccn.loc.gov/2016006834

PREFACE

There has been a resurgence in interest in lysosomes based on exciting new discoveries over the past decade. Lysosomal function was observed microscopically in the late 19th century, and lysosomes were purified in the 1950s by the group of Christian De Duve [1]. During the same period, accumulation of undigested material in cells was observed in pathological examination of tissues from patients with a variety of diseases [2–4]. With the biochemical and morphological characterization of lysosomes, the linkage of the accumulated material with these organelles led to significant insights into the functional importance of lysosomes.

In the second half of the 20th century, there were groundbreaking studies of the biology and biochemistry of lysosomes [5–9]. These studies were linked closely with rapid developments in understanding fundamental cellular biological processes such as secretion and endocytosis. As a result, an increasingly detailed picture emerged of the biogenesis of lysosomes and their functional role in digesting internalized cargo [10,11]. As understanding of lysosomal function increased, mechanism-based strategies for treating lysosomal diseases emerged. These included substrate reduction therapies (e.g., for Gaucher disease) [12,13] and enzyme replacement therapies [14].

While there continued to be advances in basic cell biology and biochemistry, as well as in new therapeutic modalities, many investigators had a sense that the exciting era of discovery in lysosome biology was ending in the early 2000s. As an example, the Gordon Conference on "Lysosomes," which for many years was one of the premier meetings on membrane traffic, changed its name to "Lysosomes and Endocytosis" in 2004.

Several related areas of investigation have blossomed over the past decade, and these have brought lysosomes back into the forefront of basic cell biology and biochemistry. One of these areas is autophagy. This process for lysosomal digestion of cytoplasmic organelles had been known for decades, but there were few handles on how to study it. With genetic studies leading to identification of key molecular components in the formation of autophagosomes and their subsequent fusion with lysosomes, it became possible to analyze this process in detail. As a result, autophagy is now recognized as playing a key role in processes including maintenance of organelle integrity, catabolism of lipid droplets, and responses to stress [15,16]. Additionally, autophagy is essential for the survival and proliferation of some cancer cells, making it a novel target for development of therapies [17,18]. Furthermore, genetic and molecular biological data accentuate the broad importance of the lysosome in aging and age-related diseases, including cardiovascular and neurodegenerative diseases, which make improving lysosome function an attractive target.

One of the most exciting recent developments has been the recognition that lysosomes are key regulators of signaling processes that regulate metabolism. The elucidation of the mTOR signaling pathways has shown that hydrolytic activity in lysosomes is used by the cell to sense nutrient status [19]. Among other activities, mTOR regulates autophagy to enhance the availability of new molecular building blocks when lysosomal production of catabolites is reduced. In another related area, it was recognized a few years ago that there is a coordinated transcriptional regulation of the genes involved in lysosome biogenesis [20,21].

Along with these basic science developments, there have been important advances in the understanding of lysosomal storage disorders and in new methods for treatment. In some cases, this is beginning to turn these devastating diseases into conditions that can be managed. At the same time, there is increasing recognition that drugs used for various purposes can interact with lysosomal processes. A dramatic example of this is the discovery of mTOR as a mechanistic target for the immunosuppressive drug rapamycin [22]. Many pharmacological drugs in widespread use can affect lysosomal function [23–26], and it is important to understand the impact of these effects.

With all of these interrelated advances in understanding of lysosome biology, it seemed worthwhile to assemble an updated and integrated book on lysosomes. There are several notable earlier books on lysosomes, and a few of them will be cited here with apologies to the authors whose contributions may have been overlooked. Eric Holtzman [27] wrote a classic monograph that is still worth reading for its historical background and insights into the role of lysosomes in biology. This was followed a few years later by a book by Brian Storrie and Robert Murphy [28]. A book by Paul Saftig [29] focused on the basic biology and function of lysosomes. There have been several excellent books on lysosomal storage disorders, including one by Fran Platt and Steven Walkley [30]. More recently, there was a book emphasizing methods for the study of lysosomes [31].

The current book is intended for a broad audience of researchers interested in multiple facets of lysosome biology. Chapters 1–7 and 12 cover fundamental roles of lysosomes in physiological processes; Chapters 8–11 discusses involvement of lysosomes in various pathological conditions; Chapters 13–20 focus on the contribution of lysosomes in various aspects of drug development, including the lysosomal pathway as a target for drug discovery, toxicity, and special pharmacokinetics attributed to lysosomal accumulation and sequestration

We thank all contributors who provided their chapters despite other pressing responsibilities. We also thank our editors for their diligent effort and David B. Iaea for the cover illustration.

We hope that the broad scope, which includes both basic science and clinical applications, can promote a productive interchange among scientists working across the spectrum of lysosomal studies and nurture drug development efforts targeting lysosome pathways. Ultimately, discovery of new drugs that could improve lysosomal function will benefit multiple therapeutics areas.

REFERENCES

[1] Novikoff AB, Beaufay H, De Duve C. Electron microscopy of lysosome rich fractions from rat liver. J Biophys Biochem Cytol 1956;2:179–184.

[2] Hers HG. alpha-Glucosidase deficiency in generalized glycogen storage disease (Pompe's disease). Biochem J 1963;86:11–16.

[3] Baudhuin P, Hers HG, Loeb H. An electron microscopic and biochemical study of type II glycogenosis. Lab Invest 1964;13:1139–1152.

[4] Kjellman B, Gamstorp I, Brun A, Ockerman PA, Palmgren B. Mannosidosis: a clinical and histopathologic study. J Pediatr 1969;75:366–373.

[5] Novikoff AB, Novikoff PM. Cytochemical contributions to differentiating GERL from the Golgi apparatus. Histochem J 1977;9:525–551.

[6] Novikoff AB. The endoplasmic reticulum: a cytochemist's view (a review). Proc Natl Acad Sci U S A 1976;73:2781–2787.

[7] Griffiths G, Hoflack B, Simons K, Mellman I, Kornfeld S. The mannose 6-phosphate receptor and the biogenesis of lysosomes. Cell 1988;52:329–341.

[8] Hasilik A, Neufeld EF. Biosynthesis of lysosomal enzymes in fibroblasts. Synthesis as precursors of higher molecular weight. J Biol Chem 1980;255:4937–4945.

[9] Kaplan A, Fischer D, Achord D, Sly W. Phosphohexosyl recognition is a general characteristic of pinocytosis of lysosomal glycosidases by human fibroblasts. J Clin Invest 1977;60:1088–1093.

[10] Kornfeld S, Mellman I. The biogenesis of lysosomes. Annu Rev Cell Biol 1989;5: 483–525.

[11] Mukherjee S, Ghosh RN, Maxfield FR. Endocytosis. Physiol Rev 1997;77:759–803.

[12] Zimran A, Elstein D. Gaucher disease and the clinical experience with substrate reduction therapy. Philos Trans R Soc Lond B Biol Sci 2003;358:961–966.

[13] Elstein D, Hollak C, Aerts JM, van Weely S, Maas M, et al. Sustained therapeutic effects of oral miglustat (Zavesca, N-butyldeoxynojirimycin, OGT 918) in type I Gaucher disease. J Inherit Metab Dis 2004;27:757–766.

[14] Desnick RJ, Schuchman EH. Enzyme replacement and enhancement therapies: lessons from lysosomal disorders. Nat Rev Genet 2002;3:954–966.

[15] Kaur J, Debnath J. Autophagy at the crossroads of catabolism and anabolism. Nat Rev Mol Cell Biol 2015;16:461–472.

[16] Mizushima N, Levine B, Cuervo AM, Klionsky DJ. Autophagy fights disease through cellular self-digestion. Nature 2008;451:1069–1075.

[17] Kenific CM, Debnath J. Cellular and metabolic functions for autophagy in cancer cells. Trends Cell Biol 2015;25:37–45.

[18] Amaravadi RK, Lippincott-Schwartz J, Yin XM, Weiss WA, Takebe N, et al. Principles and current strategies for targeting autophagy for cancer treatment. Clin Cancer Res 2011;17:654–666.

[19] Sancak Y, Bar-Peled L, Zoncu R, Markhard AL, Nada S, et al. Ragulator-Rag complex targets mTORC1 to the lysosomal surface and is necessary for its activation by amino acids. Cell 2010;141:290–303.

[20] Palmieri M, Impey S, Kang H, di Ronza A, Pelz C, et al. Characterization of the CLEAR network reveals an integrated control of cellular clearance pathways. Hum Mol Genet 2011;20:3852–3866.

[21] Sardiello M, Palmieri M, di Ronza A, Medina DL, Valenza M, et al. A gene network regulating lysosomal biogenesis and function. Science 2009;325:473–477.

[22] Abraham RT. Regulation of the mTOR signaling pathway: from laboratory bench to bedside and back again. F1000 Biol Rep 2009;1:8.

[23] Lu S, Jessen B, Strock C, Will Y. The contribution of physicochemical properties to multiple in vitro cytotoxicity endpoints. Toxicol In Vitro 2012;26:613–620.

[24] Ashoor R, Yafawi R, Jessen B, Lu S. The contribution of lysosomotropism to autophagy perturbation. PLoS One 2013;8:e82481.

[25] Khoh-Reiter S, Sokolowski SA, Jessen B, Evans M, Dalvie D, et al. Contribution of membrane trafficking perturbation to retinal toxicity. Toxicol Sci 2015;145:383–395.

[26] Slavov SH, Wilkes JG, Buzatu DA, Kruhlak NL, Willard JM, et al. Computational identification of a phospholipidosis toxicophore using (13)C and (15)N NMR-distance based fingerprints. Bioorg Med Chem 2014;22:6706–6714.

[27] Holtzman E. *Lysosomes (Cellular Organelles)*. Springer; 1989.

[28] Storrie B, Murphy RF. *Endosomes and Lysosomes: A Dynamic Relationship, Volume 1 (Advances in Cellular and Molecular Biology of Membranes and Organelles)*. Elsevier Science; 1993.

[29] Saftig P. *Lysosomes*. Springer; 2005.

[30] Platt FM, Walkley SU. *Lysosomal Disorders of the Brain: Recent Advances in Molecular and Cellular Pathogenesis and Treatment*. Oxford University Press; 2004.

[31] Platt FM, Platt N. *Lysosomes and lysosomal Diseases*. Academic Press; 2015.

LIST OF CONTRIBUTORS

Ravi K. Amaravadi, Department of Medicine, University of Pennsylvania, Philadelphia, PA, USA

Andrea Ballabio, Telethon Institute of Genetics and Medicine (TIGEM), Naples, Italy; Medical Genetics, Department of Translational Medicine, Federico II University, Naples, Italy; Department of Molecular and Human Genetics, Baylor College of Medicine, Houston, TX, USA; Jan and Dan Duncan Neurological Research Institute, Texas Children Hospital, Houston, TX, USA

Elfrida R. Benjamin, Amicus Therapeutics, Cranbury, NJ, USA

Robert E. Boyd, Amicus Therapeutics, Cranbury, NJ, USA

Thomas Burgoyne, Department of Cell Biology, UCL Institute of Ophthalmology, University College London, London, UK

Graciana Diez-Roux, Telethon Institute of Genetics and Medicine (TIGEM), Naples, Italy

Maureen Dougher, Oncology Research Unit, World Wide Research and Development, Pfizer Inc., Pearl River, NY, USA

Emily R. Eden, Department of Cell Biology, UCL Institute of Ophthalmology, University College London, London, UK

Albert De Felice, Division of Cardiovascular and Renal Products, Office of Drug Evaluation I, Center for Drug Evaluation and Research, United States Food and Drug Administration, Silver Spring, MD, USA

Clare E. Futter, Department of Cell Biology, UCL Institute of Ophthalmology, University College London, London, UK

Dominik Haddad, VIB Center for the Biology of Disease, KU Leuven, Leuven, Belgium; Laboratory of Neuronal Communication, Leuven Institute for Neurodegenerative Disease (LIND), Center for Human Genetics, KU Leuven, Leuven, Belgium; Gladstone Institutes of Neurological Disease, University of California, San Francisco, CA, USA

Abigail S. Haka, Department of Biochemistry, Weill Cornell Medical College, New York, NY, USA

Nadia Hamid, Department of Pharmaceutical Chemistry, The University of Kansas, Lawrence, KS, USA

Marja Jäättelä, Cell Death and Metabolism, Danish Cancer Society Research Center, Danish Cancer Society, Copenhagen, Denmark

Bart Jessen, Drug Safety Research and Development, Pfizer Inc., San Diego, CA, USA

Jennifer Kahler, Oncology Research Unit, World Wide Research and Development, Pfizer Inc., Pearl River, NY, USA

Tuula Kallunki, Cell Death and Metabolism, Danish Cancer Society Research Center, Danish Cancer Society, Copenhagen, Denmark

Richie Khanna, Amicus Therapeutics, Cranbury, NJ, USA

Judith Klumperman, Department of Cell Biology, University Medical Center Utrecht, Utrecht, The Netherlands

Jeffrey P. Krise, Department of Pharmaceutical Chemistry, The University of Kansas, Lawrence, KS, USA

Philip L. Leopold, Department of Chemistry, Chemical Biology, and Biomedical Engineering, Stevens Institute of Technology, Hoboken, NJ, USA

Shuyan Lu, Drug Safety Research and Development, Pfizer Inc., San Diego, CA, USA

Xiaohong Ma, Department of Medicine, University of Pennsylvania, Philadelphia, PA, USA

Michelle Mack, Oncology Research Unit, World Wide Research and Development, Pfizer Inc., Pearl River, NY, USA

Adriana R. Mantegazza, Department of Pathology and Laboratory Medicine, Children's Hospital of Philadelphia Research Institute, Philadelphia, PA, USA; Departments of Pathology and Laboratory Medicine and Physiology, Perelman School of Medicine; University of Pennsylvania, Philadelphia, PA, USA

Michael S. Marks, Department of Pathology and Laboratory Medicine, Children's Hospital of Philadelphia Research Institute, Philadelphia, PA, USA; Departments of Pathology and Laboratory Medicine and Physiology, Perelman School of Medicine, University of Pennsylvania, Philadelphia, PA, USA

Frederick R. Maxfield, Department of Biochemistry, Weill Cornell Medical College, New York, NY, USA

Quentin Mcafee, Department of Medicine, University of Pennsylvania, Philadelphia, PA, USA

Ralph A. Nixon, Center for Dementia Research, Nathan S. Kline Institute, Orangeburg, NY, USA; Department of Psychiatry, New York University Langone Medical Center, New York, NY, USA; Department of Cell Biology, New York University Langone Medical Center, New York, NY, USA

Shengfu Piao, Department of Medicine, University of Pennsylvania, Philadelphia, PA, USA

Fulvio Reggiori, Department of Cell Biology, University Medical Center Utrecht, Utrecht, The Netherlands; Department of Cell Biology, The University Medical Center Groningen, Groningen, The Netherlands

Urška Repnik, Department of Biochemistry and Molecular and Structural Biology, Jožef Stefan Institute, Ljubljana, Slovenia; Department of Biosciences, University of Oslo, Oslo, Norway

Michael Ritchie, Oncology Research Unit, World Wide Research and Development, Pfizer Inc., Pearl River, NY, USA

Puja Sapra, Oncology Research Unit, World Wide Research and Development, Pfizer Inc., Pearl River, NY, USA

Edward H. Schuchman, Department of Genetics and Genomic Sciences, Icahn School of Medicine at Mount Sinai, New York, NY, USA

Boris Shor, Oncology Research Unit, World Wide Research and Development, Pfizer Inc., Pearl River, NY, USA

Calogera M. Simonaro, Department of Genetics and Genomic Sciences, Icahn School of Medicine at Mount Sinai, New York, NY, USA

Rajesh K. Singh, Department of Biochemistry, Weill Cornell Medical College, New York, NY, USA

Greg Stevens, Drug Safety Research and Development, Pfizer Inc., San Diego, CA, USA

Matthew Sung, Oncology Research Unit, World Wide Research and Development, Pfizer Inc., Pearl River, NY, USA

Boris Turk, Department of Biochemistry and Molecular and Structural Biology, Jožef Stefan Institute, Ljubljana, Slovenia; Center of Excellence CIPKEBIP, Ljubljana, Slovenia; Faculty of Chemistry and Chemical Technology, University of Ljubljana, Ljubljana, Slovenia

Kenneth J. Valenzano, Amicus Therapeutics, Cranbury, NJ, USA

Patrik Verstreken, VIB Center for the Biology of Disease, KU Leuven, Leuven, Belgium; Laboratory of Neuronal Communication, Leuven Institute for Neurodegenerative Disease (LIND), Center for Human Genetics, KU Leuven, Leuven, Belgium; Gladstone Institutes of Neurological Disease, University of California, San Francisco, CA, USA

Yvonne Will, Drug Safety Research and Development, Pfizer Inc., Groton, CT, USA

James M. Willard, Division of Cardiovascular and Renal Products, Office of Drug Evaluation I, Center for Drug Evaluation and Research, United States Food and Drug Administration, Silver Spring, MD, USA

Su Xu, Amicus Therapeutics, Cranbury, NJ, USA

1

LYSOSOMES: AN INTRODUCTION

FREDERICK R. MAXFIELD
Department of Biochemistry, Weill Cornell Medical College, New York, NY, USA

Late endosomes and lysosomes (LE/Ly) are the main digestive organelles of eukaryotic cells. They contain a variety of enzymes and accessory proteins [1] that are capable of the hydrolysis of many biological molecules. The LE/Ly are maintained at an acidic pH, and most lysosomal enzymes are acid hydrolases with acidic pH optima. Substrates are delivered to LE/Ly either by endocytic processes (e.g., receptor-mediated endocytosis, pinocytosis, and phagocytosis) or by autophagic processes in which autophagosomes containing cellular content fuse with the LE/Ly.

The history of lysosomal studies has been discussed in several publications [2–5]. The study of lysosomal digestion began in the late 19th century with microscopic observations of cells ingesting material. In the same period, physicians began to see abnormal storage of material in pathology specimens from patients with unusual illnesses. In the 1950s and 1960s, lysosomes were purified by differential centrifugation techniques, and the linkage was established between these organelles and the microscopic observations of storage material in what then became known as lysosomal storage diseases. These discoveries were followed by contributions from many laboratories that have led to a fairly detailed understanding of the biogenesis, composition, and function of lysosomal organelles.

There have been several recent advances in our understanding of lysosomal function. In particular, we now have a much better understanding of the mechanisms underlying the regulation of lysosome formation (see Chapters 2 and 7). In addition,

Lysosomes: Biology, Diseases, and Therapeutics, First Edition.
Edited by Frederick R. Maxfield, James M. Willard and Shuyan Lu.

while autophagy has been known for many years, there have been dramatic increases in our understanding of its molecular mechanisms in the past decade (see Chapter 2).

A brief note about nomenclature may be useful. As with other organelles, the acidic digestive organelles are heterogeneous and undergo rapid changes due to various membrane trafficking processes. Several years ago, a definition was proposed [6] in which the organelles that were actively receiving new lysosomal enzymes from the trans-Golgi network would be called late endosomes, while more mature acidic digestive organelles would retain the name lysosomes. This distinction remains useful in describing two broad groups of acidic digestive organelles, but it does lead to some semantic difficulties. For example, in many cells, digestion of endocytosed proteins by lysosomal hydrolases occurs mainly in the late endosomes. Thus, "lysosomal digestion" is nearly complete before the endocytosed material can enter "lysosomes." In addition, even within the same cell there can be significant differences in the properties (size, morphology, enzyme content, substrate content, etc.) in organelles that are broadly grouped as late endosomes or lysosomes. These difficulties arise because of the intrinsically dynamic nature of these organelles, which are rapidly exchanging membrane and lumenal content with each other and with other organelles. Unfortunately, a completely accurate and descriptive terminology may not be possible.

1.1 HISTORICAL BACKGROUND

The study of lysosomes began with the microscopic observations summarized by Metchnikoff [7] in the late 19th century. He and his contemporaries reported their direct microscopic analysis on the uptake of foreign particles into cells, and the subsequent digestion of these particles. In some cases, Metchnikoff observed that ingested pieces of litmus would change from blue to red, implying an acid environment surrounding the ingested particles. Metchnikoff emphasized that aspects of phagocytic uptake were evolutionarily conserved from single cell organisms through leukocytes obtained from mammals.

One of the ongoing themes in studies of lysosomes is that there is a dynamic interplay between the study of diseases and the contributions of basic science. In parallel with the developments in the cell biology of digestion of internalized material, there were descriptions beginning in the 19th century of various storage diseases in which cellular accumulations of material were observed upon pathological examination of tissues from patients with a variety of diseases [3]. These included Gaucher, Tay–Sachs [8], and Niemann–Pick diseases among others. Many of these storage disorders were recognized to have a hereditary component, and several of them were also recognized to lead to accumulation of certain molecular species, especially lipids.

The modern era of lysosome study began with the purification of lysosomes by Christian de Duve and his colleagues in the 1950s, and a review of these discoveries presents many of the important details [5]. De Duve's laboratory sought to understand the localization of glucose-6-phosphatase in liver fractions as part of study of the mechanisms of insulin action. Among the phosphatase activities that they studied was acid phosphatase, and they noticed that its activity was latent and could only be observed when membranes were disrupted. At first they considered that this might suggest a mitochondrial localization, but improved centrifugation techniques led to separation of the acid phosphatase fraction from the mitochondria. In 1955, they published a paper describing isolation of particles that contained five enzyme activities with acidic pH optima [9]. Alex Novikoff had also been working on the distribution of various enzymes in cells. After visiting de Duve's laboratory, Novikoff carried out the first electron microscopic studies of the newly isolated organelles. To do this, he would carry fresh samples from the de Duve laboratory in Belgium to the new electron microscope available in the laboratory of Albert Claude in Paris [10]. He observed that the fractions were enriched in dense bodies about 370 nm in length, and he also noted that they resembled in many ways similar structures observed in hepatocytes [11]. Working independently, Straus [12] isolated "droplets" from rat kidneys and showed that these were enriched in acid phosphatase and other enzymes [13]. Straus also showed that similar, but larger, droplets were observed after intraperitoneal injection of egg white and reported that the injected protein could be found in these droplets. This linked the degradative organelles with the uptake of extracellular material.

After the discovery of lysosomes by de Duve and his coworkers, the idea that lysosomal defects might be the underlying cause of the storage diseases began to be considered. The clear demonstration of association of these storage disorders with lysosomes was first made by Hers [14,15] when the defect in a glycogen storage disease was shown to be a deficiency in the lysosomal enzyme acid maltase. This was followed by the association of several of the storage disorders, now classified as lysosomal storage disorders, with a deficiency in specific enzymes.

Work from Neufeld and others [16] showed that enzymes secreted by one cell could be added to the culture medium of a cell lacking a particular enzyme and correct the storage defect. It was found, however, that enzymes secreted from I-cell disease fibroblasts could not correct the enzyme deficiencies in other cells. This and other observations led to the hypothesis that there must be a tag of some type on lysosomal enzymes and there must be receptors on the surface of cells that could selectively endocytose the tagged enzymes [4]. Shortly thereafter, it was found that the uptake of β-glucuronidase could be inhibited by mannose-6-phosphate and that the uptake of the enzyme could also be blocked by pretreatment with a phosphatase [17]. Mannose-6-phosphate glycoconjugates were then identified on many lysosomal enzymes.

The cation-independent mannose-6-phosphate receptor was first isolated from bovine liver in 1981 [18], and the sequence of the human receptor was determined in 1988 [19]. It was found to be identical to an independently identified receptor for insulin-like growth factor II (IGF-II) [20]. A second mannose-6-phosphate receptor (the cation-dependent mannose-6-phosphate receptor) was identified in 1985 [21].

In 1978, Ohkuma and Poole [22], who were colleagues of de Duve, used the uptake of fluorescein–dextran into lysosomes to measure the pH of these organelles accurately based on the pH dependence of fluorescein fluorescence. They also showed that lysosomal acidification required ATP and that weak bases could increase the pH of these organelles. Most of the lysosomal hydrolases have acidic pH optima, and it seems likely that this pH dependence is protective for events that lead to disruption of lysosomes since the enzymes will have greatly reduced activity at cytoplasmic or extracellular pH.

From the 1980s to the present, there has been an explosive growth in the number of studies of various aspects of lysosome biology, and many aspects of lysosomal biogenesis and function are becoming well understood. It is particularly gratifying that understanding of the basic biochemistry and cell biology is leading to therapies for several of the lysosomal storage disorders. In addition, understanding of endocytic targeting to digestive organelles is leading to development of very selectively targeted delivery of therapeutic agents. Many of these studies are summarized in the chapters of this book.

REFERENCES

[1] Lubke T, Lobel P, Sleat DE. Proteomics of the lysosome. Biochim Biophys Acta 2009; 1793:625–635.

[2] Holtzman E. *Lysosomes*. New York: Plenum; 1989. 439 p.

[3] Platt FM, Walkley SU. *Lysosomal Disorders of Brain: Recent Advances in Molecular and Cellular Pathogenesis and Treatment*. Oxford; New York: Oxford University Press; 2004. xxvii, 447 p.p.

[4] Neufeld EF, Lim TW, Shapiro LJ. Inherited disorders of lysosomal metabolism. Annu Rev Biochem 1975;44:357–376.

[5] Bainton DF. The discovery of lysosomes. J Cell Biol 1981;91:66s–76s.

[6] Kornfeld S, Mellman I. The biogenesis of lysosomes. Annu Rev Cell Biol 1989; 5:483–525.

[7] Metchnikoff E. *Lectures on the Comparative Pathology of Inflammation*. London: Paul, Kegan, Trench & Trabner; 1893.

[8] Sachs B, Strauss I. The cell changes in amaurotic family idiocy. J Exp Med 1910; 12:685–695.

[9] De Duve C, Pressman BC, Gianetto R, Wattiaux R, Appelmans F. Tissue fractionation studies. 6. Intracellular distribution patterns of enzymes in rat-liver tissue. Biochem J 1955;60:604–617.

[10] Essner ES. A tribute to Alex B. Novikoff. J Histochem Cytochem 1987;35:937–938.

[11] Novikoff AB, Beaufay H, De Duve C. Electron microscopy of lysosome-rich fractions from rat liver. J Biophys Biochem Cytol 1956;2:179–184.

[12] Straus W. Isolation and biochemical properties of droplets from the cells of rat kidney. J Biol Chem 1954;207:745–755.

[13] Straus W. Concentration of acid phosphatase, ribonuclease, desoxyribonuclease, beta-glucuronidase, and cathepsin in droplets isolated from the kidney cells of normal rats. J Biophys Biochem Cytol 1956;2:513–521.

[14] Hers HG. alpha-Glucosidase deficiency in generalized glycogen storage disease (Pompe's disease). Biochem J 1963;86:11–16.

[15] Lejeune N, Thines-Sempoux D, Hers HG. Tissue fractionation studies. 16. Intracellular distribution and properties of alpha-glucosidases in rat liver. Biochem J 1963;86:16–21.

[16] Fratantoni JC, Hall CW, Neufeld EF. Hurler and Hunter syndromes: mutual correction of the defect in cultured fibroblasts. Science 1968;162:570–572.

[17] Kaplan A, Achord DT, Sly WS. Phosphohexosyl components of a lysosomal enzyme are recognized by pinocytosis receptors on human fibroblasts. Proc Natl Acad Sci U S A 1977;74:2026–2030.

[18] Sahagian GG, Distler J, Jourdian GW. Characterization of a membrane-associated receptor from bovine liver that binds phosphomannosyl residues of bovine testicular beta-galactosidase. Proc Natl Acad Sci U S A 1981;78:4289–4293.

[19] Oshima A, Nolan CM, Kyle JW, Grubb JH, Sly WS. The human cation-independent mannose 6-phosphate receptor. Cloning and sequence of the full-length cDNA and expression of functional receptor in COS cells. J Biol Chem 1988;263:2553–2562.

[20] Morgan DO, Edman JC, Standring DN, Fried VA, Smith MC, et al. Insulin-like growth factor II receptor as a multifunctional binding protein. Nat Geosci 1987;329:301–307.

[21] Hoflack B, Kornfeld S. Purification and characterization of a cation-dependent mannose 6-phosphate receptor from murine P388D1 macrophages and bovine liver. J Biol Chem 1985;260:12008–12014.

[22] Ohkuma S, Poole B. Fluorescence probe measurement of the intralysosomal pH in living cells and the perturbation of pH by various agents. Proc Natl Acad Sci U S A 1978;75:3327–3331.

2

LYSOSOME BIOGENESIS AND AUTOPHAGY

FULVIO REGGIORI[1,2]

Department of Cell Biology, University Medical Center Utrecht, Utrecht, The Netherlands ; Department of Cell Biology, The University Medical Center Groningen, Groningen, The Netherlands

JUDITH KLUMPERMAN[1]

Department of Cell Biology, The University Medical Center Groningen, Groningen, The Netherlands

2.1 INTRODUCTION

Lysosomes degrade biological components acquired by endocytosis, the major cellular pathway for internalization of extracellular material, and macroautophagy, the main system in the cell for turnover of organelles and large cytoplasmic protein aggregates. Endocytosis starts with the formation of an endocytic vesicle that buds from the plasma membrane and fuses with an early endosome (EE). EEs subsequently mature into late endosomes (LEs), which eventually fuse with lysosomes (Figure 2.1). Proteins not destined for degradation return to the plasma membrane directly from EEs or via specialized recycling endosomes (REs). Macroautophagy begins with the sequestration of parts of the cytoplasm by a cistern known as the phagophore or isolation membranes, which expands into a double membrane vesicle called autophagosome. The autophagosome fuses with LEs and/or lysosomes to form an amphisome or an autolysosome, respectively (Figure 2.1). Autolysosomes are generally larger and more irregularly shaped than lysosomes, with a highly variable content in which sometimes remnants of cytoplasmic components, for example ribosomes or mitochondria, are seen (Figure 2.2; [1]). While the onset of autophagy is independent of

Lysosomes: Biology, Diseases, and Therapeutics, First Edition.
Edited by Frederick R. Maxfield, James M. Willard and Shuyan Lu.

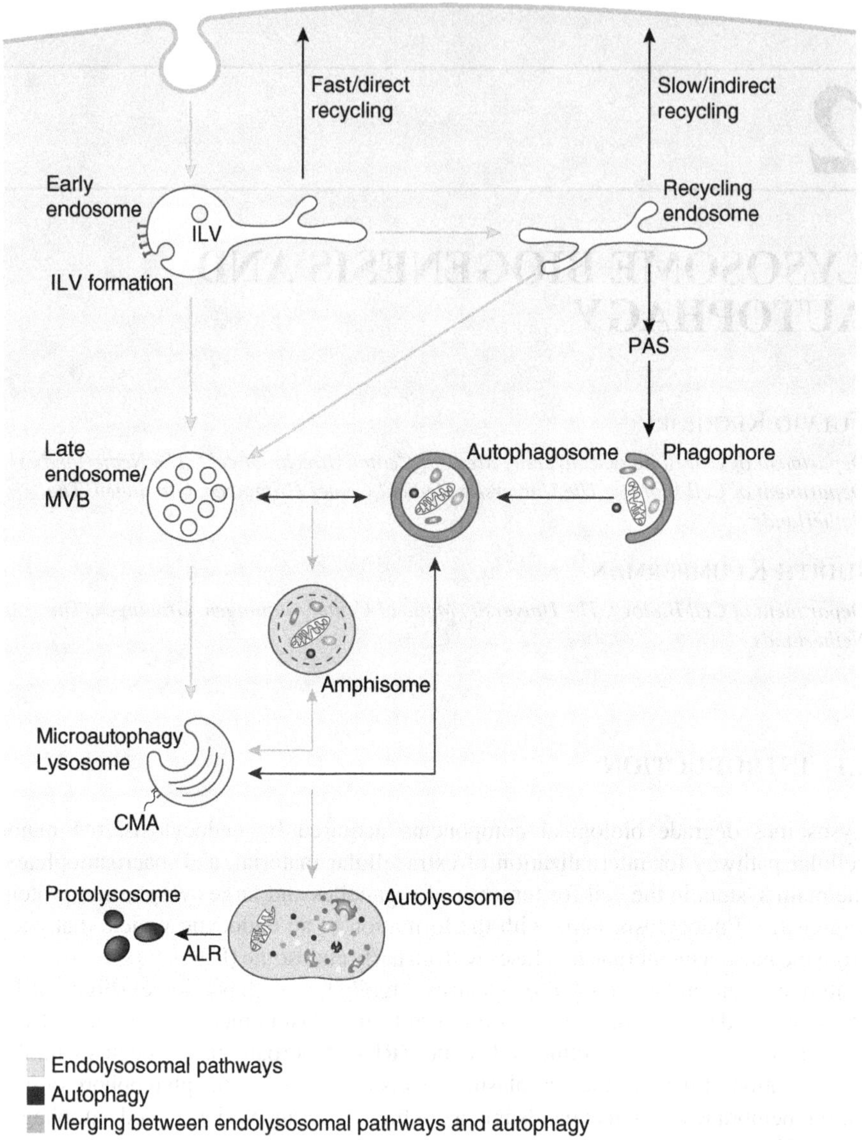

Figure 2.1 Schematic drawing depicting the endocytic and autophagy pathways to the lysosomes. ALR, autolysosome reformation; CMA, chaperone-mediated autophagy; ILV, intraluminal vesicle; MVB, multivesicular body; PAS, phagophore assembly site. (*See color plate section for the color representation of this figure.*)

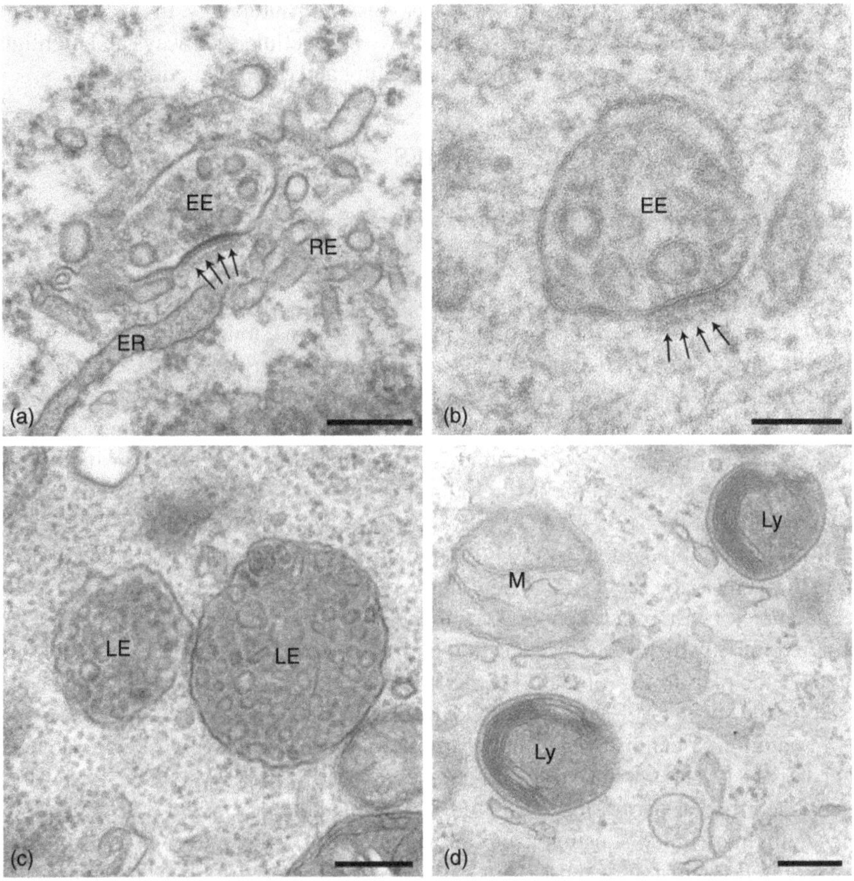

Figure 2.2 Gallery of electron micrographs providing characteristic examples of endolysosomal compartments. From (a) to (d) a sequence of early-to-late endolysosomal intermediates is shown. Arrows in (a) and (b) point to the bilayered flat clathrin coat harboring protein machinery involved in intraluminal vesicle (ILV) formation. Early endosomes (EE) and late endosomes (LE) contain an increasing number of ILVs. Lysosomes (LY) show typical membrane lamella. Pictures were taken from distinct cell models: (a) embryonic zebrafish, (b) human prostate cancer (PC3) cell, (c) activated mouse B cell, (d) HeLa cell. ER, endoplasmic reticulum; M, mitochondrion; RE, recycling endosome. Bar = 200 nm. (a–c). Courtesy of Ann de Mazière, Department of Cell Biology, University Medical Center, Utrecht, The Netherlands.

endocytosis, the formation of autophagosomes requires membranes derived from REs and completion of the autophagy process, which is characterized by the degradation and recycling to the cytoplasm of the encapsulated material, requires lysosomes. This intimate connection is the reason behind the recognition that lysosome-related disorders (LSDs) often affect both endocytosis and macroautophagy and abnormalities in lysosomal functioning become more prominent under starvation conditions. Over the past decades, important progress has been made to reveal the molecules and membrane compartments that underlie endocytosis and autophagy. In this chapter, we give an overview of these two major degradative intracellular pathways and highlight the emerging cross talks between them, in healthy and diseased conditions.

2.2 PATHWAYS TO THE LYSOSOMES

2.2.1 Biosynthetic Transport Routes to the Lysosome

Lysosomes contain more than 50 different hydrolytic enzymes involved in substrate degradation and more than 150 lysosomal membrane proteins (LMPs) involved in lysosome stability and regulation of the interior lysosomal milieu [2,3]. After synthesis in the endoplasmic reticulum (ER), lysosomal proteins traverse the Golgi complex to enter the trans-Golgi network (TGN), where the majority of the enzymes are directly sorted to the endolysosomal system. Most of the soluble lysosomal enzymes are modified in the Golgi complex with mannose-6-phosphate (M6P) residues, which in the TGN bind to a M6P receptor (MPR), that is, the 300 kDa cation-independent M6P receptor (CI-MPR) or the 46 kDa cation-dependent M6P receptor (CD-MPR) [4]. MPRs with or without their cargo [5] enter clathrin-coated vesicles that travel directly from the TGN to the EEs. This sorting step requires the heterotetrameric adaptor protein complex (AP)-1 and the Golgi-localized, γ-ear-containing, Arf-binding (GGA) protein family, which recognize specific motifs in the cytosolic tail of the MPRs and catalyze the recruitment of clathrin [6–8]. These clathrin-coated vesicles can also be used for the sorting of LMPs from TGN to EEs [9,10].

In addition, a clathrin-independent transport route has recently been discovered that delivers LMPs, but not MPRs, directly from the TGN to LE [11,12]. VPS41, originally identified as part of the yeast HOPS tethering complex, and the SNARE (soluble *N*-ethylmaleimide-sensitive factor attachment protein receptors) protein VAMP7 are required for fusion of these TGN-derived carriers with LEs, but the signals that sort LMPs into this pathway are still unresolved. Newly synthesized lysosomal hydrolases and LMPs that escape sorting in the TGN enter a default pathway to the plasma membrane, where they can subsequently be targeted to lysosomes via endocytosis. Low levels of the MPRs at the plasma membrane thereby mediate the endocytosis of these secreted hydrolases [13].

2.2.2 Endocytic Pathways to the Lysosome

There are several distinct subtypes of endocytic vesicles that bud from the plasma membrane and mediate entry into the cell. These differ by cargo and machinery proteins and include dynamin- and/or clathrin-dependent and independent pathways [14]. The commonly accepted model is that endocytic vesicles subsequently fuse with each other or directly with preexisting EEs, a network of dynamically interacting compartments at the cell's periphery [15]. RAB GTPases are small monomeric G proteins that can switch between an active (GTP-bound) and an inactive (GDP-bound) state, which provide not only the means to spatially and temporally control intracellular traffic and signaling but also organelle identity.

EE-localized RAB5 is considered the *master regulator* of the endocytic pathway, which exerts its action by attracting a variety of effectors proteins [16]. While most EEs are positive for Rab5 and its effector EEA1, subpopulations of EEs with a specific function exist, for example, EEs that lack EEA1 but contain RAB5 and APPL1 or APPL2, two RAB5 effectors that act as intermediates in signaling between the plasma membrane and the nucleus [17].

EEs have a mild pH of 5.9–6.8 and a complex structure consisting of a vacuolar body with an approximate diameter of 100–500 nm, from which tubular membrane domains emerge (Figure 2.2a and b). These tubules generate membrane carriers that fuse with either the plasma membrane (the so-called fast or direct recycling pathway), the REs (the so-called slow or indirect recycling pathway), the TGN [18,19], or the LEs [20,21]. Internalized cargo proteins destined for degradation in lysosomes, such as the epidermal growth factor receptor (EGFR) and growth hormone receptor (GHR), are retained in the EE vacuole by their incorporation into intraluminal vesicles (ILVs) that form through the budding of the endosomal limiting membrane (Figure 2.1). ILV formation starts in the EEs and continue in LEs. The process of cargo selection and ILV formation generally involves the ubiquitin-dependent ESCRT (endosomal sorting complexes required for transport) complexes, and also other mechanisms have been described (for recent review, see Ref. [22]). Hrs and STAM, two components of the ESCRT complex are localized to a characteristic bilayered, flat clathrin coat that forms patches on the EE vacuole [23]. Cargoes destined for lysosomal degradation, such as EGFR and GHR, are concentrated in this coat via interaction between ubiquitin residues appended to them and ubiquitin-binding domains in Hrs and STAM, whereas recycling proteins are not [24] implicating a role for clathrin in retention and concentration of proteins prior to their sorting into ILVs. Of note, the process of ILV formation is different from the process of microautophagy occurring at the lysosome.

EEs mature into LEs through a mechanism that involves multiple rounds of membrane fusion and fission during which the protein and lipid composition of the EEs change and they acquire more ILVs. This results in globular-shaped LEs with an approximate diameter of 250–1000 nm and a pH of ~5–6. They also contain numerous ILVs and for this reason they are often also termed multivesicular bodies (MVBs) (Figure 2.2c). During the EE to LE maturation, Rab5 is replaced by Rab7 and the phosphoinositides phosphatidylinositol-3-phosphate (PI3P) is converted into phosphatidylinositol-3,5-biphosphate [25,26]. The RAB5/RAB7 switch also requires a cytosolic complex formed by SAND1/Mon1 and Ccz1, which depends on PI3P concentrations [27] and acts as the activating guanosine exchange factor (GEF) for RAB7 [28,29]. For a detailed list of the changes accompanying the EE to LE switch, see the reviews [2,30].

LEs can fuse with other LEs (homotypic fusion) or with lysosomes (heterotypic fusion). Lysosomes are vacuoles with an approximate diameter of 200–1000 nm and a pH of 4.5–5 (Figure 2.2d). The shape and content of lysosomes are more heterogeneous (Figure 2.2d) than that of LEs and depend on the type and amount of cargo, degree of degradation and also on the cell type. Degradation of cargo starts in LEs but is optimal in lysosomes. Glycoproteins, oligosaccharides, and also lipid bilayers and DNA/RNA are degraded in lysosomes into basic metabolites, including amino acids and sugars, which are then released into the cytosol through lysosomal membrane transporters. Here they can be used for either *de novo* synthesis of cellular constituents or as a source of energy [31].

REs are generated as tubules with a 60–100 nm diameter that can extend over a distance of several micrometers, a pH of ~6.4, which form from the nonclathrin-coated areas of the main EE vacuole (Figure 2.2a). In some cells, REs associate into an endosomal recycling compartment that is typically located near the microtubule-organizing center in the perinuclear area (reviewed by Ref. [22]). REs are generally defined by the presence of RAB4 [32,33], RAB11 [34–36], the RAB11-FIP family of RAB11-interacting proteins, and the SNAREs cellubrevin/VAMP3 and syntaxin 13. Members of the large protein family of sorting nexins (SNXs) have been attributed to distinct recycling pathways emerging from REs [37]. SNXs can detect and/or induce membrane curvature and, thus, mediate the formation of tubules. REs consist of (a network of) branched tubules with multiple (clathrin-coated) buds [38], which form exits from which cargo proteins can travel to distinct cellular destinations [19]. While most pathways emerging from REs divert from the degradative track, RE can also be an intermediate station en route to the lysosomes. The adaptor protein AP-3 is associated with an RE exit that transports LMPs from REs to LEs/lysosomes [20,21]. Hence, LMPs that reach REs can via this AP3-mediated exit be transported back toward the main degradative pathway to the lysosomes.

2.2.3 Autophagy Pathways to the Lysosome

Lysosomes are not only reached by endocytosis but also by autophagic processes. Three autophagic transport pathways deliver cytoplasmic components for turnover in the lysosome: chaperone-mediated autophagy (CMA), microautophagy and macroautophagy (Figure 2.1). CMA is devoted to the transport of specific cytoplasmic proteins, while microautophagy and macroautophagy also deliver other cellular constituents, including large protein complexes, aggregates, and organelles.

Proteins destined for degradation by CMA possess a KFERQ motif or pseudo-KFERQ sequences, which are generated through regulated posttranslational modifications such as phosphorylation and acetylation [39]. Recognition of these motifs by the cytosolic chaperone HSC70 leads to substrate recruitment onto the surface of lysosomes. There, the targeted substrates bind to monomeric LAMP-2A, triggering the assembly of this LMP into a high-molecular-weight complex, which mediates translocation of the associated protein into the lysosomal lumen, an event that also requires the intralysosomal form of HSC70 [39]. CMA is thus directly dependent on the function of LMPs, providing an example for the intricate relationship between the endocytic and the autophagy pathways at the molecular level.

Microautophagy is generally described as a process that involves the selection of a specific cargo at the lysosomal limiting membrane, followed by an inward invagination and a pinching-off event, which lead to the sequestration of the cargo into the lysosome [40]. The molecular bases of microautophagy, however, still remain mysterious.

Macroautophagy, hereafter referred to as autophagy, is the most widely studied form of autophagy and characterized by the sequestration of cytoplasmic cargoes by autophagosomes (Figure 2.1). This type of autophagy is used for the degradation of a wide range of substrates, which differ in origin and size. Upon autophagy induction, the conserved factors involved in this pathway, the autophagy-related (Atg) proteins, assemble at a specialized site that has been named the *phagophore assembly site* or the *preautophagosomal structure* (PAS). The precise ultrastructure of the PAS is not known, but it is considered the organizational center of the autophagy machinery. At the PAS, the ATG proteins together with specific SNAREs, putatively mediate the biogenesis of the phagophore (Figure 2.3) by orchestrated fusion of Golgi-, RE-, and plasma membrane-derived membranes [42–44]. The ATG proteins are also involved in the subsequent elongation of the phagophore into an autophagosome, a process that requires acquisition of extra membranes. The frequent proximity and contact points between the growing phagophores and the ER have led to the hypothesis that lipids necessary for membrane elongation can directly transfer from the ER to nascent autophagosomes [45–47]. By contrast, the involvement of ER exit sites (ERES) in autophagy has underlined the possibility that vesicular traffic could also play a critical role [48,49]. Growing phagophores are also often detected adjacent to the ER–mitochondria contact sites [50,51], explaining how mitochondria could also be involved in supplying membranes, possibly by direct lipid transfer [52]. Complete autophagosomes fuse with LEs or lysosomes to enable cargo degradation (Figure 2.3). Like for the endocytic cargo, the resulting metabolites are transported into the cytoplasm by lysosomal membrane transporters. Moreover, this flux of amino acids provides a feedback mechanism to stop autophagy and restore lysosome formation through the reactivation of the mTORC1 complex.

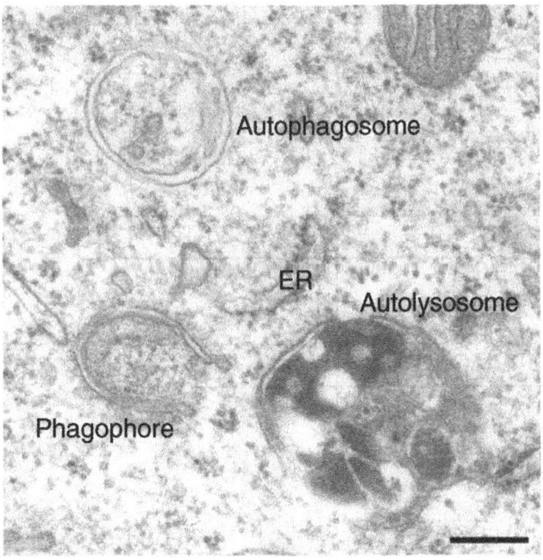

Figure 2.3 Electron micrograph showing the three main intermediates of macroautophagy: phagophore, autophagosome and autolysosome. Picture taken from Akt knockdown human prostate cell (PC3) [41]. ER, endoplasmic reticulum. Bar = 200 nm. Courtesy of Ann de Mazière, Department of Cell Biology, University Medical Center, Utrecht, The Netherlands.

2.2.4 The ATG Proteins: The Key Regulators of Autophagy

The central actors of the autophagy process are the *ATG* genes. Based on their organization in complexes and interactions, the *ATG* genes have been divided into five functional clusters that compose the core autophagy machinery [53,54] (Figure 2.4). Autophagosome biogenesis is initiated by the ULK/ATG1 complex, composed of the ULK1 or ULK2 kinase, ATG13, FIP200, and ATG101. Activation of the ULK complex and subsequent translocation from the cytosol to the PAS, possibly from REs, triggers the assembly of the rest of the autophagy machinery. Several signaling cascades regulating autophagy, including the one centered around mTOR, modulate the kinase activity of the ULK complex [55]. In particular, phosphorylation of the subunits of the ULK/ATG1 complex leads to recruitment of the autophagy-specific class III phosphatidylinositol 3-kinase (PtdIns3K) complex to the PAS [56,57]. This complex is formed by VPS34, p150, BECLIN1/ATG6, and ATG14L, and interacts with various factors such as AMBRA1 and VMP1, which

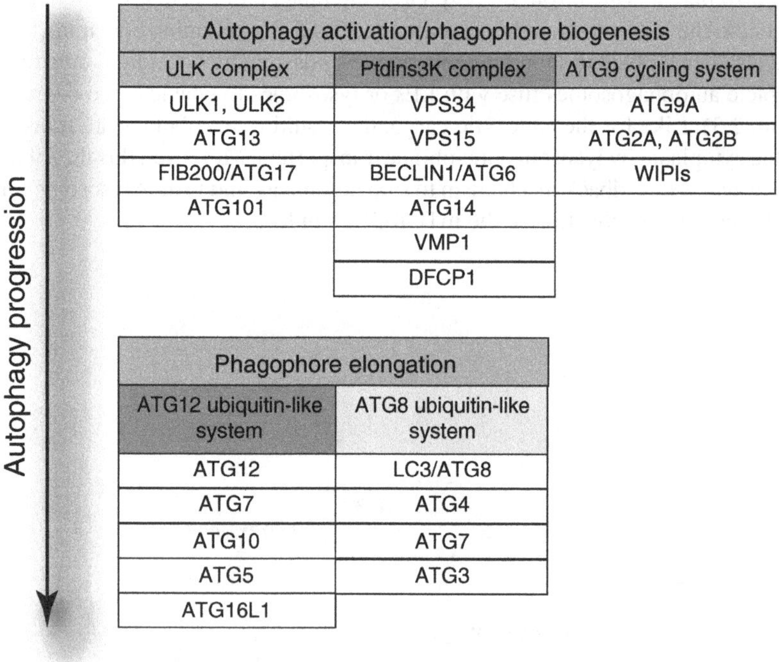

Figure 2.4 Schematic overview of the distinct Atg complexes required for autophagy. The ULK and PtsIns3K complex and the ATG9 cycling system are key in the organization of the PAS and biogenesis of the phagophore. The two ubiquitin-like systems appear to be mostly involved into the elongation of the phagophore into an autophagosome. WIPIs is a protein family with four members, that is, WIPI1-4. LC3/ATG* is composed of six proteins: LC3A–C and GABARAPL1–3. There are also four isoforms of ATG4, that is, ATG4A–D.

also regulate PtdIns3K complex localization and activity [53,54]. The generation of phosphatiylinositol-3-phosphate (PtdIns3P) at the PAS also triggers the recruitment of WIPI (WD-repeat protein interacting with phosphoinositides) proteins and DFCP1, two other proteins that mediate the formation of the phagophore and associate to membrane through binding PtdIns3P [53,54]. In this context, it is interesting to note that RAB5 can bind VPS34 and Beclin1 subunits of the PtdIns3K complex and that this interaction is relevant to the regulation of autophagy [58].

Another protein essential for the initial steps of autophagosome formation is the transmembrane protein ATG9, which together with ATG2 and members of the WIPI protein family, at least WIPI4, forms a second functional cluster [53,54]. ATG9 localizes to different compartments of the endolysosomal system, that is, REs, TGN, LE, and plasma membrane, and dynamically associates with the PAS [44,59,60]. While the precise function of ATG9 is still unknown, it has been proposed that it is required to provide some of the initial membranes that organize the PAS [61,62]. Together with the ULK and PtdIns3K complexes, ATG9 is therefore considered a key factor in generating the phagophore.

The subsequent elongation and closure of the phagophore cistern is driven by two ubiquitin-like conjugation systems: ATG7/ATG10/ATG5/ATG12 and ATG7/ATG3/ATG8. In the first, the E1-like ATG7 and the E2-like ATG10 enzymes covalently link the ubiquitin-like protein ATG12 to ATG5. The ATG12–ATG5 conjugate subsequently associates to ATG16L1 [53,54]. The ATG12–ATG5/ATG16L1 complex then recruits the second ubiquitin-like conjugation system to the PAS [53,54] by conjugating phosphatidylethanolamine (PE) to members of the LC3 (or ATG8) protein family (LC3A, LC3B, LC3C, GABARAP-L1, GABARAP-L2, and GABARAP-L3) [63]. All the members of this protein family are posttranslationally processed at their C-terminus by ATG4 to generate the nonlipidated LC3-I form. Upon autophagy induction, LC3-I is conjugated to PE on both the inner and outer membranes of the growing phagophore through the action of ATG7, the E2-like enzyme ATG3 and the ATG12–ATG5/ATG16L1 complex. The lipidated form of LC3, known as LC3-II, is a multitask factor. Together with the ATG5–ATG12/ATG16L1 complex, LC3-II is thought to form a protein coat that drives the formation of autophagosomes, possibly by deforming membranes [64]. Moreover, LC3-II appears to have fusogenic properties [65,66] that could be essential for sealing the growing phagophore and/or the subsequent fusion of autophagosomes with lysosomes [67,68]. Finally, LC3-II is involved in microtubule-dependent transport of autophagosomes toward lysosomes by interacting with microtubules and motor proteins [69].

Under starvation conditions, the autophagy pathway is a nonselective process for bulk degradation of cytoplasmic components. However, there are numerous situations in which cells can use autophagy to specifically eliminate unwanted structures, including damaged or superfluous organelles, and invading microorganisms [70]. The selective types of autophagy rely on autophagy receptors, which mediate the specific sequestration of cargo into autophagosomes. Autophagy receptors simultaneously bind to both the cargo and the pool of LC3-II present at the internal surface of the growing autophagosome [70,71]. The interaction between LC3 and the autophagy receptors is in most cases mediated through a specific amino acid sequence in the autophagy receptors: the LC3-interacting region (LIR) or Atg8-interacting motif (AIM) [70,71]. The binding of autophagy receptors to the cargo, in contrast, involves either the recognition of a specific determinant (e.g., in case of yeast Atg32 and mammalian NIX with mitochondria) or the interaction between an ubiquitin-binding domain present within the autophagy receptors (e.g., p62/SQSTM1 and NBR1) with ubiquitin moieties conjugated to the cargo.

2.3 FUSION AND FISSION BETWEEN THE ENDOLYSOSOMAL AND AUTOPHAGY PATHWAYS

Cross talk between the endocytic and autophagic pathways occurs at many levels: transcriptional regulation, protein sharing, and compartmental connections. A center stage in the regulation of these interactions is taken by the transcription factor EB (TFEB) and the mTORC1 complex. For recent reviews hereon, we refer to Ref. [72] and Chapter 7 of this book. Here we focus on the fusion and fission events between compartments of the endolysosomal system and autophagic membranes, respectively.

2.3.1 Recycling Endosomes and Autophagosome Biogenesis

An interaction between endocytic vesicles, ATG proteins, and REs occur at an early stage of autophagosome biogenesis. In particular, plasma membrane–associated ATG9 can enter clathrin-coated vesicles to travel to EEA1-positive EEs [44]. By contrast, cytoplasmic ATG16L associates with a distinct pool of clathrin-coated endocytic vesicles, which fuse between themselves and with Rab11-positive REs thus bypassing EEA1-positive EEs [43,73], which very likely are also positive for ULK1 [74]. The membrane deforming PX-BAR protein SNX18 facilitates delivery of ATG16L1 to the REs [75]. ATG9-positive recycling tubules emerging from EEA1-positive EEs can subsequently fuse with ATG16L1/RAB11 positive REs, which bring together these two proteins, a step that appears to be critical to initiate phagophore formation. The SNARE critical for the fusion of ATG9 and ATG16L1 positive membranes is VAMP3, which is present on REs. Knockdown of VAMP3 results in the accumulation of ATG9 in EEs and a decrease in autophagosome formation [44]. Several studies have shown that vesicles derived from REs are recruited to the site where autophagosomes are formed [44,74], a process that is increased upon autophagy induction [74,76]. Overexpression of the putative RabGAP TBC1D14, which associates to ULK1, impairs autophagosome formation while Rab11 depletion blocks autophagy [74]. The Rab substrate for TBC1D14 is still unknown. It binds activated RAB11 but is not a GAP for RAB11. Altogether, these data show that membrane traffic from EEs to REs is involved in autophagosome formation and that REs can probably serve as a membrane source and key regulator for the early stages of autophagosome biogenesis.

2.3.2 Autophagosome Fusion with Late Endosomes and Lysosomes

When complete, autophagosomes first fuse with LE to form amphisomes and then with lysosomes, thus generating autolysosomes [1]. In addition, lysosomes can fuse directly with an autophagosome to also form autolysosomes. The relevance of amphisomes is underlined by the fact that disruption of LE homeostasis, for example, by deleting ESCRT complex subunits, leads to the accumulation of autophagosomes and impairs the progression of autophagy [77]. Efficient fusion between lysosomes and autophagosomes requires the coordinated transport of these two organelles to the perinuclear area [78,79]. Starvation causes an increase in the intracellular pH, which induces lysosome relocalization to the perinuclear area [79,80]. Under the same conditions, mTORC1 inhibition triggers the formation of autophagosomes, which are transported to the same region of the cell by an interaction with microtubules [69].

The molecular mechanism regulating fusion of autophagosomes with lysosomes has not yet been fully established, but some key players are identified. Importantly, all the factors that have been involved so far in the RAB7-dependent fusion of LEs with lysosomes have also been shown to be essential for autolysosome formation. In both yeast and mammalian cells, impairment or deletion of Rab7 results in the accumulation of autophagosomes, indicating that this Rab GTPase is required for autophagosome–LE/lysosome fusion [81–83]. The hexameric homotypic fusion and protein sorting (HOPS) complex is a downstream effector of Rab7 involved in membrane tethering [84–86]. Recent studies in *Drosophila* and mammalian cells showed that loss of each of the six HOPS subunits results in the accumulation of autophagosomes and a block in the degradation of the autophagic cargo [84,86]. Moreover, the GEF of RAB7, the SAND1/Mon1 and Ccz1 complex, is also essential for fusion of autophagosomes with the vacuole at least in yeast [87].

More than 30 SNAREs safeguard fusion specificity and drive the fusion process in mammalian cells. Several SNAREs located in the endolysosomal system have been shown to be involved in the fusion of yeast autophagosomes with vacuoles, but until lately no SNAREs were found on autophagosomes. It was shown in both *Drosophila* and mammalian cells that upon starvation the SNARE SYNTAXIN17 redistributes in part from an ER/mitochondria onto autophagosomes [88,89]. The recruitment of SYNTAXIN17 to autophagosomes is reduced by mutations in the two glycine-rich transmembrane motifs that appear to regulate the solubility of this protein [89], but it is still unclear how the translocation of syntaxin 17 is spatially and temporarily regulated. On the autophagosome, SYNTAXIN17 binds and recruits the HOPS complex and associates with the lysosomal SNAREs SNAP-29 and VAMP8 to carry out its fusogenic function [86,88,89]. Depletion of SYNTAXIN17 results in the accumulation of autophagosomes, underscoring its importance in autolysosome biogenesis [86,88]. Interestingly, SYNTAXIN17 was originally localized to the smooth ER and implicated in the dynamics of this compartment in steroidogenic cells [90]. At least in partial agreement with this original observation, a subpopulation of SYNTAXIN17 also binds ATG14L and recruits it to ER–mitochondria contact sites, where it has been proposed to be important for autophagosome formation [51]. Whether SYNTAXIN17 is also present in isolation membranes remains a subject of debate [51,91].

Besides these classical fusion machinery proteins, more unexpected proteins are also important for the fusion between autophagosome and lysosome. For example, in patients suffering from Danon disease, the LAMP-2 deficiency causes an accumulation of autophagosomes leading to cardiomyopathy and myopathy [92]. Similarly, pancreatitis due to the depletion of LAMP-2 from a combination of alcohol exposure and endotoxemia correlates with the accumulation of autophagosomes and a relative paucity of autolysosomes [93]. How LAMP-2 participates in the fusion between autophagosomes and lysosomes is unclear.

2.3.3 Autophagic Lysosomal Reformation

Cells under sustained starvation lose their lysosomes (an event accompanied by the formation of autolysosomes) within 4 h but show a restored lysosome population after 8 h [94]. An important source for these *de novo* formed lysosomes is a process called autophagic lysosomal reformation (ALR) ([94]; Figure 2.1). The onset of ALR is accompanied by the generation of long tubules that extend in a microtubule-dependent manner from autolysosomes. Interestingly, these tubules are positive for LAMP-1 but are nonacidic and lack lysosomal enzymes, such as cathepsin D, which remains segregated in the vacuolar body of autolysosomes. The LAMP-positive tubules detach from the autolysosome through vesiculation and form globular compartments that have been named protolysosomes. It has been hypothesized that these catabolically inactive organelles are involved in the reformation of lysosomes.

The ALR process is regulated by mTOR. After prolonged starvation, the transport of metabolites from autolysosomes to the cytoplasm results in the restoration of mTOR activity, which not only attenuates autophagy but also induces the generation of the LAMP-1-positive tubules from autolysosomes [94]. For example, in *Drosophila*, the sugar transporter activity of Spinster is required to restore nutrient levels and reactivation of mTOR [95]. Interestingly, also fibroblasts from LSD patients (e.g., Pompe and Niemann–Pick diseases) show impaired mTOR reactivation and no ALR [94]. Thus, mTOR via ALR links the nutrient status of a cell to the induction or cessation of autophagy, and in addition controls the number of lysosomes. This feedback mechanism is therefore important to prevent autophagic cell death by prolonged autophagy [96] and seems to play a relevant role in the pathophysiology of LSDs.

The generation of reformation tubules from autolysosomes requires dissociation of RAB7 and the concomitant recruitment of clathrin [94,97]. This latter event is triggered by conversion of phosphatidylinositol-4-phosphate (PI4P) into phosphatidylinositol-4,5-biphosphate (PI4,5P2), which recruits the clathrin adaptor AP2 that in turn mediates the association of clathrin. The GTPase DYNAMIN2 is also recruited to the reformation tubules and its depletion prevents the fission of protolysosomes from the reformation tubules [98]. Intriguingly, PtdIns(4,5)P2, AP2, DYNAMIN2 as well as clathrin are well known for their role in clathrin-mediated endocytosis of transmembrane proteins from the plasma membrane. In normal growth conditions, these molecules are localized at the plasma membrane and not on lysosomes. It is rather unexpected, but highly interesting, that the same machinery appears to regulate different clathrin-mediated sorting events depending on the nutritional status of the cell. In contrast to clathrin-mediated endocytosis, the recruitment of AP2 to autolysosomes does not seem to require transmembrane proteins, such as LAMP-1, which contains an AP2 binding motif, but just PI(4,5)P2 [97]. This raises the question how sorting of lysosomal enzymes and LMPs is achieved during ALR. One clue is provided by the study of Sridhar et al. [99]. The synthesis of PI4P is under control of four different phosphatidylinositol 4-kinases (PI4Ks) in mammalian cells. A particular pool of one of the PI4K isoforms, PI4KIIIβ, associates with the lysosomal membrane. In PI4KIIIβ knockdown cells, the absence of PI4KIIIβ and PI4P leads to the constitutive formation of tubules from the surface of lysosomes. Notably, these tubules also contain cathepsin D and LC3 in addition to LAMP-1 and LAMP-2, and recruitment of clathrin and AP2 to lysosomes is markedly increased. These data show that in control cells PI4KIIIβ prevents tubule formation from lysosomes, but during ALR is required for retention of lysosomal enzymes in the autolysosomal lumen. How PI4P and PI4,5P2 prevent lysosomal enzymes to enter the lysosomal reformation tubules is a question that still remains unanswered.

Over time, protolysosomes formed during ALR acquire a novel set of lysosomal enzymes and become functionally active. The molecular pathways underlying the *refueling* of protolysosomes are still unknown. For example, it remains to be established whether the MPRs are required. Alternative transport routes might involve yet undiscovered pathways of lysosomal enzyme transport and/or fusion between protolysosomes and LE or lysosomes. The process of protolysosome biogenesis is seemingly reminiscent to the formation of a temporary hybrid organelle upon fusion between LEs and lysosomes, from which small, dense lysosomes can reform, possibly to avoid an excessive expansion of the lysosome volume. In this case, however, the reformed lysosomes contain lysosomal enzymes and are functionally active [100,101].

2.4 DISEASES

2.4.1 Lysosome-Related Disorders (LSDs)

To date, almost 60 different inherited LSDs are known. LSDs are caused by mutations in genes encoding for proteins that directly affect lysosomal functioning, including lysosomal hydrolases and LMPs. Specific examples of LSDs are discussed in Chapter 11. Although most LSDs have a defect in the (optimal) functioning of just one specific protein, accumulation of nondegraded material in the lysosome interior severely disturbs numerous functions of this organelle, including completion of the final stages of autophagy (i.e., autophagosome fusion and autophagosomal cargo degradation). Accordingly, a growing number of studies are highlighting that numerous LSDs display an impairment in the autophagic flux [102]. Thus, lysosomal dysfunctioning and defects in autophagy are intrinsically linked and a common feature of

LSDs. Strikingly, the pathology of most LSDs also involves neurodegeneration [103], probably because the impairment in autophagy causes an accumulation of cytoplasmic aggregates [104].

2.4.2 Lysosomes in Neurodegeneration and Its Links to Autophagy

The pathophysiological connection between the endolysosomal system and autophagy is not limited to LSDs. This is particularly relevant in neurodegenerative diseases such as Alzheimer's, Parkinson's, and Huntington's diseases, amyotrophic lateral sclerosis (ALS), and ataxias, which are caused or accompanied by an aggregation of aberrant proteins (detailed discussion in Chapter 9). This is also underlined by the observation that upregulation of autophagy can decrease the effects of toxic aggregates [105].

Recently, it was proposed that the genes such as *SNCA*, *LRRK2*, and *GBA* associated with the Lewy bodies typical of Parkinson's disease (PD) all affect lysosomal functioning [106]. In addition, forms of PD caused by mutations in *VPS35* or *ATP31A2*, a gene involved in recycling from endosomes and in acidification of the lysosome, respectively, may very likely also result from the disruption of the autophagy–endolysosomal system [106]. The direct intracerebral delivery of agents that enhance the activity of Aβ degrading enzymes is considered one of the most promising approaches for future AD treatment [107].

In Alzheimer's disease (AD), impairment of lysosomal proteolysis is among the earliest pathogenic events [108] and progression of the disease is accompanied by a continuum of pathological changes that ultimately lead to the massive accumulation of LEs, lysosomes, autolysosomes, and autophagosomes [109–112] (reviewed in Ref. [113]). Mutations in PS1, the major causative gene for early-onset familial AD, directly impair lysosome proteolysis [112,114]. An extra copy of *APP*, another gene leading to familial AD, triggers an abnormal activation of RAB5 [113]. Finally, high-dietary LDL cholesterol and overexpression of the *APOE E4* allele of APOE, a mediator of neuronal cholesterol transport and major risk factor for late onset of AD, also result in aberrant RAB5 activation [115]. The picture that emerges is that mutations causing an early onset of AD impair lysosomal proteolysis and functions, which in turn result in the accumulation of autophagosomes and/or autolysosomes with undigested material, which contribute to the pathogenesis worsening.

2.4.3 Autophagy-Related Diseases

Because of the central role of autophagy in cell and organismal physiology, hereditary genetic disorders caused by a mutation of an *ATG* gene are extremely rare. Not surprisingly, the diseases reported so far caused by a mutation in one of the *ATG* genes are either *de novo* mutations or generating a hypomorph allele.

2.4.3.1 Crohn's Disease Crohn's disease (CD) is a chronic form of inflammatory bowel disease that can affect any part of the gastrointestinal tract. It is a complex illness where the genome, microbiome, and environment determine the onset and development of the disease [116]. More than 90 distinct genomic susceptibility loci have been identified to be associated with an increased risk of CD. Those include three genes that are linked to autophagy: *NOD2*, *ATG16L1*, and *IRGM*. The coding polymorphism in *ATG16L1* (i.e., T300A) appears to lead to a decrease in selective types of autophagy resulting in altered cytokine signaling and reduced antibacterial defense [117]. The NOD2 protein is an intracellular sensor for bacteria and positively regulates autophagy in response to bacterial infections [118,119]. Cells carrying loss-of-function variants of NOD2 that have been associated with an increased susceptibility for CD display defects in antibacterial autophagy [118,119]. Similarly, the human immunity-related GTPase IRGM is required for IFN-γ-induced autophagy that controls infections [120]. The polymorphism in *IRGM* alters a binding site for miR-196 and causes deregulation of IRGM-dependent removal of intracellular pathogens [121]. Recently, additional genes linked to autophagy such as *ULK1*, *MTMR3*, and *LKKR2* have been associated to CD [122–124].

2.4.3.2 Vici Syndrome The Vici syndrome is a rare, hereditary multisystem disorder characterized by callosal agenesis, cataracts, cardiomyopathy, combined immunodeficiency, and hypopigmentation. A major causative gene for this syndrome is *EPG5* [125], the human homolog of *Caenorhabditis elegans epg-5*, which encodes a protein with a key role in autophagy in multicellular organisms [126]. EPG5 is a 2579-amino acid protein with no obvious known domains, which is involved in the late steps of autophagy because knockdown in mammalian cells leads to an accumulation of nondegradative autolysosomes, whereas overexpression accelerates autophagic catabolism [126]. Similarly, skeletal muscle tissues and fibroblasts from Vici syndrome patients display an accumulation of autophagosomes caused by an impairment in either the fusion of these carriers with lysosomes or in cargo breakdown [125]. The histopathological features are also consistent with an autophagy defect, as autophagic vacuoles containing abnormal material, such as aberrant mitochondria, are prominent [125]. This symptomology is in part recapitulated in the knockout mice [127]. Interestingly, EPG5 is probably not exclusively involved in autophagy but rather in multiple pathways terminating at the lysosome because phagolysosome formation is also defective in the absence of *epg-5* in *C. elegans* [128] and degradation of endocytosed surface receptors is impaired in EPG5 knockdown cells [127]. The mechanism of function and interacting partners of EPG5 still need to be revealed.

2.4.3.3 Neurodegeneration with Brain Iron Accumulation (NBIA) NBIA diseases comprise a set of single-gene disorders that manifest a range of neurological phenotypes, with a common feature of high iron levels in basal ganglia. Recently, it has been revealed that *de novo* mutations in *WDR45/WIPI4* are associated with NBIA [129–132]. Wdr45 is one of the four human homologs of yeast Atg18, and studies in *C. elegans*, *Drosophila melanogaster*, and human cell lines have shown that this protein is essential for autophagy and interacts with Atg2's counterparts in these organisms [133–136]. Indeed, cells from NBIA patients carrying a *WDR45* mutation display a defect in autophagy [129]. However, it remains unclear whether this defect actually causes NBIA disorders since other genes mutated in NBIA (*PANK2*, *C19orf12*, and *PLA2G6*) are not involved in autophagy and Atg18 has additional cellular functions in yeast [137–139].

2.4.3.4 Hereditary Spastic Paraparesis (HSP) Hereditary spastic paraparesis (HSP) is a group of clinically and genetically heterogeneous neurodegenerative disorders that is characterized by progressive spasticity and hyperreflexia of the lower limbs due to the axonal degeneration of the corticospinal tracts. There are hereditary forms of HSP caused by mutations in genes that are linked to autophagy. The first is a point mutation in TECPR2, which leads to a truncated version of the protein that is rapidly degraded by the proteasome [140]. While the precise function of TECPR2 is unknown, this protein interacts with all the members of the LC3 protein family, and when knocked down, autophagy is impaired [134,140]. However in patient skin fibroblasts the autophagy defect is not very prominent, possibly because TECPR2 is principally expressed in human brains, the tissue affected most by the pathology [140]. Another protein found mutated in HSP patients is spastizin (or FYVE-CENT), a PtdIns3P-binding protein that in addition to a cytosolic distribution also localizes to EEs, ER, microtubules, transport vesicles, and the midbodies formed during cell division [141–143]. Spastizin binds to the PtsIns3K complex that contains UVRAG and RUBICON, but not ATG14L, via an interaction with BECLIN1 [142,144]. This PtsIns3K complex has been implicated in the maturation of autophagosomes through its involvement in LE biogenesis [145]. In neuronal and nonneuronal cells carrying the HSP spastizin mutant forms or lacking this protein, autophagosome maturation is compromised even if the UVRAG- and Rubicon-containing PtsIns3K complex is still assembled [144]. The accumulation of immature autophagosomes is also observed in fibroblasts from patient.

2.4.3.5 Cancer Autophagy can function in both tumor suppression and tumor progression [146]. Monoallelic deletion of *BECLIN1* frequently occurs in human breast, ovarian, and prostate tumors [147,148]. In support to this, clinical studies have associated poor prognosis and aggressive tumor phenotypes with aberrant expression of Beclin1 in tumor tissue [149,150]. The sequence and the expression of other core ATG proteins such as ULK1/Atg1, ATG2B, ATG5, GABARAP/Atg8 ATG12, and ATG16L1, and autophagy-associated proteins such as UVRAG and BIF-1 have also been found to be altered in several cancers [151]. For a deeper discussion of this topic, see Chapter 13.

2.5 CONCLUDING REMARKS

It is becoming increasingly clear that the endocytic and autophagy pathways are intrinsically interconnected at many organizational and regulatory stages, both in health and disease. In principle, any defects in the endocytic pathway could also affect autophagy progression [77,152]. The next obvious challenge is to address how lysosomal functions could regulate autophagy and vice versa how autophagy could modulate lysosomal functioning [153]. For example, a block of the endolysosomal maturation may trigger autophagy to correct cellular defects [154]. Solving this type of questions will yield important insights for the development of therapeutic approaches to treat autophagy–lysosome-related diseases.

ACKNOWLEDGMENTS

We thank our many colleagues who have contributed to the work described herein and apologize to those whose work we failed to cite. We want to particularly thank our colleagues Suzanne van Dijk and Viola Oorschot for the preparation of electron microscopy samples, Ann de Mazière for providing Figures 2.2 and 2.3, and Rene Scriwanek for the figure preparation. J.K. is supported by NWO (VICI grant 918.56.611). F.R. is supported by ALW Open Program (821.02.017 and 822.02.014), DFG-NWO cooperation (DN82-303), and ZonMW VICI (016.130.606) grants.

REFERENCES

[1] Eskelinen EL, Kovacs AL. Double membranes vs. lipid bilayers, and their significance for correct identification of macroautophagic structures. Autophagy 2011;7:931–932.

[2] Saftig P, Klumperman J. Lysosome biogenesis and lysosomal membrane proteins: trafficking meets function. Nat Rev Mol Cell Biol 2009;10:623–635.

[3] Schwake M, Schroder B, Saftig P. Lysosomal membrane proteins and their central role in physiology. Traffic 2013;14:739–748.

[4] Ghosh P, Dahms NM, Kornfeld S. Mannose 6-phosphate receptors: new twists in the tale. Nat Rev Mol Cell Biol 2003;4:202–212.

[5] van Meel E, Klumperman J. TGN exit of the cation-independent mannose 6-phosphate receptor does not require acid hydrolase binding. Cell Logist 2014;4:e954441.

[6] Ghosh P, Kornfeld S. The GGA proteins: key players in protein sorting at the trans-Golgi network. Eur J Cell Biol 2004;83:257–262.

[7] Ghosh P, Kornfeld S. The cytoplasmic tail of the cation-independent mannose 6-phosphate receptor contains four binding sites for AP-1. Arch Biochem Biophys 2004;426:225–230.

[8] Bonifacino JS, Traub LM. Signals for sorting of transmembrane proteins to endosomes and lysosomes. Annu Rev Biochem 2003;72:395–447.

[9] Honing S, Griffith J, Geuze HJ, Hunziker W. The tyrosine-based lysosomal targeting signal in LAMP-1 mediates sorting into Golgi-derived clathrin-coated vesicles. EMBO J 1996;15:5230–5239.

[10] Hunziker W, Geuze HJ. Intracellular trafficking of lysosomal membrane proteins. Bioessays 1996;18:379–389.

[11] Pols MS, van Meel E, Oorschot V, ten Brink C, Fukuda M, et al. hVps41 and VAMP7 function in direct TGN to late endosome transport of lysosomal membrane proteins. Nat Commun 2013;4:1361.

[12] Swetha MG, Sriram V, Krishnan KS, Oorschot VM, ten Brink C, et al. Lysosomal membrane protein composition, acidic pH and sterol content are regulated via a light-dependent pathway in metazoan cells. Traffic 2011;12:1037–1055.

[13] Braulke T, Mach L, Hoflack B, Glossl J. Biosynthesis and endocytosis of lysosomal enzymes in human colon carcinoma SW 1116 cells: impaired internalization of plasma membrane-associated cation-independent mannose 6-phosphate receptor. Arch Biochem Biophys 1992;298:176–181.

[14] Kumari S, Mg S, Mayor S. Endocytosis unplugged: multiple ways to enter the cell. Cell Res 2010;20:256–275.

[15] Klumperman J, Raposo G. The complex ultrastructure of the endolysosomal system. Cold Spring Harb Perspect Biol 2014;6:a016857.

[16] Zeigerer A, Gilleron J, Bogorad RL, Marsico G, Nonaka H, et al. Rab5 is necessary for the biogenesis of the endolysosomal system *in vivo*. Nature 2012;485:465–470.

[17] Miaczynska M, Christoforidis S, Giner A, Shevchenko A, Uttenweiler-Joseph S, et al. APPL proteins link Rab5 to nuclear signal transduction via an endosomal compartment. Cell 2004;116:445–456.

[18] Johannes L, Wunder C. Retrograde transport: two (or more) roads diverged in an endosomal tree? Traffic 2011;12:956–962.

[19] Bonifacino JS, Rojas R. Retrograde transport from endosomes to the trans-Golgi network. Nat Rev Mol Cell Biol 2006;7:568–579.

[20] Peden AA, Oorschot V, Hesser BA, Austin CD, Scheller RH, et al. Localization of the AP-3 adaptor complex defines a novel endosomal exit site for lysosomal membrane proteins. J Cell Biol 2004;164:1065–1076.

[21] Theos AC, Tenza D, Martina JA, Hurbain I, Peden AA, et al. Functions of adaptor protein (AP)-3 and AP-1 in tyrosinase sorting from endosomes to melanosomes. Mol Biol Cell 2005;16:5356–5372.

[22] Klumperman J, Raposo G. The complex ultrastructure of the endolysosomal system. In: Schmid SL, Sorkin A, Zerial M, editors. *Endocytosis*. New York: CSH Laboratory Press; 2014. p 113–134.

[23] Sachse M, Ramm G, Strous G, Klumperman J. Endosomes: multipurpose designs for integrating housekeeping and specialized tasks. Histochem Cell Biol 2002;117:91–104.

[24] Sachse M, Urbe S, Oorschot V, Strous GJ, Klumperman J. Bilayered clathrin coats on endosomal vacuoles are involved in protein sorting toward lysosomes. Mol Biol Cell 2002;13:1313–1328.

[25] Rink J, Ghigo E, Kalaidzidis Y, Zerial M. Rab conversion as a mechanism of progression from early to late endosomes. Cell 2005;122:735–749.

[26] Gary JD, Wurmser AE, Bonangelino CJ, Weisman LS, Emr SD. Fab1p is essential for PtdIns(3)P 5-kinase activity and the maintenance of vacuolar size and membrane homeostasis. J Cell Biol 1998;143:65–79.

[27] Nordmann M, Cabrera M, Perz A, Brocker C, Ostrowicz C, et al. The Mon1–Ccz1 complex is the GEF of the late endosomal Rab7 homolog Ypt7. Curr Biol 2010; 20:1654–1659.

[28] Poteryaev D, Datta S, Ackema K, Zerial M, Spang A. Identification of the switch in early-to-late endosome transition. Cell 2010;141:497–508.

[29] Kinchen JM, Ravichandran KS. Identification of two evolutionarily conserved genes regulating processing of engulfed apoptotic cells. Nature 2010;464:778–782.

[30] Huotari J, Helenius A. Endosome maturation. EMBO J 2011;30:3481–3500.

[31] Kolter T, Sandhoff K. Lysosomal degradation of membrane lipids. FEBS Lett 2010; 584:1700–1712.

[32] van der Sluijs P, Hull M, Webster P, Male P, Goud B, et al. The small GTP-binding protein Rab4 controls an early sorting event on the endocytic pathway. Cell 1992; 70:729–740.

[33] Sheff DR, Daro EA, Hull M, Mellman I. The receptor recycling pathway contains two distinct populations of early endosomes with different sorting functions. J Cell Biol 1999;145:123–139.

[34] Ullrich O, Reinsch S, Urbe S, Zerial M, Parton RG. Rab11 regulates recycling through the pericentriolar recycling endosome. J Cell Biol 1996;135:913–924.

[35] Ren M, Xu G, Zeng J, De Lemos-Chiarandini C, Adesnik M, et al. Hydrolysis of GTP on Rab11 is required for the direct delivery of transferrin from the pericentriolar recycling compartment to the cell surface but not from sorting endosomes. Proc Natl Acad Sci U S A 1998;95:6187–6192.

[36] Wilcke M, Johannes L, Galli T, Mayau V, Goud B, et al. Rab11 regulates the compartmentalization of early endosomes required for efficient transport from early endosomes to the trans-Golgi network. J Cell Biol 2000;151:1207–1220.

[37] van Weering JR, Cullen PJ. Membrane-associated cargo recycling by tubule-based endosomal sorting. Semin Cell Dev Biol 2014;31:40–47.

[38] Stoorvogel W, Oorschot V, Geuze HJ. A novel class of clathrin-coated vesicles budding from endosomes. J Cell Biol 1996;132:21–33.

[39] Kaushik S, Cuervo AM. Chaperone-mediated autophagy: a unique way to enter the lysosome world. Trends Cell Biol 2012;22:407–417.

[40] Li WW, Li J, Bao JK. Microautophagy: lesser-known self-eating. Cell Mol Life Sci 2012;69:1125–1136.

[41] Degtyarev M, De Maziere A, Orr C, Lin J, Lee BB, et al. Akt inhibition promotes autophagy and sensitizes PTEN-null tumors to lysosomotropic agents. J Cell Biol 2008;183:101–116.

[42] Nair U, Jotwani A, Geng J, Gammoh N, Richerson D, et al. SNARE proteins are required for macroautophagy. Cell 2011;146:290–302.

[43] Moreau K, Ravikumar B, Renna M, Puri C, Rubinsztein DC. Autophagosome precursor maturation requires homotypic fusion. Cell 2011;146:303–317.

[44] Puri C, Renna M, Bento CF, Moreau K, Rubinsztein DC. Diverse autophagosome membrane sources coalesce in recycling endosomes. Cell 2013;154:1285–1299.

[45] Hayashi-Nishino M, Fujita N, Noda T, Yamaguchi A, Yoshimori T, et al. A subdomain of the endoplasmic reticulum forms a cradle for autophagosome formation. Nat Cell Biol 2009;11:1433–1437.

[46] Yla-Anttila P, Vihinen H, Jokitalo E, Eskelinen EL. 3D tomography reveals connections between the phagophore and endoplasmic reticulum. Autophagy 2009;5:1180–1185.

[47] Karanasios E, Stapleton E, Manifava M, Kaizuka T, Mizushima N, et al. Dynamic association of the ULK1 complex with omegasomes during autophagy induction. J Cell Sci 2013;126:5224–5238.

[48] Graef M, Friedman JR, Graham C, Babu M, Nunnari J. ER exit sites are physical and functional core autophagosome biogenesis components. Mol Biol Cell 2013; 24:2918–2931.

[49] Ge L, Melville D, Zhang M, Schekman R. The ER–Golgi intermediate compartment is a key membrane source for the LC3 lipidation step of autophagosome biogenesis. Elife 2013;2:e00947.

[50] Bockler S, Westermann B. Mitochondrial ER contacts are crucial for mitophagy in yeast. Dev Cell 2014;28:450–458.

[51] Hamasaki M, Furuta N, Matsuda A, Nezu A, Yamamoto A, et al. Autophagosomes form at ER–mitochondria contact sites. Nature 2013;495:389–393.

[52] Hailey DW, Rambold AS, Satpute-Krishnan P, Mitra K, Sougrat R, et al. Mitochondria supply membranes for autophagosome biogenesis during starvation. Cell 2010;141:656–667.

[53] Lamb CA, Yoshimori T, Tooze SA. The autophagosome: origins unknown, biogenesis complex. Nat Rev Mol Cell Biol 2013;14:759–774.

[54] Mizushima N, Yoshimori T, Ohsumi Y. The role of Atg proteins in autophagosome formation. Annu Rev Cell Dev Biol 2011;27:107–132.

[55] Yang Z, Klionsky DJ. Mammalian autophagy: core molecular machinery and signaling regulation. Curr Opin Cell Biol 2010;22:124–131.

[56] Jao CC, Ragusa MJ, Stanley RE, Hurley JH. A HORMA domain in Atg13 mediates PI 3-kinase recruitment in autophagy. Proc Natl Acad Sci U S A 2013;110:5486–5491.

[57] Russell RC, Tian Y, Yuan H, Park HW, Chang YY, et al. ULK1 induces autophagy by phosphorylating Beclin-1 and activating VPS34 lipid kinase. Nat Cell Biol 2013;15:741–750.

[58] Ao X, Zou L, Wu Y. Regulation of autophagy by the Rab GTPase network. Cell Death Differ 2014;21:348–358.

[59] Young ARJ, Chan EYW, Hu XW, Köchl R, Crawshaw SG, et al. Starvation and ULK1-dependent cycling of mammalian Atg9 between the TGN and endosomes. J Cell Sci 2006;119:3888–3900.

[60] Orsi A, Razi M, Dooley HC, Robinson D, Weston AE, et al. Dynamic and transient interactions of Atg9 with autophagosomes, but not membrane integration, are required for autophagy. Mol Biol Cell 2012;23:1860–1873.

[61] Mari M, Griffith J, Rieter E, Krishnappa L, Klionsky DJ, et al. An Atg9-containing compartment that functions in the early steps of autophagosome biogenesis. J Cell Biol 2010;190:1005–1022.

[62] Yamamoto H, Kakuta S, Watanabe TM, Kitamura A, Sekito T, et al. Atg9 vesicles are an important membrane source during early steps of autophagosome formation. J Cell Biol 2012;198:219–233.

[63] Shpilka T, Mizushima N, Elazar Z. Ubiquitin-like proteins and autophagy at a glance. J Cell Sci 2012;125:2343–2348.

[64] Kaufmann A, Beier V, Franquelim HG, Wollert T. Molecular mechanism of autophagic membrane-scaffold assembly and disassembly. Cell 2014;156:469–481.

[65] Weidberg H, Shpilka T, Shvets E, Abada A, Shimron F, et al. LC3 and GATE-16 N termini mediate membrane fusion processes required for autophagosome biogenesis. Dev Cell 2011;20:444–454.

[66] Nakatogawa H, Ichimura Y, Ohsumi Y. Atg8, a ubiquitin-like protein required for autophagosome formation, mediates membrane tethering and hemifusion. Cell 2007; 130:165–178.

[67] Weidberg H, Shvets E, Shpilka T, Shimron F, Shinder V, et al. LC3 and GATE-16/GABARAP subfamilies are both essential yet act differently in autophagosome biogenesis. EMBO J 2010;29:1792–1802.

[68] Manil-Segalen M, Lefebvre C, Jenzer C, Trichet M, Boulogne C, et al. The *C. elegans* LC3 acts downstream of GABARAP to degrade autophagosomes by interacting with the HOPS subunit VPS39. Dev Cell 2014;28:43–55.

[69] Monastyrska I, Rieter E, Klionsky DJ, Reggiori F. Multiple roles of the cytoskeleton in autophagy. Biol Rev Camb Philos Soc 2009;84:431–448.

[70] Reggiori F, Komatsu M, Finley K, Simonsen A. Autophagy: more than a nonselective pathway. Int J Cell Biol 2012;2012:219625.

[71] Kraft C, Peter M, Hofmann K. Selective autophagy: ubiquitin-mediated recognition and beyond. Nat Cell Biol 2010;12:836–841.

[72] Settembre C, Fraldi A, Medina DL, Ballabio A. Signals from the lysosome: a control centre for cellular clearance and energy metabolism. Nat Rev Mol Cell Biol 2013;14:283–296.

[73] Ravikumar B, Moreau K, Jahreiss L, Puri C, Rubinsztein DC. Plasma membrane contributes to the formation of pre-autophagosomal structures. Nat Cell Biol 2010;12:747–757.

[74] Longatti A, Lamb CA, Razi M, Yoshimura S, Barr FA, et al. TBC1D14 regulates autophagosome formation via Rab11- and ULK1-positive recycling endosomes. J Cell Biol 2012;197:659–675.

[75] Knaevelsrud H, Soreng K, Raiborg C, Haberg K, Rasmuson F, et al. Membrane remodeling by the PX-BAR protein SNX18 promotes autophagosome formation. J Cell Biol 2013;202:331–349.

[76] Szatmari Z, Kis V, Lippai M, Hegedus K, Farago T, et al. Rab11 facilitates cross-talk between autophagy and endosomal pathway through regulation of Hook localization. Mol Biol Cell 2014;25:522–531.

[77] Filimonenko M, Stuffers S, Raiborg C, Yamamoto A, Malerod L, et al. Functional multivesicular bodies are required for autophagic clearance of protein aggregates associated with neurodegenerative disease. J Cell Biol 2007;179:485–500.

[78] Pous C, Codogno P. Lysosome positioning coordinates mTORC1 activity and autophagy. Nat Cell Biol 2011;13:342–344.

[79] Korolchuk VI, Saiki S, Lichtenberg M, Siddiqi FH, Roberts EA, et al. Lysosomal positioning coordinates cellular nutrient responses. Nat Cell Biol 2011;13:453–460.

[80] Heuser J. Changes in lysosome shape and distribution correlated with changes in cytoplasmic pH. J Cell Biol 1989;108:855–864.

[81] Kirisako T, Baba M, Ishihara N, Miyazawa K, Ohsumi M, et al. Formation process of autophagosome is traced with Apg8/Aut7p in yeast. J Cell Biol 1999;147:435–446.

[82] Jager S, Bucci C, Tanida I, Ueno T, Kominami E, et al. Role for Rab7 in maturation of late autophagic vacuoles. J Cell Sci 2004;117:4837–4848.

[83] Gutierrez MG, Munafo DB, Beron W, Colombo MI. Rab7 is required for the normal progression of the autophagic pathway in mammalian cells. J Cell Sci 2004; 117:2687–2697.

[84] Takats S, Pircs K, Nagy P, Varga A, Karpati M, et al. Interaction of the HOPS complex with Syntaxin 17 mediates autophagosome clearance in *Drosophila*. Mol Biol Cell 2014;25:1338–1354.

[85] Rieder SE, Emr SD. A novel RING finger protein complex essential for a late step in protein transport to the yeast vacuole. Mol Biol Cell 1997;8:2307–2327.

[86] Jiang P, Nishimura T, Sakamaki Y, Itakura E, Hatta T, et al. The HOPS complex mediates autophagosome–lysosome fusion through interaction with syntaxin 17. Mol Biol Cell 2014;25:1327–1337.

[87] Wang C-W, Stromhaug PE, Shima J, Klionsky DJ. The Ccz1–Mon1 protein complex is required for the late step of multiple vacuole delivery pathways. J Biol Chem 2002;277:47917–47927.

[88] Takats S, Nagy P, Varga A, Pircs K, Karpati M, et al. Autophagosomal syntaxin 17-dependent lysosomal degradation maintains neuronal function in *Drosophila*. J Cell Biol 2013;201:531–539.

[89] Itakura E, Kishi-Itakura C, Mizushima N. The hairpin-type tail-anchored SNARE syntaxin 17 targets to autophagosomes for fusion with endosomes/lysosomes. Cell 2012;151:1256–1269.

[90] Steegmaier M, Oorschot V, Klumperman J, Scheller RH. Syntaxin 17 is abundant in steroidogenic cells and implicated in smooth endoplasmic reticulum membrane dynamics. Mol Biol Cell 2000;11:2719–2731.

[91] Hegedus K, Takats S, Kovacs AL, Juhasz G. Evolutionarily conserved role and physiological relevance of a STX17/Syx17 (syntaxin 17)-containing SNARE complex in autophagosome fusion with endosomes and lysosomes. Autophagy 2013;9:1642–1646.

[92] Saftig P, Beertsen W, Eskelinen EL. LAMP-2: a control step for phagosome and autophagosome maturation. Autophagy 2008;4:510–512.

[93] Fortunato F, Burgers H, Bergmann F, Rieger P, Buchler MW, et al. Impaired autolysosome formation correlates with Lamp-2 depletion: role of apoptosis, autophagy, and necrosis in pancreatitis. Gastroenterology 2009;137:350–360, e351–e355.

[94] Yu L, McPhee CK, Zheng L, Mardones GA, Rong Y, et al. Termination of autophagy and reformation of lysosomes regulated by mTOR. Nature 2010;465:942–946.

[95] Rong Y, McPhee CK, Deng S, Huang L, Chen L, et al. Spinster is required for autophagic lysosome reformation and mTOR reactivation following starvation. Proc Natl Acad Sci U S A 2011;108:7826–7831.

[96] Yu L, Strandberg L, Lenardo MJ. The selectivity of autophagy and its role in cell death and survival. Autophagy 2008;4:567–573.

[97] Rong Y, Liu M, Ma L, Du W, Zhang H, et al. Clathrin and phosphatidylinositol-4,5-bisphosphate regulate autophagic lysosome reformation. Nat Cell Biol 2012;14: 924–934.

[98] Schulze RJ, Weller SG, Schroeder B, Krueger EW, Chi S, et al. Lipid droplet breakdown requires dynamin 2 for vesiculation of autolysosomal tubules in hepatocytes. J Cell Biol 2013;203:315–326.

[99] Sridhar S, Patel B, Aphkhazava D, Macian F, Santambrogio L, et al. The lipid kinase PI4KIIIbeta preserves lysosomal identity. EMBO J 2013;32:324–339.

[100] Mullock BM, Bright NA, Fearon CW, Gray SR, Luzio JP. Fusion of lysosomes with late endosomes produces a hybrid organelle of intermediate density and is NSF dependent. J Cell Biol 1998;140:591–601.

[101] Bright NA, Reaves BJ, Mullock BM, Luzio JP. Dense core lysosomes can fuse with late endosomes and are re-formed from the resultant hybrid organelles. J Cell Sci 1997;110:2027–2040.

[102] Lieberman AP, Puertollano R, Raben N, Slaugenhaupt S, Walkley SU, et al. Autophagy in lysosomal storage disorders. Autophagy 2012;8:719–730.

[103] Pastores GM, Maegawa GH. Clinical neurogenetics: neuropathic lysosomal storage disorders. Neurol Clin 2013;31:1051–1071.

[104] Komatsu M, Kominami E, Tanaka K. Autophagy and neurodegeneration. Autophagy 2006;2:315–317.

[105] Tan CC, Yu JT, Tan MS, Jiang T, Zhu XC, et al. Autophagy in aging and neurodegenerative diseases: implications for pathogenesis and therapy. Neurobiol Aging 2014;35:941–957.

[106] Manzoni C, Lewis PA. Dysfunction of the autophagy/lysosomal degradation pathway is a shared feature of the genetic synucleinopathies. FASEB J 2013;27:3424–3429.

[107] Miners JS, Barua N, Kehoe PG, Gill S, Love S. Abeta-degrading enzymes: potential for treatment of Alzheimer disease. J Neuropathol Exp Neurol 2011;70:944–959.

[108] Torres M, Jimenez S, Sanchez-Varo R, Navarro V, Trujillo-Estrada L, et al. Defective lysosomal proteolysis and axonal transport are early pathogenic events that worsen with age leading to increased APP metabolism and synaptic Abeta in transgenic APP/PS1 hippocampus. Mol Neurodegener 2012;7:59.

[109] Nixon RA. Endosome function and dysfunction in Alzheimer's disease and other neurodegenerative diseases. Neurobiol Aging 2005;26:373–382.

[110] Nixon RA, Wegiel J, Kumar A, Yu WH, Peterhoff C, et al. Extensive involvement of autophagy in Alzheimer disease: an immuno-electron microscopy study. J Neuropathol Exp Neurol 2005;64:113–122.

[111] Nixon RA, Yang DS. Autophagy failure in Alzheimer's disease – locating the primary defect. Neurobiol Dis 2011;43:38–45.

[112] Lee JH, Yu WH, Kumar A, Lee S, Mohan PS, et al. Lysosomal proteolysis and autophagy require presenilin 1 and are disrupted by Alzheimer-related PS1 mutations. Cell 2010;141:1146–1158.

[113] Ihara Y, Morishima-Kawashima M, Nixon R. The ubiquitin-proteasome system and the autophagic-lysosomal system in Alzheimer disease. Cold Spring Harb Perspect Med 2012;2:a006361.

[114] Avrahami L, Eldar-Finkelman H. GSK-3 and lysosomes meet in Alzheimer's disease. Commun Integr Biol 2013;6:e25179.

[115] Ji ZS, Mullendorff K, Cheng IH, Miranda RD, Huang Y, et al. Reactivity of apolipoprotein E4 and amyloid beta peptide: lysosomal stability and neurodegeneration. J Biol Chem 2006;281:2683–2692.

[116] Marcuzzi A, Bianco AM, Girardelli M, Tommasini A, Martelossi S, et al. Genetic and functional profiling of Crohn's disease: autophagy mechanism and susceptibility to infectious diseases. Biomed Res Int 2013;2013:297501.

[117] Lassen KG, Kuballa P, Conway KL, Patel KK, Becker CE, et al. Atg16L1 T300A variant decreases selective autophagy resulting in altered cytokine signaling and decreased antibacterial defense. Proc Natl Acad Sci U S A 2014;111:7741–7746.

[118] Cooney R, Baker J, Brain O, Danis B, Pichulik T, et al. NOD2 stimulation induces autophagy in dendritic cells influencing bacterial handling and antigen presentation. Nat Med 2010;16:90–97.

[119] Travassos LH, Carneiro LA, Ramjeet M, Hussey S, Kim YG, et al. Nod1 and Nod2 direct autophagy by recruiting ATG16L1 to the plasma membrane at the site of bacterial entry. Nat Immunol 2010;11:55–62.

[120] Singh SB, Davis AS, Taylor GA, Deretic V. Human IRGM induces autophagy to eliminate intracellular mycobacteria. Science 2006;313:1438–1441.

[121] Brest P, Lapaquette P, Souidi M, Lebrigand K, Cesaro A, et al. A synonymous variant in IRGM alters a binding site for miR-196 and causes deregulation of IRGM-dependent xenophagy in Crohn's disease. Nat Genet 2011;43:242–245.

[122] Barrett JC, Hansoul S, Nicolae DL, Cho JH, Duerr RH, et al. Genome-wide association defines more than 30 distinct susceptibility loci for Crohn's disease. Nat Genet 2008;40:955–962.

[123] Henderson P, van Limbergen JE, Wilson DC, Satsangi J, Russell RK. Genetics of childhood-onset inflammatory bowel disease. Inflamm Bowel Dis 2011;17:346–361.

[124] Henckaerts L, Cleynen I, Brinar M, John JM, Van Steen K, et al. Genetic variation in the autophagy gene ULK1 and risk of Crohn's disease. Inflamm Bowel Dis 2011;17:1392–1397.

[125] Cullup T, Kho AL, Dionisi-Vici C, Brandmeier B, Smith F, et al. Recessive mutations in EPG5 cause Vici syndrome, a multisystem disorder with defective autophagy. Nat Genet 2013;45:83–87.

[126] Tian Y, Li Z, Hu W, Ren H, Tian E, et al. *C. elegans* screen identifies autophagy genes specific to multicellular organisms. Cell 2010;141:1042–1055.

[127] Zhao H, Zhao YG, Wang X, Xu L, Miao L, et al. Mice deficient in Epg5 exhibit selective neuronal vulnerability to degeneration. J Cell Biol 2013;200:731–741.

[128] Li W, Zou W, Yang Y, Chai Y, Chen B, et al. Autophagy genes function sequentially to promote apoptotic cell corpse degradation in the engulfing cell. J Cell Biol 2012;197:27–35.

[129] Saitsu H, Nishimura T, Muramatsu K, Kodera H, Kumada S, et al. De novo mutations in the autophagy gene WDR45 cause static encephalopathy of childhood with neurodegeneration in adulthood. Nat Genet 2013;45:445–449, 449e441.

[130] Haack TB, Hogarth P, Kruer MC, Gregory A, Wieland T, et al. Exome sequencing reveals de novo WDR45 mutations causing a phenotypically distinct, X-linked dominant form of NBIA. Am J Hum Genet 2012;91:1144–1149.

[131] Hayflick SJ, Kruer MC, Gregory A, Haack TB, Kurian MA, et al. β-Propeller protein-associated neurodegeneration: a new X-linked dominant disorder with brain iron accumulation. Brain 2013;136:1708–1717.

[132] Verhoeven WM, Egger JI, Koolen DA, Yntema H, Olgiati S, et al. Beta-propeller protein-associated neurodegeneration (BPAN), a rare form of NBIA: novel mutations and neuropsychiatric phenotype in three adult patients. Parkinsonism Relat Disord 2013;20:332–336.

[133] Nagy P, Hegedus K, Pircs K, Varga A, Juhasz G. Different effects of Atg2 and Atg18 mutations on Atg8a and Atg9 trafficking during starvation in *Drosophila*. FEBS Lett 2014;588:408–413.

[134] Behrends C, Sowa ME, Gygi SP, Harper JW. Network organization of the human autophagy system. Nature 2010;466:68–76.

[135] Velikkakath AK, Nishimura T, Oita E, Ishihara N, Mizushima N. Mammalian Atg2 proteins are essential for autophagosome formation and important for regulation of size and distribution of lipid droplets. Mol Biol Cell 2012;23:896–909.

[136] Lu Q, Yang P, Huang X, Hu W, Guo B, et al. The WD40 repeat PtdIns(3)P-binding protein EPG-6 regulates progression of omegasomes to autophagosomes. Dev Cell 2011;21:343–357.

[137] Michell RH, Dove SK. A protein complex that regulates PtdIns(3,5)P2 levels. EMBO J 2009;28:86–87.

[138] Krick R, Henke S, Tolstrup J, Thumm M. Dissecting the localization and function of Atg18, Atg21 and Ygr223c. Autophagy 2008;4:896–910.

[139] Efe JA, Botelho RJ, Emr SD. Atg18 regulates organelle morphology and Fab1 kinase activity independent of its membrane recruitment by phosphatidylinositol 3,5-bisphosphate. Mol Biol Cell 2007;18:4232–4244.

[140] Oz-Levi D, Ben-Zeev B, Ruzzo EK, Hitomi Y, Gelman A, et al. Mutation in TECPR2 reveals a role for autophagy in hereditary spastic paraparesis. Am J Hum Genet 2012;91:1065–1072.

[141] Murmu RP, Martin E, Rastetter A, Esteves T, Muriel MP, et al. Cellular distribution and subcellular localization of spatacsin and spastizin, two proteins involved in hereditary spastic paraplegia. Mol Cell Neurosci 2011;47:191–202.

[142] Sagona AP, Nezis IP, Pedersen NM, Liestol K, Poulton J, et al. PtdIns(3)P controls cytokinesis through KIF13A-mediated recruitment of FYVE-CENT to the midbody. Nat Cell Biol 2010;12:362–371.

[143] Hanein S, Martin E, Boukhris A, Byrne P, Goizet C, et al. Identification of the *SPG15* gene, encoding spastizin, as a frequent cause of complicated autosomal-recessive spastic paraplegia, including Kjellin syndrome. Am J Hum Genet 2008;82:992–1002.

[144] Vantaggiato C, Crimella C, Airoldi G, Polishchuk R, Bonato S, et al. Defective autophagy in spastizin mutated patients with hereditary spastic paraparesis type 15. Brain 2013;136:3119–3139.

[145] Matsunaga K, Saitoh T, Tabata K, Omori H, Satoh T, et al. Two Beclin 1-binding proteins, Atg14L and Rubicon, reciprocally regulate autophagy at different stages. Nat Cell Biol 2009;11:385–396.

[146] Choi AM, Ryter SW, Levine B. Autophagy in human health and disease. N Engl J Med 2013;368:651–662.

[147] Liang XH, Jackson S, Seaman M, Brown K, Kempkes B, et al. Induction of autophagy and inhibition of tumorigenesis by beclin 1. Nature 1999;402:672–676.

[148] Aita VM, Liang XH, Murty VV, Pincus DL, Yu W, et al. Cloning and genomic organization of beclin 1, a candidate tumor suppressor gene on chromosome 17q21. Genomics 1999;59:59–65.

[149] Koukourakis MI, Giatromanolaki A, Sivridis E, Pitiakoudis M, Gatter KC, et al. Beclin 1 over- and underexpression in colorectal cancer: distinct patterns relate to prognosis and tumour hypoxia. Br J Cancer 2010;103:1209–1214.

[150] Giatromanolaki A, Koukourakis MI, Koutsopoulos A, Chloropoulou P, Liberis V, et al. High Beclin 1 expression defines a poor prognosis in endometrial adenocarcinomas. Gynecol Oncol 2011;123:147–151.

[151] Lebovitz CB, Bortnik SB, Gorski SM. Here, there be dragons: charting autophagy-related alterations in human tumors. Clin Cancer Res 2012;18:1214–1226.

[152] Platt FM, Boland B, van der Spoel AC. The cell biology of disease: lysosomal storage disorders: the cellular impact of lysosomal dysfunction. J Cell Biol 2012;199:723–734.

[153] Shen HM, Mizushima N. At the end of the autophagic road: an emerging understanding of lysosomal functions in autophagy. Trends Biochem Sci 2014;39:61–71.

[154] Michelet X, Legouis R. Autophagy in endosomal mutants: desperately seeking to survive. Worm 2012;1:216–220.

3

MULTIVESICULAR BODIES: ROLES IN INTRACELLULAR AND INTERCELLULAR SIGNALING

EMILY R. EDEN, THOMAS BURGOYNE, AND CLARE E. FUTTER
Department of Cell Biology, UCL Institute of Ophthalmology, University College London, London, UK

3.1 INTRODUCTION

Multivesicular endosomes/bodies (MVBs) are endosomes that contain small intraluminal vesicles (ILVs) within their lumen. They were first identified in studies following the trafficking of internalized epidermal growth factor receptor (EGFR), where EGF-stimulated EGFR was found to accumulate on the ILVs of MVBs (see Figure 3.1) before delivery to the lysosome for degradation [1,2]. Proteins, such as transferrin receptor, destined for recycling to the plasma membrane, remain on the limiting membrane of the MVB before return to the cell surface [3]. MVBs were shown to undergo a gradual maturation process involving accumulation of ILVs, sorting of lysosomally directed proteins onto the ILVs and the gradual removal of recycling proteins via tubular extensions [4–6]. When all the recycling proteins have been removed, the MVB fuses directly with the lysosome and the contents are degraded [4,7]. At the time of these studies, it was already clear that MVBs would have a role in downregulating signaling through delivering activated receptor tyrosine kinases (RTKs) for degradation. The demonstration that RTKs remain active after endocytosis [8–10], suggesting the possibility of a more acute role for sorting onto ILVs of MVBs in regulating RTK signaling because sequestration on

Lysosomes: Biology, Diseases, and Therapeutics, First Edition.
Edited by Frederick R. Maxfield, James M. Willard and Shuyan Lu.

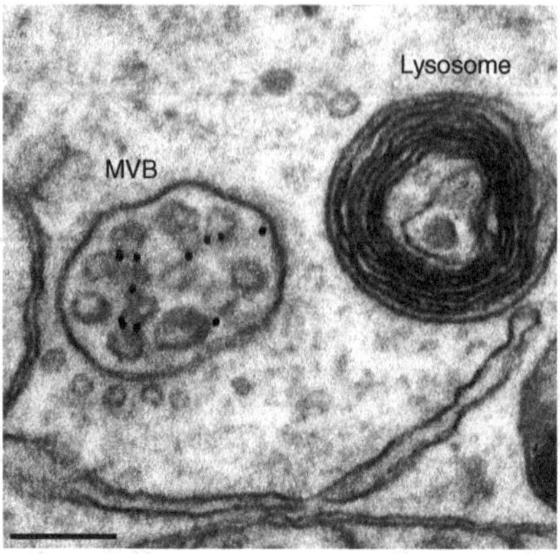

Figure 3.1 Electron micrograph of an MVB. Hela cells were stimulated with EGF in the presence of anti-EGFR 10 nm gold conjugate and prepared for transmission electron microscopy. The image shows an MVB with EGFR (gold particles) localized on the ILVs. The MVB-containing discrete ILVs are readily distinguished from the electron-dense lysosome that contains characteristic multilamellar membranous whorls. Scale bar = 200 nm.

ILVs removes the catalytic domain of the receptor from the cytosol and therefore limits its ability to interact with downstream signaling molecules.

During the last two decades, two major sets of discoveries have transformed the interest and activity in the field of MVB biology. The first is the identification of molecular mechanisms regulating ILV cargo sorting and ILV formation and the second is the discovery that fusion with the lysosome is not the only possible fate of an MVB. Identification of molecular components of the MVB sorting machinery was an important prerequisite for establishing the role of that sorting in regulating signaling. However, the machinery was initially elusive, partly because ILV formation within MVBs involves budding away from the cytosol and so has the reverse topology and is differently regulated compared with other better characterized budding events in the cell that are mediated by coat proteins such as clathrin. The best established machinery regulating both cargo sorting onto ILVs and ILV formation is the endosomal sorting complex required for transport (ESCRT) machinery, but there are multiple populations of MVBs and ILVs within them [11–13] and ESCRT-independent mechanisms of ILV formation also exist. Studies analyzing the effects of interfering with components of the ILV sorting machinery on signaling have revealed that sorting onto ILVs can have both negative and positive effects on signaling. As expected, sorting onto ILVs generally has negative effects on RTK signaling, presumably because the kinase domain and signaling proteins associated with it are sequestered from the cytoplasm. Less expected is the demonstration that inhibitors of signaling pathways can also be sequestered onto ILVs of MVBs, thereby enhancing certain signaling pathways.

A second major discovery is that fusion with the lysosome is not the only potential fate of MVBs. In specialized cell types, they are on the biosynthetic pathway of lysosome-related organelles, such as melanosomes, and in many cell types they can also fuse with the cell surface and release the ILVs, now termed exosomes, into the extracellular space. The role of exosomes in mediating intercellular communication between cells of the immune system has been well established but more recently attention has focused on the role of exosomes in mediating intercellular signaling. Release of the contents of MVBs into the extracellular space opens up the possibility of sorting onto ILVs providing a means of signaling to neighboring or even distant cells. The relationship between ILVs destined for the lysosome and those destined for release from the cell and how they are segregated is not clear.

Many studies have analyzed the effects of inhibiting endocytosis on signal transduction and revealed that, although many signaling pathways can be activated from the plasma membrane, some require endocytosis for maximal activation. In this review, we do not examine the role of endocytosis *per se* but, rather, focus on the role of MVBs. We review the ways in which sorting onto ILVs of MVBs can downregulate intracellular signaling, taking the canonical RTK, EGF receptor, as an example. We then examine the evidence for a role of sequestration of inhibitory factors on ILVs in the upregulation of signaling, focusing on the role of sequestration of glycogen synthase kinase (GSK3) in upregulating Wnt signaling. Finally, we review how sorting onto ILVs and the subsequent release of those ILVs as exosomes can promote intercellular signaling, taking long-range signaling of Notch ligands as an example.

3.2 DOWNREGULATION OF SIGNALING BY SORTING ONTO ILVs

Ligand-stimulated EGFR tyrosine kinase signaling can promote cell survival, proliferation, differentiation, or motility, depending upon the cellular context. Overexpressed EGFR or EGFR carrying activating mutations are a feature of many human cancers and so the EGFR is a target for cancer therapeutics. Ligand binding to the EGFR kinase promotes receptor dimerization, activation of the RTK activity and trans-autophosphorylation of the receptor, and generates sites for the recruitment of phosphotyrosine-binding domain (PTB)-containing components of signaling pathways. Ligand-stimulated EGFRs are endocytosed via a number of different mechanisms and a proportion of endocytosed EGFR is delivered to the lysosome via sorting onto the ILVs of MVBs.

EGF receptors are sorted onto ILVs via the ESCRT machinery, which has recently been extensively reviewed elsewhere [14–16]. It consists of a series of protein complexes, ESCRT0-III, and accessory proteins, components of which bind ubiquitinated cargo and sort them into domains on the perimeter membrane of the MVB. ESCRTs also promote membrane invagination and ultimately the budding of the ILV, which is accompanied by vacuolar protein sorting-associated protein 4 (Vps4)-mediated release of ESCRT components from the MVB-limiting membrane. Depletion of various ESCRT components has been shown to result in sustained

RTK signaling in cultured cells [17–19]. Furthermore, enhanced EGFR signaling and MAP kinase signaling has been reported in ESCRT-mutant cells in *Drosophila* tissues [20,21]. One complication of studies interfering with the ESCRT machinery is that ESCRT-mutant cells can activate a powerful apoptosis program [22,23]. However, when proapoptotic signaling is prevented, a major overproliferation is observed in ESCRT-mutant cells [24], supporting the notion that sorting onto the ILVs downregulates signaling.

As suggested earlier, a complication in the interpretation of this type of study is that ESCRT components have multiple functions in addition to sorting ubiquitinated cargo onto ILVs. These include the promotion of recycling, which has been demonstrated for certain G protein-coupled receptors [25], abscission during cytokinesis [26,27] and they also have a role in handling of microRNAs [24,28]. An alternative approach that allows the role of the ESCRT machinery in regulating sorting of EGFR onto ILVs to be distinguished from other ESCRT roles is to analyze ligand-stimulated signaling by an EGFR that is deficient in ubiquitination and so cannot engage the ESCRT machinery. This receptor is efficiently endocytosed [29], but rather than being sorted onto ILVs the receptor is recycled and signaling via the ERK MAP kinase pathway is prolonged [30].

Does sorting onto ILVs play a major/direct role in regulating signaling from the EGFR by removal of the catalytic domain of the EGFR from the cytosol or is the main role of sequestration on ILVs in signal modulation the prevention of recycling? The demonstration that some EGFR signaling pathways, such as the ERK MAP kinase pathway, are unaffected by inhibition of endocytosis [31] suggests that for these pathways inhibition of EGFR return to the plasma membrane may be the main role of the ESCRT machinery in their regulation. For other pathways, such as the B/AKT pathway, where endocytosis is required for their sustained activation [31], implying that they signal from the perimeter membrane of endosomes, sequestration on ILVs may directly regulate their potency.

Endocytosis has been shown to regulate the ability of EGFR to interact with tyrosine phosphatases that downregulate EGFR signaling by removing sites for recruitment of PTB domain-containing proteins [32]. EGFR on the perimeter membrane of MVBs comes into contact with a protein tyrosine phosphatase, PTP1B, on the cytoplasmic face of the ER via membrane contacts that form between the MVB perimeter membrane and the ER [33]. This interaction must occur before sorting of EGFR onto ILVs and so it is possible that the EGFR is already dephosphorylated before ILV targeting. PTP1B both dephosphorylates the EGFR and promotes its sorting onto ILVs, possibly through dephosphorylation of components of the ESCRT machinery [33,34]. Thus, a combination of dephosphorylation, followed by rapid sequestration onto ILVs to prevent rephosphorylation, could serve to dampen signaling at the level of the MVB (illustrated in Figure 3.2).

It is important to note that signaling proteins downstream of the EGFR kinase may be sequestered on ILVs along with the EGFR. Dephosphorylation of the EGFR before sorting onto ILVs may trigger dissociation of PTB domain-containing components of signaling pathways that associate with autophosphorylated EGFR. Alternatively sequestration of components of signaling pathways along with the EGFR

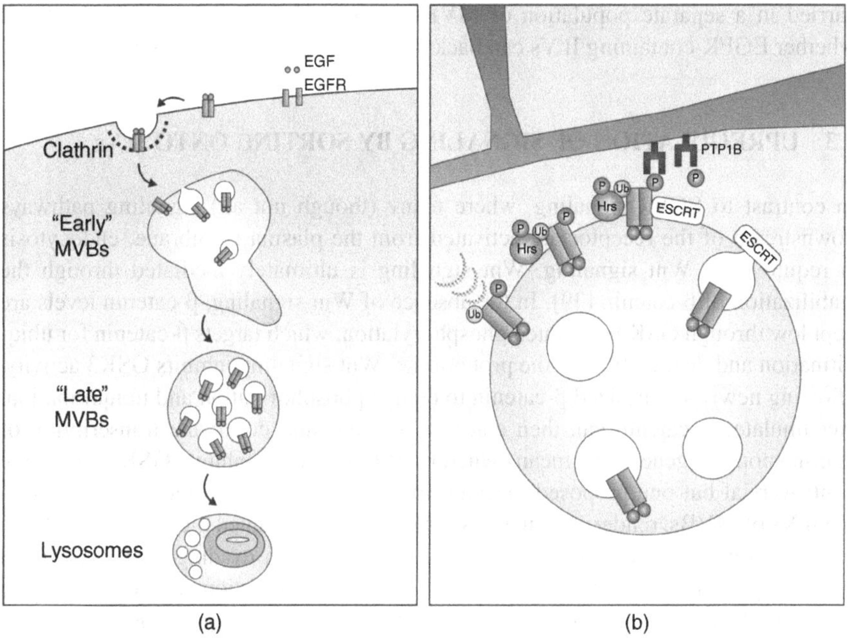

Figure 3.2 Downregulation of EGFR signaling by sorting onto ILVs. (a) EGFRs at the cell surface dimerize on ligand binding and are internalized into early endosomes where sorting of ligand–receptor complexes destined for lysosomal degradation onto ILVs begins ("early" MVBs). EGFR-containing ILVs accumulate in the maturing MVB and when all recycling proteins have been removed the "late" MVB fuses with the lysosome and the contents are degraded. (b) Signaling from the EGFR tyrosine kinase on the limiting membrane of the MVB is dampened by a combination of dephosphorylation and sequestration away from the cytosol on ILVs. Ubiquitination-dependent interaction with Hrs, and subsequent recruitment of the ESCRT machinery, concentrates EGFR on the MVB-limiting membrane where both the EGFR and ESCRTs can potentially interact with ER-localized PTP1B via direct membrane contacts between the ER and MVBs. ESCRTs then promote sequestration of EGFR on ILVs of MVBS, which ultimately fuse with the lysosome, ensuring signal termination. (*See color plate section for the color representation of this figure.*)

could contribute to downregulation of signaling. In addition, signals propagated from the EGFR may continue after the EGFR itself has been degraded or sequestered on ILVs. An ERK MAP kinase-containing signaling complex forms on the perimeter membrane of late endosomes [35], but the importance of ILV sorting in regulation of signaling from this complex is not clear.

Finally, the importance of sequestration of EGFR and downstream signaling proteins away from the cytosol in directly regulating signaling begs the question of whether or not this is an irreversible process. Back fusion of ILVs with the perimeter membrane has been demonstrated for ILVs bearing Semliki Forest virus [36] and anthrax toxin [37] in a mechanism involving Alix and the lipid, lysobisphosphatidic acid (LBPA) [38]. However, ligand-stimulated EGFR are carried in a separate population of MVBs to those that label for LBPA [11] and whether EGFR-containing ILVs can back fuse remains to be established.

3.3 UPREGULATION OF SIGNALING BY SORTING ONTO ILVs

In contrast to EGFR signaling, where many (though not all) signaling pathways downstream of the receptor are activated from the plasma membrane, endocytosis is required for Wnt signaling. Wnt signaling is ultimately mediated through the stabilization of β-catenin [39]. In the absence of Wnt signaling, β-catenin levels are kept low through GSK3-mediated phosphorylation, which targets β-catenin for ubiquitination and degradation by the proteasome. Wnt signaling inhibits GSK3 activity, allowing newly synthesized β-catenin to escape phosphorylation and ubiquitination. Accumulated β-catenin can then enter the nucleus and coactivate transcription of Wnt-responsive genes. The means whereby Wnt signaling inhibits GSK activity are controversial but one proposed mechanism involves the sequestration of GSK3 on the ILVs of MVBs, rendering it inaccessible to newly translated β-catenin [40]. Thus, sequestration of GSK3 on the ILVs results in β-catenin evading GSK3-mediated phosphorylation and resulting ubiquitination and degradation and instead being transported to the nucleus where it activates transcription of target genes.

When Wnt ligands bind to their Frizzled (Fz) receptors and low-density lipoprotein-related protein (LRP) coreceptors on the plasma membrane, disheveled (Dvl) proteins are recruited to the complex. This results in the recruitment of Axin, which, in unstimulated cells, is in an inhibitory complex with adenomatous polyposis coli (APC), casein kinase 1 (CK1) and GSK3 that targets β-catenin for degradation by facilitating the GSK3-mediated phosphorylation of β-catenin and resulting recruitment of the E3 ubiquitin ligase. Dvl polymers activate the phosphorylation of LRP5/6 by CK1, which creates docking sites for Axin, enhancing its recruitment, together with other components of the signalosome. A complex series of homo- and hetero-oligomerization events and phosphorylations by the two kinases, CK1 and GSK3, lead to the formation of a large protein complex, the Wnt signalosome. One proposed mechanism of GSK3 inhibition within this protein complex is the generation of a pseudosubstrate for GSK3 in the phosphorylated cytoplasmic domain of LRP5/6 that prevents GSK3 phosphorylating other proteins [41,42]. However, several groups have shown that the Wnt signaling complex is endocytosed. The mechanism of internalization is unclear since dependence on clathrin [43] and caveolin [44] have both been reported. A recent study indicated that internalization of the Wnt signaling complex was dependent upon dissociation of a p120-catenin/cadherin complex [45]. In all these studies, interfering with endocytosis of the Wnt signalosome, albeit in different ways, inhibited Wnt signaling.

Focusing on the role of MVBs in Wnt signaling, the finding that a proportion of GSK3 and other components of the Wnt signaling complex becomes protease resistant in digitonin-permeabilized but not Triton-permeabilized cells suggested that GSK3 became sequestered from the cytosol and immunoEM confirmed that the endocytosed protein became sequestered on the ILVs of MVBs [40]. This sequestration was inhibited by depletion of the ESCRT0 component, Hrs, or expression of dominant negative Vps4, implicating the ESCRT machinery in the sorting of GSK3 onto ILVs [40]. Importantly, Hrs depletion/Vps4 mutant expression inhibited Wnt-stimulated β-catenin stabilization, suggesting that one mechanism of GSK3 inhibition is sequestration of GSK3 away from cytosolic substrates, allowing newly synthesized β-catenin to escape phosphorylation, accumulate in the cytosol and subsequently be transported to the nucleus (illustrated in Figure 3.3). Support for the idea of sequestration of GSK3 in MVBs as a mechanism of Wnt signaling came from the recent demonstration that dissociation of the p120-catenin/cadherin complex was necessary for sequestration of GSK3 within MVBs and Wnt-stimulated β-catenin stabilization [45].

The above scheme (Figure 3.3) suggests that sequestration of GSK3 onto ILVs must be very efficient and if it is sequestered onto the same ILVs as activated EGFR, the normal destination of those MVBs would be fusion with the lysosome and degradation. However, total cellular levels of GSK3 did not change on Wnt signaling [40], suggesting either that it is only a small proportion of the GSK3 that is active and therefore sequestered onto ILVs and subsequently degraded or that the GSK3-containing ILVs can back fuse with the limiting membrane and thus escape lysosomal degradation. Alternatively, GSK3 could be sequestered in a population of MVBs distinct from those that are delivered to the lysosome and degraded.

Many questions remain about the molecular regulation of the sorting of the Wnt signaling complex on to the ILVs of MVBs. GSK3 is sorted onto ILVs in a complex with LRPs, Fx, Dvl, Axin, and CK1 in a manner that appears dependent on components of the ESCRT machinery. However, it is not known whether any components of the Wnt complex become ubiquitinated, providing a potential means of engaging the ESCRT machinery. Furthermore, the relationship between the MVBs that sequester the Wnt signaling complex and those that sequester-activated RTKs such as EGFR is not clear.

Finally, the demonstration that an inhibitory protein, such as GSK, can be sequestered on ILVs to activate signaling raises questions about the interpretation of studies where the functions of ESCRT machinery in signal regulation have been investigated. Depletion of components of the ESCRT machinery may have effects on the availability of a range of signaling regulators in addition to the one that is the target of the study.

3.4 INTERCELLULAR SIGNALING DEPENDENT ON SORTING ONTO ILVs

The first demonstration that release of ILVs into the extracellular space as exosomes was not simply a means of getting rid of unwanted material from the cell came from studies showing that exosomes from B lymphocytes could present antigen [46]. Since then the role of exosomes in the immune system has been extensively studied and more recently a plethora of roles have been attributed to exosomes, including the intercellular traffic of cell fate determining signaling molecules. These molecules are

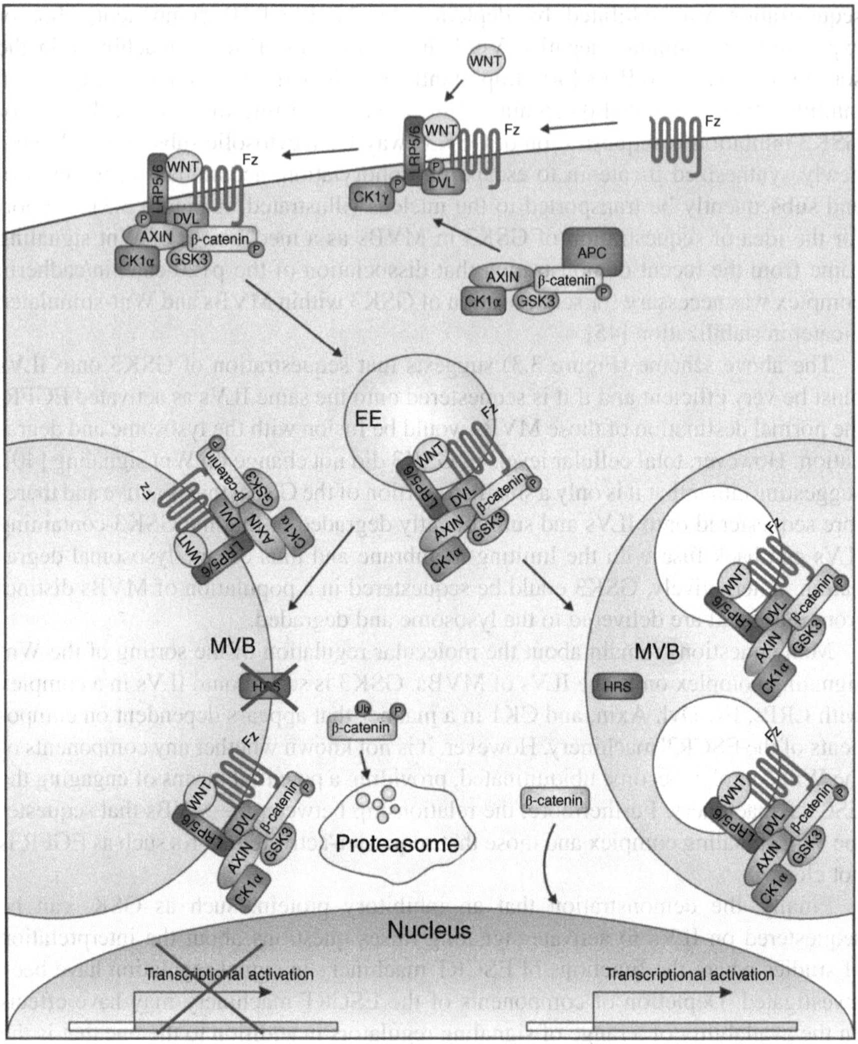

Figure 3.3 Upregulation of Wnt signaling by sorting onto ILVs. In resting cells, cytosolic β-catenin levels are kept low through the activity of an inhibitory complex that contains GSK3, which phosphorylates β-catenin, promoting its ubiquitination and proteosomal degradation. When Wnt ligands bind Frizzled receptors (Fz) and LRP coreceptors at the cell surface, a multiprotein complex is recruited that includes disheveled (Dvl), the scaffold protein, Axin, the kinases, CK1γ, and β-catenin ESCRT-dependent sequestration of the Wnt signaling complex onto the ILVs of MVBs prevents GSK3-mediated phosphorylation of newly synthesized β-catenin, allowing cytosolic β-catenin levels to rise, leading to its transport to the nucleus to activate transcription of Wnt target genes. The lack of sequestration of the Wnt signaling complex onto ILVs in the absence of Hrs renders GSK3 able to phosphorylate newly synthesized β-catenin, leading to ubiquitination and degradation, thus preventing β-catenin-mediated transcriptional activation in the nucleus. (*See color plate section for the color representation of this figure.*)

transported in the extracellular environment, forming spatial and temporal gradients, orchestrating morphogenesis during development and adult tissue homeostasis. The main families of such molecules, Hedgehog, Wnts, Notch ligands, TGF-β, EGF and FGF have all been found associated with exosomes [47]. Here we focus on the membrane-associated Notch ligands, delta/serrate/Lag2 (DSL), which until recently were thought to activate the transmembrane protein, Notch, solely via direct cell:cell contact between the signal-sending (DSL-producing) cell and the signal-receiving (Notch-expressing) cell. Notch activation triggers S2 proteolysis of the extracellular domain of Notch and then the S3 γ-secretase cleavage of the Notch intracellular domain, which can then shuttle to the nucleus, bind transcriptional cofactors, and activate the transcription of target genes. The trafficking of Notch ligands, Notch regulators, and Notch itself to promote and inhibit Notch signaling via the canonical pathway and the ligand-independent Deltex-dependent pathway is complex. This is particularly so during asymmetric cell division, when regulators of Notch signaling are partitioned unequally between daughter cells (reviewed by Furthauer and Gonzalez-Gaitan [48]). In this chapter, we focus on the canonical pathway and on the potential importance of Notch signaling via exosomes and therefore the sorting of ligand and receptor within MVBs in both the signal-sending cell and the signal-receiving cell.

A number of studies have demonstrated that endocytosis of the DSL family of Notch ligands by the signal-sending cell is necessary to activate Notch in the signal-receiving cell [49–52]. One proposal is that the DSL ligand is activated in endosomes and so must be endocytosed and recycled to plasma membrane lipid microdomains to be able to stimulate Notch [53]. Another proposal is that Delta and Notch on the signal-sending and signal-receiving cells, respectively, interact and endocytosis of Delta by the signal-sending cell "pulls" on Notch, altering its physical conformation, promoting S2 cleavage of the extracellular domain, which may be internalized with Delta in the signal-sending cell [54,55]. This creates a substrate for the subsequent S3 cleavage and release of the Notch intracellular domain for nuclear translocation and transcriptional activation in the signal-receiving cell. The finding of the Notch ligand, Delta-like Notch ligand DLL-4, associated with exosomes from tumor cells and endothelial cells suggests another possibility [56,57]. Endocytosis of Delta could be required for sorting onto ILVs before subsequent release as exosomes. One pathway of DSL ligand endocytosis requires ubiquitination and interaction with Epsin [58,59]. As described earlier, ubiquitination could allow engagement of the ESCRT machinery and sorting onto ILVs of MVBs. The formation of ILVs that are subsequently released as exosomes has been proposed to be regulated by an ESCRT-independent mechanism involving sphingomyelinase-mediated formation of the cone-shaped lipid ceramide [60], which can promote membrane curvature. However, ESCRT components are present on exosomes [61] and recently depletion of ESCRT components was shown to impact on exosome formation [62–64]. ESCRTs have frequently been found to have a negative impact on Notch signaling, but this is largely due to effects on ligand-independent Notch trafficking [65–68] where delivery of Notch to the limiting membrane of late endosomes is required for extracellular domain S2 cleavage prior to γ-secretase-mediated S3 cleavage.

In this case, cleavage and/or release of the Notch intracellular domain is inhibited by targeting onto the ILVs of MVBs. Whether the ESCRT machinery has a role in targeting of DSL ligands to ILVs for release as exosomes is not clear.

Once targeted to ILVs the MVB must move to and fuse with the plasma membrane of the signal-sending cell before release of exosomes. A number of members of the Rab family of low-molecular-weight GTPases have been implicated in this process, including Rabs 11 [69,70], 35 [71], and Rab27a and b [72]. Like most secretory events, fusion of MVBs with the plasma membrane is also regulated by calcium [69] and SNAREs (soluble NSF attachment protein receptors) [70] and a role for pH and the vacuolar ATPase [73] has also recently been reported. There appears to be some specificity in the fusion machinery, probably depending on the cell type and cargo and the specific machinery required for release of DSL-containing exosomes has not been reported.

What happens when exosomes reach the signal-receiving cell? Delta-like 4 (Dll4)-containing exosomes can elicit both stimulation and inhibition of Notch signaling in endothelial cells, promoting endothelial sprout formation and inhibiting angiogenesis, respectively [56,74]. The trafficking events that underlie this signaling upon the arrival of signal-bearing exosomes at the receiving cell are not clear. Both Dll4 and Notch are internalized in the signal-receiving cell in a manner that depends on Notch expression by the receiving cell [56]. Could this occur following fusion of exosomes at the cell surface? Microvesicles bearing the oncogenic EGFR mutant EGFRVlll have been shown to transfer oncogenic mutant receptor and signaling capacity to nonexpressing cells, suggesting fusion of exosomes at the plasma membrane [75]. Whether these microvesicles are bona fide exosomes is not clear. Exosomal fusion with the plasma membrane [76,77], macropinocytosis [78], endocytosis [79], and phagocytosis [80] have all been reported to occur upon arrival of exosomes at the receiving cell. Exosome fate may depend on cargo and cell type, and whether the function is exosome clearance, such as occurs upon macropinocytic uptake of exosomes by microglia followed by exosome degradation [78], or whether the function is antigen presentation, transfer of microRNAs, or modulation of signaling in the receiving cell. Fusion of exosomes at the signal-receiving cell surface would deliver Dll4 to the same membrane as Notch, allowing interaction in the "cis" conformation. Cis interaction between Notch and Notch ligands in the same cell could inhibit Notch signaling through competing for trans-ligand–Notch interactions or through promoting Notch endocytosis. Dll4-containing exosomes caused a marked depletion of Notch from the receiving cell surface and ultimately Notch degradation [56]. Conceivably, the selective sorting of either Notch or Notch ligand onto the ILV would allow Notch–Delta interaction in the transconformation (see Figure 3.4). If conditions within the MVB were favorable for S2 cleavage, retention of Notch on the limiting membrane could potentially allow S3 cleavage of the intracellular domain of Notch, release into the cytosol and nuclear transport. If Notch were sorted onto the ILV, then S3 cleavage would release the intracellular domain into the ILV and back fusion would be required for release into the cytosol. If exosomes do not fuse at the cell surface and the entire exosome is endocytosed, then, as shown in Figure 3.4, there could be the intriguing situation of a readymade MVB formed

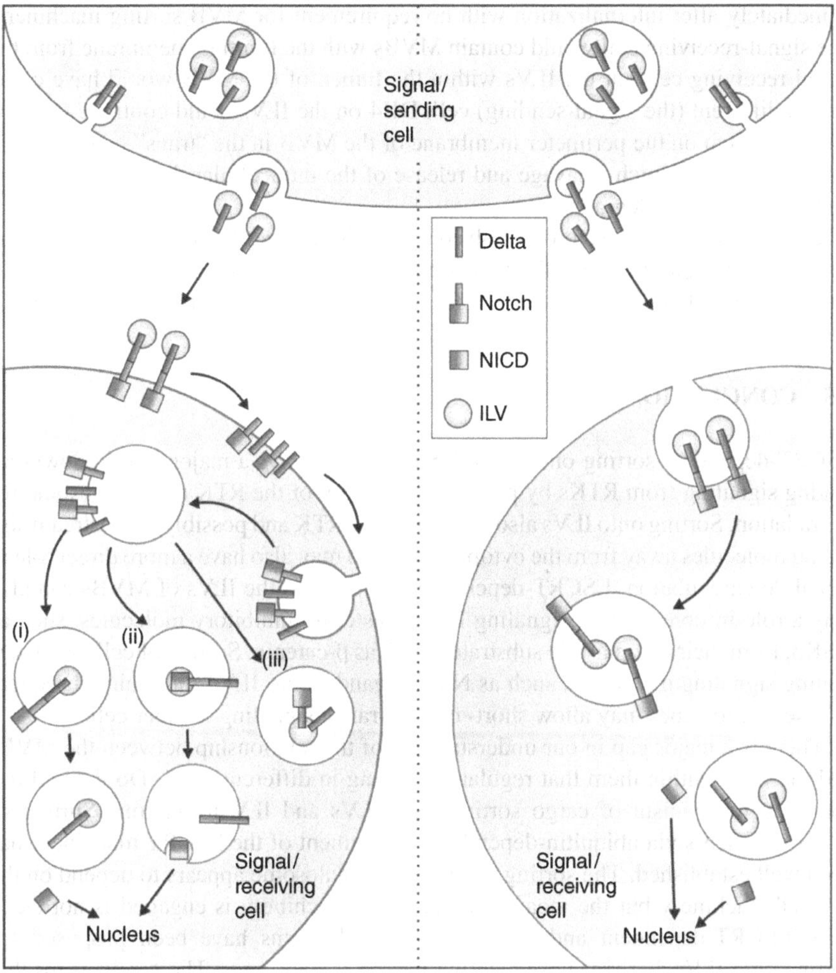

Figure 3.4 Intercellular Notch signaling mediated by ligand-bearing exosomes. Exosomes containing the Notch ligand, Delta, are released from the signal-sending cell and bind to Notch on the cell surface of the signal-receiving cell. Both Notch and Delta are endocytosed by the signal-receiving cell, but it is unclear how. In the left-hand panel, Delta-bearing exosomes fuse at the cell surface and then Delta and Notch on the same membrane are endocytosed. (i) Delta is selectively sorted onto the ILVs of MVBs where it can bind Notch on the MVB-limiting membrane. Conditions in the MVB may be favorable for S2 and then S3 cleavage releasing the Notch intracellular domain that can then traffic to the nucleus. (ii) Notch is selectively targeted to the ILV where it can bind Delta on the limiting membrane of the MVB. In this case, if S2 and S3 cleavage occurs, the intracellular domain of Notch would be sequestered within the ILV and would require back fusion of the ILV with the limiting membrane for release into the cytosol and traffic to the nucleus. (iii) Both Notch and Delta are sorted onto ILVs and this serves to target both proteins to the lysosome for degradation. In the right-hand panel, the exosomes are endocytosed forming an MVB where the limiting membrane is derived from the signal-receiving cell and the ILVs are derived from the signal-sending cell. In this case, Delta–Notch interactions could be continued after endocytosis. In all cases, the ultimate destination of the MVB in the signal-receiving cell that contains Delta and Notch, with or without the Notch intracellular domain, is likely to be the lysosome. (*See color plate section for the color representation of this figure.*)

immediately after internalization with no requirement for MVB sorting machinery. The signal-receiving cell would contain MVBs with the limiting membrane from the signal-receiving cell but the ILVs within the lumen of the MVB would have come from a different (the signal-sending) cell. Dll4 on the ILVs could continue to interact with Notch on the perimeter membrane of the MVB in the "trans" conformation leading either to Notch cleavage and release of the intracellular domain or delivery to the lysosome for degradation.

Clearly, much remains to be resolved, both with respect to the molecular mechanisms regulating sorting of cell fate determining molecules onto ILVs that are subsequently released as exosomes and the fate of those exosomes at the receiving cell.

3.5 CONCLUSION

ESCRT-dependent sorting onto the ILVS of MVBs plays a major role in downregulating signaling from RTKs by promoting delivery of the RTK to the lysosome for degradation. Sorting onto ILVs also sequesters the RTK and possibly downstream signaling molecules away from the cytoplasm and so may also have a more direct role in signal downregulation. ESCRT-depending sorting onto the ILVs of MVBs can also play a role in upregulating signaling by sequestering inhibitory molecules, such as GSK3, from their cytoplasmic substrates, such as β-catenin. Sorting of cell fate determining signaling molecules, such as Notch ligands, onto ILVs and their subsequent release as exosomes may allow short- or long-range signaling to other cells.

There is a major gap in our understanding of the relationship between the MVBs and the ILVs within them that regulate signaling in different ways. Do they all use the same mechanism of cargo sorting onto ILVs and ILV formation? Sorting of RTKs onto ILVs via ubiquitin-dependent engagement of the ESCRT machinery has been well established. The sorting of the Wnt signalosome appears to depend on the ESCRT machinery but the means whereby this machinery is engaged is not clear. Both ESCRT-dependent and -independent mechanisms have been proposed for sorting onto ILVs that are subsequently released as exosomes. The mechanisms that sort cell fate determining molecules into ILVs are not clear. Furthermore, several studies have indicated that different mechanisms of ILV formation can occur on the same membrane. If this is the case, then how are ILVs with different fates segregated? Another major question is how reversible is sorting of signaling cargo onto ILVs? This is very difficult to address experimentally but is of particular importance in cases where inhibitory molecules are sequestered from the cytoplasmic targets on ILVs. A final major question relating to ILV fate is what happens to exosomes bearing cell fate determining molecules on arrival at the signal-receiving cell?

The requirement for high-resolution techniques to visualize ILVs, the multiple roles of the ESCRT and other MVB sorting machineries, and cross talk between signaling pathways are among the challenges faced by those analyzing the role of ILV formation in regulating signaling. However, recent discoveries point to a wide-ranging role of these little vesicles in regulating cell fate and behavior.

REFERENCES

[1] Haigler HT, McKanna JA, Cohen S. Direct visualization of the binding and internalization of a ferritin conjugate of epidermal growth factor in human carcinoma cells A-431. J Cell Biol 1979;81:382–395.

[2] Miller K, Beardmore J, Kanety H, Schlessinger J, Hopkins CR. Localization of the epidermal growth factor (EGF) receptor within the endosome of EGF-stimulated epidermoid carcinoma (A431) cells. J Cell Biol 1986;102:500–509.

[3] Hopkins CR. Intracellular routing of transferrin and transferrin receptors in epidermoid carcinoma A431 cells. Cell 1983;35:321–330.

[4] Futter CE, Pearse A, Hewlett LJ, Hopkins CR. Multivesicular endosomes containing internalized EGF-EGF receptor complexes mature and then fuse directly with lysosomes. J Cell Biol 1996;132:1011–1023.

[5] Dunn KW, Maxfield FR. Delivery of ligands from sorting endosomes to late endosomes occurs by maturation of sorting endosomes. J Cell Biol 1992;117:301–310.

[6] van Deurs B, Holm PK, Kayser L, Sandvig K, Hansen SH. Multivesicular bodies in HEp-2 cells are maturing endosomes. Eur J Cell Biol 1993;61:208–224.

[7] van Deurs B, Holm PK, Kayser L, Sandvig K. Delivery to lysosomes in the human carcinoma cell line HEp-2 involves an actin filament-facilitated fusion between mature endosomes and preexisting lysosomes. Eur J Cell Biol 1995;66:309–323.

[8] Futter CE, Felder S, Schlessinger J, Ullrich A, Hopkins CR. Annexin I is phosphorylated in the multivesicular body during the processing of the epidermal growth factor receptor. J Cell Biol 1993;120:77–83.

[9] Wada I, Lai WH, Posner BI, Bergeron JJ. Association of the tyrosine phosphorylated epidermal growth factor receptor with a 55-kD tyrosine phosphorylated protein at the cell surface and in endosomes. J Cell Biol 1992;116:321–330.

[10] Nesterov A, Reshetnikova G, Vinogradova N, Nikolsky N. Functional state of the epidermal growth factor-receptor complexes during their internalization in A-431 cells. Mol Cell Biol 1990;10:5011–5014.

[11] White IJ, Bailey LM, Razi Aghakhani M, Moss SE, Futter CE. EGF stimulates annexin 1-dependent inward vesiculation in a multivesicular endosome subpopulation. Embo J 2006;25:1–12.

[12] Edgar JR, Eden ER, Futter CE. Hrs- and CD63-dependent competing mechanisms make different sized endosomal intraluminal vesicles. Traffic 2014;15:197–211.

[13] Gillooly DJ, Morrow IC, Lindsay M, Gould R, Bryant NJ, et al. Localization of phosphatidylinositol 3-phosphate in yeast and mammalian cells. Embo J 2000;19:4577–4588.

[14] Schuh AL, Audhya A. The ESCRT machinery: from the plasma membrane to endosomes and back again. Crit Rev Biochem Mol Biol 2014;49:242–261.

[15] Henne WM, Stenmark H, Emr SD. Molecular mechanisms of the membrane sculpting ESCRT pathway. Cold Spring Harb Perspect Biol 2013;5:a016766.

[16] Hanson PI, Cashikar A. Multivesicular body morphogenesis. Annu Rev Cell Dev Biol 2012;28:337–362.

[17] Babst M, Odorizzi G, Estepa EJ, Emr SD. Mammalian tumor susceptibility gene 101 (TSG101) and the yeast homologue, Vps23p, both function in late endosomal trafficking. Traffic 2000;1:248–258.

[18] Malerod L, Stuffers S, Brech A, Stenmark H. Vps22/EAP30 in ESCRT-II mediates endosomal sorting of growth factor and chemokine receptors destined for lysosomal degradation. Traffic 2007;8:1617–1629.

[19] Bache KG, Stuffers S, Malerod L, Slagsvold T, Raiborg C, et al. The ESCRT-III subunit hVps24 is required for degradation but not silencing of the epidermal growth factor receptor. Mol Biol Cell 2006;17:2513–2523.

[20] Lloyd TE, Atkinson R, Wu MN, Zhou Y, Pennetta G, et al. Hrs regulates endosome membrane invagination and tyrosine kinase receptor signaling in *Drosophila.* Cell 2002;108:261–269.

[21] Vaccari T, Rusten TE, Menut L, Nezis IP, Brech A, et al. Comparative analysis of ESCRT-I, ESCRT-II and ESCRT-III function in *Drosophila* by efficient isolation of ESCRT mutants. J Cell Sci 2009;122:2413–2423.

[22] Krempler A, Henry MD, Triplett AA, Wagner KU. Targeted deletion of the Tsg101 gene results in cell cycle arrest at G1/S and p53-independent cell death. J Biol Chem 2002;277:43216–43223.

[23] Komada M, Soriano P. Hrs, a FYVE finger protein localized to early endosomes, is implicated in vesicular traffic and required for ventral folding morphogenesis. Genes Dev 1999;13:1475–1485.

[24] Alwan HA, van Leeuwen JE. UBPY-mediated epidermal growth factor receptor (EGFR) de-ubiquitination promotes EGFR degradation. J Biol Chem 2007;282:1658–1669.

[25] Hanyaloglu AC, McCullagh E, von Zastrow M. Essential role of Hrs in a recycling mechanism mediating functional resensitization of cell signaling. Embo J 2005;24:2265–2283.

[26] Carlton JG, Martin-Serrano J. Parallels between cytokinesis and retroviral budding: a role for the ESCRT machinery. Science 2007;316:1908–1912.

[27] Mueller M, Adell MA, Teis D. Membrane abscission: first glimpse at dynamic ESCRTs. Curr Biol 2012;22:R603–605.

[28] Lee YS, Pressman S, Andress AP, Kim K, White JL, et al. Silencing by small RNAs is linked to endosomal trafficking. Nat Cell Biol 2009;11:1150–1156.

[29] Huang F, Goh LK, Sorkin A. EGF receptor ubiquitination is not necessary for its internalization. Proc Natl Acad Sci U S A 2007;104:16904–16909.

[30] Eden ER, Huang F, Sorkin A, Futter CE. The role of EGF receptor ubiquitination in regulating its intracellular traffic. Traffic 2012;13:329–337.

[31] Goh LK, Huang F, Kim W, Gygi S, Sorkin A. Multiple mechanisms collectively regulate clathrin-mediated endocytosis of the epidermal growth factor receptor. J Cell Biol 2010;189:871–883.

[32] Haj FG, Verveer PJ, Squire A, Neel BG, Bastiaens PI. Imaging sites of receptor dephosphorylation by PTP1B on the surface of the endoplasmic reticulum. Science 2002;295:1708–1711.

[33] Eden ER, White IJ, Tsapara A, Futter CE. Membrane contacts between endosomes and ER provide sites for PTP1B-EGF receptor interaction. Nat Cell Biol 2010;12:267–272.

[34] Stuible M, Abella JV, Feldhammer M, Nossov M, Sangwan V, et al. PTP1B targets the endosomal sorting machinery: dephosphorylation of regulatory sites on the endosomal sorting complex required for transport component STAM2. J Biol Chem 2010;285:23899–23907.

[35] Teis D, Wunderlich W, Huber LA. Localization of the MP1-MAPK scaffold complex to endosomes is mediated by p14 and required for signal transduction. Dev Cell 2002;3:803–814.

[36] Le Blanc I, Luyet PP, Pons V, Ferguson C, Emans N, et al. Endosome-to-cytosol transport of viral nucleocapsids. Nat Cell Biol 2005;7:653–664.

[37] Abrami L, Lindsay M, Parton RG, Leppla SH, van der Goot FG. Membrane insertion of anthrax protective antigen and cytoplasmic delivery of lethal factor occur at different stages of the endocytic pathway. J Cell Biol 2004;166:645–651.

[38] Bissig C, Lenoir M, Velluz MC, Kufareva I, Abagyan R, et al. Viral infection controlled by a calcium-dependent lipid-binding module in ALIX. Dev Cell 2013;25:364–373.

[39] Logan CY, Nusse R. The Wnt signaling pathway in development and disease. Annu Rev Cell Dev Biol 2004;20:781–810.

[40] Taelman VF, Dobrowolski R, Plouhinec JL, Fuentealba LC, Vorwald PP, et al. Wnt signaling requires sequestration of glycogen synthase kinase 3 inside multivesicular endosomes. Cell 2010;143:1136–1148.

[41] Cselenyi CS, Jernigan KK, Tahinci E, Thorne CA, Lee LA, et al. LRP6 transduces a canonical Wnt signal independently of Axin degradation by inhibiting GSK3's phosphorylation of beta-catenin. Proc Natl Acad Sci U S A 2008;105:8032–8037.

[42] Mi K, Dolan PJ, Johnson GV. The low density lipoprotein receptor-related protein 6 interacts with glycogen synthase kinase 3 and attenuates activity. J Biol Chem 2006; 281:4787–4794.

[43] Blitzer JT, Nusse R. A critical role for endocytosis in Wnt signaling. BMC Cell Biol 2006;7:28.

[44] Yamamoto H, Komekado H, Kikuchi A. Caveolin is necessary for Wnt-3a-dependent internalization of LRP6 and accumulation of beta-catenin. Dev Cell 2006;11:213–223.

[45] Vinyoles M, Del Valle-Perez B, Curto J, Vinas-Castells R, Alba-Castellon L, et al. Multivesicular GSK3 sequestration upon Wnt signaling is controlled by p120-catenin/cadherin interaction with LRP5/6. Mol Cell 2014;53:444–457.

[46] Raposo G, Nijman HW, Stoorvogel W, Liejendekker R, Harding CV, et al. B lymphocytes secrete antigen-presenting vesicles. J Exp Med 1996;183:1161–1172.

[47] Wendler F, Bota-Rabassedas N, Franch-Marro X. Cancer becomes wasteful: emerging roles of exosomes(dagger) in cell-fate determination. J Extracell Vesicles 2013;2:22390.

[48] Furthauer M, Gonzalez-Gaitan M. Endocytic regulation of notch signalling during development. Traffic 2009;10:792–802.

[49] Le Borgne R, Bardin A, Schweisguth F. The roles of receptor and ligand endocytosis in regulating Notch signaling. Development 2005;132:1751–1762.

[50] Le Borgne R, Schweisguth F. Unequal segregation of Neuralized biases Notch activation during asymmetric cell division. Dev Cell 2003;5:139–148.

[51] Le Borgne R, Schweisguth F. Notch signaling: endocytosis makes delta signal better. Curr Biol 2003;13:R273–275.

[52] Seugnet L, Simpson P, Haenlin M. Requirement for dynamin during Notch signaling in *Drosophila* neurogenesis. Dev Biol 1997;192:585–598.

[53] Heuss SF, Ndiaye-Lobry D, Six EM, Israel A, Logeat F. The intracellular region of Notch ligands Dll1 and Dll3 regulates their trafficking and signaling activity. Proc Natl Acad Sci U S A 2008;105:11212–11217.

[54] Chitnis A. Why is delta endocytosis required for effective activation of notch? Dev Dyn 2006;235:886–894.

[55] Le Borgne R. Regulation of Notch signalling by endocytosis and endosomal sorting. Curr Opin Cell Biol 2006;18:213–222.

[56] Sheldon H, Heikamp E, Turley H, Dragovic R, Thomas P, et al. New mechanism for Notch signaling to endothelium at a distance by Delta-like 4 incorporation into exosomes. Blood 2010;116:2385–2394.

[57] Ristorcelli E, Beraud E, Mathieu S, Lombardo D, Verine A. Essential role of Notch signaling in apoptosis of human pancreatic tumoral cells mediated by exosomal nanoparticles. Int J Cancer 2009;125:1016–1026.

[58] Wang W, Struhl G. *Drosophila* Epsin mediates a select endocytic pathway that DSL ligands must enter to activate Notch. Development 2004;131:5367–5380.

[59] Wang W, Struhl G. Distinct roles for Mind bomb, Neuralized and Epsin in mediating DSL endocytosis and signaling in *Drosophila*. Development 2005;132:2883–2894.

[60] Trajkovic K, Hsu C, Chiantia S, Rajendran L, Wenzel D, et al. Ceramide triggers budding of exosome vesicles into multivesicular endosomes. Science 2008;319:1244–1247.

[61] Buschow SI, Liefhebber JM, Wubbolts R, Stoorvogel W. Exosomes contain ubiquitinated proteins. Blood Cells Mol Dis 2005;35:398–403.

[62] Colombo M, Moita C, van Niel G, Kowal J, Vigneron J, et al. Analysis of ESCRT functions in exosome biogenesis, composition and secretion highlights the heterogeneity of extracellular vesicles. J Cell Sci 2013;126:5553–5565.

[63] Tamai K, Shiina M, Tanaka N, Nakano T, Yamamoto A, et al. Regulation of hepatitis C virus secretion by the Hrs-dependent exosomal pathway. Virology 2012;422:377–385.

[64] Tamai K, Tanaka N, Nakano T, Kakazu E, Kondo Y, et al. Exosome secretion of dendritic cells is regulated by Hrs, an ESCRT-0 protein. Biochem Biophys Res Commun 2010;399:384–390.

[65] Hori K, Sen A, Kirchhausen T, Artavanis-Tsakonas S. Synergy between the ESCRT-III complex and Deltex defines a ligand-independent Notch signal. J Cell Biol 2011;195:1005–1015.

[66] Yamada K, Fuwa TJ, Ayukawa T, Tanaka T, Nakamura A, et al. Roles of *Drosophila* deltex in Notch receptor endocytic trafficking and activation. Genes Cells 2011;16:261–272.

[67] Wilkin M, Tongngok P, Gensch N, Clemence S, Motoki M, et al. *Drosophila* HOPS and AP-3 complex genes are required for a Deltex-regulated activation of notch in the endosomal trafficking pathway. Dev Cell 2008;15:762–772.

[68] Hori K, Fostier M, Ito M, Fuwa TJ, Go MJ, et al. *Drosophila* deltex mediates suppressor of Hairless-independent and late-endosomal activation of Notch signaling. Development 2004;131:5527–5537.

[69] Savina A, Fader CM, Damiani MT, Colombo MI. Rab11 promotes docking and fusion of multivesicular bodies in a calcium-dependent manner. Traffic 2005;6:131–143.

[70] Koles K, Nunnari J, Korkut C, Barria R, Brewer C, et al. Mechanism of evenness interrupted (Evi)-exosome release at synaptic boutons. J Biol Chem 2012;287:16820–16834.

[71] Hsu C, Morohashi Y, Yoshimura S, Manrique-Hoyos N, Jung S, et al. Regulation of exosome secretion by Rab35 and its GTPase-activating proteins TBC1D10A-C. J Cell Biol 2010;189:223–232.

[72] Ostrowski M, Carmo NB, Krumeich S, Fanget I, Raposo G, et al. Rab27a and Rab27b control different steps of the exosome secretion pathway. Nat Cell Biol 2010;12:19–30; sup pp. 11–13.

[73] Liegeois S, Benedetto A, Garnier JM, Schwab Y, Labouesse M. The V0-ATPase mediates apical secretion of exosomes containing Hedgehog-related proteins in *Caenorhabditis elegans*. J Cell Biol 2006;173:949–961.

[74] Sharghi-Namini S, Tan E, Ong LL, Ge R, Asada HH. Dll4-containing exosomes induce capillary sprout retraction in a 3D microenvironment. Sci Rep 2014;4:4031.

[75] Al-Nedawi K, Meehan B, Micallef J, Lhotak V, May L, et al. Intercellular transfer of the oncogenic receptor EGFRvIII by microvesicles derived from tumour cells. Nat Cell Biol 2008;10:619–624.

[76] Parolini I, Federici C, Raggi C, Lugini L, Palleschi S, et al. Microenvironmental pH is a key factor for exosome traffic in tumor cells. J Biol Chem 2009;284:34211–34222.

[77] Montecalvo A, Larregina AT, Shufesky WJ, Stolz DB, Sullivan ML, et al. Mechanism of transfer of functional microRNAs between mouse dendritic cells via exosomes. Blood 2012;119:756–766.

[78] Fitzner D, Schnaars M, van Rossum D, Krishnamoorthy G, Dibaj P, et al. Selective transfer of exosomes from oligodendrocytes to microglia by macropinocytosis. J Cell Sci 2011;124:447–458.

[79] Morelli AE, Larregina AT, Shufesky WJ, Sullivan ML, Stolz DB, et al. Endocytosis, intracellular sorting, and processing of exosomes by dendritic cells. Blood 2004;104:3257–3266.

[80] Feng D, Zhao WL, Ye YY, Bai XC, Liu RQ, et al. Cellular internalization of exosomes occurs through phagocytosis. Traffic 2010;11:675–687.

4

LYSOSOMES AND MITOPHAGY

Dominik Haddad

VIB Center for the Biology of Disease, KU Leuven, Leuven, Belgium; Laboratory of Neuronal Communication, Leuven Institute for Neurodegenerative Disease (LIND), Center for Human Genetics, KU Leuven, Leuven, Belgium; Gladstone Institutes of Neurological Disease, University of California, San Francisco, CA, USA

Patrik Verstreken

VIB Center for the Biology of Disease, KU Leuven, Leuven, Belgium; Laboratory of Neuronal Communication, Leuven Institute for Neurodegenerative Disease (LIND), Center for Human Genetics, KU Leuven, Leuven, Belgium

4.1 SUMMARY

Mitochondrial autophagy or "self-eating" is a catabolic process of mitochondrial degradation by lysosomes. In contrast to nonselective autophagy (discussed in Chapter 2), mitophagy is the selective elimination of dysfunctional or unwanted mitochondria during periods of stress or during regulated development [1,2]. Given the fundamental role of mitochondria in energy production, signaling, and cellular survival, mitophagy emerged as a central mechanism in regulating mitochondrial homeostasis and quality control in health and disease [3,4].

4.2 MITOCHONDRIAL SIGNIFICANCE

Since the initial discovery of mitochondria in the 19th century, accumulating evidence continues to reshape our views of mitochondrial functions and their vital role in managing various cellular processes such as metabolic homeostasis, energy production, stress response, and cell death [5].

Lysosomes: Biology, Diseases, and Therapeutics, First Edition.
Edited by Frederick R. Maxfield, James M. Willard and Shuyan Lu.

Mitochondria are relatively small in size with a diameter of about 0.5–1 μm and a length that can be up to 5 μm [6], but depending on the cell type and cellular state this number varies significantly. Mitochondrial shape may vary from spheres to filamentous structures, but they are always composed of two enclosed membranes [6,7]. The two membranes (outer and inner) have different biochemical and physical characteristics that define their functions, with the outer membrane forming a partial selective barrier toward the cytoplasm while the inner membrane (where the major enzymatic machinery resides) forms long invaginations called "cristae" enclosing the matrix [7]. Each mitochondrion is formed by approximately 1500 distinct proteins [8]. These proteins are mainly encoded by nuclear DNA, and only 13 proteins are encoded by mitochondrial DNA [9]. Hence, specific mitochondrial functions are largely defined by the cell type these organelles reside in.

Mitochondria have numerous basic functions and they play a central role in energy metabolism through adenosine-5′-triphosphate (ATP) synthesis [10]. Mitochondria have also additional functions that are fundamental for cellular homeostasis such as calcium storage, release, and buffering [11]. Calcium has an intrinsic role in modulating mitochondrial ATP production. In addition, calcium plays an important role in controlling the function of many proteins and enzymes and therefore has a critical impact on numerous biological pathways such as neurotransmitter release, autophagy, and apoptosis [11,12]. In addition to their ability to buffer calcium, mitochondria play a central role in reactive oxygen species (ROS) homeostasis [13]. ROS (such as hydrogen peroxide, superoxide anion, and hydroxyl radical) are an essential component of different signaling pathways but can also damage the cellular machinery. Mitochondria are the main source of intracellular ROS [14].

Overall, mitochondrial dysfunction may affect several aspects of cellular life ranging from functional failures to apoptosis and cell death. Keeping mitochondria at their optimal functionality and integrity is critical for cellular survival [15,16]. Mitochondrial defects may arise from various origins such as mutated proteins or damaged mitochondrial components due to oxidative stress, all of which may affect mitochondrial structure and/or function. Consequently, if mitochondria become dysfunctional, multiple restoration mechanisms can be triggered, including mitophagy and lysosomal degradation. Here, defective mitochondria or, by virtue of mitochondrial fission, parts of mitochondria are eliminated and the mitochondrial pool remains healthy, protecting against several disorders that are associated with abnormal mitochondria [15,17] (more discussion in Chapter 9).

4.3 HISTORY OF MITOPHAGY

In 1955, Christian De Duve discovered different cytoplasmic granules containing acid hydrolases using a new scheme of cellular fractionation, where he referred to these granules as "lysosomes" to implicate their enrichment in hydrolytic enzymes [18]. Shortly thereafter, the first lysosomal engulfment of mitochondria was observed by Clark in 1957 using electron microscopy, where he described "large cytoplasmic bodies" that include "altered mitochondria" surrounded by dense membranes [19].

Similarly, Ashford and Porter in 1962 noticed upon glucagon treatment of hepatocytes promoting autophagy, a significant increase in the number of lysosomes that contain mitochondria "in various stages of breakdown or hydrolysis" [20]. However, the term "autophagy" of mitochondria was only introduced and described in 1963 [21], where mitochondria are first sequestered by membrane structures (prelysosome), then acid hydrolases are added to form an active "lysosome" and mitochondria are degraded to form a "postlysosomal body" [22]. Later, in 2005, Lemasters [23] proposed the term "mitophagy" based on similarities to the peroxisomal autophagy (pexophagy), suggesting the selective removal of cytoplasmic organelles by autophagy.

One example of selective autophagy is "pexophagy," where yeast grown on methanol as a sole source of carbon and energy has a significant increase in peroxisomal mass required for specialized metabolic pathways. Once methanol-rich media is withdrawn, selective autophagy of excess peroxisomes takes place while the turnover rate of other organelles remains unaffected [24]. This indicates that autophagy of cytoplasmic organelles is a selective process that vigorously adapts to metabolic needs. Similarly, oxidative stress may damage mitochondrial membranes and cause nonspecific pore opening, the mitochondrial permeability transition (MPT), uncoupling of oxidative phosphorylation and consequently bioenergetic failure [25]. When severe, MPT activates caspases and apoptosis [26], whereas under mild stress, MPT pore formation induces autophagy, which selectively eliminates damaged mitochondria [27]. Furthermore, starvation induces autophagy in yeast and cytoplasmic organelles degradation. However, mutation in the yeast mitochondrial outer-membrane protein Uth1p, which is the first gene identified to control autophagy, significantly inhibits mitochondrial degradation while elimination of other cytosolic components occurs normally [28]. Hence, mitophagy is a selective degradation process of damaged organelles that promotes the maintenance of a healthy pool of mitochondria and sustains the bioenergetic capacity of the cell.

4.4 MECHANISMS OF MITOPHAGY

Once mitochondria are targeted for elimination, autophagy is induced and degradation of dysfunctional or unwanted mitochondria is initiated. However, it was proposed that an early response mechanism exists, where an initial quality control pathway precedes the clearance of damaged mitochondria by autophagy. Indeed, mild stress induces the biogenesis of mitochondrial vesicles enriched for oxidized proteins and targeted for degradation by the lysosome [29]. This first line of defense operates at an early stage before the recruitment of the full blown autophagic machinery. This mechanism would be part of the quality control system that specifically removes damaged mitochondrial components via vesicular carriers and lysosomal degradation without the need to eliminate the entire organelle at once. Nonetheless, if this initial quality control intervention becomes overwhelmed and mitochondria are irreversibly damaged, autophagy is triggered and targeted mitochondria are eliminated [30].

Interestingly, an important observation that precedes mitophagy is mitochondrial fragmentation [31], a process that may in part also overlap with the production of mitochondrial vesicles destined for degradation. In order to sustain cellular needs and function, mitochondria are highly dynamic organelles undergoing constant fission and fusion events. This dynamic process helps mitochondria mix their content, preserve their integrity, and keep their physiological homeostasis balanced [32,33]. While mitochondrial length can reach up to 5 μm, autophagosomes do not exceed 1 μm causing sterical problems for their capacity to engulf mitochondria during autophagy [34]. Furthermore, when mitochondria undergo fission events, it gives rise to mitochondria with a different membrane potential spectrum, where the daughter organelle with lower membrane potential are unlikely to fuse again with "healthy mitochondria" and are then degraded by autophagy [35]. However, mitochondrial fission *per se* is not enough to induce mitophagy, suggesting the involvement of other signals that trigger mitophagy in parallel or upstream of fission [34].

Once initiated, autophagy can be divided into two major steps: induction of the general autophagic machinery (described in Chapter 2) and priming of targeted mitochondria for autophagic recognition and degradation (discussed below). Mitophagy begins with engulfment of the targeted organelle by a double-membrane structure (autophagosome), these membranes may originate from different sources such as the plasma membrane [36], Golgi [37], endoplasmic reticulum (ER) [38], and even mitochondria themselves [39]. Next, autophagosomes fuse with lysosomes for degradation and formation of the "autolysosome." More than 30 autophagy-related genes (Atg) have been discovered in yeast, and most of these are well conserved across species. Nonetheless, the exact mechanism of autophagy (as well as mitophagy) or how it is regulated at specific cellular compartments (e.g., in neurons at the synapse) is not yet fully understood and remains to be elucidated.

4.4.1 Mitophagy in Yeast

Mitophagy exists in mammalian cells [19], but the molecular mechanisms and genetic pathways were initially identified in yeast. The first genetic factor discovered to be involved in mitophagy was Uth1p, a member of the SUN family. This protein is mostly localized to the outer mitochondrial membrane (OMM) and is required for degradation of dysfunctional mitochondria during starvation [28]. Another component required for stationary-phase mitophagy is Aup1p, a member of the protein phosphatase family localized to the intermembrane space [40].

Further screening of mitophagy modulators revealed dozens of new genes. One important outcome was the identification of the 60-kDa OMM protein Atg32 that harbors its carboxy terminal in the intermembrane space while the amino terminal faces toward the cytosol [41,42]. Atg32 is required for selective autophagy during high respiration growth, but not during starvation, and is highly expressed under oxidizing conditions, reflecting its importance for mitochondrial quality control. Interestingly, under mitophagy-inducing conditions, Atg32 can interact with Atg11, known to recruit organelles into autophagosomes via Atg8 (also referred to as LC3), or directly with Atg8 through its WXXL-like Atg8-binding motif, thereby recruiting mitochondria to autophagosomes and initiating selective mitochondrial degradation [41,42] (Figure 4.1). Numerous other genes have been linked to mitophagy but

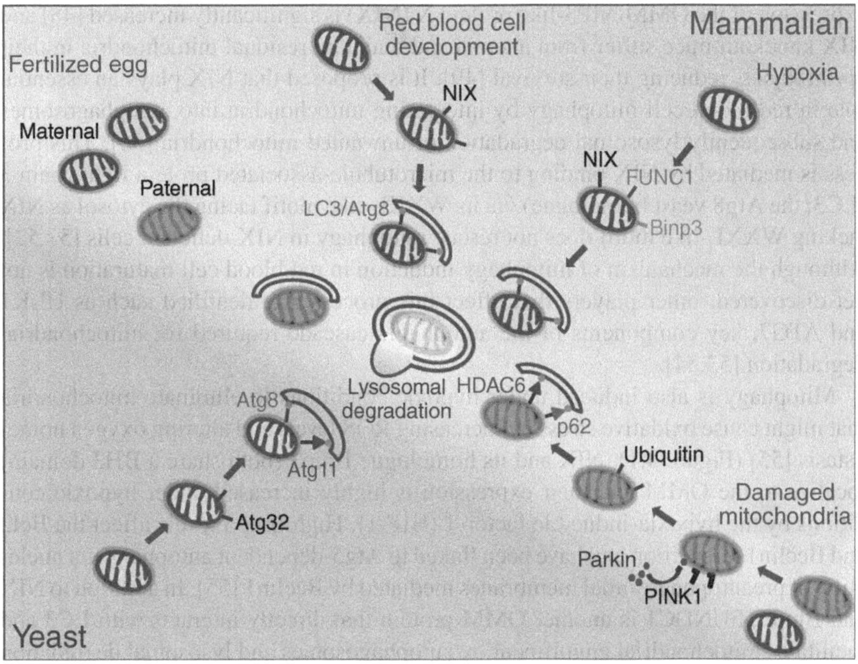

Figure 4.1 Summary of mitophagy in yeast and mammalian cells. Mitochondria require specific targeting signals to recruit the autophagic machinery and mediate autophagosome formation and lysosomal degradation. (*See color plate section for the color representation of this figure.*)

with different levels of necessity, which might indicate the existence of multiple mitophagy pathways for different purposes that remain to be uncovered [43].

4.4.2 Mitophagy in Mammals

During early stages of embryogenesis, mitophagy is needed in fertilized oocytes to eliminate paternal mitochondria [44,45]. Upon oocyte fertilization, autophagy is activated and autophagosomes are formed, causing the selective degradation of paternal mitochondria at lysosomes (Figure 4.1). When autophagy is blocked, paternal mitochondria are fragmented and fusion incompetent inhibiting their capacity to fuse with maternal mitochondria [45]. Mitochondrial DNA is mostly maternally inherited [46] and the evolutionary benefit and molecular mechanism of this selective autophagy of paternal mitochondria remains elusive.

Another type of selective mitochondrial elimination is mitophagy during red blood cell maturation [47] (Figure 4.1). As mature red blood cells (erythrocytes) lack mitochondria, autophagy efficiently eliminates those organelles throughout their maturation, and it is proposed that this elimination is necessary for the red blood cell function in oxygen transport. During the red blood cell terminal differentiation phase, expression of the OMM NIP3-like protein X (NIX) is significantly increased [48] and NIX knockout mice suffer from anemia and maintain residual mitochondria in their erythrocytes, reducing their survival [49]. It is proposed that NIX plays an essential role in red blood cell mitophagy by integrating mitochondria into autophagosomes and subsequently lysosomal degradation of unwanted mitochondria [50]. This process is mediated by NIX binding to the microtubule-associated protein light chain 3 (LC3; the Atg8 yeast homologue) via its WXXL-like motif facing the cytosol as NIX lacking WXXL-like motif does not rescue mitophagy in NIX-deficient cells [51,52]. Although the mechanism of mitophagy induction in red blood cell maturation is not yet discovered, other players that affect this process are identified such as ULK1 and ATG7, key components of the autophagic cascade required for mitochondrial degradation [53,54].

Mitophagy is also induced under hypoxic conditions to eliminate mitochondria that might cause oxidative stress by increasing ROS levels and altering oxygen homeostasis [55] (Figure 4.1). NIX and its homologue Bnip3 (both share a BH3 domain) localize to the OMM and their expression is highly increased under hypoxic conditions by the hypoxia-inducible factor-1 (HIF-1). High Bnip3 levels affect the Bcl2 and Beclin1 interaction and have been linked to Atg5-dependent autophagy via nucleation of preautophagosomal membranes mediated by Beclin1 [55]. In addition to NIX and Bnip3, FUNDC1 is another OMM protein that directly interacts with LC3 and facilitates mitochondrial engulfment by autophagosomes and lysosomal degradation under hypoxic conditions [56].

NIX is required for mitophagy in red blood cells during their differentiation; however, depolarization of mitochondria with uncouplers such as carbonilcyanide *p*-trifluoromethoxyphenylhydrazone (FCCP) rescues mitophagy in NIX-deficient cells [49]. Therefore, dissipating mitochondrial membrane potential triggers alternative mitophagy pathways that are NIX-independent. Indeed, dissipating the electrochemical gradient across the inner mitochondrial membrane using acute treatment of the mitochondrial uncoupler carbonyl cyanide *m*-chlorophenyl hydrazone (CCCP) leads to mitochondrial depolarization and autophagy initiation [57] (Figure 4.1). Under steady conditions, PINK1 a serine/threonine kinase that maintains electron transport chain function in mitochondria [58–61] is constitutively cleaved by the mitochondrial-processing protease (MPP) followed by the presenilin-associated rhomboid-like protease (PARL). But upon dissipation of the mitochondrial membrane potential by CCCP, PINK1 processing is inhibited causing its accumulation on the OMM [62–65]. PINK1 accumulation triggers the relocalization of Parkin, a cytosolic E3 ubiquitin ligase, to mitochondria [57,63]. Accumulation of PINK1 alone is sufficient to relocalize Parkin, as PINK1 targeted to other cytoplasmic components such as peroxisomes or lysosomes also recruits Parkin to the respective organelles [66]. Once autophagy is initiated, Parkin interacts with autophagy factors, for example, Ambra1, known to activate class III PI3K required for phagophore formation [67].

To selectively degrade mitochondria, these organelles are labeled with ubiquitin. Indeed, several mitochondrial targets have been identified to be ubiquitinated in a Parkin (E3 ubiquitin ligase)-dependent manner [68–71]. This process is mediated by the E2 ubiquitin conjugating enzyme Ube2a. The absence of E2 activity blocks CCCP-induced mitophagy because ubiquitin is not transferred to Parkin and to Parkin substrates, and Parkin cannot relocalize to defective mitochondria anymore [72]. Proteomic analyses identified OMM proteins such as Miro, Mitofusin, hFis1, and Tom70 [60,71,73,74] to be modified with lysine 48-linked polyubiquitin chains, suggesting a possible involvement of proteasome-dependent degradation pathways. Alternatively, other mitochondrial targets such as the OMM protein VDAC are modified with lysine 63-linked polyubiquitin chains and this is necessary for mitophagy initiation. Therefore, both types of ubiquitin linkage (via K48 and K63) are mediated by Parkin and required for mitophagy [43,70–72,75].

Following mitochondrial ubiquitination, adaptor proteins such as ubiquitin-binding adaptor protein 62 (p62/SQSTRM1) or histone deacetylase 6 (HDAC6) are required for the autophagic machinery recruitment to the targeted organelles. p62 links mitochondria to the autophagic machinery by binding both ubiquitin and LC3, which initiates the formation of autophagosomes and subsequently fusion with lysosomes for degradation [68,70].

4.5 CONCLUSION

Although mitophagy is a defined type of selective autophagy and expected to be conserved among eukaryotes, homologues of several key components involved in yeast mitophagy have not yet been characterized in mammalian systems. One potential explanation for such distinction is the purpose of mitophagy in each system. In yeast, for instance, mitophagy is involved in cell adaptation to diverse nutritional conditions, while in mammalian cells their role could vary during specific conditions such as developmental processes (e.g., maturation of red blood cells), starvation, or elimination of damaged mitochondria. Indeed, while dissipation of mitochondrial membrane potential seems a critical trigger for mitophagy in many mammalian cells, it has no effect on yeast cells. This distinction suggests that mitophagy senses mitochondrial depolarization for its initiation in mammalian cells versus growth conditions in yeast. Furthermore, mitophagy activates distinct molecular pathways depending on cell types in response to variable conditions affecting cell homeostasis. Accumulation of damaged mitochondria and defective lysosomal degradation has been linked to numerous human disorders such as neurodegeneration and cancer (discussed in Chapters 9 and 10). While some aspects of mitophagy are revealed, future discoveries will shed light on understanding the fine complexity and necessity of this system in cellular life and disease etiology.

ACKNOWLEDGMENTS

Support to the Verstreken lab is provided by an ERC starting grant (260678), FWO grants, an IUAP by BELSPO, a GOA from the Research Fund KU Leuven, the Hercules Foundation, and VIB.

REFERENCES

[1] Kroemer G, Marino G, Levine B. Autophagy and the integrated stress response. Mol Cell 2010;40:280–293.

[2] Mizushima N, Levine B. Autophagy in mammalian development and differentiation. Nat Cell Biol 2010;12:823–830.

[3] Gottlieb RA, Carreira RS. Autophagy in health and disease. 5. Mitophagy as a way of life. Am J Physiol Cell Physiol 2010;299:C203–C210.

[4] Ashrafi G, Schwarz TL. The pathways of mitophagy for quality control and clearance of mitochondria. Cell Death Differ 2013;20:31–42.

[5] Galluzzi L, Kepp O, Trojel-Hansen C, Kroemer G. Mitochondrial control of cellular life, stress, and death. Circ Res 2012;111:1198–1207.

[6] Cereghetti GM, Scorrano L. The many shapes of mitochondrial death. Oncogene 2006; 25:4717–4724.

[7] Alberts B, Johnson A, Lewis J, Raff M, Roberts K, Walter P. *Molecular Biology of the Cell.* New York: Garland Science; 2002.

[8] Lopez MF, Kristal BS, Chernokalskaya E, Lazarev A, Shestopalov AI, Bogdanova A, Robinson M. High-throughput profiling of the mitochondrial proteome using affinity fractionation and automation. Electrophoresis 2000;21:3427–3440.

[9] DiMauro S, Schon EA. Mitochondrial respiratory-chain diseases. N Engl J Med 2003; 348:2656–2668.

[10] Reichert AS, Neupert W. Mitochondriomics or what makes us breathe. Trends Genet 2004;20:555–562.

[11] Rizzuto R, Pozzan T. Microdomains of intracellular Ca2+: molecular determinants and functional consequences. Physiol Rev 2006;86:369–408.

[12] Campanella M, Pinton P, Rizzuto R. Mitochondrial Ca2+ homeostasis in health and disease. Biol Res 2004;37:653–660.

[13] Bae YS, Oh H, Rhee SG, Yoo YD. Regulation of reactive oxygen species generation in cell signaling. Mol Cells 2011;32:491–509.

[14] Jezek P, Hlavata L. Mitochondria in homeostasis of reactive oxygen species in cell, tissues, and organism. Int J Biochem Cell Biol 2005;37:2478–2503.

[15] Baker MJ, Tatsuta T, Langer T. Quality control of mitochondrial proteostasis. Cold Spring Harb Perspect Biol 2011;3:pii: a007559.

[16] Lin MT, Beal MF. Mitochondrial dysfunction and oxidative stress in neurodegenerative diseases. Nature 2006;443:787–795.

[17] Novak I. Mitophagy: a complex mechanism of mitochondrial removal. Antioxid Redox Signal 2012;17:794–802.

[18] De Duve C, Pressman BC, Gianetto R, Wattiaux R, Appelmans F. Tissue fractionation studies. 6. Intracellular distribution patterns of enzymes in rat-liver tissue. Biochem J 1955;60:604–617.

[19] Clark SL Jr. Cellular differentiation in the kidneys of newborn mice studies with the electron microscope. J Biophys Biochem Cytol 1957;3:349–362.

[20] Ashford TP, Porter KR. Cytoplasmic components in hepatic cell lysosomes. J Cell Biol 1962;12:198–202.

[21] De Duve C. *Lysosomes*. London: J. & A. Churchill; 1963.

[22] De Duve C, Wattiaux R. Functions of lysosomes. Annu Rev Physiol 1966;28:435–492.

[23] Lemasters JJ. Selective mitochondrial autophagy, or mitophagy, as a targeted defense against oxidative stress, mitochondrial dysfunction, and aging. Rejuvenation Res 2005; 8:3–5.

[24] Kiel JA, Komduur JA, van der Klei IJ, Veenhuis M. Macropexophagy in *Hansenula polymorpha*: facts and views. FEBS Lett 2003;549:1–6.

[25] Kowaltowski AJ, Castilho RF, Vercesi AE. Mitochondrial permeability transition and oxidative stress. FEBS Lett 2001;495:12–15.

[26] Kim JS, He L, Lemasters JJ. Mitochondrial permeability transition: a common pathway to necrosis and apoptosis. Biochem Biophys Res Commun 2003;304:463–470.

[27] Elmore SP, Qian T, Grissom SF, Lemasters JJ. The mitochondrial permeability transition initiates autophagy in rat hepatocytes. FASEB J 2001;15:2286–2287.

[28] Kissova I, Deffieu M, Manon S, Camougrand N. Uth1p is involved in the autophagic degradation of mitochondria. J Biol Chem 2004;279:39068–39074.

[29] Soubannier V, McLelland GL, Zunino R, Braschi E, Rippstein P, Fon EA, McBride HM. A vesicular transport pathway shuttles cargo from mitochondria to lysosomes. Curr Biol 2012;22:135–141.

[30] McLelland GL, Soubannier V, Chen CX, McBride HM, Fon EA. Parkin and PINK1 function in a vesicular trafficking pathway regulating mitochondrial quality control. EMBO J 2014;33:282–295.

[31] Gomes LC, Scorrano L. High levels of Fis1, a pro-fission mitochondrial protein, trigger autophagy. Biochim Biophys Acta 2008;1777:860–866.

[32] Chan DC. Mitochondria: dynamic organelles in disease, aging, and development. Cell 2006;125:1241–1252.

[33] Westermann B. Merging mitochondria matters: cellular role and molecular machinery of mitochondrial fusion. EMBO Rep 2002;3:527–531.

[34] Gomes LC, Scorrano L. Mitochondrial morphology in mitophagy and macroautophagy. Biochim Biophys Acta 2013;1833:205–212.

[35] Twig G, Elorza A, Molina AJ, Mohamed H, Wikstrom JD, Walzer G, Stiles L, Haigh SE, Katz S, Las G, Alroy J, Wu M, Py BF, Yuan J, Deeney JT, Corkey BE, Shirihai OS. Fission and selective fusion govern mitochondrial segregation and elimination by autophagy. EMBO J 2008;27:433–446.

[36] Ravikumar B, Moreau K, Jahreiss L, Puri C, Rubinsztein DC. Plasma membrane contributes to the formation of pre-autophagosomal structures. Nat Cell Biol 2010; 12:747–757.

[37] Kim S, Naylor SA, DiAntonio A. *Drosophila* Golgi membrane protein Ema promotes autophagosomal growth and function. Proc Natl Acad Sci U S A 2012;109: E1072–E1081.

[38] Hayashi-Nishino M, Fujita N, Noda T, Yamaguchi A, Yoshimori T, Yamamoto A. A subdomain of the endoplasmic reticulum forms a cradle for autophagosome formation. Nat Cell Biol 2009;11:1433–1437.

[39] Hailey DW, Rambold AS, Satpute-Krishnan P, Mitra K, Sougrat R, Kim PK, Lippincott-Schwartz J. Mitochondria supply membranes for autophagosome biogenesis during starvation. Cell 2010;141:656–667.

[40] Tal R, Winter G, Ecker N, Klionsky DJ, Abeliovich H. Aup1p, a yeast mitochondrial protein phosphatase homolog, is required for efficient stationary phase mitophagy and cell survival. J Biol Chem 2007;282:5617–5624.

[41] Okamoto K, Kondo-Okamoto N, Ohsumi Y. Mitochondria-anchored receptor Atg32 mediates degradation of mitochondria via selective autophagy. Dev Cell 2009;17:87–97.

[42] Kanki T, Wang K, Klionsky DJ. A genomic screen for yeast mutants defective in mitophagy. Autophagy 2010;6:278–280.

[43] Youle RJ, Narendra DP. Mechanisms of mitophagy. Nat Rev Mol Cell Biol 2011; 12:9–14.

[44] Al RS, Louvet-Vallee S, Djeddi A, Sachse M, Culetto E, Hajjar C, Boyd L, Legouis R, Galy V. Postfertilization autophagy of sperm organelles prevents paternal mitochondrial DNA transmission. Science 2011;334:1144–1147.

[45] Sato M, Sato K. Degradation of paternal mitochondria by fertilization-triggered autophagy in *C. elegans* embryos. Science 2011;334:1141–1144.

[46] Ankel-Simons F, Cummins JM. Misconceptions about mitochondria and mammalian fertilization: implications for theories on human evolution. Proc Natl Acad Sci U S A 1996;93:13859–13863.

[47] Schweers RL, Zhang J, Randall MS, Loyd MR, Li W, Dorsey FC, Kundu M, Opferman JT, Cleveland JL, Miller JL, Ney PA. NIX is required for programmed mitochondrial clearance during reticulocyte maturation. Proc Natl Acad Sci U S A 2007;104:19500–19505.

[48] Aerbajinai W, Giattina M, Lee YT, Raffeld M, Miller JL. The proapoptotic factor Nix is coexpressed with Bcl-xL during terminal erythroid differentiation. Blood 2003;102:712–717.

[49] Sandoval H, Thiagarajan P, Dasgupta SK, Schumacher A, Prchal JT, Chen M, Wang J. Essential role for Nix in autophagic maturation of erythroid cells. Nature 2008;454:232–235.

[50] Novak I, Dikic I. Autophagy receptors in developmental clearance of mitochondria. Autophagy 2011;7:301–303.

[51] Novak I, Kirkin V, McEwan DG, Zhang J, Wild P, Rozenknop A, Rogov V, Lohr F, Popovic D, Occhipinti A, Reichert AS, Terzic J, Dotsch V, Ney PA, Dikic I. Nix is a selective autophagy receptor for mitochondrial clearance. EMBO Rep 2010;11:45–51.

[52] Schwarten M, Mohrluder J, Ma P, Stoldt M, Thielmann Y, Stangler T, Hersch N, Hoffmann B, Merkel R, Willbold D. Nix directly binds to GABARAP: a possible crosstalk between apoptosis and autophagy. Autophagy 2009;5:690–698.

[53] Kundu M, Lindsten T, Yang CY, Wu J, Zhao F, Zhang J, Selak MA, Ney PA, Thompson CB. Ulk1 plays a critical role in the autophagic clearance of mitochondria and ribosomes during reticulocyte maturation. Blood 2008;112:1493–1502.

[54] Zhang J, Randall MS, Loyd MR, Dorsey FC, Kundu M, Cleveland JL, Ney PA. Mitochondrial clearance is regulated by Atg7-dependent and -independent mechanisms during reticulocyte maturation. Blood 2009;114:157–164.

[55] Zhang H, Bosch-Marce M, Shimoda LA, Tan YS, Baek JH, Wesley JB, Gonzalez FJ, Semenza GL. Mitochondrial autophagy is an HIF-1-dependent adaptive metabolic response to hypoxia. J Biol Chem 2008;283:10892–10903.

[56] Liu L, Feng D, Chen G, Chen M, Zheng Q, Song P, Ma Q, Zhu C, Wang R, Qi W, Huang L, Xue P, Li B, Wang X, Jin H, Wang J, Yang F, Liu P, Zhu Y, Sui S, Chen Q.

Mitochondrial outer-membrane protein FUNDC1 mediates hypoxia-induced mitophagy in mammalian cells. Nat Cell Biol 2012;14:177–185.

[57] Narendra D, Tanaka A, Suen DF, Youle RJ. Parkin is recruited selectively to impaired mitochondria and promotes their autophagy. J Cell Biol 2008;183:795–803.

[58] Morais VA, Verstreken P, Roethig A, Smet J, Snellinx A, Vanbrabant M, Haddad D, Frezza C, Mandemakers W, Vogt-Weisenhorn D, Van Coster R, Wurst W, Scorrano L, De Strooper B. Parkinson's disease mutations in PINK1 result in decreased Complex I activity and deficient synaptic function. EMBO Mol Med 2009;1:99–111.

[59] Gautier CA, Kitada T, Shen J. Loss of PINK1 causes mitochondrial functional defects and increased sensitivity to oxidative stress. Proc Natl Acad Sci U S A 2008; 105:11364–11369.

[60] Morais VA, Haddad D, Craessaerts K, De Bock PJ, Swerts J, Vilain S, Aerts L, Overbergh L, Grunewald A, Seibler P, Klein C, Gevaert K, Verstreken P, De Strooper B. PINK1 loss-of-function mutations affect mitochondrial complex I activity via NdufA10 ubiquinone uncoupling. Science 2014;344:203–207.

[61] Vos M, Esposito G, Edirisinghe JN, Vilain S, Haddad DM, Slabbaert JR, Van Meensel S, Schaap O, De Strooper B, Meganathan R, Morais VA, Verstreken P. Vitamin K2 is a mitochondrial electron carrier that rescues pink1 deficiency. Science 2012;336:1306–1310.

[62] Meissner C, Lorenz H, Weihofen A, Selkoe DJ, Lemberg MK. The mitochondrial intramembrane protease PARL cleaves human Pink1 to regulate Pink1 trafficking. J Neurochem 2011;117:856–867.

[63] Narendra DP, Jin SM, Tanaka A, Suen DF, Gautier CA, Shen J, Cookson MR, Youle RJ. PINK1 is selectively stabilized on impaired mitochondria to activate Parkin. PLoS Biol 2010;8:e1000298.

[64] Deas E, Plun-Favreau H, Gandhi S, Desmond H, Kjaer S, Loh SH, Renton AE, Harvey RJ, Whitworth AJ, Martins LM, Abramov AY, Wood NW. PINK1 cleavage at position A103 by the mitochondrial protease PARL. Hum Mol Genet 2011;20:867–879.

[65] Greene AW, Grenier K, Aguileta MA, Muise S, Farazifard R, Haque ME, McBride HM, Park DS, Fon EA. Mitochondrial processing peptidase regulates PINK1 processing, import and Parkin recruitment. EMBO Rep 2012;13:378–385.

[66] Lazarou M, Jin SM, Kane LA, Youle RJ. Role of PINK1 binding to the TOM complex and alternate intracellular membranes in recruitment and activation of the E3 ligase Parkin. Dev Cell 2012;22:320–333.

[67] Van Humbeeck C, Cornelissen T, Hofkens H, Mandemakers W, Gevaert K, De Strooper B, Vandenberghe W. Parkin interacts with Ambra1 to induce mitophagy. J Neurosci 2011;31:10249–10261.

[68] Lee JY, Nagano Y, Taylor JP, Lim KL, Yao TP. Disease-causing mutations in Parkin impair mitochondrial ubiquitination, aggregation, and HDAC6-dependent mitophagy. J Cell Biol 2010;189:671–679.

[69] Matsuda N, Sato S, Shiba K, Okatsu K, Saisho K, Gautier CA, Sou YS, Saiki S, Kawajiri S, Sato F, Kimura M, Komatsu M, Hattori N, Tanaka K. PINK1 stabilized by mitochondrial depolarization recruits Parkin to damaged mitochondria and activates latent Parkin for mitophagy. J Cell Biol 2010;189:211–221.

[70] Geisler S, Holmstrom KM, Skujat D, Fiesel FC, Rothfuss OC, Kahle PJ, Springer W. PINK1/Parkin-mediated mitophagy is dependent on VDAC1 and p62/SQSTM1. Nat Cell Biol 2010;12:119–131.

[71] Chan NC, Salazar AM, Pham AH, Sweredoski MJ, Kolawa NJ, Graham RL, Hess S, Chan DC. Broad activation of the ubiquitin-proteasome system by Parkin is critical for mitophagy. Hum Mol Genet 2011;20:1726–1737.

[72] Haddad DM, Vilain S, Vos M, Esposito G, Matta S, Kalscheuer VM, Craessaerts K, Leyssen M, Nascimento RM, Vianna-Morgante AM, De Strooper B, Van Esch H, Morais VA, Verstreken P. Mutations in the intellectual disability gene Ube2a cause neuronal dysfunction and impair Parkin-dependent mitophagy. Mol Cell 2013;50:831–843.

[73] Liu S, Sawada T, Lee S, Yu W, Silverio G, Alapatt P, Millan I, Shen A, Saxton W, Kanao T, Takahashi R, Hattori N, Imai Y, Lu B. Parkinson's disease-associated kinase PINK1 regulates Miro protein level and axonal transport of mitochondria. PLoS Genet 2012;8:e1002537.

[74] Ziviani E, Tao RN, Whitworth AJ. *Drosophila* Parkin requires PINK1 for mitochondrial translocation and ubiquitinates mitofusin. Proc Natl Acad Sci U S A 2010;107: 5018–5023.

[75] Lim KL, Dawson VL, Dawson TM. Parkin-mediated lysine 63-linked polyubiquitination: a link to protein inclusions formation in Parkinson's and other conformational diseases? Neurobiol Aging 2006;27:524–529.

5

LYSOSOME EXOCYTOSIS AND MEMBRANE REPAIR

Rajesh K. Singh and Abigail S. Haka
Department of Biochemistry, Weill Cornell Medical College, New York, NY, USA

5.1 INTRODUCTION

It has long been recognized that lysosomes serve as the degradative machinery within the cell where larger molecules, such as proteins and polysaccharides, are catabolized into their basic components. In addition to their role in intracellular degradation, lysosomes can be trafficked to the cell periphery where they release their lumenal contents into the extracellular space by fusing with the plasma membrane. Lysosome exocytosis occurs in a large variety of cell types and is used to regulate a number of biologically important processes. While these processes are diverse, several common themes have emerged regarding the functions of lysosome exocytosis. Thus, the role of lysosomes in cellular biology has been expanded, and, in addition to the conventional view that lysosomes are terminal compartments of the endocytic pathway, it is now accepted that lysosomes behave as regulated secretory vesicles.

5.2 FUNCTIONS OF LYSOSOME EXOCYTOSIS

Lysosomes are acidic organelles that participate not only in intracellular degradation but also in a wide variety of other cellular functions. Ca^{2+}-triggered fusion of lysosomes and other intracellular vesicles with the plasma membrane is a highly conserved phenomenon. Once considered a pathway exclusive to specialized secretory

Lysosomes: Biology, Diseases, and Therapeutics, First Edition.
Edited by Frederick R. Maxfield, James M. Willard and Shuyan Lu.

cells, we now know that Ca^{2+}-regulated exocytosis is a ubiquitous cellular process [1]. Investigations of regulated exocytosis in nonspecialized cells, such as fibroblasts and epithelial cells have identified conventional lysosomes as the major intracellular compartment capable of reacting to Ca^{2+} by fusing with the plasma membrane [2]. The capacity to sense Ca^{2+} rises and respond by lysosome exocytosis is a process present in a large number of cell types and serves a variety of effector functions [3].

5.2.1 Specialized Lysosome-Related Organelles

Lysosome-related organelle is a term used to describe the modified lysosomal vesicles found in specialized secretory cells [4–6]. The usage and definition of the term "lysosome-related organelle" can vary greatly. Some prefer to call most secretory organelles related to lysosomes "secretory lysosomes," including lytic granules of CD8-positive cytotoxic T lymphocytes (CTLs), melanosomes, platelet dense granules, azurophil granules of neutrophils, and the ruffled border secretory vesicles of osteoclasts [1,7,8]. Others use the term "lysosome-related organelle" to encompass these secretory organelles [9,10]. In this chapter, we refer to such organelles as "lysosome-related organelles." A key distinction between conventional lysosomes and lysosome-related organelles originated from the mistaken belief that conventional lysosomes were unable to undergo exocytic events. However, it is now understood that conventional lysosomes can also undergo secretion in a wide variety of cell types. Similar to conventional lysosomes, lysosome-related organelles have an acidic lumen and contain high concentrations of lysosomal acid hydrolases. However, unlike conventional lysosomes, which are multivesicular [4,6], lysosome-related organelles have a heterogeneous morphology and contain a dense core that holds specialized mediators specifically destined for secretion. Examples of such contents include melanin in melanocytes, proinflammatory mediators serotonin and histamine in mast cells, and perforin and granzyme in CTLs [11–13]. Further, lysosome-related organelles contain unique surface proteins that are specifically required for secretion and are not used during conventional lysosome exocytosis. For instance, the exocytic regulator Rab27a and its effector proteins are used by the majority of secretory cells but are not present on conventional lysosomes [6,14]. Exocytosis of lysosome-related organelles has been studied extensively and may provide insights into the molecular machinery regulating conventional lysosome exocytosis.

Exocytosis of lysosome-related organelles results in the release of their contents into the extracellular space, where they perform a number of effector functions. One of the best studied examples is the Ca^{2+}-dependent degranulation and target cell killing by CTLs [15]. CTLs function to kill tumor cells or virally infected cells by making an intimate contact with the target cell called the immunological synapse [16]. CTLs secrete contents contained in lysosome-related organelles, such as perforin and granzyme A, into the immunological synapse where these proteins elicit cytotoxic effects on the infected cell [17]. Actin depletion at the immunological synapse initiates events leading to lysosome-related organelle exocytosis, and integrins are used to seal the contact area thereby containing the secreted lysosomal contents [16,18].

Lysosome-related organelles can be exocytosed by a large variety of cell types in addition to CTLs and perform a number of other biologically important functions. Exocytosis and/or shedding of melanosome-rich packages containing melanin from the tips of melanocyte dendrites is required for their delivery to the extracellular space prior to their uptake by keratinocytes, where they protect nuclear DNA from light-induced damage in the skin [19,20]. Blood clotting relies on lysosome-related organelle exocytosis of platelet dense granules [21]. Mast cells exocytose histamine and other proinflammatory molecules from specialized secretory granules that are responsible for allergic and antiparasitic responses [22]. The exocytosis of lysosome-related organelles regulates many diverse processes and is therefore of considerable physiological importance in the context of disease and homeostasis.

5.2.2 Lysosome Exocytosis for Membrane Repair

The discovery that exocytosis of conventional lysosomes occurred in nonspecialized cells, such as fibroblasts and epithelial cells, arose during the study of cellular wound repair. Bacterial toxins, such as Streptolysin-O, create pores in the plasma membrane of cells, which must be sealed to avoid cellular necrosis or apoptosis. Ca^{2+} influx through plasma membrane wounds triggers a rapid repair response that is essential for cell survival. Calcium ionophores can induce the secretion of predocked conventional lysosomes in many different cell types [3] and pore-forming toxins in the presence of extracellular calcium cause lysosome exocytosis in a variety of cells [2]. Interfering with the exocytosis of lysosomes inhibits the ability of cells to reseal plasma membrane wounds, establishing these organelles as important mediators of plasma membrane repair [23]. Originally, it was thought that lysosomal exocytosis functioned to promote resealing of the wound solely by providing a source of new membrane. In this scenario, resealing is mediated by the addition of intracellular lysosomal membrane to the cell surface through exocytosis [24,25]. The additional membrane provided by Ca^{2+}-dependent exocytosis also results in a decrease in plasma membrane tension that is required for bilayer resealing [26–28]. However, it is now appreciated that the delivery of lysosomal contents also plays a role in wound healing. In particular, acid sphingomyelinase, a lysosomal hydrolase, promotes membrane repair by converting membrane sphingomyelin into ceramide. The generation of ceramide promotes endocytosis and clearance of membrane lesions [29]. In human patients and murine models, mutations in the genes required for lysosome exocytosis-mediated plasma membrane repair cause muscular degenerative diseases, thus highlighting the importance of understanding cellular wound healing [30–32].

Plasma membrane repair mechanisms are also important in infection as they can be hijacked to evade host defenses. After attachment of the parasite *Trypanosoma cruzi* to the cell, lysosomes surrounding the area fuse with the nascent parasitic vacuole and provide a source of membrane that promotes parasite internalization [33,34]. Inhibition of lysosome exocytosis or acid sphingomyelinase blocks the entry of *T. cruzi* into cells [33,34]. *Mycobacterium tuberculosis* infection of cells also induces lesions in the plasma membrane. Macrophages use lysosome exocytosis in an attempt to repair these lesions but *M. tuberculosis* can interfere with this repair mechanism to promote macrophage necrosis and evade host defenses [35].

5.2.3 Lysosome Exocytosis as a Source of Membrane

In addition to providing a source of new membrane during cellular wound healing, lysosome exocytosis to the nascent phagosome provides membrane to facilitate phagocytosis. This is of particular importance to promote efficient phagocytosis of large particles, allowing the phagosome to surround the particle and close [36,37]. The nervous system is also thought to rely on lysosome exocytosis to provide a source of membrane for neuron dendrite growth. Neurons are specialized cells that require extension and branching of long processes, called dendrites, in order to make connections with other neurons. Recent studies have suggested that lysosome exocytosis occurs at the tips of newly forming dendrites, which may provide a source of membrane to promote the extension of these protrusions [38].

5.2.4 Lysosome Exocytosis for Extracellular Degradation

Perhaps not surprisingly, akin to their function within the cell, lysosomes can aid degradation of moieties outside of the cell. This occurs in cases where the target to be degraded cannot be internalized by standard endocytic or phagocytic mechanisms. The best-studied example transpires during bone remodeling. Osteoclasts are multinucleated cells of hematopoietic origin that degrade the bone matrix. They secrete lysosomal contents onto areas of bone that are destined for remodeling and resorption [39]. Osteoclasts confine the area of degradation by forming an intimate contact with the region of bone to be remodeled. This contact is facilitated by a filamentous actin (F-actin) ring and adhesion molecules, called integrins, that form a sealing zone [40]. Lysosomal contents are secreted into the contact zone and are confined to this area. Lysosome exocytosis also delivers vacuolar ATPases to the sealing zone [41]. Vacuolar ATPases pump protons from the cell into the contact area, which causes acidification and greater activity of lysosomal enzymes that function more efficiently at low pH. Interestingly, the osteoclast forms elaborate membrane protrusions in the contact area called a ruffled border. As in wound repair, lysosome exocytosis is thought to provide the extra source of membrane required to form the ruffled border. The ruffled border can be extensive and provides a large surface area for the release of protons and hydrolases and the uptake of catabolized material [42]. During this process, the osteoclast is polarized, and lysosome exocytosis is thought to be a driving force to generate and maintain cell polarity [43].

Lysosome exocytosis-mediated bone remodeling plays an important role in skeletal pathologies. Recently, it was found that mice with impaired osteoclast lysosome exocytosis have lower bone mineral density, known as osteopenia, due to defective remodeling and bone formation [44]. The impaired exocytosis prevents osteoclast secretion of lysosomal enzymes required to degrade and remodel bone (in particular cathepsin K). Interestingly, osteoclasts have both lysosome-related organelles and conventional lysosomes. The lysosome-related organelles contain cathepsin K, which degrades type I collagen, the major component of bone matrix and is the main lysosomal hydrolase used to degrade bone [45]. Osteoclasts also have a second distinct population of lysosomes, which lack cathepsin K but contain cathepsin D. Upon binding of bone matrix proteins and osteoclast integrins, lysosome-related organelles fuse with the plasma membrane and initiate bone resorption [46]. However, cathepsin D, which is not thought to play a role in bone remodeling [47], is not released during the stimulatory event, suggesting that these lysosomes may serve a more conventional role in degradation within the cell [46]. Future studies are needed to examine lysosome heterogeneity in other types of secretory cells.

An additional example of lysosome exocytosis functioning to degrade extracellular species occurs during macrophage catabolism of aggregated lipoproteins. During atherosclerosis, low-density lipoproteins (LDLs) become aggregated and retained in the arterial wall [48,49]. Macrophages engage the LDL aggregates forming a stable contact zone into which lysosomal contents are secreted [50] (Figure 5.1). Again, lysosomal enzymes are confined to the area of degradation by local actin polymerization that restricts passage of molecules from the extracellular compartment to the outside environment [51]. As in osteoclasts, vacuolar ATPases pump protons into the extracellular compartment, and an acidic pH allows activity of lysosomal acid hydrolases delivered by exocytosis. Akin to the ruffled border in osteoclasts, extensive membrane ruffling is observed in macrophage contact areas with aggregated LDL

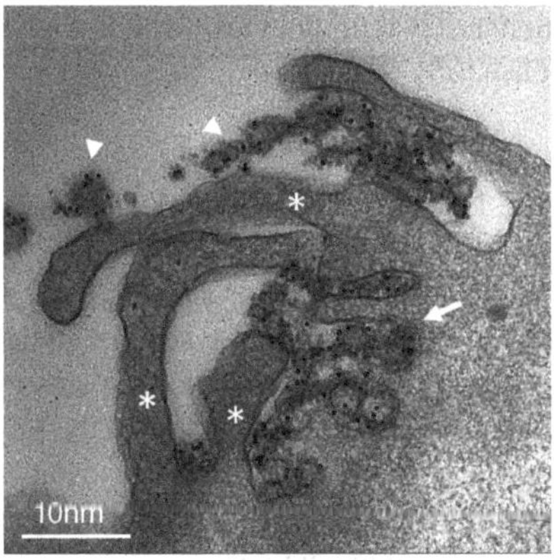

Figure 5.1 Lysosome exocytosis is required for extracellular degradation. Electron micrograph of a macrophage digesting aggregated LDL, labeled with colloidal gold, sequestered in deep membrane invaginations (arrow). During atherogenesis, macrophages make an extracellular degradative compartment to catabolize aggregated LDL. Formation of the compartment is likely aided by extensive plasma membrane ruffling at contact sites (asterisks). Lysosomal enzymes are delivered to the contact area through exocytosis and the compartment is acidified by plasma membrane vacuolar ATPase, allowing activity of lysosomal enzymes. This process promotes extracellular degradation of aggregated LDL prior to uptake of partially digested material and ensuing foam cell formation. Portions of aggregated LDL that have not been sequestered by the macrophage can also be seen (arrowheads). Image courtesy of I. Grosheva and F.R. Maxfield.

(Figure 5.1). The membrane required to form this ruffled border may be provided by lysosome exocytosis. The interaction results in the hydrolysis of LDL cholesteryl esters and transfer of free cholesterol to the macrophage with subsequent foam cell formation, a key event in atherogenesis.

5.2.5 Lysosome Exocytosis and Delivery of Proteins to the Cell Surface

In the same way that lysosome exocytosis delivers vacuolar ATPases to the ruffled border in osteoclasts and macrophages, in dendritic cells, fusion of an intracellular compartment with the plasma membrane delivers major histocompatibility complex class II (MHCII) to their surface [52]. These MHC class II compartments (MIICs) are thought to be either a lysosome-related organelle or a conventional late endosome/lysosome (discussed in depth in Chapter 12). Dendritic cells sample the local environment by endocytosis of antigens, in an effort to identify pathogens and initiate an immune response. MHCII, which is present in dendritic cell MIICs, is important in antigen presentation. Endocytosed antigens are delivered to lysosomes but are only partially degraded [53] because dendritic cell lysosomes are less acidic than lysosomes in most cells, thus dampening the activity of many lysosomal hydrolases [54]. Partially degraded peptides are then loaded onto MHCII, and this complex is presented on the cell surface. Delivery of the MHCII–antigen complex occurs by fusion of tubular MIICs with the plasma membrane during the process of dendritic cell maturation [52]. Cell surface presentation is important for recognition by other leukocytes, such as T lymphocytes, which then become activated and initiate an immune response [55].

5.3 MECHANISMS OF LYSOSOME EXOCYTOSIS

Despite its importance in a wide range of processes, current understanding of the molecular machinery regulating conventional lysosome exocytosis is limited. However, recent studies have begun to unravel the mechanisms governing exocytosis of specialized lysosome-related organelles, and many of the key proteins involved are now known. Consideration of proteins involved in lysosome-related organelle exocytosis may provide important insights into the regulatory machinery likely to be involved in more general lysosome exocytosis.

There are two main types of exocytosis that occur within the cell. The first is regulated exocytosis. In this process, cells exhibit polarized exocytosis secondary to an extracellular stimulatory signal. The regulated exocytosis of lysosomes involves integration of major cytoarchitectural changes with precisely timed and often targeted vesicular transport to the extracellular space. The second type of exocytosis is constitutive exocytosis. Molecules such as growth factors and cytokines that are required in the extracellular space for tissue homeostasis can be secreted constitutively. Such molecules are synthesized and trafficked through the trans-Golgi network and packaged into vesicles distinct from lysosomes and lysosome-related organelles, which then fuse with the plasma membrane in a constant manner [56]. In unstimulated fibroblasts, epithelial cells, and macrophages, constitutive lysosome exocytosis occurs at a low frequency, though the rate is slightly higher in macrophages, for reasons that are not known [2]. This is because in general, lysosome exocytosis is a regulated process that requires a specific stimulus to induce exocytosis [4,57]. Very little is known regarding the purpose of constitutive exocytosis of lysosomes. Therefore, we limit our discussion to stimulus-induced exocytosis of lysosomes.

In resting cells, levels of intracellular calcium are strictly maintained. However, upon stimulation, cell surface receptor-mediated signaling leads to an increase in intracellular calcium. This stimulates regulated exocytosis of lysosomes (Figure 5.2). While lysosome exocytosis is predominately regulated by calcium, other factors can also stimulate exocytosis, such as the depletion of membrane cholesterol [58].

5.3.1 Maturation of Lysosomes and Lysosome-Related Organelles

While the steps regulating lysosome maturation have been the subject of much study [6,59], the requirement for maturation of lysosomes prior to exocytosis is

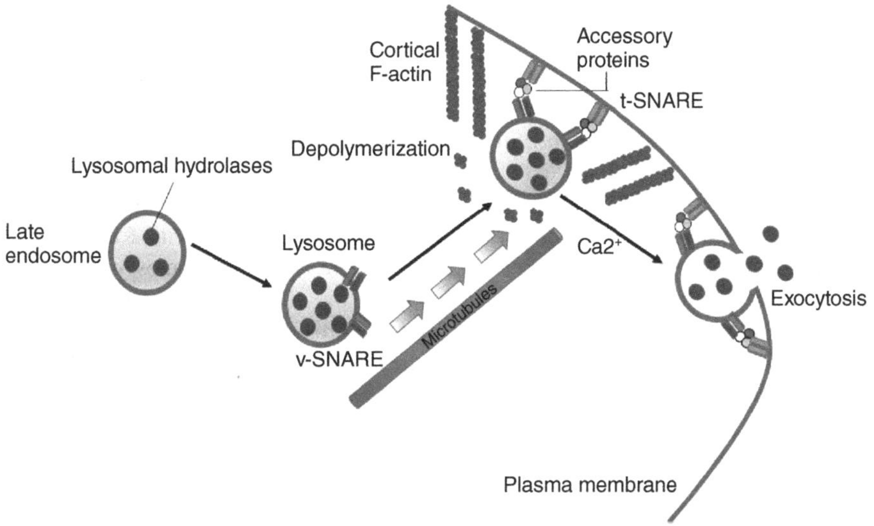

Figure 5.2 Schematic of mechanisms involved in lysosome exocytosis. Lysosomes acquire lysosomal hydrolases from late endosomes. Signaling through ligation of cell surface receptors induces microtubule-dependent movement of lysosomes toward the plasma membrane. Local depolymerization of cortical actin can promote docking of lysosomes at the plasma membrane in some systems. While in other cell types, tethering of lysosomes to cortical actin appears to be necessary for exocytosis to occur. Engagement of vesicular membrane-associated and target membrane-associated SNAREs occurs with the help of accessory proteins, such as calcium sensing synaptotagmins. The binding of SNAREs makes lysosome fusion with the plasma membrane more energetically favorable and influx of calcium stimulates exocytosis to occur. This releases lysosomal hydrolases and other secretory products into the extracellular environment where they mediate various effector functions. (*See color plate section for the color representation of this figure.*)

not well understood. By contrast, the maturation of lysosome-related organelles prior to fusion with the plasma membrane has been studied in detail (discussed in depth in Chapter 12). A critical maturation step is the fusion of exocytic vesicles with immature lysosome-related organelles [60]. Exocytic vesicles are formed from the fusion of late endosomes and recycling endosomes [61] and deliver proteins necessary for fusion with the plasma membrane [60]. In mast cells and melanosomes, lysosome-related organelles maturation can be visualized by the progressive buildup of secretory cargo in these organelles [11,62,63].

5.3.2 Transport of Lysosomes to the Plasma Membrane

An early step in the process of lysosome exocytosis is their transport to the plasma membrane. Rab GTPases are a family of evolutionary conserved small GTPases that are important regulators of intracellular trafficking [56,64]. Members of the Rab GTPase family have been shown to control regulated exocytosis of lysosome-related organelles that are secreted in response to external stimuli. Rab GTPases are typically present on the lysosome-related organelle, and upon stimulation of the cell, they become activated and bind to a discrete set of effector proteins. Rab effector proteins are involved in both the transport of lysosomes to the plasma membrane via microtubules and actin as well as their docking at the membrane prior to exocytosis.

5.3.2.1 Microtubules in Lysosome Transport Lysosomes and secretory vesicles in many cell types are transported within the cell through interaction with the microtubule network [33,65,66]. The spatial targeting of lysosomes is regulated through association with the microtubule motor proteins kinesin and dynein. Kinesins and dyneins can bind to microtubules by their heavy chains and to lysosomes by their light chains [67]. The heavy chains are responsible for the motor activity and move processively along microtubules in a manner that requires ATP hydrolysis [67]. Microtubules emerge from a structure called the microtubule-organizing center (MTOC), which is the major site of microtubule nucleation within the cell [68]. Kinesin heavy chains move along microtubules toward the plus end (the end distal to the MTOC), while dynein heavy chains move toward the minus end (the end proximal to the MTOC), transporting lysosomes with them. Lysosome dispersion within the cell and anterograde/retrograde transport of lysosomes toward the cell periphery/interior is therefore dependent on kinesin and dynein.

Several studies have attempted to characterize the interaction of kinesin/dynein light chains with lysosomes. Kinesin light chain interaction with lysosomes is thought to occur due to binding to the lysosomal proteins SifA, kinesin-interacting protein and the small GTPase Arl8 in its GTP-bound active form [69]. Dynein light chains can interact with Rab7-interacting lysosomal protein (RILP), which binds the small GTPase Rab7 in its active GTP-bound form in fibroblast-like Rat2 cells [70]. In this way, active Rab7 and RILP regulate lysosome minus end transport by recruiting dynein motors to lysosomes [71]. Dynein is known to interact with the accessory protein dynactin, which is required for minus end transport. Overexpression of the dynamitin (p50) subunit of dynactin disrupts dynein/dynactin complex formation and redistributes lysosomes to the cell periphery in HeLa cells [72]. Therefore, it seems that under resting conditions in the cell, this "tug of war" between kinesin and dynein regulates lysosome distribution within the cell.

In CTLs, the MTOC polarizes toward the cell to be killed, so minus end transport allows lysosome-related organelles to be trafficked toward the target cell [73]. This function requires Rab7 and the effector RILP, which interact with dynein and couple lysosome-related organelles to minus end transport [74]. The adaptor protein AP-3 has also been implicated in promoting minus end microtubule-dependent transport of lysosome-related organelles in CTLs [75], and consistent with this, depletion of AP-3 in HEK293T cells causes lysosomes to accumulate at the cell periphery [76]. However, AP-3 does not localize to lysosome-related organelles; therefore, it is likely that AP-3 does not interact directly with the dynein/dynactin complex but rather regulates the sorting of cargoes important for minus end transport [75,76].

Exogenous factors can also influence lysosome distribution and transport. Cholesterol is known to regulate lysosome distribution, and its depletion redistributes lysosomes to the cell periphery [58]. However, the role of microtubules in this process has not been evaluated. Cholesterol regulation of organelle distribution was recently suggested to be mediated by cholesterol binding to the Rab7 effector Orp1-like, in late endosomes [77], and this may also be true for lysosomes. In live cell experiments, macrophage lysosomes make tubular structures that extend radially and are dependent on microtubules [78,79]. Formation of tubular lysosomes can be stimulated by exogenous factors, such as lipopolysaccharide and phorbol-12-myristate 13-acetate, and their formation is dependent on dynein, kinesin, Rab7, RILP, FYCO1, Arl8, and SKIP [80,81]. Tubular lysosomes were also found to be motile more often than punctuate lysosomes [80], suggesting a link between microtubule-dependent motility of lysosomes and their morphology. Interestingly, in some cell types, microtubules appear to be less important in lysosome secretion [82]. This perhaps suggests that in certain circumstances, lysosomes in close proximity to the plasma membrane do not require microtubule transport and can be selectively exocytosed.

Rab7 is a protein known to be associated with the lysosome [71]. Rab7 function has been characterized in the endocytic pathway, where it promotes late endosome/lysosome fusion. However, it also induces microtubule-based motility of lysosomes within the cell. It is presumably through this aspect of Rab7 function that it plays a role in lysosome-related organelles exocytosis in CTLs [74]. This is supported by the fact that in CTLs, as discussed above, Rab7 interacts with the effector RILP that recruits the microtubule motor protein dynein [74]. Rab7 also likely plays a role in lysosome-related organelle exocytosis from osteoclasts [43,83]. Further, Rab7-GTP is known to interact with Rac1, so it has been suggested that this interaction may promote switching of microtubule to actin-based transport of lysosomes [84].

5.3.2.2 Cortical Actin in Lysosome Exocytosis The role of cortical actin in exocytosis has been debated in recent years. Some studies suggest that F-actin is required to tether lysosome-related organelles close to the plasma membrane prior to exocytosis while others find that F-actin creates a barrier that requires remodeling prior to exocytosis. In melanocytes, cortical actin seems to play an important role in the peripheral distribution of melanosomes prior to exocytosis. Microtubule-based transport of melanosomes is followed by their peripheral tethering to cortical actin due to the actions of the Rab27a/Melanophilin/Myosin Va complex [85,86]. This is a requirement for pigment transfer as deficiency of any protein in this complex leads to pigment dilution in the skin of mice and humans [87–92]. Similarly, melanosome secretion from retinal pigment epithelium, the color producing cell in the eye, requires a Rab27a/MyRIP/Myosin VIIa complex to perform tethering to cortical actin prior to exocytosis [93]. Melanocytes are not the only cell types to use this tethering function. In neural PC12 cells, Rab27a and MyRIP link secretory granules to F-actin and this interaction has been proposed to control the motion of granules to exocytic sites [94].

While some lysosome-related organelles show a requirement for peripheral tethering to cortical actin prior to exocytosis, there is also evidence that cortical actin acts as a barrier to lysosome secretion and must be remodeled to allow exocytosis to occur (Figure 5.2). Actin depolymerization prior to stimulation with a calcium ionophore increased apical lysosome exocytosis from Madin–Darby canine kidney epithelial cells [82]. Depolymerization of F-actin also increased lysosome exocytosis from NRK cells in response to calcium, while F-actin stabilization using phalloidin reduced lysosome exocytosis in this system [95]. Further, entry of *T. cruzi* into cells, which is dependent on lysosome exocytosis, is enhanced under conditions where actin is depolymerized [33]. F-actin depolymerization prior to stimulation also increases granule exocytosis from mast cells [66,96,97]. These studies support a barrier function for F-actin in exocytosis, in which actin polymerization slows the transport of lysosome-related organelles [98,99]. Recently, the secretory machinery itself was proposed to remodel F-actin in the local environment of secretory granules, thereby promoting exocytosis. The Rab27 effector synaptotagmin-like protein 1 (Slp1) was found to recruit RhoA-GAP GMIP and inhibit RhoA signaling on secretory granules. In neutrophils, this promotes remodeling of F-actin around secretory granules allowing them to traverse cortical actin prior to exocytosis in neutrophils [99]. Further studies are needed to determine the mechanisms by which the secretory machinery in other cell types is able to overcome this F-actin barrier prior to exocytosis as well as if and when actin remodeling is necessary.

5.3.3 Tethering of Lysosomes to the Plasma Membrane

Currently, the Rab GTPases regulating lysosome exocytosis remain largely uncharacterized, though their role in exocytosis of lysosome-related organelles has been intensively studied (Table 5.1). Several "secretory" Rab GTPases have been identified, including Rab27 and Rab3, which are related through amino acid sequence similarity. Many studies have focused on the role of Rab27 and Rab3 as they are expressed in a variety of different cell types. Rab27a regulates exocytosis of lysosome-related organelles from CD8 positive CTLs [12]. While Rab27a-deficient *ashen* CTLs polarize correctly to the site of contact with the target cell, lysosome-related organelles do not dock with the plasma membrane, suggesting a role for Rab27a in this process. CD4-positive T helper lymphocytes typically use mechanisms distinct from CTLs to kill target cells, but there have been reports

TABLE 5.1 Summary of Secretion Machinery Used by Different Cell Types for Exocytosis of Various Secretory Organelles

Cell Type	Organelle	Proteins Regulating Secretion	References
Fibroblast	Lysosome	Synaptotagmin VII	[100]
		SNAP-23	[101]
		Syntaxin4	[101]
		VAMP-7	[101]
Osteoclast	Lysosome-related organelle	Synaptotagmin VII	[39]
		Rab7	[37]
$CD4^+$ T cells	Secretory granule	Rab27a	[102]
$CD8^+$ T cells (CTLs)	Lysosome-related organelle	Rab27a	[102,103]
		Munc13-4	[104]
		Slp1	[105]
		Slp2	[105,106]
		Slp3	[107]
		Kinesin-1	[107]
		Syntaxin 11	[108,109]
		VAMP-8	[110,111]
		Vti1b	[111]
		VAMP-2	[112]
		AP-3	[71]
		Rab7	[70]
		RILP	[70]
NK cells	Lysosome-related organelle	Rab27a	[57,94,102]
		Munc13-4	[57,113]
		Syntaxin 11	[108,109]
		VAMP-4	[114]
		VAMP-7	[114]
Mast cell	Secretory granule	Rab27a/b	[93,115]
		Munc13-4	[116–118]
		Doc2α	[118]
		Rab3d	[119]
		Synaptotagmin II	[120]
		VAMP-8	[121–123]
		VAMP-7	[121]
		SNAP-23	[121]
		Syntaxin-4	[121]
Neutrophil	Azurophilic granule	Rab27a	[124]
		Rab27b	[125]
		Munc13-4	[126]
		Slp1	[126]
		Syntaxin 11	[108]
Neutrophil	Specific granule Tertiary granule	Rab27a	[127]

(*continued*)

TABLE 5.1 ***(Continued)***

Cell Type	Organelle	Proteins Regulating Secretion	References
Platelet	Dense granule	Rab27a/b	[128]
		Syntaxin 11	[129]
		Munc18b	[130]
		SNAP-23	[131]
		Slp1	[132]
		Munc13-4	[133]
Platelet	α-Granule	Syntaxin 11	[129]
		Munc18b	[130]
Platelet	Lysosome	SNAP-23	[134]
		Syntaxin 11	[129]
		Munc18b	[130]
Melanocyte	Melanosome	Rab27a	[135]
		Melanophilin	[81]
		Myosin Va	[82]
Spermatid	Acrosome	Rab27a	[136]
		Rab3a	[136,137]
		RIM	[137]
		Munc13-1	[137]

of CD4-positive lysosome-related organelle-mediated cytotoxicity in response to certain pathogens [103,138,139]. Rab27a is important for lysosome-related organelle exocytosis in CD4-positive T lymphocytes as well [140]. Mast cells use both Rab27a and its related isoform Rab27b in conjunction with Munc13-4 to promote secretory granule exocytosis [97,102,115]. In Rab27a/b double knockout mast cells, granules exhibited more prominent long-range movements than in wild-type cells. This was inhibited by nocodazole treatment, which disrupts microtubules, suggesting that Rab27 may promote plasma membrane docking of granules and in its absence, granules remain associated with microtubules [102]. Using Munc13-4 (FQL>AAA) and Munc13-4 (Δ280-285) mutants that specifically disrupt the Munc13-4/Rab27 association, it was shown that this interaction is required for stimulation-induced granule docking at the plasma membrane [115].

Rab27 and Munc13-4 have also been found to play a role in exocytic events in NK cells, platelets, and neutrophils [98,116,124–128,133,141–143]. While the rudimentary details about how Rab27/Munc13-4 function to promote exocytosis in NK cells and platelets are known, the exact mechanisms remain to be determined. It is likely that in these cell types, Rab27/Munc13-4 functions in much the same way that they do in mast cells. In addition to Rab27 and Munc13-4, recent studies have suggested a role for the effector Slp1 in neutrophil granule exocytosis. In neutrophils isolated from Slp1 knockout mice or in cells treated with Slp1 blocking antibodies, azurophilic granule secretion is inhibited [99,126]. Slp1 can bind to plasma membrane phospholipids, particularly phosphoinositide 3,4,5 triphosphate (PIP_3).

This may suggest that Slp1 promotes exocytosis by recruiting Rab27-positive granules to membrane microdomains rich in PIP_3 and allowing them to dock there prior to calcium-dependent exocytosis. Consistent with this, HL-60 neutrophil-like cells that were downregulated for Slp1 displayed increased LAMP-3-positive granule velocity and displacement, suggestive of greater microtubule-based transport and decreased granule docking at the plasma membrane [126].

Rab3, which exists in four isoforms (Rab3a, 3b, 3c, and 3d), is related to Rab27 in terms of structure [64]. Rab3a has been implicated in the regulation of exocytosis from sperm cells called spermatids [136]. Spermatids contain one large acidic dense-core granule that is related to a lysosome, called an acrosome. Upon contact with an oocyte during the process of fertilization, the acrosome is exocytosed [144]. This releases digestive enzymes onto the oocyte and helps the spermatid to break through the hard shell of the oocyte [145]. This process is similar to that described previously for osteoclast bone resorption and macrophage degradation of aggregated LDL. Acrosome exocytosis is regulated by both Rab27 and Rab3. A sequential model of action has been proposed in which Rab27 recruits a guanine nucleotide exchange factor for Rab3 that allows Rab3a to become activated [119]. Rab3d is thought to regulate exocytosis from other granulocyte cell types, such as eosinophils [146] and mast cells [101,110,134,147], though much less is known about their regulation.

5.3.4 Lysosome Fusion with the Plasma Membrane

Recent work has demonstrated a key role for *S*oluble *N*SF *A*ttachment *RE*ceptor (SNARE) proteins in membrane fusion events within the cell. SNARE proteins are expressed on each of the membranes destined to fuse. They each contain at least one α-helical domain that can bind to specific α-helical domains of other SNARE proteins. These helices form a specific and stable four-helix bundle and induce energetically favorable conditions to allow membrane fusion events to take place (Figure 5.2). The vesicular SNARE (v-SNARE) VAMP-7 and the target SNAREs (t-SNAREs) syntaxin 4 and SNAP-23 are involved in lysosome exocytosis in response to wounding in NRK cells [112]. In platelets, SNAP-23, syntaxin 2, and syntaxin 4 were found to be involved in exocytosis of conventional lysosomes [114]. For the exocytosis of lysosome-related organelles, additional SNAREs have been implicated. These include VAMP-2 and VAMP-8 for lysosome-related organelles in CTLs [121,122], VAMP-4 and VAMP-7 in NK cells [129], VAMP-7 and VAMP-8 in mast cell granule exocytosis [108,109], and Syntaxin 11 in platelets, neutrophils, NK cells, and CTLs [148–150] (Table 5.1). Recent work has indicated that lytic granules in NK cells do not fuse completely with the plasma membrane [100]. This type of exocytosis is often referred to as “kiss-and-run” and is also known to occur in the context of adrenal chromaffin cell granule release [151,152]. Similarly, in NK cells, “kiss-and-run” exocytosis may promote reclamation of lytic granule membrane and proteins; however, the factors influencing incomplete fusion of lysosome-related organelles with the plasma membrane are not well understood.

5.3.5 Calcium-Dependent Exocytosis

Lysosome exocytosis in many cell types has been shown to be calcium dependent. The protein Synaptotagmin VII (Syt VII) is thought to function as a calcium sensor for the exocytosis of lysosomes. Syt VII is important for calcium-dependent exocytosis of lysosomes in fibroblasts [104] and mediates cell invasion by *T. cruzi* that is dependent on calcium-regulated exocytosis of lysosomes [105]. It is also required for plasma membrane repair in response to scratch-induced wounding of NRK cells [23] and to repair *M. tuberculosis*-induced plasma membrane lesions [35]. In addition, Syt VII regulates calcium-dependent lysosome exocytosis at the rear of the cell. This can facilitate cell detachment and efficient migration of T cells and neutrophils [106], lysosome exocytosis in osteoclasts [44], and to nascent phagosomes in macrophages [37]. Lysosomes are the major vesicular compartment to exocytose in response to calcium in astrocytes [107], which is important for ATP release [111], and Syt VII is required for lysosome exocytosis as a source of membrane to promote neurite outgrowth [38]. The relationship between Syt VII and SNARE proteins in lysosome exocytosis has not been fully characterized, but one study found that calcium influx into wounded NRK cells stimulates formation of a complex consisting of SNAP-23/Syntaxin 4/VAMP7, which interacts with Syt VII [112]. This would allow Syt VII to confer calcium sensitivity to the SNARE fusion machinery (Figure 5.2). Another regulator of lysosome exocytosis is transcription factor EB, which can promote the fusion of lysosomes with the plasma membrane. The overexpression of transcription factor EB raises intracellular Ca^{2+} levels through the activation of the lysosomal Ca^{2+} channel MCOLN1 [113]. Interestingly, a recent study found a pool of lysosomes that can be exocytosed in response to cholesterol depletion in the plasma membrane independently of Syt VII and that this exocytosis occurred more efficiently in the absence of Ca^{2+} [58]. This suggests the existence of a distinct pool of lysosomes that may utilize different secretory proteins to undergo exocytosis in a calcium-independent manner.

5.4 CONCLUSION

Lysosome exocytosis plays an important role in diverse biological processes such as plasma membrane repair following mechanical or bacterial cellular damage, immune responses, parasite entry into cells, bone remodeling, phagocytosis of large particles, and extracellular degradation of disease-associated substrates. While significant progress has been made in understanding how lysosome exocytosis is regulated in various physiological systems, there is much yet to be learned. Parallels can be drawn between the exocytic regulation of conventional lysosomes and lysosome-related organelles. This is likely to serve as a template to guide future studies attempting to understand lysosome exocytosis in different cell types and systems. While our current understanding of the secretory machinery regulating lysosome exocytosis is limited, it is clear that elucidation of such mechanisms will have direct significance for human health. Discoveries may offer new strategies and drug targets for the prevention and treatment of a wide array of human diseases.

ACKNOWLEDGMENTS

We would like to thank Frederick Maxfield for helpful discussions and feedback that contributed to the chapter. We would also like to thank Leona Cohen-Gould at the EM and Histology Core at Weill Cornell Medical College for her generous help and support in generating the original image in Figure 5.1. We apologize to those whose work we omitted.

REFERENCES

[1] Idone V, Tam C, Andrews NW. Two-way traffic on the road to plasma membrane repair. Trends Cell Biol 2008;18:552–559.

[2] Rodriguez A, Webster P, Ortego J, Andrews NW. Lysosomes behave as Ca2+-regulated exocytic vesicles in fibroblasts and epithelial cells. J Cell Biol 1997;137:93–104.

[3] Jaiswal JK, Andrews NW, Simon SM. Membrane proximal lysosomes are the major vesicles responsible for calcium-dependent exocytosis in nonsecretory cells. J Cell Biol 2002;159:625–635.

[4] Blott EJ, Griffiths GM. Secretory lysosomes. Nat Rev Mol Cell Biol 2002;3:122–131.

[5] Griffiths GM. Secretory lysosomes – a special mechanism of regulated secretion in haemopoietic cells. Trends Cell Biol 1996;6:329–332.

[6] Luzio JP, Hackmann Y, Dieckmann NM, Griffiths GM. The biogenesis of lysosomes and lysosome-related organelles. Cold Spring Harb Perspect Biol 2014;6:a016840.

[7] Page LJ, Darmon AJ, Uellner R, Griffiths GM. L is for lytic granules: lysosomes that kill. Biochim Biophys Acta 1998;1401:146–156.

[8] Griffiths G. What's special about secretory lysosomes? Semin Cell Dev Biol 2002;13:279–284.

[9] Marks MS, Heijnen HF, Raposo G. Lysosome-related organelles: unusual compartments become mainstream. Curr Opin Cell Biol 2013;25:495–505.

[10] Harris DA, Kim K, Nakahara K, Vasquez-Doorman C, Carthew RW. Cargo sorting to lysosome-related organelles regulates siRNA-mediated gene silencing. J Cell Biol 2011;194:77–87.

[11] Raposo G, Tenza D, Mecheri S, Peronet R, Bonnerot C, et al. Accumulation of major histocompatibility complex class II molecules in mast cell secretory granules and their release upon degranulation. Mol Biol Cell 1997;8:2631–2645.

[12] Stinchcombe JC, Barral DC, Mules EH, Booth S, Hume AN, et al. Rab27a is required for regulated secretion in cytotoxic T lymphocytes. J Cell Biol 2001;152:825–834.

[13] Wasmeier C, Romao M, Plowright L, Bennett DC, Raposo G, et al. Rab38 and Rab32 control post-Golgi trafficking of melanogenic enzymes. J Cell Biol 2006;175:271–281.

[14] Fukuda M. Rab27 effectors, pleiotropic regulators in secretory pathways. Traffic 2013;14:949–963.

[15] Barry M, Bleackley RC. Cytotoxic T lymphocytes: all roads lead to death. Nat Rev Immunol 2002;2:401–409.

[16] Grakoui A, Bromley SK, Sumen C, Davis MM, Shaw AS, et al. The immunological synapse: a molecular machine controlling T cell activation. Science 1999;285:221–227.

[17] Trapani JA, Smyth MJ. Functional significance of the perforin/granzyme cell death pathway. Nat Rev Immunol 2002;2:735–747.

[18] Ritter AT, Asano Y, Stinchcombe JC, Dieckmann NMG, Chen B, et al. Actin depletion initiates events leading to granule secretion at the immunological synapse. Immunity 2015;42:864–876.

[19] Tarafder AK, Bolasco G, Correia MS, Pereira FJ, Iannone L, et al. Rab11b mediates melanin transfer between donor melanocytes and acceptor keratinocytes via coupled exo/endocytosis. J Invest Dermatol 2014;134:1056–1066.

[20] Wu XS, Masedunskas A, Weigert R, Copeland NG, Jenkins NA, et al. Melanoregulin regulates a shedding mechanism that drives melanosome transfer from melanocytes to keratinocytes. Proc Natl Acad Sci U S A 2012;109:E2101–E2109.

[21] Fitch-Tewfik JL, Flaumenhaft R. Platelet granule exocytosis: a comparison with chromaffin cells. Front Endocrinol (Lausanne) 2013;4:77.

[22] Abraham SN, St John AL. Mast cell-orchestrated immunity to pathogens. Nat Rev Immunol 2010;10:440–452.

[23] Reddy A, Caler E, Andrews NW. Plasma membrane repair is mediated by Ca2+-regulated exocytosis of lysosomes. Molecular Biology of the Cell 2001;12:266a.

[24] McNeil PL. Repairing a torn cell surface: make way, lysosomes to the rescue. J Cell Sci 2002;115:873–879.

[25] McNeil PL, Vogel SS, Miyake K, Terasaki M. Patching plasma membrane disruptions with cytoplasmic membrane. J Cell Sci 2000;113:1891–1902.

[26] Dai J, Sheetz MP. Regulation of endocytosis, exocytosis, and shape by membrane tension. Cold Spring Harbor Symp Quant Biol 1995;60:567–571.

[27] Togo T, Alderton JM, Bi GQ, Steinhardt RA. The mechanism of facilitated cell membrane resealing. J Cell Sci 1999;112:719–731.

[28] Togo T, Krasieva TB, Steinhardt RA. A decrease in membrane tension precedes successful cell-membrane repair. Mol Biol Cell 2000;11:4339–4346.

[29] Tam C, Idone V, Devlin C, Fernandes MC, Flannery A, et al. Exocytosis of acid sphingomyelinase by wounded cells promotes endocytosis and plasma membrane repair. J Cell Biol 2010;189:1027–1038.

[30] Chakrabarti S, Kobayashi KS, Flavell RA, Marks CB, Miyake K, et al. Impaired membrane resealing and autoimmune myositis in synaptotagmin VII-deficient mice. J Cell Biol 2003;162:543–549.

[31] Bansal D, Miyake K, Vogel SS, Groh S, Chen CC, et al. Defective membrane repair in dysferlin-deficient muscular dystrophy. Nature 2003;423:168–172.

[32] Liu J, Aoki M, Illa I, Wu C, Fardeau M, et al. Dysferlin, a novel skeletal muscle gene, is mutated in Miyoshi myopathy and limb girdle muscular dystrophy. Nat Genet 1998;20:31–36.

[33] Tardieux I, Webster P, Ravesloot J, Boron W, Lunn JA, et al. Lysosome recruitment and fusion are early events required for trypanosome invasion of mammalian cells. Cell 1992;71:1117–1130.

[34] Rodriguez A, Samoff E, Rioult MG, Chung A, Andrews NW. Host cell invasion by trypanosomes requires lysosomes and microtubule/kinesin-mediated transport. J Cell Biol 1996;134:349–362.

[35] Divangahi M, Chen M, Gan H, Desjardins D, Hickman TT, et al. Mycobacterium tuberculosis evades macrophage defenses by inhibiting plasma membrane repair. Nat Immunol 2009;10:899–906.

[36] Braun V, Fraisier V, Raposo G, Hurbain I, Sibarita JB, et al. TI-VAMP/VAMP7 is required for optimal phagocytosis of opsonised particles in macrophages. EMBO J 2004;23:4166–4176.

[37] Czibener C, Sherer NM, Becker SM, Pypaert M, Hui E, et al. Ca2+ and synaptotagmin VII-dependent delivery of lysosomal membrane to nascent phagosomes. J Cell Biol 2006;174:997–1007.

[38] Arantes RM, Andrews NW. A role for synaptotagmin VII-regulated exocytosis of lysosomes in neurite outgrowth from primary sympathetic neurons. J Neurosci 2006; 26:4630–4637.

[39] Lacombe J, Karsenty G, Ferron M. Regulation of lysosome biogenesis and functions in osteoclasts. Cell Cycle 2013;12:2744–2752.

[40] Matsumoto T, Nagase Y, Hirose J, Tokuyama N, Yasui T, et al. Regulation of bone resorption and sealing zone formation in osteoclasts occurs through protein kinase B-mediated microtubule stabilization. J Bone Miner Res 2013;28:1191–1202.

[41] Yang DQ, Feng S, Chen W, Zhao H, Paulson C, et al. V-ATPase subunit ATP6AP1 (Ac45) regulates osteoclast differentiation, extracellular acidification, lysosomal trafficking, and protease exocytosis in osteoclast-mediated bone resorption. J Bone Miner Res 2012;27:1695–1707.

[42] Stenbeck G. Formation and function of the ruffled border in osteoclasts. Semin Cell Dev Biol 2002;13:285–292.

[43] Zhao H, Laitala-Leinonen T, Parikka V, Vaananen HK. Downregulation of small GTPase Rab7 impairs osteoclast polarization and bone resorption. J Biol Chem 2001; 276:39295–39302.

[44] Zhao H, Ito Y, Chappel J, Andrews NW, Teitelbaum SL, et al. Synaptotagmin VII regulates bone remodeling by modulating osteoclast and osteoblast secretion. Dev Cell 2008;14:914–925.

[45] Saftig P, Hunziker E, Wehmeyer O, Jones S, Boyde A, et al. Impaired osteoclastic bone resorption leads to osteopetrosis in cathepsin-K-deficient mice. Proc Natl Acad Sci U S A 1998;95:13453–13458.

[46] van Meel E, Boonen M, Zhao H, Oorschot V, Ross FP, et al. Disruption of the Man-6-P targeting pathway in mice impairs osteoclast secretory lysosome biogenesis. Traffic 2011;12:912–924.

[47] Zhao H. Membrane trafficking in osteoblasts and osteoclasts: new avenues for understanding and treating skeletal diseases. Traffic 2012;13:1307–1314.

[48] Smith EB, Massie IB, Alexander KM. The release of an immobilized lipoprotein fraction from atherosclerotic lesions by incubation with plasmin. Atherosclerosis 1976;25:71–84.

[49] Tabas I, Williams KJ, Boren J. Subendothelial lipoprotein retention as the initiating process in atherosclerosis: update and therapeutic implications. Circulation 2007; 116:1832–1844.

[50] Haka AS, Grosheva I, Chiang E, Buxbaum AR, Baird BA, et al. Macrophages create an acidic extracellular hydrolytic compartment to digest aggregated lipoproteins. Mol Biol Cell 2009;20:4932–4940.

[51] Grosheva I, Haka AS, Qin C, Pierini LM, Maxfield FR. Aggregated LDL in contact with macrophages induces local increases in free cholesterol levels that regulate local actin polymerization. Arterioscler Thromb Vasc Biol 2009;29:1615–1621.

[52] Chow A, Toomre D, Garrett W, Mellman I. Dendritic cell maturation triggers retrograde MHC class II transport from lysosomes to the plasma membrane. Nature 2002;418:988–994.

[53] Savina A, Jancic C, Hugues S, Guermonprez P, Vargas P, et al. NOX2 controls phagosomal pH to regulate antigen processing during crosspresentation by dendritic cells. Cell 2006;126:205–218.

[54] Jancic C, Savina A, Wasmeier C, Tolmachova T, El-Benna J, et al. Rab27a regulates phagosomal pH and NADPH oxidase recruitment to dendritic cell phagosomes. Nat Cell Biol 2007;9:367–378.

[55] Bousso P. T-cell activation by dendritic cells in the lymph node: lessons from the movies. Nat Rev Immunol 2008;8:675–684.

[56] Stenmark H. Rab GTPases as coordinators of vesicle traffic. Nat Rev Mol Cell Biol 2009;10:513–525.

[57] Carta S, Lavieri R, Rubartelli A. Different members of the IL-1 family come out in different ways: DAMPs vs. cytokines? Front Immunol 2013;4:123.

[58] Hissa B, Duarte JG, Kelles LF, Santos FP, del Puerto HL, et al. Membrane cholesterol regulates lysosome-plasma membrane fusion events and modulates *Trypanosoma cruzi* invasion of host cells. PLoS Neglected Trop Dis 2012;6:e1583.

[59] Luzio JP, Pryor PR, Bright NA. Lysosomes: fusion and function. Nat Rev Mol Cell Biol 2007;8:622–632.

[60] Menager MM, Menasche G, Romao M, Knapnougel P, Ho CH, et al. Secretory cytotoxic granule maturation and exocytosis require the effector protein hMunc13-4. Nat Immunol 2007;8:257–267.

[61] van der Sluijs P, Zibouche M, van Kerkhof P. Late steps in secretory lysosome exocytosis in cytotoxic lymphocytes. Front Immunol 2013;4:359.

[62] Wasmeier C, Hume AN, Bolasco G, Seabra MC. Melanosomes at a glance. J Cell Sci 2008;121:3995–3999.

[63] Wernersson S, Pejler G. Mast cell secretory granules: armed for battle. Nat Rev Immunol 2014;14:478–494.

[64] Pereira-Leal JB, Seabra MC. Evolution of the Rab family of small GTP-binding proteins. J Mol Biol 2001;313:889–901.

[65] Carta S, Tassi S, Semino C, Fossati G, Mascagni P, et al. Histone deacetylase inhibitors prevent exocytosis of interleukin-1beta-containing secretory lysosomes: role of microtubules. Blood 2006;108:1618–1626.

[66] Martin-Verdeaux S, Pombo I, Iannascoli B, Roa M, Varin-Blank N, et al. Evidence of a role for Munc18-2 and microtubules in mast cell granule exocytosis. J Cell Sci 2003;116:325–334.

[67] Schliwa M, Woehlke G. Molecular motors. Nature 2003;422:759–765.

[68] Tucker JB. Spatial organization of microtubule-organizing centers and microtubules. J Cell Biol 1984;99:55s–62s.

[69] Rosa-Ferreira C, Munro S. Arl8 and SKIP act together to link lysosomes to kinesin-1. Dev Cell 2011;21:1171–1178.

[70] Tan SC, Scherer J, Vallee RB. Recruitment of dynein to late endosomes and lysosomes through light intermediate chains. Mol Biol Cell 2011;22:467–477.

[71] Jordens I, Fernandez-Borja M, Marsman M, Dusseljee S, Janssen L, et al. The Rab7 effector protein RILP controls lysosomal transport by inducing the recruitment of dynein–dynactin motors. Curr Biol 2001;11:1680–1685.

[72] Burkhardt JK, Echeverri CJ, Nilsson T, Vallee RB. Overexpression of the dynamitin (p50) subunit of the dynactin complex disrupts dynein-dependent maintenance of membrane organelle distribution. J Cell Biol 1997;139:469–484.

[73] Stinchcombe JC, Majorovits E, Bossi G, Fuller S, Griffiths GM. Centrosome polarization delivers secretory granules to the immunological synapse. Nature 2006;443: 462–465.

[74] Daniele T, Hackmann Y, Ritter AT, Wenham M, Booth S, et al. A role for Rab7 in the movement of secretory granules in cytotoxic T lymphocytes. Traffic 2011;12:902–911.

[75] Clark RH, Stinchcombe JC, Day A, Blott E, Booth S, et al. Adaptor protein 3-dependent microtubule-mediated movement of lytic granules to the immunological synapse. Nat Immunol 2003;4:1111–1120.

[76] Ivan V, Martinez-Sanchez E, Sima LE, Oorschot V, Klumperman J, et al. AP-3 and Rabip4′ coordinately regulate spatial distribution of lysosomes. PLoS ONE 2012; 7:e48142.

[77] van der Kant R, Fish A, Janssen L, Janssen H, Krom S, et al. Late endosomal transport and tethering are coupled processes controlled by RILP and the cholesterol sensor ORP1L. J Cell Sci 2013;126:3462–3474.

[78] Swanson JA, Locke A, Ansel P, Hollenbeck PJ. Radial movement of lysosomes along microtubules in permeabilized macrophages. J Cell Sci 1992;103(Pt 1):201–209.

[79] Swanson J, Bushnell A, Silverstein SC. Tubular lysosome morphology and distribution within macrophages depend on the integrity of cytoplasmic microtubules. Proc Natl Acad Sci U S A 1987;84:1921–1925.

[80] Mrakovic A, Kay JG, Furuya W, Brumell JH, Botelho RJ. Rab7 and Arl8 GTPases are necessary for lysosome tubulation in macrophages. Traffic 2012;13:1667–1679.

[81] Hollenbeck PJ, Swanson JA. Radial extension of macrophage tubular lysosomes supported by kinesin. Nature 1990;346:864–866.

[82] Xu J, Toops KA, Diaz F, Carvajal-Gonzalez JM, Gravotta D, et al. Mechanism of polarized lysosome exocytosis in epithelial cells. J Cell Sci 2012;125:5937–5943.

[83] DeSelm CJ, Miller BC, Zou W, Beatty WL, van Meel E, et al. Autophagy proteins regulate the secretory component of osteoclastic bone resorption. Dev Cell 2011; 21:966–974.

[84] Sun Y, Buki KG, Ettala O, Vaaraniemi JP, Vaananen HK. Possible role of direct Rac1-Rab7 interaction in ruffled border formation of osteoclasts. J Biol Chem 2005; 280:32356–32361.

[85] Hume AN, Collinson LM, Hopkins CR, Strom M, Barral DC, et al. The leaden gene product is required with Rab27a to recruit myosin Va to melanosomes in melanocytes. Traffic 2002;3:193–202.

[86] Wu XS, Rao K, Zhang H, Wang F, Sellers JR, et al. Identification of an organelle receptor for myosin-Va. Nat Cell Biol 2002;4:271–278.

[87] Matesic LE, Yip R, Reuss AE, Swing DA, O'Sullivan TN, et al. Mutations in Mlph, encoding a member of the Rab effector family, cause the melanosome transport defects observed in leaden mice. Proc Natl Acad Sci U S A 2001;98:10238–10243.

[88] Wilson SM, Yip R, Swing DA, O'Sullivan TN, Zhang Y, et al. A mutation in Rab27a causes the vesicle transport defects observed in ashen mice. Proc Natl Acad Sci U S A 2000;97:7933–7938.

[89] Pastural E, Barrat FJ, Dufourcq-Lagelouse R, Certain S, Sanal O, et al. Griscelli disease maps to chromosome 15q21 and is associated with mutations in the myosin-Va gene. Nat Genet 1997;16:289–292.

[90] Menasche G, Pastural E, Feldmann J, Certain S, Ersoy F, et al. Mutations in RAB27A cause Griscelli syndrome associated with haemophagocytic syndrome. Nat Genet 2000;25:173–176.

[91] Mercer JA, Seperack PK, Strobel MC, Copeland NG, Jenkins NA. Novel myosin heavy chain encoded by murine dilute coat colour locus. Nature 1991;349:709–713.

[92] Menasche G, Ho CH, Sanal O, Feldmann J, Tezcan I, et al. Griscelli syndrome restricted to hypopigmentation results from a melanophilin defect (GS3) or a MYO5A F-exon deletion (GS1). J Clin Invest 2003;112:450–456.

[93] Lopes VS, Ramalho JS, Owen DM, Karl MO, Strauss O, et al. The ternary Rab27a-Myrip-Myosin VIIa complex regulates melanosome motility in the retinal pigment epithelium. Traffic 2007;8:486–499.

[94] Desnos C, Schonn JS, Huet S, Tran VS, El-Amraoui A, et al. Rab27A and its effector MyRIP link secretory granules to F-actin and control their motion towards release sites. J Cell Biol 2003;163:559–570.

[95] Rodriguez A, Martinez I, Chung A, Berlot CH, Andrews NW. cAMP regulates Ca2+-dependent exocytosis of lysosomes and lysosome-mediated cell invasion by trypanosomes. J Biol Chem 1999;274:16754–16759.

[96] Foger N, Jenckel A, Orinska Z, Lee KH, Chan AC, et al. Differential regulation of mast cell degranulation versus cytokine secretion by the actin regulatory proteins Coronin1a and Coronin1b. J Exp Med 2011;208:1777–1787.

[97] Singh RK, Mizuno K, Wasmeier C, Wavre-Shapton ST, Recchi C, et al. Distinct and opposing roles for Rab27a/Mlph/MyoVa and Rab27b/Munc13-4 in mast cell secretion. FEBS J 2013;280:892–903.

[98] Liu D, Meckel T, Long EO. Distinct role of rab27a in granule movement at the plasma membrane and in the cytosol of NK cells. PLoS ONE 2010;5:e12870.

[99] Johnson JL, Monfregola J, Napolitano G, Kiosses WB, Catz SD. Vesicular trafficking through cortical actin during exocytosis is regulated by the Rab27a effector JFC1/Slp1 and the RhoA-GTPase-activating protein Gem-interacting protein. Mol Biol Cell 2012;23:1902–1916.

[100] Liu D, Martina JA, Wu XS, Hammer JA 3rd, Long EO. Two modes of lytic granule fusion during degranulation by natural killer cells. Immunol Cell Biol 2011;89: 728–738.

[101] Roa M, Paumet F, Le Mao J, David B, Blank U. Involvement of the ras-like GTPase rab3d in RBL-2H3 mast cell exocytosis following stimulation via high affinity IgE receptors (Fc epsilonRI). J Immunol 1997;159:2815–2823.

[102] Mizuno K, Tolmachova T, Ushakov DS, Romao M, Abrink M, et al. Rab27b regulates mast cell granule dynamics and secretion. Traffic 2007;8:883–892.

[103] Yasukawa M, Ohminami H, Yakushijin Y, Arai J, Hasegawa A, et al. Fas-independent cytotoxicity mediated by human CD4+ CTL directed against herpes simplex virus-infected cells. J Immunol 1999;162:6100–6106.

[104] Martinez I, Chakrabarti S, Hellevik T, Morehead J, Fowler K, et al. Synaptotagmin VII regulates Ca(2+)-dependent exocytosis of lysosomes in fibroblasts. J Cell Biol 2000;148:1141–1149.

[105] Caler EV, Chakrabarti S, Fowler KT, Rao S, Andrews NW. The exocytosis-regulatory protein synaptotagmin VII mediates cell invasion by *Trypanosoma cruzi*. J Exp Med 2001;193:1097–1104.

[106] Colvin RA, Means TK, Diefenbach TJ, Moita LF, Friday RP, et al. Synaptotagmin-mediated vesicle fusion regulates cell migration. Nat Immunol 2010;11:495–502.

[107] Li D, Ropert N, Koulakoff A, Giaume C, Oheim M. Lysosomes are the major vesicular compartment undergoing Ca2+-regulated exocytosis from cortical astrocytes. J Neurosci 2008;28:7648–7658.

[108] Sander LE, Frank SP, Bolat S, Blank U, Galli T, et al. Vesicle associated membrane protein (VAMP)-7 and VAMP-8, but not VAMP-2 or VAMP-3, are required for activation-induced degranulation of mature human mast cells. Eur J Immunol 2008; 38:855–863.

[109] Tiwari N, Wang CC, Brochetta C, Ke G, Vita F, et al. VAMP-8 segregates mast cell-preformed mediator exocytosis from cytokine trafficking pathways. Blood 2008;111:3665–3674.

[110] Pombo I, Martin-Verdeaux S, Iannascoli B, Le Mao J, Deriano L, et al. IgE receptor type I-dependent regulation of a Rab3D-associated kinase: a possible link in the calcium-dependent assembly of SNARE complexes. J Biol Chem 2001;276: 42893–42900.

[111] Zhang Z, Chen G, Zhou W, Song A, Xu T, et al. Regulated ATP release from astrocytes through lysosome exocytosis. Nat Cell Biol 2007;9:945–953.

[112] Rao SK, Huynh C, Proux-Gillardeaux V, Galli T, Andrews NW. Identification of SNAREs involved in synaptotagmin VII-regulated lysosomal exocytosis. J Biol Chem 2004;279:20471–20479.

[113] Medina DL, Fraldi A, Bouche V, Annunziata F, Mansueto G, et al. Transcriptional activation of lysosomal exocytosis promotes cellular clearance. Dev Cell 2011;21:421–430.

[114] Chen D, Lemons PP, Schraw T, Whiteheart SW. Molecular mechanisms of platelet exocytosis: role of SNAP-23 and syntaxin 2 and 4 in lysosome release. Blood 2000;96:1782–1788.

[115] Elstak ED, Neeft M, Nehme NT, Voortman J, Cheung M, et al. The munc13-4-rab27 complex is specifically required for tethering secretory lysosomes at the plasma membrane. Blood 2011;118:1570–1578.

[116] Wood SM, Meeths M, Chiang SC, Bechensteen AG, Boelens JJ, et al. Different NK cell-activating receptors preferentially recruit Rab27a or Munc13-4 to perforin-containing granules for cytotoxicity. Blood 2009;114:4117–4127.

[117] Feldmann J, Callebaut I, Raposo G, Certain S, Bacq D, et al. Munc13-4 is essential for cytolytic granules fusion and is mutated in a form of familial hemophagocytic lymphohistiocytosis (FHL3). Cell 2003;115:461–473.

[118] Holt O, Kanno E, Bossi G, Booth S, Daniele T, et al. Slp1 and Slp2-a localize to the plasma membrane of CTL and contribute to secretion from the immunological synapse. Traffic 2008;9:446–457.

[119] Bustos MA, Lucchesi O, Ruete MC, Mayorga LS, Tomes CN. Rab27 and Rab3 sequentially regulate human sperm dense-core granule exocytosis. Proc Natl Acad Sci U S A 2012;109:E2057–E2066.

[120] Menasche G, Menager MM, Lefebvre JM, Deutsch E, Athman R, et al. A newly identified isoform of Slp2a associates with Rab27a in cytotoxic T cells and participates to cytotoxic granule secretion. Blood 2008;112:5052–5062.

[121] Loo LS, Hwang LA, Ong YM, Tay HS, Wang CC, et al. A role for endobrevin/VAMP8 in CTL lytic granule exocytosis. Eur J Immunol 2009;39:3520–3528.

[122] Matti U, Pattu V, Halimani M, Schirra C, Krause E, et al. Synaptobrevin2 is the v-SNARE required for cytotoxic T-lymphocyte lytic granule fusion. Nat Commun 2013;4:1439.

[123] Kurowska M, Goudin N, Nehme NT, Court M, Garin J, et al. Terminal transport of lytic granules to the immune synapse is mediated by the kinesin-1/Slp3/Rab27a complex. Blood 2012;119:3879–3889.

[124] Munafo DB, Johnson JL, Ellis BA, Rutschmann S, Beutler B, et al. Rab27a is a key component of the secretory machinery of azurophilic granules in granulocytes. Biochem J 2007;402:229–239.

[125] Johnson JL, Brzezinska AA, Tolmachova T, Munafo DB, Ellis BA, et al. Rab27a and Rab27b regulate neutrophil azurophilic granule exocytosis and NADPH oxidase activity by independent mechanisms. Traffic 2010;11:533–547.

[126] Brzezinska AA, Johnson JL, Munafo DB, Crozat K, Beutler B, et al. The Rab27a effectors JFC1/Slp1 and Munc13-4 regulate exocytosis of neutrophil granules. Traffic 2008;9:2151–2164.

[127] Herrero-Turrion MJ, Calafat J, Janssen H, Fukuda M, Mollinedo F. Rab27a regulates exocytosis of tertiary and specific granules in human neutrophils. J Immunol 2008;181:3793–3803.

[128] Tolmachova T, Abrink M, Futter CE, Authi KS, Seabra MC. Rab27b regulates number and secretion of platelet dense granules. Proceed Natl Acad Sci U S A 2007; 104:5872–5877.

[129] Krzewski K, Gil-Krzewska A, Watts J, Stern JN, Strominger JL. VAMP4- and VAMP7-expressing vesicles are both required for cytotoxic granule exocytosis in NK cells. Eur J Immunol 2011;41:3323–3329.

[130] Dressel R, Elsner L, Novota P, Kanwar N, Fischer von Mollard G. The exocytosis of lytic granules is impaired in Vti1b- or Vamp8-deficient CTL leading to a reduced cytotoxic activity following antigen-specific activation. J Immunol 2010;185:1005–1014.

[131] Marcenaro S, Gallo F, Martini S, Santoro A, Griffiths GM, et al. Analysis of natural killer-cell function in familial hemophagocytic lymphohistiocytosis (FHL): defective CD107a surface expression heralds Munc13-4 defect and discriminates between genetic subtypes of the disease. Blood 2006;108:2316–2323.

[132] Neeft M, Wieffer M, de Jong AS, Negroiu G, Metz CH, et al. Munc13-4 is an effector of rab27a and controls secretion of lysosomes in hematopoietic cells. Mol Biol Cell 2005;16:731–741.

[133] Shirakawa R, Higashi T, Tabuchi A, Yoshioka A, Nishioka H, et al. Munc13-4 is a GTP-Rab27-binding protein regulating dense core granule secretion in platelets. J Biol Chem 2004;279:10730–10737.

[134] Carroll K, Ray K, Helm B, Carey E. Differential expression of Rab3 isoforms in high- and low-secreting mast cell lines. Eur J Cell Biol 2001;80:295–302.

[135] Higashio H, Nishimura N, Ishizaki H, Miyoshi J, Orita S, et al. Doc2 alpha and Munc13-4 regulate Ca(2+)-dependent secretory lysosome exocytosis in mast cells. J Immunol 2008;180:4774–4784.

[136] Michaut M, Tomes CN, De Blas G, Yunes R, Mayorga LS. Calcium-triggered acrosomal exocytosis in human spermatozoa requires the coordinated activation of Rab3A and N-ethylmaleimide-sensitive factor. Proc Natl Acad Sci U S A 2000;97:9996–10001.

[137] Melicoff E, Sansores-Garcia L, Gomez A, Moreira DC, Datta P, et al. Synaptotagmin-2 controls regulated exocytosis but not other secretory responses of mast cells. J Biol Chem 2009;284:19445–19451.

[138] Appay V, Zaunders JJ, Papagno L, Sutton J, Jaramillo A, et al. Characterization of CD4(+) CTLs ex vivo. J Immunol 2002;168:5954–5958.

[139] Ochoa MT, Stenger S, Sieling PA, Thoma-Uszynski S, Sabet S, et al. T-cell release of granulysin contributes to host defense in leprosy. Nat Med 2001;7:174–179.

[140] Haddad EK, Wu X, Hammer JA 3rd, Henkart PA. Defective granule exocytosis in Rab27a-deficient lymphocytes from Ashen mice. J Cell Biol 2001;152:835–842.

[141] Singh RK, Liao W, Tracey-White D, Recchi C, Tolmachova T, et al. Rab27a-mediated protease release regulates neutrophil recruitment by allowing uropod detachment. J Cell Sci 2012;125:1652–1656.

[142] Pivot-Pajot C, Varoqueaux F, de Saint BG, Bourgoin SG. Munc13-4 regulates granule secretion in human neutrophils. J Immunol 2008;180:6786–6797.

[143] Johnson JL, Hong H, Monfregola J, Kiosses WB, Catz SD. Munc13-4 restricts motility of Rab27a-expressing vesicles to facilitate lipopolysaccharide-induced priming of exocytosis in neutrophils. J Biol Chem 2011;286:5647–5656.

[144] Breitbart H, Rubinstein S, Lax Y. Regulatory mechanisms in acrosomal exocytosis. Rev Reprod 1997;2:165–174.

[145] Mayorga LS, Tomes CN, Belmonte SA. Acrosomal exocytosis, a special type of regulated secretion. IUBMB Life 2007;59:286–292.

[146] Nishio H, Suda T, Sawada K, Miyamoto T, Koike T, et al. Molecular cloning of cDNA encoding human Rab3D whose expression is upregulated with myeloid differentiation. Biochim Biophys Acta 1999;1444:283–290.

[147] Tuvim MJ, Adachi R, Chocano JF, Moore RH, Lampert RM, et al. Rab3D, a small GTPase, is localized on mast cell secretory granules and translocates to the plasma membrane upon exocytosis. Am J Respir Cell Mol Biol 1999;20:79–89.

[148] Ye S, Karim ZA, Al Hawas R, Pessin JE, Filipovich AH, et al. Syntaxin-11, but not syntaxin-2 or syntaxin-4, is required for platelet secretion. Blood 2012;120:2484–2492.

[149] D'Orlando O, Zhao F, Kasper B, Orinska Z, Muller J, et al. Syntaxin 11 is required for NK and CD8+ T-cell cytotoxicity and neutrophil degranulation. Eur J Immunol 2013;43:194–208.

[150] Bryceson YT, Rudd E, Zheng C, Edner J, Ma D, et al. Defective cytotoxic lymphocyte degranulation in syntaxin-11 deficient familial hemophagocytic lymphohistiocytosis 4 (FHL4) patients. Blood 2007;110:1906–1915.

[151] Henkel AW, Kang G, Kornhuber J. A common molecular machinery for exocytosis and the 'kiss-and-run' mechanism in chromaffin cells is controlled by phosphorylation. J Cell Sci 2001;114:4613–4620.

[152] Wang N, Lee AK, Yan L, Simpson MR, Tse A, et al. Granule matrix property and rapid "kiss-and-run" exocytosis contribute to the different kinetics of catecholamine release from carotid glomus and adrenal chromaffin cells at matched quantal size. Can J Physiol Pharmacol 2012;90:791–801.

6

ROLE OF LYSOSOMES IN LIPID METABOLISM

Frederick R. Maxfield
Department of Biochemistry, Weill Cornell Medical College, New York, NY, USA

6.1 INTRODUCTION

Many of the acid hydrolases in late endosomes and lysosomes (LE/Ly) participate in the breakdown of lipids. Lipids are delivered to lysosomes by endocytic uptake of lipoproteins, as components of the lipid bilayer involved in vesicle transport, by autophagy, and by cytoplasmic lipid transport proteins. Lipids can leave lysosomes by vesicle transport, by export to cytoplasmic lipid transport proteins, or by lysosome secretion. For several reasons, specialized mechanisms are required for hydrolysis of lipids in LE/Ly. First, the membrane integrity of the LE/Ly must be maintained even as some of the lipid components that were once part of its limiting bilayer are hydrolyzed, so the outer limiting membrane must be protected from hydrolysis even as extensive lipid hydrolysis is taking place. Second, many of the lipid components are highly insoluble in water, so mechanisms must be employed to allow enzymatic access to cleavage sites. In some cases (e.g., cholesterol), special mechanisms are required to solubilize the hydrophobic products of hydrolysis. The mechanisms used by LE/Ly to hydrolyze lipids and some of the disease states that occur when these mechanisms fail are discussed in this chapter. The role of lysosomes in responding to nutrient status is also discussed.

Lysosomes: Biology, Diseases, and Therapeutics, First Edition.
Edited by Frederick R. Maxfield, James M. Willard and Shuyan Lu.

6.2 ENDOCYTIC UPTAKE OF LIPOPROTEINS

Probably the best characterized example of lysosomal digestion of lipids is the hydrolysis of cholesteryl esters that are present in the core of low-density lipoprotein (LDL) particles taken into the cell by receptor-mediated endocytosis [1–3] (Figure 6.1). LDL receptors are expressed on the surface of most nucleated mammalian cells, and they bind LDL particles, which are derived from larger lipoproteins, very low-density lipoproteins (VLDLs), that are secreted from the liver. The VLDL is acted on by

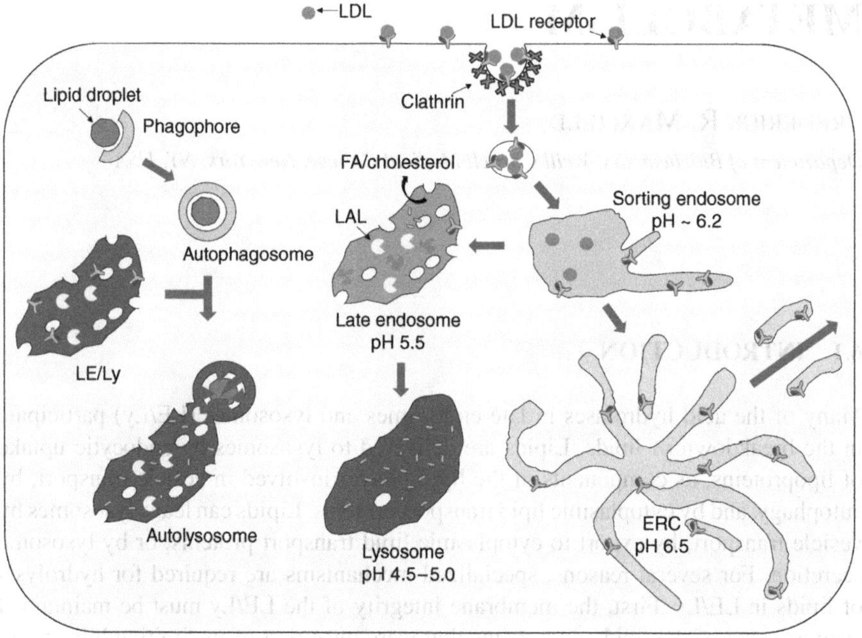

Figure 6.1 Schematic diagram of lipoprotein and lipid droplet degradation by lysosomes. Low-density lipoproteins (LDLs) bind to receptors on the surface of cells and are rapidly internalized via clathrin-coated pits. The contents of the clathrin-coated pits are delivered to sorting endosomes, which are transient organelles with an internal pH of about 6.2. The low pH causes LDL to dissociate from its receptor, and the empty receptors are rapidly exported from the sorting endosome in narrow diameter tubules. The receptors return to the cell surface either directly or after passing through an intermediate organelle, the endocytic recycling compartment (ERC). The sorting endosomes mature into late endosomes, which are somewhat more acidic and have lysosomal enzymes that have been delivered from the trans-Golgi network. Membrane invaginations at the surface of the late endosomes create internal vesicles that are enriched in BMP. Lysosomal enzymes, which are activated by the low pH, and these digest the internalized LDL. Membrane lipids in the bilayer are mainly digested on the internal vesicles, which protects the limiting membrane against leakage. Lipid droplets can be surrounded by phagophores to create an autophagosome, which fuses with late endosomes or lysosomes to create a digestive autolysosome. Figure created by David B. Iaea. (*See color plate section for the color representation of this figure.*)

a nonlysosomal enzyme, lipoprotein lipase, which is held on the surface of certain endothelial cells by a GPI-anchored receptor, GPIHBP1 [4–6]. The lipase hydrolyzes the triacylglycerides in the core of VLDL, leaving smaller cholesteryl ester-rich LDL particles that circulate throughout the body but not the central nervous system. LDL receptors, which bind the circulating LDL, are internalized via clathrin-coated pits and delivered to early sorting endosomes, which maintain an internal pH of about 6.2 [7,8]. This pH is sufficiently acidic to cause dissociation of LDL from its receptor, and the LDL receptors then are free to recycle while the LDL remains in the early sorting endosome as it matures into a late endosome (LE) [7,9]. Unless retained in the sorting endosomes by a signal, most membrane components are returned to the cell surface. This is accomplished by an iterative process in which newly formed vesicles that are pinched off from the surface continue to fuse with a sorting endosome for several minutes [7,10]. During this time, narrow diameter tubules, which have a high surface-to-volume ratio, repeatedly pinch off from the sorting endosome and deliver the membrane components (and a small amount of the volume) back to the surface either directly or through an intermediate compartment called the endocytic recycling compartment (ERC) [10–12].

In the LE, the LDL begins to encounter lysosomal enzymes that have been delivered from the trans-Golgi network [13]. The lysosomal proteases digest the protein components, and a single enzyme, lysosomal acid lipase, digests the cholesteryl esters and triacylglycerides in the core of lipoproteins [14]. This releases cholesterol and a fatty acid from the cholesteryl esters and glycerol and fatty acids from the triacylglycerides. (Deficiencies in lysosomal acid lipase lead to Wollman disease or cholesteryl ester storage disease [14].) Cholesterol is highly insoluble in water, so it needs to be bound to transport proteins when it is not in a bilayer. In the LE/Ly, a soluble protein called NPC2 serves this role [15]. It now appears that NPC2 delivers the cholesterol to an LE/Ly membrane protein, NPC1, which helps to shuttle the cholesterol to the limiting membrane of the LE/Ly [16–18]. The itinerary of cholesterol after insertion into the limiting membrane is uncertain, but it may be transported by one or more cytoplasmic sterol-binding proteins and delivered to other organelles [19,20]. Homozygous defects in either NPC1 or NPC2 cause Niemann–Pick C disease, which is characterized by abnormal accumulation of cholesterol and other lipids in LE/Ly. β-Cyclodextrins can chelate cholesterol and facilitate its exchange among membranes. When delivered into the LE/Ly, β-cyclodextrins can bypass the requirement for both NPC1 and NPC2, suggesting that delivery of sterol to the limiting membrane of the LE/Ly is the essential role of these proteins [20]. Clinical trials for treatment of NPC patients with β-cyclodextrins have been initiated [21].

Some LE are also described as multivesicular bodies (MVBs) because their lumen is filled with small vesicles and inward budding tubules. The internal vesicles are formed by the invagination of the limiting membrane into the lumen of the LE under the control of a set of proteins called the endosomal sorting complexes required for transport (ESCRT) complex [22]. These inward invaginations are enriched in proteins that had been mono-ubiquitinated, and their inclusion in these structures helps prevent their recycling. The internal vesicles are highly enriched in an unusual lipid, bis(monoacylglycero)phosphate (BMP) (also called lysobisphosphatidic acid in some older publications), a negatively charged glycerophospholipid with an unusual sn-1;sn-1′ structural configuration, that is present at very low levels elsewhere in the cell [23,24]. Under some circumstances, these internal vesicles containing BMP could hold some of the cholesterol that is released by the action of lysosomal acid lipase, but typically these vesicles become enriched in ceramide following hydrolysis of sphingomyelin (SM) by acid sphingomyelinase. The ceramide can displace cholesterol from lipid bilayers [25], so normally these internal membranes would be expected to have low cholesterol content.

In peripheral cells (i.e., outside the central nervous system), the cholesterol derived from LDL uptake provides some of the cholesterol used by cells for incorporation into lipid bilayers. Excess cholesterol is esterified by acyl-coA-cholesterol acyltransferase (ACAT), an ER-associated enzyme, and the cholesteryl esters are stored in cytoplasmic lipid droplets [26]. Some cholesterol is effluxed from the plasma membrane, mainly by delivery to high-density lipoproteins (HDLs) [27]. Small amounts of cholesterol are converted into oxysterols, which play regulatory roles in cells by binding to oxysterol-binding proteins and to some nuclear receptors [28,29].

Most cells can also synthesize cholesterol, so the total cholesterol in a cell is a balance between uptake and synthesis versus export and conversion to oxysterols (or bile salts in the liver). The major regulator of total cell cholesterol is the SREBP2 transcriptional regulator [30]. SREBP2 is a membrane protein with two transmembrane domains. In conditions of high cholesterol, SREBP2 is retained in the ER by binding to SCAP and Insig [31]. When cholesterol levels are reduced, SCAP and SREBP2 are released from Insig, and they are transported to the Golgi apparatus via COP-II transport vesicles. In the Golgi, SREBP2 is cleaved sequentially by two endoproteases, releasing a cytoplasmic fragment that is translocated into the nucleus, where it activates transcription of several genes involved in the synthesis of cholesterol. SREBP2 also activates the transcription of the LDL receptor, so it can simultaneously increase cholesterol uptake and synthesis by cells.

Two important cell types for lipid metabolism are hepatocytes and macrophages. Hepatocytes are the key cells in mammals for regulation of lipoproteins and cholesterol. Dietary cholesterol is packaged in intestinal cells into large lipoproteins called chylomicrons that contain ApoB48 and other apolipoproteins, including ApoE. These particles are enriched in triacylglycerides and to a lesser extent cholesteryl esters. After being acted on by lipoprotein lipase, the chylomicron remnants are internalized by hepatocytes by binding to a variety of receptors [32,33], and they are hydrolyzed in LE/Ly as described earlier. Some of the fatty acids and cholesterol are reesterified and packaged into VLDL particles, which are large lipoproteins containing ApoB100 and other apolipoproteins including ApoE. The core of the VLDL is also mainly triacylglycerides with some cholesteryl esters. Similar to chylomicrons, the VLDL is acted on by lipoprotein lipase, hydrolyzing the triacylglycerides and leaving a cholesteryl ester-enriched LDL particle, which can be taken up by many cell types as described earlier. When overall cholesterol levels are in balance, LDL receptors on the hepatocyte play a major role in clearing LDL from the circulation by endocytosis into hepatocytes. Since cholesterol is not catabolized, the main pathway for clearance from the body is conversion to bile salts in hepatocytes and excretion [34]. In conditions of excess LDL cholesterol, the hepatic LDL receptors are downregulated, and LDL levels in the circulation rise due to reduced hepatic uptake.

Recently, it has been found that a circulating protein, proprotein convertase subtilisin/kexin type 9 (PCSK9), promotes downregulation of LDL receptors in certain tissues, including liver [35]. PCSK9 binds to the LDL receptor and redirects it from its normal recycling itinerary toward lysosomal degradation by a mechanism that is not well understood at present. Antibody therapies to inactivate PCSK9 have recently been approved for treatment of elevated LDL in some individuals [36].

Some LDL percolates through the endothelium, especially in areas of high shear stress, and this LDL can accumulate in the wall of blood vessels. Macrophages normally clear this LDL and digest it by lysosomal hydrolases. Excess cholesterol in the macrophages is effluxed to HDL and can return to the liver. Under conditions of elevated LDL cholesterol, the cellular cholesterol load in the macrophages is beyond the capacity of the efflux system, and the excess cholesterol is esterified by ACAT and stored as lipid droplets. Under conditions of extreme cholesterol loading, the macrophages become filled with lipid droplets that are enriched in cholesteryl esters, and these macrophage foam cells can eventually die as a consequence of the lipid loading [37–39].

6.3 LIPID METABOLISM IN LATE ENDOSOMES AND LYSOSOMES

Lipids are brought into LE/Ly by vesicle transport, and LE/Ly are a significant site for the catabolism of these lipids. It should be noted, however, that breakdown and remodeling of lipids occur in many membranes of the cell, and this contributes to the variation in lipid composition among membrane organelles [40]. Much of our knowledge about the various lipid degrading enzymes in LE/Ly was initiated by studies of patients with lysosomal storage diseases, as discussed later in this section. Most lipid degradation products that are formed in LE/Ly are subsequently used by the cell to create new lipids. For example, in human fibroblasts, it was found that 90% of the sphingoid base incorporated into glucosylceramide was recycled after degradation of sphingolipids, and only 10% was newly synthesized [41]. Fatty acids released from LE/Ly following lipid hydrolysis are also used for energy production.

Hydrolysis of membrane lipids in lysosomes presents a special problem because it is essential to maintain the integrity of the limiting membrane while allowing efficient degradation of lipids delivered to the LE/Ly. Protection of the limiting membrane is achieved by having a dense enzyme-resistant glycocalyx composed of the glycoconjugates of lysosomal membrane proteins – the lysosomal integral membrane proteins (LIMPs) and lysosome-associated membrane proteins (LAMPs) [42,43]. The invaginated membranes and internal vesicles formed by the ESCRT complexes lack this glycocalyx, and they are sites of lipid hydrolysis as well as the degradation of integral membrane proteins. The special properties of these internal vesicles have been described in detail in several reviews [43–45].

Glycerolipids can be cleaved by several lysosomal phospholipases, including isoenzymes of acid phospholipase A2 (group XV PLA2), which have broad specificity and some preference for phosphatidylcholine and phosphatidylethanolamine

[43,46]. Sphingomyelin (SM) and glycosphingolipids (GSLs), which have phosphorylcholine or glycoconjugates attached to the hydroxyl group of ceramide, have more complex degradation mechanisms [44,45]. The degradation of the GSLs begins with stepwise removal of monosaccharides from the terminal ends of the glycoconjugates. These ends of the glycoconjugates protrude into the aqueous phase and are easily reached by the acid hydrolases in the LE/Ly. However, the last few sugars (i.e., those closest to the membrane) can only be removed with the assistance of sphingolipid activator proteins (saposins or SAPs) that solubilize the glycolipids and present them to acid hydrolases. Four saposins, SAP A–D, are derived by proteolytic hydrolysis of a precursor protein, prosaposin, which is delivered to LE/Ly by the mannose-6-phoshate receptor system. Another soluble activator protein, the GM2-activator protein (GM2-AP), is essential for the degradation of ganglioside GM2 by β-hexosaminidase A [47]. GM2-AP selectively extracts gangliosides such as GM1 and GM2 from membranes and presents them in the aqueous phase to the active sites of acid hydrolases. The GM2-AP is an essential component of the enzymatic complex. Lack of GM2-AP leads to a lysosomal storage disorder with accumulation of the GM2 ganglioside. GM2-AP and other activator proteins are sometimes called "liftases" because they have no enzymatic activity of their own, but they pull the lipid substrates out of the bilayer and present them to the hydrolases that cleave them [48].

SAPs A–D are small, highly homologous, protease-resistant proteins that are soluble in the lumen of LE/Ly. They are homologous to other lipid-binding proteins, including some of the pore-forming lipid-binding proteins released by bacteria. The saposins have positively charged surfaces that can interact with negatively charged membranes. One of the key features of the internal vesicles in LE is that they are highly enriched in BMP, which retains a negative charge even in the acid environment inside LE/Ly [49,50]. It has been estimated that more than half of the lipids in some of the internal vesicles are BMP [51]. This negatively charged lipid provides a favorable docking platform for the saposins [49]. Although all four of the saposins are able to extract lipids from membrane bilayers, they differ in their specificity for presenting sphingolipids to acid hydrolases as seen by the differences in the lipids that accumulate when different saposins are defective in human lysosomal storage diseases [45]. Degradation of sphingomyelin by acid sphingomyelinase does not require an activator protein because acid sphingomyelinase has a SAP homology domain, which carries out the same function [45].

The important role of the anionic lipid BMP has implications for understanding some drug-induced effects on sphingolipid breakdown (see Chapter 20). It is known that many amphiphilic weak bases can cause accumulation of phospholipids, sphingolipids, and cholesterol in LE/Ly [52]. Because the uncharged species of a weak base can cross a membrane by passive diffusion much more rapidly than the positively charged form, weak bases will accumulate in acidic organelles such as LE/Ly since it is the uncharged species that equilibrates across the membrane (see Chapter 18). In acidic organelles, the fraction of molecules in the protonated, positively charged form is increased. An amphiphilic weak base (i.e., one with a hydrophobic portion in addition to the titratable weak base – usually an amine) will also partition into membrane bilayers. When these positively charged weak bases associate with the internal vesicles in LE/Ly, they neutralize the negative charge imparted by BMP. This causes the saposins and acid sphingomyelinase to dissociate from the internal vesicles in the LE/Ly. When released from the internal vesicles, these proteins are more susceptible to proteolytic degradation, further reducing their activity [43].

An interesting aspect of this is the pH dependence of enzymes involved in sphingolipid hydrolysis. The pH of endolysosomal organelles drops from early sorting endosomes (pH $\approx$ 6.2) to LEs (pH $\approx$ 5.5) to lysosomes (pH 5.0 or below) [7,53]. These pH changes alter the activity of enzymes involved in sphingolipid catabolism. The enzymes acid sphingomyelinase and glucosylceramide-β-glucosidase, which produce ceramide from sphingomyelin and glucosylceramide, respectively, have pH optima near pH 5.5 [45]. This favors ceramide production at the pH found in LEs. In contrast, acid ceramidase, which converts ceramide into sphingosine, has a pH optimum near 4.0, and at pH 5.5 it actually favors the reverse reaction converting sphingosine to ceramide. These reactions all keep ceramide relatively high in the LE/MVBs. The ceramide competes with cholesterol for sites in the lipid bilayer because both of these lipids have small headgroups that need to be shielded from the aqueous phase [25]. High ceramide on internal membranes in MVBs may help to deliver cholesterol to NPC2 by reducing the stability of the cholesterol in the bilayer (i.e., increasing its chemical activity coefficient) [49,54]. Sphingosine has a positive charge and would neutralize the negative charges from BMP as discussed above. This would promote dissociation of saposins and acid sphingomyelinase from internal membranes and slow down sphingolipid catabolism.

As described in Chapter 11, many lysosomal storage diseases are associated with abnormal lipid accumulation. It has been noted that several of these disorders with different primary deficiencies lead to a similar spectrum of lipid accumulation [55]. Thus, several disorders lead to accumulation of sphingolipids and cholesterol in the storage organelles. This similar lipid profile arising from different primary causes can be understood in terms of the general properties of lipid biophysics and the biochemistry of lipid degradation. For example, Niemann–Pick disease type C arises from recessive mutations in genes encoding NPC1 and NPC2 [56]. As expected, cholesterol levels are increased in the LE/Ly of cells with homozygous defects in either the *NPC1* or *NPC2* genes. However, these cells also accumulate sphingomyelin, other sphingolipids, and BMP. In neuronal cells, the accumulation of gangliosides is particularly evident [57]. Studies of the biophysics of membranes have demonstrated that sphingomyelin and cholesterol have an energetically favorable association in lipid bilayers [58], so the increased cholesterol may stabilize sphingomyelin and slow its degradation by acid sphingomyelinase. Ceramide, the product of sphingomyelinase, would displace cholesterol from membranes, so the cholesterol:sphingomyelin interactions would be mutually reinforcing [25]. In fact, when acid sphingomyelinase levels are increased in the LE/Ly of Niemann–Pick C mutant cells, the cholesterol storage in these organelles is greatly reduced [59]. The same cholesterol:sphingolipid interactions would lead to increased cholesterol content in LE/Ly of cells with defects in acid sphingomyelinase or in GSL hydrolysis.

6.4 AUTOPHAGY AND LYSOSOMAL LIPID TURNOVER

Autophagy, which is discussed in Chapter 2, plays important roles in lipid degradation in lysosomes. In this process, portions of the cytoplasm, including organelles, are engulfed by membranes to form sealed autophagosomes. Lysosomes fuse with the autophagosomes to form autophagolysosomes, which digest the contents of the autophagosomes. As with other catabolic processes in lysosomes, the breakdown products are then released back into the cell and can be reused.

Autophagy may play an important role in the synthesis of BMP. The biosynthetic route to production of BMP is unknown, but potential precursors include phosphatidylglycerol (PG) and cardiolipin [49]. Studies with mutant cells lacking the ability to synthesize PG in the ER strongly implicate this as the main precursor for BMP. Cardiolipin is found almost exclusively in mitochondria. There is not a major lipid vesicle transport pathway directly from the ER directly to the LE/Ly, and there is no vesicle transport from mitochondria to LE/Ly. Thus, it is not clear how either PG or cardiolipin would be delivered to LE/Ly in sufficient abundance to serve as an adequate precursor pool for production of BMP. One possibility is that these lipids are delivered to lysosomes by autophagy of ER segments or of mitochondria (mitophagy) [49]. Another possibility is that PG is transported from the ER to LE/Ly at membrane contact sites, which have been observed between these organelles and are recognized to play important roles in many aspects of lipid transport between organelles.

In the past few years, it has been shown that autophagy of lipid droplets and lysosomal breakdown of the cholesteryl ester and triacylglycerides in their cores is an important component of the catabolism of lipid droplets. It had long been thought that breakdown of the neutral lipids in the core of lipid droplets was carried out exclusively by cytoplasmic enzymes with pH optima near 7 [60]. Recent studies have shown that a significant amount of the catabolism of lipid droplets in hepatocytes [60] and in macrophages [61,62] follows sequestration of the lipid droplets in autophagosomes and subsequent fusion with lysosomes. The cholesteryl esters and triacylglycerides are then hydrolyzed by lysosomal acid lipase, the enzyme that is also responsible for breakdown of these lipids when delivered to LE/Ly in the core of lipoproteins.

The balance between formation and breakdown of lipid droplets in hepatocytes is important for the overall metabolic regulation of the organism. In the presence of high glucose after a meal, fatty acids are synthesized and stored as TGs in lipid droplets. During periods of fasting, these lipid droplets are utilized to provide fatty acids for gluconeogenesis to produce glucose, which is the main energy source for the brain. Thus, the formation and breakdown of these lipid droplets in hepatocytes must be under tight regulatory control, which is only partially understood at present. Inhibition of autophagy in hepatocytes by 3-methyladenine led to a significant increase in stored TGs in the cells [60]. There was a further increase upon inhibition of all lipolysis, indicating that autophagy was only one component of TG turnover. The association of lipid droplets with autophagic membranes could be observed by both optical and electron microscopies [60]. Treatment with rapamycin, which releases the inhibition of autophagy by mTOR, leads to a decrease in lipid droplets and in the amount of TG in the cells. Starvation increased the association of lipid droplets with autophagosomes, indicating that the rates of autophagic flux of lipid droplets can be metabolically regulated.

Macrophages are another cell type in which the formation and breakdown of lipid droplets play an important role in normal physiology and in disease [39]. Macrophages ingest cholesterol deposited in lipoproteins in the wall of blood vessels. Some of this cholesterol is exported to HDLs and returned to the liver, where it can be converted into bile salts or repackaged into lipoproteins. Excess cholesterol in the macrophages is esterified by ACAT and stored in lipid droplets. As in the liver, the prevailing view had long been that the cholesteryl ester in lipid droplets was hydrolyzed by neutral hydrolases in the cytoplasm, although the relative importance of specific neutral cholesteryl ester hydrolases has remained uncertain. Recently, it was shown that autophagy of lipid droplets and lysosomal hydrolysis of their cholesteryl esters by lysosomal acid lipase plays a significant role in releasing cholesterol contained in lipid droplets [61,62]. The cholesterol can then be shuttled out of the lysosomes by NPC2 and NPC1 as described earlier and delivered to various cellular compartments including the plasma membrane, where it can be exported from the cell onto HDL carriers.

6.5 LYSOSOMAL LIPID HYDROLYSIS AND METABOLIC REGULATION

Recent studies have started to elucidate mechanistic links between nutrient sensing and the regulation of lysosomal lipid catabolism (Chapter 7). Transcriptional factors such as transcription factor EB (TFEB; see Chapter 7) regulate lysosome biogenesis and autophagy, including the autophagy of lipid droplets. Under conditions of nutrient abundance, the mechanistic target of rapamycin complex 1 (mTORC1) is activated at the lysosomal surface, and this stimulates lipogenesis through the SREBP transcription factors and adipogenesis through the PPARγ transcription factors. The activation of mTORC1 also inhibits autophagy. Conversely, low levels of amino acids in the lysosomes cause inactivation of mTORC1, with decreases of lipid synthesis and increased autophagy. The stimulated autophagy includes lipophagy, leading to hydrolysis of triacylglycerides by lysosomal acid lipase in autophagolysosomes. This provides a source of fatty acids for both energy production and synthesis of phospholipids.

TFEB is a member of the microphthalmia-associated transcription factor (MITF) family. Under the condition of nutrient abundance, it is mainly held in the cytoplasm due to phosphorylation of key serine residues by active mTORC1. When nutrient levels are reduced, mTORC1 is inactivated, and the lack of phosphorylation allows TFEB to enter the nucleus. TFEB stimulates the synthesis of proteins involved in lysosome biogenesis and autophagy. The main regulators of mTORC1 activation are amino acids in the lumen of the organelles. It is unclear whether lipid catabolism in lysosomes can also influence mTORC1 signaling directly. In addition, it remains unclear how lipophagy is regulated beyond the transcriptional regulation discussed here. It seems likely that molecules recruited to the surface of lipid droplets can regulate lipophagy, but such regulatory mechanisms are not well characterized at present.

6.6 SUMMARY

There has been enormous progress in elucidating mechanisms of lipid catabolism in LE/Ly. Many of the important pathways of lysosomal lipid metabolism have been elucidated by analysis of the metabolic defects in inherited lysosomal storage disorders. Basic biochemical and genetic studies of cells from these patients have revealed surprisingly complex and interrelated mechanisms for lipid breakdown. One unique feature of lipid breakdown in LE/Ly is the important role for internal vesicle membranes that are enriched in BMP, a lipid that is found almost exclusively in these organelles. Restricting lipid hydrolysis mainly to these internal membranes helps to preserve the integrity of the limiting membrane. Often, the accumulation of one lipid, due to a defective hydrolytic enzyme, leads to accumulation of several other lipids and to expansion of the internal membranes. These accumulations can be understood in terms of the biophysical association of certain lipids such as cholesterol and sphingolipids, which will mutually stabilize each other in the bilayer in a way that partially protects them from extraction. In recent years, it has become apparent that autophagy plays an important role in the turnover of lipids in lipid droplets. This means that lysosomal turnover is important in the digestion of cholesteryl esters and triacylglycerides in the core of the droplets. It seems likely that there will be further discoveries linking the biology of the LE/Ly to regulatory mechanisms governing lipid synthesis and degradation.

REFERENCES

[1] Brown MS, Goldstein JL. A receptor-mediated pathway for cholesterol homeostasis. Science 1986;232(4746):34–47. PubMed PMID: 3513311.

[2] Goldstein JL, Brown MS. The LDL receptor. Arterioscler Thromb Vasc Biol 2009;29(4): 431–438. DOI: 10.1161/ATVBAHA.108.179564. PubMed PMID: 19299327; PubMed Central PMCID: PMC2740366.

[3] Goldstein JL, Brown MS. A century of cholesterol and coronaries: from plaques to genes to statins. Cell 2015;161(1):161–172. DOI: 10.1016/j.cell.2015.01.036. PubMed PMID: 25815993.

[4] Beigneux AP, Weinstein MM, Davies BS, Gin P, Bensadoun A, Fong LG, et al. GPIHBP1 and lipolysis: an update. Curr Opin Lipidol 2009;20(3):211–216. DOI: 10.1097/MOL.0b013e32832ac026. PubMed PMID: 19369870; PubMed Central PMCID: PMC2810420.

[5] Beigneux AP, Davies BS, Bensadoun A, Fong LG, Young SG. GPIHBP1, a GPI-anchored protein required for the lipolytic processing of triglyceride-rich lipoproteins. J Lipid Res 2009;50(Suppl):S57–S62. DOI: 10.1194/jlr.R800030-JLR200PubMed PMID: 18854402; PubMed Central PMCID: PMC2674691.

[6] Goulbourne CN, Gin P, Tatar A, Nobumori C, Hoenger A, Jiang H, et al. The GPIHBP1-LPL complex is responsible for the margination of triglyceride-rich lipoproteins in capillaries. Cell Metab 2014;19(5):849–860. DOI: 10.1016/j.cmet.2014.01.017. PubMed PMID: 24726386; PubMed Central PMCID: PMC4143151.

[7] Mukherjee S, Ghosh RN, Maxfield FR. Endocytosis. Physiol Rev 1997;77(3):759–803. PubMed PMID: 9234965.

[8] Presley JF, Mayor S, McGraw TE, Dunn KW, Maxfield FR. Bafilomycin A1 treatment retards transferrin receptor recycling more than bulk membrane recycling. J Biol Chem 1997;272(21):13929–13936. PubMed PMID: 9153255.

[9] Dunn KW, Maxfield FR. Delivery of ligands from sorting endosomes to late endosomes occurs by maturation of sorting endosomes. J Cell Biol 1992;117(2):301–310PubMed PMID: 1560027; PubMed Central PMCID: PMC2289412.

[10] Dunn KW, McGraw TE, Maxfield FR. Iterative fractionation of recycling receptors from lysosomally destined ligands in an early sorting endosome. J Cell Biol 1989;109 (6 Pt 2):3303–3314PubMed PMID: 2600137; PubMed Central PMCID: PMC2115921.

[11] Mayor S, Presley JF, Maxfield FR. Sorting of membrane components from endosomes and subsequent recycling to the cell surface occurs by a bulk flow process. J Cell Biol 1993;121(6):1257–1269PubMed PMID: 8509447; PubMed Central PMCID: PMC2119709.

[12] Hao M, Maxfield FR. Characterization of rapid membrane internalization and recycling. J Biol Chem 2000;275(20):15279–15286. PubMed PMID: 10809763.

[13] Braulke T, Bonifacino JS. Sorting of lysosomal proteins. Biochim Biophys Acta 2009; 1793(4):605–614. DOI: 10.1016/j.bbamcr.2008.10.016. PubMed PMID: 19046998.

[14] Du H, Sheriff S, Bezerra J, Leonova T, Grabowski GA. Molecular and enzymatic analyses of lysosomal acid lipase in cholesteryl ester storage disease. Mol Genet Metab 1998;64(2):126–134. DOI: 10.1006/mgme.1998.2707. PubMed PMID: 9705237.

[15] Friedland N, Liou HL, Lobel P, Stock AM. Structure of a cholesterol-binding protein deficient in Niemann–Pick type C2 disease. Proc Natl Acad Sci U S A 2003;100(5): 2512–2517. DOI: 10.1073/pnas.0437840100. PubMed PMID: 12591954; PubMed Central PMCID: PMC151372.

[16] Wang ML, Motamed M, Infante RE, Abi-Mosleh L, Kwon HJ, Brown MS, et al. Identification of surface residues on Niemann–Pick C2 essential for hydrophobic handoff of cholesterol to NPC1 in lysosomes. Cell Metab 2010;12(2):166–173. DOI: 10.1016/ j.cmet.2010.05.016PubMed PMID: 20674861; PubMed Central PMCID: PMC3034247.

[17] Infante RE, Wang ML, Radhakrishnan A, Kwon HJ, Brown MS, Goldstein JL. NPC2 facilitates bidirectional transfer of cholesterol between NPC1 and lipid bilayers, a step in cholesterol egress from lysosomes. Proc Natl Acad Sci U S A 2008; 105(40):15287–15292. DOI: 10.1073/pnas.0807328105. PubMed PMID: 18772377; PubMed Central PMCID: PMC2563079.

[18] Deffieu MS, Pfeffer SR. Niemann–Pick type C 1 function requires lumenal domain residues that mediate cholesterol-dependent NPC2 binding. Proc Natl Acad Sci U S A 2011;108(47):18932–18936. DOI: 10.1073/pnas.1110439108PubMed PMID: 22065-762; PubMed Central PMCID: PMC3223457.

[19] Du X, Kumar J, Ferguson C, Schulz TA, Ong YS, Hong W, et al. A role for oxysterol-binding protein-related protein 5 in endosomal cholesterol trafficking. J Cell Biol 2011;192(1):121–135doi: 10.1083/jcb.201004142. PubMed PMID: 21220512; PubMed Central PMCID: PMCPMC3019559.

[20] Rosenbaum AI, Maxfield FR. Niemann–Pick type C disease: molecular mechanisms and potential therapeutic approaches. J Neurochem 2011;116(5):789–795. DOI: 10.1111/j.1471-4159.2010.06976.x. PubMed PMID: 20807315; PubMed Central PMCID: PMCPMC3008286.

[21] Ottinger EA, Kao ML, Carrillo-Carrasco N, Yanjanin N, Shankar RK, Janssen M, et al. Collaborative development of 2-hydroxypropyl-beta-cyclodextrin for the treatment of Niemann–Pick type C1 disease. Curr Top Med Chem 2014;14(3):330–339PubMed PMID: 24283970; PubMed Central PMCID: PMCPMC4048128.

[22] Henne WM, Buchkovich NJ, Emr SD. The ESCRT pathway. Dev Cell 2011;21(1):77–91. DOI: 10.1016/j.devcel.2011.05.015. PubMed PMID: 21763610.

[23] Kobayashi T, Beuchat MH, Lindsay M, Frias S, Palmiter RD, Sakuraba H, et al. Late endosomal membranes rich in lysobisphosphatidic acid regulate cholesterol transport. Nat Cell Biol 1999;1(2):113–118. DOI: 10.1038/10084. PubMed PMID: 10559883.

[24] Kobayashi T, Stang E, Fang KS, de Moerloose P, Parton RG, Gruenberg J. A lipid associated with the antiphospholipid syndrome regulates endosome structure and function. Nature 1998;392(6672):193–197. DOI: 10.1038/32440. PubMed PMID: 9515966.

[25] Megha LE. Ceramide selectively displaces cholesterol from ordered lipid domains (rafts): implications for lipid raft structure and function. J Biol Chem 2004;279(11):9997–10004. DOI: 10.1074/jbc.M309992200. PubMed PMID: 14699154.

[26] Chang TY, Chang CC, Ohgami N, Yamauchi Y. Cholesterol sensing, trafficking, and esterification. Annu Rev Cell Dev Biol 2006;22:129–157. DOI: 10.1146/annurev.cellbio.22.010305.104656. PubMed PMID: 16753029.

[27] Rosenson RS, Brewer HB Jr, Davidson WS, Fayad ZA, Fuster V, Goldstein J, et al. Cholesterol efflux and atheroprotection: advancing the concept of reverse cholesterol transport. Circulation 2012;125(15):1905–1919. DOI: 10.1161/CIRCULATIONAHA.111.066589. PubMed PMID: 22508840; PubMed Central PMCID: PMC4159082.

[28] Olkkonen VM. Macrophage oxysterols and their binding proteins: roles in atherosclerosis. Curr Opin Lipidol 2012;23(5):462–470. DOI: 10.1097/MOL.0b013e328356dba0. PubMed PMID: 22814702.

[29] Spann NJ, Glass CK. Sterols and oxysterols in immune cell function. Nat Immunol 2013;14(9):893–900. DOI: 10.1038/ni.2681. PubMed PMID: 23959186.

[30] Brown MS, Goldstein JL. A proteolytic pathway that controls the cholesterol content of membranes, cells, and blood. Proc Natl Acad Sci U S A 1999;96(20):11041–11048. PubMed PMID: 10500120; PubMed Central PMCID: PMC34238.

[31] Radhakrishnan A, Goldstein JL, McDonald JG, Brown MS. Switch-like control of SREBP-2 transport triggered by small changes in ER cholesterol: a delicate balance. Cell Metab 2008;8(6):512–521. DOI: 10.1016/j.cmet.2008.10.008. PubMed PMID: 19041766; PubMed Central PMCID: PMC2652870.

[32] Havel RJ. Receptor and non-receptor mediated uptake of chylomicron remnants by the liver. Atherosclerosis 1998;141(Suppl 1):S1–S7. PubMed PMID: 9888635.

[33] Mahley RW, Ji ZS. Remnant lipoprotein metabolism: key pathways involving cell-surface heparan sulfate proteoglycans and apolipoprotein E. J Lipid Res 1999;40(1):1–16. PubMed PMID: 9869645.

[34] Dawson PA. Chapter 12 - Bile acid metabolism. In: McLeod NDRS, editor. *Biochemistry of Lipids, Lipoproteins and Membranes*. Sixth ed. Boston: Elsevier; 2016. p 359–389.

[35] Wang Y, Huang Y, Hobbs HH, Cohen JC. Molecular characterization of proprotein convertase subtilisin/kexin type 9-mediated degradation of the LDLR. J Lipid Res 2012;53(9):1932–1943. DOI: 10.1194/jlr.M028563. PubMed PMID: 22764087; PubMed Central PMCID: PMCPMC3413232.

[36] Giunzioni I, Tavori H. New developments in atherosclerosis: clinical potential of PCSK9 inhibition. Vasc Health Risk Manag 2015;11:493–501. DOI: 10.2147/VHRM.S74692. PubMed PMID: 26345307; PubMed Central PMCID: PMCPMC4554462.

[37] Moore KJ, Tabas I. Macrophages in the pathogenesis of atherosclerosis. Cell 2011;145(3):341–355. DOI: 10.1016/j.cell.2011.04.005. PubMed PMID: 21529710; PubMed Central PMCID: PMC3111065.

[38] Tabas I, Garcia-Cardena G, Owens GK. Recent insights into the cellular biology of atherosclerosis. J Cell Biol 2015;209(1):13–22. DOI: 10.1083/jcb.201412052. PubMed PMID: 25869663; PubMed Central PMCID: PMC4395483.

[39] Maxfield FR, Tabas I. Role of cholesterol and lipid organization in disease. Nature 2005;438(7068):612–621. DOI: 10.1038/nature04399. PubMed PMID: 16319881.

[40] Maxfield FR, Menon AK. Chapter 14 - Intramembrane and intermembrane lipid transport. In: McLeod NDRS, editor. *Biochemistry of Lipids, Lipoproteins and Membranes*. 6th ed. Boston: Elsevier; 2016. p 415–436.

[41] Gillard BK, Clement RG, Marcus DM. Variations among cell lines in the synthesis of sphingolipids in de novo and recycling pathways. Glycobiology 1998;8(9):885–890. PubMed PMID: 9675221.

[42] Saftig P, Klumperman J. Lysosome biogenesis and lysosomal membrane proteins: trafficking meets function. Nat Rev Mol Cell Biol 2009;10(9):623–635. http://www.nature.com/nrm/journal/v10/n9/suppinfo/nrm2745_S1.html.

[43] Schulze H, Kolter T, Sandhoff K. Principles of lysosomal membrane degradation: cellular topology and biochemistry of lysosomal lipid degradation. Biochim Biophys Acta 2009;1793(4):674–683. DOI: 10.1016/j.bbamcr.2008.09.020. PubMed PMID: 19014978.

[44] Schulze H, Sandhoff K. Sphingolipids and lysosomal pathologies. Biochim Biophys Acta 2014;1841(5):799–810. DOI: 10.1016/j.bbalip.2013.10.015. PubMed PMID: 24184515.

[45] Kolter T, Sandhoff K. Lysosomal degradation of membrane lipids. FEBS Lett 2010; 584(9):1700–1712. DOI: 10.1016/j.febslet.2009.10.021. PubMed PMID: 19836391.

[46] Stahelin RV. Chapter 8 - Phospholipid catabolism. In: McLeod NDRS, editor. *Biochemistry of Lipids, Lipoproteins and Membranes*. 6th ed. Boston: Elsevier; 2016. p 237–257.

[47] Conzelmann E, Sandhoff K. Purification and characterization of an activator protein for the degradation of glycolipids GM2 and GA2 by hexosaminidase A. Hoppe Seylers Z Physiol Chem 1979;360(12):1837–1849. PubMed PMID: 527942.

[48] Wendeler M, Werth N, Maier T, Schwarzmann G, Kolter T, Schoeniger M, et al. The enzyme-binding region of human GM2-activator protein. FEBS J 2006;273(5):982–991. DOI: 10.1111/j.1742-4658.2006.05126.x.

[49] Gallala HD, Sandhoff K. Biological function of the cellular lipid BMP-BMP as a key activator for cholesterol sorting and membrane digestion. Neurochem Res 2011; 36(9):1594–1600. DOI: 10.1007/s11064-010-0337-6. PubMed PMID: 21136156.

[50] Wilkening G, Linke T, Uhlhorn-Dierks G, Sandhoff K. Degradation of membrane-bound ganglioside GM1. Stimulation by bis(monoacylglycero)phosphate and the activator proteins SAP-B and GM2-AP. J Biol Chem 2000;275(46):35814–35819. DOI: 10.1074/jbc.M006568200. PubMed PMID: 10942779.

[51] Kobayashi T, Beuchat MH, Chevallier J, Makino A, Mayran N, Escola JM, et al. Separation and characterization of late endosomal membrane domains. J Biol Chem 2002;277(35):32157–32164. DOI: 10.1074/jbc.M202838200. PubMed PMID: 12065580.

[52] Shayman JA, Abe A. Drug induced phospholipidosis: an acquired lysosomal storage disorder. Biochim Biophys Acta 2013;1831(3):602–611. DOI: 10.1016/j.bbalip.2012.08.013. PubMed PMID: 22960355; PubMed Central PMCID: PMCPMC3528828.

[53] Maxfield FR, McGraw TE. Endocytic recycling. Nat Rev Mol Cell Biol 2004; 5(2):121–132. DOI: 10.1038/nrm1315. PubMed PMID: 15040445.

[54] Lange Y, Ye J, Steck TL. Activation mobilizes the cholesterol in the late endosomes-lysosomes of Niemann–Pick type C cells. PLoS One 2012;7(1):e30051doi: 10.1371/journal.pone.0030051. PubMed PMID: 22276143; PubMed Central PMCID: PMCPMC3262792.

[55] Pagano RE, Puri V, Dominguez M, Marks DL. Membrane traffic in sphingolipid storage diseases. Traffic 2000;1(11):807–815. PubMed PMID: 11208071.

[56] Patterson MC, Hendriksz CJ, Walterfang M, Sedel F, Vanier MT, Wijburg F, et al. Recommendations for the diagnosis and management of Niemann–Pick disease type C: an update. Mol Genet Metab 2012;106(3):330–344. DOI: 10.1016/j.ymgme.2012.03.012. PubMed PMID: 22572546.

[57] Walkley SU, Vanier MT. Secondary lipid accumulation in lysosomal disease. Biochim Biophys Acta 2009;1793(4):726–736. DOI: 10.1016/j.bbamcr.2008.11.014. PubMed PMID: 19111580; PubMed Central PMCID: PMCPMC4382014.

[58] Radhakrishnan A, McConnell H. Condensed complexes in vesicles containing cholesterol and phospholipids. Proc Natl Acad Sci U S A 2005;102(36):12662–12666. DOI: 10.1073/pnas.0506043102. PubMed PMID: 16120676; PubMed Central PMCID: PMCPMC1200296.

[59] Devlin C, Pipalia NH, Liao X, Schuchman EH, Maxfield FR, Tabas I. Improvement in lipid and protein trafficking in Niemann–Pick C1 cells by correction of a secondary enzyme defect. Traffic 2010;11(5):601–615. DOI: 10.1111/j.1600-0854.2010.01046.x. PubMed PMID: 20412078; PubMed Central PMCID: PMCPMC3000737.

[60] Singh R, Kaushik S, Wang Y, Xiang Y, Novak I, Komatsu M, et al. Autophagy regulates lipid metabolism. Nature 2009;458(7242):1131–1135. DOI: 10.1038/nature07976. PubMed PMID: 19339967; PubMed Central PMCID: PMCPMC2676208.

[61] Ouimet M, Franklin V, Mak E, Liao X, Tabas I, Marcel YL. Autophagy regulates cholesterol efflux from macrophage foam cells via lysosomal acid lipase. Cell Metab 2011; 13(6):655–667. DOI: 10.1016/j.cmet.2011.03.023. PubMed PMID: 21641547; PubMed Central PMCID: PMCPMC3257518.

[62] Ouimet M, Marcel YL. Regulation of lipid droplet cholesterol efflux from macrophage foam cells. Arterioscler Thromb Vasc Biol 2012;32(3):575–581. DOI: 10.1161/ATVBAHA.111.240705. PubMed PMID: 22207731.

7

TFEB, MASTER REGULATOR OF CELLULAR CLEARANCE

GRACIANA DIEZ-ROUX

Telethon Institute of Genetics and Medicine (TIGEM), Naples, Italy

ANDREA BALLABIO

Telethon Institute of Genetics and Medicine (TIGEM), Naples, Italy; Medical Genetics, Department of Translational Medicine, Federico II University, Naples, Italy; Department of Molecular and Human Genetics, Baylor College of Medicine, Houston, TX, USA; Jan and Dan Duncan Neurological Research Institute, Texas Children Hospital, Houston, TX, USA

7.1 LYSOSOME

Lysosome is the main digestive compartment of the cell. This organelle degrades a wide variety of structurally diverse substances, such as proteins, glycosaminoglycans (GAGs), nucleic acids, oligosaccharides, and complex lipids into their building blocks, which are either recycled in biosynthetic pathways or further degraded to generate energy (Figure 7.1). Numerous functions, including the turnover of cellular proteins, autophagy, cholesterol homeostasis, downregulation of surface receptors, release of endocytosed nutrients, inactivation of pathogenic organisms, repair of the plasma membrane, and bone remodeling, have been shown to depend on normal lysosomal function. The lysosome is delimited by a single-layer lipid membrane. An acidic internal pH is maintained by an ATP-dependent proton pump. Lysosomal proteins include acidic hydrolases, soluble accessory proteins, and lysosomal membrane proteins, for most of which the precise function is unknown. Lysosomal biogenesis involves maturation of the early endosome to form a multivesicular body (i.e., late

Lysosomes: Biology, Diseases, and Therapeutics, First Edition.
Edited by Frederick R. Maxfield, James M. Willard and Shuyan Lu.

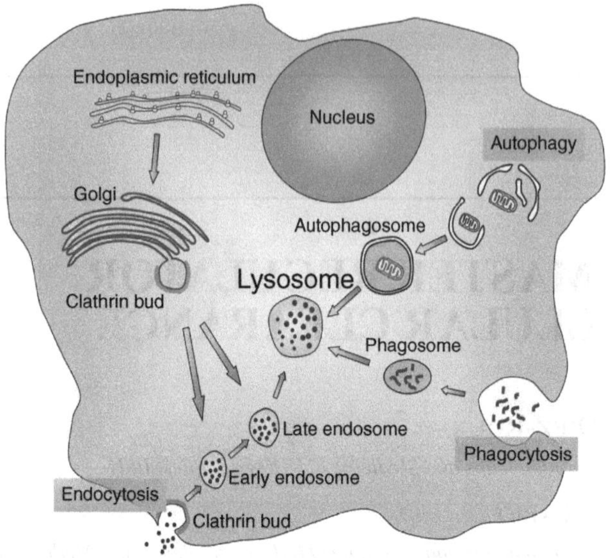

Figure 7.1 Central role of lysosomes in key cellular processes. (*See color plate section for the color representation of this figure.*)

endosome) followed by fusion to lysosomes and by lysosome reformation [1]. Lysosomal hydrolases are imported into lysosomes by a mannose-6-phosphate (M6P) trafficking pathway that involves binding to the M6P receptors. Other proteins are transported to lysosomes by alternative receptors such as the lysosomal integral membrane protein LIMP-2 and sortilin. Cellular and foreign material to be degraded can reach the lysosome via endocytosis, phagocytosis, autophagy, or direct transport. Autophagy is a lysosomal degradation pathway with a crucial role in health and disease. A growing number of intracellular substrates such as polyubiquitinated proteins, protein aggregates, and dysfunctional organelles are degraded, and their components recycled, by autophagy. During autophagy, a double-membrane vesicle (autophagosome AV) originates in the proximity of the ER and targets cytosolic cargoes to lysosomes. A crucial step of the autophagic pathway is fusion of the autophagosome with the lysosome. A schematic view of lysosome biogenesis and of the autophagic pathway is depicted in Figure 7.1.

7.2 THE TRANSCRIPTIONAL REGULATION OF LYSOSOMAL FUNCTION

The cell has developed strategies to facilitate the coordination of complex functions, such as cellular clearance, by concentrating these processes in specialized organelles. As the degradative requirements of the cell vary depending upon tissue type, age, and environmental conditions, it was reasonable to expect the existence of systems that allow regulation of lysosomal function. This hypothesis led to the discovery of a gene

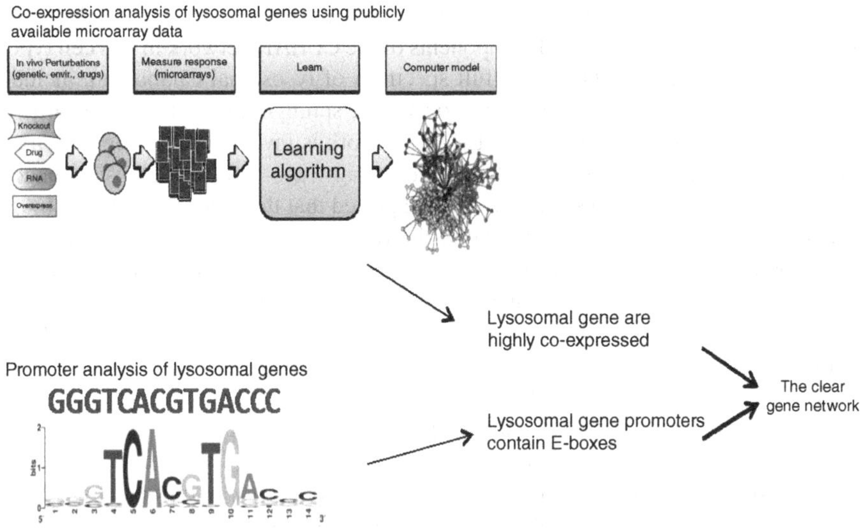

Figure 7.2 Systems biology approach used to discover the CLEAR network. The clear network is inferred by integrating the coexpression analysis of microarray data and promoter analysis. Pattern discovery analysis of the promoter regions of the known lysosomal genes resulted in the identification of a palindromic 10-base preferentially located at 200 bp from the transcription start site. The CLEAR consensus sequence overlaps that of the E-box (CANNTG), a known target site for basic helix–loop–helix transcription factors. (*See color plate section for the color representation of this figure.*)

regulatory network (CLEAR: Coordinated Lysosomal Enhancement And Regulation) that controls lysosomal biogenesis and function [2]. The coordinated regulation of lysosomal genes was discovered using both coexpression and promoter analyses. The latter resulted in the identification of a palindromic sequence of 10 bases, preferentially located at 200 bp from the transcription start site. This sequence, named the CLEAR consensus sequence, overlaps with that of the E-box, a known target for basis helix–loop–helix transcription factors (Figure 7.2).

Members of the microphthalmia–transcription factor E (MiT/TFE) subfamily of bHLH factors were found to bind sequences identical to the CLEAR consensus. The MiT/TFE subfamily is composed of four members in humans: MITF, TFE3, TFEB, and TFEC [3]. All members of the family share a similar structure and are capable of homo- or heterodimerizing with each other in order to bind to DNA and activate transcription. MiTF and TFEC have a restricted pattern of expression in specific cell types (MiTF: melanocytes, osteoclasts, mast cells, macrophages, NK cells, B cells and heart; TFEC, cells of myeloid origin), while TFEB and TFE3 have a ubiquitous pattern of expression [3–7].

TFEB binds to CLEAR sites and positively regulates the expression of lysosomal genes. TFEB overexpression in cells increases the number of lysosomes and enhances the degradative capabilities of the cells, as a result of an induction of lysosomal gene expression [2]. A detailed analysis on TFEB target genes in HeLa cells, using promoter, transcriptome, and ChIP-seq analyses, identified 471 TFEB direct targets representing the essential components of the CLEAR network in this cell type, which include genes regulating the full spectrum of lysosomal catabolism, including degradation of proteins, glycosaminoglycans, sphingolipids, and glycogen [8]. This analysis also revealed that the CLEAR network is involved in the regulation of lysosome-associated processes, including autophagy, exo- and endocytosis, phagocytosis, and immune response. These data suggested that the function of the CLEAR network, and TFEB, may extend beyond the regulation of lysosomal function and impact other important cellular processes related to cellular catabolism [8]. Recent studies suggest that the CLEAR network may be remarkably different in different tissues and cell types. For instance, in liver, TFEB regulates genes involved in lipid catabolism [9].

TFEB overexpression was also shown to induce autophagy by increasing the number of autophagosomes and enhancing lysosome-to-autophagosome fusion and degradation of long-lived proteins, which are known autophagy substrates [10]. TFEB also has a direct impact on lysosomal exocytosis, a process by which lysosomes secrete their content into the extracellular matrix by fusing with the plasma membrane through a Ca^{2+}-regulated mechanism [11,12]. TFEB regulates different aspects of this process including docking of lysosomes to the plasma membrane and fusion of lysosomes to the plasma membrane via the calcium channel mucolipin-1 [13].

Other members of the MiT/TFE family, such as TFE3, have been shown to have a function similar to that of TFEB (and similar regulation as described in the following section). In fact, both TFE3 and MITF have been shown to increase the expression of lysosomal and autophagic genes and to stimulate lysosomal biogenesis and autophagy [14,15]. More studies are necessary to determine the functional redundancy of this family of transcription factors and the physiological relevance of each.

7.3 TFEB SUBCELLULAR REGULATION IS REGULATED BY ITS PHOSPHORYLATION

In most cell types, TFEB is located in the cytoplasm under normal nutrient conditions. However, in response to certain stimuli, such as starvation, TFEB translocates from the cytoplasm to the nucleus [2,16]. The translocation of TFEB to the nucleus is mediated by the phosphorylation status of two particular serines, Ser142 and Ser211. When these two serines are phosphorylated, TFEB is located predominantly in the cytoplasm, while when they are unphosphorylated it is mainly nuclear. Mutations of either of these serines into alanines, preventing their phosphorylation, result in a nuclear, and constitutively active, TFEB [16–19]. ERK2 kinase was found to phosphorylate TFEB S142 and consequently its inhibition leads to TFEB nuclear translocation [16]. The mTOR complex 1 (mTORC1) kinase complex is also able to phosphorylate TFEB and plays a critical role in determining TFEB subcellular localization and activity [17–20]. mTORC1 phosphorylates both TFEB S142 and S211 [16,19]. When S211 is phosphorylated, TFEB binds to several isoforms of the 14-3-3 protein family and this binding retains TFEB in the cytoplasm [17,18]. This mechanism is also shared by other members of the bHLH family of transcription factors, such as MITF and TFE3, which are closely related to TFEB [7,18].

Interestingly, the transcriptional response of TFEB to nutrients is mediated by an autoregulatory feedback loop in which TFEB regulates its own expression in a starvation-dependent manner [9]. Thus, the regulation of TFEB activity by nutrients involves a rapid, phosphorylation-dependent, post-transcriptional switch, which is responsible for the nuclear translocation of TFEB and a transcriptional autoregulatory component for a slower, more sustained, response.

While the role of the kinases that phosphorylate TFEB has been studied by several groups [17–19,21], the phosphatase(s) involved in its dephosphorylation have remained elusive for some time. Presumably, kinases and phosphatases would have an opposite effect of TFEB subcellular localization, the former promoting a cytoplasmic, while the latter a nuclear localization. A high content (HC) screening of a phosphatase siRNA library using a cellular assay based on cytoplasm-to-nucleus shuttling of TFEB during starvation was used to identify phosphatases that dephosphorylate TFEB, thus promoting its nuclear translocation. This approach led to the identification of calcineurin, a calcium-dependent serine–threonine phosphatase [22], which binds and dephosphorylates TFEB [23]. Calcineurin inhibition, either by pharmacologic or genetic approaches, suppresses TFEB activity even during starvation. Conversely, calcineurin overexpression and constitutive activation have the opposite effect. Interestingly, the effect of calcineurin on TFEB subcellular localization overrides that of mTORC1, as the inhibition of both calcineurin and mTOR activities leads to a cytoplasmic localization of TFEB [23].

7.4 A LYSOSOME-TO-NUCLEUS SIGNALING MECHANISM

The discovery that the mTORC1 kinase complex, a master growth regulator that is stimulated by nutrients and growth factors, exerts its activity on the lysosomal surface has shed new light on the role of the lysosome in signaling processes [24,25]. The lysosomal localization of mTORC1 suggests that the lysosome acts as a link between the regulation of anabolic and catabolic processes. Growth factors, hormones, amino acids, glucose, stress, and oxygen are the major activators of mTORC1, which in turn positively regulates protein, mRNA and lipid biosynthesis, and ATP production [24,26,27]. When nutrients are present, mTORC1 is activated, promotes anabolic process and inhibits autophagy through the direct phosphorylation and inhibition of the kinase complex ULK1–Atg13–FIP200 (unc-51-like kinase 1/mammalian autophagy-related gene 13/focal adhesion kinase family-interacting protein of 200 kDa) [28–30], which is required to induce autophagosome biogenesis [31,32]. When mTORC1 is inhibited, by either starvation or drugs, autophagy is stimulated. mTORC1 is part of a complex signaling machinery that sits on the lysosomal surface. This machinery includes the Rag GTPases proteins, which regulate mTORC1 subcellular localization [26] and Rheb GTPases, responsible for mTORC1 activation [26]. Rag GTPases are activated by nutrients, in particular, amino acids, and are responsible for the translocation of mTORC1 to the lysosomal surface. Upon starvation, the tuberous sclerosis complex (TSC1/2) maintains Rheb in its GDP-bound status, thereby blocking the activation of mTORC1 [33]. mTORC1 is further regulated by additional lysosome-associated proteins that include Ragulator, which is responsible for both the subcellular localization of the Rags and regulation of their nucleotide-binding state [25]. The sensing of amino acids by the lysosomal proton pump v-ATPase is necessary to activate the Ragulator complex that exerts a GEF activity toward the Rag proteins, thus promoting the translocation of mTORC1 to the lysosomal surface [25]. Additional GTPase-activating protein (GAP) complexes, such as GATOR1 and folliculin-FNIP1/2, promote the GTPase activity of Rag proteins [34,35]. More recently, an additional component of the lysosomal amino acid machinery that controls mTORC1 has been described, SLC38A9, a member of the solute carrier (SLC) family. SLC38A9 is a transmembrane protein that associates with the lysosomal Rag-Ragulator complex. SLC38A9 overexpression renders mTORC1 insensitive to amino acid depletion, thus active during starvation [36,37].

Interestingly, the cytoplasmic form of TFEB is found both free in the cytosol and anchored to the lysosomal surface, where it interacts with mTORC1 [19]. Cellular conditions that alter lysosomal function such as stress and lysosomal inhibition lead to mTORC1 inactivation and induce TFEB nuclear translocation [17–19]. Thus, the lysosome controls its own biogenesis by sending a signal to the nucleus via TFEB.

Lysosomal Ca^{2+} release plays a crucial role in the activation of calcineurin, the phosphatase that dephosphorylates TFEB [23]. This process is mediated by the calcium channel mucolipin-1 (MCOLN1), also known as transient receptor potential calcium channel MucoLipin subfamily member 1 (TRPML1), which is localized on the lysosomal surface. Inhibition or downregulation of MCOLN1 significantly reduced cytoplasm-to-nucleus shuttling of TFEB after starvation [23]. Interestingly, the induction of autophagy and lysosomal biogenesis via TFEB requires MCOLN1-mediated calcineurin activation, thus linking lysosomal calcium signaling to autophagy regulation [23]. Together these findings identify the lysosome as a central hub for the signaling pathways that regulate cellular homeostasis (Figure 7.3).

7.5 TFEB AND CELLULAR CLEARANCE IN HUMAN DISEASE

Lysosomal dysfunction is a major contributor to the pathogenesis of a significant number of human diseases, including neurodegenerative disorders, disorders of lipid and glucose metabolism, infectious diseases, bone diseases, and disorders of the immune system. Even aging has been associated to decline in lysosomal function [38]. Therefore, boosting lysosomal activity and pathways related to cellular clearance could have a significant impact on a number of human conditions. Therapeutic strategies aimed at rescuing and/or enhancing lysosomal and autophagic function may impact lysosomal storage disorders (LSDs) as well as common neurodegenerative diseases.

The function of TFEB, and the associated CLEAR network, has a direct impact on each step associated with cellular clearance: regulation of cargo delivery of substrates

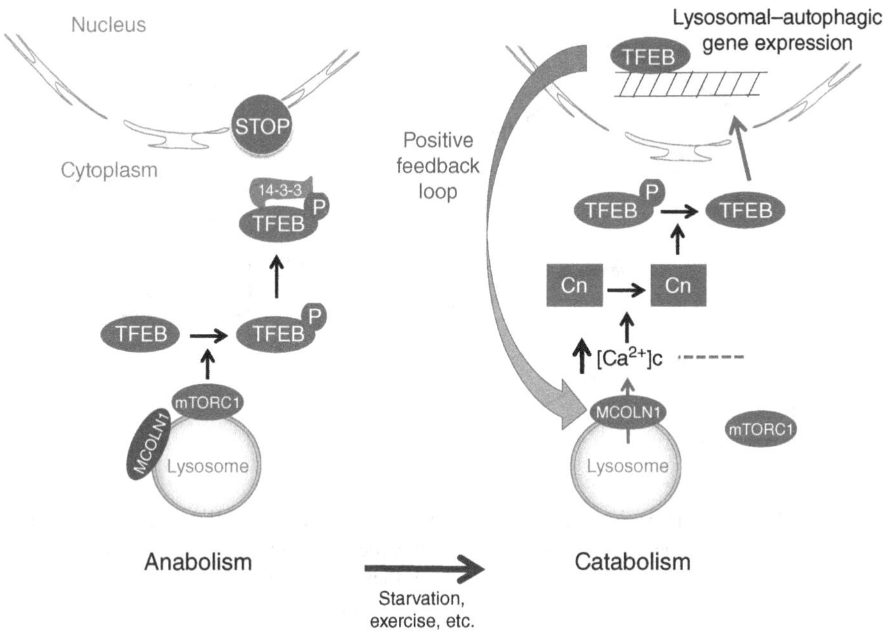

Figure 7.3 Model depicting Ca^{2+}-mediated regulation of TFEB. This figure illustrates how transcription factor TFEB is induced by starvation and mediates the starvation response. TFEB, in adequate nutrition condition, is phosphorylated by mTORC1 on the lysosomal surface. This keeps TFEB inactive by cytosolic sequestration. During starvation, mTORC1 is released from the lysosomal surface and becomes inactive while a calcium-dependent serine–threonine phosphatase, calcineurin, dephosphorylates TFEB. Thus, TFEB can no longer be phosphorylated by mTORC1 and its dephosphorylation promotes its nuclear translocation, where it induces its own transcription. Therefore, starvation regulates TFEB activity through a dual mechanism that involves a posttranslational modification (phosphorylation) and a transcriptional autoregulatory loop. Once in the nucleus, TFEB regulates the expression of genes involved in the lysosomal–autophagy pathway. (*See color plate section for the color representation of this figure.*)

by inducing autophagosome biogenesis and autophagosome–lysosome fusion, promotion of substrate degradation by increasing the levels of lysosomal enzymes, induction of lysosomal exocytosis with the consequent release of storage material outside the cells (Figure 7.4). The major impact that TFEB has on cellular clearance identifies this pathway as a possible therapeutic target for a variety of diseases (Figure 7.5).

7.5.1 Lysosomal Storage Disorders

LSDs are characterized by progressive accumulation of undigested macromolecules within the cell due to lysosomal dysfunction. LSDs include nearly 50 different inherited disorders, each sharing a genetic defect that renders the lysosomal system

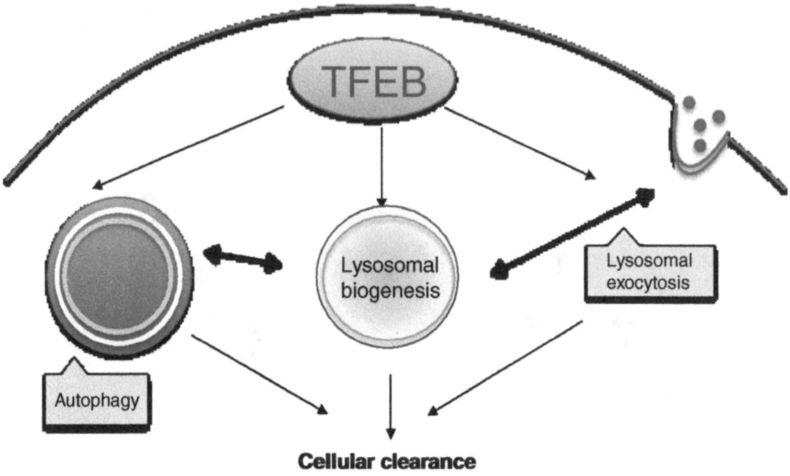

Figure 7.4 TFEB activation on cellular mechanisms that lead to cellular clearance. TFEB controls lysosomal biogenesis, autophagosome biogenesis, autophagosome–lysosome fusion, and lysosomal exocytosis. The concerted action of these processes leads to cellular clearance. (*See color plate section for the color representation of this figure.*)

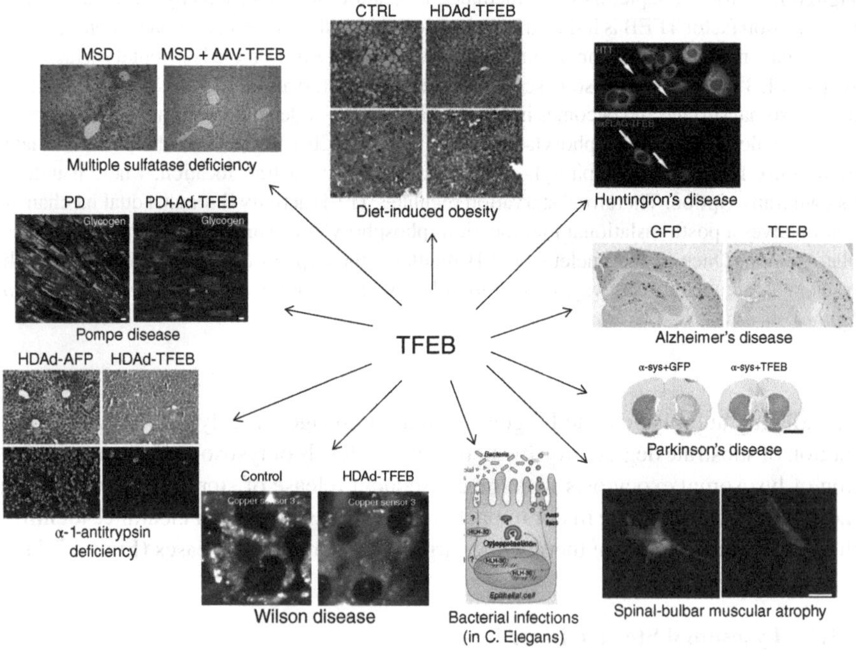

Figure 7.5 Diseases that respond to TFEB-mediated clearance. The major impact that TFEB has on cellular clearance identifies this pathway as a possible therapeutic target for a variety of diseases. (*See color plate section for the color representation of this figure.*)

dysfunctional and unable to degrade specific materials normally processed within the cell. As a consequence, many tissues and organ systems are affected, including brain, viscera, bone, and cartilage, with early-onset central nervous system (CNS) dysfunction predominating. The clinical features of these disorders vary widely. Most are fatal within the first two decades of life following many years of worsening disease. The progressive nature of phenotype development is one of the hallmarks of LSDs.

Viral-mediated gene transfer of TFEB in mouse models of LSDs, including multiple sulfatase deficiency (MSD) and Pompe disease, resulted in clearance lysosomal storage, in spite of the differences in the nature of accumulating substrates (e.g., glycogen in Pompe disease and GAG accumulation in MSD and MPSIIIA) [13,39]. Other secondary effects of storage, such as inflammation and apoptosis, were also reduced. The proposed mechanism of cellular clearance is through the induction of lysosomal exocytosis, as a way to empty the cells from the unwanted storage of undigested substrates. Interestingly, the clearance effect mediated by TFEB required a functional autophagic pathway [39].

7.5.2 Neurodegenerative Disorders

A series of experimental evidences indicate that lysosomal and autophagy dysfunction are associated to the disease pathogenesis of common neurodegenerative disorders such as Parkinson's disease (PD), Alzheimer's disease (AD), and Huntington's disease (HD) [40,41]. Therefore, several studies addressed whether boosting lysosomal and autophagic function through TFEB overexpression could arrest or delay disease progression in several models of common neurodegenerative disorders.

TFEB overexpression resulted in beneficial effects in two mouse models of PD, one pharmacologically induced [42] and the other obtained by α-synuclein overexpression. In both models, AAV-mediated TFEB overexpression had beneficial effects, significantly reducing death of dopaminergic neurons [42,43]. Clearance of α-synuclein oligomers and significant neuroprotection was also observed [43]. Interestingly, inhibition of TFEB expression, by the overexpression of a microRNA known to target TFEB, exacerbated the toxicity of α-synuclein [43].

TFEB was also shown to effectively reduce neurofibrillary tangle pathology and improve behavioral and synaptic defects as well as neurodegeneration in a mouse model of AD, induced by the overexpression of tau. TFEB was shown to specifically target hyperphosphorylated and misfolded tau species [44], and not influence the levels of wild-type soluble tau. Similar results were also obtained in a mouse model of HD, in which TFEB was shown to be the downstream effector of the ability of PGC1α to promote cellular clearance and decrease neurotoxicity in this disease model [45].

Finally, TFEB function was found to be inhibited in a mouse model of X-linked spinal and bulbar muscular atrophy (SBMA), due to polyglutamine-expanded androgen receptor (polyQ-AR). PolyQ-AR was shown to interfere with TFEB function, while normal AR induces TFEB activation. TFEB overexpression rescued metabolic and autophagic flux defects in SBMA cell lines [46].

The fact that TFEB overexpression has such a strong impact on the phenotype of a wide variety of neurodegenerative diseases further strengthens the link between neurodegenerative disorders and lysosomal-autophagic dysfunction.

7.5.3 Metabolic Syndrome

There is accumulating evidence that the lysosomal-autophagic pathway directly influences lipid metabolism. In fact, in a process known as macrolipophagy [47,48], lipid droplets are transported to lysosomes via autophagic vesicles, where they are hydrolyzed into free fatty acids (FFAs) and glycerol. The link is further strengthened by evidence indicating that autophagy levels in the liver directly influence the phenotype of genetically induced obese mice and that anomalies in lysosomal lipases, such as seen in Wolman disease, can lead to severe intracellular fat accumulation. Upon starvation, the organism induces a starvation response that includes the degradation of lipids to produce energy. One of the immediate responders to starvation is TFEB, which translocates to the nucleus upon nutrient deprivation and activates the expression of target genes.

In the liver, TFEB exerts an integrated control on cellular lipid metabolism by activating genes involved in several steps of lipid breakdown, such as lipophagy, fatty acid oxidation, and ketogenesis [9]. This effect is mediated by a TFEB-dependent activation of PGC1α and PPARα, key mediators of the lipid catabolism during the starvation response [9,49,50]. Interestingly, when mice overexpress TFEB, they are leaner and show increased levels of lipid catabolism as well as increased energy expenditure. Most importantly, when TFEB was overexpressed in mouse models of obesity, both genetically and diet induced, it rescued the effects of metabolic syndrome [9].

7.5.4 Cancer, Inborn Errors of Metabolism, Immunity, and Longevity

The activation of the autophagic/lysosomal pathway has also been shown to have beneficial effects in other disorders and disease conditions. For instance, the overexpression of TFEB in the liver of a mouse model of α1-antitrypsin (ATZ) deficiency resulted in the clearance of mutated protein aggregates and rescue of liver fibrosis phenotype. Once again in this case, the increase in ATZ degradation was accompanied by an increase in the autophagic flux, decreased liver fibrosis, and decreased apoptosis, which are the key features of the disease [51].

Using alternative animal models, such as *Caenorhabditis elegans*, TFEB was shown to have a direct effect on the transcription of cytoprotective, antimicrobial, and autophagy genes, suggesting that TFEB has an important role in host response to infection and innate immunity [52].

Gene fusions caused by chromosomal translocations that involve either *TFE3* or *TFEB* are associated with a specific type of renal cell carcinoma (RCC); however, the molecular mechanisms underlying the renal-specific tumorigenesis of these genes remain largely unclear. Determining which pathways are most important in TFEB/TFE3-dependent oncogenesis will be critical in discovering the most promising therapeutic targets.

Finally, of particular interest is the recent finding that TFEB promotes longevity in *C. elegans* [53], suggesting that modulation of lysosomal function may even prevent aging and prolong lifespan [52].

To date, therapeutic strategies for disorders due to intracellular accumulation of storage material have been tailored to the specific gene or disease, significantly increasing the costs of preclinical investigation. The global enhancement of cellular clearance as the basis for a new therapeutic approach may overcome this major limitation and set the basis to develop therapies for a large number of disorders. More need to be understood on this mechanism to be able to fine-tune its activity, to understand potential unwanted effects, and maximize its benefits.

REFERENCES

[1] Luzio JP, Pryor PR, Bright NA. Lysosomes: fusion and function. Nat Rev Mol Cell Biol 2007;8:622–632.

[2] Sardiello M, Palmieri M, di Ronza A, Medina DL, Valenza M, et al. A gene network regulating lysosomal biogenesis and function. Science 2009;325:473–477.

[3] Steingrimsson E, Copeland NG, Jenkins NA. Melanocytes and the microphthalmia transcription factor network. Ann Rev Genet 2004;38:365–411.

[4] Massari ME, Murre C. Helix-loop-helix proteins: regulators of transcription in eukaryotic organisms. Mol Cell Biol 2000;20:429–440.

[5] Meadows NA, Sharma SM, Faulkner GJ, Ostrowski MC, Hume DA, et al. The expression of Clcn7 and Ostm1 in osteoclasts is coregulated by microphthalmia transcription factor. J Biol Chem 2007;282:1891–1904.

[6] Rehli M, Lichanska A, Cassady AI, Ostrowski MC, Hume DA. TFEC is a macrophage-restricted member of the microphthalmia-TFE subfamily of basic helix-loop-helix leucine zipper transcription factors. J Immunol 1999;162:1559–1565.

[7] Martina JA, Diab HI, Li H, Puertollano R. Novel roles for the MiTF/TFE family of transcription factors in organelle biogenesis, nutrient sensing, and energy homeostasis. Cell Mol Life Sci 2014;71:2483–2497.

[8] Palmieri M, Impey S, Kang H, di Ronza A, Pelz C, et al. Characterization of the CLEAR network reveals an integrated control of cellular clearance pathways. Hum Mol Genet 2011;20:3852–3866.

[9] Settembre C, De Cegli R, Mansueto G, Saha PK, Vetrini F, et al. TFEB controls cellular lipid metabolism through a starvation-induced autoregulatory loop. Nat Cell Biol 2013;15:647–658.

[10] Cuervo AM. Cell biology. Autophagy's top chef. Science 2011;332:1392–1393.

[11] Chavez RA, Miller SG, Moore HP. A biosynthetic regulated secretory pathway in constitutive secretory cells. J Cell Biol 1996;133:1177–1191.

[12] Coorssen JR, Schmitt H, Almers W. Ca2+ triggers massive exocytosis in Chinese hamster ovary cells. EMBO J 1996;15:3787–3791.

[13] Medina DL, Fraldi A, Bouche V, Annunziata F, Mansueto G, et al. Transcriptional activation of lysosomal exocytosis promotes cellular clearance. Dev Cell 2011;21:421–430.

[14] Martina JA, Diab HI, Lishu L, Jeong AL, Patange S, et al. The nutrient-responsive transcription factor TFE3 promotes autophagy, lysosomal biogenesis, and clearance of cellular debris. Sci Signal 2014;7:ra9.

[15] Ploper D, Taelman VF, Robert L, Perez BS, Titz B, et al. MITF drives endolysosomal biogenesis and potentiates Wnt signaling in melanoma cells. Proc Natl Acad Sci U S A 2015;112:E420–E429.

[16] Settembre C, Di Malta C, Polito VA, Garcia Arencibia M, Vetrini F, et al. TFEB links autophagy to lysosomal biogenesis. Science 2011;332:1429–1433.

[17] Martina JA, Chen Y, Gucek M, Puertollano R. MTORC1 functions as a transcriptional regulator of autophagy by preventing nuclear transport of TFEB. Autophagy 2012;8:903–914.

[18] Roczniak-Ferguson A, Petit CS, Froehlich F, Qian S, Ky J, et al. The transcription factor TFEB links mTORC1 signaling to transcriptional control of lysosome homeostasis. Sci Signaling 2012;5:ra42.

[19] Settembre C, Zoncu R, Medina DL, Vetrini F, Erdin S, et al. A lysosome-to-nucleus signalling mechanism senses and regulates the lysosome via mTOR and TFEB. EMBO J 2012;31:1095–1108.

[20] Pena-Llopis S, Vega-Rubin-de-Celis S, Schwartz JC, Wolff NC, Tran TA, et al. Regulation of TFEB and V-ATPases by mTORC1. EMBO J 2011;30:3242–3258.

[21] Ferron M, Settembre C, Shimazu J, Lacombe J, Kato S, et al. A RANKL-PKCbeta-TFEB signaling cascade is necessary for lysosomal biogenesis in osteoclasts. Genes Dev 2013;27:955–969.

[22] Stewart AA, Ingebritsen TS, Manalan A, Klee CB, Cohen P. Discovery of a Ca2+- and calmodulin-dependent protein phosphatase: probable identity with calcineurin (CaM-BP80). FEBS Lett 1982;137:80–84.

[23] Medina DL, Di Paola S, Peluso I, Armani A, De Stefani D, et al. Lysosomal calcium signalling regulates autophagy through calcineurin and TFEB. Nat Cell Biol 2015;17:288–299.

[24] Sancak Y, Bar-Peled L, Zoncu R, Markhard AL, Nada S, et al. Ragulator-Rag complex targets mTORC1 to the lysosomal surface and is necessary for its activation by amino acids. Cell 2010;141:290–303.

[25] Zoncu R, Bar-Peled L, Efeyan A, Wang S, Sancak Y, et al. mTORC1 senses lysosomal amino acids through an inside-out mechanism that requires the vacuolar H(+)-ATPase. Science 2011;334:678–683.

[26] Laplante M, Sabatini DM. mTOR signaling in growth control and disease. Cell 2012;149:274–293.

[27] Efeyan A, Zoncu R, Chang S, Gumper I, Snitkin H, et al. Regulation of mTORC1 by the Rag GTPases is necessary for neonatal autophagy and survival. Nature 2012;493:679–683.

[28] Ganley IG, Lam du H, Wang J, Ding X, Chen S, et al. ULK1.ATG13.FIP200 complex mediates mTOR signaling and is essential for autophagy. J Biol Chem 2009;284: 12297–12305.

[29] Hosokawa N, Hara T, Kaizuka T, Kishi C, Takamura A, et al. Nutrient-dependent mTORC1 association with the ULK1-Atg13-FIP200 complex required for autophagy. Mol Biol Cell 2009;20:1981–1991.

[30] Jung CH, Jun CB, Ro SH, Kim YM, Otto NM, et al. ULK-Atg13-FIP200 complexes mediate mTOR signaling to the autophagy machinery. Mol Biol Cell 2009; 20:1992–2003.

[31] Chan EY, Kir S, Tooze SA. siRNA screening of the kinome identifies ULK1 as a multidomain modulator of autophagy. J Biol Chem 2007;282:25464–25474.

[32] Hara T, Takamura A, Kishi C, Iemura S, Natsume T, et al. FIP200, a ULK-interacting protein, is required for autophagosome formation in mammalian cells. J Cell Biol 2008;181:497–510.

[33] Inoki K, Li Y, Xu T, Guan KL. Rheb GTPase is a direct target of TSC2 GAP activity and regulates mTOR signaling. Genes Dev 2003;17:1829–1834.

[34] Tsun ZY, Bar-Peled L, Chantranupong L, Zoncu R, Wang T, et al. The folliculin tumor suppressor is a GAP for the RagC/D GTPases that signal amino acid levels to mTORC1. Mol Cell 2013;52:495–505.

[35] Bar-Peled L, Chantranupong L, Cherniack AD, Chen WW, Ottina KA, et al. A tumor suppressor complex with GAP activity for the Rag GTPases that signal amino acid sufficiency to mTORC1. Science 2013;340:1100–1106.

[36] Rebsamen M, Pochini L, Stasyk T, de Araujo ME, Galluccio M, et al. SLC38A9 is a component of the lysosomal amino acid sensing machinery that controls mTORC1. Nature 2015;519:477–481.

[37] Wang S, Tsun ZY, Wolfson RL, Shen K, Wyant GA, et al. Metabolism. Lysosomal amino acid transporter SLC38A9 signals arginine sufficiency to mTORC1. Science 2015;347:188–194.

[38] Rubinsztein DC, Marino G, Kroemer G. Autophagy and aging. Cell 2011;146:682–695.

[39] Spampanato C, Feeney E, Li L, Cardone M, Lim JA, et al. Transcription factor EB (TFEB) is a new therapeutic target for Pompe disease. EMBO Mol Med 2013;5:691–706.

[40] Wong E, Cuervo AM. Autophagy gone awry in neurodegenerative diseases. Nat Neurosci 2010;13:805–811.

[41] Harris H, Rubinsztein DC. Control of autophagy as a therapy for neurodegenerative disease. Nat Rev Neurol 2012;8:108–117.

[42] Dehay B, Bove J, Rodriguez-Muela N, Perier C, Recasens A, et al. Pathogenic lysosomal depletion in Parkinson's disease. J Neurosci 2010;30:12535–12544.

[43] Decressac M, Mattsson B, Weikop P, Lundblad M, Jakobsson J, et al. TFEB-mediated autophagy rescues midbrain dopamine neurons from alpha-synuclein toxicity. Proc Nat Acad Sci U S A 2013;110:E1817–E1826.

[44] Polito VA, Li H, Martini-Stoica H, Wang B, Yang L, et al. Selective clearance of aberrant tau proteins and rescue of neurotoxicity by transcription factor EB. EMBO Mol Med 2014;6:1142–1160.

[45] Tsunemi T, Ashe TD, Morrison BE, Soriano KR, Au J, et al. PGC-1alpha rescues Huntington's disease proteotoxicity by preventing oxidative stress and promoting TFEB function. Science Translational Medicine 2012;4:142ra197.

[46] Cortes CJ, Miranda HC, Frankowski H, Batlevi Y, Young JE, et al. Polyglutamine-expanded androgen receptor interferes with TFEB to elicit autophagy defects in SBMA. Nat Neurosci 2014;17:1180–1189.

[47] Singh R, Cuervo AM. Autophagy in the cellular energetic balance. Cell Metab 2011; 13:495–504.

[48] Singh R, Kaushik S, Wang Y, Xiang Y, Novak I, et al. Autophagy regulates lipid metabolism. Nature 2009;458:1131–1135.

[49] Finck BN, Kelly DP. PGC-1 coactivators: inducible regulators of energy metabolism in health and disease. J Clin Invest 2006;116:615–622.

[50] Kersten S, Desvergne B, Wahli W. Roles of PPARs in health and disease. Nature 2000;405:421–424.

[51] Pastore N, Blomenkamp K, Annunziata F, Piccolo P, Mithbaokar P, et al. Gene transfer of master autophagy regulator TFEB results in clearance of toxic protein and correction of hepatic disease in alpha-1-anti-trypsin deficiency. EMBO Mol Med 2013;5:397–412.

[52] Visvikis O, Ihuegbu N, Labed SA, Luhachack LG, Alves AM, et al. Innate host defense requires TFEB-mediated transcription of cytoprotective and antimicrobial genes. Immunity 2014;40:896–909.

[53] Lapierre LR, De Magalhaes Filho CD, McQuary PR, Chu CC, Visvikis O, et al. The TFEB orthologue HLH-30 regulates autophagy and modulates longevity in *Caenorhabditis elegans*. Nat Commun 2013;4:2267.

8

LYSOSOMAL MEMBRANE PERMEABILIZATION IN CELL DEATH

URŠKA REPNIK

Department of Biochemistry and Molecular and Structural Biology, Jožef Stefan Institute, Ljubljana, Slovenia; Department of Biosciences, University of Oslo, Oslo, Norway

BORIS TURK

Department of Biochemistry and Molecular and Structural Biology, Jožef Stefan Institute, Ljubljana, Slovenia; Center of Excellence CIPKEBIP, Ljubljana, Slovenia; Faculty of Chemistry and Chemical Technology, University of Ljubljana, Ljubljana, Slovenia

8.1 INTRODUCTION

The organelles of the endocytic pathway comprise early endosomes, late endosomes, lysosomes, and hybrid organelles between the last two. As explained in the earlier chapters, pH decreases and degradative potential increases along the endocytic pathway. This supports the function of the late endocytic compartments (hereafter referred to as lysosomes), which is degradation of cargo, delivered either by endocytosis, phagocytosis, or autophagy [1]. However, more than 60 different hydrolases packed in vesicles lined by a single phospholipid bilayer pose a threat if released into the cytosol as realized already by de Duve [2–4] when characterizing these organelles. Nevertheless, the lysosomal membrane seems sturdy and lysosomal membrane permeabilization (LMP) takes place either as a result of a direct insult to lysosomes or in response to stress signaling [4]. In either case, LMP interferes with the degradative function of the endocytic pathway. Moreover, in the event of considerable permeabilization, the luminal hydrolases are released into the cytosol. On the one hand,

Lysosomes: Biology, Diseases, and Therapeutics, First Edition.
Edited by Frederick R. Maxfield, James M. Willard and Shuyan Lu.

these hydrolases can generate mediators that signal programmed cell death; on the other hand, the extensive LMP can lead to widespread hydrolysis in the cytoplasm, which causes nonprogrammed cell necrosis. Herein, LMP is discussed in the context of cell death together with the agents that cause it. In addition to exogenous compounds that directly destabilize lysosomal membrane, the endogenous signaling events that lead to LMP are described. Finally, cell death signaling downstream of LMP is summarized.

8.2 CELL DEATH MODALITIES

In multicellular organisms, cell death process has evolved into an array of regulated signaling pathways that can respond to external and internal signals and limit their potentially harmful consequences for the sake of preserving the homeostasis of the organism. Based on the ultrastructure, cell death is apoptotic or nonapoptotic. Apoptotic cells show chromatin and cytoplasm condensation, membrane blebbing, and apoptotic body formation. Nonapoptotic cell death is characterized by oncosis (dilatation of the ER and the nuclear envelope, swelling of the cytosol), which leads to the loss of plasma membrane integrity. Based on biochemical criteria, additional types of cell death were identified. At the highest level of classification, cell death can be programmed and nonprogrammed, also termed accidental. While nonprogrammed cell death is a consequence of a catastrophic event and can only be prevented with the removal of the stressor, programmed cell death is "characterized" with a specific biochemical pathway [3,4]. Apoptosis and programmed necrosis represent two morphologically distinct types of programmed cell death.

Biochemically, apoptosis is characterized by activation of the executioner caspases-3 and -7. Depending on the upstream signaling, apoptosis can be extrinsic or intrinsic, but the principles of the activation of the pathway-specific "initiator caspases" are similar. Extrinsic apoptosis is initiated upon binding of cognate ligands (TNF-α, TRAIL, FasL, or CD95L) to death receptors on the plasma membrane. This leads to receptor conformational changes and the assembly of the multiprotein platform termed death-inducing signaling complex (DISC), which mediates the initiator caspase-8 activation. Activated caspase-8 can proteolytically activate executioner caspases or recruits the intrinsic apoptotic pathway via the cleavage of the BCL-2 protein family member BID. In the intrinsic pathway, the critical event is mitochondrial outer membrane permeabilization (MOMP). The integrity of the mitochondrial outer membrane (MOM) is regulated by BCL-2 proteins [5,6]. MOMP enables translocation of proapoptotic factors into the cytosol. Among them, cytochrome c and cytosol-resident apoptotic peptidase activating factor (APAF1) form the apoptosome, which serves as a platform for the activation of the initiator capsase-9. Activated caspase-9 then proteolytically activates the executioner caspases-3 and -7, which degrade several cellular components thereby causing cell demise [7,8].

Programmed necrosis includes necroptosis and pyroptosis. Similarly to the extrinsic apoptosis, necroptosis is initiated by ligation of the same death receptors;

however, the multiprotein complex that is assembled is different and thereby it mediates a different biological response. In addition to death receptors, necroptosis can also be initiated by pattern-recognition receptors, which include toll-like receptors (TLRs) either on the plasma membrane or in the endosomal limiting membrane. The executioner mechanisms of necroptosis are complex and include generation of ROS and activation of hydrolytic enzymes, which collectively affect mitochondrial, lysosomal, and plasma membrane integrity [9]. Pyroptosis is a proinflammatory cell death pathway and characteristic for cells that can express caspase-1, most notably macrophages. Pyroptosis stimuli, which include viral DNA and RNA, intracellular bacteria, and crystalline material accumulated in endosomes, activate NOD-like receptors (NLRs). This leads to the assembly of the inflammasome, which serves as a platform for caspase-1 activation. Caspase-1 activation depends on potassium or chloride ion efflux and calcium mobilization; however, they alone are not sufficient for caspase-1 activation. Cell demise in pyroptosis seems to be largely mediated by the proteolytic activity of caspase-1, which either directly degrades structural and functional elements or proteolytically activates other hydrolases [10].

8.3 LYSOSOMAL MEMBRANE PERMEABILIZATION (LMP) AND CELL DEATH

In the context of cell death, LMP commonly refers to the permeabilization of the limiting membrane of lysosomes, late endosomes, and hybrid organelles between these two, as they all harbor hydrolases, which can promote cell death [11]. LMP can contribute to different cell death pathways. Actually, the same exogenous compound can initiate different cell death modalities depending on the intensity of the stimulus and cell death signaling machinery of the affected cell. In general, it was postulated that limited LMP initiates signaling that leads to or amplifies apoptosis, whereas extensive LMP results in general proteolysis/hydrolysis and causes nonprogrammed (accidental) necrosis [12–16]. The intensity of the stimulus determines the extent of the LMP and its reversibility. Minor destabilization would enable the loss of ion gradients, including the proton gradient, which would cause the loss of acidic pH in the lumen of the endosomes and lysosomes [17]. This would in turn affect the activity of luminal hydrolases whose stability and/or activity depends on acidic environment. If minor destabilization is transient, then the degradative potential of endosomes could be restored and the cell saved from more severe consequences, which can take place if luminal hydrolases are released into the cytosol. However, this requires larger membrane lesions to allow the translocation of lysosomal hydrolases (25–80 kDa) into the cytosol. It was recently shown that upon such considerable destabilization of the lysosomal membrane cellular and tissue homeostasis depends on the sequestration of damaged lysosomes by autophagy [18].

Signaling events both upstream and downstream of LMP are poorly understood. As described below proteolytic activation of proapoptotic messengers and generalized proteolysis seem the most important mechanism through which LMP promotes cell death. An indirect effect of LMP and consequent loss of lysosomal degradative capacity is also the interruption of the autophagic flux. By sequestering bulk cytoplasm or specifically damaged organelles, autophagy removes damaged organelles and provides metabolites that can fuel glycolysis. This aspect of autophagy may influence the ability of cells to overcome the stress and avoid cell death [15,19].

8.3.1 Mechanisms of LMP

8.3.1.1 Characteristics of the Lysosomal Membranes The inner leaflet of the lysosomal membrane is lined by a thick glycocalyx, which is composed of the carbohydrate parts of lysosomal integral membrane proteins (LIMPs) and lysosome-associated membrane proteins (LAMPs) and protects the limiting membrane from the degradative enzymes [20]. Compared to the plasma membrane, the lysosomal limiting membrane contains less cholesterol and sphingolipids, but the content of cholesterol is still higher than that in the intracellular membranes of other organelles [21]. However, the lipid composition of the limiting membrane differs considerably from the intralysosomal membrane of intraluminal vesicles, which start forming by ESCRT-dependent sorting already early in the endocytic pathway [22] and represent the platform for membrane degradation [23]. The intralysosomal membranes are characterized by the increased content of negatively charged lipid bis(monoacylglycero)phosphate (BMP) and to a lesser extent phosphatidylinositol and dolichol phosphate [23]. These negative charges facilitate the adhesion of the soluble positively charged hydrolases and activator proteins [24]. Cholesterol is almost absent from intralysosomal membranes and sphingomyelin is progressively degraded to ceramide by the acid sphingomyelinase (SMase) [23]. Changes in its lipid composition can sensitize the lysosomal membrane to LMP [25]. Sphingomyelin and cholesterol form liquid ordered phase, which represents the so-called membrane rafts and give the membrane stability [21]. SMase, which degrades sphingosine to ceramide, sensitizes lysosomes to damage, including oxidative injury, by reducing sphingomyelin content [26]. Accumulation of ceramide perturbs the mechanical integrity of membranes, which can lead to increased membrane permeability or even membrane rupture [27]. On the other hand, ceramidase can degrade ceramide to sphingosine, which can also affect membrane permeability by rigidifying lipid acyl chains [28] and was shown to be associated with increased risk for LMP [29]. In addition, sphingosine would act as a cationic amphiphile and induce lipidosis by neutralizing the negative charge of BMP and displacing lysosomal proteins from membranes [23]. In contrast to pH 4.5, at pH 5.5, ceramide catalyzes the reverse reaction, which maintains low level of sphingosine in endosomes [30]. Another factor suggested to participate in sphingomyelin metabolism is Hsp70, which was found to exhibit a protective effect on the lysosomal membrane that was associated with its role as an essential cofactor for lysosomal sphingomyelin metabolism [31,32]. Depletion of cholesterol by adding methyl cyclodextrin to cell culture sensitized lysosomes to osmotic stress or membranolytic action exerted by Gly-Phe-β-naphthylamide [33]. This is possibly due to decreased mechanical rigidity of the membrane as well as increased permeability to water, potassium ions, and protons [21,34,35]. Moreover, increased content of cholesterol was identified as a protective factor against LMP [36].

8.3.1.2 Selective Targeting of the Lysosomal Membrane Selective permeabilization of the lysosomal membrane entails lysosomal localization of the causative agents or the susceptible factor. The former applies mostly to compounds applied externally (exogenous factors) and the latter to internal signaling. Selective accumulation of exogenous compounds can be achieved by several routes. Membrane-nonpermeable compounds can be taken up by endocytosis and transferred along the endocytic pathway to accumulate in its late compartments. Membrane-permeable compounds can diffuse throughout the cell, but in order to be retained and to accumulate inside the endocytic organelles, they need to be modified in the lumen in such a way to become impermeable. Lysosomotropic amines are thus made impermeable by protonation in the acidic lysosomal lumen [37], whereas esterified compounds are retained in the lysosomes by de-esterification carried out by luminal hydrolases [38,39]. However, additional enzymatic modifications could be involved in other types of compounds.

Mediators that can transmit cellular stress signals into LMP are mostly localized in the limiting membrane and their modification destabilizes membrane in such a way that the membrane permeability increases. Lysosomal membranes are particularly prone to ROS-mediated damage, because of the lysosomal iron content, which can produce intralysosomal ROS in the Fenton reaction [40]. The initial burst of ROS may be generated inside lysosomes [41] or outside, for example, by destabilized mitochondria [42]. However, hardly any of these accumulation mechanisms is entirely specific for lysosomes, and, therefore, one has to be aware that LMP might not be the only consequence of the treatment with a particular LMP-inducing compound [43].

The two main mechanisms of LMP include direct membranolytic activity and osmotic lysis (Figure 8.1).

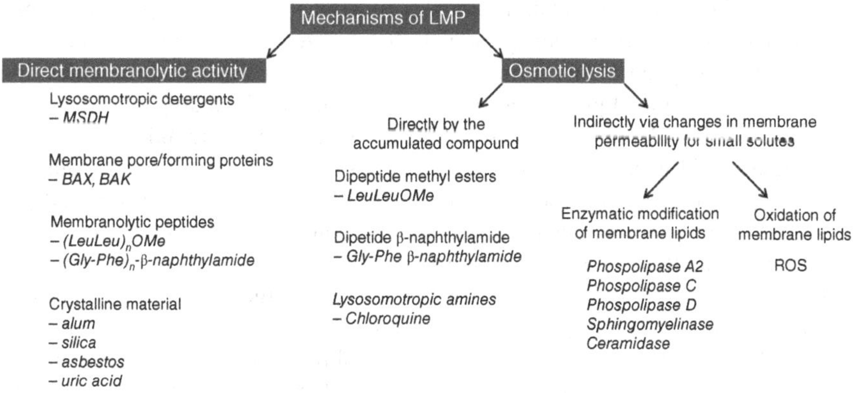

Figure 8.1 Mechanisms of LMP.

8.3.1.3 Direct Membrane Lysis Direct membranolytic activity is usually associated with detergents. These amphiphilic molecules partition between the water phase and the phospholipid bilayer. With their increasing concentration, a point is reached at which the membranes are solubilized and mixed micelles consisting of membrane lipids and detergent molecules are formed [44]. Lysosomotropic amines with linear-chain hydrocarbon "tails" induce LMP due to detergent activity. They include *O*-methyl-serine dodecylamide hydrochloride (MSDH) [13] and *N*-dodecylimidazole [45–47]. Lipofuscin, which is generated in lysosomes by oxidative polymerization of protein and lipid residues, may be an endogenous compound with membranolytic activity on lysosomal membranes [48]. In particular, long-lived postmitotic cells are prone to accumulation of this nondegradable compound, as they cannot dilute it by cell division [49].

Apart from lipid detergents, also peptides can exert membranolytic activity, as demonstrated by antimicrobial peptides, in particular, amphipathic α-helical peptides. Initially, these accumulate at the bilayer surface like a carpet. In the next step, peptide accumulation leads to thinning of the bilayer, which in turn leads to conditions allowing a localized collapse of the lipid bilayer by the formation of discrete pores or simply a detergent-like disintegration of the bilayer structure [50]. Upon intralysosomal oligomerization also certain dipeptide methyl esters and β-naphthylamides exert membranolytic properties [51].

8.3.1.4 Osmotic Lysis If lysosomal membrane permeability to otherwise nonpermeable solutes increases, the net influx of solutes will increase the osmolarity in the lumen and cause an influx of water, which may lead to osmotic lysis. Such increased permeability could be mediated by increased transient water defects of the membrane. These act as mobile free volumes that can carry small molecules and ions across the membrane. Their frequency can be increased by the modification of membrane lipids [52] or the incorporation of external perturbants [53,54]. Another example is membrane lipids that can be modified by ROS-mediated lipid peroxidation or enzymatically [52]. Lipid peroxidation oxidizes functional groups in lipid tails, which causes their conformational change so that the oxidized tails bend toward the water phase and the oxygen atoms form hydrogen bonds with water and the polar lipid headgroup. Membrane lipids are also modified by lipases, including cytosolic phospholipase 2 (cPLA2), phospholipase C (PLC), SMase, and ceramidase [28,55,56]. Arachidonic acid generated by cPLA2 and phosphatide acid generated mainly by phospholipase D both increase lysosomal permeability to potassium ions and protons [56,57]. In addition, detergents at sublytic concentrations can act as external perturbants that affect membrane permeability [58].

Increased permeability for small solutes and even larger molecules can also be mediated by pore-forming proteins. These proteins are long enough to transverse the length of the hydrophobic core of the membrane. Notable examples include complement membrane attack complex [59], perforin [60], and BAX and BAK molecules [61].

Rather than increasing osmotic pressure across the lysosomal membrane by changing permeability to small solutes, compounds themselves can directly build up osmotic pressure. This is true for compounds that accumulate inside lysosomes after being modified to become impermeable. Lysosomotropic amines, such as chloroquine or ammonium chloride, mediate the so-called "proton sponge effect," which is underlined with the rise in pH and the increase in the lysosomal volume. However, the osmotic lysis is not commonly observed [47,62]. This is in contrast to dipeptide methyl esters, such as LeuLeuOMe, or dipeptide-β-naphthylamides, such as Gly-Phe-β-naphthylamide, which were among the first compounds that were shown to destabilize lysosomal membranes. However, they are also oligomerized by lysosomal enzymes, in particular, cathepsin C, and comparison with lysosomotropic amines indicates that efficient permeabilization of lysosomal membranes by dipeptide methyl esters or β-naphthylamides is mediated largely by detergent-like properties of oligopeptides [38,39,51].

8.3.2 Upstream of LMP: Direct Insult Versus Molecular Signaling

Historically, the role of LMP in cell death signaling has been demonstrated with exogenous compounds, which directly target the lysosomal membrane or osmotic stability (Figure 8.1). Among best-studied compounds that induce apoptosis via LMP are dipeptide methyl esters [63–65]. LeuLeuOMe has even been tested in clinical trials for *ex vivo* purging of perforin-positive T cells in order to prevent a graft-versus-host disease after allogeneic bone marrow transplantation [66,67]. Another group of agents that can induce LMP-triggered apoptosis are lysosomotropic detergents [13,46,47]. However, when added to cells, these compounds can destabilize the plasma membrane and impair its barrier function causing necrotic cell death even below the critical micelle concentration [13,45,68,69]. Natural compounds include quinolone antibiotics ciprofloxacin and norfloxacin. Norfloxacin triggers ROS-mediated LMP after UV light activation [70]. Similarly, aminoglycoside antibiotic gentamicin [41] and covalent conjugates between the photosensitizer chlorin(e6) and hyperbranched poly(ether-ester) (HPEE) [71] also trigger LMP by local generation of ROS. These compounds present candidates for therapeutic agents that would induce cell death specifically in targeted cells, for example, cancer cells [72], as also discussed in Chapter 10.

Crystalline materials, such as aluminum hydroxide (alum), silicon dioxide (silica), uric acid, and asbestos, permeabilize phagolysosomal membranes and induce pyroptosis. However, the uptake of these small crystals is limited to cells that can phagocytose these particulate compounds, most notably macrophages [73,74]. The exact mechanism of phagosomal membrane destabilization is not known, but it was shown that ROS are generated by a NADPH oxidase upon phagocytosis of asbestos or silica particles [74]. The current understanding of the immunogenic effect of alum adjuvants is that it mediates cell death, which transiently induces the release of endogenous danger signals and proinflammatory cytokines. These in turn activate antigen-presenting cells and thereby increase the immune response to the antigen [75]. To date, there is no report that LMP could initiate necroptosis.

As mentioned, LMP can occur also as a consequence of molecular signaling that was initiated outside the endocytic pathway (Figure 8.2). The paramount question

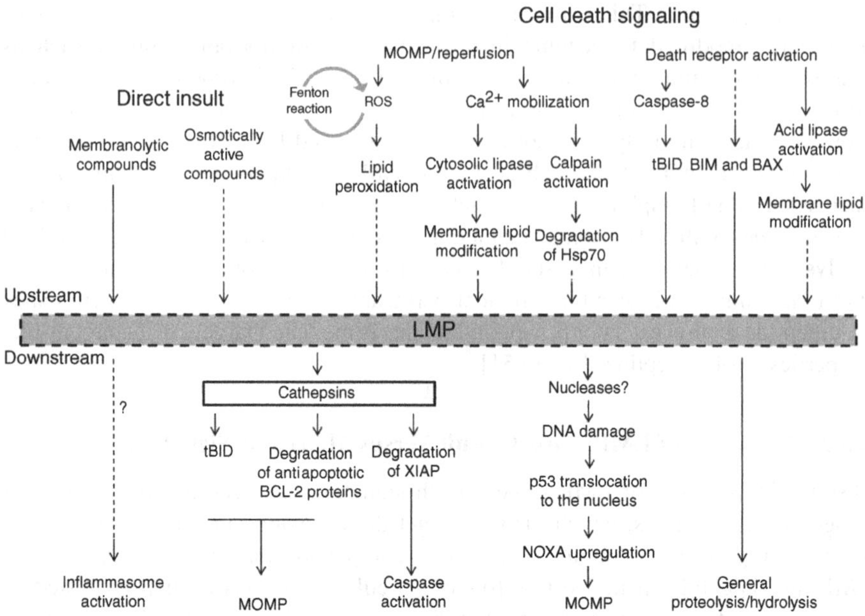

Figure 8.2 Molecular events upstream and downstream of LMP. Most exogenous compounds cause LMP as a consequence of direct targeting of the lysosomal membrane or osmotic stability. In contrast, LMP can also occur in consequence of cell death signaling. The most likely factors that mediate LMP downstream of MOMP activation are ROS and increased intracellular calcium concentration linked to activation of several lipases, which collectively affect the osmotic stability of lysosomes. Downstream of LMP, the best-studied effector mechanism is the proteolytic activity of lysosomal cathepsins that translocate to the cytosol. In the case of moderate LMP, cell response is dominated by the proteolytic activation of BID and degradation of antiapoptotic BCL-2 proteins and caspase inhibitors. Extensive LMP can cause general proteolysis and thereby necrosis.

is whether LMP occurs early or late in the signaling cascade, so to say upstream or downstream of the point that irreversibly commits the cell to death. Most data indicate that LMP, unless a consequence of a direct insult on lysosomal membranes, amplifies cell death signaling beyond the point of no return represented by mitochondria destabilization, in apoptosis, pyroptosis, or necroptosis, and contributes to the execution phase of cell death [9,76–80].

The critical regulators of mitochondrial integrity are the BCL-2 proteins and it is the interactions between its antiapoptotic and proapoptotic members that regulate the integrity of outer mitochondrial membrane in response to several stress modalities. The effector proteins BAX and BAK can homooligomerize in the outer mitochondrial membrane and thereby form pores through which proapoptotic agents, including cytochrome c, are released into the cytosol. Stress affects the transcriptional and posttranscriptional modulation of BH3-only proapoptotic proteins, which can sequester and neutralize the antiapoptotic proteins. There are two models to explain the activation of the effector molecules BAX and BAK. According to the "direct model," BAX and BAK activation depends on the interaction with the so-called BH3-only activators BID and BIM [5]. In contrast, "indirect model" suggests that the activity of BAX and BAK is independent of the association with BH3-only activators and instead the BH3-only proteins sequester antiapopotic BCL-2 proteins, which inhibit BAX and BAK, which in turn releases BAX and BAK activity [81].

Some studies suggest that BCL-2 proteins may also regulate the integrity of the lysosomal membrane. It was reported that BAX can permeabilize the lysosomal membrane after treatment with STS [82] and TRAIL. The latter was shown to induce a recruitment of multifunctional sorting protein phosphofurin acidic cluster sorting protein-2 (PACS-2) on death receptor-5 (DR5)-positive endosomes, which then in turn recruits BIM and BAX [83]. In addition, caspase-8 generated tBID was reported to be able to permeabilize lysosomal membranes on its own. Phosphatidic acid facilitates the insertion of tBID deeply into lipid bilayers, where it undergoes homooligomerization and triggers the formation of highly curved nonbilayer lipid phases [84].

LMP upstream of MOMP was observed also early after TNF treatment. It was suggested to be mediated by sphingosine–cathepsin B interaction [85], caspase-8-mediated cleavage of caspase-9 [86], or increased sphingosine generation [29]. Other studies suggested that also for TNF-α LMP occurs downstream of MOMP and depends on mitochondrial ROS generation [78,79]. Similarly, late LMP that occurs after caspase activation was shown to take place in extrinsic apoptotic pathways initiated by ligation of TRAIL or Fas receptors [77,87,88]. In fibroblasts and monocytes exposed to etoposide, ultraviolet light, FasL, or interleukin-3 (IL-3) deprivation, which represent stimuli that initiate intrinsic apoptotic pathway, lysosomes were not directly perforated by Bax/Bak but by effectors downstream of the apoptosome. Neither Bax nor Bak was localized to the lysosomes [80]. Probably the most common mediator of LMP is ROS that have been generated in consequence to mitochondrial destabilization and caspase activation. Critical for the generation of mitochondrial ROS is caspase-3-mediated cleavage of the p75 subunit of complex I of the electron transport chain [89].

In the execution phase of cell death, including apoptosis, necroptosis, and pyroptosis, cytosolic Ca^{2+} concentration increases [9,90,91]. This activates cytosolic lipases, which can modify lipids in lysosomal membranes and thereby change their permeability for ions and small metabolites, leading to osmotic lysis of lysosomes [92]. Moreover, increased ceramide generation may lead to aggregation and assembly of NADPH subunits, whereby ROS are generated [93]. Increased calcium concentration was shown to generate H_2O_2 and thereby oxidative stress [94]. Ca^{2+}-mediated activation of cPLA2 has also been involved in lysosomal membrane destabilization induced by heavy metals [95] or 17-β-estradiol [96].

In neurons exposed to ischemia, LMP is postulated to occur with combined action of calpains and ROS. Immediately after ischemia, cytosolic Ca^{2+} concentration increases and activates μ-calpains. Upon reperfusion, hydroxyl radicals generated within lysosomes through the Fenton reaction generate an oxidated lipid 4-hydroxy-2-nonenal (HNE). This can in turn carbonylate proteins, including

Hsp70.1. Carbonylated Hsp70.1 can be cleaved by μ-calpains. From the point of Hsp70.1 inactivation, several mechanisms were proposed to contribute to cell death [97]. Hypothetically, lysosomal membrane is destabilized by the increase in sphingomyelin, whose degradation depends on Hsp70.1 enhanced activity of acid SMase [98].

In addition, proteasome inhibitors [99] and microtubule stabilizing agents [100] have been shown to mediate cell death via LMP, although detailed molecular signaling has not been established.

For many compounds, the evidence that they can induce LMP is often not compelling, even more challenging is the evaluation of the critical role of the putative LMP for cell death. As mentioned earlier, targeting of LMP-inducing compounds is not entirely specific for the endocytic pathway and potential effects on other cell organelles should be investigated. When LMP is the result of complex molecular signaling, the cause and the consequence relationship is considerably more complex and therefore extremely challenging to decipher [101].

8.3.3 Signaling Downstream of LMP

In general, limited damage to lysosomes triggers apoptosis and an extensive one results in necrosis [12–16]. In both cases, the signaling seems to be mediated by lysosomal hydrolases and/or ROS. However, extensive LMP, which may occur after an intense direct insult of lysosomes or in the execution phase of different cell death modalities, leads to general hydrolysis and contributes to cell demise [13,102,103] (Figure 8.2).

8.3.3.1 Lysosomal Cathepsins Lysosomal cathepsins, among them cysteine cathepsins and aspartic cathepsin D, are the most common effector molecules associated with cell death signaling downstream of LMP [16]. These enzymes are the major lysosomal proteases and among the smallest lysosomal hydrolases (25–30 kDa) that often translocate to the cytosol during LMP. Cysteine cathepsins are members of papain-like cysteine proteases. In humans, 11 cysteine cathepsins have been found, including cathepsins B, C (dipeptidyl peptidase I, DPPI), F, H, K, L, O, S, V, W, and X. They all share the same core structure and are all monomers of 30 kDa with the exception of cathepsin C, which is a homotetramer. Cysteine cathepsins are synthesized as inactive precursors and require acidic pH for proteolytic processing to active forms, which usually takes place in late endosomes and lysosomes. The enzymes are delivered to these organelles by the mannose 6-phosphate receptor-mediated protein sorting. Cathepsins have a broad specificity for substrates, consistent with their role of recycling enzymes, and exhibit considerable redundancy [104]. In contrast to the aspartic cathepsin D, the catalytic mechanism of cysteine cathepsins is not dependent on acidic pH. However, they are structurally unstable at neutral pH, which makes their activity short lived. There are though considerable differences between individual members in their half-life at neutral pH, ranging from a few minutes in the case of CatL to a few hours for CatS [105–108]. Their activity can be prolonged by binding to various ligands, including substrates, glycosaminoglycans (GAGs), and even inhibitors that may under special circumstances serve as cathepsin "reservoirs" [106,107,109,110]. The endogenous inhibitors include cystatin C in the extracellular space and stefins and serpins in the cytosol [16,111]. Interestingly, transcription factor Stat3 upregulates cathepsins B and L and downregulates their cytosolic inhibitor serpin SpiA2, which collectively promotes LMP-mediated apoptosis in mammary epithelium [112]. The translocation of cysteine cathepsins from lysosomes to the cytosol has been demonstrated by enzymatic activity or immunodetection in the cytosolic fractions of cell extracts, or by immunolocalization in fixed cells [113]. Their involvement in cell death signaling has also been demonstrated by using specific synthetic inhibitors, gene knockdown or knockout systems [64,99,114].

8.3.3.2 Lysosomal Apoptotic Pathway Collectively, the proteolytic cell death signaling by cathepsins is referred to as the lysosomal apoptotic pathway (Figure 8.3). Initially in a cell-free system, it was established that cysteine cathepsins trigger apoptosis by a cleavage of the BID molecule rather than by a direct cleavage of executioner caspases-3 and -7 [63,115]. In cell experiments, it was shown that LMP triggered by hydroxychloroquine (HCQ) or LeuLeuOMe does not suffice to trigger caspase activation and that Bax/Bak-dependent MOMP is a critical step of LMP-induced cell death [64,116]. The BID protein belongs to proapoptotic BH3-only BCL-2 proteins. It may act as a direct activator of pore-forming effector molecules BAX and BAK or/and indirectly promote their activity by sequestering antiapoptotic BCL-2 proteins [6]. It contains an unstructured, the so-called bait loop, which is cleaved by several proteases associated with cell death, including initiator caspase-8 [117], granzyme B [118], and calpain [119]. Cysteine cathepsins can cleave mouse BID at several sites, most notably at Arg65 [63]. These exact cleavage sites are not entirely preserved in the human BID; however, similar cleavage fragments are generated [64]. Cathepsin D also cleaves mouse BID in the bait loop, however at Trp48 [120]. A truncated form tBID then translocates to the mitochondrial membrane via specific targeting mediated by cardiolipin [121]. In the next step, tBID recruits soluble BAX and assists homooligomerization of BAX or the membrane-resident BAK into pores spanning the MOM. Upstream of MOMP, cathepsins not only proteolytically activate tBID but also proteolytically inactivate or degrade antiapoptotic BCL-2 homologues BCL-2, BCL-x_L, and MIC-1, and thereby promote the activity of BAX and BAK proteins, which target the integrity of mitochondria [64]. In this way, LMP is transmitted to MOMP, which is critical for signaling in the intrinsic apoptotic pathway (Figure 8.3). In addition to cytochrome c, other proapoptotic factors are translocated through the pores in the MOM, among them SMAC/DIABLO that antagonizes and the serine protease HtrA2/Omi that degrades the inhibitors of apoptosis proteins (IAPs) [7], thereby promoting caspase activity. Similarly, cysteine cathepsins can degrade XIAP [64], thereby simultaneously promoting apoptosis upstream and downstream of MOMP. Interestingly, tBID pathway is independent of p53 and so cancer cells with mutant p53 remain sensitive for the lysosomal apoptotic pathway [122], which offers an opportunity for therapeutic approaches [72].

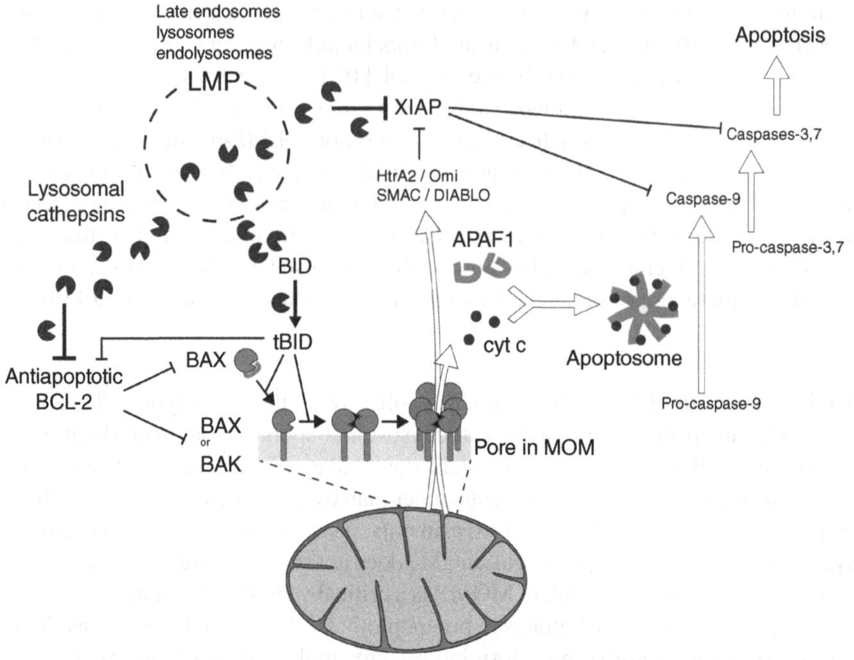

Figure 8.3 Lysosomal apoptotic pathway. Upon LMP, lysosomal cysteine cathepsins and aspartic cathepsin D can be translocated to the cytosol, where they promote apoptotic signaling upstream and downstream of MOMP by proteolytic activation of BID and degradation of antiapoptotic BCL-2 proteins and caspase inhibitor X-linked inhibitor of apoptosis (XIAP) (bold arrows). Truncated BID (tBID) transmits LMP to MOMP via recruitment of BAX and homooligomerization of BAX and BAK in the outer mitochondrial membrane (regular arrows) and/or sequestration of antiapoptotic BCL-2 proteins. Downstream of MOMP (light arrows), cytochrome c, which is translocated from the intermembrane space of mitochondria into the cytosol, mediates the assembly of the apoptosome, which serves as a platform for the activation of the initiator caspase-9. Caspase-9 then activates executioner caspases-3 and -7. Additional proapoptotic factors are translocated through the pores in the MOM, including SMAC/DIABLO that antagonizes and the serine protease HtrA2/Omi that degrades the inhibitors of apoptotic proteins (IAPs), thereby promoting caspase activity.

8.3.3.3 ***Alternative Signaling*** For many compounds that induce cell death via LMP, the signaling downstream of LMP is not understood. In some cases, it was demonstrated that cell death is largely independent of lysosomal cathepsins [123,124]. In others, lysosomal cathepsins even seem to have a protective role [125], which may be linked to the protective role of autophagy. Moreover, even when lysosomal cathepsins seem to be involved, there may be considerable differences in the execution phase in particular with regard to caspase activation. Caspases may be activated, yet cell death cannot be prevented with caspase inhibitors [124,126], which suggests that the cell has experienced severe caspase-independent injuries that cause metabolic imbalance [3]. Alternatively, caspases may not get activated at all, which suggests that cell death in such cases is nonapoptotic [127–129].

There are few mechanistic reports on BID-independent apoptotic signaling pathways downstream of LMP. Oxidative stress induced by H_2O_2 or exogenous iron loading induced LMP that leads to an increase in oxidative DNA damage and p53-dependent Noxa expression, which in turn induces MOMP and eventual activation of apoptosis signaling [130]. The involvement of p53 nuclear translocation has also been shown upon oxysterol (oxidized derivatives of cholesterol)-mediated LMP and cysteine cathepsin translocation to the cytosol [131].

Compared to apoptosis, pyroptosis has been established rather recently and thus much less studied. In pyroptosis that is seemingly initiated by LMP, the mechanism of NLRP inflammasome activation is not understood and the role of cysteine cathepsins seems controversial [101]. LMP and translocation of cysteine cathepsins into the cytosol either appears essential [73,132,133], or LMP was observed to occur downstream of inflammasome activation [103]. So far, ROS generated by phagosomal NADPH have been identified as the only direct link between LMP and inflammasome activation [74].

8.4 CONCLUSION

The map of cell death signaling has gained complexity with regard to the diversity of distinct pathways and the cross talk between them. Lysosomes harbor numerous hydrolases, which can promote cell death if released into the cytosol, similarly to apoptotic factors contained in the intermembrane space of mitochondria. The potential of lysosomes to generate ROS seems to be another important factor involved in both LMP itself and its downstream cell death signaling. LMP, which is a consequence of a direct insult on lysosomes, can initiate lysosomal pathway of apoptosis, necrosis, or pyroptosis. In addition, LMP that occurs downstream of MOMP amplifies cell death signaling likely in any cell death modality, including extrinsic and intrinsic apoptosis, pyroptosis, and necroptosis. In the context of the role of LMP in cell death, our understanding in particular of the effectors that signal cell death downstream of LMP is modest. At present, proteolytic regulation of the BCL-2 proteins by cysteine cathepsins and aspartic cathepsin D seems the predominant mechanism [134]; however, experimental data suggest that it is not the only one. Understanding the role of LMP in cell death and the factors regulating it holds promise for an alternative approach for drug delivery, therapeutic induction of cell death in cancer cells, or its prevention in ischemic tissue.

ACKNOWLEDGMENTS

The work was supported by grants P1–0140, J1–4121, and J1–3602 from the Slovene Research Agency to B.T.

REFERENCES

[1] Huotari J, Helenius A. Endosome maturation. EMBO J 2011;30:3481–3500.

[2] de Duve C. Cellular pathology. J Sci Med Lille 1959;77:437–443.

[3] Kroemer G, Galluzzi L, Vandenabeele P, Abrams J, Alnemri ES, et al. Classification of cell death: recommendations of the Nomenclature Committee on Cell Death 2009. Cell Death Differ 2009;16:3–11.

[4] Leist M, Jaattela M. Four deaths and a funeral: from caspases to alternative mechanisms. Nat Rev Mol Cell Biol 2001;2:589–598.

[5] Chipuk JE, Moldoveanu T, Llambi F, Parsons MJ, Green DR. The BCL-2 family reunion. Mol Cell 2010;37:299–310.

[6] Czabotar PE, Lessene G, Strasser A, Adams JM. Control of apoptosis by the BCL-2 protein family: implications for physiology and therapy. Nat Rev Mol Cell Biol 2014;15:49–63.

[7] Riedl SJ, Salvesen GS. The apoptosome: signalling platform of cell death. Nat Rev Mol Cell Biol 2007;8:405–413.

[8] Taylor RC, Cullen SP, Martin SJ. Apoptosis: controlled demolition at the cellular level. Nat Rev Mol Cell Biol 2008;9:231–241.

[9] Vandenabeele P, Galluzzi L, Vanden Berghe T, Kroemer G. Molecular mechanisms of necroptosis: an ordered cellular explosion. Nat Rev Mol Cell Biol 2010;11:700–714.

[10] Bergsbaken T, Fink SL, Cookson BT. Pyroptosis: host cell death and inflammation. Nat Rev Microbiol 2009;7:99–109.

[11] Repnik U, Hafner Cesen M, Turk B. The endolysosomal system in cell death and survival. Cold Spring Harb Perspect Biol 2013;5:a008755.

[12] Boya P, Kroemer G. Lysosomal membrane permeabilization in cell death. Oncogene 2008;27:6434–6451.

[13] Li W, Yuan X, Nordgren G, Dalen H, Dubowchik GM, et al. Induction of cell death by the lysosomotropic detergent MSDH. FEBS Lett 2000;470:35–39.

[14] Kagedal K, Zhao M, Svensson I, Brunk UT. Sphingosine-induced apoptosis is dependent on lysosomal proteases. Biochem J 2001;359:335–343.

[15] Repnik U, Stoka V, Turk V, Turk B. Lysosomes and lysosomal cathepsins in cell death. Biochim Biophys Acta 2012;1824:22–33.

[16] Turk B, Turk V. Lysosomes as "suicide bags" in cell death: myth or reality? J Biol Chem 2009;284:21783–21787.

[17] Steinberg BE, Huynh KK, Brodovitch A, Jabs S, Stauber T, et al. A cation counterflux supports lysosomal acidification. J Cell Biol 2010;189:1171–1186.

[18] Maejima I, Takahashi A, Omori H, Kimura T, Takabatake Y, et al. Autophagy sequesters damaged lysosomes to control lysosomal biogenesis and kidney injury. EMBO J 2013;32:2336–2347.

[19] Wang Y, Singh R, Massey AC, Kane SS, Kaushik S, et al. Loss of macroautophagy promotes or prevents fibroblast apoptosis depending on the death stimulus. J Biol Chem 2008;283:4766–4777.

[20] Carlsson SR, Roth J, Piller F, Fukuda M. Isolation and characterization of human lysosomal membrane glycoproteins, h-lamp-1 and h-lamp-2. Major sialoglycoproteins carrying polylactosaminoglycan. J Biol Chem 1988;263:18911–18919.

[21] van Meer G, Voelker DR, Feigenson GW. Membrane lipids: where they are and how they behave. Nat Rev Mol Cell Biol 2008;9:112–124.

[22] Raiborg C, Stenmark H. The ESCRT machinery in endosomal sorting of ubiquitylated membrane proteins. Nature 2009;458:445–452.

[23] Schulze H, Kolter T, Sandhoff K. Principles of lysosomal membrane degradation: cellular topology and biochemistry of lysosomal lipid degradation. Biochim Biophys Acta 2009;1793:674–683.

[24] Gallala HD, Sandhoff K. Biological function of the cellular lipid BMP-BMP as a key activator for cholesterol sorting and membrane digestion. Neurochem Res 2011; 36:1594–1600.

[25] Johansson AC, Appelqvist H, Nilsson C, Kagedal K, Roberg K, et al. Regulation of apoptosis-associated lysosomal membrane permeabilization. Apoptosis 2010;15: 527–540.

[26] Zager RA, Burkhart KM, Johnson A. Sphingomyelinase and membrane sphingomyelin content: determinants of Proximal tubule cell susceptibility to injury. J Am Soc Nephrol 2000;11:894–902.

[27] Lopez-Montero I, Monroy F, Velez M, Devaux PF. Ceramide: from lateral segregation to mechanical stress. Biochim Biophys Acta 2010;1798:1348–1356.

[28] Goni FM, Alonso A. Biophysics of sphingolipids I. Membrane properties of sphingosine, ceramides and other simple sphingolipids. Biochim Biophys Acta 2006;1758: 1902–1921.

[29] Ullio C, Casas J, Brunk UT, Sala G, Fabrias G, et al. Sphingosine mediates TNF-alpha-induced lysosomal membrane permeabilization and ensuing programmed cell death in hepatoma cells. J Lipid Res 2012;53:1134–1143.

[30] Okino N, He X, Gatt S, Sandhoff K, Ito M, et al. The reverse activity of human acid ceramidase. J Biol Chem 2003;278:29948–29953.

[31] Kirkegaard T, Roth AG, Petersen NH, Mahalka AK, Olsen OD, et al. Hsp70 stabilizes lysosomes and reverts Niemann–Pick disease-associated lysosomal pathology. Nature 2010;463:549–553.

[32] Gyrd-Hansen M, Nylandsted J, Jaattela M. Heat shock protein 70 promotes cancer cell viability by safeguarding lysosomal integrity. Cell Cycle 2004;3:1484–1485.

[33] Jadot M, Andrianaivo F, Dubois F, Wattiaux R. Effects of methylcyclodextrin on lysosomes. Eur J Biochem 2001;268:1392–1399.

[34] Deng D, Jiang N, Hao SJ, Sun H, Zhang GJ. Loss of membrane cholesterol influences lysosomal permeability to potassium ions and protons. Biochim Biophys Acta 2009;1788:470–476.

[35] Hao SJ, Hou JF, Jiang N, Zhang GJ. Loss of membrane cholesterol affects lysosomal osmotic stability. Gen Physiol Biophys 2008;27:278–283.

[36] Appelqvist H, Nilsson C, Garner B, Brown AJ, Kagedal K, et al. Attenuation of the lysosomal death pathway by lysosomal cholesterol accumulation. Am J Pathol 2011;178:629–639.

[37] de Duve C, de Barsy T, Poole B, Trouet A, Tulkens P, et al. Commentary. Lysosomotropic agents. Biochem Pharmacol 1974;23:2495–2531.

[38] Goldman R, Kaplan A. Rupture of rat liver lysosomes mediated by L-amino acid esters. Biochim Biophys Acta 1973;318:205–216.

[39] Reeves JP. Accumulation of amino acids by lysosomes incubated with amino acid methyl esters. J Biol Chem 1979;254:8914–8921.

[40] Kurz T, Terman A, Gustafsson B, Brunk UT. Lysosomes in iron metabolism, ageing and apoptosis. Histochem Cell Biol 2008;129:389–406.

[41] Denamur S, Tyteca D, Marchand-Brynaert J, Van Bambeke F, Tulkens PM, et al. Role of oxidative stress in lysosomal membrane permeabilization and apoptosis induced by gentamicin, an aminoglycoside antibiotic. Free Radic Biol Med 2011;51:1656–1665.

[42] Turrens JF. Mitochondrial formation of reactive oxygen species. J Physiol 2003;552:335–344.

[43] Hafner Cesen M, Repnik U, Turk V, Turk B. Siramesine triggers cell death through destabilisation of mitochondria, but not lysosomes. Cell Death Dis 2013;4:e818.

[44] Helenius A, Simons K. Solubilization of membranes by detergents. Biochim Biophys Acta 1975;415:29–79.

[45] Boyer MJ, Horn I, Firestone RA, Steele-Norwood D, Tannock IF. pH dependent cytotoxicity of N-dodecylimidazole: a compound that acquires detergent properties under acidic conditions. Br J Cancer 1993;67:81–87.

[46] Miller DK, Griffiths E, Lenard J, Firestone RA. Cell killing by lysosomotropic detergents. J Cell Biol 1983;97:1841–1851.

[47] Wilson PD, Firestone RA, Lenard J. The role of lysosomal enzymes in killing of mammalian cells by the lysosomotropic detergent N-dodecylimidazole. J Cell Biol 1987;104:1223–1229.

[48] Schutt F, Bergmann M, Kopitz J, Holz FG. Detergent-like effects of the lipofuscin retinoid component A2-E in retinal pigment epithelial cells. Ophthalmologe 2002; 99:861–865.

[49] Terman A, Brunk UT. Lipofuscin. Int J Biochem Cell Biol 2004;36:1400–1404.

[50] Sato H, Feix JB. Peptide-membrane interactions and mechanisms of membrane destruction by amphipathic alpha-helical antimicrobial peptides. Biochim Biophys Acta 2006;1758:1245–1256.

[51] Thiele DL, Lipsky PE. Mechanism of L-leucyl-L-leucine methyl ester-mediated killing of cytotoxic lymphocytes: dependence on a lysosomal thiol protease, dipeptidyl peptidase I, that is enriched in these cells. Proc Natl Acad Sci U S A 1990;87:83–87.

[52] Wong-Ekkabut J, Xu Z, Triampo W, Tang IM, Tieleman DP, et al. Effect of lipid peroxidation on the properties of lipid bilayers: a molecular dynamics study. Biophys J 2007;93:4225–4236.

[53] Deamer DW, Bramhall J. Permeability of lipid bilayers to water and ionic solutes. Chem Phys Lipids 1986;40:167–188.

[54] Hu JS, Li YB, Wang JW, Sun L, Zhang GJ. Mechanism of lysophosphatidylcholine-induced lysosome destabilization. J Membr Biol 2007;215:27–35.

[55] Wang X, Zhao HF, Zhang GJ. Mechanism of cytosol phospholipase C and sphingomyelinase-induced lysosome destabilization. Biochimie 2006;88:913–922.

[56] Zhang G, Yi YP, Zhang GJ. Effects of arachidonic acid on the lysosomal ion permeability and osmotic stability. J Bioenerg Biomembr 2006;38:75–82.

[57] Yi YP, Wang X, Zhang G, Fu TS, Zhang GJ. Phosphatidic acid osmotically destabilizes lysosomes through increased permeability to K+ and H+. Gen Physiol Biophys 2006;25:149–160.

[58] Ahyayauch H, Bennouna M, Alonso A, Goni FM. Detergent effects on membranes at subsolubilizing concentrations: transmembrane lipid motion, bilayer permeabilization, and vesicle lysis/reassembly are independent phenomena. Langmuir 2010;26:7307–7313.

[59] Wang Y, Bjes ES, Esser AF. Molecular aspects of complement-mediated bacterial killing. Periplasmic conversion of C9 from a protoxin to a toxin. J Biol Chem 2000; 275:4687–4692.

[60] Lopez JA, Susanto O, Jenkins MR, Lukoyanova N, Sutton VR, et al. Perforin forms transient pores on the target cell plasma membrane to facilitate rapid access of granzymes during killer cell attack. Blood 2013;121:2659–2668.

[61] Westphal D, Kluck RM, Dewson G. Building blocks of the apoptotic pore: how Bax and Bak are activated and oligomerize during apoptosis. Cell Death Differ 2014;21:196–205.

[62] Maclean KH, Dorsey FC, Cleveland JL, Kastan MB. Targeting lysosomal degradation induces p53-dependent cell death and prevents cancer in mouse models of lymphomagenesis. J Clin Invest 2008;118:79–88.

[63] Cirman T, Oresic K, Mazovec GD, Turk V, Reed JC, et al. Selective disruption of lysosomes in HeLa cells triggers apoptosis mediated by cleavage of Bid by multiple papain-like lysosomal cathepsins. J Biol Chem 2004;279:3578–3587.

[64] Droga-Mazovec G, Bojic L, Petelin A, Ivanova S, Romih R, et al. Cysteine cathepsins trigger caspase-dependent cell death through cleavage of bid and antiapoptotic Bcl-2 homologues. J Biol Chem 2008;283:19140–19150.

[65] Uchimoto T, Nohara H, Kamehara R, Iwamura M, Watanabe N, et al. Mechanism of apoptosis induced by a lysosomotropic agent, L-Leucyl-L-Leucine methyl ester. Apoptosis 1999;4:357–362.

[66] Filicko-O'Hara J, Grosso D, Flomenberg PR, Friedman TM, Brunner J, et al. Antiviral responses following L-leucyl-L-leucine methyl ester (LLME)-treated lymphocyte infusions: graft-versus-infection without graft-versus-host disease. Biol Blood Marrow Transplant 2009;15:1609–1619.

[67] Friedman TM, Filicko-O'Hara J, Mookerjee B, Wagner JL, Grosso DA, et al. T cell repertoire complexity is conserved after LLME treatment of donor lymphocyte infusions. Biol Blood Marrow Transplant 2007;13:1439–1447.

[68] Aranzazu Partearroyo M, Ostolaza H, Goni FM, Barbera-Guillem E. Surfactant-induced cell toxicity and cell lysis. A study using B16 melanoma cells. Biochem Pharmacol 1990;40:1323–1328.

[69] Forster S, Scarlett L, Lloyd JB. The effect of lysosomotropic detergents on the permeability properties of the lysosome membrane. Biochim Biophys Acta 1987; 924:452–457.

[70] Boya P, Andreau K, Poncet D, Zamzami N, Perfettini JL, et al. Lysosomal membrane permeabilization induces cell death in a mitochondrion-dependent fashion. J Exp Med 2003;197:1323–1334.

[71] Li P, Zhou G, Zhu X, Li G, Yan P, et al. Photodynamic therapy with hyperbranched poly(ether-ester) chlorin(e6) nanoparticles on human tongue carcinoma CAL-27 cells. Photodiagnosis Photodyn Ther 2012;9:76–82.

[72] Fehrenbacher N, Jaattela M. Lysosomes as targets for cancer therapy. Cancer Res 2005;65:2993–2995.

[73] Hornung V, Bauernfeind F, Halle A, Samstad EO, Kono H, et al. Silica crystals and aluminum salts activate the NALP3 inflammasome through phagosomal destabilization. Nat Immunol 2008;9:847–856.

[74] Dostert C, Petrilli V, Van Bruggen R, Steele C, Mossman BT, et al. Innate immune activation through Nalp3 inflammasome sensing of asbestos and silica. Science 2008;320:674–677.

[75] Kool M, Soullie T, van Nimwegen M, Willart MA, Muskens F, et al. Alum adjuvant boosts adaptive immunity by inducing uric acid and activating inflammatory dendritic cells. J Exp Med 2008;205:869–882.

[76] Vanden Berghe T, Vanlangenakker N, Parthoens E, Deckers W, Devos M, et al. Necroptosis, necrosis and secondary necrosis converge on similar cellular disintegration features. Cell Death Differ 2010;17:922–930.

[77] Spes A, Sobotic B, Turk V, Turk B. Cysteine cathepsins are not critical for TRAIL- and CD95-induced apoptosis in several human cancer cell lines. Biol Chem 2012;393:1417–1431.

[78] Klaric M, Tao S, Stoka V, Turk B, Turk V. Cysteine cathepsins are not critical for TNF-alpha-induced cell death in T98G and U937 cells. Biochim Biophys Acta 2009;1794:1372–1377.

[79] Huai J, Vogtle FN, Jockel L, Li Y, Kiefer T, et al. TNF-alpha-induced lysosomal membrane permeability is downstream of MOMP and triggered by caspase-mediated NDUFS1 cleavage and ROS formation. J Cell Sci 2013;126:4015–4025.

[80] Oberle C, Huai J, Reinheckel T, Tacke M, Rassner M, et al. Lysosomal membrane permeabilization and cathepsin release is a Bax/Bak-dependent, amplifying event of apoptosis in fibroblasts and monocytes. Cell Death Differ 2010;17:1167–1178.

[81] Willis SN, Fletcher JI, Kaufmann T, van Delft MF, Chen L, et al. Apoptosis initiated when BH3 ligands engage multiple Bcl-2 homologs, not Bax or Bak. Science 2007;315:856–859.

[82] Kagedal K, Johansson AC, Johansson U, Heimlich G, Roberg K, et al. Lysosomal membrane permeabilization during apoptosis--involvement of Bax? Int J Exp Pathol 2005;86:309–321.

[83] Werneburg NW, Bronk SF, Guicciardi ME, Thomas L, Dikeakos JD, et al. Tumor necrosis factor-related apoptosis-inducing ligand (TRAIL) protein-induced lysosomal translocation of proapoptotic effectors is mediated by phosphofurin acidic cluster sorting protein-2 (PACS-2). J Biol Chem 2012;287:24427–24437.

[84] Zhao K, Zhou H, Zhao X, Wolff DW, Tu Y, et al. Phosphatidic acid mediates the targeting of tBid to induce lysosomal membrane permeabilization and apoptosis. J Lipid Res 2012;53:2102–2114.

[85] Werneburg NW, Guicciardi ME, Bronk SF, Gores GJ. Tumor necrosis factor-alpha-associated lysosomal permeabilization is cathepsin B dependent. Am J Physiol Gastrointest Liver Physiol 2002;283:G947–G956.

[86] Gyrd-Hansen M, Farkas T, Fehrenbacher N, Bastholm L, Hoyer-Hansen M, et al. Apoptosome-independent activation of the lysosomal cell death pathway by caspase-9. Mol Cell Biol 2006;26:7880–7891.

[87] Bojic L, Petelin A, Stoka V, Reinheckel T, Peters C, et al. Cysteine cathepsins are not involved in Fas/CD95 signalling in primary skin fibroblasts. FEBS Lett 2007; 581:5185–5190.

[88] Wattiaux R, Wattiaux-de Coninck S, Thirion J, Gasingirwa MC, Jadot M. Lysosomes and Fas-mediated liver cell death. Biochem J 2007;403:89–95.

[89] Ricci JE, Munoz-Pinedo C, Fitzgerald P, Bailly-Maitre B, Perkins GA, et al. Disruption of mitochondrial function during apoptosis is mediated by caspase cleavage of the p75 subunit of complex I of the electron transport chain. Cell 2004;117:773–786.

[90] Boehning D, Patterson RL, Sedaghat L, Glebova NO, Kurosaki T, et al. Cytochrome c binds to inositol (1,4,5) trisphosphate receptors, amplifying calcium-dependent apoptosis. Nat Cell Biol 2003;5:1051–1061.

[91] Mattson MP, Chan SL. Calcium orchestrates apoptosis. Nat Cell Biol 2003;5: 1041–1043.

[92] Orrenius S, Zhivotovsky B, Nicotera P. Regulation of cell death: the calcium-apoptosis link. Nat Rev Mol Cell Biol 2003;4:552–565.

[93] Li PL, Zhang Y. Cross talk between ceramide and redox signaling: implications for endothelial dysfunction and renal disease. Handb Exp Pharmacol 2013;216:171–197.

[94] Tripathi A, Chaube SK. High cytosolic free calcium level signals apoptosis through mitochondria-caspase mediated pathway in rat eggs cultured *in vitro*. Apoptosis 2012;17:439–448.

[95] Marchi B, Burlando B, Moore MN, Viarengo A. Mercury- and copper-induced lysosomal membrane destabilisation depends on [Ca2+]i dependent phospholipase A2 activation. Aquat Toxicol 2004;66:197–204.

[96] Burlando B, Marchi B, Panfoli I, Viarengo A. Essential role of Ca2+ -dependent phospholipase A2 in estradiol-induced lysosome activation. Am J Physiol Cell Physiol 2002;283:C1461–C1468.

[97] Yamashima T. Hsp70.1 and related lysosomal factors for necrotic neuronal death. J Neurochem 2012;120:477–494.

[98] Petersen NH, Kirkegaard T, Olsen OD, Jaattela M. Connecting Hsp70, sphingolipid metabolism and lysosomal stability. Cell Cycle 2010;9:2305–2309.

[99] Berndtsson M, Beaujouin M, Rickardson L, Havelka AM, Larsson R, et al. Induction of the lysosomal apoptosis pathway by inhibitors of the ubiquitin-proteasome system. Int J Cancer 2009;124:1463–1469.

[100] Broker LE, Huisman C, Span SW, Rodriguez JA, Kruyt FA, et al. Cathepsin B mediates caspase-independent cell death induced by microtubule stabilizing agents in non-small cell lung cancer cells. Cancer Res 2004;64:27–30.

[101] Repnik U, Hafner Cesen M, Turk B. Lysosomal membrane permeabilization in cell death: concepts and challenges. Mitochondrion 2014;19(Pt A):49–57.

[102] Messner B, Ploner C, Laufer G, Bernhard D. Cadmium activates a programmed, lysosomal membrane permeabilization-dependent necrosis pathway. Toxicol Lett 2012;212:268–275.

[103] Lima H Jr, Jacobson LS, Goldberg MF, Chandran K, Diaz-Griffero F, et al. Role of lysosome rupture in controlling Nlrp3 signaling and necrotic cell death. Cell Cycle 2013;12:1868–1878.

[104] Turk V, Stoka V, Vasiljeva O, Renko M, Sun T, et al. Cysteine cathepsins: from structure, function and regulation to new frontiers. Biochim Biophys Acta 2012;1824:68–88.

[105] Turk B, Dolenc I, Lenarcic B, Krizaj I, Turk V, et al. Acidic pH as a physiological regulator of human cathepsin L activity. Eur J Biochem 1999;259:926–932.

[106] Turk B, Dolenc I, Turk V, Bieth JG. Kinetics of the pH-induced inactivation of human cathepsin L. Biochemistry 1993;32:375–380.

[107] Turk B, Bieth JG, Bjork I, Dolenc I, Turk D, et al. Regulation of the activity of lysosomal cysteine proteinases by pH-induced inactivation and/or endogenous protein inhibitors, cystatins. Biol Chem Hoppe Seyler 1995;376:225–230.

[108] Kirschke H, Wiederanders B, Brömme D, Rinne A. Cathepsin S from bovine spleen. Purification, distribution, intracellular localization and action on proteins. Biochem J 1989;264:467–473.

[109] Almeida PC, Nantes IL, Chagas JR, Rizzi CC, Faljoni-Alario A, et al. Cathepsin B activity regulation. Heparin-like glycosaminogylcans protect human cathepsin B from alkaline pH-induced inactivation. J Biol Chem 2001;276:944–951.

[110] Almeida PC, Nantes IL, Rizzi CC, Judice WA, Chagas JR, et al. Cysteine proteinase activity regulation. A possible role of heparin and heparin-like glycosaminoglycans. J Biol Chem 1999;274:30433–30438.

[111] Turk B, Turk D, Salvesen GS. Regulating cysteine protease activity: essential role of protease inhibitors as guardians and regulators. Curr Pharm Des 2002;8:1623–1637.

[112] Kreuzaler PA, Staniszewska AD, Li W, Omidvar N, Kedjouar B, et al. Stat3 controls lysosomal-mediated cell death *in vivo*. Nat Cell Biol 2011;13:303–309.

[113] Ivanova S, Repnik U, Bojic L, Petelin A, Turk V, et al. Lysosomes in apoptosis. Methods Enzymol 2008;442:183–199.

[114] Vasiljeva O, Korovin M, Gajda M, Brodoefel H, Bojic L, et al. Reduced tumour cell proliferation and delayed development of high-grade mammary carcinomas in cathepsin B-deficient mice. Oncogene 2008;27:4191–4199.

[115] Stoka V, Turk B, Schendel SL, Kim TH, Cirman T, et al. Lysosomal protease pathways to apoptosis. Cleavage of bid, not pro-caspases, is the most likely route. J Biol Chem 2001;276:3149–3157.

[116] Boya P, Gonzalez-Polo RA, Poncet D, Andreau K, Vieira HL, et al. Mitochondrial membrane permeabilization is a critical step of lysosome-initiated apoptosis induced by hydroxychloroquine. Oncogene 2003;22:3927–3936.

[117] Li H, Zhu H, Xu CJ, Yuan J. Cleavage of BID by caspase 8 mediates the mitochondrial damage in the Fas pathway of apoptosis. Cell 1998;94:491–501.

[118] Pinkoski MJ, Waterhouse NJ, Heibein JA, Wolf BB, Kuwana T, et al. Granzyme B-mediated apoptosis proceeds predominantly through a Bcl-2-inhibitable mitochondrial pathway. J Biol Chem 2001;276:12060–12067.

[119] Gil-Parrado S, Fernandez-Montalvan A, Assfalg-Machleidt I, Popp O, Bestvater F, et al. Ionomycin-activated calpain triggers apoptosis. A probable role for Bcl-2 family members. J Biol Chem 2002;277:27217–27226.

[120] Appelqvist H, Johansson AC, Linderoth E, Johansson U, Antonsson B, et al. Lysosome-mediated apoptosis is associated with cathepsin D-specific processing of bid at Phe24, Trp48, and Phe183. Ann Clin Lab Sci 2012;42:231–242.

[121] Lutter M, Fang M, Luo X, Nishijima M, Xie X, et al. Cardiolipin provides specificity for targeting of tBid to mitochondria. Nat Cell Biol 2000;2:754–761.

[122] Erdal H, Berndtsson M, Castro J, Brunk U, Shoshan MC, et al. Induction of lysosomal membrane permeabilization by compounds that activate p53-independent apoptosis. Proc Natl Acad Sci U S A 2005;102:192–197.

[123] Melo FR, Lundequist A, Calounova G, Wernersson S, Pejler G. Lysosomal membrane permeabilization induces cell death in human mast cells. Scand J Immunol 2011;74:354–362.

[124] Rammer P, Groth-Pedersen L, Kirkegaard T, Daugaard M, Rytter A, et al. BAMLET activates a lysosomal cell death program in cancer cells. Mol Cancer Ther 2010;9:24–32.

[125] Marques C, Oliveira CS, Alves S, Chaves SR, Coutinho OP, et al. Acetate-induced apoptosis in colorectal carcinoma cells involves lysosomal membrane permeabilization and cathepsin D release. Cell Death Dis 2013;4:e507.

[126] Mediavilla-Varela M, Pacheco FJ, Almaguel F, Perez J, Sahakian E, et al. Docetaxel-induced prostate cancer cell death involves concomitant activation of caspase and lysosomal pathways and is attenuated by LEDGF/p75. Mol Cancer 2009;8:68.

[127] Hornick JR, Vangveravong S, Spitzer D, Abate C, Berardi F, et al. Lysosomal membrane permeabilization is an early event in Sigma-2 receptor ligand mediated cell death in pancreatic cancer. J Exp Clin Cancer Res 2012;31:41.

[128] Mena S, Rodriguez ML, Ponsoda X, Estrela JM, Jaattela M, et al. Pterostilbene-induced tumor cytotoxicity: a lysosomal membrane permeabilization-dependent mechanism. PLoS One 2012;7:e44524.

[129] Villalpando Rodriguez GE, Torriglia A. Calpain 1 induce lysosomal permeabilization by cleavage of lysosomal associated membrane protein 2. Biochim Biophys Acta 2013;1833:2244–2253.

[130] Eno CO, Zhao G, Venkatanarayan A, Wang B, Flores ER, et al. Noxa couples lysosomal membrane permeabilization and apoptosis during oxidative stress. Free Radic Biol Med 2013;65:26–37.

[131] Li W, Laskar A, Sultana N, Osman E, Ghosh M, et al. Cell death induced by 7-oxysterols via lysosomal and mitochondrial pathways is p53-dependent. Free Radic Biol Med 2012;53:2054–2061.

[132] Duewell P, Kono H, Rayner KJ, Sirois CM, Vladimer G, et al. NLRP3 inflammasomes are required for atherogenesis and activated by cholesterol crystals. Nature 2010; 464:1357–1361.

[133] Tseng WA, Thein T, Kinnunen K, Lashkari K, Gregory MS, et al. NLRP3 inflammasome activation in retinal pigment epithelial cells by lysosomal destabilization: implications for age-related macular degeneration. Invest Ophthalmol Vis Sci 2013;54:110–120.

[134] Turk B, Turk D, Turk V. Protease signalling: the cutting edge. EMBO J 2012;31: 1630–1643.

9

THE LYSOSOME IN AGING-RELATED NEURODEGENERATIVE DISEASES

RALPH A. NIXON
Departments of Psychiatry and Cell Biology, New York University Langone Medical Center, New York, NY, USA; Center for Dementia Research, Nathan S. Kline Institute, Orangeburg, NY, USA

9.1 INTRODUCTION

The efficiency with which lysosomes receive, hydrolyze, and recycle substrates is a vital aspect of cell homeostasis that becomes increasingly important during aging. Substrate delivery to lysosomes occurs through several routes, including endocytosis, macroautophagy, chaperone-mediated autophagy (CMA), and macroautophagy, as summarized in Figure 9.1 and extensively reviewed elsewhere. As the only step common to these pathways, lysosomes are vital determinants of the cell's success in maintaining quality control over its components and may be the compartment most vulnerable to failure in neurodegenerative disease states [1]. Because postmitotic cells such as neurons and cardiac myocytes have fewer options than dividing cells for removing unwanted materials, they are particularly reliant on the lysosomal system. Neurons are further challenged by their extreme geometric asymmetry that places many compartments that sequester substrates at long distances from lysosomes, the degradative compartments that are located mainly in cell bodies. Beyond these intrinsic limitations, cellular aging superimposes additional challenges to lysosome efficiency. Indeed, declining efficiency of lysosomal-related degradation is believed to be a cardinal mechanism underlying aging at the cell and organismal levels and one

Lysosomes: Biology, Diseases, and Therapeutics, First Edition.
Edited by Frederick R. Maxfield, James M. Willard and Shuyan Lu.

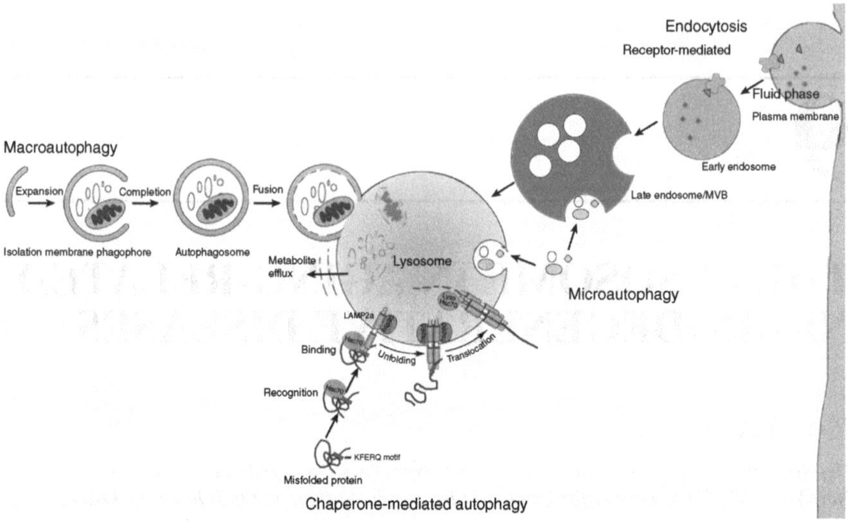

Figure 9.1 Major routes of substrate delivery to lysosomes. (a) Macroautophagy is characterized by the sequestration of structures targeted for degradation into double-membrane vesicles called autophagosomes. Fully formed autophagosomes may first fuse with late endosomes to form an amphisome before fusion with hydrolase-filled lysosomes, which causes degradation of the inner limiting membrane releasing the hydrolases into the lumen of the created autolysosome. Introduction of a fully activated proton pump (V-ATPase) induces full acidification of the autolysosomal lumen necessary to activate acid hydrolases for optimal digestion of substrates. The resulting metabolites are transported into the cytoplasm and used for synthesis of new macromolecules or as a source of energy. (b) During chaperone-mediated autophagy, proteins carrying the pentapeptide KFERQ-like sequence are recognized by the Hsc70 chaperone, which then associates with the integral lysosome membrane protein LAMP-2A, triggering its oligomerization. This event leads to the translocation of the bound protein into the lysosome interior through a process that requires Hsc70. (c) Microautophagy involves "bulk" or chaperone-mediated internalization and degradation of cytoplasmic substrates into late endosome/MVB or lysosomal compartments by a process of membrane invagination followed by membrane scission to release the cargo into the lysosomal lumen for degradation. (d) Heterophagy involves the lysosomal degradation of plasma membrane components and exogenous substrates after they are internalized by bulk or receptor-mediated endocytosis. After selected proteins are sorted to different cellular destinations or recycled to the plasma membrane, proteins targeted for degradation are trafficked to late endosomes/MVB, which fuse with a lysosome or with autophagosomes to effect degradation. (*See color plate section for the color representation of this figure.*)

that promotes the emergence of late-onset neurodegenerative diseases characterized by proteinopathy.

In this chapter, lysosomal system aging is discussed as a key factor in determining the onset and progression of neurodegenerative diseases in late life. After briefly considering the dysfunction of lysosomes in nonneural tissues in the context of aging, we review information on four late age-onset neurodegenerative diseases: Alzheimer's disease (AD), Parkinson's disease (PD) and related disorders, Lewy body disease, and frontotemporal dementia (FD). In each of these disorders, lysosomal dysfunction contributes critically to the functional decline and loss of neurons and to development of the signature proteinopathies characterized in each of these disorders.

9.2 LYSOSOME FUNCTION IN AGING ORGANISMS

In familial forms of many common late age-onset degenerative diseases, pathogenic misfolded proteins are produced continuously from an early age and often many decades before the protein accumulates abnormally in the neuron or induces evident cellular damage. This observation suggests that additional essential disease-promoting factors emerge during the aging process. A likely vital contribution of cellular aging to degenerative disease in late adulthood is diminished proteostasis [2] and particularly a decline in the efficiency of the lysosomal system, which is a well-established pathological hallmark of cellular aging [3–5]. At the organismal and cellular levels, aging also involves a progressive loss of physiological integrity through multiple pathways [4], including shortening of telomeres [6], accumulation of extrachromosomal DNA [7] and oxygen-free radicals [8], and an altered balance of cell cycle regulating factors [9] as well as alterations of insulin/mTOR signaling, protein secretion, and epigenetic gene regulation [10]. In many of these aspects of cell aging, progressive lysosomal degradative deficits can be implicated as either a primary driving force or as a direct consequence of another aging-related impairment. The close relationship between lysosomes and cellular aging is underscored by evidence that the dozen or so experimental manipulations known to extend lifespan in a wide range of organisms share the property of increasing the activity or efficiency of autophagy and lysosomal function [2,11]. Conversely, autophagy impairment can induce a range of senescence-related changes in certain cells [12–14].

The activity of mTOR, a molecular rheostat that balances autophagy and protein synthetic activities, is particularly critical in mediating lifespan extension and is influenced by a broad array of signaling pathways, environmental and disease factors [15,16]. Although mTOR and autophagy induction, and not lysosomes *per se*, are usually the focus in longevity studies, mTOR activity and lysosome biogenesis are cross-regulated by transcription factors, such as TFEB and TFE3, which modulate the expression of genes encoding most lysosomal components and additional components of the autophagosome machinery [2,17]. The upregulation of autophagy flux via mTOR inhibition in lifespan extension models, therefore, includes major changes in lysosomal function. Indeed, the lysosome and specifically the activity of the lysosomal proton pump, V-ATPase, has recently been singled out in yeast as a critical determinant of longevity [18–20] related to its control of lysosomal pH and thus the hydrolytic capacity of lysosomes for clearing damaged mitochondria and modified substrates generated during cellular aging. The mechanism involves the amino acid sensing function of the lysosomal vacuole, which, in turn, regulates mTOR activity [20]. During Ras-induced senescence, lysosomes spatially link mTOR and autophagy in a compartment known as the TOR-autophagy spatial coupling compartment [21].

Although a need for greater lysosome (and autophagy) capacity is expected in cells as they age, evidence suggests that lysosome system efficiency and capacity for clearing compromised organelles decline with age, which is considered fundamental to cellular aging. Mitochondrial turnover is entirely dependent on autophagy (mitophagy) and adequate lysosomal hydrolytic capacity. Mitochondria in aging cells become enlarged and less efficiently sequestered by mitophagy [22]. These enlarged mitochondria have reduced inner membrane potential and produce ATP less efficiently, making them more vulnerable to acute damage, which can then trigger mitochondrial membrane permeabilization and apoptosis or necrosis. In younger cells, an efficient autophagic turnover of dysfunctional or damaged mitochondria counteracts these structural changes. Conversely, in certain disease states, such as PD discussed later, the expression of specific genes that regulate mitophagy is decreased and compounds aging-related inefficiencies. Extensive nuclear changes that accompany cell senescence, the state of irreversible cell cycle arrest, include the remodeling of chromatin and progressive loss of total histones, which are, in part, lysosome-dependent processes. The autophagic-lysosomal degradation of chromatin released from nuclei is believed to drive senescence by making cell cycle reentry less likely [10]. It could be speculated that impaired lysosomal function, as seen in AD for example, promotes the abortive attempts of neurons to reenter the cell cycle, which is associated with neurodegeneration in this disease [23].

Beyond effects on specific organelles, cellular aging involves an increasing and cumulative oxidative damage to cytosolic and membrane-associated proteins and a rise in translational errors leading to larger synthesis of defective proteins [24–27]. Many of the covalent modifications of proteins common in aging such as oxidation, glycation, phosphorylation, deamidation, carbonyl modification, and misfolding [28] reduce the proteolytic susceptibility of the proteins to proteolysis [29]. Free oxygen radicals, which accumulate with age, cross-link proteins through isopeptide bonds and confer additional resistance to proteolysis. Moreover, disease-related gene mutations promote misfolding and aggregation of specific pathogenic proteins or their metabolites and thereby alter proteolytic susceptibility [30]. The cell's primary line of defense against these mounting aging- and disease-related substrate modifications are mainly two proteolytic clearance systems, the ubiquitin proteasome system (UPS) and the autophagic pathway [26,31–33]. These two systems act somewhat interdependently as suggested by the shared use of overlapping families of adaptor proteins such as ubiquitin and p62, which earmark substrates for elimination [34] and the tendency of lysosomal degradation to become upregulated when proteasome activity is inhibited [35,36].

The increased need for lysosomal degradation to clear the growing burden of damaged constituents in aging cells is, unfortunately, not matched by an increased degradative capacity because overall proteolysis actually declines with age in a wide range of organisms and tissues. This decline is greatest for long-lived proteins, which are most often lysosomal substrates, while declines in activity of the ubiquitin proteasome pathway may be less marked [37,38], except in certain pathological states [39]. A third proteolytic capability of cells, the calpain–calpastatin system [40,41], may selectively initiate turnover of certain proteins, but its capacity for only carrying out limited proteolysis results in the generation of protein fragments that must also be cleared by lysosomes and the UPS. Notably, the activities of calpains increase in most aging tissues [41,42], which may further contribute to an overburdening of the two main clearance systems.

Besides effects on the proteolytic substrates, cell aging may also adversely impact lysosomes directly. One manifestation of this impact, seen especially in postmitotic cells, is an accumulation of the age pigment [43] lipofuscin within lysosomal compartments, which is a hallmark of cell aging [44]. Lipofuscin content of cells is linearly correlated with age in various organisms [45], although environmental factors can independently affect net lipofuscin accumulation in a given organism [46]. Lipofuscin is an autofluorescent polymeric pigment composed of aldehyde cross-linked protein fragments, oxidized lipids, carbohydrates, and trace amount of metals, especially iron [47,48]. It is formed by the iron-catalyzed peroxidation of decomposed lipids principally from vesicular organelles such as autophagocytosed mitochondria and its presence reflects the impaired intralysosomal degradation of these autophagic substrates under conditions of growing oxidative stress within aging cells. Loss of lysosomal enzyme activity, as seen experimentally using inhibitors of protease or acidification, increases the opportunity for lipid peroxidation, and accelerates lipofuscin generation as well as other cardinal manifestations of brain aging [5,49–51]. The influence of lipopigment accumulation on lysosomal function was initially thought to be negligible, but further study has documented lipofuscin-dependent decreases in activities of lysosomal cysteine proteases [48,52] and, interestingly, also the proteasome [36].

Although the accumulation of lipofuscin in aging cells suggests dysfunctional hydrolysis within lysosomes, reliable information on the hydrolytic efficacy of lysosomes in living cells is still fragmentary and a better understanding is hampered by the limitations of available methods to assess lysosomal activity *in vivo*. Hydrolase activities measured *in vitro* are a poor reflection of the total activities within lysosomes in living cells. It is, therefore, difficult to interpret observations that the *in vitro* activities of cathepsin D and certain lysosomal cysteine proteases may be higher in aging rat liver [53] or in rat brain [54,55] and that *in vitro* β-galactosidase activity may be elevated in aging human fibroblasts. Compensatory increases in cathepsin protein expression in compromised lysosomes may yield higher activity *in vitro* even though the *in situ* activity is lowered because of defective lysosomal acidification [56] or other luminal lysosomal changes. Age-related changes in the intralysosomal pH, levels of endogenous lysosomal cysteine protease inhibitors (e.g., cystatins B and C) [57], membrane stability [58], or in myriad other possible conditions within the intraluminal environment of the lysosome, may well explain the overall decreased lysosomal activity in aging cells, although these possibilities are yet to be fully investigated.

Another clear indication of the diminished function of aging lysosomes is the striking decline in the activity of CMA seen in many cell types and tissues [59,60]. Underlying this decline are age-related changes in lysosomal membrane lipid composition that alter the dynamics and stability of LAMP-2A in lysosomes [60–62] without changing transcription, synthesis, and lysosomal targeting of the LAMP-2A protein during lysosome biogenesis. The loss of LAMP-2A by degradation in the lysosomal lumen reduces the ability of the lysosomes to bind and translocate substrate proteins into the lumen [62]. Age-dependent declines in CMA may be promoted by increased oxidative stress and attempts to eliminate oxidatively damaged proteins [63]. That these CMA deficits accelerate other aspects of the aging process is shown by the improvement in the health span of aged mice [64] when an exogenous copy of LAMP-2A is expressed in the liver.

9.3 LYSOSOMES AND DISEASES OF LATE AGE ONSET

In light of the adverse impact of aging on lysosome function, it is not surprising that diseases that manifest at an old age would have a lysosomal component. Less well appreciated is that in some of these disorders, the disease-related mechanism itself helps to cripple the lysosomal system and also contributes significantly to primary pathogenesis. This is particularly an emerging theme in recent conceptualizations of neurodegenerative disease mechanisms. We first briefly consider a few examples of lysosome-aging interactions in nonneural disorders and later discuss major neurodegenerative disorders that are most prevalent at advanced age in the following sections.

9.3.1 Cardiovascular Disease

Lipofuscin accumulates robustly within aging cardiac myocytes and other long-lived postmitotic cells, like neurons, because it is neither degraded nor exocytosed efficiently [5]. Along with lipofuscin accumulation, reflecting incomplete degradation of autophagocytosed material, myocardial aging is characterized by the abnormal accretion of lysosomal "waste" material, especially defective enlarged mitochondria. Arising by oxidative damage and/or altered fission, these enlarged mitochondria are believed to be autophagocytosed less efficiently than smaller ones, leading to the progressive accumulation characteristic "giant" mitochondria in aged cardiac myocytes [22].

Normal arteries express very low levels of cathepsins, but these proteases become abundantly expressed and secreted in atherosclerotic vessels of elderly individuals [65–67] stimulated in part by increased release of inflammatory cytokines [65,66,68]. Cysteine cathepsins K, L, and S contribute especially to the formation and progression of the atherosclerotic plaque and may influence its stability and thus thrombogenic potential [69]. The effect of cathepsins in atherosclerotic lesions is balanced by the expression of endogenous cathepsin inhibitors, the cystatins, which normally buffer protease activity, but the levels of cystatin C are reduced in atherosclerotic lesions [70]. This situation likely contributes to increased degradation of the extracellular matrix of the vessel, degradation of the elastic lamina, and formation of larger plaques [71]. Lysosomal membrane permeabilization (LMP)-dependent apoptosis can be initiated in macrophages by lipids, including the oxidized lipoprotein particles and oxysterols accumulating in lysosomes of the affected vessels [72] [73].

Age-related macular degeneration (AMD), the leading cause of central vision loss in the elderly in developed countries, is a progressive degenerative disease involving multiple genetic and environmental factors impinging mainly on the retinal pigment epithelium (RPE). AMD is a disease of aging, rising in incidence exponentially 8- to 10-fold between ages 50 and 90 (https://nei.nih.gov/eyedata/amd). Interestingly, the variant B mutation in cystatin C, a widely expressed lysosomal protease inhibitor is associated with both AMD and Alzheimer's disease [74] and is one of several parallels that have been noted in these two disorders [75,76]. Links were recently suggested between mutations in the Npc1 and Npc2 gene that cause Niemann–Pick type C (NPC), a lysosomal storage disease, and retinal degeneration associated with marked AMD-like lipofuscin accumulation within the RPE [77]. Age, genetics, diet, smoking, and many cardiovascular factors cause aging-related functional changes in the RPE, choroid, and neural retina, and there is increasing evidence that long-term oxidative stress, impaired autophagy clearance, and inflammasome-mediated inflammation in the RPE are early pathogenic events. Each RPE cell phagocytoses and digests the material produced by 30–50 overlying photoreceptor cells, which shed 10% of their mass daily. The intense and continual phagocytic activity of RPE cells results in the progressive accumulation of lipofuscin and other indigestible products in their lysosomal compartments [78]. Unlike lipofuscins found in other tissues, which are composed mainly of protein, RPE lipofuscin consists predominantly of lipid-bisretinoids (LBs) and only 2% protein [79]. Once *N*-retinylidene-*N*-ethanolamine (A2E) and other less abundant bisretinoids are formed, they are refractory to all known lysosomal hydrolases and progressively accumulate in the lysosomes of RPE cells [80,81]. The concept that LB accumulation causes retinal degeneration is supported by evidence showing that excessive LBs are toxic for cultured RPE cells [82,83], photoreceptors overlying A2E-laden RPE become more susceptible to degeneration [84] and in Stargardt's disease, excessive accumulation of LBs precedes macular neurodegeneration [85]. Mice carrying null mutations in Abca4 and Rdh8, two genes linked to familial forms of AMD, develop blindness, basal laminar deposits, and focal accumulations of extracellular debris between the RPE and the Bruch membrane (drusen) [86]. Notably, beta cyclodextrins (β-CDs), which bind LBs and protect them against oxidation, markedly reduce the LB content of RPE in the Abca4-Rdh8 double knockout (DKO) mice model of AMD [87].

Lysosomes may be a convergence point of multiple factors contributing to AMD pathogenesis. Iron-rich lysosomes are sensitive to oxidative stress, and diffusion of peroxides into this compartment causes lipofuscin to form in normal aging postmitotic cells. Iron contributes to the pathogenesis of AMD as a source of free radicals that damage the tissue [88,89]. Iron levels increase normally with age in the neural retina [90] [91,92]. When the greater abundance of iron observed in vulnerable of AMD-affected eyes [93,94] is mimicked in an AMD mouse model, the mice exhibit age-dependent retinal iron accumulation and retinal degeneration with AMD features [95,96]. When the ARPE-19 cell model of AMD is exposed to increased iron to mimic the iron accumulation with age, these cells display markedly decreased phagocytosis activity, interrupted cathepsin D processing, and reduced cathepsin D activity [97]. As previously discussed, increased iron-catalyzed free radical generation within lysosomes generates lipid peroxidation products, such as 4-hydroxynonenal (HNE) and malondialdehyde (MDA), known to increase lipofuscinogenesis and reduce autophagy activity in RPE cells [98] by directly inhibiting autophagic sequestration of cytoplasmic material and by inactivating lysosomal cysteine proteases via covalent binding to the active site, thereby creating proteolysis-resistant substrates that become competitive protease inhibitors [99]. In addition, A2-E is a potent inhibitor of the ATPase proton pump [100] and its accumulation increases lysosomal pH and this inhibits lysosomal hydrolases. In support of the importance of this effect, several approaches used to reacidify lysosomes, including stimulating beta-adrenergic, A2A adenosine, and D5 dopamine receptors, and targeting acidic nanoparticles to RPE lysosomes, have been shown to lower lysosomal pH and improve degradation of outer segments within compromised RPE cells from aged ABCA4(−/−) mice and other AMD models [101].

Finally, a number of AMD-associated insults may disrupt the integrity of RPE lysosomes. For example, the lipofuscin component A2E has been shown to permeabilize RPE lysosomes via its detergent properties [102]. Destabilization of RPE lysosomes induces NLRP3 inflammasome activation, which may contribute to AMD pathology through the release of the proinflammatory cytokine IL-1β and through caspase-1-mediated cell death, known as "pyroptosis" [103].

9.4 LYSOSOMES IN AGING-RELATED NEURODEGENERATIVE DISEASES

Varying degrees of lysosomal-related pathology have been reported in most major adult-onset neurodegenerative diseases, including AD [104,105], PD [106], amyotrophic lateral sclerosis (ALS) [107,108], Huntington's disease (HD) [109], and several others [109]. Historically, the possible pathogenic significance of this lysosomal pathology has been largely dismissed as generalized secondary or end-stage responses of the lysosomal system to neuronal compromise. More recently, however, with the emerging appreciation of lysosome and autophagy involvement in neurodegenerative disease mechanisms, the concern is now increasing that modest AV/lysosome changes in a disease tend to be overinterpreted as pathogenically meaningful without performing adequate quantitative analytical assessments. Where extensive quantitative analyses of this pathology have been carried out, as in AD, for example, the findings strongly point to a disease process significantly involving lysosomes, which has revealed important clues to the underlying pathobiology.

Compelling genetic evidence now links mutation-driven lysosome dysfunction to the development of familial forms of multiple late-onset neurodegenerative diseases [1,110,111] and to the mechanism of neuronal cell death itself [112,113]. In a growing array of aging-related neurodegeneration diseases, the suspected pathogenic protein is often an integral or associated regulatory component of the lysosome or the larger lysosomal network (e.g., autophagy or endosomal lysosomal pathway). In many of these cases, the disease mutation has also been shown to disrupt a critical function of the lysosomal system. Equally compelling supporting evidence is the growing number of instances in which the same lysosomal system gene can cause either a neurodegenerative disease of late adult onset or a congenital lysosomal storage disorder (LSD), depending on the mutation or mutant gene dosage. In neurodegenerative diseases characterized by signature patterns of proteinopathy, the pathogenic protein is very frequently a known substrate of lysosomal mediated pathways (e.g., macroautophagy or CMA) and its appearance as inclusions or aggregates is related in part to a lysosomal effect induced by the protein. Finally, the striking therapeutic effects achieved by modulating autophagy or lysosome efficacy in mouse models of major late age-onset neurodegenerative diseases underscore the significance of lysosomal system dysfunction as a pathogenic mechanism and as a target for future therapies. We consider four major neurodegenerative diseases that develop with increased incidence in old age (AD, PD, diffuse Lewy body diseases, and FD) and their earlier onset familial variants as follows.

9.4.1 Alzheimer's Disease (AD)

AD, the most common dementia and a disease involving widespread neuronal cell death, increases exponentially in incidence after the age of 65, underscoring the close relationship of pathogenesis to brain aging. Autosomal-dominant mutations in one of three genes presenilin 1 (PS1), presenilin 2 (PS2), and amyloid precursor protein (APP) (APP gene duplication in the case of Down syndrome), account for less than 5% of AD cases. Risk in the remaining 95% of AD cases is influenced substantially by the epsilon 4 allele of APOE and more modestly by a growing number of other genes. AD is defined by the coexistence of two neuropathological features: intraneuronal aggregates of the microtubule-associated protein tau (neurofibrillary tangles) and extracellular lesions known as "senile or neuritic plaques," which are foci of degenerating or dystrophic axons and dendrites associated with deposits of beta-amyloid peptide (Aβ), a cleavage product of APP able to form amyloid fibrils and toxic Aβ oligomers. Equally characteristic of AD pathology, however, are changes in the endosomal lysosomal system, beginning with very early appearing disease-specific enlargement of early endosomes and increased lysosome biogenesis and subsequently autophagy failure evidenced by profuse accumulation of autophagic vacuoles (AVs), especially autolysosomes, in affected neurons. Lysosomal dysfunction and enormously swollen dystrophic neurites containing mainly AVs are widespread in affected brain regions and are highly characteristic of neuropathology in AD compared to that in the several other adult-onset degenerative diseases sufficiently studied. Such accumulations of autophagic and endocytic "waste" in autolysosomes are, however, reminiscent of those seen in some primary LSDs [114]. Notably, LSDs (e.g., NPC and mucopolysaccharidosis type IIB [115,116]) are among the few disorders in which neurofibrillary tangles, a hallmark lesion of AD, develop in some cases, along with other features of AD pathology [117–120]. In NPC, for example, highly disease-selective abnormalities of endosomes, elevated levels of the beta-site APP-cleaving enzyme (BACE) and

BACE-cleaved carboxyl-terminal fragments of APP (βCTFs), and modest extracellular Aβ deposition arise in the context of endosomal trafficking of cholesterol and accumulation of unesterified cholesterol in late endosomes/lysosomes. Interestingly, allele-selective influences of apolipoprotein E (APOE) genotype on pathology are also seen in both NPC and AD [121].

Converging lines of evidence point to defective autolysosomal proteolysis is being principally responsible for disruption of autophagy in AD. Enlarged substrate-laden autolysosomes are the principal organelles in giant neuritic swellings [104], and similar neuritic dystrophy and selective accumulation of AVs can be reproduced by blocking lysosomal proteolysis pharmacologically or genetically *in vitro* or *in vivo* (reviewed in Reference [122]). The morphological and functional deficits in AD models that overexpress mutant APP develop lysosomal deficits, not unlike those in the well-studied presenilin-1 loss-of-function models although different mechanisms are likely involved (see below). The endosomal–lysosomal dysfunction in these mouse models includes impaired axonal transport of selected cargoes, increased amyloidogenesis, reduced amyloid clearance, and lowered survival of neurons [122,123]. The pathologic significance of autophagy failure in AD is further supported by evidence that selectively enhancing *lysosomal* proteolysis substantially ameliorates AD-related pathologies (e.g., Aβ deposition, tau pathology), synaptic dysfunction, and cognitive deficits in AD models [124–126]. This range of therapeutic effects has been seen *in vivo* in mouse AD models in which the endogenous lysosomal cysteine protease inhibitors, cystatin C [124] or cystatin B, were deleted to increase cysteine protease activation or cathepsin B was overexpressing or pharmacologically modulated to enhance lysosomal proteolysis [124,126–130]. Rapamycin administration prior to development of AD pathology and presumably before lysosomal failure delays the onset and diminishes the severity of the AD phenotype while administration after the appearance of pathology has little beneficial effect highlighting the importance of lysosomal failure [127]. Although the mTOR inhibition actions of rapamycin to induce autophagy may be a major factor in the preventative effects, the additional effects of activating the transcriptional factors that regulate lysosomal biogenesis, for example, TFEB, TFE3, are also likely to be important.

Although the known functions of the genes causing early-onset AD argue strongly for a pathogenic role of APP and its metabolites in AD, the three autosomal-dominant causative AD genes and many of the other genes that confer increased AD risk also have direct or close functional relationships to the endosomal–lysosomal system, which is notably also the organellar system most active in generating and clearing APP metabolites [123]. Emerging evidence indicates that, independently of toxic effects mediated by Aβ, the alterations of the key genes and many of the newly identified genetic risk factors adversely alter lysosomal network function. Given that endosomal–lysosomal and autophagic pathways are strongly implicated in both the generation and clearance of Aβ and thus, the negative effects of AD genes on APP processing and on lysosomal system function are mechanistically intertwined [123]. Key genetic risk factors and proteins linked to AD pathogenesis impair lysosome function either directly (e.g., presenilins) or by altering processes upstream in lysosomal pathways (e.g., APP), which requires aged and failing lysosomes to deal with a substantially greater burden of less digestible substrates, including Aβ, which can further corrupt lysosome function.

The most direct genetic link between lysosomal dysfunction and AD pathogenesis are mutations of PS1, the most common cause of early-onset familial AD [131]. PS1, a ubiquitous protein with 9 transmembrane domains, exists in the endoplasmic reticulum (ER) as a 65 kDa holoprotein and elsewhere as a cleaved two-chain catalytic subunit of the gamma (γ)-secretase enzyme complex that mediates intramembranous cleavage of well over 25 different substrates [132,133]. Notable among its actions is generation of Aβ from a carboxyl-terminal fragment of APP generated by BACE-1 (ß-APP cleaving enzyme) [134]. PS1 mutations in AD confer loss of function to γ-secretase [135] resulting in reduced levels of Aβ peptides [136,137] and often, though not invariably, a modest increase in the proportion of a toxic 42 amino acid peptide (Aß42) prone to greater aggregation relative to a less toxic 40 amino acid form (Aß40) [133,135,138]. Although PS1 pathogenicity in AD is often ascribed to this minor shift in Aβ42 and 40 and the promotion of Aβ oligomerization, the evidence for this mechanism is correlational rather than direct. Additional or alternative explanations for pathogenicity of PS1 mutations have also been sought [139] involving alternative actions of PS1 as a component of γ-secretase [140] or via its secretase-independent roles as a holoprotein, which include disruption of its roles in lysosomal acidification [141], Wnt signaling [142], and cellular calcium regulation [143–145].

Mutations of PS1, which accelerate AD onset, considerably exacerbate lysosomal system pathology in early-onset AD or mouse models of AD [146] and similar lysosomal pathology can be induced by deleting PS1 [141,147,148]. Exploring the basis for these effects, Lee et al. [141] demonstrated that PS1 holoprotein acts as a chaperone in the ER for the V-ATPase V0a1 subunit, a six-pass transmembrane protein (TMEM) core subunit of the proton pump principally responsible for full acidification of lysosomes. While glycosylation at one candidate site on the V0a1 subunit did not influence V-ATPase stability [149], recent studies (Lee, unpublished) have shown that glycosylation failure at a second V0a1 site, an event normally facilitated by PS1 holoprotein, leads to V0a1 protein instability and increased V-ATPase degradation by ERAD before it can be delivered to lysosomes as earlier proposed [141]. In diverse cell types lacking PS1 and in fibroblasts from patients with AD-causing PS1 mutations, the V0a1 subunit of the V-ATPase is poorly glycosylated [150] and is associated with reduced assembly and function of the V-ATPase leading to a failure to fully acidify lysosomes, which is a requirement for lysosomal protease activation and autophagy [141,151] (Lee et al., 2014, submitted). The role of PS1 in lysosome acidification has been confirmed in various systems [151–155], and recent evidence has shown that normalizing lysosomal acidification fully reverses defective autophagy and lysosome function in PS1-deficient cells [149].

The failure of lysosomal acidification under conditions of PS1 loss of function has additional untoward effects on the signaling roles of lysosomes and lysosome trafficking. Loss of PS1 holoprotein function has previously been implicated in mediating γ-secretase-independent effects of AD-causing PS1 mutations (or PS1 deletion) that dysregulate calcium homeostasis at the ER via multiple mechanisms [156–158].

Lysosome mechanisms, however, have recently emerged as an additional important factor driving calcium dysregulation. Although ER contains the largest stores of cellular calcium [159], lysosomes are the second most abundant calcium store [160]. Abnormal efflux of calcium from lysosomes [149,161] via the TRPML1 calcium channel is a major contribution to cytosolic calcium elevations in PS1-deficient cells (Lee et al., submitted) and arises as a direct consequence of defective lysosomal acidification. Notably, reversing calcium efflux abnormalities alone does not rescue either acidification or autophagy deficits, whereas correcting lysosomal pH rescues calcium and autophagy deficits indicating that lysosomal calcium dysregulation is a secondary consequence of elevated pH. Although the lysosomal calcium efflux abnormality in PS1-deficient cells does not drive autophagy/lysosomal abnormalities, it does cause activation of calpains [162] and calcium effectors, such as protein kinases (e.g., cdk5, ERK 1/2), which potentially mediate tau hyperphosphorylation [162]. In addition, signaling that regulates endolysosomal compartments is specifically impeded when lysosomal proteolysis is inhibited by blocking acidification or directly inhibiting cathepsins, promoting the selective accumulation of these organelles within axon swellings [163]. Local cytoskeletal protein hyperphosphorylation within dystrophic neurites also seen under these conditions [163] may involve calcium release from accumulating deacidified autolysosomes.

Mutations of APP and APP gene duplication in the context of Down syndrome cause AD in which severe autophagy neuropathology develops with a late age onset, as seen in all forms of AD and in mouse models. In TgCRND8 mice overexpressing human APP with two FAD mutations [126,164], florid autophagic-lysosomal pathology includes strikingly enlarged autolysosomes containing incompletely digested autophagic materials, including APP metabolites, membranous structures, and a lipopigment component. Stimulating lysosomal proteolytic efficiency in TgCRND8 mice by deleting an endogenous inhibitor of lysosomal cysteine proteases (cystatin B) prevents development of these pathological features, decreases extracellular amyloid deposition, and ameliorates learning and memory deficits [126], supporting the pathogenic significance of autophagic-lysosomal dysfunction, and specifically deficient lysosomal proteolysis in AD. Similar therapeutic effects have also been reported in another AD model of β-amyloidosis after overexpressing cathepsin B or deleting cystatin C [124,130].

One likely contributor to the pathogenic actions of APP in all forms of AD, besides Aβ, is the effect of the β-cleaved carboxyl-terminal fragment (βCTF) of APP on the endosomal–lysosomal system. The earliest disease-specific pathologic change in sporadic AD appearing before amyloid is deposited in the neocortex is the enlargement of Rab5- and Rab7-positive endosomes [165,166], and upregulated transcription of genes related to endocytosis, such as Rab5, Rab7, and Rab4 [167]. Recruitment to endosomes of the proteins encoded by these genes promotes fusion and abnormal enlargement of early and late endosomes [105,165], pathological acceleration of endocytosis, and diverse additional deficits demonstrated in mouse models [165,167,168].

The same pathological endosomal response mediated by βCTF [168] is seen in beginning decades before classical AD neuropathology in Down syndrome (trisomy 21), a cause of early-onset AD, attributed to an extra copy of App on the trisomic region of chromosome 21 [169]. The downstream consequences of this abnormal endocytic response include disrupted neurotrophin signaling, cholinergic neuronal degenerative changes [170], and increased apoptotic signaling [171]. The acceleration of endocytosis in cells of individuals with Down syndrome causes increased protein and lipid accumulation in endosomes and slowed lysosomal degradation of endocytic cargoes [165] as well as additional impairments of proteolysis within lysosomes (Jiang et al., submitted). Rab5 activation and the endosomal phenotype are exacerbated by both FAD mutations of APP [171] [172] and by inheritance of the ε4 allele of APOE, an endocytic transporter of cholesterol and a major genetic risk factor for late-onset AD [166] and by dietary cholesterol itself, another suspected AD risk factor. By upregulating endocytosis, elevated dietary cholesterol alone or overexpression of its receptor APOE (particularly APOE4) elevate βCTF levels and also lead to increased delivery of Aβ1-42 to lysosomes [173,174]. Expression in a hAPP mouse model of AD of an APOE ε4 allele, but not the APOE ε3 allele, increases levels of intracellular Aβ in lysosomes, altering their function and causing neurodegeneration of hippocampal CA1, entorhinal, and septal neurons [175]. As in several AD mouse models, various lipid storage materials (e.g., cholesterol, sphingolipids) accumulate in endolysosomal compartments within cells from individuals with Down syndrome. In several settings, these lipid accumulations have been shown to promote lysosome deacidification [176–178] and to alter the processing and clearance of APP and its metabolites [179]. Notably, chronic administration of cyclodextrin, an agent that promotes clearance of cholesterol from lysosomes, reduces amyloidogenesis and improves cognitive function in a PS/APP mouse model [180]. Interestingly, among a very few additional pathological conditions in which AD-like endosomal pathology develop is NPC, a disorder of cholesterol homeostasis. Although less well characterized, a growing number of AD risk genes for late-onset AD (e.g., BINI, CLU, PICALM, ABCA7, MS4A4A) identified through genome-wide screens and association studies have direct links to endocytic regulation in part because of the actions on these processes, APP processing [181,182]. Thus, in sporadic AD and under conditions of increased genetic and environmental AD risk, accelerated endocytosis mediated by Rab5 and βCTF seems to be a common pathway in AD leading to increased delivery of substrates into degradative compartments.

Aβ peptide can be sequestered or generated from APP during autophagy [183] although it is normally then degraded in lysosomes [184–186]. Although less well studied as "Aβ degrading proteases" than the zinc metallopeptidase family [187,188], cathepsins are now considered an important route for Aβ/amyloid clearance [123,125,130] and human neurons may be particularly dependent on this mechanism [189]. Among its various processing routes, tau also undergoes lysosomal degradation by CMA upon hsc70 recognition of one of the two targeting motifs in its C-terminus [190]. Mutant tau variants, however, abnormally bind to LAMP-2A and are only partially internalized. The portion of the protein that gains entry into the lysosomal lumen is trimmed, resulting in the formation of smaller amyloidogenic tau fragments at the lysosomal membrane that oligomerize directly at the surface of lysosomes and may disrupt lysosomal membrane integrity and block CMA of other substrates [190].

Slowed degradation of Aβ and other substrates within lysosomes encourages further oxidation of substrates [191] and additional lysosomal dysfunction. In those inheriting the ApoE4 risk allele, a unique proteolytic product of ApoE4, and not ApoE3 or ApoE2, may be generated in lysosomes, which yields a "molten globule" structure that induces reactive intermediates and destabilizes lysosomal membranes leading to lysosomal leakage and apoptosis [174]. Also, the expression of ApoE4, but not the ApoE3 allele, increases levels of intracellular A-beta peptide (Aβ), enlarges lysosomes and alters their morphology in a mouse AD model and causes neurodegeneration of neurons typically vulnerable in AD [175]. Similarly, overexpression of human Aβ 42, but not Aβ 40, in *Drosophila* neurons induces age-related autophagic/lysosomal dysfunction and neurotoxicity [192] believed to arise from lysosomal membrane destabilization mediated directly by Aβ [193,194] or by incompletely degraded oxidized autophagic substrates [191]. Neurons cannot dilute by cell division any toxic protein buildup and are, therefore, particularly vulnerable to potentially toxic, mutant, oxidized, and aggregated proteins and peptide fragments. Accumulation of these substrates in lysosomes, promoted by increasing hydrolytic dysfunction, are conditions that may increase membrane permeability and release of hydrolases into the cytoplasm, even from otherwise intact lysosomes [195]. A close connection between lysosomal network dysfunction and mechanisms of neurodegeneration is well documented [114].

9.4.2 Parkinson's Disease and Related Disorders

In *PD*, the death of neurons involves mainly the dopaminergic population in the substantia nigra. These neurons develop signature Lewy body inclusions containing fibrillar aggregates of α-synuclein (SNCA), a cytosolic protein that dynamically interacts with membranes [196] and influences vesicle behavior, neurotransmission, and synaptic plasticity [197]. Point mutations of the α-synuclein gene (PARK1 locus) are causative for an autosomal-dominant form of PD [198–203], and single nucleotide polymorphisms (SNPs) in the *SNCA* gene increase PD susceptibility [204]. Additional families with multiplication (duplication or triplication) of its allele (*PARK4* locus) [205,206] suggest that even increased levels of normal α-synuclein may be causal for PD or other synucleinopathies. Besides SNCA, causative genes for PD include autosomal-dominant mutations in leucine-rich repeat kinase (LRRK2) or glucocerebrosidase (GBA) or autosomal recessive mutations of ATP13a2, VPS35, UCH-L1 (ubiquitin carboxy-terminal hydrolase-L1 Parkin), PINK1, and DJ-1. For these last three recessive genetic disorders, the relationship to the synucleinopathies is unclear since neuropathological reports are limited for PINK1- and PARK7-related PD cases [207] and LBs are infrequent in the few PARK2 cases studied. Nevertheless, PARK2 and PINK1 are known to regulate the autophagy of mitochondria (mitophagy) [208].

Although they have relevant pathogenic effects on the lysosomal system and potentially on other processes, most of the genes mutated in PD have an impact on lysosomal α-synuclein clearance leading to its accumulation. α-Synuclein undergoes various types of posttranslational modification [209–213], affecting its turnover and oligomerization into putatively neurotoxic oligomers [196,210,214–216] that may be released to propagate the pathological process in a prion-like manner from one neuron to neighboring neurons [200,217] possibly involving the endosomal–lysosomal pathway [218]. α-Synuclein may be released by calcium-dependent exocytosis [219,220], a phenomenon exacerbated by inhibition of lysosome function [221].

Most α-synuclein, especially overexpressed and misfolded aggregate-prone forms, is degraded by macroautophagy, which is upregulated when CMA, the other main degenerative pathway, becomes compromised in disease states [222,223]. Abnormal α-synuclein accumulation implicates lysosomal dysfunction as a primary or secondary consequence of α-synuclein mishandling, which is reflected by an accumulation of autolysosomes when either mutant or wild-type α-synuclein is overexpressed [223,224]. Moreover, in rat PC12 cells, expression of mutant α-synuclein can decrease lysosomal acidification [225] and slow lysosomal protein turnover [226]. Interestingly, the characteristic α-synuclein aggregation of PD is also a feature of Sanfilippo syndrome, a congenital LSD caused by α-*N*-acetylglucosaminidase (NAGLU) gene mutations. An SNP in NAGLU was also associated with PD risk [227]. The pathogenic importance of the lysosomal dysfunction induced by α-synuclein overexpression can also be appreciated from the neuroprotection provided by viral-mediated overexpression of transcription factor EB (TFEB), LAMP2a, or Beclin-1 in rodent models of PD [223,228,229], which upregulate different aspects of autophagy, including the lysosomal clearance steps.

Several other PD-related genetic factors have also been shown to disrupt lysosomal function and the general clearance of autophagic substrates. Loss-of-function mutations of the lysosomal acid β-glucosidase gene (GBA) in the homozygous state cause Gaucher disease, a congenital LSD leading to lysosomal accumulation of glucocerebrosidase and reduced lysosomal content of ceramide. In heterozygous carriers who are not affected developmentally by Gaucher disease, the same mutations greatly increase the risk of developing PD [230,231]. A number of recent studies have highlighted a reciprocal relationship between α-synuclein and GCase. Synuclein levels markedly increase in the presence of disease-associated mutations [232] due to preferential resistance to lysosomal proteolysis [233] that may involve direct interaction between α-synuclein and GCase [234]. The compromised lysosomal degradation and neurodegeneration seen after deleting the gene encoding GCase in mice can be largely prevented by overexpressing wild-type but not mutant forms of GCase [235].

A common property linking pathogenic actions of some mutant proteins in PD, as well as several other neurodegenerative diseases, is their ability to disrupt CMA, which is dysfunctional in both familial [226,236–238] and sporadic [236,239] PD. CMA-targeting motifs are present in the majority of PD-related proteins, and the two most commonly mutated proteins in patients with familial PD, α-synuclein and LRRK2, are known to be degraded in lysosomes via CMA [226,236–239] as is a third PD-related gene product, UCH-L1 [240]. Despite their normal chaperoned delivery to the lysosomal membrane via cytosolic hsc70, pathogenic mutant variants of α-synuclein, LRRK2, and UCH-L1 fail to reach the lysosomal lumen to be

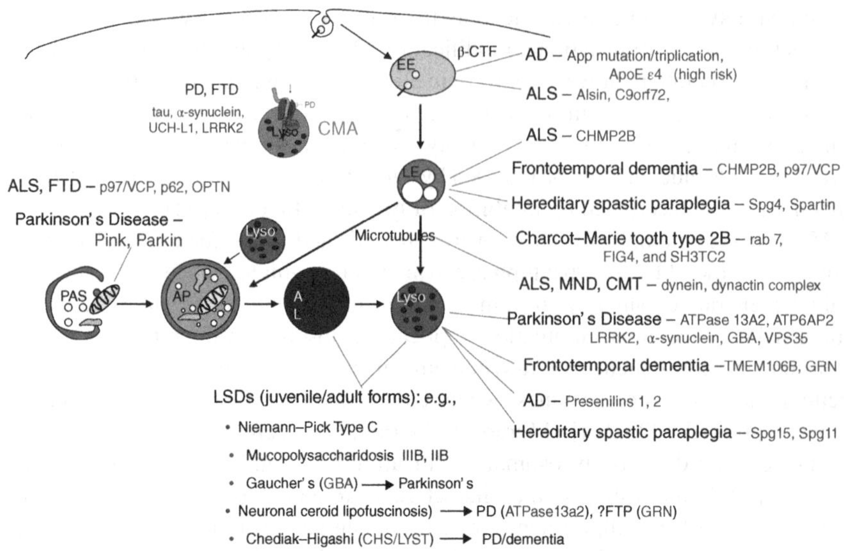

Figure 9.2 Genetic evidence strongly implicates the lysosomal network in the pathogenesis of neurodegenerative disease. The diagram identifies selected neurological disorders in which the pathogenic gene encodes a protein that plays a vital role in lysosomal network functioning. Mutations of the indicated genes are causative for familial forms of each disease and have also been shown to disrupt lysosomal function directly or through malfunctions of another compartment in the lysosomal network. Because of an extensive cross-talk among vesicular compartments comprising the lysosomal network, a defect in any component of the various pathways to lysosomes can potentially impede the efficiency of lysosomal digestion and signaling. (*See color plate section for the color representation of this figure.*)

degraded by CMA [226,236] due to their aberrant association with LAMP-2A [226,236,240]. LRRK2 mutant proteins show enhanced lysosomal binding in the presence of other CMA substrates, which then interferes with the proper organization of the active CMA translocon (Figure 9.2). The toxic interactions of α-synuclein and LRRK2 mutants with the CMA transporter preclude not only their own degradation but also inhibit the degradation of other CMA substrates and α-synuclein in particular [226,236,240]. Wild-type α-synuclein modified by dopamine seems to cause similar CMA dysfunction suggesting that this form of α-synuclein toxicity applies to sporadic and familial forms of PD [241] and may partly explain the role of environmental or cellular stressors (e.g., pesticides, oxidative stress) in developmental PD [242]. Even in the absence of noticeable posttranslational modifications, an increase in the cellular levels of either α-synuclein [226] or LRRK2 [236] beyond a tolerable threshold, has very similar inhibitory effects on CMA activity. In fact, these two proteins seem to potentiate each other's toxic effect on CMA [236].

The most common autosomal-dominant form of PD and a familial variant that closely resembles sporadic PD clinically [243] involves mutations of *LRRK2*. Similar to SNCA, increased activity of this kinase associated with aggregate formation has been a prevailing hypothesis for mutant LRRK2-mediated toxicity [244–246], although only a few *LRRK2* mutations have been shown to enhance LRRK2 kinase activity [244,247]. The range of pathogenic mutations identified along the molecule have suggested other pathogenic effects of misfolded LRRK2, including disruption of proteostasis [248]. Neurons ectopically expressing mutant LRRK2 develop a PD-like cellular phenotype and degenerate at a rate predicted from the accumulated level of diffuse mutant LRRK2 [249]. Additional findings suggest that LRRK2 levels are more important than kinase activity per se in predicting toxicity [249]. Overexpressing the C-terminus of Hsp70 interacting protein or downregulating Hsp90, both of which interact with LRRK2, reduces mutant LRRK2-induced toxicity in a kinase-independent manner. Interaction with either of these proteins is associated with lower steady-state levels of mutant LRRK2 [250,251].

α-Synuclein may play a role in LRRK2-mediated toxicity by modulating LRRK2 levels. α-Synuclein accumulates in cells with elevated LRRK2 levels independently of its kinase activity [252,253], and, in PD patients that have synuclein pathology due to LRRK2 mutations [254], symptoms occur earlier if certain genetic variants in the synuclein gene [255] are present along with the LRRK2 mutation. In neurons ectopically expressing mutant LRRK, these levels are significantly lower in the absence of α-synuclein. Because α-synuclein buildup disrupts protein homeostasis pathways [226,256,257] and lowering α-synuclein levels enhances them [257], synuclein accumulation may impede mutant LRRK2 clearance and promote its toxicity. However, the increase in α-synuclein by LRRK2 is probably insufficient to fully explain the range of mutant LRRK2 toxic effects, which include dysfunction in vesicular trafficking, neurotransmitter release, cytoskeletal dynamics, protein degradation [258], and mitochondrial dynamics [259,260]. The range of possible *LRRK2* functions is expanding [261], and roles in autophagy and endosomal trafficking (via GTPase regulation) have been proposed as particularly relevant to pathogenesis [262,263]. Recently, three LRRK2 interactors were identified that are risk factors for PD, including the small GTPase Rab7 (a causative gene in Charcot–Marie–Tooth neuropathy), cyclin G-associated kinase, and BCL2-associated athanogene (BAG5) [264]. Altered *LRRK2* activity due to PD-related mutations has been linked to defects in endosomal–lysosomal trafficking [263] [265], lysosomal pH and calcium regulation [266] and CMA [236]. LRRK2 deletion in mice or siRNA knockdown in human cells prominently alters markers of autophagy and the lysosomal pathway [267,268], suggesting a possible physiological role in regulating autophagy [266,269], although the mechanisms are unclear. Interestingly, LRRK2 inhibition has been shown to stimulate macroautophagy, suggesting that two putative pathogenic LRRK2-related mechanisms may be interlinked [268].

Kufor-Rakeb syndrome, a neurodegenerative dementia, associated with early-onset parkinsonism [270] is caused by mutations in the P5-type ATPase 13a2 gene (ATP13a2) within the PARK9 PD susceptibility locus. ATP13a2 is a lysosomal protein involved in heavy metal transport and possibly local lipid dynamics during vesicle formation and membrane fusion [271] and *in vitro* studies show that mutant proteins do not traffic correctly to the lysosome [272,273]. Loss of ATP13a2 in cellular and animal models results in the accumulation of autophagosomes and lysosomes with reduced proteolytic activity possibly due to defective lysosomal acidification [272,274,275]. Although the presence of α-synuclein aggregates or LB has not yet been established in human cases [271], either knockout and knockdown of ATP13a2 causes an accumulation of insoluble α-synuclein [276] [275,277], along with NCL-like neuropathology [278], and induces neurotoxicity. Interestingly, in this context, loss of ATP13A2 function is also associated with the LSD, neuronal ceroid lipofuscinosis (NCL) in a strain of dogs and in at least one family [277,279,280]. Overexpression of ATP13A2 suppressed the toxicity of overexpressed α-synuclein in animal and neuronal PD models [281].

Mutations in vacuolar protein sorting35 (VPS35) cause a parkinsonian syndrome clinically similar to sporadic PD [282,283]. VPS35 is part of the retromer complex mediating transport of endosomes back to the trans-Golgi and in the sorting of receptors for hydrolases, thus affecting vacuole/lysosomal function and biogenesis. Loss of function in VPS35 induces vacuolar/lysosomal abnormalities [284] including accelerating MPR turnover, impeding CatD maturation, and inducing α-synuclein accumulation [285]. Manipulating LRRK2 and a candidate gene for the PARK16 locus, RAB7L1, alters endosomal trafficking, linking to the previously reported function of VPS35 [265].

Hereditary spastic paraplegias (HSPs) are clinically and genetically heterogeneous neurological disorders characterized by length-dependent axonopathy of corticospinal motor neurons, resulting in progressive spasticity and weakness of the legs. Two nearly identical HSP gene variants, SPG15 and SPG11, exhibit distinctive features of early-onset parkinsonism, cognitive impairment, white matter changes, retinal abnormalities, and lens opacities. Although several different pathogenic mechanisms have been suggested for these two autosomal recessive disorders [286–288], pathogenic alterations in lysosomes or autophagy are favored in recent studies [287,288]. Recent studies [289] show that the SPG15 protein spastizin and SPG11 protein spatacsin play pivotal roles in autophagic lysosome reformation (ALR), a pathway generating new lysosomes. Loss of spastizin or spatacsin results in depletion of free lysosomes and accumulation of autolysosomes. Depletion without affecting fusion of autophagosomes and lysosomes, as previously suggested [287], or altering the acidic environment of autolysosome. However, ALR via lysosome tubulation is significantly impaired, causing accumulation of autolysosomes [287] [288]

A mouse model of PD resulting from the exposure of the neurotoxin 1-methyl-4-phenyl-1,2,3,6-tetrahydropyridine (MPTP) shows accumulation of autophagosomes and an early decrease in lysosome number in dopaminergic neurons as a result of lysosomal membrane destabilization and cytosolic release of cathepsins [290]. Mitochondrial dysfunction and oxidative stress induced by the neurotoxin MPTP is characteristic of this well-characterized neurotoxin model of PD [291,292]. Lysosome number drops in dopaminergic neurons very early after MPTP exposure, related in part to LMP and in part due to Bax internalization within lysosomes [293]. Effects of cytosolic release of cathepsins are partially prevented by cathepsin inhibitors. These events and the secondary accumulation of autophagosomes and neurodegeneration are attenuated by genetic or pharmacological activation of TFEB or rapamycin, both of which increase lysosomal biogenesis [290] in addition to inducing autophagy.

9.4.3 Diffuse Lewy Body Disease (DLB)

Another relatively common synucleinopathy is characterized by adult-onset dementia, parkinsonism, and the presence of Lewy bodies (LBs) throughout the CNS (central nervous system) and PNS (peripheral nervous system) and not infrequently varying degrees of amyloid pathology [294]. Except for a few cases of familial aggregation, DLB is generally considered a sporadic disorder. While the genetics of AD and PD are generally not overlapping, a small number of reports suggest an involvement of APOE (an AD risk factor) and GBA (a PD causative factor) as risk alleles for the development of DLB [295–297]. In addition, two additional PD-related genes were recently identified as DLB risk factors: SNCA involving a distinct risk haplotype from PD and SCARB2, or lysosomal integral membrane protein 2 (LIMP2), which is required for endolysosomal biogenesis and maintenance [298] and acts as a lysosomal receptor for GBA targeting [299]. While only a weak risk factor for PD [283], LIMP2 seems to be a relatively strong risk factor for LBD. Localization of LRRK2 to enlarged endolysosomal compartment seen in PD is also seen in LBD [300].

9.4.4 Frontotemporal Lobar Degeneration (FTLD)

FTLD is the second most common dementia in people younger than 65 years and 25% of FTLD cases appear after this age [301]. A family history is positive in about 30–50% of FTLD cases [302,303]. In addition to cognitive impairment, motor dysfunction is often detected and may meet criteria for ALS in some of these cases [304], reflecting mounting evidence that ALS and FTLD are at opposite ends of a single disease continuum. The pathological hallmarks of the major subtype, FTLD-TDP, are intracellular ubiquitin and TAR DNA-binding protein (TDP)-43 positive inclusions [305,306], which distinguish this common variant from other FTLD types, such as FTLD-tau, FTLD-FUS (fused in sarcoma) [307] or FTLD-DPR [308]. TDP-43, a DNA- and RNA-binding protein involved in transcription and splicing, is hyperphosphorylated, proteolytically processed, and frequently mislocalized to the cytoplasm in FTLD-TDP [305,306,309]. Autosomal recessive loss-of-function mutations in the progranulin (GRN) gene [310,311], which severely reduce tissue GRN levels [312,313], account for up to 20% of familial FTLD-TDP cases [314,315] and are only a rare cause of ALS [316]. Additional missense mutations reduce functional GRN by impairing secretion or inducing premature degradation of the misfolded protein [317–319]. Mutations of C9orf72, which induce DNA-repeat expansion, account for another 20% of familial FTD TDP and 5–7% of the apparently sporadic FTLD cases. Mutations of C9orf72, unlike GRN, are also a common cause of fALS.

Endolysosomal mechanisms are increasingly suspected in FTLD-TDP pathogenesis. As with synucleinopathies, the same gene (GRN) has been shown to cause either FTLD or LSD. Although rare so far, homozygous mutations of GRN were identified in two siblings with dementia associated with an adult-onset form of NCL, a family of childhood neurodegenerative diseases characterized by massive storage of lipofuscin [320] and caused by recessive mutations in genes mainly involved in lysosomal processing (PPT1, TPP1, CLN3, CLN5, CLN6, MFSD8, CLN8, CTSD, or DNAJC5) [305,320,321]. The clinical picture in these siblings, which included retinopathy, seizures, and ataxia, contrasts with the FTLD-TDP phenotype due to heterozygous GRN mutations, where behavioral and cognitive impairments are rarely accompanied by these additional features. Interestingly, progranulin-deficient mice exhibit [313,322] pathobiochemical features of both GRN-associated FTLD-TDP and NCL. Supporting the potential pathological overlap of FTLD and NCL are further observations that mice lacking cathepsin D, Ctsd (−/−), a model for NCL, show elevations of pathologically phosphorylated TDP-43 similar to those in mice lacking Grn. Accompanying the neuronal storage of abnormal autofluorescent lipopigment and characteristic ultrastructural features of NCL in GRN-deleted mice are elevated levels of the NCL storage components saposin D and subunit c of mitochondrial ATP synthase (SCMAS).

GRN-deleted mice have increased levels of TMEM 106B, a lysosomal protein known as a risk factor for FTLD-TDP pathology [323–325] and a genetic factor that confers risk for cognitive impairment in ALS controls *(Ryan Vass)*. TMEM106B variants that promote FTLD-TDP increase levels of TMEM106B mRNA and protein [326], which can be demonstrated in GRN mutation carriers [324,327]. Moreover, microRNA-132, which represses TMEM106B expression [328], is halved in FTLD-TDP, thus promoting pathologically high levels of TMEM106B. Evidence suggests that TMEM106B and progranulin mechanisms are interrelated. Endogenous neuronal TMEM106B colocalizes with progranulin in LAMP-1-positive late endolysosomes [323–325] and TMEM106B overexpression increases intracellular levels of progranulin. TMEM106B risk genotype correlates with age at onset in GRN(_) FTLD-TDP [329], and both protein and mRNA for TMEM106B are increased in expression and abnormally localized in neuronal processes in GRN(_)FTLD-TDP cases.

The TMEM106B interactors, VPS11 and VPS13D, are known to participate in endocytic vesicle maturation and in delivery to lysosomes [330–334] and molecular interactions of TMEM106B with these and other endocytosis-related proteins may contribute to the delayed degradation of endocytic cargoes when TMEM106B is overexpressed [323]. Overexpression of TMEM106B in neurons inhibits lysosome transport [323,324] and leads to accumulation of enlarged lysosomes in the cell soma. Reduced TMEM106B not only increases the percentage of actively transported lysosomes but also impairs acidification of endolysosomes and mannose-6-phosphate-receptor trafficking [335].

Further suggesting the involvement of lysosome in FTLD [336,337] are mutations in the gene encoding valosin-containing protein (VCP, also known as p97) [338,339], a conserved and highly abundant multifunctional protein and member of the class II AAA ATPAse family (ATPases associated with diverse cellular activities) [340]. VCP is responsible for IBMPFD, a disorder characterized by disabling muscle weakness, IBM-inclusion body myopathy, osteolytic bone lesions (Paget's disease), and FD [339]. VCP is considered an FTLD-related gene, due to its characteristic language and/or behavioral dysfunction and FTLD-TDP D type pathology, which features prominent ubiquitinated inclusions in multiple tissues. VCP mutations are also responsible for autosomal-dominant fALS in an Italian family in which other genetic causes of ALS were excluded in small percentages of other fALS cases [341]. VCP regulates endolysosomal sorting of endocytosed ubiquitinated cargoes [342] and VCP mutation is known to disrupt selective autophagy. VCP depletion or the expression of IBMPFD-linked mutants of VCP in cell models causes immature autophagosomes containing ubiquitinated substrates to accumulate [336,343]. Mutations of CHMP2B (loss-of-function charged multivesicular body protein – 2B), causing frontotemporal dementia linked to chromosome 3 (FTD3), disrupt ESCRT machinery, impair amphisome formation, and lead to accumulation of ubiquitin-positive aggregates in neurons.

C9orf72 expansion can cause FTLD-TDP, classical ALS, or combination of both symptom complexes [344]. TDP-43 accumulates in neuronal and oligodendroglial inclusions within various brain regions. Also observed specifically in *C9orf72* expansion cases are ubiquitin/p62-positive, TDP-43 negative neuronal inclusions, which contain dipeptide-repeat (DPR) proteins generated by unconventional repeat-associated translation of *C9orf72* transcripts [344,345]. C9orf72 interacts with proteostasis molecules actin and ubiquilin-2, the latter involved in binding and transporting ubiquitinated cargo to the proteasome and autophagosome [346,347]. Notably, mutations of UBQLN2 are present in rare cases of ALS with FTLD [347], and *SQSTM1* (p62) mutations have now been found in both ALS and FTLD [348–350]. Similarly, rare ALS-associated mutations of another selective autophagy receptor, OPTN, cause this protein to accumulate in TDP-43-positive neuronal inclusions [351]. Collectively, these observations imply that a cluster of risk genes for FTLD (and ALS), including C9orf72, may engage trafficking of cargoes for delivery to the autophagosome and/or proteasome. In C9orf72 expansion disease, nuclear RNA foci also form throughout the brain [344,345]. Wild-type C9orf72 interacts with ALS-related heterogeneous nuclear ribonucleoproteins, hnRNPA2/B1 and hnRNPA1, which shuttle continuously between the nucleus and cytoplasm and accumulate in cytoplasmic SGs during cellular stress [352], as, for example, under conditions of proteasome inhibition, raising the possibility that C9orf72 interference with cargo delivery could trigger this response.

C9orf72 is believed to regulate aspects of endosomal and autophagic vacuole trafficking [345]. Consistent with its predicted activity as a Rab guanine exchange factor activity capable of activating Rab5, C9orf72 in neurons colocalizes with Rab proteins (Rab1, Rab5, Rab7, and Rab11) and with LC3-positive vesicles implicated in endocytic transport and autophagy and this colocalization increases in ALS patient motor neurons. Depletion of C9orf72 inhibits internalization of TrkB receptor and Shiga toxin transport from the plasma membrane to Golgi apparatus. Ectopic overexpression of C9orf72 activates autophagy and its silencing attenuates autophagy induction suggesting a role in regulating autophagy activation in response to ROS, a well-established stimulus for autophagy induction and also for C9orf72 expression.

Genetic linkage in FTLD families with autosomal-dominant disinhibition, dementia, parkinsonism, and amyotrophy was first demonstrated on chromosome 17q21 (FTDP-17) [353] and traced to missense or splice site mutations of the tau (MAPT) gene, which disrupt the normal equimol proportions of tau isoforms with either three or four repeat microtubule binding domains and drive aggregation into ubiquitinated tau inclusions within neurons and, in some cases, glia [354,355]. Tau has a CMA-targeting motif and, as in other neurodegenerative diseases involving misfolded and aggregated pathogenic proteins, mutant tau corrupts both its own lysosomal luminal delivery but that of other substrates [190,226,236]. Lysosome dysfunction, however, is unlikely to account for the entire phenotype in this disease, given the number of other toxic gain of function effects of mutant tau on calcium homeostasis and calpain activation, seen in this FTLD variant [162].

9.5 CONCLUSION

The past 5 years has witnessed a remarkable convergence of genetic and molecular-biological data underscoring the broad importance of the lysosome and its interacting compartments in the pathogenesis of late age-onset neurodegenerative diseases. In view of the unique properties of neurons among other cell types, the heavy reliance on efficient lysosomal clearance mechanisms is not surprising. Equally unsurprising is the vulnerability to the lysosomal system impairments in postmitotic cells such as neurons expected to survive over the lifetime of the organism. Lysosomal vulnerability is magnified by aging, which adds further progressive impediments to efficient substrate delivery and clearance at a period in life when increasing damage to proteins and membranes may require an even greater capacity for quality control and clearance. Understanding the critical role evidently played by the lysosomal system in cellular aging and organismal lifespan is an important frontier of biology holding many clues to neurodegenerative disease development and therapeutic strategies. Beyond influences of aging, increasing numbers of genes responsible for forms of AD, PD, and possibly other proteopathies have been shown to have specific and often direct roles in regulating lysosomal function and causative gene mutations in these disorders are known to corrupt these functions. It may now be reasonable to view the realm of degenerative lysosomal disorders as a continuum across the lifespan [114] with congenital disorders involving severe lysosomal dysfunction at one end and at the other end, adult and late age-onset disorders involving milder, though still disease-selective lysosomal dysfunction, which emerge as cellular aging adds a final set of insults to the lysosome. Challenges lie ahead in identifying how to target the lysosomal network effectively as a disease therapy; nevertheless, recent advances in this field are a clear positive sign that the next decade will bring exciting progress toward this goal.

ACKNOWLEDGMENTS

We are grateful to Nicole Gogel for assistance with manuscript preparation and to Corrinne Peterhoff for assistance with figure preparation and artwork. Studies from our laboratories have been supported by the National Institute on Aging and the Alzheimer's Association.

REFERENCES

[1] Nixon RA. The role of autophagy in neurodegenerative disease. Nat Med 2013;19: 983–997.

[2] Vilchez D, Simic MS, Dillin A. Proteostasis and aging of stem cells. Trends Cell Biol 2014;24:161–170.

[3] Lynch G, Bi X. Lysosomes and brain aging in mammals. Neurochem Res 2003;28:1725–1734.

[4] Cuervo AM, Dice JF. When lysosomes get old. Exp Gerontol 2000;35:119–131.

[5] Terman A, Kurz T, Gustafsson B, Brunk UT. The involvement of lysosomes in myocardial aging and disease. Curr Cardiol Rev 2008;4:107–115.

[6] Wright WE, Brasiskyte D, Piatyszek MA, Shay JW. Experimental elongation of telomeres extends the lifespan of immortal x normal cell hybrids. EMBO J 1996; 15:1734–1741.

[7] Sinclair DA, Guarente L. Extrachromosomal rDNA circles – a cause of aging in yeast. Cell 1997;91:1033–1042.

[8] Harman D. The biologic clock: the mitochondria? J Am Geriatr Soc 1972;20:145–147.

[9] Afshari C, Barret J. Molecular genetics of *in vitro* cellular senescence. In: Holbrrok N, Martin G, Lockshin R, editors. *Cellular Aging and Cell Death*. New York: Wiley-Liss, Inc.; 1996. p 109–121.

[10] Ivanov A, Pawlikowski J, Manoharan I, van Tuyn J, Nelson DM, et al. Lysosome-mediated processing of chromatin in senescence. J Cell Biol 2013;202:129–143.

[11] Chin RM, Fu X, Pai MY, Vergnes L, Hwang H, et al. The metabolite alpha-ketoglutarate extends lifespan by inhibiting ATP synthase and TOR. Nature 2014;510:397–401.

[12] Rubinsztein DC, Marino G, Kroemer G. Autophagy and aging. Cell 2011;146:682–695.

[13] Markaki M, Tavernarakis N. Metabolic control by target of rapamycin and autophagy during ageing – a mini-review. Gerontology 2013;59:340–348.

[14] Kang HT, Lee KB, Kim SY, Choi HR, Park SC. Autophagy impairment induces premature senescence in primary human fibroblasts. PLoS One 2011;6:e23367.

[15] Kenyon CJ. The genetics of ageing. Nature 2010;464:504–512.

[16] Zoncu R, Efeyan A, Sabatini DM. mTOR: from growth signal integration to cancer, diabetes and ageing. Nat Rev Mol Cell Biol 2011;12:21–35.

[17] Lapierre LR, De Magalhaes Filho CD, McQuary PR, Chu CC, Visvikis O, et al. The TFEB orthologue HLH-30 regulates autophagy and modulates longevity in *Caenorhabditis elegans*. Nat Commun 2013;4:2267.

[18] Ruckenstuhl C, Netzberger C, Entfellner I, Carmona-Gutierrez D, Kickenweiz T, et al. Lifespan extension by methionine restriction requires autophagy-dependent vacuolar acidification. PLoS Genet 2014;10:e1004347.

[19] Stephan J, Franke J, Ehrenhofer-Murray AE. Chemical genetic screen in fission yeast reveals roles for vacuolar acidification, mitochondrial fission, and cellular GMP levels in lifespan extension. Aging Cell 2013;12:574–583.

[20] Hughes AL, Gottschling DE. An early age increase in vacuolar pH limits mitochondrial function and lifespan in yeast. Nature 2012;492:261–265.

[21] Narita M, Young AR, Arakawa S, Samarajiwa SA, Nakashima T, et al. Spatial coupling of mTOR and autophagy augments secretory phenotypes. Science 2011;332:966–970.

[22] Terman A, Dalen H, Eaton JW, Neuzil J, Brunk UT. Mitochondrial recycling and aging of cardiac myocytes: the role of autophagocytosis. Exp Gerontol 2003;38:863–876.

[23] Herrup K. The contributions of unscheduled neuronal cell cycle events to the death of neurons in Alzheimer's disease. Front Biosci (Elite Ed) 2012;4:2101–2109.

[24] Terman A. Garbage catastrophe theory of aging: imperfect removal of oxidative damage? Redox Rep 2001;6:15–26.

[25] Sohal RS, Mockett RJ, Orr WC. Mechanisms of aging: an appraisal of the oxidative stress hypothesis. Free Radic Biol Med 2002;33:575–586.

[26] Levine B, Klionsky DJ. Development by self-digestion; molecular mechanisms and biological functions of autophagy. Dev Cell 2004;6:463–477.

[27] Roy AK, Oh T, Rivera O, Mubiru J, Song CS, et al. Impacts of transcriptional regulation on aging and senescence. Ageing Res Rev 2002;1:367–380.

[28] Gafni A. Structural modifications of proteins during aging. J Am Geriatr Soc 1997; 45:871–880.

[29] Sukharev SA, Pleshakova OV, Moshnikova AB, Sadovnikov VB, Gaziev AI. Age- and radiation-dependent changes in carbonyl content, susceptibility to proteolysis, and antigenicity of soluble rat liver proteins. Comp Biochem Physiol B Biochem Mol Biol 1997;116:333–338.

[30] Soto C. Unfolding the role of protein misfolding in neurodegenerative diseases. Nat Rev Neurosci 2003;4:49–60.

[31] Grune T, Shringarpure R, Sitte N, Davies K. Age-related changes in protein oxidation and proteolysis in mammalian cells. J Gerontol A Biol Sci Med Sci 2001;56:B459–467.

[32] Goldberg AL. Protein degradation and protection against misfolded or damaged proteins. Nature 2003;426:895–899.

[33] Berke SJ, Paulson HL. Protein aggregation and the ubiquitin proteasome pathway: gaining the UPPer hand on neurodegeneration. Curr Opin Genet Dev 2003;13:253–261.

[34] Dikic I, Johansen T, Kirkin V. Selective autophagy in cancer development and therapy. Cancer Res 2010;70:3431–3434.

[35] Fortun J, Dunn WA Jr, Joy S, Li J, Notterpek L. Emerging role for autophagy in the removal of aggresomes in Schwann cells. J Neurosci 2003;23:10672–10680.

[36] Sitte N, Huber M, Grune T, Ladhoff A, Doecke WD, et al. Proteasome inhibition by lipofuscin/ceroid during postmitotic aging of fibroblasts. FASEB J 2000;14:1490–1498.

[37] DeMartino GN, Slaughter CA. The proteasome, a novel protease regulated by multiple mechanisms. J Biol Chem 1999;274:22123–22126.

[38] Shibatani T, Ward WF. Effect of age and food restriction on alkaline protease activity in rat liver. J Gerontol A Biol Sci Med Sci 1996;51:B175–178.

[39] Shang F, Gong X, Palmer HJ, Nowell TR Jr, Taylor A. Age-related decline in ubiquitin conjugation in response to oxidative stress in the lens. Exp Eye Res 1997;64:21–30.

[40] Carafoli E, Molinari M. Calpain: a protease in search of a function? Biochem Biophys Res Commun 1998;247:193–203.

[41] Glaser T, Schwarz-Benmeir N, Barnoy S, Barak S, Eshhar Z, et al. Calpain (Ca(2+)-dependent thiol protease) in erythrocytes of young and old individuals. Proc Natl Acad Sci U S A 1994;91:7879–7883.

[42] Saito K, Elce JS, Hamos JE, Nixon RA. Widespread activation of calcium-activated neutral proteinase (calpain) in the brain in Alzheimer disease: a potential molecular basis for neuronal degeneration. Proc Natl Acad Sci U S A 1993;90:2628–2632.

[43] Jolly RD, Dalefield RR, Palmer DN. Ceroid, lipofuscin and the ceroid-lipofuscinoses (Batten disease). J Inherit Metab Dis 1993;16:280–283.

[44] Terman A, Brunk U. Lipofuscin. Int J Biochem Cell Biol 2004;36:1400–1404.

[45] Strehler BL, Mark DD, Mildvan AS. GEE MV: rate and magnitude of age pigment accumulation in the human myocardium. J Gerontol 1959;14:430–439.

[46] Sheehy MR. Quantitative comparison of in situ lipofuscin concentration with soluble autofluorescence intensity in the crustacean brain. Exp Gerontol 1996;31:421–432.

[47] Brunk UT, Terman A. Lipofuscin: mechanisms of age-related accumulation and influence on cell function. Free Radic Biol Med 2002;33:611–619.

[48] Seehafer SS, Pearce DA. You say lipofuscin, we say ceroid: defining autofluorescent storage material. Neurobiol Aging 2006;27:576–588.

[49] Bednarski E, Ribak CE, Lynch G. Suppression of cathepsins B and L causes a proliferation of lysosomes and the formation of meganeurites in hippocampus. J Neurosci 1997;17:4006–4021.

[50] Terman A, Brunk UT. Ceroid/lipofuscin formation in cultured human fibroblasts: the role of oxidative stress and lysosomal proteolysis. Mech Ageing Dev 1998; 104:277–291.

[51] Ivy GO, Schottler F, Wenzel J, Baudry M, Lynch G. Inhibitors of lysosomal enzymes: accumulation of lipofuscin-like dense bodies in the brain. Science 1984;226:985–987.

[52] Amano T, Nakanishi H, Kondo T, Tanaka T, Oka M, et al. Age-related changes in cellular localization and enzymatic activities of cathepsins B, L and D in the rat trigeminal ganglion neuron. Mech Ageing Dev 1995;83:133–141.

[53] Keppler D, Walter R, Perez C, Sierra F. Increased expression of mature cathepsin B in aging rat liver. Cell Tissue Res 2000;302:181–188.

[54] Nakanishi H, Tominaga K, Amano T, Hirotsu I, Inoue T, et al. Age-related changes in activities and localizations of cathepsins D, E, B, and L in the rat brain tissues. Experimental Neurology 1994;126:119–128.

[55] Porta E, Llesuy S, Monserrat AJ, Benavides S, Travacio M. Changes in cathepsin B and lipofuscin during development and aging in rat brain and heart. Gerontology 1995;41(Suppl 2):81–93.

[56] Xue X, Wang LR, Sato Y, Jiang Y, Berg M, et al. Single-walled carbon nanotubes alleviate autophagic/lysosomal defects in primary glia from a mouse model of Alzheimer's disease. Nano Lett 2014;14:5110–5117.

[57] Kay P, Yang YC, Hiscott P, Gray D, Maminishkis A, et al. Age-related changes of cystatin C expression and polarized secretion by retinal pigment epithelium: potential age-related macular degeneration links. Invest Ophthalmol Vis Sci 2014;55:926–934.

[58] Nakamura Y, Takeda M, Suzuki H, Morita H, Tada K, et al. Lysosome instability in aged rat brain. Neurosci Lett 1989;97:215–220.

[59] Cuervo AM, Dice JF. Age-related decline in chaperone-mediated autophagy. J Biol Chem 2000;275:31505–31513.

[60] Kiffin R, Kaushik S, Zeng M, Bandyopadhyay U, Zhang C, et al. Altered dynamics of the lysosomal receptor for chaperone-mediated autophagy with age. J Cell Sci 2007;120:782–791.

[61] Kaushik S, Massey A, Cuervo AM. Lysosome membrane lipid microdomains: novel regulators of chaperone-mediated autophagy. EMBO J 2006;25:3921–3933.

[62] Rodriguez-Navarro JA, Kaushik S, Koga H, Dall'Armi C, Shui G, et al. Inhibitory effect of dietary lipids on chaperone-mediated autophagy. Proc Natl Acad Sci U S A 2012;109:E705–714.

[63] Massey AC, Kaushik S, Sovak G, Kiffin R, Cuervo AM. Consequences of the selective blockage of chaperone-mediated autophagy. Proc Nat Acad Sci 2006;103:5805–5810.

[64] Zhang C, Cuervo AM. Restoration of chaperone-mediated autophagy in aging liver improves cellular maintenance and hepatic function. Nat Med 2008;14:959–965.

[65] Sukhova GK, Shi GP, Simon DI, Chapman HA, Libby P. Expression of the elastolytic cathepsins S and K in human atheroma and regulation of their production in smooth muscle cells. J Clin Invest 1998;102:576–583.

[66] Liu J, Sukhova GK, Yang JT, Sun J, Ma L, et al. Cathepsin L expression and regulation in human abdominal aortic aneurysm, atherosclerosis, and vascular cells. Atherosclerosis 2006;184:302–311.

[67] Li W, Yuan XM. Increased expression and translocation of lysosomal cathepsins contribute to macrophage apoptosis in atherogenesis. Ann N Y Acad Sci 2004; 1030:427–433.

[68] Cheng XW, Kuzuya M, Nakamura K, Di Q, Liu Z, et al. Localization of cysteine protease, cathepsin S, to the surface of vascular smooth muscle cells by association with integrin alphanubeta3. Am J Pathol 2006;168:685–694.

[69] Rodgers KJ, Watkins DJ, Miller AL, Chan PY, Karanam S, et al. Destabilizing role of cathepsin S in murine atherosclerotic plaques. Arterioscler Thromb Vasc Biol 2006;26:851–856.

[70] Shi GP, Sukhova GK, Grubb A, Ducharme A, Rhode LH, et al. Cystatin C deficiency in human atherosclerosis and aortic aneurysms. J Clin Invest 1999;104:1191–1197.

[71] Bengtsson E, To F, Grubb A, Hakansson K, Wittgren L, et al. Absence of the protease inhibitor cystatin C in inflammatory cells results in larger plaque area in plaque regression of apoE-deficient mice. Atherosclerosis 2005;180:45–53.

[72] Yuan XM, Li W, Brunk UT, Dalen H, Chang YH, et al. Lysosomal destabilization during macrophage damage induced by cholesterol oxidation products. Free Radic Biol Med 2000;28:208–218.

[73] Appelqvist H, Waster P, Kagedal K, Ollinger K. The lysosome: from waste bag to potential therapeutic target. J Mol Cell Biol 2013;5:214–226.

[74] Paraoan L, Hiscott P, Gosden C, Grierson I. Cystatin C in macular and neuronal degenerations: implications for mechanism(s) of age-related macular degeneration. Vision Res 2010;50:737–742.

[75] Tatton W, Chen D, Chalmers-Redman R, Wheeler L, Nixon R, et al. Hypothesis for a common basis for neuroprotection in glaucoma and Alzheimer's disease: anti-apoptosis by alpha-2-adrenergic receptor activation. Surv Ophthalmol 2003;48(Suppl 1):25–37.

[76] Ohno-Matsui K. Parallel findings in age-related macular degeneration and Alzheimer's disease. Prog Retin Eye Res 2011;30:217–238.

[77] Claudepierre T, Paques M, Simonutti M, Buard I, Sahel J, et al. Lack of Niemann–Pick type C1 induces age-related degeneration in the mouse retina. Mol Cell Neurosci 2010; 43:164–176.

[78] Sparrow JR, Boulton M. RPE lipofuscin and its role in retinal pathobiology. Exp Eye Res 2005;80:595–606.

[79] Ng KP, Gugiu B, Renganathan K, Davies MW, Gu X, et al. Retinal pigment epithelium lipofuscin proteomics. Mol Cell Proteomics 2008;7:1397–1405.

[80] Wu Y, Zhou J, Fishkin N, Rittmann BE, Sparrow JR. Enzymatic degradation of A2E, a retinal pigment epithelial lipofuscin bisretinoid. J Am Chem Soc 2011;133: 849–857.

[81] Sparrow JR, Kim SR, Cuervo AM, Bandhyopadhyayand U. A2E, a pigment of RPE lipofuscin, is generated from the precursor, A2PE by a lysosomal enzyme activity. Adv Exp Med Biol 2008;613:393–398.

[82] Schutt F, Davies S, Kopitz J, Holz FG, Boulton ME. Photodamage to human RPE cells by A2-E, a retinoid component of lipofuscin. Invest Ophthalmol Vis Sci 2000; 41:2303–2308.

[83] Sparrow JR, Cai B. Blue light-induced apoptosis of A2E-containing RPE: involvement of caspase-3 and protection by Bcl-2. Invest Ophthalmol Vis Sci 2001;42:1356–1362.

[84] Holz FG, Bellman C, Staudt S, Schutt F, Volcker HE. Fundus autofluorescence and development of geographic atrophy in age-related macular degeneration. Invest Ophthalmol Vis Sci 2001;42:1051–1056.

[85] Allikmets R, Singh N, Sun H, Shroyer NF, Hutchinson A, et al. A photoreceptor cell-specific ATP-binding transporter gene (ABCR) is mutated in recessive Stargardt macular dystrophy. Nat Genet 1997;15:236–246.

[86] Maeda A, Maeda T, Golczak M, Palczewski K. Retinopathy in mice induced by disrupted all-trans-retinal clearance. J Biol Chem 2008;283:26684–26693.

[87] Nociari MM, Lehmann GL, Perez Bay AE, Radu RA, Jiang Z, et al. Beta cyclodextrins bind, stabilize, and remove lipofuscin bisretinoids from retinal pigment epithelium. Proc Natl Acad Sci U S A 2014;111:E1402–1408.

[88] He X, Hahn P, Iacovelli J, Wong R, King C, et al. Iron homeostasis and toxicity in retinal degeneration. Prog Retin Eye Res 2007;26:649–673.

[89] Wong RW, Richa DC, Hahn P, Green WR, Dunaief JL. Iron toxicity as a potential factor in AMD. Retina 2007;27:997–1003.

[90] Chen H, Liu B, Lukas TJ, Suyeoka G, Wu G, et al. Changes in iron-regulatory proteins in the aged rodent neural retina. Neurobiol Aging 2009;30:1865–1876.

[91] Hahn P, Ying GS, Beard J, Dunaief JL. Iron levels in human retina: sex difference and increase with age. Neuroreport 2006;17:1803–1806.

[92] Sergeant C, Vesvres MH, Pravikoff MS, Deves G, Yefimova M, et al. Trace metals in human retina during aging period. Metal Ion Biol Med 2004;8:491–495.

[93] Hahn P, Milam AH, Dunaief JL. Maculas affected by age-related macular degeneration contain increased chelatable iron in the retinal pigment epithelium and Bruch's membrane. Arch Ophthalmol 2003;121:1099–1105.

[94] Dentchev T, Hahn P, Dunaief JL. Strong labeling for iron and the iron-handling proteins ferritin and ferroportin in the photoreceptor layer in age-related macular degeneration. Arch Ophthalmol 2005;123:1745–1746.

[95] Hadziahmetovic M, Dentchev T, Song Y, Haddad N, He X, et al. Ceruloplasmin/hephaestin knockout mice model morphologic and molecular features of AMD. Invest Ophthalmol Vis Sci 2008;49:2728–2736.

[96] Hahn P, Qian Y, Dentchev T, Chen L, Beard J, et al. Disruption of ceruloplasmin and hephaestin in mice causes retinal iron overload and retinal degeneration with features of age-related macular degeneration. Proc Natl Acad Sci U S A 2004;101:13850–13855.

[97] Chen H, Lukas TJ, Du N, Suyeoka G, Neufeld AH. Dysfunction of the retinal pigment epithelium with age: increased iron decreases phagocytosis and lysosomal activity. Invest Ophthalmol Vis Sci 2009;50:1895–1902.

[98] Kurz T, Eaton JW, Brunk UT. Redox activity within the lysosomal compartment: implications for aging and apoptosis. Antioxid Redox Signal 2010;13:511–523.

[99] Krohne TU, Kaemmerer E, Holz FG, Kopitz J. Lipid peroxidation products reduce lysosomal protease activities in human retinal pigment epithelial cells via two different mechanisms of action. Exp Eye Res 2010;90:261–266.

[100] Bergmann M, Schutt F, Holz FG, Kopitz J. Inhibition of the ATP-driven proton pump in RPE lysosomes by the major lipofuscin fluorophore A2-E may contribute to the pathogenesis of age-related macular degeneration. FASEB J 2004;18:562–564.

[101] Guha S, Coffey EE, Lu W, Lim JC, Beckel JM, et al. Approaches for detecting lysosomal alkalinization and impaired degradation in fresh and cultured RPE cells: evidence for a role in retinal degenerations. Exp Eye Res 2014;126:68–76.

[102] Schutt F, Bergmann M, Holz FG, Kopitz J. Isolation of intact lysosomes from human RPE cells and effects of A2-E on the integrity of the lysosomal and other cellular membranes. Graefes Arch Clin Exp Ophthalmol 2002;240:983–988.

[103] Tseng WA, Thein T, Kinnunen K, Lashkari K, Gregory MS, et al. NLRP3 inflammasome activation in retinal pigment epithelial cells by lysosomal destabilization: implications for age-related macular degeneration. Invest Ophthalmol Vis Sci 2013;54:110–120.

[104] Nixon RA, Wegiel J, Kumar A, Yu WH, Peterhoff C, et al. Extensive involvement of autophagy in Alzheimer disease: an immuno-electron microscopy study. J Neuropathol Exp Neurol 2005;64:113–122.

[105] Cataldo AM, Barnett JL, Pieroni C, Nixon RA. Increased neuronal endocytosis and protease delivery to early endosomes in sporadic Alzheimer's disease: neuropathologic evidence for a mechanism of increased beta-amyloidogenesis. J Neurosci 1997; 17:6142–6151.

[106] Anglade P, Vyas S, Javoy-Agid F, Herrero MT, Michel PP, et al. Apoptosis and autophagy in nigral neurons of patients with Parkinson's disease. Histol Histopathol 1997;12:25–31.

[107] Nakano I, Shibata T, Uesaka Y. On the possibility of autolysosomal processing of skein-like inclusions. Electron microscopic observation in a case of amyotrophic lateral sclerosis. J Neurol Sci 1993;120:54–59.

[108] Sasaki S. Autophagy in spinal cord motor neurons in sporadic amyotrophic lateral sclerosis. J Neuropathol Exp Neurol 2011;70:349–359.

[109] Rudnicki DD, Pletnikova O, Vonsattel JP, Ross CA, Margolis RL. A comparison of huntington disease and huntington disease-like 2 neuropathology. J Neuropathol Exp Neurol 2008;67:366–374.

[110] Harris H, Rubinsztein DC. Control of autophagy as a therapy for neurodegenerative disease. Nat Rev Neurol 2011;8:108–117.

[111] Cuervo AM, Wong E. Chaperone-mediated autophagy: roles in disease and aging. Cell Res 2014;24:92–104.

[112] Nixon RA, Yang D. Autophagy and neuronal cell death in neurological disorders. Cold Spring Harb Perspect Biol 2012;4:a008839.

[113] Boya P. Lysosomal function and dysfunction: mechanism and disease. Antioxid Redox Signal 2012;17:766–774.

[114] Nixon RA, Yang DS, Lee JH. Neurodegenerative lysosomal disorders: a continuum from development to late age. Autophagy 2008;4:590–599.

[115] Ohmi K, Kudo LC, Ryazantsev S, Zhao HZ, Karsten SL, et al. Sanfilippo syndrome type B, a lysosomal storage disease, is also a tauopathy. Proc Natl Acad Sci U S A 2009; 106:8332–8337.

[116] Ryazantsev S, Yu WH, Zhao HZ, Neufeld EF, Ohmi K. Lysosomal accumulation of SCMAS (subunit c of mitochondrial ATP synthase) in neurons of the mouse model of mucopolysaccharidosis III B. Mol Genet Metab 2007;90:393–401.

[117] Chevrier M, Brakch N, Lesueur C, Genty D, Ramdani Y, et al. Autophagosome maturation is impaired in Fabry disease. Autophagy 2010;6:589–599.

[118] Fukuda T, Ewan L, Bauer M, Mattaliano RJ, Zaal K, et al. Dysfunction of endocytic and autophagic pathways in a lysosomal storage disease. Ann Neurol 2006;59:700–708.

[119] Koike M, Shibata M, Waguri S, Yoshimura K, Tanida I, et al. Participation of autophagy in storage of lysosomes in neurons from mouse models of neuronal ceroid-lipofuscinoses (Batten disease). Am J Pathol 2005;167:1713–1728.

[120] Settembre C, Fraldi A, Rubinsztein DC, Ballabio A. Lysosomal storage diseases as disorders of autophagy. Autophagy 2008;4:113–114.

[121] Nixon RA. Niemann–Pick Type C disease and Alzheimer's disease: the APP-endosome connection fattens up. Am J Pathol 2004;164:757–761.

[122] Nixon RA, Yang DS. Autophagy failure in Alzheimer's disease-locating the primary defect. Neurobiol Dis 2011;43:38–45.

[123] Nixon RA. Autophagy, amyloidogenesis and Alzheimer disease. J Cell Sci 2007; 120:4081–4091.

[124] Sun B, Zhou Y, Halabisky B, Lo I, Cho SH, et al. Cystatin C-cathepsin B axis regulates amyloid beta levels and associated neuronal deficits in an animal model of Alzheimer's disease. Neuron 2008;60:247–257.

[125] Butler D, Hwang J, Estick C, Nishiyama A, Kumar SS, et al. Protective effects of positive lysosomal modulation in Alzheimer's disease transgenic mouse models. PLoS One 2011;6:e20501.

[126] Yang DS, Stavrides P, Mohan PS, Kaushik S, Kumar A, et al. Reversal of autophagy dysfunction in the TgCRND8 mouse model of Alzheimer's disease ameliorates amyloid pathologies and memory deficits. Brain 2011;134:258–277.

[127] Majumder S, Richardson A, Strong R, Oddo S. Inducing autophagy by rapamycin before, but not after, the formation of plaques and tangles ameliorates cognitive deficits. PLoS One 2011;6:e25416.

[128] Spilman P, Podlutskaya N, Hart MJ, Debnath J, Gorostiza O, et al. Inhibition of mTOR by rapamycin abolishes cognitive deficits and reduces amyloid-beta levels in a mouse model of Alzheimer's disease. PLoS One 2010;5:e9979.

[129] Greco SJ, Bryan KJ, Sarkar S, Zhu X, Smith MA, et al. Leptin reduces pathology and improves memory in a transgenic mouse model of Alzheimer's disease. J Alzheimers Dis 2010;19:1155–1167.

[130] Mueller-Steiner S, Zhou Y, Arai H, Roberson ED, Sun B, et al. Antiamyloidogenic and neuroprotective functions of cathepsin B: implications for Alzheimer's disease. Neuron 2006;51:703–714.

[131] Sherrington R, Rogaev EI, Liang Y, Rogaeva EA, Levesque G, et al. Cloning of a gene bearing missense mutations in early-onset familial Alzheimer's disease. Nature 1995;375:754–760.

[132] Citron M, Westaway D, Xia W, Carlson G, Diehl T, et al. Mutant presenilins of Alzheimer's disease increase production of 42- residue amyloid beta-protein in both transfected cells and transgenic mice. Nat Med 1997;3:67–72.

[133] De Strooper B, Saftig P, Craessaerts K, Vanderstichele H, Guhde G, et al. Deficiency of presenilin-1 inhibits the normal cleavage of amyloid precursor protein. Nature 1998;391:387–390.

[134] Cai H, Wang Y, McCarthy D, Wen H, Borchelt DR, et al. BACE1 is the major beta-secretase for generation of Abeta peptides by neurons. Nat Neurosci 2001;4:233–234.

[135] Shioi J, Georgakopoulos A, Mehta P, Kouchi Z, Litterst CM, et al. FAD mutants unable to increase neurotoxic Abeta 42 suggest that mutation effects on neurodegeneration may be independent of effects on Abeta. J Neurochem 2007;101:674–681.

[136] Bentahir M, Nyabi O, Verhamme J, Tolia A, Horre K, et al. Presenilin clinical mutations can affect γ-secretase activity by different mechanisms. Journal of Neurochemistry 2006;96:732–742.

[137] Kumar-Singh S, Theuns J, Van Broeck B, Pirici D, Vennekens K, et al. Mean age-of-onset of familial Alzheimer disease caused by presenilin mutations correlates with both increased Abeta42 and decreased Abeta40. Hum Mutat 2006;27:686–695.

[138] Steiner H, Duff K, Capell A, Romig H, Grim MG, et al. A Loss of function mutation of presenilin-2 interferes with amyloid beta-peptide production and notch signaling. J Biol Chem 1999;274:28669–28673.

[139] Pimplikar SW, Nixon RA, Robakis NK, Shen J, Tsai LH. Amyloid-independent mechanisms in Alzheimer's disease pathogenesis. J Neurosci 2010;30:14946–14954.

[140] Shen J, Kelleher RJ 3rd. The presenilin hypothesis of Alzheimer's disease: evidence for a loss-of-function pathogenic mechanism. Proc Natl Acad Sci U S A 2007;104:403–409.

[141] Lee JH, Yu WH, Kumar A, Lee S, Mohan PS, et al. Lysosomal proteolysis and autophagy require presenilin 1 and are disrupted by Alzheimer-related PS1 mutations. Cell 2010;141:1146–1158.

[142] Kang DE, Soriano S, Frosch MP, Collins T, Naruse S, et al. Presenilin 1 facilitates the constitutive turnover of beta-catenin: differential activity of Alzheimer's disease-linked PS1 mutants in the beta-catenin-signaling pathway. J Neurosci 1999;19:4229–4237.

[143] Tu H, Nelson O, Bezprozvanny A, Wang Z, Lee SF, et al. Presenilins form ER Ca2+ leak channels, a function disrupted by familial Alzheimer's disease-linked mutations. Cell 2006;126:981–993.

[144] Zampese E, Fasolato C, Kipanyula MJ, Bortolozzi M, Pozzan T, et al. Presenilin 2 modulates endoplasmic reticulum (ER)-mitochondria interactions and Ca2+ cross-talk. Proc Natl Acad Sci U S A 2011;108:2777–2782.

[145] Shilling D, Mak DO, Kang DE, Foskett JK. Lack of evidence for presenilins as endoplasmic reticulum Ca2+ leak channels. J Biol Chem 2012;287:10933–10944.

[146] Cataldo AM, Peterhoff CM, Schmidt SD, Terio NB, Duff K, et al. Presenilin mutations in familial Alzheimer disease and transgenic mouse models accelerate neuronal lysosomal pathology. J Neuropathol Exp Neurol 2004;63:821–830.

[147] Esselens C, Oorschot V, Baert V, Raemaekers T, Spittaels K, et al. Presenilin 1 mediates the turnover of telencephalin in hippocampal neurons via an autophagic degradative pathway. J Cell Biol 2004;166:1041–1054.

[148] Wilson CA, Murphy DD, Giasson BI, Zhang B, Trojanowski JQ, et al. Degradative organelles containing mislocalized α- and β-synuclein proliferate in presenilin-1 null neurons. J Cell Biol 2004;165:335–346.

[149] Coen K, Flannagan R, Baron S, Carraro-Lacroix L, Wang D, et al. Lysosomal calcium homeostasis defects, not proton pump defects, cause endo-lysosomal dysfunction in PSEN-deficient cells. J Cell Biol 2012;198:23–35.

[150] Lee J-H, Wolfe D, McBrayer M, Haslett L, Kumar A, et al. *Presenilin Maintains Lysosomal Calcium Homeostasis by Regulating v-ATPase-Mediated Lysosomal Acidification.* San Diego, CA, USA: Society for Neuroscience; 2013.

[151] Wolfe DM, Lee JH, Kumar A, Lee S, Orenstein SJ, et al. Autophagy failure in Alzheimer's disease and the role of defective lysosomal acidification. Eur J Neurosci 2013;37:1949–1961.

[152] Dobrowolski R, Vick P, Ploper D, Gumper I, Snitkin H, et al. Presenilin deficiency or lysosomal inhibition enhances Wnt signaling through relocalization of GSK3 to the late-endosomal compartment. Cell Rep 2012;2:1316–1328.

[153] Torres M, Jimenez S, Sanchez-Varo R, Navarro V, Trujillo-Estrada L, et al. Defective lysosomal proteolysis and axonal transport are early pathogenic events that worsen with age leading to increased APP metabolism and synaptic Abeta in transgenic APP/PS1 hippocampus. Mol Neurodegener 2012;7:59.

[154] Avrahami L, Farfara D, Shaham-Kol M, Vassar R, Frenkel D, et al. Inhibition of GSK-3 ameliorates beta-amyloid(A-beta) pathology and restores lysosomal acidification and mTOR activity in the Alzheimer's disease mouse model: *in vivo* and *in vitro* studies. J Biol Chem 2013;288:1295–1306.

[155] Sawa A. Neuronal cell death in Down's syndrome. J Neural Transm Suppl 1999;57: 87–97.

[156] LaFerla FM. Calcium dyshomeostasis and intracellular signalling in Alzheimer's disease. Nat Rev Neurosci 2002;3:862–872.

[157] Mattson MP. ER calcium and Alzheimer's disease: in a state of flux. Sci Signal 2010;3:pe10.

[158] Stutzmann GE, Caccamo A, LaFerla FM, Parker I. Dysregulated IP3 signaling in cortical neurons of knock-in mice expressing an Alzheimer's-linked mutation in presenilin1 results in exaggerated Ca2+ signals and altered membrane excitability. J Neurosci 2004;24:508–513.

[159] Alvarez J, Montero M, Garcia-Sancho J. Subcellular Ca(2+) dynamics. News Physiol Sci 1999;14:161–168.

[160] Lloyd-Evans E, Platt FM. Lysosomal Ca(2+) homeostasis: role in pathogenesis of lysosomal storage diseases. Cell Calcium 2011;50:200–205.

[161] Neely Kayala KM, Dickinson GD, Minassian A, Walls KC, Green KN, et al. Presenilin-null cells have altered two-pore calcium channel expression and lysosomal calcium: implications for lysosomal function. Brain Res 2012;1489:8–16.

[162] Rao MV, McBrayer MK, Campbell J, Kumar A, Hashim A, et al. Specific calpain inhibition by calpastatin prevents tauopathy and neurodegeneration and restores normal lifespan in tau P301L mice. J Neurosci 2014;34:9222–9234.

[163] Lee S, Sato Y, Nixon RA. Lysosomal proteolysis inhibition selectively disrupts axonal transport of degradative organelles and causes an Alzheimer's-like axonal dystrophy. J Neurosci 2011;31:7817–7830.

[164] Chishti MA, Yang DS, Janus C, Phinney AL, Horne P, et al. Early-onset amyloid deposition and cognitive deficits in transgenic mice expressing a double mutant form of amyloid precursor protein 695. J Biol Chem 2001;276:21562–21570.

[165] Cataldo AM, Mathews PM, Boiteau AB, Hassinger LC, Peterhoff CM, et al. Down syndrome fibroblast model of Alzheimer-related endosome pathology. Accelerated endocytosis promotes late endocytic defects. Am J Pathol 2008;173:370–384.

[166] Cataldo AM, Peterhoff CM, Troncoso JC, Gomez-Isla T, Hyman BT, et al. Endocytic pathway abnormalities precede amyloid beta deposition in sporadic Alzheimer's disease and Down syndrome: differential effects of APOE genotype and presenilin mutations. Am J Pathol 2000;157:277–286.

[167] Ginsberg SD, Alldred MJ, Counts SE, Cataldo AM, Neve RL, et al. Microarray analysis of hippocampal CA1 neurons implicates early endosomal dysfunction during Alzheimer's disease progression. Biol Psychiatry 2010;68:885–893.

[168] Jiang Y, Mullaney KA, Peterhoff CM, Che S, Schmidt SD, et al. Alzheimer's-related endosome dysfunction in Down syndrome is Abeta-independent but requires APP and is reversed by BACE-1 inhibition. Proc Natl Acad Sci U S A 2010;107:1630–1635.

[169] Cataldo AM, Petanceska S, Peterhoff CM, Terio NB, Epstein CJ, et al. App gene dosage modulates endosomal abnormalities of Alzheimer's disease in a segmental trisomy 16 mouse model of down syndrome. J Neurosci 2003;23:6788–6792.

[170] Salehi A, Delcroix JD, Belichenko PV, Zhan K, Wu C, et al. Increased App expression in a mouse model of Down's syndrome disrupts NGF transport and causes cholinergic neuron degeneration. Neuron 2006;51:29–42.

[171] Laifenfeld D, Patzek LJ, McPhie DL, Chen Y, Levites Y, et al. Rab5 mediates an amyloid precursor protein signaling pathway that leads to apoptosis. J Neurosci 2007;27:7141–7153.

[172] Cataldo A, Rebeck GW, Ghetri B, Hulette C, Lippa C, et al. Endocytic disturbances distinguish among subtypes of Alzheimer's disease and related disorders. Ann Neurol 2001;50:661–665.

[173] Cossec JC, Simon A, Marquer C, Moldrich RX, Leterrier C, et al. Clathrin-dependent APP endocytosis and Abeta secretion are highly sensitive to the level of plasma membrane cholesterol. Biochim Biophys Acta 2010;1801:846–852.

[174] Ji ZS, Mullendorff K, Cheng IH, Miranda RD, Huang Y, et al. Reactivity of apolipoprotein E4 and amyloid beta peptide: lysosomal stability and neurodegeneration. J Biol Chem 2006;281:2683–2692.

[175] Belinson H, Lev D, Masliah E, Michaelson DM. Activation of the amyloid cascade in apolipoprotein E4 transgenic mice induces lysosomal activation and neurodegeneration resulting in marked cognitive deficits. J Neurosci 2008;28:4690–4701.

[176] Karaca I, Tamboli IY, Glebov K, Richter J, Fell LH, et al. Deficiency of sphingosine-1-phosphate lyase impairs lysosomal metabolism of the amyloid precursor protein. J Biol Chem 2014;289:16761–16772.

[177] Kimura N, Yanagisawa K. Endosomal accumulation of GM1 ganglioside-bound amyloid beta-protein in neurons of aged monkey brains. Neuroreport 2007;18:1669–1673.

[178] Grimm MO, Rothhaar TL, Hartmann T. The role of APP proteolytic processing in lipid metabolism. Exp Brain Res 2012;217:365–375.

[179] Tamboli IY, Hampel H, Tien NT, Tolksdorf K, Breiden B, et al. Sphingolipid storage affects autophagic metabolism of the amyloid precursor protein and promotes Abeta generation. J Neurosci 2011;31:1837–1849.

[180] Yao J, Ho D, Calingasan NY, Pipalia NH, Lin MT, et al. Neuroprotection by cyclodextrin in cell and mouse models of Alzheimer disease. J Exp Med 2012;209:2501–2513.

[181] Tosto G, Reitz C. Genome-wide association studies in Alzheimer's disease: a review. Curr Neurol Neurosci Rep 2013;13:381.

[182] Chouraki V, Seshadri S. Genetics of Alzheimer's disease. Adv Genet 2014;87:245–294.

[183] Yu WH, Cuervo AM, Kumar A, Peterhoff CM, Schmidt SD, et al. Macroautophagy – a novel β-amyloid peptide-generating pathway activated in Alzheimer's disease. J Cell Biol 2005;171:87–98.

[184] Bahr BA, Bendiske J. The neuropathogenic contributions of lysosomal dysfunction. J Neurochem 2002;83:481–489.

[185] Florez-McClure ML, Hohsfield LA, Fonte G, Bealor MT, Link CD. Decreased insulin-receptor signaling promotes the autophagic degradation of beta-amyloid peptide in C. elegans. Autophagy 2007;3:569–580.

[186] Heinrich M, Wickel M, Schneider-Brachert W, Sandberg C. Gahr J, et al. Cathepsin D targeted by acid sphingomyelinase-derived ceramide EMBO J 1999;18:5252–5263.

[187] Eckman EA, Eckman CB. Abeta-degrading enzymes: modulators of Alzheimer's disease pathogenesis and targets for therapeutic intervention. Biochem Soc Trans 2005;33:1101–1105.

[188] Guenette SY. Mechanisms of Abeta clearance and catabolism. Neuromolecular Med 2003;4:147–160.

[189] LeBlanc AC, Goodyer CG. Role of endoplasmic reticulum, endosomal–lysosomal compartments, and microtubules in amyloid precursor protein metabolism of human neurons. J Neurochem 1999;72:1832–1842.

[190] Wang Y, Martinez-Vicente M, Kruger U, Kaushik S, Wong E, et al. Tau fragmentation, aggregation and clearance: the dual role of lysosomal processing. Hum Mol Genet 2009;18:4153–4170.

[191] Kurz T, Terman A, Gustafsson B, Brunk UT. Lysosomes and oxidative stress in aging and apoptosis. Biochim Biophys Acta 2008;1780:1291–1303.

[192] Ling D, Song HJ, Garza D, Neufeld TP, Salvaterra PM. Abeta42-induced neurodegeneration via an age-dependent autophagic-lysosomal injury in *Drosophila*. PLoS One 2009;4:e4201.

[193] Glabe C. Intracellular mechanisms of amyloid accumulation and pathogenesis in Alzheimer's disease. J Mol Neurosci 2001;17:137–145.

[194] Yang AJ, Chandswangbhuvana D, Margol L, Glabe CG. Loss of endosomal/lysosomal membrane impermeability is an early event in amyloid Abeta1-42 pathogenesis. J Neurosci Res 1998;52:691–698.

[195] Kroemer G, Martin SJ. Caspase-independent cell death. Nat Med 2005;11:725–730.

[196] Auluck PK, Caraveo G, Lindquist S. alpha-Synuclein: membrane interactions and toxicity in Parkinson's disease. Annu Rev Cell Dev Biol 2010;26:211–233.

[197] Bellani S, Sousa VL, Ronzitti G, Valtorta F, Meldolesi J, et al. The regulation of synaptic function by alpha-synuclein. Commun Integr Biol 2010;3:106–109.

[198] Polymeropoulos MH, Lavedan C, Leroy E, Ide SE, Dehejia A, et al. Mutation in the alpha-synuclein gene identified in families with Parkinson's disease. Science 1997; 276:2045–2047.

[199] Athanassiadou A, Voutsinas G, Psiouri L, Leroy E, Polymeropoulos MH, et al. Genetic analysis of families with Parkinson disease that carry the Ala53Thr mutation in the gene encoding alpha-synuclein. Am J Hum Genet 1999;65:555–558.

[200] Puschmann A, Wszolek ZK, Farrer M, Gustafson L, Widner H, et al. Alpha-synuclein multiplications with parkinsonism, dementia or progressive myoclonus? Parkinsonism Relat Disord 2009;15:390–392.

[201] Spira PJ, Sharpe DM, Halliday G, Cavanagh J, Nicholson GA. Clinical and pathological features of a Parkinsonian syndrome in a family with an Ala53Thr alpha-synuclein mutation. Ann Neurol 2001;49:313–319.

[202] Ki CS, Stavrou EF, Davanos N, Lee WY, Chung EJ, et al. The Ala53Thr mutation in the alpha-synuclein gene in a Korean family with Parkinson disease. Clin Genet 2007; 71:471–473.

[203] Choi JM, Woo MS, Ma HI, Kang SY, Sung YH, et al. Analysis of PARK genes in a Korean cohort of early-onset Parkinson disease. Neurogenetics 2008;9:263–269.

[204] Simon-Sanchez J, Schulte C, Bras JM, Sharma M, Gibbs JR, et al. Genome-wide association study reveals genetic risk underlying Parkinson's disease. Nat Genet 2009; 41:1308–1312.

[205] Singleton AB, Farrer M, Johnson J, Singleton A, Hague S, et al. alpha-Synuclein locus triplication causes Parkinson's disease. Science 2003;302:841.

[206] Chartier-Harlin MC, Kachergus J, Roumier C, Mouroux V, Douay X, et al. Alpha-synuclein locus duplication as a cause of familial Parkinson's disease. Lancet 2004; 364:1167–1169.

[207] Houlden H, Singleton AB. The genetics and neuropathology of Parkinson's disease. Acta Neuropathol 2012;124:325–338.

[208] Deas E, Wood NW, Plun-Favreau H. Mitophagy and Parkinson's disease: the PINK1-parkin link. Biochim Biophys Acta 2011;1813:623–633.

[209] Giasson BI, Duda JE, Murray IV, Chen Q, Souza JM, et al. Oxidative damage linked to neurodegeneration by selective alpha-synuclein nitration in synucleinopathy lesions 723. Science 2000;290:985–989.

[210] Fujiwara H, Hasegawa M, Dohmae N, Kawashima A, Masliah E, et al. alpha-Synuclein is phosphorylated in synucleinopathy lesions. Nat Cell Biol 2002;4:160–164.

[211] Tofaris GK, Razzaq A, Ghetti B, Lilley KS, Spillantini MG. Ubiquitination of alpha-synuclein in Lewy bodies is a pathological event not associated with impairment of proteasome function. J Biol Chem 2003;278:44405–44411.

[212] Anderson JP, Walker DE, Goldstein JM, de Laat R, Banducci K, et al. Phosphorylation of Ser-129 is the dominant pathological modification of alpha-synuclein in familial and sporadic Lewy body disease. J Biol Chem 2006;281:29739–29752.

[213] Dorval V, Fraser PE. Small ubiquitin-like modifier (SUMO) modification of natively unfolded proteins tau and alpha-synuclein. J Biol Chem 2006;281:9919–9924.

[214] Norris EH, Giasson BI, Ischiropoulos H, Lee VM. Effects of oxidative and nitrative challenges on alpha-synuclein fibrillogenesis involve distinct mechanisms of protein modifications. J Biol Chem 2003;278:27230–27240.

[215] Winner B, Jappelli R, Maji SK, Desplats PA, Boyer L, et al. *In vivo* demonstration that alpha-synuclein oligomers are toxic. Proc Natl Acad Sci U S A 2011;108:4194–4199.

[216] Cremades N, Cohen SI, Deas E, Abramov AY, Chen AY, et al. Direct observation of the interconversion of normal and toxic forms of alpha-synuclein. Cell 2012; 149:1048–1059.

[217] Hansen C, Angot E, Bergstrom AL, Steiner JA, Pieri L, et al. alpha-Synuclein propagates from mouse brain to grafted dopaminergic neurons and seeds aggregation in cultured human cells. J Clin Invest 2011;121:715–725.

[218] Sigurdsson EM. Tau immunotherapy and imaging. Neurodegener Dis 2014;12: 103–106.

[219] Lee HJ, Patel S, Lee SJ. Intravesicular localization and exocytosis of alpha-synuclein and its aggregates. J Neurosci 2005;25:6016–6024.

[220] Emmanouilidou E, Melachroinou K, Roumeliotis T, Garbis SD, Ntzouni M, et al. Cell-produced alpha-synuclein is secreted in a calcium-dependent manner by exosomes and impacts neuronal survival. J Neurosci 2010;30:6838–6851.

[221] Alvarez-Erviti L, Seow Y, Schapira AH, Gardiner C, Sargent IL, et al. Lysosomal dysfunction increases exosome-mediated alpha-synuclein release and transmission. Neurobiol Dis 2011;42:360–367.

[222] Ebrahimi-Fakhari D, Cantuti-Castelvetri I, Fan Z, Rockenstein E, Masliah E, et al. Distinct roles *in vivo* for the ubiquitin-proteasome system and the autophagy-lysosomal pathway in the degradation of alpha-synuclein. J Neurosci 2011;31:14508–14520.

[223] Spencer B, Potkar R, Trejo M, Rockenstein E, Patrick C, et al. Beclin 1 gene transfer activates autophagy and ameliorates the neurodegenerative pathology in alpha-synuclein models of Parkinson's and Lewy body diseases. J Neurosci 2009;29:13578–13588.

[224] Yu WH, Dorado B, Figueroa HY, Wang L, Planel E, et al. Metabolic activity determines efficacy of macroautophagic clearance of pathological oligomeric alpha-synuclein. Am J Pathol 2009;175:736–747.

[225] Stefanis L, Larsen KE, Rideout HJ, Sulzer D, Greene LA. Expression of A53T mutant but not wild-type alpha-synuclein in PC12 cells induces alterations of the ubiquitin-dependent degradation system, loss of dopamine release, and autophagic cell death. J Neurosci 2001;21:9549–9560.

[226] Cuervo AM, Stefanis L, Fredenburg R, Lansbury PT, Sulzer D. Impaired degradation of mutant α-synuclein by chaperone-mediated autophagy. Science 2004;305:1292–1295.

[227] Winder-Rhodes SE, Garcia-Reitbock P, Ban M, Evans JR, Jacques TS, et al. Genetic and pathological links between Parkinson's disease and the lysosomal disorder Sanfilippo syndrome. Mov Disord 2012;27:312–315.

[228] Decressac M, Mattsson B, Weikop P, Lundblad M, Jakobsson J, et al. TFEB-mediated autophagy rescues midbrain dopamine neurons from alpha-synuclein toxicity. Proc Natl Acad Sci U S A 2013;110:E1817–1826.

[229] Xilouri M, Brekk OR, Landeck N, Pitychoutis PM, Papasilekas T, et al. Boosting chaperone-mediated autophagy *in vivo* mitigates alpha-synuclein-induced neurodegeneration. Brain 2013;136:2130–2146.

[230] Westbroek W, Gustafson AM, Sidransky E. Exploring the link between glucocerebrosidase mutations and parkinsonism. Trends Mol Med 2011;17:485–493.

[231] Sidransky E, Lopez G. The link between the GBA gene and parkinsonism. Lancet Neurol 2012;11:986–998.

[232] Cullen V, Sardi SP, Ng J, Xu YH, Sun Y, et al. Acid beta-glucosidase mutants linked to Gaucher disease, Parkinson disease, and Lewy body dementia alter alpha-synuclein processing. Ann Neurol 2011;69:940–953.

[233] Mazzulli JR, Xu YH, Sun Y, Knight AL, McLean PJ, et al. Gaucher disease glucocerebrosidase and alpha-synuclein form a bidirectional pathogenic loop in synucleinopathies. Cell 2011;146:37–52.

[234] Yap TL, Gruschus JM, Velayati A, Westbroek W, Goldin E, et al. Alpha-synuclein interacts with glucocerebrosidase providing a molecular link between Parkinson and Gaucher diseases. J Biol Chem 2011;286:28080–28088.

[235] Sardi SP, Clarke J, Kinnecom C, Tamsett TJ, Li L, et al. CNS expression of glucocerebrosidase corrects alpha-synuclein pathology and memory in a mouse model of Gaucher-related synucleinopathy. Proc Natl Acad Sci U S A 2011;108:12101–12106.

[236] Orenstein SJ, Kuo SH, Tasset I, Arias E, Koga H, et al. Interplay of LRRK2 with chaperone-mediated autophagy. Nat Neurosci 2013;16:394–406.

[237] Mak SK, McCormack AL, Manning-Bog AB, Cuervo AM, Di Monte DA. Lysosomal degradation of alpha-synuclein *in vivo*. J Biol Chem 2010;285:13621–13629.

[238] Vogiatzi T, Xilouri M, Vekrellis K, Stefanis L. Wild type alpha-synuclein is degraded by chaperone-mediated autophagy and macroautophagy in neuronal cells. J Biol Chem 2008;283:23542–23556.

[239] Martinez-Vicente M, Talloczy Z, Kaushik S, Massey AC, Mazzulli J, et al. Dopamine-modified alpha-synuclein blocks chaperone-mediated autophagy. J Clin Invest 2008; 118:777–788.

[240] Kabuta T, Furuta A, Aoki S, Furuta K, Wada K. Aberrant interaction between Parkinson disease-associated mutant UCH-L1 and the lysosomal receptor for chaperone-mediated autophagy. J Biol Chem 2008;283:23731–23738.

[241] Xilouri M, Vogiatzi T, Vekrellis K, Park D, Stefanis L. Aberrant alpha-synuclein confers toxicity to neurons in part through inhibition of chaperone-mediated autophagy. PLoS One 2009;4:e5515.

[242] Oueslati A, Fournier M, Lashuel HA. Role of post-translational modifications in modulating the structure, function and toxicity of alpha-synuclein: implications for Parkinson's disease pathogenesis and therapies. Prog Brain Res 2010;183:115–145.

[243] Rudenko IN, Chia R, Cookson MR. Is inhibition of kinase activity the only therapeutic strategy for LRRK2-associated Parkinson's disease? BMC Med 2012;10:20.

[244] Greggio E, Jain S, Kingsbury A, Bandopadhyay R, Lewis P, et al. Kinase activity is required for the toxic effects of mutant LRRK2/dardarin. Neurobiol Dis 2006;23: 329–341.

[245] MacLeod D, Dowman J, Hammond R, Leete T, Inoue K, et al. The familial Parkinsonism gene LRRK2 regulates neurite process morphology. Neuron 2006;52:587–593.

[246] Lee BH, Lee MJ, Park S, Oh DC, Elsasser S, et al. Enhancement of proteasome activity by a small-molecule inhibitor of USP14. Nature 2010;467:179–184.

[247] West AB, Moore DJ, Choi C, Andrabi SA, Li X, et al. Parkinson's disease-associated mutations in LRRK2 link enhanced GTP-binding and kinase activities to neuronal toxicity. Hum Mol Genet 2007;16:223–232.

[248] Gidalevitz T, Ben-Zvi A, Ho KH, Brignull HR, Morimoto RI. Progressive disruption of cellular protein folding in models of polyglutamine diseases. Science 2006; 311:1471–1474.

[249] Skibinski G, Nakamura K, Cookson MR, Finkbeiner S. Mutant LRRK2 toxicity in neurons depends on LRRK2 levels and synuclein but not kinase activity or inclusion bodies. J Neurosci 2014;34:418–433.

[250] Wang L, Xie C, Greggio E, Parisiadou L, Shim H, et al. The chaperone activity of heat shock protein 90 is critical for maintaining the stability of leucine-rich repeat kinase 2. J Neurosci 2008;28:3384–3391.

[251] Ko HS, Bailey R, Smith WW, Liu Z, Shin JH, et al. CHIP regulates leucine-rich repeat kinase-2 ubiquitination, degradation, and toxicity. Proc Natl Acad Sci U S A 2009;106:2897–2902.

[252] Lin X, Parisiadou L, Gu XL, Wang L, Shim H, et al. Leucine-rich repeat kinase 2 regulates the progression of neuropathology induced by Parkinson's-disease-related mutant alpha-synuclein. Neuron 2009;64:807–827.

[253] Sanchez-Danes A, Richaud-Patin Y, Carballo-Carbajal I, Jimenez-Delgado S, Caig C, et al. Disease-specific phenotypes in dopamine neurons from human iPS-based models of genetic and sporadic Parkinson's disease. EMBO Mol Med 2012;4:380–395.

[254] Biskup S, West AB. Zeroing in on LRRK2-linked pathogenic mechanisms in Parkinson's disease. Biochim Biophys Acta 2009;1792:625–633.

[255] Botta-Orfila T, Ezquerra M, Pastor P, Fernandez-Santiago R, Pont-Sunyer C, et al. Age at onset in LRRK2-associated PD is modified by SNCA variants. J Mol Neurosci 2012;48:245–247.

[256] Snyder H, Mensah K, Theisler C, Lee J, Matouschek A, et al. Aggregated and monomeric alpha-synuclein bind to the S6' proteasomal protein and inhibit proteasomal function. J Biol Chem 2003;278:11753–11759.

[257] Fornai F, Schluter OM, Lenzi P, Gesi M, Ruffoli R, et al. Parkinson-like syndrome induced by continuous MPTP infusion: convergent roles of the ubiquitin-proteasome system and alpha-synuclein. Proc Natl Acad Sci U S A 2005;102:3413–3418.

[258] Greggio E, Bisaglia M, Civiero L, Bubacco L. Leucine-rich repeat kinase 2 and alpha-synuclein: intersecting pathways in the pathogenesis of Parkinson's disease? Mol Neurodegener 2011;6:6.

[259] Nakamura K, Nemani VM, Azarbal F, Skibinski G, Levy JM, et al. Direct membrane association drives mitochondrial fission by the Parkinson disease-associated protein alpha-synuclein. J Biol Chem 2011;286:20710–20726.

[260] Wang X, Yan MH, Fujioka H, Liu J, Wilson-Delfosse A, et al. LRRK2 regulates mitochondrial dynamics and function through direct interaction with DLP1. Hum Mol Genet 2012;21:1931–1944.

[261] Manzoni C. LRRK2 and autophagy: a common pathway for disease. Biochem Soc Trans 2012;40:1147–1151.

[262] Xiong Y, Coombes CE, Kilaru A, Li X, Gitler AD, et al. GTPase activity plays a key role in the pathobiology of LRRK2. PLoS Genet 2010;6:e1000902.

[263] Dodson MW, Zhang T, Jiang C, Chen S, Guo M. Roles of the *Drosophila* LRRK2 homolog in Rab7-dependent lysosomal positioning. Hum Mol Genet 2012;21:1350–1363.

[264] Beilina A, Rudenko IN, Kaganovich A, Civiero L, Chau H, et al. Unbiased screen for interactors of leucine-rich repeat kinase 2 supports a common pathway for sporadic and familial Parkinson disease. Proc Natl Acad Sci U S A 2014;111:2626–2631.

[265] MacLeod DA, Rhinn H, Kuwahara T, Zolin A, Di Paolo G, et al. RAB7L1 interacts with LRRK2 to modify intraneuronal protein sorting and Parkinson's disease risk. Neuron 2013;77:425–439.

[266] Gomez-Suaga P, Luzon-Toro B, Churamani D, Zhang L, Bloor-Young D, et al. Leucine-rich repeat kinase 2 regulates autophagy through a calcium-dependent pathway involving NAADP. Hum Mol Genet 2012;21:511–525.

[267] Tong Y, Giaime E, Yamaguchi H, Ichimura T, Liu Y, et al. Loss of leucine-rich repeat kinase 2 causes age-dependent bi-phasic alterations of the autophagy pathway. Mol Neurodegener 2012;7:2.

[268] Manzoni C, Mamais A, Dihanich S, McGoldrick P, Devine MJ, et al. Pathogenic Parkinson's disease mutations across the functional domains of LRRK2 alter the autophagic/lysosomal response to starvation. Biochem Biophys Res Commun 2013;441:862–866.

[269] Alegre-Abarrategui J, Christian H, Lufino MM, Mutihac R, Venda LL, et al. LRRK2 regulates autophagic activity and localizes to specific membrane microdomains in a novel human genomic reporter cellular model. Hum Mol Genet 2009;18:4022–4034.

[270] Ramirez A, Heimbach A, Grundemann J, Stiller B, Hampshire D, et al. Hereditary parkinsonism with dementia is caused by mutations in ATP13A2, encoding a lysosomal type 5 P-type ATPase. Nat Genet 2006;38:1184–1191.

[271] van Veen S, Sorensen DM, Holemans T, Holen HW, Palmgren MG, et al. Cellular function and pathological role of ATP13A2 and related P-type transport ATPases in Parkinson's disease and other neurological disorders. Front Mol Neurosci 2014;7:48.

[272] Park JS, Mehta P, Cooper AA, Veivers D, Heimbach A, et al. Pathogenic effects of novel mutations in the P-type ATPase ATP13A2 (PARK9) causing Kufor-Rakeb syndrome, a form of early-onset parkinsonism. Hum Mutat 2011;32:956–964.

[273] Podhajska A, Musso A, Trancikova A, Stafa K, Moser R, et al. Common pathogenic effects of missense mutations in the P-type ATPase ATP13A2 (PARK9) associated with early-onset parkinsonism. PLoS One 2012;7:e39942.

[274] Dehay B, Ramirez A, Martinez-Vicente M, Perier C, Canron MH, et al. Loss of P-type ATPase ATP13A2/PARK9 function induces general lysosomal deficiency and leads to Parkinson disease neurodegeneration. Proc Nat Acad Sci 2012;109:9611–9616.

[275] Usenovic M, Tresse E, Mazzulli JR, Taylor JP, Krainc D. Deficiency of ATP13A2 leads to lysosomal dysfunction, alpha-synuclein accumulation, and neurotoxicity. J Neurosci 2012;32:4240–4246.

[276] Yeger-Lotem E, Riva L, Su LJ, Gitler AD, Cashikar AG, et al. Bridging high-throughput genetic and transcriptional data reveals cellular responses to alpha-synuclein toxicity. Nat Genet 2009;41:316–323.

[277] Schultheis PJ, Fleming SM, Clippinger AK, Lewis J, Tsunemi T, et al. Atp13a2-deficient mice exhibit neuronal ceroid lipofuscinosis, limited alpha-synuclein accumulation and age-dependent sensorimotor deficits. Hum Mol Genet 2013;22:2067–2082.

[278] Matsui H, Uemura N, Yamakado H, Takeda S, Takahashi R. Exploring the pathogenetic mechanisms underlying Parkinson's disease in medaka fish. J Parkinsons Dis 2014;4:301–310.

[279] Farias FH, Zeng R, Johnson GS, Wininger FA, Taylor JF, et al. A truncating mutation in ATP13A2 is responsible for adult-onset neuronal ceroid lipofuscinosis in Tibetan terriers. Neurobiol Dis 2011;42:468–474.

[280] Bras J, Verloes A, Schneider SA, Mole SE, Guerreiro RJ. Mutation of the parkinsonism gene ATP13A2 causes neuronal ceroid-lipofuscinosis. Hum Mol Genet 2012; 21:2646–2650.

[281] Gitler AD, Chesi A, Geddie ML, Strathearn KE, Hamamichi S, et al. Alpha-synuclein is part of a diverse and highly conserved interaction network that includes PARK9 and manganese toxicity. Nat Genet 2009;41:308–315.

[282] Vilarino-Guell C, Wider C, Ross OA, Dachsel JC, Kachergus JM, et al. VPS35 mutations in Parkinson disease. Am J Hum Genet 2011;89:162–167.

[283] Zimprich A, Benet-Pages A, Struhal W, Graf E, Eck SH, et al. A mutation in VPS35, encoding a subunit of the retromer complex, causes late-onset Parkinson disease. Am J Hum Genet 2011;89:168–175.

[284] Arighi CN, Hartnell LM, Aguilar RC, Haft CR, Bonifacino JS. Role of the mammalian retromer in sorting of the cation-independent mannose 6-phosphate receptor. J Cell Biol 2004;165:123–133.

[285] Miura E, Hasegawa T, Konno M, Suzuki M, Sugeno N, et al. VPS35 dysfunction impairs lysosomal degradation of alpha-synuclein and exacerbates neurotoxicity in a *Drosophila* model of Parkinson's disease. Neurobiol Dis 2014;71:1–13.

[286] Voeltz GK, Prinz WA, Shibata Y, Rist JM, Rapoport TA. A class of membrane proteins shaping the tubular endoplasmic reticulum. Cell 2006;124:573–586.

[287] Hu J, Shibata Y, Voss C, Shemesh T, Li Z, et al. Membrane proteins of the endoplasmic reticulum induce high-curvature tubules. Science 2008;319:1247–1250.

[288] Park SH, Zhu PP, Parker RL, Blackstone C. Hereditary spastic paraplegia proteins REEP1, spastin, and atlastin-1 coordinate microtubule interactions with the tubular ER network. J Clin Invest 2010;120:1097–1110.

[289] Chang J, Lee S, Blackstone C. Spastic paraplegia proteins spastizin and spatacsin mediate autophagic lysosome reformation. J Clin Invest 2014;124:5249–5262.

[290] Dehay B, Bove J, Rodriguez-Muela N, Perier C, Recasens A, et al. Pathogenic lysosomal depletion in Parkinson's disease. J Neurosci 2010;30:12535–12544.

[291] Yokoyama H, Kuroiwa H, Yano R, Araki T. Targeting reactive oxygen species, reactive nitrogen species and inflammation in MPTP neurotoxicity and Parkinson's disease. Neurol Sci 2008;29:293–301.

[292] Bezard E, Przedborski S. A tale on animal models of Parkinson's disease. Mov Disord 2011;26:993–1002.

[293] Bove J, Martinez-Vicente M, Dehay B, Perier C, Recasens A, et al. BAX channel activity mediates lysosomal disruption linked to Parkinson disease. Autophagy 2014;10:889–900.

[294] Marui W, Iseki E, Kato M, Akatsu H, Kosaka K. Pathological entity of dementia with Lewy bodies and its differentiation from Alzheimer's disease. Acta Neuropathol 2004;108:121–128.

[295] Clark LN, Kartsaklis LA, Wolf Gilbert R, Dorado B, Ross BM, et al. Association of glucocerebrosidase mutations with dementia with Lewy bodies. Arch Neurol 2009;66:578–583.

[296] Kobayashi S, Tateno M, Park TW, Utsumi K, Sohma H, et al. Apolipoprotein E4 frequencies in a Japanese population with Alzheimer's disease and dementia with Lewy bodies. PLoS One 2011;6:e18569.

[297] Hardy J, Crook R, Prihar G, Roberts G, Raghavan R, et al. Senile dementia of the Lewy body type has an apolipoprotein E epsilon 4 allele frequency intermediate between controls and Alzheimer's disease. Neurosci Lett 1994;182:1–2.

[298] Coutinho MF, Prata MJ, Alves S. A shortcut to the lysosome: the mannose-6-phosphate-independent pathway. Mol Genet Metab 2012;107:257–266.

[299] Gaspar P, Kallemeijn WW, Strijland A, Scheij S, Van Eijk M, et al. Action myoclonus-renal failure syndrome: diagnostic applications of activity-based probes and lipid analysis. J Lipid Res 2014;55:138–145.

[300] Higashi S, Moore DJ, Yamamoto R, Minegishi M, Sato K, et al. Abnormal localization of leucine-rich repeat kinase 2 to the endosomal–lysosomal compartment in Lewy body disease. J Neuropathol Exp Neurol 2009;68:994–1005.

[301] Onyike CU, Diehl-Schmid J. The epidemiology of frontotemporal dementia. Int Rev Psychiatry 2013;25:130–137.

[302] Seelaar H, Rohrer JD, Pijnenburg YA, Fox NC, van Swieten JC. Clinical, genetic and pathological heterogeneity of frontotemporal dementia: a review. J Neurol Neurosurg Psychiatry 2011;82:476–486.

[303] Bennion Callister J, Pickering-Brown SM. Pathogenesis/genetics of frontotemporal dementia and how it relates to ALS. Exp Neurol 2014;262:84–90.

[304] Burrell JR, Kiernan MC, Vucic S, Hodges JR. Motor neuron dysfunction in frontotemporal dementia. Brain 2011;134:2582–2594.

[305] Arai T, Hasegawa M, Akiyama H, Ikeda K, Nonaka T, et al. TDP-43 is a component of ubiquitin-positive tau-negative inclusions in frontotemporal lobar degeneration and amyotrophic lateral sclerosis. Biochem Biophys Res Commun 2006;351:602–611.

[306] Neumann M, Sampathu DM, Kwong LK, Truax AC, Micsenyi MC, et al. Ubiquitinated TDP-43 in frontotemporal lobar degeneration and amyotrophic lateral sclerosis. Science 2006;314:130–133.

[307] Mackenzie IR, Neumann M, Bigio EH, Cairns NJ, Alafuzoff I, et al. Nomenclature and nosology for neuropathologic subtypes of frontotemporal lobar degeneration: an update. Acta Neuropathol 2010;119:1–4.

[308] Mori K, Weng SM, Arzberger T, May S, Rentzsch K, et al. The C9orf72 GGGGCC repeat is translated into aggregating dipeptide-repeat proteins in FTLD/ALS. Science 2013;339:1335–1338.

[309] Janssens J, Van Broeckhoven C. Pathological mechanisms underlying TDP-43 driven neurodegeneration in FTLD-ALS spectrum disorders. Hum Mol Genet 2013; 22:R77–87.

[310] Baker M, Mackenzie IR, Pickering-Brown SM, Gass J, Rademakers R, et al. Mutations in progranulin cause tau-negative frontotemporal dementia linked to chromosome 17. Nature 2006;442:916–919.

[311] Cruts M, Gijselinck I, van der Zee J, Engelborghs S, Wils H, et al. Null mutations in progranulin cause ubiquitin-positive frontotemporal dementia linked to chromosome 17q21. Nature 2006;442:920–924.

[312] Finch N, Baker M, Crook R, Swanson K, Kuntz K, et al. Plasma progranulin levels predict progranulin mutation status in frontotemporal dementia patients and asymptomatic family members. Brain 2009;132:583–591.

[313] Ghidoni R, Benussi L, Glionna M, Franzoni M, Binetti G. Low plasma progranulin levels predict progranulin mutations in frontotemporal lobar degeneration. Neurology 2008;71:1235–1239.

[314] Gass J, Cannon A, Mackenzie IR, Boeve B, Baker M, et al. Mutations in progranulin are a major cause of ubiquitin-positive frontotemporal lobar degeneration. Hum Mol Genet 2006;15:2988–3001.

[315] Gijselinck I, Van Broeckhoven C, Cruts M. Granulin mutations associated with frontotemporal lobar degeneration and related disorders: an update. Hum Mutat 2008; 29:1373–1386.

[316] Schymick JC, Talbot K, Traynor BJ. Genetics of sporadic amyotrophic lateral sclerosis. Hum Mol Genet 2007;16 Spec No. 2:R233–R242.

[317] Mukherjee O, Wang J, Gitcho M, Chakraverty S, Taylor-Reinwald L, et al. Molecular characterization of novel progranulin (GRN) mutations in frontotemporal dementia. Hum Mutat 2008;29:512–521.

[318] van der Zee J, Le Ber I, Maurer-Stroh S, Engelborghs S, Gijselinck I, et al. Mutations other than null mutations producing a pathogenic loss of progranulin in frontotemporal dementia. Hum Mutat 2007;28:416.

[319] Wang J, Van Damme P, Cruchaga C, Gitcho MA, Vidal JM, et al. Pathogenic cysteine mutations affect progranulin function and production of mature granulins. J Neurochem 2010;112:1305–1315.

[320] Smith KR, Damiano J, Franceschetti S, Carpenter S, Canafoglia L, et al. Strikingly different clinicopathological phenotypes determined by progranulin-mutation dosage. Am J Hum Genet 2012;90:1102–1107.

[321] Amritraj A, Wang Y, Revett TJ, Vergote D, Westaway D, et al. Role of cathepsin D in U18666A-induced neuronal cell death: potential implication in Niemann–Pick type C disease pathogenesis. J Biol Chem 2013;288:3136–3152.

[322] Ghoshal N, Dearborn JT, Wozniak DF, Cairns NJ. Core features of frontotemporal dementia recapitulated in progranulin knockout mice. Neurobiol Dis 2012;45:395–408.

[323] Brady OA, Zheng Y, Murphy K, Huang M, Hu F. The frontotemporal lobar degeneration risk factor, TMEM106B, regulates lysosomal morphology and function. Hum Mol Genet 2013;22:685–695.

[324] Chen-Plotkin AS, Unger TL, Gallagher MD, Bill E, Kwong LK, et al. TMEM106B, the risk gene for frontotemporal dementia, is regulated by the microRNA-132/212 cluster and affects progranulin pathways. J Neurosci 2012;32:11213–11227.

[325] Lang CM, Fellerer K, Schwenk BM, Kuhn PH, Kremmer E, et al. Membrane orientation and subcellular localization of transmembrane protein 106B (TMEM106B), a major risk factor for frontotemporal lobar degeneration. J Biol Chem 2012;287:19355–19365.

[326] Van Deerlin VM, Sleiman PM, Martinez-Lage M, Chen-Plotkin A, Wang LS, et al. Common variants at 7p21 are associated with frontotemporal lobar degeneration with TDP-43 inclusions. Nat Genet 2010;42:234–239.

[327] Finch N, Carrasquillo MM, Baker M, Rutherford NJ, Coppola G, et al. TMEM106B regulates progranulin levels and the penetrance of FTLD in GRN mutation carriers. Neurology 2011;76:467–474.

[328] Gotzl JK, Mori K, Damme M, Fellerer K, Tahirovic S, et al. Common pathobiochemical hallmarks of progranulin-associated frontotemporal lobar degeneration and neuronal ceroid lipofuscinosis. Acta Neuropathol 2014;127:845–860.

[329] Cruchaga C, Graff C, Chiang HH, Wang J, Hinrichs AL, et al. Association of TMEM106B gene polymorphism with age at onset in granulin mutation carriers and plasma granulin protein levels. Arch Neurol 2011;68:581–586.

[330] Bankaitis VA, Johnson LM, Emr SD. Isolation of yeast mutants defective in protein targeting to the vacuole. Proc Natl Acad Sci U S A 1986;83:9075–9079.

[331] Banta LM, Robinson JS, Klionsky DJ, Emr SD. Organelle assembly in yeast: characterization of yeast mutants defective in vacuolar biogenesis and protein sorting. J Cell Biol 1988;107:1369–1383.

[332] Rieder SE, Emr SD. A novel RING finger protein complex essential for a late step in protein transport to the yeast vacuole. Mol Biol Cell 1997;8:2307–2327.

[333] Rampoldi L, Dobson-Stone C, Rubio JP, Danek A, Chalmers RM, et al. A conserved sorting-associated protein is mutant in chorea-acanthocytosis. Nat Genet 2001; 28:119–120.

[334] Park JS, Neiman AM. VPS13 regulates membrane morphogenesis during sporulation in *Saccharomyces cerevisiae*. J Cell Sci 2012;125:3004–3011.

[335] Stagi M, Klein ZA, Gould TJ, Bewersdorf J, Strittmatter SM. Lysosome size, motility and stress response regulated by fronto-temporal dementia modifier TMEM106B. Mol Cell Neurosci 2014;61:226–240.

[336] Tresse E, Salomons FA, Vesa J, Bott LC, Kimonis V, et al. VCP/p97 is essential for maturation of ubiquitin-containing autophagosomes and this function is impaired by mutations that cause IBMPFD. Autophagy 2010;6:217–227.

[337] Urwin H, Authier A, Nielsen JE, Metcalf D, Powell C, et al. Disruption of endocytic trafficking in frontotemporal dementia with CHMP2B mutations. Hum Mol Genet 2010;19:2228–2238.

[338] Watts GD, Thomasova D, Ramdeen SK, Fulchiero EC, Mehta SG, et al. Novel VCP mutations in inclusion body myopathy associated with Paget disease of bone and frontotemporal dementia. Clin Genet 2007;72:420–426.

[339] Watts GD, Wymer J, Kovach MJ, Mehta SG, Mumm S, et al. Inclusion body myopathy associated with Paget disease of bone and frontotemporal dementia is caused by mutant valosin-containing protein. Nat Genet 2004;36:377–381.

[340] Muller JM, Deinhardt K, Rosewell I, Warren G, Shima DT. Targeted deletion of p97 (VCP/CDC48) in mouse results in early embryonic lethality. Biochem Biophys Res Commun 2007;354:459–465.

[341] Johnson JO, Mandrioli J, Benatar M, Abramzon Y, Van Deerlin VM, et al. Exome sequencing reveals VCP mutations as a cause of familial ALS. Neuron 2010;68: 857–864.

[342] Ritz D, Vuk M, Kirchner P, Bug M, Schutz S, et al. Endolysosomal sorting of ubiquitylated caveolin-1 is regulated by VCP and UBXD1 and impaired by VCP disease mutations. Nat Cell Biol 2011;13:1116–1123.

[343] Ju JS, Fuentealba RA, Miller SE, Jackson E, Piwnica-Worms D, et al. Valosin-containing protein (VCP) is required for autophagy and is disrupted in VCP disease. J Cell Biol 2009;187:875–888.

[344] Mackenzie IR, Frick P, Neumann M. The neuropathology associated with repeat expansions in the C9ORF72 gene. Acta Neuropathol 2014;127:347–357.

[345] Farg MA, Sundaramoorthy V, Sultana JM, Yang S, Atkinson RA, et al. C9ORF72, implicated in amytrophic lateral sclerosis and frontotemporal dementia, regulates endosomal trafficking. Hum Mol Genet 2014;23:3579–3595.

[346] Brettschneider J, Van Deerlin VM, Robinson JL, Kwong L, Lee EB, et al. Pattern of ubiquilin pathology in ALS and FTLD indicates presence of C9ORF72 hexanucleotide expansion. Acta Neuropathol 2012;123:825–839.

[347] Deng HX, Chen W, Hong ST, Boycott KM, Gorrie GH, et al. Mutations in UBQLN2 cause dominant X-linked juvenile and adult-onset ALS and ALS/dementia. Nature 2011;477:211–215.

[348] Hirano M, Nakamura Y, Saigoh K, Sakamoto H, Ueno S, et al. Mutations in the gene encoding p62 in Japanese patients with amyotrophic lateral sclerosis. Neurology 2013;80:458–463.

[349] Le Ber I, Camuzat A, Guerreiro R, Bouya-Ahmed K, Bras J, et al. SQSTM1 mutations in French patients with frontotemporal dementia or frontotemporal dementia with amyotrophic lateral sclerosis. JAMA Neurol 2013;70:1403–1410.

[350] Rubino E, Rainero I, Chio A, Rogaeva E, Galimberti D, et al. SQSTM1 mutations in frontotemporal lobar degeneration and amyotrophic lateral sclerosis. Neurology 2012;79:1556–1562.

[351] Ito H, Nakamura M, Komure O, Ayaki T, Wate R, et al. Clinicopathologic study on an ALS family with a heterozygous E478G optineurin mutation. Acta Neuropathol 2011;122:223–229.

[352] Guil S, Long JC, Caceres JF. hnRNP A1 relocalization to the stress granules reflects a role in the stress response. Mol Cell Biol 2006;26:5744–5758.

[353] Hutton M, Lendon CL, Rizzu P, Baker M, Froelich S, et al. Association of missense and 5'-splice-site mutations in tau with the inherited dementia FTDP-17. Nature 1998;393:702–705.

[354] Grover A, Houlden H, Baker M, Adamson J, Lewis J, et al. 5' splice site mutations in tau associated with the inherited dementia FTDP-17 affect a stem-loop structure that regulates alternative splicing of exon 10. J Biol Chem 1999;274:15134–15143.

[355] Hutton M. Missense and splice site mutations in tau associated with FTDP-17: multiple pathogenic mechanisms 1023. Neurology 2001;56:S21–S25.

10

LYSOSOME AND CANCER

Marja Jäättelä and Tuula Kallunki

Cell Death and Metabolism, Danish Cancer Society Research Center, Danish Cancer Society, Copenhagen, Denmark

10.1 INTRODUCTION

Cancer development and progression induces fundamental changes in lysosomes, which in turn imposes profound effects in tumor invasion and metastasis and in sensitivity to anticancer treatments. Enhanced lysosomal hydrolase activity, lysosome trafficking, and extracellular secretion of lysosomal hydrolases (e.g., cathepsins and heparanase) are important hallmarks of invasive cancer and are associated with its metastatic capacity. On the other hand, lysosomal membrane permeabilization (LMP) that often leads to lysosomal cell death (LCD) may be useful as an anticancer treatment, provided that it is specially targeted toward cancer cells. In this chapter, we present the most central cancer-associated alterations in the function and composition of lysosomes and discuss their importance for tumorigenesis and the possibilities they may provide for the development of anticancer treatments.

10.2 LYSOSOMAL FUNCTION AND ITS IMPORTANCE FOR CANCER DEVELOPMENT AND PROGRESSION

Lysosomes are membrane-bounded acidic organelles that are filled with powerful hydrolytic enzymes. From their functional point of view, lysosomes can be described as cellular recycling centers [1–3]. They receive material via endocytosis, autophagy,

Lysosomes: Biology, Diseases, and Therapeutics, First Edition.
Edited by Frederick R. Maxfield, James M. Willard and Shuyan Lu.

and phagocytosis, and their most basal function in cancer, as well as in healthy organisms, is to degrade and destruct larger, unnecessary, worn out, and sometimes even harmful cellular and extracellular components for the production of energy and building material. Accordingly, lysosomes compose an important source of cellular nutrients and building blocks for rapidly dividing cancer cells.

Lysosomes are also contributing to cellular fate by functioning as specific endpoints of well-controlled degradation programs. For example, endocytic downregulation of the epidermal growth factor receptor (EGFR/ErbB1) tyrosine kinase is an important and tightly regulated process that is executed in lysosomes [4,5]. Its impairment is associated to cancer, since the defective downregulation of EGFR can lead to increased, uncontrolled signaling. In many cancers, EGFR signaling is facilitated and one of the mechanisms contributing to this is its cancer-associated increased recycling and escape from lysosomal degradation [5]. Another example of lysosomes contributing to cell fate is the lysosome-mediated processing of chromatin that can contribute to cellular senescence, proliferation arrest, and tumor suppression [6]. Thus, processes that affect lysosomal function may have profound and multiple effects on the cell fate.

In addition to downregulating important cellular processes and supporting cell survival and growth by providing nutrients and building blocks, lysosomes and lysosomal hydrolases are also involved in tissue remodeling as for mammary gland involution [7,8], a process that can contribute to cancer progression. Moreover, invading cancer cells can harness the cellular clearance mechanism called lysosomal exocytosis to empty their digestive contents to the extracellular space [9–11]. Upon secretion to extracellular space, lysosomal hydrolases can induce and facilitate invasion. Numerous studies have shown that specific inhibition of lysosomal hydrolase activity can significantly delay invasion and metastasis in various *in vivo* model systems as reviewed earlier by us [11].

Lysosomes are central contributors of cancer development and progression. The active involvement of lysosomes in these processes makes them attractive anticancer therapy targets. Thus, understanding the lysosomal function and its alteration during cancer development and progression can assist in the development of novel lysosome based anticancer therapies.

10.3 CANCER-INDUCED CHANGES IN LYSOSOMAL FUNCTION

Cancer-induced changes in lysosomal function are summarized in Figure 10.1.

10.3.1 Increased Activity of Lysosomal Enzymes

Lysosomes contain over 50 hydrolases including proteinases, glycosidases, phosphatases, sulfatases, nucleases, and lipases. Since early 1950s, several studies have reported increased activities of various lysosomal enzymes in solid tumors in comparison to their tissue of origin [12]. Especially, increased biogenesis and activity

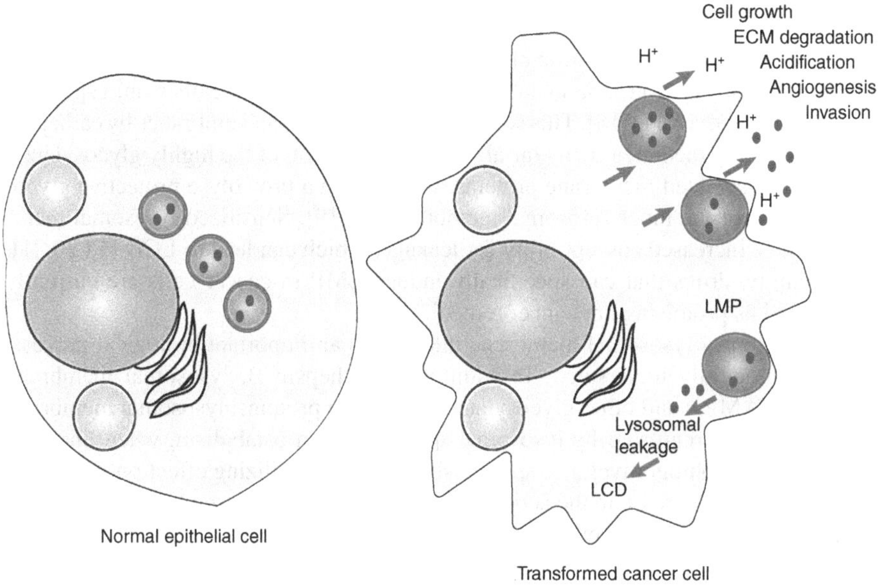

Figure 10.1 Cancer development induces changes in the lysosomal function. Normal, healthy cells undergo dramatic changes that affect their lysosomal function upon transformation to cancer cells. Cancer development and progression induces lysosomal biogenesis increasing the expression of various lysosomal hydrolases (gray). It alters the lysosomal membrane integrity, sensitizing them to LMP and lysosomal leakage that can lead to LCD. It additionally increases the size of lysosomes and alters their distribution in a manner where normally mostly perinuclear lysosomes adapt pericellular locations close to the invasive cellular protrusions at the cell membrane, which allows them to secrete or exocytose their hydrolytic contents by a process called "lysosomal exocytosis." Upon reaching the extracellular space, lysosomal hydrolases can induce cell growth, extracellular matrix degradation, invasion, angiogenesis, and extracellular acidification. (*See color plate section for the color representation of this figure.*)

of cathepsins, heparanase, acid ceramidase, and vacuolar-type ATPase have been implicated in the progression of different types of cancers and linked to metastatic disease and poor prognosis [11]. The increased activity of lysosomal hydrolases, such as cathepsins and heparanase, is mostly resulting from their cancer-induced increased gene expression and their pharmacological or genetic inhibition can hinder invasion and metastasis in *in vivo* model systems [11]. Of these enzymes, cysteine cathepsin B is the most studied and thus more is known about its regulation in cancer in comparison to the others. Oncogenic Ras, Src, and ErbB2 can increase cysteine cathepsin mRNA and protein expression leading to increased hydrolase activity and increased invasion [13–16]. Moreover, increased cathepsin B activity can sensitize cancer cells to LMP and cathepsin B is also one of the most central mediators of LCD [17].

10.3.2 Altered Lysosome Membrane Permeability

Oncogene-driven transformation alters lysosomal membranes and sensitizes them to LMP. The sensitization is associated with increased cysteine cathepsin, especially cathepsin B, activity [11,18]. This sensitization of lysosomal membranes by cathepsin B is most likely mediated by its intralysosomal digestion of the highly glycosylated, lysosome-associated membrane proteins, which form a proteolyse protective glycocalyx shield of the inner lysosomal membrane [14,19]. Sensitized lysosomal membranes have increased susceptibility for leakage, which can lead to LCD [18,20,21]. Accordingly, drugs that can specifically induce LMP in cancer cells are currently considered as promising anticancer remedies.

Regulation of lysosomal membrane integrity is an important biological process, which is not fully understood. In addition to cathepsin B, lysosomal membrane proteins (LAMPs) and other glycosylated membrane proteins, lysosomal membrane integrity is also regulated by lysosomal sphingomyelin catabolism, where increase in the level of sphingomyelin or sphingosine has a destabilizing effect on lysosomal membranes and increase in the activity of acid sphingomyelinase (ASM) has a stabilizing effect [21]. Thus, a promising cancer therapeutic remedy could, for example, involve ASM inhibition that would lead to the induction of LMP and LCD in cancer cells [22]. LMP can also be induced by the inhibition of sphingosine kinase 1, an enzyme that converts sphingosine into sphingosine-1-phosphate [23]. Interestingly, cathepsin B can cleave sphingosine kinase 1 leading to its degradation [24], which is likely to contribute to the oncogene-induced destabilization of lysosomal membranes.

10.3.3 Increased Lysosome Size

Lysosome size varies between 0.1 and 1.2 μm. Tumor cells have larger lysosomes than noncancerous cells [9,14]. Lysosome size is regulated by a balance between vesicle fusion and fission and can be reversibly altered by, for example, acidifying the extracellular environment [9]. Acidification of the environment results in larger lysosomal vesicles. This occurs more frequently in highly metastatic human breast cancer cells than in less metastatic breast cancer cells or mammary epithelial cells [9]. Increase in the intracellular Ca^{2+} can trigger homotypic lysosomal fusion that can lead to increase in the lysosome size. Similarly, deficient lysosome fission can give rise to larger lysosomes. The absence of the Chediak–Higashi gene CHS1/Lyst gives rise to enlarged lysosomes in a disease called Chediak–Higashi syndrome most likely due to decreased rate of lysosome fission [25]. It is not fully known what causes the increased lysosome sizes in malignant cancer cells beyond increased extracellular acidification. Larger lysosomes have shown to be more sensitive for breakage [26]. Some LMP-inducing anticancer drugs increase lysosomal volume prior to inducing cell death. For example, treatment of HeLa human cervical cancer cells with vincristine, a widely used microtubule-destabilizing anticancer drug, induces increased lysosome volume and LCD [27]. Similar observations have been reported already 40 years ago from studies where mouse mammary carcinoma cells were treated with radiotherapy or chemotherapy (cyclophosphamide) resulting in increased lysosome size and contents followed by degradative changes in the tumor cells [28,29].

10.3.4 Altered Lysosome Trafficking – Increased Lysosomal Exocytosis

Cancer cells have altered lysosomal trafficking [9,14,30]. Oncogenic activation often induces pericellular distribution of lysosomes leading to their accumulation to the close vicinity of cell membrane at the invasive cellular protrusions, for example, invadosomes. This is likely to be promoted by extracellular acidosis, which is typical for malignant cancer cells [31]. In noncancerous cells, lysosomal distribution alters according to the nutritional status of the cells [32]. Normally, lysosomes cluster around the nucleus in nutrient starvation conditions, and they move along microtubules to a more peripheral location in nutrient-rich conditions [32].

In cancer cells, pericellular distribution of lysosomes enables them to exocytose their highly hydrolytic contents into the extracellular space allowing the release of lysosomal proteases and further acidification of the extracellular milieu. In turn, the acidified extracellular milieu provides optimal conditions for lysosomal hydrolases. There is a vast amount of evidence indicating that the lysosomal hydrolases (e.g., cathepsins and heparanase) have an important extracellular role in cancer progression and are contributing to cancer invasion and metastasis [11,33,34].

10.4 CANCER-INDUCED CHANGES IN LYSOSOME COMPOSITION

Cancer-induced changes in lysosomal function are closely linked to changes in its composition, as summarized in Figure 10.2. We describe in detail the most important molecular changes affecting lysosomal components in connection to cancer in the following sections.

10.4.1 Changes in Lysosomal Hydrolases

10.4.1.1 Cathepsins Cathepsins comprise three subgroups: aspartate cathepsins (cathepsins D and E), cysteine cathepsins (cathepsins B, C, F, H, K, L, O, S, V/U/L2, W, X/Z/Y), and serine cathepsins (cathepsins A and G). Cathepsins, like other lysosomal hydrolases, have optimal pH at 4.5 or lower. However, cathepsins can also function in higher pH, which in some cases can change their functional parameters such as substrate specificity. For example, cathepsin B, which is primarily an exopeptidase in acidic lysosomal environment, will gain endopeptidase activity in neutral pH [35,36]. Cysteine cathepsins are overexpressed in variety of tumor cells and tumor associated cells such as fibroblasts, macrophages, osteoclasts, neutrophils, mast cells, myoepithelial cells, T lymphocytes and endothelial cells, however the diverse expression of cathepsins, tumor type and stage dependently, suggests for variable roles for cathepsins during progression of different types of cancers [30]. Increased expression, activity, and secretion of cathepsins can enhance tumor growth, angiogenesis, invasion, and metastasis. Cathepsins, especially cathepsins B, L, and D, are often overexpressed in cancer. In addition to cancer cells, cathepsin overexpression can also be detected in tumor-associated macrophages, leucocytes, fibroblasts, osteoclasts, and in myoepitehlial and endothelial cells [11,30]. All cathepsin-targeted cancer therapeutics aim at inhibiting its extracellular activity.

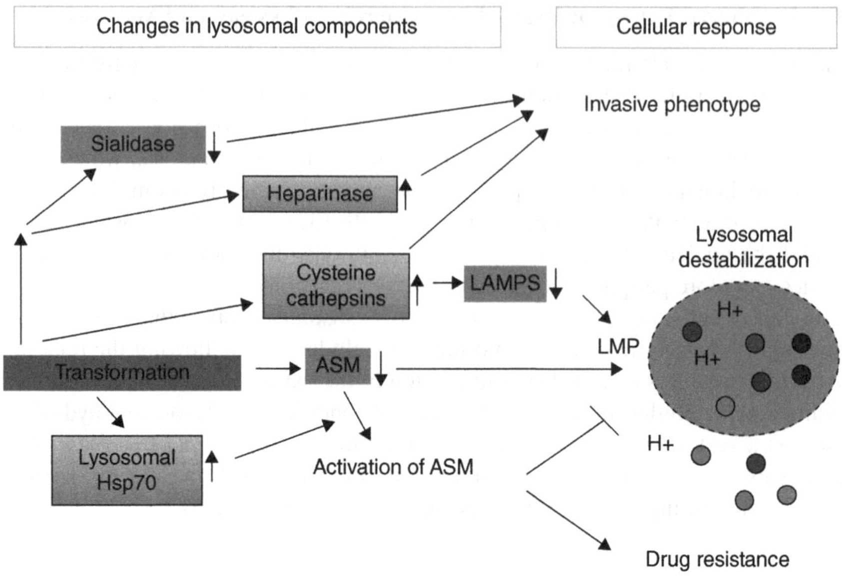

Figure 10.2 Cancer induces changes in lysosomal composition. These include changes in the expression levels and activity of several lysosomal hydrolases. Some of these changes are "protumor" processes contributing to tumor growth, invasion, metastasis, and drug resistance, and some of them are "antitumor" processes that sensitize lysosomes to LMP and LCD. (*See color plate section for the color representation of this figure.*)

10.4.1.1.1 Cathepsin D Cathepsin D is an aspartate cathepsin that belongs to the pepsin superfamily of proteases [37]. Increased levels of cathepsin D in human neoplastic tissues was first time reported almost 30 years ago [38]. Since then, numerous studies have demonstrated increased cathepsin D expression in almost any type of solid tumors, including neuroblastoma, glioma, melanoma, breast tumors, ovarian and endometrial tumors, head and neck tumors, thyroid tumors, pancreatic tumors, liver tumors, gastric, bladder, lung and colorectal carcinomas, and prostate tumors in comparison to their tissue of origin [39,40]. Thus, several studies have identified positive correlation between cathepsin D levels and tumor size or grade, tumor aggressiveness, metastasis, prognosis, degree of chemoresistance, and risk of recurrence [40,41]. Cathepsin D and especially its pro-form have been reported to act as growth-stimulating autocrine growth factors for various cancer cells, as well as for stromal cells. This function of cathepsin D is independent of its enzymatic activity [39,41]. Activated cathepsin D can also promote invasion and metastasis upon secretion to extracellular space and cathepsin D and its propeptide can induce activation of the MAPK pathway and expression of various genes to enhance the secretion of different cytokines [41–43]. The role of the protease inactive cathepsin D pro-form in cancer progression challenges the development of cathepsin D targeting anticancer drugs, since small-molecule enzyme inhibitors of cathepsin D are not likely to be very effective.

10.4.1.1.2 Cathepsin B Cathepsin B is a cysteine cathepsin and belongs to the papain superfamily of proteases. In human colorectal and breast cancers, the expression of cathepsin B is increased in cancer cells as well as in the tumor-associated cells such as macrophages and stromal fibroblasts [11,44,45]. Increased expression and secretion of cathepsin B have also been reported in brain, lung, and prostate cancers. In general, increased cathepsin B protein levels detected in immunohistochemical staining correlate well with higher mRNA levels in cancer cells and tissues, indicating a transcriptional upregulation of cathepsin B gene (*CTSB*) in various cancers [46]. In tumors, cathepsin B expression is predominantly increased in the tumor cells at the invasive front of the tumors [47,48]. Overexpression of oncogenic Ras [14–16] or ErbB2 [13,14] can induce cathepsin B expression and activity. The ErbB2-induced expression of *CTSB* is transcriptionally regulated, however independently of the lysosomal biogenesis regulator transcription factor EB (TFEB) [13,49]. Instead, ErbB2 induces the myeloid zinc finger transcription factor 1 (MZF1), which binds to an ErbB2-inducible enhancer element in the first intron of *CTSB* gene *in vivo*. The ErbB2-mediated activity of MZF1 is regulated via an ErbB2-activated signaling network consisting of PAK4, 5 and 6 cdc42bpβ, PKCα, and MAPK-ERK2 serine–threonine kinases as well as TGFβRI and 2 [13]. Interestingly, MAPK-ERK2 activity, which is essential for the ErbB2-induced activation of MZF1 and expression of *CTSB*, can inhibit the nuclear import of TFEB [50], further supporting the observations that ErbB2 induces CTSB expression independently of TFEB [13]. In malignant, highly invasive tumors overexpressed cathepsin B is often secreted into the extracellular space where it associates with the tumor cell surface [16,51,52].

The active role of cathepsin B in cancer is strongly supported by murine models. In the RIP1-Tag2/RT2 (RT2) pancreatic islet cell carcinogenesis model mice develop multiple tumors in their pancreatic islets due to overexpression of oncogenic SV40 T-antigen in insulin producing ß-cells. These mice express high levels of cathepsin B [33,53] and its genetic inactivation results in impaired tumor formation, angiogenesis, and invasiveness [53]. The suggested mechanism of how cathepsin B contributes to increased tumor formation involves secretion of cathepsin B from tumor cells and tumor-associated macrophages and cathepsin B-mediated direct cleavage of the epithelial transmembrane protein E-cadherin as well as activation of the urokinase plasminogen activator by cleavage of its propeptide [53,54]. In addition to activating plasmin and various matrix metalloproteases, cathepsin B can also directly degrade extracellular matrix by cleaving laminin, type VI collagen, and tenascin-C [55–57]. Similarly, *Ctsb* deficiency significantly delays the onset and growth of primary mammary tumors as well as formation of lung metastases in murine mammary tumor virus polyoma middle-T antigen (PyMT) transgenic mice and cathepsin B overexpression in these mice promotes mammary tumor growth and metastasis [58,59]. In the PyMT model, the expression of cathepsin B in the actual cancer cells rather than in the stromal cells promotes tumor progression [60].

Since cathepsin B is overexpressed in various cancers and several studies show beneficial effects of its inhibition by RNA interference or pharmacological compounds *in vitro* and its genetic or pharmacological inhibition *in vivo* in mouse studies, it is believed that targeting cathepsin B could have significant therapeutic effect [61].

However, even though many compounds that can inhibit cathepsin B have been isolated and developed, none of them have shown to be potential enough for clinics [61].

10.4.1.1.3 Cathepsin L Cathepsin L is ubiquitously expressed cysteine cathepsin that is often overexpressed in human tumors [45]. It is capable of cleaving various extracellular matrix proteins such as laminin, fibronectin, and collagens as well as E-cadherin and proheparanase [53,62–65]. Several studies conclude that cathepsin L plays an important role in cancer cell invasion and migration by decreasing cell–cell adhesion and increasing extracellular matrix degradation [30,66]. Studies elucidating the role of cathepsin L in angiogenesis have contradictory results, suggesting that it is either angiogenesis promoting or not involved at all [65]. For example, in glioma patients, cathepsin L expression is not associated significantly with new vasculature of astrocytic tumors [67] but a cathepsin L-specific inhibitor NapSul-Ile-Trp-CHO (NSITC) displays potent antiangiogenic activity *in vitro* in chick chorioallantoic membrane and mouse Matrigel models of angiogenesis [68]. Crossing of an RT2 transgenic mice modeling pancreatic carcinoma with cathepsin L knockout mice showed decreased tumor growth and invasiveness when compared to the corresponding cathepsin B knockout mice. However, these mice showed no signs of reduction in the angiogenic switch on the contrary to the corresponding cathepsin B knockout mice [53]. It is to be noted that these pancreatic tumors exhibited increased expression of several cathepsins and thus it is possible that some other cathepsins were able to compensate the loss of cathepsin L [53].

In colorectal cancer, nuclear staining of cathepsin L increases and cytoplasmic staining decreases with advancing tumor stage and is associated with poor prognosis [69]. Indeed, a short form of cathepsin L that lacks the signal peptide needed for lysosomal targeting has been found in nucleus. There it cleaves histone H3 and specific transcription factors, such as the CCAAT-displacement protein/cut homeobox (CDP/Cux), altering gene expression patterns and leading into delay in cell cycle progression and contributing to oncogenic transformation [70–72]. Interestingly, Ras transformation can increase both the production of short nuclear cathepsin L isoforms and the cleavage of CDP/Cux [72].

Inhibition of cathepsin L can reverse cancer cell's resistance to various chemotherapeutic drugs. This is suggested to occur due to cathepsin L inhibition-mediated stabilization and increased availability of cytoplasmic and nuclear targets including estrogen receptor-α, Bcr-Abl, topoisomerase-II α, histone deacetylase 1, and the androgen receptor [68,73].

However, contradicting to the beneficial role of cathepsin L inhibition in many other types of cancers, inactivation of cathepsin L, for example, in mouse intestinal epithelia promotes tumor progression [74]. Similarly, in mouse skin cancer models, genetic inactivation of cathepsin L results in increased tumorigenesis [75,76]. Indeed, cathepsin L has been found to be essential for the negative regulation of the growth factor and growth factor receptor recycling in keratinocytes. Thus, its inhibition will result in increased receptor levels at the plasma membrane as well as increased levels of growth factors [77]. Moreover, transgenic overexpression of serine protease inhibitor hurpin, also known as serpin, a specific inhibitor of cathepsin L, mainly expressed in skin epidermis, leads to increased skin cancer progression. Hurpin overexpressing mouse skin undergoes less apoptosis than corresponding wild-type skin upon exposure to UVB radiation [78], suggesting for a possible involvement of cathepsin L in apoptosis regulation. Due to recent reports indicating that cathepsin L deficiency can promote tumor progression, its usefulness as therapeutic target has become questionable.

10.4.1.1.4 Cathepsin K Unlike cysteine cathepsins B and L that are expressed ubiquitously, expression of cysteine cathepsin K is mainly restricted to skeleton and particularly in osteoclasts, where it is the predominant cysteine cathepsin [79]. It has a central role in osteoclastic resorption and inactivating mutants of the human cathepsin K gene (CTSK) result in pycnodysostosis characterized by short stature, skull deformities, and skeletal abnormalities [80]. Cathepsin K is expressed in breast and prostate carcinomas and its expression is enhanced in bone metastases of these cancers when compared to corresponding primary tumors [81–83]. Several cathepsin K inhibitors (e.g., CKI, AFG-495, balicatib, and odanacatib) are developed to treat osteoporosis-associated bone loss [79,84,85]. Of these, for example, CKI can reduce breast cancer–induced osteolysis and skeletal tumor burden in a mouse experimental skeletal metastasis model [82]. However, most successful of cathepsin K inhibitors, thus far, is odanacatib that is currently on phase III trial as a treatment to reduce the risk of breast cancer–induced bone metastasis (http://www.cancer.gov/drugdictionary). Interestingly, cathepsin K is also expressed in melanoma where its expression correlates with advanced metastatic disease. Indeed, pharmacological inhibition of cathepsin K with a rather unspecific cathepsin K inhibitor Boc-I decreases melanoma cell invasion through the Matrigel basement membrane matrix [86].

10.4.1.1.5 Additional Cysteine Cathepsins The expression of cysteine cathepsins F, H, S, V (also known as L2 or U), and Z (also known as Y or X) is also increased in cancer. In addition to cathepsins B and L, increase in the expression of cathepsins K and V has been observed in breast carcinomas [83,87]. Furthermore, cathepsin V is overexpressed in renal, ovarian, colorectal, and squamous cell carcinomas, cathepsin S and Z expression is increased in prostate carcinoma, and expression of cathepsin Z is increased in gastric cancers [88–90].

In a mouse RT2 model of pancreatic carcinoma, pancreatic tumors exhibit increased expression of cathepsins C, S, H, and Z in addition to cathepsins B and L and treatment of mice with a broad-spectrum pharmacological cell-permeable cathepsin inhibitor led to delayed proliferation, angiogenic switch, tumor vascularity, and invasion [91]. Genetic inactivation of cathepsin S in this model system resulted in impaired tumor formation and angiogenesis, similar to the inactivation of cathepsin B [53]. In the RT2 model, the absence of cathepsin B, L, or S resulted in delayed tumor invasion [53]. E-cadherin, which is a suppressor of invasion, was identified as a target for cathepsins B, L, and S but not C in these tumors, and E-cadherin levels were well maintained in knockout RT2 pancreatic carcinomas compared to the corresponding RT2 wild-type tumors [53].

Increased expression of cathepsins B and Z is also detectable on the surface of the cancer cells in the polyoma middle-T oncogene (MMTV-PyMT) induced mammary breast carcinomas [58]. Genetic inactivation of cathepsin B or Z slightly delays the appearance of tumors as well as lung metastases [92,93], but their combined inactivation results in significant delay in tumor burden and reduction in the number and size of metastases [92]. Cathepsin S, which is mainly expressed by endothelial cells, can function in angiogenesis by degrading antiangiogenic, type IV collagen-derived peptides and generating proangiogenic peptides by cleavage of laminin-5 [94].

10.4.1.2 Deregulation of Cystatins: Endogenous Cathepsin Inhibitors The increased activity of cysteine cathepsins is often connected to the cancer-associated deregulation of cystatins, which are endogenous cysteine cathepsin inhibitors [95]. There are four main cystatins (A, B, C, and M). They are epigenetically silenced in breast, brain, lung, and pancreatic cancers, linking them to the increased activity of cysteine cathepsins in these tissues [96]. The functional importance of cystatins in cancer is demonstrated, for example, by an experimental human fibrosarcoma lung metastasis model in which overexpression of cystatin C reduces metastasis [97] and by *in vitro* invasion assays where overexpression of cystatin M or S can suppress breast cancer and melanoma cell invasion [95,96].

10.4.1.3 Pharmacological Targeting of Cysteine Cathepsins Of all the cysteine cathepsins, only inhibitors of cathepsin K, the osteoclast-specific cysteine cathepsin, have reached the clinics. In most cases, often more than one cysteine cathepsin is upregulated upon oncogene-driven cell transformation *in vitro*, in preclinical murine cancer models as well as in human cancer. Their highly overlapping substrate specificities and prooncogenic functions in tumorigenesis suggest that their combined inhibition might be required for efficient anticancer therapy. In line with this, the combined inactivation of *Ctsb* and *Ctsz* in PyMT mice [92,93], treatment of mice with a cell-permeable, broad-spectrum cysteine cathepsin inhibitor JPM-OEt was more efficient than single depletions of *Ctsb*, *Ctsl*, or *Ctss* in reducing angiogenic switching in the murine RT2 pancreatic carcinoma model [33,53]. On the contrary, JPM-OEt is not efficient in the PyMT murine mammary cancer model and fails to reduce metastasis in a breast cancer bone metastasis model in which the non-cell-permeable cathepsin B inhibitor, CA-074, is effective [98,99]. This may be due to differences in the bioavailability of these inhibitors in different tissues [99]. Another possible explanation is that different cysteine cathepsins may still have unexplored opposing roles in some cancers. However, the currently available preclinical data is encouraging and supports the continuation of the development of broad-spectrum cysteine cathepsin inhibitors as cancer therapeutics [100]. However, more basic research on the function and localization of individual cathepsins is needed to identify the optimal treatment modalities.

10.4.1.4 Heparanase Heparanase is an endo-β-D-glucuronidase and the most important mammalian enzyme degrading heparan sulfate. Heparanase is processed in lysosomes and late endosomes and in addition to these it is also active in extracellular matrix. Cleavage of heparan sulfate side chains from the heparan sulfate proteoglycans by heparanase leads to the disassembly of the extracellular matrix and is associated with tissue remodeling, cell movement, cancer progression, metastasis, and angiogenesis [34]. Heparan sulfate binds and sequesters several growth factors, cytokines, and chemokines into the extracellular matrix and cell surface and its cleavage will therefore also increase their release and tissue availability [101,102]. Heparanase expression is induced in nearly all human cancers [103,104] and increased heparanase activity correlates with the metastatic potential of tumor cells, contributes to remodeling of the extracellular matrix and is associated with neovascularization [105,106]. Cancer cells can secrete heparanase in response to suitable stimulus such as ATP, ADP, and adenosine nucleotides with similar kinetics than cathepsin D [107]. Heparanase is also found in the body fluids of patients with metastatic cancer, indicating its secretion *in vivo* [108,109].

Heparan and its derivatives, sulfated oligoshaccarides suramin and suramin analogues, have potent anti-metastatic activity in mouse models of metastatic cancers [110–112]. Heparin itself is a close mimic of heparan sulfate and a natural heparanase inhibitor. However, it is not clinically usable due to its potent anticoagulant activity. Thus, heparin modifications have been largely used for the development of efficient heparanase inhibitors [113]. These include SST0001, which is currently in Phase I/II clinical trial for myeloma treatment [113]. Another heparin derivative is M402 that is also currently in Phase I/II clinical trial in combination with gemcitabine against metastatic pancreatic cancer [113]. The PI-88, a heparanase inhibitor with potent antiangiogenic activity [114,115], is currently on Phase III clinical trial as "the adjuvant treatment of subjects with hepatitis virus-related hepatocellular carcinoma after surgical resection" (http://www.cancer.gov/drugdictionary). Moreover, an additional heparan sulfate mimetic PG545 that is a potent heparanase and angiogenesis inhibitor has shown to be promising when tested in preclinical murine cancer models [116] and is now on Phase I clinical trial (http://www.cancer.gov/drugdictionary). All of these heparanase inhibitors aim at inhibiting the extracellular heparanase activity.

10.4.1.5 Sialidase 1/Neuraminidase 1 Aberrant glycosylation is one of the typical features of cancer cells. Changes in sialylation have been connected with the cancer cell invasiveness and metastatic potential [117]. There are four types of mammalian sialidases (Neu1, Neu2, Neu3, and Neu4). Of these, Neu1 is the predominant sialidase found in lysosomes [118]. When total sialidase activity is generally increased in cancer, Neu1 activity is decreased and, for example, in rat 3Y1 fibroblasts transformed with src or v-fos and mouse adenocarcinoma colon 26 cells, Neu1 expression correlates negatively with the metastatic potential [117]. Ectopic expression of Neu1 in human colon cancer HT-29 cells significantly decreases and its depletion increases their potential to metastasize into liver in a mouse xenograft model. The metastasis-suppressing function of Neu1 is mediated via sialylation of β4 integrin, resulting in its decreased phosphorylation and signaling [119]. Moreover, Neu1 is a negative regulator of lysosomal exocytosis, thus decreased Neu1 levels may contribute to increased lysosomal exocytosis supporting its antimetastatic activity [120]. Excessive lysosomal exocytosis in Neu1-deficient cells can be inhibited by silencing of lysosome-associated membrane protein 1 (LAMP1), suggesting that oversialylated

LAMP1 is a key mediator of excessive exocytosis of lysosomal proteases in these cells [120]. Increased sialylation of LAMP1 and lysosome-associated membrane protein 2 (LAMP2) can be observed in some human cancers, and, for example, in colorectal cancer it strongly correlates with the metastatic potential [121]. Because sialylated LAMPs are essential for the lysosomal membrane stability [122], the excessive sialylation of LAMPs may protect lysosomal membranes from destabilization and thus enhance cancer cell survival. The fact that its inactivation promotes cancer progression makes sialidase a challenging target for anticancer treatment.

10.4.2 Changes in the Lysosomal Membrane Proteins

10.4.2.1 Lysosomal Membrane Proteins LAMP1 and LAMP2 The most abundant and the most studied lysosomal membrane proteins are LAMP1 and LAMP2. Several less studied lysosomal membrane proteins exist whose role or changes in cancer are not established and thus they are not discussed here [3,19,123]. LAMP1 and 2 are type-1 transmembrane proteins containing a large and heavily glycosylated luminal domain and a short C-terminal cytosolic domain. Together they comprise over 50% of the lysosomal membrane proteins and are essential for the maintenance of the structural integrity of the lysosomal membranes [14,19].

Oncogenic transformation of fibroblasts decreases the levels of LAMPs on their lysosomal membranes thus increasing their sensitivity to LMP and LCD [14]. The glycosylated tails of LAMPs form a sugar coat called glycocalyx to the inner surface of the limiting membrane protecting the lysosomal membrane from degradation by lysosomal hydrolases [19]. LAMPs have shown to exhibit overlapping and distinct functions in lysosomal trafficking and exocytosis, chaperone-mediated autophagy, autophagosome–lysosome fusion, and cholesterol transport [19,120,124], all of which may be affected when LAMP levels are altered.

10.4.2.2 V-ATPase Vacuolar-type H^+ ATPase (V-ATPase) is a proton pump that is associated with the membranes of endosomes, lysosomes, and secretory vesicles, and it is also found at the plasma membranes of many cell types. V-ATPase is important for the acidification of lysosomes and endosomes is also involved in the regulation of the pH of the cytosol and possibly also of the extracellular space [125]. Its proper function is necessary for the maintenance of the highly acidic pH of the lysosomal lumen and for the maintenance of the calcium homeostasis in lysosomes [126,127].

In spite of the cancer-associated increase in acid production, the entire V-ATPase complex does not seem to be upregulated in cancer cells. Instead, the increased expression of specific subunits has been observed and is suggested to contribute to altered pump activity. Thus, overexpression of the gene encoding for the V1 subunit C (*ATP6V1C1*) in oral squamous carcinoma cells can enhance the pump activity by promoting the assembly of V0 and V1 subcomplexes [128]. On the other hand, the V0 subunit C (ATP6V0C) is upregulated in cisplatin-resistant cells [129]. The expression pattern of the subunit isoforms can affect the subcellular localization of the V-ATPase. The appearance of the V0 a3 isoform favors plasma membrane localization of this type of V-ATPase [130], whereas the depletion of a3 isoform in highly invasive breast cancer or melanoma cells results in decreased cytosolic pH and reduced invasiveness [131,132], thus making it less favorable for cathepsins to function on the cleavage of the extracellular matrix to enhance invasion. The expression of V-ATPases at the cell membrane is normally low but is, for example, prominently increased in highly invasive MDA-MB-231 breast cancer cells in comparison to the less invasive MCF7 breast cancer cells [131,133], possibly through increased lysosomal exocytosis. Inhibition of V-ATPase with bafilomycin A1, a specific V-ATPase inhibitor, reduces the *in vitro* invasiveness of the MDA-MB-231 cells [131].

Several different V-ATPase inhibitors exist and they are known to induce programmed cell death, revert multidrug-resistant phenotype and induce cell cycle arrest in cancer cells and decrease mouse xenograft tumor growth [125,134,135]. Perhaps the most interesting of these are the benzolactone enamides that may even discriminate between different V-ATPases and thus could be good basic compounds for further development [136,137]. Moreover, benzolactone enamide RTA203 and taxol can synergistically kill H1155 cancer cells at low doses [138], supporting the possible use of benzolactone enamide-based compounds in combinatory therapy. These very potent V-ATPase inhibitors are however quite toxic, which limits their use in the clinic. Due to this reason, well-tolerated inhibitors of H^+/K^+ ATPase, which is an enzyme involved in proton secretion in the stomach (e.g., omeprazole and esomeprazole), have been suggested to be utilized instead. These are capable of inhibiting V-ATPase and inducing LCD *in vitro*, a notion that has been supported by some promising *in vivo* results [139–141].

10.5 MOLECULAR CHANGES INVOLVING LYSOSOMAL INTEGRITY

Oncogene-induced, decreased expression of LAMPs and increased expression and activity of cysteine cathepsins, especially cathepsin B, can contribute to LMP by sensitizing lysosomal membranes for leakage and rupture, as presented earlier in this chapter. Moreover, cytoskeletal/cytosolic proteins, such as cytoskeleton-associated motor proteins, protect lysosomes from LMP [142], and disruption of cytoskeleton and cellular trafficking by microtubule-destabilizing drugs induces LMP [27,143]. There are additional, mainly cytosolic molecules that may contribute to lysosomal stability [18,144,145]. One of these is the heat shock protein 70 (Hsp70) that can promote cell survival by inhibiting LMP [146]. This occurs via Hsp70 binding to the endolysosomal anionic phospholipid bis(monoacylglycero) phosphate (BMP), which is an essential cofactor for lysosomal sphingomyelin metabolism and has an important role in the maintenance of lysosomal integrity [147].

10.5.1 Cancer-Associated Changes in Lysosomal Sphingolipid Metabolism

Sphingolipids modulate diverse physiological and pathological processes including cell growth, cell death, autophagy, and angiogenesis [148]. Cancer-associated dysregulation of sphingolipid metabolism contributes to cancer progression and resistance

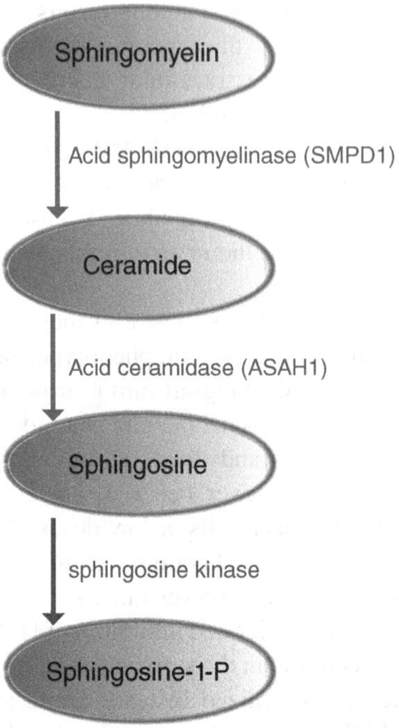

Figure 10.3 Lysosomal sphingolipid catabolism. Sphingomyelin is an abundant structural lipid of biological membranes. Acid sphingomyelinase catalyzes sphingomyelin into ceramide, which is a membrane lipid that can be associated with lipid rafts and function as a signaling molecule regulating cell death and autophagy. Acid ceramidase converts ceramide into sphingosine, which is a soluble detergent that can also function as a signaling molecule. Finally, sphingosine kinase phosphorylates sphingosine turning it into sphingosine-1-phosphate, a soluble signaling molecule that can bind to cell surface receptor promoting cell survival and autophagy.

to chemotherapeutics [149]. Sphingomyelinases and ceramidases are key enzymes of sphingolipid catabolism (Figure 10.3). Human cells express several forms of these enzymes with varying optimal pH. Here, we present only their lysosomal forms that can function in low pH, that is, ASM and acid ceramidase (AC). They regulate the formation and degradation of ceramide, one of the most studied sphingolipids. The vast interest to ceramide in cancer owes to its capacity to induce programmed cell death and inhibit cell proliferation [150]. Genetic inactivation of acid sphingomyelinase, which is the major sphingomyelinase in lysosomes, protects mice against various cellular stresses including ionizing radiation and treatment with cisplatin, assumably due to deficient ceramide production, supporting the apoptosis-promoting role of ceramide [151].

Sphingolipids and sphingolipid metabolites such as ceramide regulate cell survival with diverse mechanisms that are not completely understood. The proapoptotic activity of ASM is mostly attributed to its localization to the plasma membrane in response to various stresses [152,153]. This induces generation of ceramide at the plasma membrane contributing to the formation of ordered membrane domains that can enhance death receptor function and lead to the induction of apoptosis [151–153]. ASM activity is generally considered to negatively regulate cancer progression, mainly due to the proapoptotic properties of ceramide [154,155]. Contrasting this, lysosomal ASM has an opposite effect on cell survival, since its activity is necessary for the maintenance of the lysosomal membrane integrity [22,144,147]. Another mechanism how ASM can positively contribute to cancer progression is by ceramide-mediated activation of pre-pro cathepsin D resulting in increased activity of cathepsin D [156]. Interestingly, a database search reveals a highly significant cancer-associated reduction in the mRNA level of *SMPD1* encoding for ASM in multiple studies of gastrointestinal, hepatocellular, salivary gland, renal, and head and neck carcinomas. A modest but significant increase in *SMPD1* expression was found in single studies of hairy cell leukemia and smoldering myeloma (www.oncomine.org). The decreased *SMPD1* expression in several cancers may contribute to the increased sensitivity of transformed cells to LCD pathways. On the other hand, the *SMPD1* mRNA levels do not necessarily reflect the levels and activity of ASM protein, which are further contributed by its stability, binding with BMP and association with Hsp70 [146,147,157].

Lysosomal ceramide is degraded by AC to produce sphingosine, which diffuses rapidly from lysosomes to cytosol and is converted to sphingosine-1-phosphate [158, 159]. Aberrant expression of AC has been attributed to several human cancers [160]. Inhibition of AC activity is expected to lead to increased ceramide levels, which can stimulate programmed cell death. This has been verified with different cancer and primary cell lines using various AC inhibitors such as *N*-oleoylethanolamine, B13, and De-MAPP [160,161]. Although this data suggests that AC inhibition could be a good cancer therapy strategy, it is to be noticed that the specificity of many of these inhibitors toward AC in respect to, for example, other ceramidases remains to be verified. In addition to survival, sphingolipid metabolism is also involved in other cellular processes that are important in cancer including proliferation, cell cycle progression, and inflammation, which have increased its interest as a potential cancer therapy target [159].

10.5.2 Targeting Lysosomal Membrane Integrity

Cancer cells frequently escape from spontaneous and therapy-induced apoptosis due to mutational defects in the caspase-dependent apoptosis pathways [162]. Thus, identification and utilization of caspase-independent cellular death programs to selectively kill cancer cells have become increasingly important [17,163,164]. Cancer development and progression involves dramatic changes in the composition and integrity of lysosomal membranes, which provides intriguing possibilities for drug-induced LMP and LCD [14,17,22,165]. Several established anticancer agents induce LCD [17]. These include DNA-damaging drugs camptothecin, cisplatin, and etoposide [14,166,167]; microtubule targeting drugs discodermolide, docetaxel, epothilone B, paclitaxel, and vincristine [27,143,168]; V-ATPase inhibitors omeprazole and bafilomycin A1 [139,169,170] among many others [17]. Interestingly, in all of these cases, LCD can be mediated by single cathepsins B, D, or L, or their different combinations [17].

Recently, lysosomotropic detergents that specifically destabilize lysosomal membranes in cancer cells have evolved as highly promising, potential anticancer remedies [17,21]. These include siramesine [22,171,172], BAMLET [173], hydroxychloroquine [174], LeuLeuOMe [175], sphingosine [176], mefloquine [177], and SU11652 and sunitinib [178]. Recent work with siramesine has led to the identification of a group of compounds, a subset of cationic amphiphilic drugs, which specifically sensitize lysosomal membranes of cancer cells by a mechanism that involves direct inhibition of ASM [22]. Interestingly, these compounds involve some feasible antidepressants and antiallergy remedies [22,179]. In addition to ASM inhibition, their detergent-like activity and ability to induce the production of reactive oxygen species is likely to further contribute to LMP and LCD [22]. Very exciting and promising properties associated with these LCD-inducing agents are that they can also efficiently kill multidrug-resistant cancer cells [17,22].

10.6 CONCLUSION

Lysosomes are packed with powerful hydrolases that upon release from the lysosomal compartment can degrade the environment that they encounter. This imposes opposing outcomes to cancer cells depending on where the lysosomes "leak," thus challenging the design of the lysosome targeting therapies. In principle, lysosomal content that encounters the cytosol will most likely induce cell death and destruction of the cancer cell. On the contrary, leakage of lysosomal hydrolases into the extracellular space can facilitate cancer cell invasion by degrading the extracellular matrix. Thus, these two options provide two different possibilities to intervene. Consequently, two types of promising anticancer therapy strategies involving lysosomes are currently being explored: inhibition of extracellular lysosomal hydrolase activities and induction of lysosomal permeabilization and leakage within cancer cells.

Long-term research on cysteine cathepsins and heparanase has led to the development of several highly promising anticancer agents with a strategy to inhibit invasion and metastasis by inhibiting the activity of these proteases [11,113,180]. Many of these are already in clinical trials. Moreover, recent identification of several commonly used anticancer remedies as compounds that can induce cancer-specific LMP [17] together with the identification of a widely used subset of cationic amphiphilic drugs including tricyclic antidepressants, antihistamines, and calcium channel blockers as direct ASM inhibitors capable of destabilizing lysosomal membranes in cancer cells [22] has rapidly provided many novel and feasible possibilities to intervene cancer progression by targeting lysosomes in cancer cells. Especially, important and promising are the observations indicating that they can target apoptosis-deficient cancer cells as well as those that have become multidrug resistant [17,22,178]. Since cathepsins are important mediators of both invasion and LCD, more research is needed to dissect those cancers that would mostly benefit from cathepsin inhibition from those that would need full cathepsin activity for the execution of cell death. One possible option could be to combine both processes: the induction of cell death and the inhibition of invasion. This could, for example, be achieved by combining LCD-inducing treatments with treatments that inhibit lysosomal exocytosis. This would require more research on lysosomal exocytosis and its regulation in cancer.

REFERENCES

[1] Kolter T, Sandhoff K. Lysosomal degradation of membrane lipids. FEBS Lett 2010; 584:1700–1712.

[2] Pryor PR, Luzio JP. Delivery of endocytosed membrane proteins to the lysosome. Biochim Biophys Acta 2009;1793:615–624.

[3] Saftig P, Klumperman J. Lysosome biogenesis and lysosomal membrane proteins: trafficking meets function. Nat Rev Mol Cell Biol 2009;10:623–635.

[4] Grandal MV, Madshus IH. Epidermal growth factor receptor and cancer: control of oncogenic signalling by endocytosis. J Cell Mol Med 2008;12:1527–1534.

[5] Mellman I, Yarden Y. Endocytosis and cancer. Cold Spring Harb Perspect Biol 2013; 5:a016949.

[6] Ivanov A, Pawlikowski J, Manoharan I, van Tuyn J, Nelson DM, et al. Lysosome-mediated processing of chromatin in senescence. J Cell Biol 2013;202:129–143.

[7] Kreuzaler PA, Staniszewska AD, Li W, Omidvar N, Kedjouar B, et al. Stat3 controls lysosomal-mediated cell death *in vivo*. Nat Cell Biol 2011;13:303–309.

[8] Watson CJ, Kreuzaler PA. The role of cathepsins in involution and breast cancer. J Mammary Gland Biol Neoplasia 2009;14:171–179.

[9] Glunde K, Guggino SE, Solaiyappan M, Pathak AP, Ichikawa Y, et al. Extracellular acidification alters lysosomal trafficking in human breast cancer cells. Neoplasia 2003;5:533–545.

[10] Liu Y, Zhou Y, Zhu K. Inhibition of glioma cell lysosome exocytosis inhibits glioma invasion. PLoS One 2012;7:e45910.

[11] Kallunki T, Olsen OD, Jaattela M. Cancer-associated lysosomal changes: friends or foes? Oncogene 2013;32:1995–2004.

[12] Allison AC. Lysosomes in cancer cells. J Clin Pathol Suppl (R Coll Pathol) 1974; 7:43–50.

[13] Rafn B, Nielsen CF, Andersen SH, Szyniarowski P, Corcelle-Termeau E, et al. ErbB2-driven breast cancer cell invasion depends on a complex signaling network activating myeloid zinc finger-1-dependent cathepsin B expression. Mol Cell 2012; 45:764–776.

[14] Fehrenbacher N, Bastholm L, Kirkegaard-Sorensen T, Rafn B, Bottzauw T, et al. Sensitization to the lysosomal cell death pathway by oncogene-induced down-regulation of lysosome-associated membrane proteins 1 and 2. Cancer Res 2008;68:6623–6633.

[15] Kim K, Cai J, Shuja S, Kuo T, Murnane MJ. Presence of activated ras correlates with increased cysteine proteinase activities in human colorectal carcinomas. Int J Cancer 1998;79:324–333.

[16] Sloane BF, Moin K, Sameni M, Tait LR, Rozhin J, et al. Membrane association of cathepsin B can be induced by transfection of human breast epithelial cells with c-Ha-ras oncogene. J Cell Sci 1994;107(Pt 2):373–384.

[17] Groth-Pedersen L, Jaattela M. Combating apoptosis and multidrug resistant cancers by targeting lysosomes. Cancer Lett 2013;332:265–274.

[18] Boya P, Kroemer G. Lysosomal membrane permeabilization in cell death. Oncogene 2008;27:6434–6451.

[19] Eskelinen EL. Roles of LAMP-1 and LAMP-2 in lysosome biogenesis and autophagy. Mol Aspects Med 2006;27:495–502.

[20] Fehrenbacher N, Jaattela M. Lysosomes as targets for cancer therapy. Cancer Res 2005;65:2993–2995.

[21] Aits S, Jaattela M. Lysosomal cell death at a glance. J Cell Sci 2013;126:1905–1912.

[22] Petersen NH, Olsen OD, Groth-Pedersen L, Ellegaard AM, Bilgin M, et al. Transformation-associated changes in sphingolipid metabolism sensitize cells to lysosomal cell death induced by inhibitors of acid sphingomyelinase. Cancer Cell 2013; 24:379–393.

[23] Mora R, Dokic I, Kees T, Huber CM, Keitel D, et al. Sphingolipid rheostat alterations related to transformation can be exploited for specific induction of lysosomal cell death in murine and human glioma. Glia 2010;58:1364–1383.

[24] Taha TA, Kitatani K, Bielawski J, Cho W, Hannun YA, et al. Tumor necrosis factor induces the loss of sphingosine kinase-1 by a cathepsin B-dependent mechanism. J Biol Chem 2005;280:17196–17202.

[25] Durchfort N, Verhoef S, Vaughn MB, Shrestha R, Adam D, et al. The enlarged lysosomes in beige j cells result from decreased lysosome fission and not increased lysosome fusion. Traffic 2012;13:108–119.

[26] Ono K, Kim SO, Han J. Susceptibility of lysosomes to rupture is a determinant for plasma membrane disruption in tumor necrosis factor alpha-induced cell death. Mol Cell Biol 2003;23:665–676.

[27] Groth-Pedersen L, Ostenfeld MS, Hoyer-Hansen M, Nylandsted J, Jaattela M. Vincristine induces dramatic lysosomal changes and sensitizes cancer cells to lysosome-destabilizing siramesine. Cancer Res 2007;67:2217–2225.

[28] Brandes D, Anton E. The role of lysosomes in cellular lytic processes. 3. Electron histochemical changes in mammary tumors after treatment with cytoxan and vitamin A. Lab Invest 1966;15:987–1006.

[29] Brandes D, Sloan KW, Anton E, Bloedorn F. The effect of x-irradiation on the lysosomes of mouse mammary gland carcinomas. Cancer Res 1967;27:731–746.

[30] Mohamed MM, Sloane BF. Cysteine cathepsins: multifunctional enzymes in cancer. Nat Rev Cancer 2006;6:764–775.

[31] Chiche J, Brahimi-Horn MC, Pouyssegur J. Tumour hypoxia induces a metabolic shift causing acidosis: a common feature in cancer. J Cell Mol Med 2010;14:771–794.

[32] Korolchuk VI, Saiki S, Lichtenberg M, Siddiqi FH, Roberts EA, et al. Lysosomal positioning coordinates cellular nutrient responses. Nat Cell Biol 2011;13:453–460.

[33] Joyce JA, Hanahan D. Multiple roles for cysteine cathepsins in cancer. Cell Cycle 2004;3:1516–1619.

[34] Li JP. Heparin, heparan sulfate and heparanase in cancer: remedy for metastasis? Anticancer Agents Med Chem 2008;8:64–76.

[35] Nagler DK, Storer AC, Portaro FC, Carmona E, Juliano L, et al. Major increase in endopeptidase activity of human cathepsin B upon removal of occluding loop contacts. Biochemistry 1997;36:12608–12615.

[36] Almeida PC, Nantes IL, Chagas JR, Rizzi CC, Faljoni-Alario A, et al. Cathepsin B activity regulation. Heparin-like glycosaminogylcans protect human cathepsin B from alkaline pH-induced inactivation. J Biol Chem 2001;276:944–951.

[37] Zaidi N, Maurer A, Nieke S, Kalbacher H. Cathepsin D: a cellular roadmap. Biochem Biophys Res Commun 2008;376:5–9.

[38] Reid WA, Valler MJ, Kay J. Immunolocalization of cathepsin D in normal and neoplastic human tissues. J Clin Pathol 1986;39:1323–1330.

[39] Liaudet-Coopman E, Beaujouin M, Derocq D, Garcia M, Glondu-Lassis M, et al. Cathepsin D: newly discovered functions of a long-standing aspartic protease in cancer and apoptosis. Cancer Lett 2006;237:167–179.

[40] Leto G, Tumminello FM, Crescimanno M, Flandina C, Gebbia N. Cathepsin D expression levels in nongynecological solid tumors: clinical and therapeutic implications. Clin Exp Metastasis 2004;21:91–106.

[41] Benes P, Vetvicka V, Fusek M. Cathepsin D – many functions of one aspartic protease. Crit Rev Oncol Hematol 2008;68:12–28.

[42] Benes P, Vashishta A, Saraswat-Ohri S, Fusek M, Pospisilova S, et al. Effect of procathepsin D activation peptide on gene expression of breast cancer cells. Cancer Lett 2006;239:46–54.

[43] Leto G, Sepporta MV, Crescimanno M, Flandina C, Tumminello FM. Cathepsin L in metastatic bone disease: therapeutic implications. Biol Chem 2010;391:655–664.

[44] Mason SD, Joyce JA. Proteolytic networks in cancer. Trends Cell Biol 2011;21: 228–237.

[45] Jedeszko C, Sloane BF. Cysteine cathepsins in human cancer. Biol Chem 2004; 385:1017–1027.

[46] Yan S, Sloane BF. Molecular regulation of human cathepsin B: implication in pathologies. Biol Chem 2003;384:845–854.

[47] Sinha AA, Gleason DF, Deleon OF, Wilson MJ, Sloane BF. Localization of a biotinylated cathepsin B oligonucleotide probe in human prostate including invasive cells and invasive edges by in situ hybridization. Anat Rec 1993;235:233–240.

[48] Demchik LL, Sameni M, Nelson K, Mikkelsen T, Sloane BF. Cathepsin B and glioma invasion. Int J Dev Neurosci 1999;17:483–494.

[49] Rafn B, Kallunki T. A way to invade: a story of ErbB2 and lysosomes. Cell Cycle 2012;11:2415–2416.

[50] Settembre C, Ballabio A. TFEB regulates autophagy: an integrated coordination of cellular degradation and recycling processes. Autophagy 2011;7:1379–1381.

[51] Frosch BA, Berquin I, Emmert-Buck MR, Moin K, Sloane BF. Molecular regulation, membrane association and secretion of tumor cathepsin B. APMIS 1999;107:28–37.

[52] Sameni M, Elliott E, Ziegler G, Fortgens PH, Dennison C, et al. Cathepsin B and D are localized at the surface of human breast cancer cells. Pathol Oncol Res 1995;1:43–53.

[53] Gocheva V, Zeng W, Ke D, Klimstra D, Reinheckel T, et al. Distinct roles for cysteine cathepsin genes in multistage tumorigenesis. Genes Dev 2006;20:543–556.

[54] Gocheva V, Chen X, Peters C, Reinheckel T, Joyce JA. Deletion of cathepsin H perturbs angiogenic switching, vascularization and growth of tumors in a mouse model of pancreatic islet cell cancer. Biol Chem 2010;391:937–945.

[55] Mai J, Sameni M, Mikkelsen T, Sloane BF. Degradation of extracellular matrix protein tenascin-C by cathepsin B: an interaction involved in the progression of gliomas. Biol Chem 2002;383:1407–1413.

[56] Buck MR, Karustis DG, Day NA, Honn KV, Sloane BF. Degradation of extracellular-matrix proteins by human cathepsin B from normal and tumour tissues. Biochem J 1992;282(Pt 1):273–278.

[57] Kobayashi H, Schmitt M, Goretzki L, Chucholowski N, Calvete J, et al. Cathepsin B efficiently activates the soluble and the tumor cell receptor-bound form of the proenzyme urokinase-type plasminogen activator (Pro-uPA). J Biol Chem 1991;266:5147–5152.

[58] Vasiljeva O, Papazoglou A, Kruger A, Brodoefel H, Korovin M, et al. Tumor cell-derived and macrophage-derived cathepsin B promotes progression and lung metastasis of mammary cancer. Cancer Res 2006;66:5242–5250.

[59] Sevenich L, Werner F, Gajda M, Schurigt U, Sieber C, et al. Transgenic expression of human cathepsin B promotes progression and metastasis of polyoma-middle-T-induced breast cancer in mice. Oncogene 2011;30:54–64.

[60] Bengsch F, Buck A, Gunther SC, Seiz JR, Tacke M, et al. Cell type-dependent pathogenic functions of overexpressed human cathepsin B in murine breast cancer progression. Oncogene 2013;33:4474–4484.

[61] Gondi CS, Rao JS. Cathepsin B as a cancer target. Expert Opin Ther Targets 2013; 17:281–291.

[62] Abboud-Jarrous G, Atzmon R, Peretz T, Palermo C, Gadea BB, et al. Cathepsin L is responsible for processing and activation of proheparanase through multiple cleavages of a linker segment. J Biol Chem 2008;283:18167–18176.

[63] Corvo I, Cancela M, Cappetta M, Pi-Denis N, Tort JF, et al. The major cathepsin L secreted by the invasive juvenile *Fasciola hepatica* prefers proline in the S2 subsite and can cleave collagen. Mol Biochem Parasitol 2009;167:41–47.

[64] Ishidoh K, Kominami E. Procathepsin L degrades extracellular matrix proteins in the presence of glycosaminoglycans *in vitro*. Biochem Biophys Res Commun 1995; 217:624–631.

[65] Lankelma JM, Voorend DM, Barwari T, Koetsveld J, Van der Spek AH, et al. Cathepsin L, target in cancer treatment? Life Sci 2010;86:225–233.

[66] Gocheva V, Joyce JA. Cysteine cathepsins and the cutting edge of cancer invasion. Cell Cycle 2007;6:60–64.

[67] Strojnik T, Kavalar R, Trinkaus M, Lah TT. Cathepsin L in glioma progression: comparison with cathepsin B. Cancer Detect Prev 2005;29:448–455.

[68] Rebbaa A, Chu F, Sudha T, Gallati C, Dier U, et al. The anti-angiogenic activity of NSITC, a specific cathepsin L inhibitor. Anticancer Res 2009;29:4473–4481.

[69] Sullivan S, Tosetto M, Kevans D, Coss A, Wang L, et al. Localization of nuclear cathepsin L and its association with disease progression and poor outcome in colorectal cancer. Int J Cancer 2009;125:54–61.

[70] Ceru S, Konjar S, Maher K, Repnik U, Krizaj I, et al. Stefin B interacts with histones and cathepsin L in the nucleus. J Biol Chem 2010;285:10078–10086.

[71] Duncan EM, Muratore-Schroeder TL, Cook RG, Garcia BA, Shabanowitz J, et al. Cathepsin L proteolytically processes histone H3 during mouse embryonic stem cell differentiation. Cell 2008;135:284–294.

[72] Goulet B, Baruch A, Moon NS, Poirier M, Sansregret LL, et al. A cathepsin L isoform that is devoid of a signal peptide localizes to the nucleus in S phase and processes the CDP/Cux transcription factor. Mol Cell 2004;14:207–219.

[73] Zheng X, Chu F, Chou PM, Gallati C, Dier U, et al. Cathepsin L inhibition suppresses drug resistance *in vitro* and *in vivo*: a putative mechanism. Am J Physiol Cell Physiol 2009;296:C65–74.

[74] Boudreau F, Lussier CR, Mongrain S, Darsigny M, Drouin JL, et al. Loss of cathepsin L activity promotes claudin-1 overexpression and intestinal neoplasia. FASEB J 2007;21:3853–3865.

[75] Dennemarker J, Lohmuller T, Mayerle J, Tacke M, Lerch MM, et al. Deficiency for the cysteine protease cathepsin L promotes tumor progression in mouse epidermis. Oncogene 2010;29:1611–1621.

[76] Benavides F, Perez C, Blando J, Contreras O, Shen J, et al. Protective role of cathepsin L in mouse skin carcinogenesis. Mol Carcinog 2011;51:352–361.

[77] Reinheckel T, Hagemann S, Dollwet-Mack S, Martinez E, Lohmuller T, et al. The lysosomal cysteine protease cathepsin L regulates keratinocyte proliferation by control of growth factor recycling. J Cell Sci 2005;118:3387–3395.

[78] Walz M, Kellermann S, Bylaite M, Andree B, Ruther U, et al. Expression of the human Cathepsin L inhibitor hurpin in mice: skin alterations and increased carcinogenesis. Exp Dermatol 2007;16:715–723.

[79] Grabowskal U, Chambers TJ, Shiroo M. Recent developments in cathepsin K inhibitor design. Curr Opin Drug Discov Devel 2005;8:619–630.

[80] Gelb BD, Shi GP, Chapman HA, Desnick RJ. Pycnodysostosis, a lysosomal disease caused by cathepsin K deficiency. Science 1996;273:1236–1238.

[81] Brubaker KD, Vessella RL, True LD, Thomas R, Corey E. Cathepsin K mRNA and protein expression in prostate cancer progression. J Bone Miner Res 2003;18: 222–230.

[82] Le Gall C, Bellahcene A, Bonnelye E, Gasser JA, Castronovo V, et al. A cathepsin K inhibitor reduces breast cancer induced osteolysis and skeletal tumor burden. Cancer Res 2007;67:9894–9902.

[83] Littlewood-Evans AJ, Bilbe G, Bowler WB, Farley D, Wlodarski B, et al. The osteoclast-associated protease cathepsin K is expressed in human breast carcinoma. Cancer Res 1997;57:5386–5390.

[84] Lewiecki EM. Odanacatib, a cathepsin K inhibitor for the treatment of osteoporosis and other skeletal disorders associated with excessive bone remodeling. IDrugs 2009;12:799–809.

[85] Podgorski I. Future of anticathepsin K drugs: dual therapy for skeletal disease and atherosclerosis? Future Med Chem 2009;1:21–34.

[86] Quintanilla-Dieck MJ, Codriansky K, Keady M, Bhawan J, Runger TM. Cathepsin K in melanoma invasion. J Invest Dermatol 2008;128:2281–2288.

[87] Santamaria I, Velasco G, Cazorla M, Fueyo A, Campo E, et al. Cathepsin L2, a novel human cysteine proteinase produced by breast and colorectal carcinomas. Cancer Res 1998;58:1624–1630.

[88] Krueger S, Kalinski T, Hundertmark T, Wex T, Kuster D, et al. Up-regulation of cathepsin X in *Helicobacter pylori* gastritis and gastric cancer. J Pathol 2005;207:32–42.

[89] Fernandez PL, Farre X, Nadal A, Fernandez E, Peiro N, et al. Expression of cathepsins B and S in the progression of prostate carcinoma. Int J Cancer 2001;95:51–55.

[90] Nagler DK, Kruger S, Kellner A, Ziomek E, Menard R, et al. Up-regulation of cathepsin X in prostate cancer and prostatic intraepithelial neoplasia. Prostate 2004;60:109–119.

[91] Joyce JA, Baruch A, Chehade K, Meyer-Morse N, Giraudo E, et al. Cathepsin cysteine proteases are effectors of invasive growth and angiogenesis during multistage tumorigenesis. Cancer Cell 2004;5:443–453.

[92] Sevenich L, Schurigt U, Sachse K, Gajda M, Werner F, et al. Synergistic antitumor effects of combined cathepsin B and cathepsin Z deficiencies on breast cancer progression and metastasis in mice. Proc Natl Acad Sci U S A 2010;107:2497–2502.

[93] Vasiljeva O, Korovin M, Gajda M, Brodoefel H, Bojic L, et al. Reduced tumour cell proliferation and delayed development of high-grade mammary carcinomas in cathepsin B-deficient mice. Oncogene 2008;27:4191–4199.

[94] Wang B, Sun J, Kitamoto S, Yang M, Grubb A, et al. Cathepsin S controls angiogenesis and tumor growth via matrix-derived angiogenic factors. J Biol Chem 2006;281: 6020–6029.

[95] Cox JL. Cystatins and cancer. Front Biosci (Landmark Ed) 2009;14:463–474.

[96] Rivenbark AG, Coleman WB. Epigenetic regulation of cystatins in cancer. Front Biosci (Landmark Ed) 2009;14:453–462.

[97] Kopitz C, Anton M, Gansbacher B, Kruger A. Reduction of experimental human fibrosarcoma lung metastasis in mice by adenovirus-mediated cystatin C overexpression in the host. Cancer Res 2005;65:8608–8612.

[98] Withana NP, Blum G, Sameni M, Slaney C, Anbalagan A, et al. Cathepsin B inhibition limits bone metastasis in breast cancer. Cancer Res 2012;72:1199–1209.

[99] Schurigt U, Sevenich L, Vannier C, Gajda M, Schwinde A, et al. Trial of the cysteine cathepsin inhibitor JPM-OEt on early and advanced mammary cancer stages in the MMTV-PyMT-transgenic mouse model. Biol Chem 2008;389:1067–1074.

[100] Palermo C, Joyce JA. Cysteine cathepsin proteases as pharmacological targets in cancer. Trends Pharmacol Sci 2008;29:22–28.

[101] Bernfield M, Gotte M, Park PW, Reizes O, Fitzgerald ML, et al. Functions of cell surface heparan sulfate proteoglycans. Annu Rev Biochem 1999;68:729–777.

[102] Folkman J, Klagsbrun M, Sasse J, Wadzinski M, Ingber D, et al. A heparin-binding angiogenic protein – basic fibroblast growth factor – is stored within basement membrane. Am J Pathol 1988;130:393–400.

[103] Friedmann Y, Vlodavsky I, Aingorn H, Aviv A, Peretz T, et al. Expression of heparanase in normal, dysplastic, and neoplastic human colonic mucosa and stroma. Evidence for its role in colonic tumorigenesis. Am J Pathol 2000;157:1167–1175.

[104] Ilan N, Elkin M, Vlodavsky I. Regulation, function and clinical significance of heparanase in cancer metastasis and angiogenesis. Int J Biochem Cell Biol 2006;38: 2018–2039.

[105] Parish CR, Freeman C, Hulett MD. Heparanase: a key enzyme involved in cell invasion. Biochim Biophys Acta 2001;1471:M99–108.

[106] Vlodavsky I, Friedmann Y. Molecular properties and involvement of heparanase in cancer metastasis and angiogenesis. J Clin Invest 2001;108:341–347.

[107] Shafat I, Vlodavsky I, Ilan N. Characterization of mechanisms involved in secretion of active heparanase. J Biol Chem 2006;281:23804–23811.

[108] Davidson B, Shafat I, Risberg B, Ilan N, Trope CG, et al. Heparanase expression correlates with poor survival in metastatic ovarian carcinoma. Gynecol Oncol 2007; 104:311–319.

[109] Quiros RM, Rao G, Plate J, Harris JE, Brunn GJ, et al. Elevated serum heparanase-1 levels in patients with pancreatic carcinoma are associated with poor survival. Cancer 2006;106:532–540.

[110] Marchetti D, Reiland J, Erwin B, Roy M. Inhibition of heparanase activity and heparanase-induced angiogenesis by suramin analogues. Int J Cancer 2003;104: 167–174.

[111] Vlodavsky I, Mohsen M, Lider O, Svahn CM, Ekre HP, et al. Inhibition of tumor metastasis by heparanase inhibiting species of heparin. Invasion Metastasis 1994;14:290–302.

[112] Norrby K. Low-molecular-weight heparins and angiogenesis. APMIS 2006;114: 79–102.

[113] Pisano C, Vlodavsky I, Ilan N, Zunino F. The potential of heparanase as a therapeutic target in cancer. Biochem Pharmacol 2014;89:12–19.

[114] Basche M, Gustafson DL, Holden SN, O'Bryant CL, Gore L, et al. A phase I biological and pharmacologic study of the heparanase inhibitor PI-88 in patients with advanced solid tumors. Clin Cancer Res 2006;12:5471–5480.

[115] Liu CJ, Lee PH, Lin DY, Wu CC, Jeng LB, et al. Heparanase inhibitor PI-88 as adjuvant therapy for hepatocellular carcinoma after curative resection: a randomized phase II trial for safety and optimal dosage. J Hepatol 2009;50:958–968.

[116] Dredge K, Hammond E, Handley P, Gonda TJ, Smith MT, et al. PG545, a dual heparanase and angiogenesis inhibitor, induces potent anti-tumour and anti-metastatic efficacy in preclinical models. Br J Cancer 2011;104:635–642.

[117] Miyagi T. Aberrant expression of sialidase and cancer progression. Proc Jpn Acad Ser B Phys Biol Sci 2008;84:407–418.

[118] Bonten E, van der Spoel A, Fornerod M, Grosveld G, d'Azzo A. Characterization of human lysosomal neuraminidase defines the molecular basis of the metabolic storage disorder sialidosis. Genes Dev 1996;10:3156–3169.

[119] Uemura T, Shiozaki K, Yamaguchi K, Miyazaki S, Satomi S, et al. Contribution of sialidase NEU1 to suppression of metastasis of human colon cancer cells through desialylation of integrin beta4. Oncogene 2009;28:1218–1229.

[120] Yogalingam G, Bonten EJ, van de Vlekkert D, Hu H, Moshiach S, et al. Neuraminidase 1 is a negative regulator of lysosomal exocytosis. Dev Cell 2008;15:74–86.

[121] Saitoh O, Wang WC, Lotan R, Fukuda M. Differential glycosylation and cell surface expression of lysosomal membrane glycoproteins in sublines of a human colon cancer exhibiting distinct metastatic potentials. J Biol Chem 1992;267:5700–5711.

[122] Eskelinen EL, Tanaka Y, Saftig P. At the acidic edge: emerging functions for lysosomal membrane proteins. Trends Cell Biol 2003;13:137–145.

[123] Honing S, Griffith J, Geuze HJ, Hunziker W. The tyrosine-based lysosomal targeting signal in lamp-1 mediates sorting into Golgi-derived clathrin-coated vesicles. EMBO J 1996;15:5230–5239.

[124] Schneede A, Schmidt CK, Holtta-Vuori M, Heeren J, Willenborg M, et al. Role for LAMP-2 in endosomal cholesterol transport. J Cell Mol Med 2011;15:280–295.

[125] Hernandez A, Serrano G, Herrera-Palau R, Perez-Castineira JR, Serrano A. Intraorganellar acidification by V-ATPases: a target in cell proliferation and cancer therapy. Recent Pat Anticancer Drug Discov 2010;5:88–98.

[126] Mindell JA. Lysosomal acidification mechanisms. Annu Rev Physiol 2012;74:69–86.

[127] Morgan AJ, Platt FM, Lloyd-Evans E, Galione A. Molecular mechanisms of endolysosomal Ca2+ signalling in health and disease. Biochem J 2011;439:349–374.

[128] Otero-Rey EM, Somoza-Martin M, Barros-Angueira F, Garcia-Garcia A. Intracellular pH regulation in oral squamous cell carcinoma is mediated by increased V-ATPase activity via over-expression of the ATP6V1C1 gene. Oral Oncol 2008;44:193–199.

[129] Torigoe T, Izumi H, Ishiguchi H, Uramoto H, Murakami T, et al. Enhanced expression of the human vacuolar H+-ATPase c subunit gene (ATP6L) in response to anticancer agents. J Biol Chem 2002;277:36534–36543.

[130] Toei M, Saum R, Forgac M. Regulation and isoform function of the V-ATPases. Biochemistry 2010;49:4715–4723.

[131] Hinton A, Sennoune SR, Bond S, Fang M, Reuveni M, et al. Function of a subunit isoforms of the V-ATPase in pH homeostasis and *in vitro* invasion of MDA-MB231 human breast cancer cells. J Biol Chem 2009;284:16400–16408.

[132] Nishisho T, Hata K, Nakanishi M, Morita Y, Sun-Wada GH, et al. The a3 isoform vacuolar type H(+)-ATPase promotes distant metastasis in the mouse B16 melanoma cells. Mol Cancer Res 2011;9:845–855.

[133] Sennoune SR, Bakunts K, Martinez GM, Chua-Tuan JL, Kebir Y, et al. Vacuolar H+-ATPase in human breast cancer cells with distinct metastatic potential: distribution and functional activity. Am J Physiol Cell Physiol 2004;286:C1443–1452.

[134] Perez-Sayans M, Somoza-Martin JM, Barros-Angueira F, Rey JM, Garcia-Garcia A. V-ATPase inhibitors and implication in cancer treatment. Cancer Treat Rev 2009; 35:707–713.

[135] Spugnini EP, Citro G, Fais S. Proton pump inhibitors as anti vacuolar-ATPases drugs: a novel anticancer strategy. J Exp Clin Cancer Res 2010;29:44.

[136] Sugimoto Y, Konoki K, Murata M, Matsushita M, Kanazawa H, et al. Design, synthesis, and biological evaluation of fluorinated analogues of salicylihalamide. J Med Chem 2009;52:798–806.

[137] Huss M, Wieczorek H. Inhibitors of V-ATPases: old and new players. J Exp Biol 2009; 212:341–346.

[138] Whitehurst AW, Bodemann BO, Cardenas J, Ferguson D, Girard L, et al. Synthetic lethal screen identification of chemosensitizer loci in cancer cells. Nature 2007;446:815–819.

[139] Scaringi L, Cornacchione P, Ayroldi E, Corazzi L, Capodicasa E, et al. Omeprazole induces apoptosis in jurkat cells. Int J Immunopathol Pharmacol 2004;17:331–342.

[140] Luciani F, Spada M, De Milito A, Molinari A, Rivoltini L, et al. Effect of proton pump inhibitor pretreatment on resistance of solid tumors to cytotoxic drugs. J Natl Cancer Inst 2004;96:1702–1713.

[141] De Milito A, Canese R, Marino ML, Borghi M, Iero M, et al. pH-dependent antitumor activity of proton pump inhibitors against human melanoma is mediated by inhibition of tumor acidity. Int J Cancer 2010;127:207–219.

[142] Groth-Pedersen L, Aits S, Corcelle-Termeau E, Petersen NH, Nylandsted J, et al. Identification of cytoskeleton-associated proteins essential for lysosomal stability and survival of human cancer cells. PLoS One 2012;7:e45381.

[143] Broker LE, Huisman C, Span SW, Rodriguez JA, Kruyt FA, et al. Cathepsin B mediates caspase-independent cell death induced by microtubule stabilizing agents in non-small cell lung cancer cells. Cancer Res 2004;64:27–30.

[144] Kirkegaard T, Jaattela M. Lysosomal involvement in cell death and cancer. Biochim Biophys Acta 2009;1793:746–754.

[145] Kroemer G, Jaattela M. Lysosomes and autophagy in cell death control. Nat Rev Cancer 2005;5:886–897.

[146] Nylandsted J, Gyrd-Hansen M, Danielewicz A, Fehrenbacher N, Lademann U, et al. Heat shock protein 70 promotes cell survival by inhibiting lysosomal membrane permeabilization. J Exp Med 2004;200:425–435.

[147] Kirkegaard T, Roth AG, Petersen NH, Mahalka AK, Olsen OD, et al. Hsp70 stabilizes lysosomes and reverts Niemann–Pick disease-associated lysosomal pathology. Nature 2010;463:549–553.

[148] Kolter T. A view on sphingolipids and disease. Chem Phys Lipids 2011;164:590–606.

[149] Ryland LK, Fox TE, Liu X, Loughran TP, Kester M. Dysregulation of sphingolipid metabolism in cancer. Cancer Biol Ther 2011;11:138–149.

[150] Saddoughi SA, Ogretmen B. Diverse functions of ceramide in cancer cell death and proliferation. Adv Cancer Res 2013;117:37–58.

[151] Jenkins RW, Canals D, Hannun YA. Roles and regulation of secretory and lysosomal acid sphingomyelinase. Cell Signal 2009;21:836–846.

[152] Carpinteiro A, Dumitru C, Schenck M, Gulbins E. Ceramide-induced cell death in malignant cells. Cancer Lett 2008;264:1–10.

[153] Dumitru CA, Gulbins E. TRAIL activates acid sphingomyelinase via a redox mechanism and releases ceramide to trigger apoptosis. Oncogene 2006;25:5612–5625.

[154] Ogretmen B, Hannun YA. Biologically active sphingolipids in cancer pathogenesis and treatment. Nat Rev Cancer 2004;4:604–616.

[155] Canals D, Perry DM, Jenkins RW, Hannun YA. Drug targeting of sphingolipid metabolism: sphingomyelinases and ceramidases. Br J Pharmacol 2011;163:694–712.

[156] Heinrich M, Wickel M, Schneider-Brachert W, Sandberg C, Gahr J, et al. Cathepsin D targeted by acid sphingomyelinase-derived ceramide. EMBO J 1999;18:5252–5263.

[157] Hantschel M, Pfister K, Jordan A, Scholz R, Andreesen R, et al. Hsp70 plasma membrane expression on primary tumor biopsy material and bone marrow of leukemic patients. Cell Stress Chaperones 2000;5:438–442.

[158] Shida D, Takabe K, Kapitonov D, Milstien S, Spiegel S. Targeting SphK1 as a new strategy against cancer. Curr Drug Targets 2008;9:662–673.

[159] Ozbayraktar FB, Ulgen KO. Molecular facets of sphingolipids: mediators of diseases. Biotechnol J 2009;4:1028–1041.

[160] Park JH, Schuchman EH. Acid ceramidase and human disease. Biochim Biophys Acta 2006;1758:2133–2138.

[161] Liu X, Zeidan YH, Elojeimy S, Holman DH, El-Zawahry AM, et al. Involvement of sphingolipids in apoptin-induced cell killing. Mol Ther 2006;14:627–636.

[162] Hanahan D, Weinberg RA. Hallmarks of cancer: the next generation. Cell 2011;144: 646–674.

[163] Leist M, Jaattela M. Four deaths and a funeral: from caspases to alternative mechanisms. Nat Rev Mol Cell Biol 2001;2:589–598.

[164] Cesen MH, Pegan K, Spes A, Turk B. Lysosomal pathways to cell death and their therapeutic applications. Exp Cell Res 2012;318:1245–1251.

[165] Fehrenbacher N, Gyrd-Hansen M, Poulsen B, Felbor U, Kallunki T, et al. Sensitization to the lysosomal cell death pathway upon immortalization and transformation. Cancer Res 2004;64:5301–5310.

[166] Paquet C, Sane AT, Beauchemin M, Bertrand R. Caspase- and mitochondrial dysfunction-dependent mechanisms of lysosomal leakage and cathepsin B activation in DNA damage-induced apoptosis. Leukemia 2005;19:784–791.

[167] Emert-Sedlak L, Shangary S, Rabinovitz A, Miranda MB, Delach SM, et al. Involvement of cathepsin D in chemotherapy-induced cytochrome c release, caspase activation, and cell death. Mol Cancer Ther 2005;4:733–742.

[168] Mediavilla-Varela M, Pacheco FJ, Almaguel F, Perez J, Sahakian E, et al. Docetaxel-induced prostate cancer cell death involves concomitant activation of caspase and lysosomal pathways and is attenuated by LEDGF/p75. Mol Cancer 2009;8:68.

[169] De Milito A, Iessi E, Logozzi M, Lozupone F, Spada M, et al. Proton pump inhibitors induce apoptosis of human B-cell tumors through a caspase-independent mechanism involving reactive oxygen species. Cancer Res 2007;67:5408–5417.

[170] Nakashima S, Hiraku Y, Tada-Oikawa S, Hishita T, Gabazza EC, et al. Vacuolar H+-ATPase inhibitor induces apoptosis via lysosomal dysfunction in the human gastric cancer cell line MKN-1. J Biochem 2003;134:359–364.

[171] Ostenfeld MS, Fehrenbacher N, Hoyer-Hansen M, Thomsen C, Farkas T, et al. Effective tumor cell death by sigma-2 receptor ligand siramesine involves lysosomal leakage and oxidative stress. Cancer Res 2005;65:8975–8983.

[172] Ostenfeld MS, Hoyer-Hansen M, Bastholm L, Fehrenbacher N, Olsen OD, et al. Anticancer agent siramesine is a lysosomotropic detergent that induces cytoprotective autophagosome accumulation. Autophagy 2008;4:487–499.

[173] Rammer P, Groth-Pedersen L, Kirkegaard T, Daugaard M, Rytter A, et al. BAMLET activates a lysosomal cell death program in cancer cells. Mol Cancer Ther 2010;9:24–32.

[174] Boya P, Gonzalez-Polo RA, Poncet D, Andreau K, Vieira HL, et al. Mitochondrial membrane permeabilization is a critical step of lysosome-initiated apoptosis induced by hydroxychloroquine. Oncogene 2003;22:3927–3936.

[175] Cirman T, Oresic K, Mazovec GD, Turk V, Reed JC, et al. Selective disruption of lysosomes in HeLa cells triggers apoptosis mediated by cleavage of Bid by multiple papain-like lysosomal cathepsins. J Biol Chem 2004;279:3578–3587.

[176] Kagedal K, Zhao M, Svensson I, Brunk UT. Sphingosine-induced apoptosis is dependent on lysosomal proteases. Biochem J 2001;359:335–343.

[177] Sukhai MA, Prabha S, Hurren R, Rutledge AC, Lee AY, et al. Lysosomal disruption preferentially targets acute myeloid leukemia cells and progenitors. J Clin Invest 2013;123:315–328.

[178] Ellegaard AM, Groth-Pedersen L, Oorschot V, Klumperman J, Kirkegaard T, et al. Sunitinib and SU11652 inhibit acid sphingomyelinase, destabilize lysosomes, and inhibit multidrug resistance. Mol Cancer Ther 2013;12:2018–2030.

[179] Jahchan NS, Dudley JT, Mazur PK, Flores N, Yang D, et al. A drug repositioning approach identifies tricyclic antidepressants as inhibitors of small cell lung cancer and other neuroendocrine tumors. Cancer Discov 2013;3:1364–1377.

[180] Appelqvist H, Waster P, Kagedal K, Ollinger K. The lysosome: from waste bag to potential therapeutic target. J Mol Cell Biol 2013;5:214–226.

11

THE GENETICS OF SPHINGOLIPID HYDROLASES AND SPHINGOLIPID STORAGE DISEASES

EDWARD H. SCHUCHMAN AND CALOGERA M. SIMONARO

Department of Genetics and Genomic Sciences, Icahn School of Medicine at Mount Sinai, New York, NY, USA

11.1 INTRODUCTION AND OVERVIEW

Sphingolipids are a diverse group of over 100 bioactive lipids involved in many aspects of cellular function (see Ref. [1–4] for recent reviews). They share a common sphingosine (2-amino-4-octadecene-1,3-diol) backbone, which may be linked via an amide bond to a fatty acid (forming ceramide) or phosphorylated to form sphingosine-1-phosphate. Ceramide itself also may be modified to form glycosphingolipids, sphingomyelin, or ceramide-1-phosphate.

The metabolism of sphingolipids is complex and carefully regulated [5–7]. Abnormal metabolism can have profound effects on cellular function, leading to enhanced cell death, proliferation, and/or abnormal cell differentiation. In general, the synthesis and breakdown of sphingolipids occurs in distinct cellular compartments. For example, numerous enzymes contribute to the de novo synthesis of sphingolipids, mostly found in the endoplasmic reticulum and Golgi apparatus. In contrast, the breakdown of sphingolipids occurs primarily within endosomes and lysosomes by a series of hydrolytic enzymes that function optimally at acidic pH. Hydrolysis of sphingolipids also may occur in nonendosomal/lysosomal compartments, for example, by enzymes located at the plasma membrane or within mitochondria. Overall, over 100 enzymes participate in the synthesis or breakdown of sphingolipids.

Lysosomes: Biology, Diseases, and Therapeutics, First Edition.
Edited by Frederick R. Maxfield, James M. Willard and Shuyan Lu.

Historically, most information concerning the importance of sphingolipids in human disease stems from studies of the sphingolipid storage diseases [8]. These disorders are each due to deficient function of one or more of the hydrolytic enzymes found within the endosomal/lysosomal system, resulting in sphingolipid accumulation [9]. Each of these diseases is inherited as a Mendelian trait (recessive or X-linked), the cDNAs/genes encoding the respective enzymes have been isolated, and many mutations causing the human diseases have been identified. In addition, genetic abnormalities in several of these enzymes/genes have now been linked to more common diseases (e.g., Parkinson's disease and beta-glucocerebrosidase) [10]. Table 11.1 lists the sphingolipid storage diseases and the respective enzymes involved.

This review focuses on the genetics of sphingolipid storage diseases and related hydrolytic enzymes. Each of the diseases/enzymes will be briefly reviewed, with an emphasis on the importance of genetic mutations in disease.

11.2 ACID CERAMIDASE DEFICIENCY: FARBER DISEASE

As with other lysosomal storage disorders, patients with Farber disease (also known as Farber Lipogranulomatosis) may exhibit a wide variation of phenotypes ranging from early-onset neurological forms to attenuated forms that lack neurological involvement. A "triad" of clinical manifestations is typical of Farber disease, including painful and progressively deformed joints, subcutaneous nodules, particularly near joints, and progressive hoarseness due to laryngeal involvement [11]. About half of the reported patients have neurological deficits [12], but this is likely to be overrepresented since these patients are on the severe end of the spectrum and the most likely to be diagnosed. Moreover, moderate and attenuated cases primarily present as idiopathic juvenile arthritis and are seen by pediatric rheumatologists who often have little or no knowledge of the disease.

Acid ceramidase (EC 3.5.1.23) was first identified and purified in the 1960s. It was recognized as the enzyme deficient in Farber disease in the 1970s [13]. The full-length cDNAs and genes encoding AC have been isolated from humans, mice, and other species. The human cDNA encodes a 395-amino-acid precursor polypeptide, and the gene is located on chromosomal region 8p22. Of note, the newly synthesized AC precursor is inactive and only assumes an active form after N-glycosylation and the internal cleavage of a peptide bond, resulting in alpha and beta subunits. Interestingly, this cleavage event is carried out by AC itself, and it is therefore considered a "self-regulating" enzyme. In addition, AC associates with other lipid hydrolases within lysosomes and other cell compartments, including acid sphingomyelinase (ASM) (see Section 11.3). The importance of this multienzyme complex on AC activity, as well as on sphingolipid metabolism in general, is not clearly understood. Finally, full AC activity, at least in lysosomes, depends on the expression of a sphingolipid activator protein (SAP-D), which is expressed by a distinct gene (see Section 11.9).

Acid ceramidase is an "amidase" that hydrolyzes the amide bond between sphingosine and the fatty acid in ceramide. Ceramides are a heterogenous group of lipids

TABLE 11.1 Genetics of Human Sphingolipid Hydrolases and Diseases

Sphingolipid Hydrolase/ Activator	Disease	Primary Accumulating Lipid	Gene Designation	Gene Location
Acid ceramidase	Farber disease	Ceramide	ASAH1	8p22
Acid sphingomyelinase	Niemann–Pick (types A and B)	Sphingomyelin	SMPD1	11p15.4-p15.1
Beta-glucocerebrosidase	Gaucher disease	Glucosylceramide	GBA1	1q21
Beta-galactocerebrosidase	Krabbe disease	Galactosylceramide	GALC	14q31
Arylsulfatase A	Metachromatic leukodystrophy	Sulfatide	ARSA	22q13.31-qter
Alpha-galactosidase A	Fabry disease	Globotriosylceramide	GLA	Xq22.1
Beta-galactosidase	GM1 gangliosidosis	GM1 ganglioside	GLB1	3p22.31
Hexosaminidase	GM2 gangliosidosis			
Hexosaminidase A	Tay–Sachs disease	GM2 ganglioside	HEXA	15q24
Hexosaminidase B	Sandhoff disease	GM2 ganglioside	HEXB	5q13
GM2 activator	Tay–Sachs A/B variant	GM2 ganglioside	GMA2	5q31.1
Prosaposin	Variable	Variable	PSAP	10q21-q22

defined by the length of their fatty acid chains. While the substrate specificity of AC is not entirely clear, ceramides with fatty acid chains shorter than 12 carbons cannot be efficiently hydrolyzed by the enzyme *in vitro*. *In vivo*, it is likely that AC can recognize and cleave ceramides of varying chain lengths. The ceramides that accumulate in Farber disease patients are primarily found within lysosomes, leading to the original designation of AC as a lysosomal enzyme. AC also has maximal activity at lysosomal (acidic) pH, although it has been localized to sites other than the lysosomes as well. It is therefore not entirely clear to what extent the biological activity of AC depends on the acidic environment, and in fact a "reverse" ceramidase reaction has been attributed to the enzyme at more neutral pH [14]. Numerous *in vitro* assays have been developed for AC, and some can be used to diagnose Farber disease in cultured cells from suspected patients, or prenatally using amniocytes or cultured chorionic villi. However, due to the very hydrophobic nature of ceramides, it remains one of the more difficult diagnostic enzyme assays to carry out.

To date, only about 100 Farber disease patients have been reported in the literature [15]. Several cases of hydrops fetalis due to acid ceramidase deficiency also have been reported. Complete inactivation of the AC gene in mice leads to very early embryonic lethality (four-cell stage), suggesting that severe mutations in the human *ASAH1* gene will lead to embryonic lethality as well [16]. Of note, mouse embryos lacking AC can be rescued by the addition of recombinant AC (obtained from overexpressing Chinese hamster ovary [CHO] cells) to the culture media. Moreover, addition of recombinant AC to the culture media during *in vitro* fertilization (mouse, bovine) enhances embryo production and quality, resulting in more live births [17].

Thus, AC is an essential enzyme required for very early mammalian development. One explanation for the lethal phenotype in mice is that in the absence of this enzymatic activity, early-stage embryos (four cells in mice) undergo apoptosis due to ceramide accumulation. It should be recognized, however, that an additional consequence of such AC "loss-of-function" during embryogenesis is reduced production of sphingosine, which in turn is converted to S1P, an important mitogenic/proliferative lipid. Thus, AC mutations also may reduce S1P production, contributing to the lethality as well. Indeed, embryos lacking AC can be partially rescued by S1P.

As noted earlier, the first relationship between AC and human disease arose from studies on Farber disease. To date, over 25 mutations have been described in the *ASAH1* gene leading to Farber disease (e.g., see Ref. [15] for review and Ref. [18]). Most are point mutations, although several splice variants leading to small deletions have been described as well. To date, no patients have been found homozygous for a complete loss-of-function mutation, likely because such individuals do not survive embryonic development. Farber disease is inherited as an autosomal recessive disorder, and to develop the disorder mutations must be inherited on both *ASAH1* alleles. No reports of clinical findings in the parents of affected patients (obligate carriers) have been published; however, such individuals are not usually followed by clinicians and follow-up analysis is lacking.

Recently, a "knock-in" mouse model of Farber disease has been reported where a specific human *ASAH1* mutation (P361R) was expressed [19]. These animals exhibited massive ceramide storage in tissues, a shortened lifespan, and an early-onset inflammatory profile typical of Farber disease patients. In addition, a conditional AC knockout mouse has been developed, where AC can be inactivated at various stages of development or in specific tissues [20].

In addition to Farber disease, a second rare genetic disorder – spinal muscular atrophy with myoclonic epilepsy (SMA-PME) – was found to be due to AC deficiency and mutations in the *ASAH1* gene [21,22]. These patients exhibit massive ceramide storage in motor neurons of the spinal cord and markedly reduced AC activity in cultured cells, but have an otherwise distinct phenotype from Farber disease patients. The molecular basis of this phenotypic heterogeneity is not understood, but SMA-PME clearly represents a new sphingolipid storage disease that might benefit from AC-mediate therapies.

Other than Farber disease and SMA-PME, which are due to the deficiency of AC, constitutive overexpression of the AC gene occurs in many types of cancer [23,24]. This may promote tumorigenesis and/or metastasis due to overproduction of S1P or may lead to chemoresistance due to an enhanced ability to hydrolyze proapoptotic ceramide [2,25]. The molecular basis of AC overexpression in cancer has not been studied extensively, but at least in some cases it appears to be due to elevated transcription. Gain-of-function mutations also could explain this finding. A recent report also showed that the *ASAH1* gene was downregulated in a Chinese population of schizophrenic patients and was associated with two single nucleotide polymorphisms in this gene [26]. Due to the central role that ceramide plays in cell signaling, AC-related diseases may be more common than currently known since the enzyme is essential for maintaining the proper balance between this lipid, sphingosine, and S1P.

Finally, it must be noted that several other ceramidase activities have been described in humans, and at least four other genes have been cloned [27]. These ceramidases are defined by their unique pH, subcellular location, and other biological properties. These enzymes also play important roles in sphingolipid metabolism and/or signaling and are likely to contribute to human disease pathogenesis. The biology of the other ceramidases has been reviewed extensively elsewhere [27].

11.3 ACID SPHINGOMYELINASE DEFICIENCY: TYPES A AND B NIEMANN–PICK DISEASE

Types A and B Niemann–Pick disease (NPD) result from mutations in the gene (*SMPD1*) encoding ASM (EC 3.1.4.12) [28,29]. Patients with Type A NPD present in early infancy with hepatosplenomegaly and failure to thrive and develop a rapidly degenerative neurological phenotype that leads to death by 3 years of age. In contrast, patients with Type B NPD have a later-onset disease and usually lack neurological findings. Such individuals also frequently present with hepatosplenomegaly and may have pulmonary involvement as well. Dyslipidemia (high LDL-C, low HDL-C, and high triglycerides) is also common. Patients with Type B NPD may survive into adulthood, although their lifespan is generally shortened by the disease. The cause of death is unknown but may occur from complications related to liver disease, trauma-induced bleeding episodes due to thrombocytopenia, early-onset cardiovascular disease, or pulmonary disease.

ASM was found to be deficient in Type A and B NPD patients in the late 1960s. Although the enzyme had been discovered several years earlier in rat tissues, it was not substantially purified until the late 1980s. The cDNAs and genes encoding human and mouse ASM were cloned in the early 1990s. The human gene is located on chromosome 11p15.1. Of interest, among genes encoding lysosomal enzymes, it is the only one for which genomic imprinting has been described. This form of genetic regulation is generally reserved for genes with important developmental roles.

Over 100 mutations have been found in the *SMPD1* gene leading to Types A or B NPD (http://www.hgmd.org). Of note, several of these mutations are common within specific regions or ethnic populations, allowing genetic screening programs to be established. Homozygosity for these mutations is required to develop NPD, and it is inherited as an autosomal recessive trait. However, heterozygous carriers of *SMPD1* mutations may develop dyslipidemia and late-onset features of the disorder. In part, the development of a phenotype in the carrier individuals may be related to their imprinting status (i.e., whether the mutations are inherited from the mother or father) [30]. Unlike acid ceramidase and several other sphingolipid hydrolases, ASM does not require an exogenous activator protein to achieve full function. Thus, mutations in the SAP gene (see Section 11.10) do not affect ASM activity or cause Types A and B NPD. Of interest, the lysosomal lipid, bis(monacyl) phosphate, may activate the enzyme *in vitro* and *in vivo*.

In addition to ASM, several other sphingomyelinases (nonacidic) exist in mammalian cells [31]. These are defined by their unique subcellular location or other biochemical properties. At least four human genes have been isolated encoding these sphingomyelinases. Definitive proof that mutations in the *SMPD1* gene were solely responsible for Types A and B NPD came from studies in the knockout mouse model (ASMKO), where neutral and other sphingomyelinase activities were normal, despite the fact that these animals completely lacked ASM and developed NPD-like pathology and clinical features [32]. In addition, these early studies in the ASMKO mice showed that the previously identified zinc-activated, serum form of ASM was encoded by the same *SMPD1* gene as lysosomal ASM [33]. It is now known that ASM is, in fact, a zinc-requiring enzyme. The lysosomal form of the enzyme has high levels of zinc bound and thus does not require exogenous zinc when measured *in vitro*. In contrast, the serum form of the enzyme requires exogenous zinc in the assay systems routinely used.

Another early finding in the ASMKO mouse was that ceramide, the product of sphingomyelin hydrolysis by ASM, was also elevated in tissues of these animals. While initially a surprising result, it is likely that the elevated ceramide derives from breakdown of the accumulating sphingomyelin in NPD tissues by other sphingomyelinases. Sphingomyelin levels may be up to 30-fold above normal in the ASMKO mice and ceramide two- to fivefold above normal. Most recently, sphingosine levels also were found to be high in tissues from ASMKO mice, probably due to breakdown of the accumulating ceramide by ceramidases (Schuchman et al., unpublished observations). Of interest, the degree of sphingosine accumulation in these animals (>20-fold above normal) cannot be explained by ceramide alone, perhaps suggesting some defect in converting sphingosine into S1P.

The ASMKO mice develop a phenotype intermediate between Types A and B NPD. They die prematurely at ~6–8 months, secondary to neurological disease. They have been extensively used to study the pathobiology of this disease and to develop/evaluate new therapies. In particular, enzyme replacement therapy (ERT) using recombinant ASM produced in CHO cells has been developed in the mouse model [34], and a phase 1 clinical trial in adult patients with Type B NPD has been completed (http://clinicaltrials.gov). Of interest, in the ASMKO mice, dose-related toxicity was observed following ERT (which was not seen in healthy animals), presumably due to rapid breakdown of the accumulating sphingomyelin to ceramide. Toxicity could be prevented by pretreating the animals with low doses of enzyme ("debulking" the accumulating sphingomyelin), followed by higher dose maintenance therapy. Based on these findings, a second repeat dosing (phase 1b) clinical was carried out in adult Type B NPD patients using a dose escalation scheme to debulk the accumulating sphingomyelin. The findings revealed that the recombinant ASM could be safely administered to these patients, resulting in significant clinical improvements. A phase 2/3 clinical trial is now planned in adult patients and a phase 2 study in children.

Beginning in the late 1990s, the ASMKO mice also were extensively used to investigate the role of ASM in cell signaling. This is the subject of several reviews and will not be focused on in this chapter. However, it is now clear that ASM plays an important role in stress-induced ceramide generation and apoptosis and that this occurs through hydrolysis of sphingomyelin at the cell surface within lipid rafts [35]. How ASM, which has an acidic pH optimum *in vitro* and functions within lysosomes, participates in these processes has been the subject of much debate. It is now known that after the induction of stress or developmental signals, ASM may be rapidly translocated to lipid rafts at the cell surface. The mechanism underlying this rapid translocation is not entirely clear, but phosphorylation within the ER/Golgi may be responsible in part. It is also known that although ASM requires an acidic pH for optimal activity *in vitro*, *in vivo*, it is capable of efficiently hydrolyzing sphingomyelin at physiological pH. This highlights the complex biology of this and other sphingolipid hydrolases, particularly in terms of pH, subcellular location, and other properties.

These observations on the ASMKO mice have also highlighted the important role of ASM in human disease, beyond NPD [36–38]. Elevated ASM activity has now been detected in numerous common diseases, including diabetes, Alzheimer disease, cardiovascular diseases, pulmonary diseases, depression, and others. These will also be reviewed elsewhere and not discussed further here. At the present time, it is not known if the elevated ASM found in these diseases is due to an activating genetic mutation, enhanced transcription, and/or posttranslational modification of the enzyme.

11.4 BETA-GLUCOCEREBROSIDASE DEFICIENCY: GAUCHER DISEASE

Gaucher disease, due to the deficient activity of acid beta-glucocerebrosidase (also known as beta-glucosidase [E.C.#3.2.1.45]), is one of the most common sphingolipid storage disorders [128,129]. There are three clinical subtypes of Gaucher disease, including an early-onset, neurological form (Type 2), an intermediate form (Type 3), and a late-onset, nonneurological form (Type 1) [39,40]. All Gaucher patients present with hepatosplenomegaly, but the progression of the disease and survival are markedly different. For example, Type 3 children may only survive until 2–3 years of age, while Type 1 patients may survive into adulthood. If left untreated, the hematological and bone lesions in Type 1 patients progress with age, seriously limiting quality of life. A massively enlarged spleen and frequent bone fractures may become common, requiring splenectomy, joint replacement, or other surgical interventions [41,42]. Bleeding is also a common presenting feature of Gaucher disease due to thrombocytopenia. Pulmonary involvement may similarly occur. Of note, neoplastic disorders, particularly of the blood forming organs, are more common in Gaucher disease patients than in the general population. These include chronic lymphocytic leukemia, multiple myeloma, and Hodgkin and non-Hodgkin lymphoma. The molecular basis of this increased cancer risk remains unknown [43].

All Gaucher patients are characterized by the presence of "Gaucher" cells in various organs, lipid-filled cells derived from the monocyte/macrophage system. Gaucher cells have a distinctive appearance by light microscopy and can be readily distinguished from cells of other sphingolipid storage disorders. Gaucher cells are distributed throughout the body wherever macrophages reside. The largest numbers are found in the spleen, sinusoids of the liver, bone marrow, and lymph nodes.

The major accumulating lipid in Gaucher disease is *N*-acyl-sphingosyl-1-0-β-D-glucoside (glucosylceramide, also known as glucosylcerebroside; GC). GC is widely distributed in mammalian tissues in small quantities as a metabolic intermediate in the synthesis and degradation of complex glycosphingolipids, such as gangliosides or globoside. The deacylated form of GC, glucosylsphingosine, is not only normally found in tissues but also accumulates in Gaucher disease cells. GC levels in plasma may be 2- to 20-fold above normal.

The major site of GC turnover in cells is lysosome. Beta-glucocerebrosidase is normally found in this organelle and functions optimally at an acidic pH. As with the other lipid hydrolases, *in vitro* detection of this enzymatic activity can be difficult, and there have been various factors (lipids, protein, and sugars) shown to influence the assay systems. *In vivo*, sphingolipid activator-derived peptides, particularly saposin (SAP) C, appear to be required for full enzymatic activity (see Section 11.10). Nonacidic glucocerebrosidases also have been described and are encoded by distinct genes. Whether these activities can be modifying factors for the Gaucher phenotype has been the subject of much discussion but remains unclear.

The gene encoding beta-glucocerebrosidase (*GBA1*) has been mapped to chromosome 1q21. It is ~7 kb and contains 11 exons. Of note, a 5 kb "pseudogene" is located immediately downstream from the *GBA1* gene. This pseudogene has maintained a high degree of homology with the functional gene and is transcribed. However, the nucleotide changes within the pseudogene sequence prevent the occurrence of a long open reading frame, precluding the possibility of expressing an active beta-glucocerebrosidase [44]. Full-length cDNAs from many species also have been obtained that can be used to express the active enzyme.

Numerous mutations have been found in the *GBA1* gene causing Gaucher disease [45]. As described for NPD above, several common mutations are found in specific populations, and DNA-based screening is now routinely carried out in the Ashkenazi Jewish population [46]. Estimates of the Gaucher disease carrier frequency in this population range from ~1:10 to 1:25. Some of the mutations also can be used to predict the occurrence of severe or mild Gaucher phenotypes, assisting in family planning and genetic counseling.

Recently, mutations in the *GBA1* gene also have been associated with the occurrence of Parkinson's disease, although the precise relationship between the *GBA1* gene, glucocerebrosidase activity, and Parkinson's disease remains uncertain [47]. One of the first clues suggesting this linkage came from the small number of Gaucher patients who exhibited Parkinson-like features, while stronger associations later came from Gaucher relatives who were carriers for *GBA1* mutations. Currently, heterozygous *GBA1* mutations are considered to be one of the most common genetic risk factor for Parkinson's disease [48].

A complete knockout mouse model for Gaucher was constructed and had an early neonatal lethal phenotype, dying within 34 days after birth. These animals had massive GC storage, and while the cause of death was not entirely clear, skin abnormalities were prevalent. Several knock-in models also have been constructed, and some exhibited longer survival and later-onset GC storage in macrophages, more closely resembling Gaucher disease patients [49,50].

Gaucher disease was the first sphingolipid storage disease for which ERT was developed in the early 1990s, initially using beta-glucocerebrosidase purified from placentae (Ceredase) and subsequently using recombinant enzyme produced in CHO cells (Cerezyme) [51,52]. For nonneurological patients, ERT was life-changing, resulting in reduction in spleen size and correction of the hematological abnormalities. Bone disease has been much less responsive to ERT and remains a challenge in the treatment of Gaucher disease, as does treatment of the brain disease [51]. Since the introduction of Cerezyme, several other recombinant forms of GC also have become available, and in general similar clinical responses have been observed. Another treatment strategy currently available for some Gaucher patients uses small-molecule inhibitors of GC synthesis (Miglustat) to slow the accumulation of this lipid [53,54]. The target of this inhibitor is the enzyme glucosylceramide synthase.

Of note, treatment of Gaucher patients by either ERT or with Miglustat should elevate ceramide and could potentially result in ceramide-mediated toxicity. To date, however, such toxicity has not been observed in animal models or patients. There are several factors that might explain this observation. First, compared to sphingomyelin in NPD, GC is generally found in small quantities in cells and is not a major structural component of cell membranes. Thus, accumulation of GC in Gaucher cells is more exclusively found in the lysosomes than NPD and does not disrupt plasma membrane structure and signaling to the same extent as sphingomyelin. In addition, ceramide produced by ASM (compared to GC) is likely to remain within the cell membrane. Finally, the ERT dose typically used to treat Gaucher patients (1.6 mg/kg, every 2 weeks) may be below the threshold needed to induce ceramide-mediated toxicity.

11.5 GALACTOCEREBROSIDASE DEFICIENCY: KRABBE DISEASE/GLOBOID CELL LEUKODYSTROPHY

Infantile globoid cell leukodystrophy (GLD), also known as Krabbe disease, is a rapidly progressive, fatal neurological disease of infants [55]. The disease usually begins between 3 and 6 months of life with symptoms that may include irritability or hypersensitivity to stimuli. Patients rapidly exhibit severe mental and motor deterioration and rarely survive beyond the second year [56]. Clinical manifestations in classical GLD patients are limited to the nervous system. Blindness and deafness are common, and peripheral neuropathy is almost always detectable. Patients with later-onset forms of the disease, including adult-onset forms, may present with blindness, spastic paraparesis, and dementia [57,58].

The presence of numerous multinucleated globoid cells, almost total loss of myelin and oligodendrocytes, and astrocytic gliosis in the white matter are the morphologic basis of the disease. "Globoid cells" in Krabbe disease are enlarged macrophages that contain undigested galactosylceramide and other lipids. Demyelination, axonal degeneration, fibrosis, and histiocytic infiltration are also common in the peripheral nervous system. Consistent with the myelin loss, the white matter is severely depleted of all lipids, particularly glycolipids. Under normal circumstances, galactosylceramide is present almost exclusively in the myelin sheath. The ratio of galactosylceramide to sulfatide (3-0 sulfogalactosylceramide) is abnormally high in GLD.

The genetic cause of Krabbe disease is mutations in the gene encoding galactocerebrosidase beta-galactosidase (GALC; E.C. #3.2.1.46). This lysosomal enzyme normally degrades galactosylceramide to ceramide and galactose. It is postulated that accumulation of a toxic metabolite, psychosine (galactosylsphingosine), which is also a substrate for the missing enzyme, leads to early destruction of the oligodendrocytes. The total brain content of galactosylceramide is not increased in patients with Krabbe disease, supporting this hypothesis. It should be noted that mammalian tissues contain two genetically distinct lysosomal beta-galactosidases, GALC and GM1-ganglioside beta-galactosidase (see Section 11.8). These two enzymes have overlapping substrate specificities *in vitro*, which can present complications when attempting to make the enzymatic diagnosis of Krabbe disease [59].

The human *GALC* gene has been mapped to chromosome 14q24.3-32.1, and the genomic and/or full-length cDNA sequences have been cloned from multiple species. The human cDNA encodes a 669 amino acid polypeptide with six potential N-glycosylation sites. The fully glycosylated precursor polypeptide is ~85 kDa and is processed into ~50 and 30 kDa subunits. To date, over 70 mutations in the *GALC* gene have been identified in infantile and late-onset GALC-deficient patients (http://www.hgmd.org). Most of these mutations are "private," occurring in individual families, but some may be commonly found in specific regions or ethnicities. Among these, the most common mutation occurs in patients of European ancestry and deletes a large segment of the *GALC* gene. Several polymorphisms also have been found in the *GALC* gene, and some of these may alter amino acids and thus potentially affect the function of the GALC polypeptide, although not enough to cause GLD.

In general, carriers of *GALC* mutations are considered clinically normal, although one interesting report did show an association between the risk for primary open-angle glaucoma and a heterozygous deletion of the *GALC* gene [60]. Another interesting recent paper also showed that GALC contributed to the maintenance of a functional hematopoietic stem cell niche, suggesting that genetic alterations in the *GALC* gene may be associated with hematological conditions as well. Indeed, hematopoietic stem cells retrieved from Twitcher mice (GALC deficient) had defects in their frequency, proliferation, clonogenic potential, and engraftment.

As noted above, in 1972, Miyatake and Suzuki formulated a hypothesis to explain the unusually rapid and complete destruction of oligodendrocytes in GLD. It was known at that time that psychosine (galactosyl sphingosine), due to its free amino group, was highly cytotoxic and that this lipid served as a substrate for GALC but not GM1 beta-galactosidase. In the central nervous system (CNS), psychosine is primarily formed within oligodendrocytes, the site of myelin synthesis, and this may explain the selective destruction of these cells in GLD brains. Indeed, in the white matter of brains from Krabbe patients, psychosine levels may be elevated up to 100-fold, while galacotosylceramide levels are either normal or only elevated slightly.

GALC deficiency occurs in a number of naturally occurring animal models, including mice, sheep, several breeds of dogs, and the rhesus monkey. The mouse model, in particular (also known as the "Twitcher" mouse), has been exploited for numerous therapeutic studies [61]. Affected mice develop clinical signs at ~20 days, with stunted growth, twitching, and hind leg weakness. By 40 days, they reach a near-terminal stage. Twitcher mice have been treated by ERT, gene therapies, BMT (bone marrow transplantation), and other modalities [62–64]. BMT, in particular, may have positive therapeutic effects; however, all of the treated mice eventually still died of characteristic pathological findings. Thus, this and the other available therapies may slow the disease progress to some degree but not prevent it.

Since 2006, New York State in the United States has screened all newborns for the occurrence of Krabbe disease. Pilot screening programs are now underway in other states and parts of the world. This screening is accomplished by GALC activity assays on dried blood spots, followed by DNA analyses. The rationale for this screening program is that the early identification of infants with Krabbe disease will permit early BMT and other therapeutic interventions, improving the quality of life [56]. However, since BMT has not been curative in the animal models, screening for severe neurological disorders such as this remains controversial.

11.6 ARYLSULFATASE A DEFICIENCY: METACHROMATIC LEUKODYSTROPHY

Metachromatic leukodystrophy (MLD) is an autosomal recessively inherited disorder in which desulfation of 3-0-sulfogalactosylsphingosine or 3-0-sulfogalactosyl glycolipids is deficient. Among the accumulating glycolipids in MLD are sulfatide (galactosylceramide-3-sulfate), lysosulfatide (i.e., similar to sulfatide but lacking the fatty acid in the ceramide moiety), lactosylceramide-3-sulfate, and others. These sulfated glycolipids normally occur in the myelin sheaths in the central and peripheral nervous systems, and, to a lesser extent, in visceral organs.

The clinical onset and severity of MLD is highly variable [65]. The late infantile form is usually recognized in the second year of life and fatal in early childhood. The juvenile form presents between age four and puberty; the adult form may occur at any age after puberty. In both of the later-onset variants, gait disturbance and mental regression are the earliest signs. In the childhood variants, other common signs are blindness, loss of speech, quadriparesis, peripheral neuropathy, and seizures [66].

In 1963, Austin et al. reported the deficiency of arylsulfatase A (ASA, E.C. 3.1.6.8) activity in MLD, also known as sulfatide sulfatase, and mutations in this gene (*ARSA*) account for the vast majority of MLD patients. There are also two other minor, genetic causes of MLD. One is due to mutations in the gene (*PSAP*, see Section 11.9) encoding the SAP, saposin B. Mutations in the saposin B encoding region of the *PSAP* gene lead to deficient ASA activity, and a clinical picture indistinguishable from MLD. In fact, these observations in MLD patients provided the first *in vivo* proof that such SAPs were required for *in vivo* activity of this and other sphingolipid hydrolases. In addition to ASA and SAP B mutations, patients with multiple sulfatase deficiency, due to mutations in the gene encoding sulfatase-modifying factor 1 (*SUMF-1*), also have a reduction in ASA and other sulfatase activities and develop features of MLD.

As with the other sphingolipid hydrolases, full-length genomic and cDNA sequences encoding ASA have been isolated from several species. The human *ARSA* gene has been mapped to chromosomal region 22q13. It is a remarkably simple gene and encompasses only 3.2 kb divided into eight exons. As noted above, almost all cases of MLD are due to mutations in this gene, except for the rare cases of *PSAP* or *SUMF-1* mutations. The human *ARSA* gene encodes a 507 amino acid polypeptide precursor that undergoes posttranslational modifications much like other lysosomal enzymes (e.g., N-glycosylation). Also, as with all known eukaryotic sulfatases, a formylglycine residue is substituted for a cysteine in human ASA. The formylglycine residue is generated by oxidation of the thiol group to an aldehyde by the enzyme SUMF-1. Mutations in the gene encoding this cysteine-modifying enzyme lead to the synthesis of inactive sulfatases and are the cause for multiple sulfatase deficiency [67].

Over 100 MLD-causing mutations have been described in the *ARSA* gene. The most common mutation in patients with late-infantile MLD is a G to A transition that eliminates the donor splice site at the start of intron 2 [68]. This causes a loss of all ASA enzymatic activity, producing a severe early-onset phenotype. In contrast, in adults with MLD the most frequent mutation is a C to T transition that results in substitution of a leucine for proline in amino acid residue 426. Patients with this mutation have a small amount of residual enzyme activity and therefore exhibit late-onset MLD. Common mutations in specific populations have also been described.

Notably, there also are two mutations that cause "pseudodeficiency" of ASA activity. The term "pseudodeficiency" reflects the fact that these mutations cause severe reduction of ASA activity *in vitro* but do not result in MLD [69]. One of these mutations alters an N-glycosylation site, but this has no effect on enzyme function. The other affects a polyadenylation site in the mRNA. The pseudodeficiency alleles may be found in ~3–8% of the normal population and, therefore, pose a diagnostic challenge since the artificial *in vitro* assay systems may reveal very low ASA activity, even though the individual is highly unlikely to develop MLD.

Alterations in the *ARSA* gene also have been linked with the occurrence of several common diseases. For example, as early as 1991, Kappler et al. found that among a small set of patients with multiple sclerosis (MS) there was a higher incidence of the ASA pseudodeficiency alleles than in the general population [70]. In addition, variants of the ASA protein and/or gene have been associated with depression and alcoholism. MLD itself also has been associated with several psychiatric disorders, and there is at least one report where polymorphisms in the genes encoding the serotonin and dopamine pathways affected the MLD phenotype [71].

Other than humans, naturally occurring mutations causing MLD do not occur in other species. ASA knockout mice have been constructed that exhibit a complete loss of ASA activity and sulfatide storage in multiple organs [67]. While these animals exhibit some neurological deficits, they do not develop the predicted severe phenotype of MLD and live a near-normal lifespan. They also do not develop demyelinating disease. Thus, these observations in the mouse suggest the existence of compensating pathways in this species that do not exist in humans, perhaps providing a link to novel therapies. Several therapies have been evaluated in the MLD knockout mouse model, including BMT, ERT, and gene therapies [72,73]. BMT also has been extensively used in MLD patients. While it is not effective in infantile cases, BMT is often recommended for the later-onset cases [74,75]. However, the data supporting an improved clinical outcome in the late-onset MLD cases are fragmented and incomplete, and it is unclear whether the benefit outweighs the potential risks.

11.7 ALPHA-GALACTOSIDASE A DEFICIENCY: FABRY DISEASE

Fabry disease is inherited as an X-linked recessive trait, the only such X-linked disease among the sphingolipid storage disorders [76,77]. Therefore, all males ("hemizygotes") carrying mutations in the Fabry gene (*GLA*) develop symptoms of the disease although the severity may vary depending on the nature of the individual mutation. Female heterozygotes who inherit one copy of the mutant gene on one of their X chromosomes may also exhibit features of the disease depending on the pattern of X-inactivation ("Lyonization") that occurs in their individual tissues.

The gene responsible for Fabry disease encodes an enzyme, alpha-galactosidase A (E.C.#3.2.1.22), that is required to hydrolyze glycosphingolipids with terminal alpha-galactosyl moieties, predominately globotriosylceramide, and to a lesser extent galabiosylceramide and blood group B substances. Affected males have extensive deposition of these lipids in body fluids and in the lysosomes of endothelial, perithelial, and smooth muscle cells of blood vessels. Deposition also occurs in ganglion cells, and in many cell types in the heart, kidney, eyes, and most other tissues [78].

Clinical manifestations in classically affected hemizygotes with no detectable alpha-galactosidase A activity include the onset during childhood or adolescence of pain and paresthesias in the extremities, angiokeratomas in the skin and mucous membranes, and hypohidrosis. Corneal and lenticular opacities also are early findings. With increasing age, proteinuria, hyposthenuria, and lymphedema appear. Severe renal impairment leads to hypertension and uremia. Death usually occurs in adulthood from renal failure or from cardiac or cerebrovascular disease. Atypical variant males with residual alpha-galactosidase A activity may be asymptomatic or have late-onset disease. Heterozygous females may have an attenuated form of the disease depending on the pattern of X-inactivation. They usually are asymptomatic although can present with clinical disease as severe as hemizygous males.

As noted above, in Fabry disease glycosphingolipids with terminal alpha-D-galactosyl residues are not properly broken down due to the alpha-galactosidase A enzyme defect, and these lipids therefore accumulate in cells. The predominant accumulating lipid is globotriaosylceramide (Gal(a1-4)Gal(a1-4)Glc(b1-1′)Cer). Since alpha-galactosidase A is mostly found in lysosomes, the major site of lipid accumulation is this organelle. Elevated globotriaosylceramide is found in most tissues but predominantly in kidney.

Other minor accumulating lipids include galabiosylceramide (Gala1-4)(Gal(b1-1′) Cer) and the blood group B and P glycosphingolipids [79]. In human erythrocytes, there are two neutral glycosphingolipids with terminal alpha galactosyl residues that inhibit blood group B-specific hemagglutination. The structure of these blood group B glycosphingolipids has been determined, and they are found at high levels in Fabry patients. This raises an interesting and unique point about Fabry disease. Fabry hemizygotes and heterozygotes who have blood groups B and AB accumulate four glycosphingolipids, while those who are blood group A or O only accumulate two (globotriaosylceramide and galabiosylceramide). A fifth neutral glycosphingolipid that can accumulate in Fabry is the P1 blood group antigen, which also has terminal alpha-galactosyl residues. The role of these individual blood types on the Fabry phenotype is an interesting area of ongoing research.

In humans, there are two alpha-galactosidases (alpha-galactosidases A and B). Fabry disease is caused by mutations in the gene (*GLA*) encoding the "A" type [80]. However, when measuring alpha-galactosidase activity in blood from Fabry patients *in vitro*, classic hemizygotes may exhibit up to 25% of normal activity due to the presence of the "B" form. Alpha-galactosidase B is encoded by a distinct gene (*NAGA*) and is now known to be an alpha-*N*-acetylgalactosaminidase that recognizes these moieties in glycoproteins, complex carbohydrates, and other molecules. Because the artificial substrates routinely used to measure alpha-galactosidase activity do not readily distinguish these enzymes, substantial misdiagnosis can occur by simple enzyme testing.

The human gene encoding alpha-galactosidase A resides on chromosomal region Xq22. It encompasses ~12 kb and contains seven exons. The full-length human cDNA contains a 1290 bp open reading frame that encodes an unglycosylated precursor polypeptide of ~48 kDa. As with most lysosomal hydrolases, there are several N-glycosylation sites in the predicted polypeptide sequence and the oligosaccharide chains undergo mannose-6-phosphate modifications, which facilitates targeting of the enzyme to lysosomes. The mature, fully glycosylated alpha-galactosidase A in lysosomes is ~51 kDa. Genomic and cDNA sequences encoding alpha-galactosidases have now been isolated from many species. An unusual feature of most human alpha-galactosidase A cDNAs is the lack of a 3′ untranslated region. The polyadenylation signal sequence is often actually found in the coding region, 12 bp from the termination codon, followed by a poly(A) tract.

The full-length cDNA encoding human alpha-galactosidase A had been used to express and purify the recombinant enzyme from CHO cells. This enzyme was extensively characterized and used in the Fabry disease mouse model to evaluate ERT. Based on these results and after extensive clinical testing, a recombinant enzyme drug (Fabrazyme) was approved for use in human Fabry disease patients in 2003. This represents the second sphingolipid storage disorder for which ERT became available (Gaucher disease was the first). ERT in Fabry patients reduces pain, proteinuria, and endothelial cell storage of glycolipids and improves the quality of life for Fabry patients [81]. Since no naturally occurring animal model of Fabry disease has been found, the preclinical evaluation of ERT was performed in a knockout mouse model. These animals exhibit age-dependent glycosphingolipid storage, but few clinical findings of Fabry disease. Since the approval of Fabrazyme several additional recombinant alpha-galactosidases also have become available and are used for ERT in Fabry patients (e.g., Replagal). The comparative efficacy and appropriate dosing of these enzymes is a subject of ongoing debate, but suffice it to say clinical benefit is provided by each of the enzymes currently being used.

Numerous mutations have been found in the *GLA* gene causing Fabry disease, spread throughout the enzyme-coding region [82]. Several regulatory (promoter) and splice site mutations also have been found. As noted above, the presence of one *GLA* mutation in hemizygous males will cause disease, but the severity will depend on the effect of this mutation on the residual alpha-galactosidase A activity. In female heterozygotes, because of X-inactivation some may develop disease [83]. It should be noted that such "Fabry" females carrying single mutations will be essentially "mosaics" for alpha-galactosidase A, with some cells expressing the mutant allele and others the normal allele (due to random inactivation of one X chromosome). Because the enzyme can be released from cells at low levels (as with all lysosomal hydrolases), and then taken up by neighboring cells, "cross-correction" of the lipid storage is possible. Aside from Fabry disease, there has been some evidence that mutations in the *GLA* gene also may be a previously unrecognized cause of stroke. Indeed, newborn screening programs have revealed that in some populations the incidence of *GLA* mutations is unexpectedly high, suggesting that this gene/disease association should be studied further as a potential risk factor for this and other common disorders of the vasculature [84]. In the context of newborn screening, it is also important to recognize that many individuals with *GLA* gene mutations may not develop clinical disease or perhaps may develop disease late in life. Thus, the very high estimates of Fabry disease derived from these ongoing screening efforts may overestimate the number of actual clinical cases.

11.8 BETA-GALACTOSIDASE DEFICIENCY: GM1 GANGLIOSIDOSIS

An inherited deficiency of lysosomal acid beta-galactosidase (E.C. 3.2.1.23) is expressed as two clinically distinct diseases, GM1 gangliosidosis and Morquio B disease [85–87]. GM1 gangliosidosis affects both neurological and somatic tissues, occurring mainly in early infancy (type 1). Developmental arrest is usually observed a few months after birth, followed by progressive neurologic deterioration and generalized rigospasticity with sensorimotor and psycho/intellectual dysfunction. As with many of the other neurodegenerative lysosomal storage diseases, cherry-red maculae are common, as is facial dysmorphism, hepatosplenomegaly, and generalized skeletal dysplasia. Later-onset, juvenile, and adult forms of beta-galactosidase deficiency have also been described (types 2 and 3, respectively). They are observed as progressive neurologic diseases in childhood or in young adults. Dysmorphic skeletal changes are less prominent or absent in these clinical forms. A protracted course, mainly presenting as dystonia, is the major neurological manifestation in adults with GM1 gangliosidosis.

Morquio B disease is clinically expressed as a generalized skeletal dysplasia with corneal clouding, resulting in short stature, pectus carinatum, platyspondylia, odontoid hypoplasia, kyphoscoliosis, and genu valgum. There is no CNS involvement, although spinal cord compression may occur at late stages of the disease. Intelligence is normal and hepatosplenomegaly is not present.

In humans, two lysosomal enzymes are known to be responsible for the hydrolysis of terminal beta-linked galactose residues at acidic pH in various glycoconjugates. One is the enzyme being discussed here, beta-galactosidase (EC 3.2.1.23), and the other is galactocerebrosidase (EC 3.2.1.46), whose primary substrates include galactosylceramide, galactosylsphingosine, and other lipids. This is the enzyme deficient in GLD (Krabbe disease) and was discussed in the Section 11.5. Beta-galactosidase activity is severely deficient in cells and tissues from patients with both GM1 gangliosidosis and Morquio B disease, and both diseases are now known to be due to genetic mutations in the same gene (*GLB1*) [88]. The primary substrates for beta-galactosidase include galactose-containing oligosaccharides (e.g., keratan sulfate) and GM1 ganglioside. The asialo derivative of GM1, GA1 ganglioside, may also be a substrate for this enzyme.

As expected, tissues from patients with beta-galactosidase deficiency exhibit a broad spectrum of accumulating galactose-containing macromolecules, including GM1, GA1, and keratan sulfate. Gangliosides are normal components of plasma membranes, concentrated in neuronal membranes, especially in the regions of nerve endings and dendrites. GM1 is the major ganglioside in the brains of vertebrates. Gangliosides display a broad spectrum of interactions and may act as binding molecules for toxins and hormones, and are also involved in cell differentiation, cell–cell interactions, and cell signaling.

In the brains of patients with GM1 gangliosidosis, storage of GM1 is the most prominent observation [89]. The accumulating GM1 has the same fatty acid composition, sugar composition and sequence, and glycosidic linkages as normal GM1. Visceral organs also show storage of GM1 ganglioside but to a lesser degree. Also to a lesser degree, GA1 ganglioside may accumulate in the brain of GM1 gangliosidosis patients, as do glycosylceramide, lactosylceramide, and GM2 ganglioside. White matter shows chemical manifestations of myelin breakdown, include low proteolipid protein, low total lipid, and the presence of esterified cholesterol [90].

Other than gangliosides, the other major storage molecule in beta-galactosidase-deficient patients is keratan sulfate [91]. This glycosaminoglycan normally exists in a proteoglycan linked with chondroitin sulfate. It is found primarily within connective tissues, particularly cartilage, explaining the skeletal dysplasia typical of many beta-galactosidase-deficient patients. After proteolysis and release from the proteoglycan, free keratan sulfate chains are hydrolyzed by a series of exoenzymes, including beta-galactosidase.

It should be noted that another protein, referred to as protective protein/cathepsin A (PPCA), has been associated with both beta-galactosidase and another lysosomal hydrolase, neuraminidase, and stabilizes and in some cases activates the enzymes [92]. A genetic deficiency of this protective protein results in combined deficiency of beta-galactosidase and neuraminidase (galactosialidosis). The gene encoding PPCA is distinct from both of these enzymes and will not be discussed further here.

The gene encoding beta-galactosidase (*GLB1*) is located in chromosomal region 3p21.33. Genes encoding this enzyme also have been isolated from mice and several other species. The full-length human *GLB1* cDNA encodes a 677 amino acid polypeptide with beta-galactosidase activity. Mutations in the *GLB1* gene are responsible for both GM1 gangliosidosis and Morquio B disease. As with other lysosomal diseases, most are point mutations altering single amino acids, although small deletions, splice site mutations, and other alterations also have been identified. To date, no correlation has been made between the location or type of *GLB1* mutations and the occurrence of GM1 gangliosidosis or Morquio B disease.

The pathogenesis of beta-galactosidase deficiency is complex. GM1 ganglioside stimulates neurite outgrowth, affects neuronal differentiation, and enhances the action of nerve growth factor [93]. Golgi and electron microscopic studies of cortical neurons from this and several other neurological lysosomal diseases exhibit large neural processes (meganeurites), which may be explained by GM1 accumulation. The extent of meganeurite development is related to the onset, severity, and clinical course of the disease. In other studies, significant changes in neuronal connectivity were found in the cerebral cortex of animals with beta-galactosidase deficiency, and cholinergic function was altered as well. Phosphoinositol-specific phospholipase C and adenyl cyclase activities were altered in the membranes of cerebral cortex from GM1 gangliosidosis cats. In addition, the beta subunit of cholera toxin, which binds specifically to GM1 in the outer leaflet of the cell membrane, was found to induce an increase of calcium and manganese influx in N18 cells, and this has been associated with activation of an L-type voltage-dependent calcium channel. This channel has important implications for neural development, and its reliance on GM1 ganglioside is also likely related to the pathogenesis of GM1 storage. These and other data suggest various morphologic and metabolic aberrations occur in the brains of GM1 gangliosidosis patients and animals, and can be linked to the biology of GM1.

Naturally occurring animal models of GM1 gangliosidosis have been reported in cats, dogs, sheep, and calves. The cat model in particular has been studied in some detail, and is a neurologic disorder with clinical, morphological, biochemical, and genetic similarities to human GM1 gangliosidosis [94]. Affected kittens appear normal at birth, but tremors of the head and hind limbs are first noted at 2–3 months of age, followed by generalized dysmetria and spastic quadriplegia. GM1 ganglioside and other galactose-containing oligosaccharides accumulate in the tissues of affected animals. A genetic knockout of beta-galactosidase activity also has been generated in mice by gene targeting [95]. Progressive and severe neurodegeneration occurs in these animals, and GM1 and GA1 gangliosides accumulate in the brain. Unlike patients, GA1 ganglioside may accumulate in beta-galactosidase deficiency mice to a greater extent than GM1, which may be explained by the fact that there is a unique neuraminidase in mice that converts GM1 to GA1. Skeletal dysplasia, which is characteristic of patients with beta-galactosidase deficiency, does not occur in the mice. This also is due to the unique metabolism of keratan sulfate in mice compared to humans.

11.9 HEXOSAMINIDASE A AND B DEFICIENCY: GM2 GANGLIOSIDOSES

The GM2 gangliosidoses are a group of genetic disorders caused by excessive accumulation of GM2 ganglioside and related glycolipids in lysosomes, mainly of neuronal cells [96]. The general cause of GM2 gangliosidosis is deficient activity of the enzyme beta-hexosaminidase (E.C. 3.2.1.30). The enzymatic hydrolysis of GM2 ganglioside by this enzyme requires that it be complexed with a substrate-specific cofactor, the GM2 activator. There are two isozymes of beta-hexosaminidase, Hex A, which is composed of alpha and beta subunits, and Hex B, which is composed of two beta subunits. Hex A and B can only act on the GM2/GM2 activator complex.

Defects in any of three genes can lead to GM2 gangliosidosis, *HEXA*, which encodes the alpha subunit of Hex A, *HEXB*, which encodes the beta subunit of Hex A and Hex B, or *GM2A*, which encodes the GM2 activator protein [97]. Three clinical forms of GM2 gangliosidosis occur from mutations in these genes: Tay–Sachs disease and variants, resulting from mutations of the *HEXA* gene, and resulting in deficient activity of Hex A; Sandhoff disease and variants, resulting from mutations of the *HEXB* gene, and resulting in deficient activity of both Hex A and Hex B, and GM2 activator deficiency, caused by mutations in the *GM2A* gene and characterized by normal Hex A and Hex B, but the inability to form a functional ganglioside GM2/GM2 activator complex.

GM2 gangliosidosis can present with a wide spectrum of severity, ranging from infantile-onset, rapidly progressive neurodegenerative disease culminating in death before age 4 years (classical Tay–Sachs and Sandhoff diseases and GM2 activator deficiency), to adult-onset, slowly progressive neurological conditions compatible with long survival with little or no effect on intellect [98–101]. The clinical phenotypes of the acute, infantile form of any of the three genetic GM2 gangliosidosis (*HEXA*, *HEXB*, or *GM2A* mutations) are essentially indistinguishable. Affected infants generally appear normal at birth. The earliest signs of disease are often mild motor weakness, beginning at 3–5 months of age. An exaggerated startle response is also commonly observed at this early stage. Soon after, regression and loss of already acquired mental and motor skills becomes obvious. The disease is rapidly progressive and leads to death in early childhood.

As noted above, later-onset forms of GM2 gangliosidosis may also occur. The clinical phenotype varies widely among these patients, and onset can occur at any time from the late infantile period to adults. In general, in the later-onset cases involvement of the deeper brain structures are more prominent compared to the overwhelming generalized gray matter involvement in the infantile form. Manifestations include dystonia, other extrapyramidal signs, such as ataxia, choreoathetoid movements, signs of spinocerebellar degeneration, and ALS-like motor neuron involvement. Psychotic manifestations are not uncommon as well. It should also be noted that in a small number of later-onset cases, mental capacity is well preserved, although severe dysarthria often masks the preserved intelligence.

Similar to most lysosomal glycosidases, beta-hexosaminidase hydrolyzes a broad spectrum of substrates. It is specific for only the terminal nonreducing sugar (GlcNac or GalNac) in beta linkage. The primary substrates for Hex A and Hex B are glycoproteins, oligosaccharides, glycosaminoglycans, and glycolipids, including the ganglioside GM2 when complexed with the GM2 activator protein. It should be noted that a minor isoform composed of two alpha subunits (Hex S) also has been identified, but the biochemical and clinical significance of this isoform remains unclear. Both the alpha and beta subunits of beta-hexosaminidase possess an active site, although dimerization is required for activity. The substrate specificity of the two subunits differs; however, the beta subunit prefers neutral, water-soluble substrates, while the alpha subunit also hydrolyzes negatively charged substrates such as GM2 gangliosides or glycosaminoglycans. The fact that mutations in the *HEXA* gene, in which the Hex B isozyme (composed of two beta subunits) is normal, lead to accumulated GM2 ganglioside demonstrates that the preferred substrate for the alpha subunit is GM2. However, as noted above, the Hex A isozyme can only hydrolyze GM2 if complexed with the GM2 activator protein.

Mutations in the alpha subunit encoding the *HEXA* gene are therefore primarily characterized by accumulation of GM2 ganglioside. Indeed, GM2 ganglioside is the primary neuronal storage material in all three genetic causes, hence their grouping as GM2 gangliosidoses. *HEXA* mutations are responsible for classical Tay–Sachs disease and the later-onset Tay–Sachs variants. The pathology of classical Tay–Sachs disease has been described extensively. The brain is atrophic during the early stage, but usually increases over time until death. Histological analysis shows classical neuronal storage disease in essentially all neurons of the CNS and the peripheral nervous system as well, including swollen retinal ganglion cells. The membranous storage bodies characteristic of Tay–Sachs disease upon electron microscopic examination are composed of cholesterol, phospholipid, and GM2 ganglioside. Increased apoptotic neurons have also been observed. The neuronal pathology in the acute or subacute forms of Tay–Sachs disease is generally less severe than that in the infantile form and tends to be more prominent in the hypothalamus, cerebellum, brain stem, and spinal cord.

In the Sandhoff disease variants, due to mutations in the *HEXB* gene, the pathological and clinical findings are very similar to those of Tay–Sachs. However, yellowing of the cerebral cortex and deeper structures has been documented in Sandhoff disease, possibly due to accumulated asialoganglioside. Also, additional accumulation of sphingoglycolipids with a terminal hexosamine residue and fragments of undigested glycoprotein in systemic organs has been found. Patients with *GM2A* mutations are indistinguishable from those with *HEXA* or *HEXB* mutations.

The *HEXA* gene, encoding the alpha subunit, is located within the chromosomal region 15q23-q24. It is ~35 kb in length and contains 14 exons. The *HEXB* gene, encoding the beta subunit, is located at chromosomal region 5q13 and is ~45 kb in length that is divided into 14 exons. Analysis of these genes strongly suggests that they arose from a common ancestor. They share a very common exon structure, and in both genes the promoter activity resides within ~150 bp of the initiating ATG codon. Overall, the alpha and beta subunits share ~57% amino acid identity. The *GM2A* gene maps to chromosomal region 5q31.1-31.3. In addition, an evolutionarily related pseudogene, *GM2AP*, maps to chromosome 3. The *GM2A* gene is ~16 kb and contains four exons.

Numerous mutations have been described in the *HEXA* gene responsible for Tay–Sachs disease and its variants. Infantile Tay–Sachs disease occurs most commonly among individuals of Ashkenazi Jewish ancestry (carrier frequency ~1:25), and there are two mutations that are the most frequent; one is a 4 bp insertion in exon 11, and the second is a donor splice site mutation in intron 11 [102]. Both of these mutations result in severely deficient or absent *HEXA* mRNA expression, leading to absent Hex A and Hex S activities. Another common HEXA mutation causing infantile Tay–Sachs disease occurs in French Canadian patients and is due to a deletion that extends from ~2 kb upstream of the 5′ end of the gene into intron 1 [103,104]. This also results in the absence of Hex A mRNA and protein. Many other mutations in the *HEXA* gene have been described causing classical Tay–Sachs disease, effecting protein processing, catalytic activity, and/or mRNA processing. In addition, mutations causing the later-onset forms are known, and generally result in expression of *HEXA* mRNA and residual enzyme activity. An important *HEXA* mutation is also known as the B1 variant, in which there is normal activity toward the artificial substrates generally used to measure beta-hexosaminidase activity, but no activity of the mutant enzyme toward GM2 ganglioside. Individuals with this mutation pose a unique diagnostic challenge since they will appear enzymatically normal but develop severe disease. Many mutations in the *HEXB* gene causing Sandhoff disease and its variants also have been described. A much small number of mutations in the *GM2A* gene are known. As noted above, the fact that mutations in this gene caused a severe clinical disease provided the first proof of the physiological significance of the GM2/GM2 activator complex [105,106].

Naturally occurring mutations causing GM2 gangliosidosis have been described in dogs, cats, and pigs [107–109]. In all species, abundant neuronal storage is the major finding. Visceral storage is only found in the feline model, however. Meganeurite formation, which has been observed in human GM2 gangliosidosis patients, occurs in dog and cat models. A number of groups also have independently generated murine models of Tay–Sachs, Sandhoff, and GM2 activator deficiency in mice using gene targeting strategies. In general, knockout of the *HEXA* gene in mice (Tay–Sachs disease) results in a much milder neurological disease than predicted from the human cases [110]. Mutations in the *HEXB* gene in mice results in a more severe phenotype, with onset of clinical disease at ~3 months associated with excessive neuronal GM2 storage [130]. Mutations in the *GM2A* gene of mice also resulted in a much milder disease than predicted [111].

In part, the distinction between the mouse models and the human disorder are due to the distinct degradation of GM2 ganglioside by these species [112]. In humans, GM2 is degraded nearly exclusively by Hex A in collaboration with the activator protein to yield the ganglioside GM3. In contrast, in mice GM2 can be degraded by two different pathways. One is identical to the human pathway, and the second is essentially unique to the mouse and is initiated by sialidase acting on GM2 to yield GA2 ganglioside. The GA2 is then degraded by either Hex A with activator protein, or to a lesser extent by Hex B. This explains the mild phenotype due to alpha subunit mutations (*HEXA*), since the Hex B can act on the GA2 produced by sialidase. In contrast, *HEXB* mutations result in deficiency of both enzymes and a complete block of degradation.

11.10 SPHINGOLIPID ACTIVATOR PROTEINS

As discussed in the earlier sections, the lysosomal degradation of sphingolipids requires small nonenzymatic glycoproteins, referred to as "SAPs" [113,114]. These include the saposin proteins (SAPs A–D, and GM2 activator protein). There are two genes responsible for these proteins. One is *GMA2*, which encodes the GM2 activator protein and is located on chromosome 5 [115], and the other is *PSAP*, located on chromosome 10 and encoding the saposin precursor protein and the individual SAPs A–D, which are derived from proteolytic cleavage of the precursor [116]. The GM2 activator protein and resulting mutations (responsible for the "A/B" Tay–Sachs variant) were discussed in the previous section and will not be discussed further here.

The *PSAP* gene encodes the SAP precursor protein (prosaposin), with a total of 524 amino acids and 5 N-glycosylation sites [116]. There are four homologous domains located within the precursor protein, each of ~80 amino acids. A major portion of the newly synthesized precursor is exported to the cell surface and then imported into the lysosomal compartment, where it is processed to the mature SAP A–D proteins. Notably, unlike most lysosomal proteins the prosaposin protein is transported to the lysosome via its interaction with the alternative receptor, sortilin [117,118]. The occurrence of the nonprocessed SAP precursor in body fluids and its neurotrophic properties indicates that its function may not be restricted to being the precursor of the individual SAPs. For example, prosaposin has been found in milk, serum, and seminal fluid, and treatment of cell lines with PSAP increased cell survival and was antiapoptotic. Indeed, serum prosaposin levels are increased in patients with advanced prostate cancer. Of interest is whether these effects of PSAP on cell survival are a direct effect of the protein itself, or rather the effect on the individual processed SAPs on the activation of the sphingolipid hydrolases.

In vitro, SAP-A stimulates beta-glucocerebrosidase (the enzyme deficient in Gaucher disease) and beta-galactocerebrosidase (the enzyme deficient in Krabbe disease) activities in the presence of detergents. SAP-B is a nonspecific glycolipid binding protein that stimulates the hydrolysis of ~20 or more glycolipids by different enzymes *in vitro*, including the hydrolysis of sulfatide by ASA (the enzyme deficient in MLD). SAP-C stimulates the *in vitro* activities of glucosyl- and galactosylcerebrosidases, as well as sphingomyelin by ASM (the enzyme deficient in Types A and B NPD) [119]. SAP-D is required for the degradation of ceramide by AC (the enzyme deficient in Farber disease [120].

Analysis of mutations in the genes encoding the *PSAP* protein have provided important insights into the role of the SAP proteins for the *in vivo* hydrolysis of sphingolipids and their relationship to human disease. Of note, a SAP precursor deficiency was found in a child who died at 16 weeks and who was homozygous for a mutation in the initiation codon of the *PSAP* gene. The complete absence of the precursor protein and four resultant SAPs led to a generalized accumulation of ceramide, glucosylceramide, galactosylceramide, sulfatide, lactosylceramide, digalactosylceramide, and other sphingolipids. The symptoms of the disease resembled that of type 2 Gaucher disease. Other point mutations within the region of the *PSAP* gene encoding SAP B caused accumulation of sulfatide and some other sphingolipids, and a clinical course resembling that of juvenile MLD. Mutations in the SAP-C region only led to accumulation of glucosylceramide, similar to juvenile Gaucher disease [121]. No sphingomyelin storage was observed in individuals with SAP-C mutations, and this has been explained by the fact that there is a SAP-like domain within ASM that compensates for the loss of SAP-C in these patients. Patients with mutations in the SAP-A region of the *PSAP* gene resemble those with Krabbe disease (GLD), and accumulate galactosylcerebroside within the CNS and peripheral nervous system [122]. There are no human patients with SAP-D mutations, but the requirement by AC for SAP-D has been confirmed *in vivo* in mutant mice, who accumulate ceramides with alpha-hydroxylated fatty acids mainly in the brain and kidney, and develop Purkinje cell loss and defects in the urinary system [123,124].

A mouse model of SAP precursor deficiency has also been constructed by gene targeting, and the clinical and biochemical features resembled that of the human disease [125]. Mice developed massive sphingolipid storage and neurological disease by ~20 days, and died by ~35–38 days. Several knock-in models expressing specific human mutations have also been constructed, confirming the *in vivo* importance of these proteins [126,127].

REFERENCES

[1] Chen H, Tran J-TA, Brush RS, Saadi A, Rahman AK, et al. Ceramide signaling in retinal degeneration. Adv Exp Med Biol 2012;723:553–558.

[2] Furuya H, Shimizu Y, Kawamori T. Sphingolipids in cancer. Cancer Metastasis Rev 2011;30:567–576.

[3] Petrache I, Petrusca DN, Bowler RP, Kamocki K. Involvement of ceramide in cell death responses in the pulmonary circulation. Proc Am Thorac Soc 2011;8:492–496.

[4] Podbielska M, Krotkiewski H, Hogan EL. Signaling and regulatory functions of bioactive sphingolipids as therapeutic targets in multiple sclerosis. Neurochem Res 2012;37:1154–1169.

[5] Mullen TD, Hannun YA, Obeid LM. Ceramide synthases at the centre of sphingolipid metabolism and biology. Biochem J 2012;441:789–802.

[6] Wennekes T, van den Berg RJBHN, Boot RG, van der Marel GA, Overkleeft HS, et al. Glycosphingolipids – nature, function, and pharmacological modulation. Angew Chem Int Ed Engl 2009;48:8848–8869.

[7] Worgall TS. Sphingolipid synthetic pathways are major regulators of lipid homeostasis. Adv Exp Med Biol 2011;721:139–148.

[8] Schulze H, Sandhoff K. Lysosomal lipid storage diseases. Cold Spring Harb Perspect Biol 2011;3:a004804.

[9] Cox TM, Cachon-Gonzalez MB. The cellular pathology of lysosomal diseases. J Pathol 2012;226:241–254.

[10] Young SA, Mina JG, Denny PW, Smith TK. Sphingolipid and ceramide homeostasis: potential therapeutic targets. Biochem Res Int 2012;2012:248135–248135.

[11] Farber S, Cohen J, Uzman LL. Lipogranulomatosis; a new lipo-glycoprotein storage disease. J Mt Sinai Hosp N Y 1957;24:816–837.

[12] Ahmad A, Mazhar AU, Anwar M. Farber disease: a rare neurodegenerative disorder. J Coll Physicians Surg Pak 2009;19:67–68.

[13] Sugita M, Dulaney JT, Moser HW. Ceramidase deficiency in Farber's disease (lipogranulomatosis). Science 1972;178:1100–1102.

[14] Okino N, He X, Gatt S, Sandhoff K, Ito M, et al. The reverse activity of human acid ceramidase. J Biol Chem 2003;278:29948–29953.

[15] Park J-H, Schuchman EH. Acid ceramidase and human disease. Biochim Biophys Acta 2006;1758:2133–2138.

[16] Li C-M, Park J-H, Simonaro CM, He X, Gordon RE, et al. Insertional mutagenesis of the mouse acid ceramidase gene leads to early embryonic lethality in homozygotes and progressive lipid storage disease in heterozygotes. Genomics 2002;79:218–224.

[17] Eliyahu E, Shtraizent N, Martinuzzi K, Barritt J, He X, et al. Acid ceramidase improves the quality of oocytes and embryos and the outcome of *in vitro* fertilization. FASEB J 2010;24:1229–1238.

[18] Bashyam MD, Chaudhary AK, Kiran M, Reddy V, Nagarajaram HA, et al. Molecular analyses of novel ASAH1 mutations causing Farber lipogranulomatosis: analyses of exonic splicing enhancer inactivating mutation. Clin Genet 2014;86:530–538.

[19] Alayoubi AM, Wang JC, Au BC, Carpentier S, Garcia V, et al. Systemic ceramide accumulation leads to severe and varied pathological consequences. EMBO Mol Med 2013;5:827–842.

[20] Eliyahu E, Shtraizent N, Shalgi R, Schuchman EH. Construction of conditional acid ceramidase knockout mice and *in vivo* effects on oocyte development and fertility. Cell Physiol Biochem 2012;30:735–748.

[21] Zhou J, Tawk M, Tiziano FD, Veillet J, Bayes M, et al. Spinal muscular atrophy associated with progressive myoclonic epilepsy is caused by mutations in ASAH1. Am J Hum Genet 2012;91:5–14.

[22] Dyment DA, Sell E, Vanstone MR, Smith AC, Garandeau D, et al. Evidence for clinical, genetic and biochemical variability in spinal muscular atrophy with progressive myoclonic epilepsy. Clin Genet 2014;86:558–563.

[23] Park J-H, Eliyahu E, Narla G, DiFeo A, Martignetti JA, et al. KLF6 is one transcription factor involved in regulating acid ceramidase gene expression. Biochim Biophys Acta 2005;1732:82–87.

[24] Beckham TH, Lu P, Cheng JC, Zhao D, Turner LS, et al. Acid ceramidase-mediated production of sphingosine 1-phosphate promotes prostate cancer invasion through upregulation of cathepsin B. Int J Cancer 2012;131:2034–2043.

[25] Flowers M, Fabrias G, Delgado A, Casas J, Abad JL, et al. C6-ceramide and targeted inhibition of acid ceramidase induce synergistic decreases in breast cancer cell growth. Breast Cancer Res Treat 2012;133:447–458.

[26] Zhang H, Li D, Su Y, Jiang S, Xu Y, et al. Identification of the N-acylsphingosine amidohydrolase 1 gene (ASAH1) for susceptibility to schizophrenia in a Han Chinese population. World J Biol Psychiatry 2012;13:106–113.

[27] Mao C, Obeid LM. Ceramidases: regulators of cellular responses mediated by ceramide, sphingosine, and sphingosine-1-phosphate. Biochim Biophys Acta 2008; 1781:424–434.

[28] McGovern MMSE. In: Pagon RABT, Dolan CR, Stephens K, Adam MP, editors. *Acid Sphingomyelinase Deficiency*. Seattle (WA): University of Washington; 2009.

[29] Schuchman EH. The pathogenesis and treatment of acid sphingomyelinase-deficient Niemann–Pick disease. Int J Clin Pharmacol Ther 2009;47(Suppl 1):48–57.

[30] Simonaro CM, Park JH, Eliyahu E, Shtraizent N, McGovern MM, et al. Imprinting at the SMPD1 locus: implications for acid sphingomyelinase-deficient Niemann–Pick disease. Am J Hum Genet 2006;78:865–870.

[31] Clarke CJ, Wu BX, Hannun YA. The neutral sphingomyelinase family: identifying biochemical connections. Adv Enzyme Regul 2011;51:51–58.

[32] Horinouchi K, Erlich S, Perl DP, Ferlinz K, Bisgaier CL, et al. Acid sphingomyelinase deficient mice: a model of types A and B Niemann–Pick disease. Nat Genet 1995; 10:288–293.

[33] Schissel SL, Keesler GA, Schuchman EH, Williams KJ, Tabas I. The cellular trafficking and zinc dependence of secretory and lysosomal sphingomyelinase, two products of the acid sphingomyelinase gene. J Biol Chem 1998;273:18250–18259.

[34] Miranda SR, He X, Simonaro CM, Gatt S, Dagan A, et al. Infusion of recombinant human acid sphingomyelinase into Niemann–Pick disease mice leads to visceral, but not neurological, correction of the pathophysiology. FASEB J 2000;14:1988–1995.

[35] Zeidan YH, Hannun YA. The acid sphingomyelinase/ceramide pathway: biomedical significance and mechanisms of regulation. Curr Mol Med 2010;10:454–466.

[36] Becker KA, Grassme H, Zhang Y, Gulbins E. Ceramide in *Pseudomonas aeruginosa* infections and cystic fibrosis. Cell Physiol Biochem 2010;26:57–66.

[37] Bikman BT, Summers SA. Sphingolipids and hepatic steatosis. Adv Exp Med Biol 2011;721:87–97.

[38] Truman J-P, Al Gadban MM, Smith KJ, Hammad SM. Acid sphingomyelinase in macrophage biology. Cell Mol Life Sci 2011;68:3293–3305.

[39] Lo SM, Choi M, Liu J, Jain D, Boot RG, et al. Phenotype diversity in type 1 Gaucher disease: discovering the genetic basis of Gaucher disease/hematologic malignancy phenotype by individual genome analysis. Blood 2012;119:4731–4740.

[40] Pastores GM. Neuropathic Gaucher disease. Wien Med Wochenschr 2010;160: 605–608.

[41] Mikosch P, Hughes D. An overview on bone manifestations in Gaucher disease. Wien Med Wochenschr 2010;160:609–624.

[42] Goker-Alpan O. Therapeutic approaches to bone pathology in Gaucher disease: past, present and future. Mol Genet Metab 2011;104:438–447.

[43] Choy FYM, Campbell TN. Gaucher disease and cancer: concept and controversy. Int J Cell Biol 2011;2011:150450–150450.

[44] Brown JT, Lahey C, Laosinchai-Wolf W, Hadd AG. Polymorphisms in the glucocerebrosidase gene and pseudogene urge caution in clinical analysis of Gaucher disease allele c.1448T>C (L444P). BMC Med Genet 2006;7:69–69.

[45] Hruska KS, LaMarca ME, Scott CR, Sidransky E. Gaucher disease: mutation and polymorphism spectrum in the glucocerebrosidase gene (GBA). Hum Mutat 2008; 29:567–583.

[46] Scott SA, Edelmann L, Liu L, Luo M, Desnick RJ, et al. Experience with carrier screening and prenatal diagnosis for 16 Ashkenazi Jewish genetic diseases. Hum Mutat 2010;31:1240–1250.

[47] Kinghorn KJ. Pathological looping in the synucleinopathies: investigating the link between Parkinson's disease and Gaucher disease. Dis Model Mech 2011;4:713–715.

[48] Anheim M, Elbaz A, Lesage S, Durr A, Condroyer C, et al. Penetrance of Parkinson disease in glucocerebrosidase gene mutation carriers. Neurology 2012;78:417–420.

[49] Farfel-Becker T, Vitner EB, Futerman AH. Animal models for Gaucher disease research. Dis Model Mech 2011;4:746–752.

[50] Sardi SP, Singh P, Cheng SH, Shihabuddin LS, Schlossmacher MG. Mutant GBA1 expression and synucleinopathy risk: first insights from cellular and mouse models. Neurodegener Dis 2012;10:195–202.

[51] Hughes DA, Pastores GM. The pathophysiology of GD – current understanding and rationale for existing and emerging therapeutic approaches. Wien Med Wochenschr 2010;160:594–599.

[52] Lachmann RH. Enzyme replacement therapy for lysosomal storage diseases. Curr Opin Pediatr 2011;23:588–593.

[53] Abian O, Alfonso P, Velazquez-Campoy A, Giraldo P, Pocovi M, et al. Therapeutic strategies for Gaucher disease: miglustat (NB-DNJ) as a pharmacological chaperone for glucocerebrosidase and the different thermostability of velaglucerase alfa and imiglucerase. Mol Pharm 2011;8:2390–2397.

[54] Benito JM, Garcia Fernandez JM, Ortiz Mellet C. Pharmacological chaperone therapy for Gaucher disease: a patent review. Expert Opin Ther Pat 2011;21:885–903.

[55] Kohlschutter A, Eichler F. Childhood leukodystrophies: a clinical perspective. Expert Rev Neurother 2011;11:1485–1496.

[56] Duffner PK, Barczykowski A, Jalal K, Yan L, Kay DM, et al. Early infantile Krabbe disease: results of the World-Wide Krabbe Registry. Pediatr Neurol 2011;45:141–148.

[57] Malandrini A, D'Eramo C, Palmeri S, Gaudiano C, Gambelli S, et al. Peripheral neuropathy in late-onset Krabbe disease: report of three cases. Neurol Sci 2013;34:79–83.

[58] Perlman SJ, Mar S. Leukodystrophies. Adv Exp Med Biol 2012;724:154–171.

[59] Sakai N. Pathogenesis of leukodystrophy for Krabbe disease: molecular mechanism and clinical treatment. Brain Dev 2009;31:485–487.

[60] Liu Y, Gibson J, Wheeler J, Kwee LC, Santiago-Turla CM, et al. GALC deletions increase the risk of primary open-angle glaucoma: the role of Mendelian variants in complex disease. PLoS One 2011;6:e27134.

[61] Luzi P, Rafi MA, Zaka M, Curtis M, Vanier MT, et al. Generation of a mouse with low galactocerebrosidase activity by gene targeting: a new model of globoid cell leukodystrophy (Krabbe disease). Mol Genet Metab 2001;73:211–223.

[62] Neri M, Ricca A, di Girolamo I, Alcala'-Franco B, Cavazzin C, et al. Neural stem cell gene therapy ameliorates pathology and function in a mouse model of globoid cell leukodystrophy. Stem Cells 2011;29:1559–1571.

[63] Reddy AS, Kim JH, Hawkins-Salsbury JA, Macauley SL, Tracy ET, et al. Bone marrow transplantation augments the effect of brain- and spinal cord-directed adeno-associated virus 2/5 gene therapy by altering inflammation in the murine model of globoid-cell leukodystrophy. J Neurosci 2011;31:9945–9957.

[64] Lin D-S, Hsiao C-D, Liau I, Lin S-P, Chiang M-F, et al. CNS-targeted AAV5 gene transfer results in global dispersal of vector and prevention of morphological and function deterioration in CNS of globoid cell leukodystrophy mouse model. Mol Genet Metab 2011;103:367–377.

[65] Renaud DL. Lysosomal disorders associated with leukoencephalopathy. Seminars in neurology. 2012;32(1):51–4.

[66] Kehrer C, Blumenstock G, Gieselmann V, Krageloh-Mann I. The natural course of gross motor deterioration in metachromatic leukodystrophy. Dev Med Child Neurol 2011;53:850–855.

[67] Gieselmann V, Zlotogora J, Harris A, Wenger DA, Morris CP. Molecular genetics of metachromatic leukodystrophy. Hum Mutat 1994;4:233–242.

[68] Lugowska A, Poninska J, Krajewski P, Broda G, Ploski R. Population carrier rates of pathogenic ARSA gene mutations: is metachromatic leukodystrophy underdiagnosed? PLoS One 2011;6:e20218.

[69] Tinsa F, Caillaud C, Vanier MT, Bousnina D, Boussetta K, et al. An unusual homozygous arylsulfatase: a pseudodeficiency in a metachromatic leukodystrophy Tunisian patient. J Child Neurol 2010;25:82–86.

[70] Baronica KB, Mlinac K, Ozretic D, Vladic A, Bognar SK. Arylsulfatase a gene polymorphisms in relapsing remitting multiple sclerosis: genotype-phenotype correlation and estimation of disease progression. Coll Antropol 2011;35(Suppl 1):11–16.

[71] Kumperscak HG, Dolzan V, Videtic A, Plesnicar BK. Polymorphisms in genes encoding the serotonin and dopamine pathways in two sisters with metachromatic leukodystrophy. J Int Med Res 2008;36:1123–1128.

[72] Matthes F, Stroobants S, Gerlach D, Wohlenberg C, Wessig C, et al. Efficacy of enzyme replacement therapy in an aggravated mouse model of metachromatic leukodystrophy declines with age. Hum Mol Genet 2012;21:2599–2609.

[73] Stroobants S, Gerlach D, Matthes F, Hartmann D, Fogh J, et al. Intracerebroventricular enzyme infusion corrects central nervous system pathology and dysfunction in a mouse model of metachromatic leukodystrophy. Hum Mol Genet 2011;20:2760–2769.

[74] Solders G, Celsing G, Hagenfeldt L, Ljungman P, Isberg B, et al. Improved peripheral nerve conduction, EEG and verbal IQ after bone marrow transplantation for adult metachromatic leukodystrophy. Bone Marrow Transplant 1998;22:1119–1122.

[75] Gassas A, Raiman J, White L, Schechter T, Clarke J, et al. Long-term adaptive functioning outcomes of children with inherited metabolic and genetic diseases treated with hematopoietic stem cell transplantation in a single large pediatric center: parents' perspective. J Pediatr Hematol Oncol 2011;33:216–220.

[76] Tarabuso AL. Fabry disease. Skinmed 2011;9:173–177.

[77] Toyooka K. Fabry disease. Curr Opin Neurol 2011;24:463–468.

[78] Aerts JM, Boot RG, van Eijk M, Groener J, Bijl N, et al. Glycosphingolipids and insulin resistance. Adv Exp Med Biol 2011;721:99–9119.

[79] Linthorst GE, Folman CC, Aerts JM, Hollak CE. Blood group does not correlate with disease severity in patients with Fabry disease (alpha-galactosidase A deficiency). Blood Cells Mol Dis 2003;31:324–326.

[80] Tomasic IB, Metcalf MC, Guce AI, Clark NE, Garman SC. Interconversion of the specificities of human lysosomal enzymes associated with Fabry and Schindler diseases. J Biol Chem 2010;285:21560–21566.

[81] Rozenfeld P, Neumann PM. Treatment of Fabry disease: current and emerging strategies. Curr Pharm Biotechnol 2011;12:916–922.

[82] Schaefer E, Mehta A, Gal A. Genotype and phenotype in Fabry disease: analysis of the Fabry Outcome Survey. Acta Paediatr Suppl 2005;94:87–92; discussion 79.

[83] Bouwman MG, Rombach SM, Schenk E, Sweeb A, Wijburg FA, et al. Prevalence of symptoms in female Fabry disease patients: a case-control survey. J Inherit Metab Dis 2012;35:891–898.

[84] Spačil Z, Elliott S, Reeber SL, Gelb MH, Scott CR, et al. Comparative triplex tandem mass spectrometry assays of lysosomal enzyme activities in dried blood spots using fast liquid chromatography: application to newborn screening of Pompe, Fabry, and Hurler diseases. Anal Chem 2011;83:4822–4828.

[85] Brunetti-Pierri N, Scaglia F. GM1 gangliosidosis: review of clinical, molecular, and therapeutic aspects. Mol Genet Metab 2008;94:391–396.

[86] Morita M, Saito S, Ikeda K, Ohno K, Sugawara K, et al. Structural bases of GM1 gangliosidosis and Morquio B disease. J Hum Genet 2009;54:510–515.

[87] Caciotti A, Garman SC, Rivera-Colon Y, Procopio E, Catarzi S, et al. GM1 gangliosidosis and Morquio B disease: an update on genetic alterations and clinical findings. Biochim Biophys Acta 2011;1812:782–790.

[88] Hofer D, Paul K, Fantur K, Beck M, Burger F, et al. GM1 gangliosidosis and Morquio B disease: expression analysis of missense mutations affecting the catalytic site of acid beta-galactosidase. Hum Mutat 2009;30:1214–1221.

[89] Nada R, Gupta K, Lal SB, Vasishta RK. An autopsy case of infantile GM1 gangliosidosis with adrenal calcification. Metab Brain Dis 2011;26:307–310.

[90] Steenweg ME, Vanderver A, Blaser S, Bizzi A, de Koning TJ, et al. Magnetic resonance imaging pattern recognition in hypomyelinating disorders. Brain 2010;133:2971–2982.

[91] Ohto U, Usui K, Ochi T, Yuki K, Satow Y, et al. Crystal structure of human beta-galactosidase: structural basis of Gm1 gangliosidosis and morquio B diseases. J Biol Chem 2012;287:1801–1812.

[92] Tatano Y, Takeuchi N, Kuwahara J, Sakuraba H, Takahashi T, et al. Elastogenesis in cultured dermal fibroblasts from patients with lysosomal beta-galactosidase, protective protein/cathepsin A and neuraminidase-1 deficiencies. J Med Invest 2006;53:103–112.

[93] van der Voorn JP, Kamphorst W, van der Knaap MS, Powers JM. The leukoencephalopathy of infantile GM1 gangliosidosis: oligodendrocytic loss and axonal dysfunction. Acta Neuropathol 2004;107:539–545.

[94] Claro E, Wallace MA, Fain JN, Nair BG, Patel TB, et al. Altered phosphoinositide-specific phospholipase C and adenylyl cyclase in brain cortical membranes of cats with GM1 and GM2 gangliosidosis. Brain Res Mol Brain Res 1991;11:265–271.

[95] Itoh M, Matsuda J, Suzuki O, Ogura A, Oshima A, et al. Development of lysosomal storage in mice with targeted disruption of the beta-galactosidase gene: a model of human G(M1)-gangliosidosis. Brain Dev 2001;23:379–384.

[96] Maegawa G. GM2 gangliosidosis: the prototype of lysosomal storage disorders. Dev Med Child Neurol 2012;54:104–105.

[97] Cordeiro P, Hechtman P, Kaplan F. The GM2 gangliosidoses databases: allelic variation at the HEXA, HEXB, and GM2A gene loci. Genet Med 2000;2:319–327.

[98] Bley AE, Giannikopoulos OA, Hayden D, Kubilus K, Tifft CJ, et al. Natural history of infantile G(M2) gangliosidosis. Pediatrics 2011;128:e1233–1241.

[99] Smith NJ, Winstone AM, Stellitano L, Cox TM, Verity CM. GM2 gangliosidosis in a UK study of children with progressive neurodegeneration: 73 cases reviewed. Dev Med Child Neurol 2012;54:176–182.

[100] Jain A, Kohli A, Sachan D. Infantile Sandhoff's disease with peripheral neuropathy. Pediatr Neurol 2010;42:459–461.

[101] Osher E, Fattal-Valevski A, Sagie L, Urshanski N, Amir-Levi Y, et al. Pyrimethamine increases beta-hexosaminidase A activity in patients with Late Onset Tay Sachs. Mol Genet Metab 2011;102:356–363.

[102] Frisch A, Colombo R, Michaelovsky E, Karpati M, Goldman B, et al. Origin and spread of the 1278insTATC mutation causing Tay–Sachs disease in Ashkenazi Jews: genetic drift as a robust and parsimonious hypothesis. Hum Genet 2004;114:366–376.

[103] De Braekeleer M, Hechtman P, Andermann E, Kaplan F. The French Canadian Tay–Sachs disease deletion mutation: identification of probable founders. Hum Genet 1992; 89:83–87.

[104] Martin DC, Mark BL, Triggs-Raine BL, Natowicz MR. Evaluation of the risk for Tay–Sachs disease in individuals of French Canadian ancestry living in New England. Clin Chem 2007;53:392–398.

[105] Peleg L, Meltzer F, Karpati M, Goldman B. GM2 gangliosidosis B1 variant: biochemical and molecular characterization of hexosaminidase A. Biochem Mol Med 1995; 54:126–132.

[106] Tanaka A, Hoang LTN, Nishi Y, Maniwa S, Oka M, et al. Different attenuated phenotypes of GM2 gangliosidosis variant B in Japanese patients with HEXA mutations at codon 499, and five novel mutations responsible for infantile acute form. J Hum Genet 2003;48:571–574.

[107] Yamato O, Matsuki N, Satoh H, Inaba M, Ono K, et al. Sandhoff disease in a golden retriever dog. J Inherit Metab Dis 2002;25:319–320.

[108] Kosanke SD, Pierce KR, Read WK. Morphogenesis of light and electron microscopic lesions in porcine GM2-gangliosidosis. Vet Pathol 1979;16:6–17.

[109] Bradbury AM, Morrison NE, Hwang M, Cox NR, Baker HJ, et al. Neurodegenerative lysosomal storage disease in European Burmese cats with hexosaminidase beta-subunit deficiency. Mol Genet Metab 2009;97:53–59.

[110] Yamanaka S, Johnson MD, Grinberg A, Westphal H, Crawley JN, et al. Targeted disruption of the Hexa gene results in mice with biochemical and pathologic features of Tay–Sachs disease. Proc Natl Acad Sci U S A 1994;91:9975–9979.

[111] Phaneuf D, Wakamatsu N, Huang JQ, Borowski A, Peterson AC, et al. Dramatically different phenotypes in mouse models of human Tay–Sachs and Sandhoff diseases. Hum Mol Genet 1996;5:1–14.

[112] Baek RC, Martin DR, Cox NR, Seyfried TN. Comparative analysis of brain lipids in mice, cats, and humans with Sandhoff disease. Lipids 2009;44:197–205.

[113] Sandhoff K, Kolter T. Processing of sphingolipid activator proteins and the topology of lysosomal digestion. Acta Biochim Pol 1998;45:373–384.

[114] Kolter T, Sandhoff K. Lysosomal degradation of membrane lipids. FEBS Lett 2010; 584:1700–1712.

[115] Wendeler M, Lemm T, Weisgerber J, Hoernschemeyer J, Bartelsen O, et al. Expression of recombinant human GM2-activator protein in insect cells: purification and characterization by mass spectrometry. Protein Expr Purif 2003;27:259–266.

[116] Misasi R, Hozumi I, Inuzuka T, Capozzi A, Mattei V, et al. Biochemistry and neurobiology of prosaposin: a potential therapeutic neuro-effector. Cent Nerv Syst Agents Med Chem 2009;9:119–131.

[117] Wahe A, Kasmapour B, Schmaderer C, Liebl D, Sandhoff K, et al. Golgi-to-phagosome transport of acid sphingomyelinase and prosaposin is mediated by sortilin. J Cell Sci 2010;123:2502–2511.

[118] Yuan L, Morales CR. Prosaposin sorting is mediated by oligomerization. Exp Cell Res 2011;317:2456–2467.

[119] Kolzer M, Ferlinz K, Bartelsen O, Hoops SL, Lang F, et al. Functional characterization of the postulated intramolecular sphingolipid activator protein domain of human acid sphingomyelinase. Biol Chem 2004;385:1193–1195.

[120] Linke T, Wilkening G, Sadeghlar F, Mozcall H, Bernardo K, et al. Interfacial regulation of acid ceramidase activity. Stimulation of ceramide degradation by lysosomal lipids and sphingolipid activator proteins. J Biol Chem 2001;276:5760–5768.

[121] Vaccaro AM, Motta M, Tatti M, Scarpa S, Masuelli L, et al. Saposin C mutations in Gaucher disease patients resulting in lysosomal lipid accumulation, saposin C deficiency, but normal prosaposin processing and sorting. Hum Mol Genet 2010;19: 2987–2997.

[122] Grossi S, Regis S, Rosano C, Corsolini F, Uziel G, et al. Molecular analysis of ARSA and PSAP genes in twenty-one Italian patients with metachromatic leukodystrophy: identification and functional characterization of 11 novel ARSA alleles. Hum Mutat 2008;29:220–230.

[123] Oya Y, Nakayasu H, Fujita N, Suzuki K, Suzuki K. Pathological study of mice with total deficiency of sphingolipid activator proteins (SAP knockout mice). Acta Neuropathol 1998;96:29–40.

[124] Tohyama J, Oya Y, Ezoe T, Vanier MT, Nakayasu H, et al. Ceramide accumulation is associated with increased apoptotic cell death in cultured fibroblasts of sphingolipid activator protein-deficient mouse but not in fibroblasts of patients with Farber disease. J Inherit Metab Dis 1999;22:649–662.

[125] Matsuda J, Yoneshige A, Suzuki K. The function of sphingolipids in the nervous system: lessons learnt from mouse models of specific sphingolipid activator protein deficiencies. J Neurochem 2007;103(Suppl 1):32–38.

[126] Sun Y, Witte DP, Ran H, Zamzow M, Barnes S, et al. Neurological deficits and glycosphingolipid accumulation in saposin B deficient mice. Hum Mol Genet 2008;17: 2345–2356.

[127] Sun Y, Ran H, Zamzow M, Kitatani K, Skelton MR, et al. Specific saposin C deficiency: CNS impairment and acid beta-glucosidase effects in the mouse. Hum Mol Genet 2010; 19:634–647.

[128] Campbell TN, Choy FY. Gaucher disease and the synucleinopathies: refining the relationship. Orphanet J Rare Dis 2012;7:12.

[129] Lee JY, Lee BH, Kim GH, Jung CW, Lee J, et al. Clinical and genetic characteristics of Gaucher disease according to phenotypic subgroups. Korean J Pediatr 2012;55:48–53.

[130] Gulinello M, Chen F, Dobrenis K. Early deficits in motor coordination and cognitive dysfunction in a mouse model of the neurodegenerative lysosomal storage disorder, Sandhoff disease. Behav Brain Res 2008;193:315–319.

12

LYSOSOME-RELATED ORGANELLES: MODIFICATIONS OF THE LYSOSOME PARADIGM

Adriana R. Mantegazza and Michael S. Marks

Department of Pathology and Laboratory Medicine, Children's Hospital of Philadelphia Research Institute, Philadelphia, PA, USA; Departments of Pathology and Laboratory Medicine and Physiology, Perelman School of Medicine, University of Pennsylvania, Philadelphia, PA, USA

Lysosomes are essential organelles for the physiological maintenance and viability of all metazoan cells, with conserved functions in nutrient and ion homeostasis, cell signaling, macromolecule turnover, and plasma membrane repair among others. The complex physiology of multicellular organisms requires certain cell types to fulfill needs that exploit some features of lysosomes but that are specialized for specific functions – such as cell-mediated cytolysis by cytotoxic T lymphocytes (CTLs) or regulated secretion of clotting factors, surfactants, or other specialized cargoes. Many such cells have evolved to generate either modified forms of lysosomes or completely distinct membranous structures to fulfill these functions. Such organelles are called *lysosome-related organelles (LROs)*, and are broadly defined as cell type-specific compartments that share features of lysosomal or endosomal organelles but are functionally and morphologically distinct. Since LROs are required for the physiology of the organism but not for the viability of the cell that generates them, heritable defects in the formation, composition, motility, or secretion of LROs often result in disease. In this chapter, we briefly describe the variety of LROs that have been defined to date in mammals and model organisms, relate recent insights into the biogenesis and

Lysosomes: Biology, Diseases, and Therapeutics, First Edition.
Edited by Frederick R. Maxfield, James M. Willard and Shuyan Lu.

secretion of a few model LROs and how these insights relate to disease, and end with a focus on LROs of the immune system and perspectives for future investigation.

12.1 DIFFERENCES BETWEEN LROs AND SECRETORY GRANULES

LROs have been rather difficult to precisely define, even by experts in the field. It is generally agreed that all LROs contain essential functional elements of endolysosomal compartments, and the late endosomal/lysosomal tetraspanin CD63 has been raised as a potential common component of most, if not all, LROs (e.g., see Refs [1–3]). This distinguishes LROs from classical regulated secretory granules, such as the dense core granules of chromaffin cells or pancreatic β cells, in which the functional components are derived exclusively from the secretory pathway. However, the degree to which critical LRO components derive from the secretory pathway varies. At one end of the spectrum, all protein components of melanosomes that have been studied to date pass through endosomal compartments prior to arrival at their destination. At the other end, the long von Willebrand factor (vWF) polymers of Weibel–Palade bodies (WPBs) assemble and bud from the Golgi and are joined by endosome-derived factors such as the membrane proteins CD63 and P-selectin [4]. Most LROs are secretory organelles that release their contents upon cellular stimulation; indeed, LROs have also been referred to as "secretory lysosomes" [5] based largely on the behavior of LROs that double as the host cell's lysosome, such as lytic granules of CTLs and natural killer (NK) cells. However, this term belies the fact that many LROs coexist with bona fide lysosomes in their host cells and that bona fide lysosomes can also be secreted under certain conditions [6,7]. Moreover, not all LROs are secreted. For example, notochord vacuoles function to provide structural rigidity to the cells of the inner notochord layer [8], NOX2-containing LROs fuse with nascent phagosomes [9,10], melanosomes in ocular pigment cells are retained intracellularly, and it is not yet clear whether epidermal melanocytes secrete melanin or whether their melanosome-enriched tips are engulfed by neighboring keratinocytes [11,12]. Here we consider LROs as tissue-specific membranous structures in which the origins of a substantial fraction of their components can be traced to the endolysosomal system. Table 12.1 provides a list of compartments that conform to this definition and that have been referred to as LROs or secretory lysosomes in the literature. Electron micrographs of examples of these LROs are shown in Figure 12.1.

12.2 PHYSIOLOGICAL FUNCTIONS OF LROs

LROs fulfill a variety of functions within metazoans. Many LROs are found in hematopoietic cells (Table 12.1 and Figure 12.1), and contribute to important functions in innate or adaptive immunity. For example, the lytic granules of CTLs or NK cells store pore-forming components such as perforin and cytolytic agents such as granzymes; signal-dependent release of these components into the synapse between a CTL or NK cell and a target cell lead to target cell permeabilization and death [14].

TABLE 12.1 LROs and Their Host Cells, Functions, and Cargoes

LRO	Cell Type	Physiologic Function	Sample Cargoes
Vertebrates			
Melanosome	Melanocytes, eye pigment cells (retinal pigment epithelia, others)	Melanin production and storage for pigmentation	Tyrosinase, tyrosinase-related proteins, OCA2, PMEL
Weibel–Palade body	Endothelial cells	Storage and release of hemostatic and proinflammatory factors	von Willebrand factor, P-selectin
Cytolytic granule	Cytotoxic T cells, NK cells	Targeted destruction of virally infected or cancerous cells	Perforin, granzymes, lysosomal enzymes
Dense granule	Platelets	Storage and release of small molecules for platelet adhesion	Serotonin, calcium, ADP, ATP, polyphosphates
α-Granule	Platelets	Storage and release of proteins for platelet hemostasis, inflammation, angiogenesis, others	von Willebrand factor, platelet factor-4, fibrinogen
Lamellar body	Alveolar type II epithelial cells	Synthesis, storage, and secretion of surfactant	Surfactant proteins A-D, peroxiredoxin 6, ABCA3
Basophilic granule	Mast cells, basophils	Storage and secretion of histamines	Histamine
Azurophilic (primary) granules	Neutrophils, eosinophils	Storage and release into phagosomes of microbicidal compounds	Myeloperoxidase, elastase, cathepsin G, lysozyme, defensins
Specific (secondary) granules	Neutrophils	Storage and release of microbicidal compounds and antioxidant activity	NADPH oxidase, gelatinase, lactoferrin, lysozyme
Gelatinase (tertiary) granules	Neutrophils	Storage and release of microbicidal compounds	Gelatinase
Osteoclast secretory lysosome	Osteoclasts	Bone digestion	Cathepsin K, vATPase, tartrate acid-resistant phosphatase

(continued)

TABLE 12.1 (*Continued*)

LRO	Cell Type	Physiologic Function	Sample Cargoes
Acrosome	Sperm cells	Egg penetration during fertilization	Acrosin, matrix metalloproteinases
Notochord vacuole	Notochord inner cells	Structural rigidity of the notochord	LAMP1, vATPase, ?
MHC class II compartment	Activated dendritic cells, B lymphocytes	Antigen processing for presentation to T lymphocytes	MHC class II, HLA-DM, cathepsin S
NOX2+ inhibitory lysosome	Conventional dendritic cells	Neutralization of phagosomes for cross-presentation	NADPH oxidase complex
IRF7 signaling lysosome	Plasmacytoid dendritic cells	Toll-like receptor signaling for type I interferon production	TRAF3, IRF7, TLR7, TLR9
Phagosome	Macrophages, neutrophils, dendritic cells, others	Pathogen destruction, antigen presentation	Variable
Invertebrates			
Gut granules *C. elegans*	Intestinal cells (coelomocytes)	Storage compartment?	Zinc, MRP4, other ABC transporters, UMP synthase-1
Eye pigment granules (*D. melanogaster*)	Ocular pigment cells	Light absorption during vision	ABC transporters, pigments

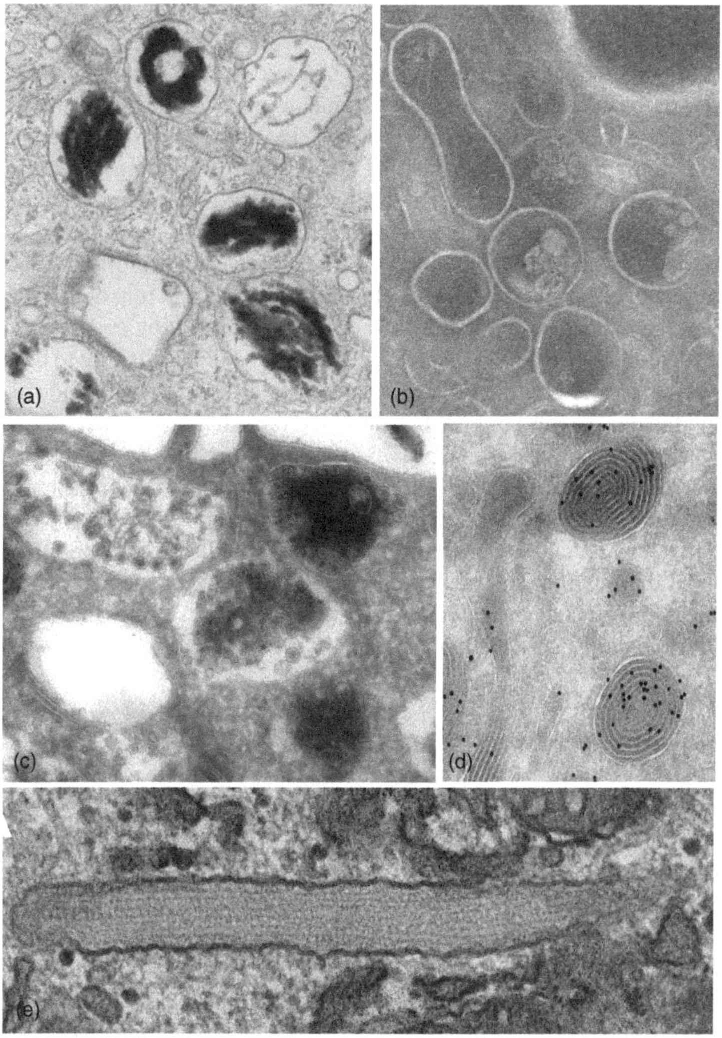

Figure 12.1 Examples of LRO ultrastructure. Shown are transmission electron micrographs (courtesy of G. Raposo, Insitut Curie and CNRS, Paris, France) of a few LROs described in this chapter. (a) Melanosomes in a human melanoma cell line; (b) azurophilic granules from an eosinophil; (c) cytolytic granules in a cytotoxic T cell; (d) MIIC from dendritic cells (immunogold labeled for MHC-II); (e) a Weibel–Palade body in a human umbilical vein endothelial cell. Raposo et al. [13]. Copyright 2007, Reproduced with permission of Elsevier.

Similarly, the azurophilic granules of eosinophils and the basophilic granules of basophils and mast cells store effectors that are released upon cell stimulation to damage extracellular microbes and activate proinflammatory pathways. Components delivered to maturing phagosomes upon fusion with inhibitory NOX2 compartments in dendritic cells (DCs) or with a number of morphologically distinct granules in neutrophils destroy internalized microbes and protect phagocytosed antigens from degradation [15,16]. Stimulus-dependent secretion of a variety of LROs from platelets (α granules and dense granules) and endothelial cells (WPBs) function in hemostasis by providing factors that stimulate platelet adherence and thrombus formation at sites of blood vessel damage [17,18]. Additional LROs in hematopoietic cells serve nonsecretory roles, including the antigen processing compartments of professional antigen presenting cells [19] and interferon regulatory factor 7 (IRF7) signaling compartments in plasmacytoid DCs [20].

LROs are also found in nonhematopoietic cell types and fulfill unusual roles in physiology. Melanosomes in pigment cells of the eye and skin function in the synthesis and storage of melanin pigments to support photoprotection and visual development [21]. Lamellar bodies within alveolar type 2 epithelial cells (not to be confused with the lamellar bodies of the skin, which are distinct structures) function in the synthesis, storage, and secretion of surfactant to facilitate lung plasticity [22]. Recently, vacuoles within vertebrate notochords that provide turgidity for structural support have been shown to have features of LROs [8], as do acrosomes in sperm – which function in the penetration of the egg during fertilization [23]. LROs have also been described in invertebrates, including the pigment granules within pigment cells of the compound eyes of insects such as *Drosophila melanogaster* and the gut granules within intestinal cells of *Caenorhabditis elegans*. The latter serve as interesting model systems for the biogenesis and maintenance of LROs.

12.3 LRO BIOGENESIS

LROs fall into two distinct subsets: those that correspond to traditional lysosomes that have been functionally "accessorized" by the expression of tissue-specific constituents and those that coexist with traditional lysosomes within their cell types [24]. Some LROs, such as cytolytic granules in CTLs and NK cells, are emerging as a hybrid of these two subsets, in which traditional lysosomes that harbor lytic granule components are further matured upon cellular stimulation by fusion with other endocytic organelles [25]. The formation of nascent LROs in cells that generate them has been a topic of interest for some time, in large part due to heritable diseases in which the biogenesis of LRO subsets is impaired such as Chediak–Higashi syndrome (CHS), Gray Platelet syndrome (GPS), and Hermansky–Pudlak syndrome (HPS). Here, we illustrate several examples in which our understanding of LRO biogenesis is substantially mature and in which the products of the genes that are defective in these diseases participate (Figure 12.2).

12.3.1 Chediak–Higashi Syndrome and Gray Platelet Syndrome

CHS is characterized by oculocutaneous albinism due to melanosome malformation, bleeding diathesis due to platelet storage pool deficiency, severe immunodeficiency due at least in part to defective cytolytic granules and neutropenia, and frequent neurological defects [26–28]. Numerous cell types from CHS patients harbor enlarged

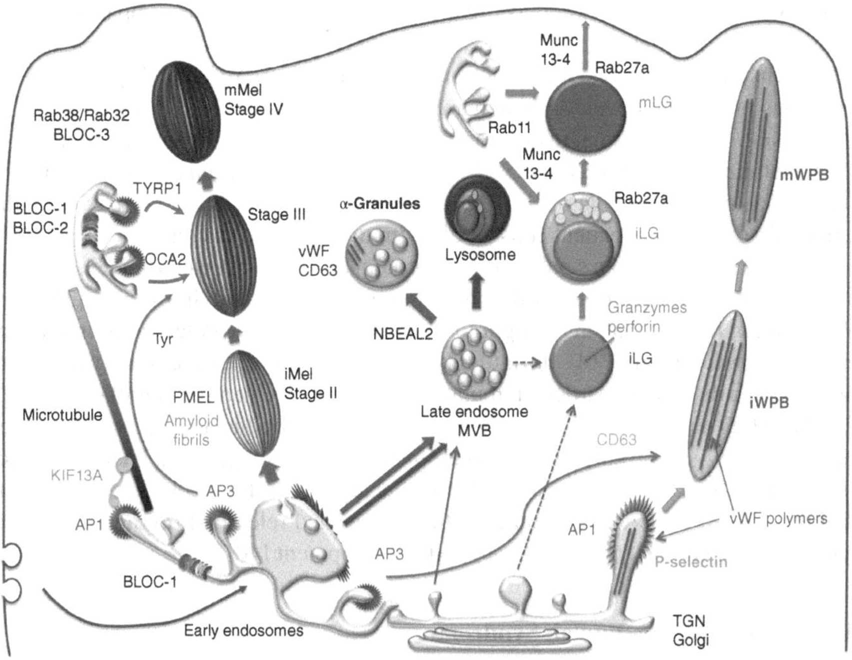

Figure 12.2 Model for biogenesis of four vertebrate LROs. Shown are models for the biogenesis of immature (iMel) and mature (mMel) melanosomes (left, brown and gold), platelet α granules (pink), lysosomes (violet), CTL LGs (gray), and WPBs (right, blue) relative to endosomal and biosynthetic organelles. Golgi, *trans*-Golgi network (TGN), early endosomes, late endosomes/multivesicular bodies (MVBs), and lysosomes are indicated. Key cargo molecules discussed in the text are noted in the same color as the LRO, and effectors involved in biogenetic steps are labeled in black text. Arrows indicate relevant trafficking pathways. Left, immature melanosomes (iMel) emerge from vacuolar domains of early endosomes, and mature by cargo delivery from tubulovesicular domains of early endosomes through AP-1-coated or AP-3-coated vesicles; recycling endosomal domains associated with KIF13A and AP-1 migrate along microtubules toward maturing melanosomes for delivery of some cargoes as indicated. BLOC-1 facilitates tubule-mediated transport; BLOC-2, BLOC-3, RAB32, and RAB38 likely function downstream. Center left, platelet α granules derive in an NBEAL2-dependent process from late endosomes within megakaryocytes, and receive both biosynthetic and endocytic cargoes. Late endosomes in the same cells also fuse with lysosomes to deliver other cargoes. Center right, in CTLs and NK cells, immature LGs (iLGs) also derive by fusion of late endosomes with dense core structures containing perforin and granzymes. iLGs then fuse with recycling endosome-derived structures upon stimulation by target cells to form mature LGs (mLGs). MUNC13-4 controls the fusion of RAB11-containing exocytic compartments with RAB27A-containing iLG in a process that does not require a physical interaction between MUNC13-4 and RAB27A. In the final step, MUNC13-4 and RAB27A cooperate in the docking of lytic granules to the plasma membrane, to allow for granule content secretion. Right, vWF forms long fibrous polymers in the TGN of endothelial cells. Nascent immature WPBs (iWPBs) then bud off from the TGN encasing the vWF fibers likely with membrane cargoes such as P-selectin. Other cargoes, such as CD63, are then delivered from early endosomes in an AP-3-dependent manner. Marks et al. [24]. Copyright 2013, Reproduced with permission of Elsevier. (*See color plate section for the color representation of this figure.*)

lysosomes and/or LROs, whereas organelles of earlier endocytic compartments are comparatively normal [29,30]. The defective gene in CHS and its *beige* mouse model, *CHS1*, encodes a 3801-residue cytoplasmic protein called LYST [31,32]. How LYST functions is still unknown. Although some data suggest that LYST positively regulates fission from lysosomal fusion intermediates [33], the majority of data support the view that mammalian LYST and its homologue in *Dictyostelium discoideum*, LvsB, function as negative regulators of lysosome or LRO fusion [29,30,34], such that the enlarged organelles in patient cells result from uncontrolled fusion.

LYST is a member of a small family of proteins with a conserved BEACH domain, a pleckstrin homology domain that might engage phosphatidylinositides [35], and WD domain repeats that are thought to form an interface for protein:protein interactions (reviewed in Ref. [36]). Intriguingly, mutations in several other family members also result in disease. In particular, mutations in the family member neurobeachin-like 2 (NBEAL2) result in GPS, a syndrome characterized by the absence of α granules in platelets [37–39]. Analyses of a mouse model of GPS suggest that NBEAL2 functions during α granule formation within megakaryocytes (Figure 12.2, center panel) and that loss of NBEAL2 correlates with uncontrolled release of α granule contents and concomitant defects in megakaryocyte development [40].

12.3.2 Hermansky–Pudlak Syndrome

HPS is a genetically heterogeneous group of disorders characterized in all cases by oculocutaneous albinism due to melanosome malformation in pigment cells and by bleeding diathesis primarily due to an absence of dense granules in platelets. Some HPS isoforms are also associated with a lethal lung fibrosis and/or immunodeficiency [41–43]. The ten genes that have been identified to date as defective in HPS patients encode subunits of four multisubunit protein complexes – adaptor protein (AP)-3 and Biogenesis of lysosome-related organelles complex (BLOC) -1, -2 and -3 – that are thought to regulate membrane trafficking steps required for the maturation of LROs, particularly, melanosomes, platelet dense granules, and alveolar lamellar bodies (reviewed in Refs [24,43–46]). Mutations in orthologous genes and several additional subunits of the same complexes result in a similar disorder in mice, and orthologous complexes regulate the biogenesis of gut granules in *C. elegans* [47–49] and ocular pigment granules in *D. melanogaster* [50–58]. In addition, mutations in vacuolar protein sorting 33A (Vps33a) – encoding a SNARE-interacting Sec1/Mun18 (SM) family member and a subunit of the homotypic fusion and protein sorting (HOPS) tethering complex – and in *Rab38* – encoding a tissue-restricted RAB GTPase – underlie a similar disorder in rodents [59–63]. Mutations in genes encoding additional HOPS subunits and orthologues of RAB38 underlie LRO defects in invertebrates [47,49,64–67].

The products of all of these genes appear to regulate the delivery of cargoes required for the maturation of LROs such as melanosomes. AP-3 is a member of the heterotetrameric adaptor protein complexes that facilitate protein sorting into transport carriers for post-Golgi trafficking in eukaryotes, and AP-3 specifically facilitates protein sorting to lysosomes in cells that lack LROs [68,69]. The yeast

HOPS complex facilitates homotypic fusion of the lysosome-like vacuole and fusion of the vacuole with transport vesicles, and mammalian HOPS is thought to function at multiple endocytic fusion events [70]. BLOC-3 has recently been shown to function as a guanine nucleotide exchange factor (GEF) for RAB38 and the highly related RAB32 GTPase [71]. The molecular functions of BLOC-1 and BLOC-2 are less well understood, and how any of these proteins function during LRO biogenesis is just beginning to unfold. Their function during melanosome biogenesis is best understood (Figure 12.2, left panel) and is described in the following sections.

12.3.3 Melanosome Biogenesis

The pigment-storing organelles within melanocytes and developing retinal pigment epithelia (RPE) have emerged as excellent models for LRO biogenesis due to an easily identifiable morphology, a wealth of genetic variants, excellent cultured cell models, and well-characterized cargoes [46,72]. Melanosomes in mammalian cells that make predominantly black and brown eumelanins mature through four distinct morphological stages. The first two stages lack pigment but are characterized by development of a fibrillar matrix upon which melanins deposit in later stages [73]. Stage I melanosomes are accessible to the endocytic pathway and correspond to vacuolar domains of early endosomes with flat clathrin lattices on their cytoplasmic face and few intraluminal vesicles [74] but differ from early endosomes in other cell types by their hyperacidification and content of nascent amyloid fibrils that emerge from the intraluminal vesicles [73–76]. These fibrils, composed of the pigment cell-specific premelanosome (PMEL) protein (reviewed in Ref. [77]), elongate and assemble into sheets during maturation of ellipsoidal stage II melanosomes [76]. As stage I melanosomes mature into stage II, they become inaccessible to endocytic material [74]; how this occurs concomitant with maturation of early endosomes to endocytic-accessible late endosomes is still not understood, but likely reflects fission of a common precursor. This is suggested by the finding that distinct classes of intraluminal vesicles [78] are associated with distinct proteolytic fragments of PMEL destined either for lysosome-bound late endosomes or stage II melanosomes [2].

Intriguingly, PMEL is synthesized as a transmembrane glycoprotein that becomes proteolytically processed en route to and/or within stage I melanosomes [75,79–82] to a lumenal fragment that has biophysical characteristics of amyloid [79,83]. PMEL was the first described mammalian member of a family of "functional amyloid" proteins that exploit the amyloid fold – commonly associated with neurodegenerative diseases such as Alzheimer, Parkinson, and Creutzfeldt–Jakob diseases – for physiological advantage [84]. A number of distinct domains within PMEL have amyloidogenic capacity *in vitro* [85–89], but N-terminal domains appear to be most critical *in vivo* [85,90]. How PMEL adopts an amyloid fold within melanosome precursors and averts pathology is a subject of intensive investigation. Indeed, certain *PMEL* gene mutations in animal hypopigmentation models are associated with amyloid forms that are more detrimental than inactivating mutations for pigment accumulation and perhaps for cell viability [91–95], suggesting that the process and/or timing of amyloid formation is exquisitely controlled.

12.3.4 HPS and Melanosome Maturation

Maturation of nonpigmented stage II melanosomes to pigmented stage III and IV melanosomes requires the delivery of melanogenic enzymes (such as Tyrosinase [TYR] and Tyrosinase-related protein-1 [TYRP1]), and membrane transporters (such as OCA2 and ATP7A) that modify the lumenal environment to favor melanogenesis (Figure 12.2, left panel). The delivery of these cargoes occurs by vesicular transport from distinct domains of early endosomes and is regulated in part by BLOC-1, BLOC-2, AP-3, and the RAB32 and RAB38 GTPases that are defective in HPS models (reviewed in Ref. [46]). Except for the major cohort of TYR, most cargoes analyzed to date absolutely require BLOC-1 for their export from vacuolar domains of early endosomes toward melanosomes [96–99]. BLOC-1-dependent transport occurs through tubular elements of the recycling endosome pathway [100] that are "pulled" along microtubules by the kinesin-3 motor, KIF13A [101]. Thus, depletion of KIF13A results in hypopigmentation and entrapment of BLOC-1-dependent cargoes in enlarged early endosomes [100]. Recruitment of KIF13A and sorting of many of these cargoes toward melanosomes requires AP-1, another member of the clathrin-associated tetrameric adaptor family, which appears to play a dual role of sorting adaptor for many melanosomal cargoes and recruiter of KIF13A [100].

How BLOC-1 functions in this pathway is not clear. BLOC-1 consists of eight subunits, of which three and five, respectively, are encoded by genes that are defective in HPS and its mouse models. A cohort of BLOC-1 associates with AP-3 [102], and in neurons, BLOC-1 cooperates with AP-3 in the sorting of a number of transmembrane cargoes into synaptic vesicles [103–105], but BLOC-1 functions to sort both AP-3-dependent and -independent cargoes in melanocytes [97–99,106]. BLOC-1 interacts physically with a number of additional effectors that might facilitate sorting, formation, or stabilization of transport carriers, or eventual fusion of transport carriers with melanosomes. These include the SNARE proteins syntaxin-13 and SNAP-25 [107–110], actin regulatory proteins such as Arp2/3 and WASH [111], phosphatidylinositol 4-kinase IIα [112], and a number of other candidates identified by quantitative mass spectrometry analyses [113]. Moreover, a BLOC-1-like complex in the yeast *Saccharomyces cerevisiae* is responsible for recruiting the GTPase-activating protein for the master early endosomal Rab GTPase, RAB5 [114], suggesting that mammalian BLOC-1 might participate in RAB exchange to promote the formation of transport carriers. A low-resolution rotary shadowed structure of recombinant BLOC-1 suggests that the complex adopts a curvilinear shape [115], as expected for a protein that induces or stabilizes membrane curvature on tubular transport intermediates [116].

BLOC-2 functions downstream of BLOC-1 and regulates the melanosomal delivery of a similar set of cargoes [97,102,117–119]. BLOC-2 binds a cohort of BLOC-1 in cells and, similar to BLOC-1, decorates tubular elements of the early endosomal system [102]. Moreover, BLOC-2 appears to be an effector of RAB32 and RAB38, with which it partially colocalizes on melanosomes or structures closely apposed to melanosomes [120]. How BLOC-2 functions is not yet known. BLOC-2 consists of three large subunits, all of which are products of genes that are defective in HPS variants and mouse models. The only clear structural domains in any of them are a WD40 repeat in the HPS5 subunit [121] and a clathrin binding box that is capable of engaging clathrin *in vitro* [122], but the functional relevance of these features is not clear. When overexpressed in HeLa cells, the HPS6 subunit interacts with a component of the dynein/dynactin complex [123], but whether this reflects a function of intact BLOC-2 is not yet clear.

The genes that are disrupted in HPS are all ubiquitously expressed (albeit perhaps at different levels in different cell types). How are then tissue-specific effects observed? The answer likely lies at least in part in the collaboration of BLOC-1, -2, -3 and AP-3 with tissue-restricted Rab GTPases. In particular, RAB32 and RAB38 play important roles in the biogenesis of melanosomes and other LROs. These two Rab proteins localize in part to melanosomes and in part to presumed transport intermediates in melanocytes [63,71,120], as well as to subsets of lamellar bodies in alveolar type II cells [124] and of dense granule-like compartments in a megakaryocytoid cell line [125]. TYR and TYRP1 fail to be properly delivered to maturing melanosomes upon simultaneous depletion of these two highly homologous RAB proteins in immortalized mouse melanocytes and a human melanoma cell line [63,71] or in RAB38-depleted RPE [61]. BLOC-3 – composed of two subunits, HPS1 and HPS4, that are targeted by mutation in human and mouse HPS models – was recently shown to function as a GEF i.e., activator) for these two RAB proteins [71]. Accordingly, similar phenotypes were observed in a BLOC-3-depleted human melanoma cell line [71] and in RPE and interfollicular skin melanocytes in mouse HPS1 and HPS4 models [126]. However, the lack of a similar phenotype in choroidal melanocytes or hair follicle melanocytes from these same models [127] suggests that either an additional GEF must exist for RAB32 and RAB38 in these cells or the requirements for RAB32/38 activation in these cells are more limited.

What do RAB32 and RAB38 do? In epidermal melanocytes, these two proteins play partially redundant roles in cargo transport, although unique effects on some cargoes have been observed specifically upon RAB32 knockdown [120]. Recent evidence supports the view that both RAB GTPases, in their GTP-bound form, physically engage AP-1, AP-3, and BLOC-2, suggesting an important role throughout cargo transfer to melanocytes [120], and to the myosin motor, Myosin Vc [128]. The latter effector apparently plays contrasting roles in melanosome biogenesis, as knockdown of Myosin Vc with inhibitory RNA depletes cargoes such as TYRP1 from melanosomes but paradoxically enhances pigmentation. Part of the latter effect seems to reflect a reduction in release of melanin that occurs constitutively in the melanoma cell line used in this study [128], perhaps reflecting melanin exocytosis. An additional effector of RAB32 and RAB38 is VPS9-ankyrin-repeat protein (VARP), a large scaffolding protein that also functions as a GEF for RAB21 and engages the retromer complex and the closed conformation of the vSNARE, VAMP7/TI-VAMP, through distinct binding sites [129–134]. Extinguished expression of VARP, VAMP7, or retromer impairs melanosome protein transport [133,135,136]. The RAB32/38 binding site is dispensable for VARP recruitment to endosomes and for endosomal recycling in nonmelanocytic cells [129] but is necessary for melanosomal protein transport [136], suggesting that VARP plays a specialized role in melanocytes.

12.3.5 HPS and the Biogenesis of Other LROs

As predicted from the symptoms of HPS, BLOC-1, -2, and -3, AP-3 and RAB32/RAB38 have important functions in the generation of other LROs besides melanosomes. In patients and mouse models of all forms of HPS, platelets lack detectable dense granules by whole mount electron microscopy and have reduced storage pools of serotonin and adenine nucleotides [137–142]. In addition, platelets from mouse HPS models and two HPS case studies show impaired secretion of α granules and lysosomes, as well as of protein disulfide isomerase from the so-called "T granules," to low doses of agonists [143–145]; the α granule and T granule secretion defects are secondary to impaired signaling due to loss of secreted dense granule contents, but BLOC-1 and AP-3 likely regulate cargo delivery to lysosomes since lysosome secretion from platelets lacking these complexes is intrinsically impaired [143]. The result of all of these defects in HPS is impaired platelet aggregation at sites of vascular damage [143,145,146], correlating with excessive bleeding in patients (reviewed in Ref. [147]). RAB38-deficient rats also present with bleeding diathesis due to absent dense granules [62,148], and depletion of RAB38 or RAB32 from a megakaryocytoid cell line resulted in reduced size and mobility of dense granule-like compartments [125]. The cargoes that localize to dense granules and that might be regulated by these components are not yet known but likely include transmembrane transporters that pump calcium, adenine nucleotides, serotonin, and polyphosphates into the dense granule lumen. Among the potential cargoes are the vesicular serotonin transporter VMAT2 [125], the ABC transporter MRP4 [149], and a member of the sugar–nucleotide transporter family, SLC35D3, which is depleted in HPS model platelets [150] and which is the target of mutation in a mouse dense granule storage deficiency [151].

HPS patients that lack BLOC-3 or AP-3 also present with a lethal lung fibrosis [147], and mouse models lacking many of the HPS complexes individually or in combination have a distinct lung disorder [152,153] and heightened susceptibility to lung irritants [154,155]. This appears to result from defects in the formation of lamellar bodies, the LROs of alveolar type II cells in which surfactant is synthesized and stored, and correlates with lung disorders in HPS model mice [156–159]. In addition, RAB38-deficient rats present with progressive and severe lung fibrosis like BLOC-3-deficient HPS1 and HPS4 patients [62]. Consistently, RAB38 is expressed in alveolar type II epithelial cells (AT2) [160] and RAB38-deficient cells harbor abnormal surfactant levels and abnormally enlarged lamellar bodies [124,161]. The cargoes that are transported to lamellar bodies in an HPS-dependent manner are not yet identified.

12.3.6 HPS and Neurosecretory Granule Biogenesis

Although not considered traditional LROs, certain synaptic vesicles in neuronal subpopulations are also affected by HPS mutations. AP-3 regulates the formation of synaptic vesicles from early endosomes in PC12 cells [162]. Neurons express unique AP-3 β and μ chains, and only the AP-3B form that harbors these subunits functions in synaptic vesicle recycling [163]; hence *mocha* mice that lack the single AP-3 δ chain or the neuron-specific μ3B chain have neurological phenotypes, but HPS type 2 (HPS2) patients and *pearl* mice with mutations in nonneuronal AP3B1 do not [164,165]. Neuronal AP-3 deficiency leads to an imbalance in the composition of synaptic vesicles in neurons, characterized by a depletion of cargoes such as the zinc transporter ZnT3, the glutamate transporters Vglut1 and VGAT, the chloride channel ClC3, SNAREs such as VAMP2, and enzymes such as phosphatidylinositol (4) kinase IIα [103,166–168]. Some of the same cargoes are also depleted in neurons from BLOC-1-deficient mice [103,105,116], particularly in certain neuronal subpopulations [104]. In particular, BLOC-1 mutations interfere with sorting to neurites and nerve terminals [169]. These mutations are also associated with altered neurotransmitter release and behavioral abnormalities in mice [170–172], and polymorphisms in BLOC-1 subunit genes are associated with increased risk for schizophrenia in humans (reviewed in Ref. [173]).

12.3.7 Weibel–Palade Body Biogenesis

HPS, CHS, and GPS clearly impact the formation of LROs for which cargoes are derived exclusively from the endocytic pathway. However, not all LROs appear to be formed via these pathways. A particularly well-characterized example is the formation of WPBs in endothelial cells [174] (Figure 12.2, right panel). The primary function of these elongated "cigar"-shaped organelles is to assemble vWF into large polymers or fibers for agonist-induced secretion; the fibers then serve as a mechanism to recruit platelets and clotting factors to sites of endothelial cell damage [175–177]. vWF polymers assemble in the Golgi complex following proteolytic maturation of pro-vWF to the mature form [178,179]. Unlike PMEL in melanocytes, vWF does not adopt an amyloid fold, but rather assembles into regularly spaced helices that can extend up to 3–5 μm in length and that coalesce into paracrystalline structures [180–182]. An exciting recent study showed that the length of the helices is determined by the degree to which Golgi stacks assemble into the Golgi ribbon in endothelial cells; "quanta" of vWF polymers within individual Golgi stacks are assembled into longer polymers in the *trans*-Golgi network (TGN) within a Golgi ribbon structure but remain as smaller quanta if ribbon formation is blocked [183]. The increased length of vWF polymers is required for optimum capture of platelets at sites of vascular damage [183,184].

WPB precursors bud from the TGN in a process that requires clathrin and its adaptor AP-1 in an unconventional manner [185]. Rather than regulating cargo budding into clathrin-coated vesicles, AP-1 and clathrin nearly completely coat the nascent WPBs [185] (Figure 12.2, right panel). This likely provides a structural support for the organelles during their formation or maturation but is not required to maintain the shape of fully formed WPBs [185]. Once segregated from the Golgi, WPBs mature by recruitment of additional membrane cargoes, such as P-selectin/CD62p and the tetraspanin CD63, from the endosomal system [186]. Interestingly, recruitment of CD63 requires AP-3, whereas the recruitment of P-selectin does not and relies on targeting determinants in several topological domains [186]. P-selectin is thought to be retained within WPBs through a direct interaction of the lumenal domain with vWF [187]. The RAB GTPase, RAB27A, is recruited to mature WPBs in a manner that is dependent on the vWF cargo [188], where it facilitates vWF secretion in response to secretagogues and regulates the degree of vWF polymerization released into the extracellular space [189–191].

Smaller vWF polymers, P-selectin and the RAB27A-related GTPase, RAB27B, are also major components of α granules in platelets, which have a distinct morphology [192]. Whether this morphology reflects differing levels of vWF biosynthesis, fractured Golgi elements in the megakaryocyte precursors, or another regulatory step is not yet understood.

12.4 LRO MOTILITY, DOCKING, AND SECRETION

A number of diseases result from defects in LRO motility or secretion. Griscelli syndrome types 1, 2, and 3 (GS1, GS2, and GS3) reflect defects in the intracellular transport of selected LROs due to mutations in the genes encoding Myosin VA, RAB27A, and MLPH (Melanophilin), respectively [193]. A feature of each of these diseases is cutaneous albinism due to defective positioning of mature melanosomes within melanocytes and consequent failure to transfer melanosomes to keratinocytes [194–201]. In melanocytes, RAB27A forms a tripartite complex with the linker protein, MLPH, and the processive actin motor, Myosin VA, that links melanosomes to the peripheral actin cytoskeleton. This linkage captures melanosomes from bidirectional movement along microtubules, in which minus-end-directed motions predominate; thus, loss of any of these components results in accumulation of melanosomes in the pericentriolar region [202,203]. Myosin VA in this complex functions either as a tether for melanosomes on peripheral actin [202,203] or to mediate long-range movements toward the cell periphery [204]. The same RAB27A/MLPH/Myosin VA complex impedes spontaneous secretion of immature WPBs from endothelial cells [190]. In RPE, RAB27A and MLPH associate with a distinct motor, Myosin VIIA, to facilitate motility of melanosomes into the apical processes that interdigitate with the outer segments of photoreceptors [205,206].

GS3 (MLPH deficiency) is associated primarily with hypopigmentation, but patients with GS1 and GS2 suffer from a number of additional symptoms. Defective Myosin VA in GS1 is most often associated with neurological impairment due to important roles of this myosin motor in neurons [207], particularly in the development of dendritic spines [208]. The neuronal function of Myosin VA does not require association with MLPH; consequently, some GS1 patients and *dilute* mouse models with particular Myosin VA mutations present either with pigmentation or neurologic defects, but not both [209,210].

GS2 (RAB27A deficiency) is not associated with neurological defects but rather with immune defects and often with a hyperinflammatory syndrome referred to as hemophagocytic lymphohistiocytosis (HLH) [199]. The latter is primarily due to defects in cytolytic granule mobilization within CTLs [211,212]. When bound to GTP, RAB27A engages numerous effectors that play roles in the maturation of cytolytic granules and their positioning toward the immunological synapse (IS) in CTLs and NK cells [213,214]. These are further discussed in the following text in the context of cytolytic granule maturation and secretion. RAB27A and its effectors are also required for proper positioning and exocytosis of numerous other LROs and non-LRO compartments, including mast cell granules [215], azurophilic granules [216,217], tertiary and specific granules of neutrophils [218,219], and even insulin-containing dense core secretory granules [220]. RAB27A and its homologue, RAB27B, also regulate distinct steps in the release of exosomes [221], and RAB27B controls dense granule release in platelets [222].

Other aspects in the regulation of LRO secretion have been discovered in the context of hyperproliferative disorders of the immune system like that observed in GS2 and CHS patients. These are discussed as follows in the broader context of LROs in adaptive and innate immunity.

12.5 LROs AND IMMUNITY TO PATHOGENS

LROs are particularly enriched within cells of the adaptive and innate immune systems, and diseases associated with their formation, positioning, or secretion lead to immunodeficiency and often to hyperproliferative immune disorders. Investigations into the etiology of these disorders has extended our understanding of general principles of LRO biogenesis and secretion, and raise interesting questions regarding how LROs might be defined.

12.5.1 Cytolytic Granules

Directed secretion of the contents of CTL and NK cell cytolytic granules into the synapse with target cells is necessary for antiviral immunity, cancer immunosurveillance, and downregulation of immune responses. Hence, defects in cytolytic granule biogenesis or secretion lead to heightened susceptibility to viral infection and cancer, and often to the typically fatal hyperinflammatory disorder HLH. HLH is characterized by an exacerbated yet ineffective immune response [223,224] in which augmented cytokine secretion by CTLs and NK cells recruits and activates phagocytes; these cells then produce additional cytokines, leading to a "cytokine storm" that sustains inflammation and harms infiltrated tissues [225]. HLH is often observed in GS2 and CHS patients [199,226], but surprisingly only rarely in HPS2 patients, despite impaired antiviral immunity [227].

Impaired cytolytic activity by CTLs in CHS, HPS2, and GS2 does not seem to be due to abnormal composition of these granules. Perforin and granzyme are correctly localized to the lytic granules in AP-3-deficient CTLs [228] (but lysosomal pools of perforin are reduced in NK cells, suggesting possible defects in NK cell differentiation or in selective trafficking to granules [229]). The defect within CTLs in CHS, HPS2, and GS2 resides primarily in the secretion of the granule contents. In HPS2, granules are unable to efficiently engage and move along microtubules for positioning at the IS and for subsequent polarized content secretion [228]. Movement of the granules along microtubules is driven by the minus-end-directed motor, dynein [230]. Once the granules are polarized toward the microtubule organizing center (MTOC), the MTOC moves toward the IS at the plasma membrane, ferrying the associated cytolytic granules with it [230,231]. This step requires RAB27A in complex with its effector, Slp3, and the kinesin-1 microtubule motor [232]. During this process, the perforin-containing lytic granules fuse with polarized exocytic vesicles that carry the recycling endosome-associated RAB11A [233] (Figure 12.2, center panel). Fusion requires the RAB27A effector MUNC13-4 [233], which is discussed later, and enables functional maturation of the perforin-containing granules for exocytosis. Thus, the final transport step and docking at the IS are blocked in GS2 CTLs [199,233]. Motility of effector-associated granules is also blocked in CHS [29]. The polarization of the MTOC to the IS is tightly controlled by the strength of T-cell signaling, at least in part via activation of the LCK tyrosine kinase [234], and requires the dynein motor [230,235].

12.5.2 Familial Hemophagocytic Lymphohistiocytosis and Cytolytic Granule Secretion

While HLH is observed together with pigmentation defects in CHS and GS2, HLH can also manifest as a primary immunodeficiency disorder with mild or absent symptoms of oculocutaneous albinism. In this case, HLH is due to mutations in other genes involved in the exocytosis of lytic granules and comprises a group of diseases called familial HLH (FHL) [236]. The affected gene in FHL type 2 (FHL2) encodes perforin, a cytolytic granule cargo that is the main effector of cytolytic activity in target cells. By contrast, the affected genes in FHL types 3, 4, and 5 encode components of the cytolytic granule secretory machinery: MUNC13-4 (defective in FHL3), Syntaxin-11 (STX11; defective in FHL4) and syntaxin-binding protein-2 (STXBP2, also referred to as MUNC18-2 in humans or Munc18-b in mice; defective in FHL5). FHL types 3–5 are also associated with bleeding diathesis, which is caused by impairment in agonist-stimulated release of α granules, dense granules, and lysosomes from platelets [237–240]. This indicates that cytolytic granules and platelet LROs and lysosomes share mechanisms of exocytosis.

The RAB27A effector MUNC13-4 plays at least two distinct roles in cytolytic granule release. First, as described above, MUNC13-4 controls the fusion of RAB11-containing exocytic compartments with RAB27A-containing "immature" lytic granules in a process that does not require a physical interaction between MUNC13-4 and RAB27A [233]. Second, MUNC13-4 and RAB27A cooperate in the docking of lytic granules to the plasma membrane, a necessary step in secretion [241,242]. MUNC13-4 also binds directly to the SNARE complex containing STX11 [237], and thus likely facilitates SNARE complex formation during exocytosis.

STX11 is a member of the syntaxin family of SNARE proteins. SNAREs on opposing membranes form a stable four-helix bundle that brings the membranes into close apposition and thus drives their fusion [243]. STX11 is an unusual target SNARE (tSNARE) protein in that it lacks a transmembrane helix, and instead is tethered to the membrane by a lipid anchor [244,245]. In FHL4 patients (with mutations in STX11), lytic granule secretion in CTLs and NK cells is blocked, indicating that an STX11-containing tSNARE complex mediates a fusion step required for LRO exocytosis [246]. Recent findings suggest that STX11 functions during the final exocytic step [247], but the localization of STX11 to intracellular structures that lack RAB27A in CTLs and NK cells [246,248], and the finding that CTL-dependent killing under certain circumstances in FHL4 patients is intact [248] raises the possibility that STX11 might additionally – or instead – function in the fusion of RAB11- and RAB27A-containing compartments to form mature granules.

The CTL deficiencies observed in FHL5 patients with mutations in STXBP2 are similar to those of FHL4. Indeed, the MUNC18-2 product of the STXBP2 gene is a member of the Sec1-Munc18 family of syntaxin "chaperones" that stabilizes STX11 and facilitates its SNARE complex formation [249]. In platelets, Munc18b facilitates the formation of an STX11 complex with the Qbc SNARE SNAP-23 on the plasma membrane and the R-SNARE VAMP8 on granule membranes to mediate fusion and granule content release [237,239]. MUNC18-2 in CTLs and NK cells likely facilitates the formation of a similar complex to facilitate either maturation or exocytosis of lytic granules. Intriguingly, a recent report identified a dominant negative mutant of STXBP2 in an FHL5 patient that bound tightly to STX11 itself but impaired formation of a four-helix bundle SNARE complex, suggesting that SNARE complex assembly is indeed a key function for MUNC18-2 [250].

12.5.3 Azurophilic Granules

HPS2 patients suffer from recurrent bacterial infections, in part due to chronic neutropenia [251]. The basis for neutropenia is not completely understood, but it has been associated with impaired trafficking of elastase to azurophilic granules in the absence of AP-3 [229,252,253]. Elastase is required for normal differentiation of myeloid progenitors to mature neutrophils [254]. Consequently, HPS2 patients show arrested maturation of neutrophil precursors at the promyelocyte stage [253]. However, neutrophil counts increase during infections, and recurrent infections persist even upon treatment of patients with granulocyte-colony stimulating factor to increase neutrophil numbers [255–257]. This suggests the presence of other immune dysfunctions in these patients. Azurophilic granules are also required for neutrophil bactericidal activity, and defective fusion of these granules with phagosomes has been observed in CHS [258].

12.5.4 NADPH Oxidase-Containing LROs

Chronic granulomatous disease (CGD) is an immunodeficiency disorder in which pigmentation is normal (and hence there is no defect in melanosome formation) but in which a number of immune system LROs fail to function properly. Immunodeficiency is evidenced by recurrent respiratory, intestinal, and cutaneous bacterial infections and delayed clearance of viruses [251]. These immune defects are attributed to impaired activity of the "phagocyte-specific" NADPH oxidase (NOX2) in phagocytes, including conventional DCs. The $gp91^{phox}$ subunit of NOX2 is stored in "specific granules" in neutrophils or in less well-defined compartments in macrophages that fuse with nascent phagosomes following phagocytosis of large particles such as bacteria; there it generates reactive oxygen species that are thought to damage the phagocytosed cargoes [259]. In DCs, gp91^{phox} is also stored within LROs that are mobilized to fuse with nascent phagosomes [10,260] (Figure 12.3, left). These LROs have been called "inhibitory LROs" due to their ability to limit acidification when recruited to phagosomes in phagocytes. The reactive oxygen species generated by NOX2 capture protons in phagosomes and therefore counteract the vacuolar ATPase activity in DCs, preventing acidification. This activity limits the degradation of bacterial antigens within phagosomes to preserve antigenic peptides [10]. DCs from patients and mouse models of CGD show increased phagosomal acidification and impaired antigen cross-presentation due to the defective recruitment of NOX2 to phagosomes, perhaps contributing to impaired immune responses to infections [10,260]. Recruitment of inhibitory LROs to phagosomes is mediated by RAB27A [261]; delayed recruitment of NOX2 to nascent phagosomes in DCs in the RAB27A-deficient *ashen* mouse model of GS2 leads to increased acidification and consequently increased protein degradation and impaired antigen cross-presentation to naïve CD8+ T cells [261].

12.5.5 IRF7-Signaling LROs and Type I Interferon Induction

Plasmacytoid DCs (pDCs) are a specialized DC subset that functions largely in the production of type-I interferon (IFN) and plays an important role during viral infections. The pDCs recognize viral DNA at least in part through endosomal toll-like receptors (TLRs) including TLR7 and TLR9 [262]. TLR9 is activated by proteolytic processing within endosomes and then signals through the adaptor protein MyD88 via two bifurcating pathways. One pathway leads to NF-κB activation and the production of proinflammatory cytokines. The other pathway requires activation of the adaptor protein TNF receptor-associated factor 3 (TRAF3), and leads to phosphorylation and activation of the transcription factor IRF7 to stimulate production of type I IFN [263]. Interestingly, these two signaling cascades are initiated in different compartments [20] (Figure 12.3, right). NF-κB activation is initiated from a VAMP3-containing early endosomal compartment that lacks the lysosomal-associated membrane protein 2 (LAMP2). By contrast, TRAF3 is activated on a distinct LRO-like compartment that contains LAMP2 and to which IRF7 is recruited [20]. Trafficking of TLR9 from endosomes to this IRF7 signaling LRO requires AP-3 [20] and also likely BLOC-1 and BLOC-2 [264]. A potential cargo of this pathway is SLC15A4 [264], a transporter recently shown to facilitate cargo escape from phagosomes in conventional DCs [265]. The defective production of type-I IFN by pDCs in the absence of AP-3 might contribute to the recurrent viral infections observed in HPS2 patients.

12.5.6 MIICs: LROs or Conventional Late Endosome/Lysosomes?

The T-cell receptor on CD4+ lymphocytes recognizes antigenic peptides bound to major histocompatibility complex class II (MHC-II) molecules on the surface of

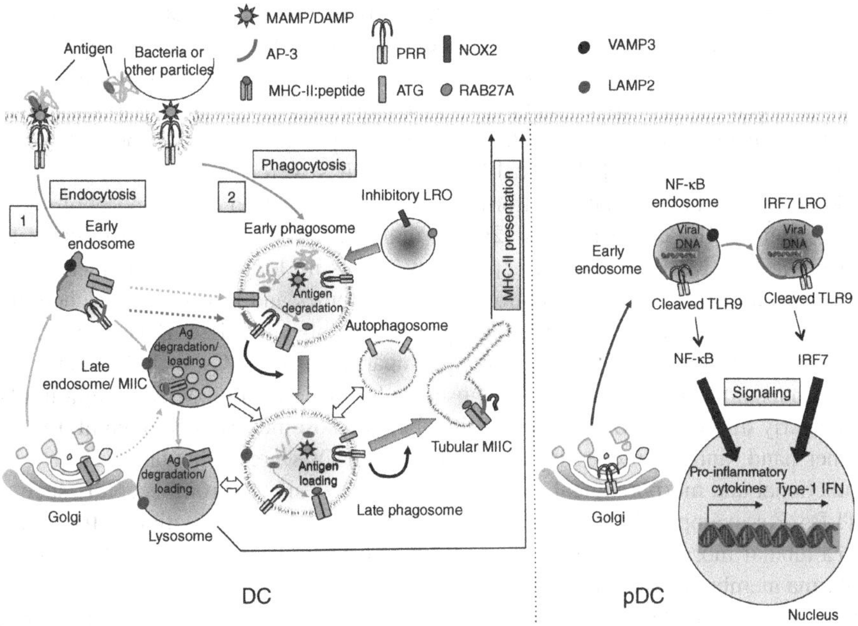

Figure 12.3 Model for LRO function in conventional DCs and pDCs. Left, LROs and phagosome maturation in DCs. 1. Antigens captured by endocytosis or macropinocytosis enter the endolysosomal system and are degraded to smaller peptides. Peptides encounter MHC-II molecules as endosomes mature to late endosomes/MIICs and lysosomes. Antigen loading mostly occurs within late endosomes/MIICs and lysosomes that are enriched in the regulatory component HLA-DM. From these compartments, peptide-loaded MHC-II molecules are delivered to the cell surface (black straight arrow) for presentation to T cells. 2. Particulate antigens captured by phagocytosis are degraded in phagosomes as the phagosomes mature. Maturation is achieved by the acquisition of content (including MHC-II) from early and late endosomes/MIICs and lysosomes by both direct fusion (open arrows) and vesicular transport (turquoise arrows; dashed arrows indicate possible pathways for MHC-II transport that are not yet confirmed). Antigen is loaded onto MHC-II predominantly in late phagosomes. Phagosome maturation is supported by autophagy (ATG; light violet) proteins, which might derive from autophagosomes themselves or independently from the cytosol, and by signaling from pattern recognition receptors (PRRs, black arrow) such as TLRs. PRRs such as TLR4 are delivered to phagosomes from early endosomes in an AP-3-dependent manner (mauve dashed arrow). From late phagosomes, peptide-loaded MHC-II is delivered to the cell surface either directly (not shown) or via an intermediate tubular MHC-II storage compartment for presentation to T cells (black straight arrow). Delivery to or from this compartment appears to be regulated by PRR signaling from the phagosome (dashed black arrows). Right, IRF7 LROs and type I IFN signaling in pDCs. TLR9 is trafficked from the biosynthetic pathway to VAMP-3-positive early endosomes. Here TLR9 is cleaved and becomes competent for signaling to activate proinflammatory cytokine expression through NF-κB (NF-κB endosome). Cleaved TLR9 is then targeted in an AP-3-dependent manner to a LAMP2-positive LRO (IRF7 LRO) harboring the adaptor TRAF3. Here TLR9 signals through IRF7 to induce the transcription of type-1 interferon genes. Adapted from Traffic 14:135–152, Mantegazza AR, Magalhaes JG, Amigorena S, Marks MS, Presentation of phagocytosed antigen by MHC class I and II, 2013 and from Science 329:1530–1534, Sasai M, Linehan MM, Iwasaki A, Bifurcation of toll-like receptor 9 signaling by adaptor protein 3, 2010. (*See color plate section for the color representation of this figure.*)

antigen presenting cells, which include DCs, B lymphocytes, activated macrophages, and a few other cell types [266]. In Epstein–Barr virus transformed B lymphocytes and conventional DCs, MHC-II accumulates in multivesicular and multilaminar compartments called MIICs that harbor lysosomal proteins such as CD63, LAMP1, and cathepsins [267–269] as well as the antigen processing chaperones HLA-DM and HLA-DO [270,271] (Figure 12.3, left). MIICs were historically described as LROs due to the unusually uniform multilamellar electron dense morphology of the "later" MIIC compartments [267] (Figure 12.1), which could be induced by expression of MHC-II alone [272]. However, it is no longer clear whether MIICs are true LROs or merely conventional late endosomes and lysosomes enriched in a special cargo. Indeed, antigen processing and assembly with MHC-II has been shown to occur in different MHC-II-containing compartments in APCs, including early and late endosomes, lysosomes, and phagosomes [273]. Therefore, MIICs might not be as clearly differentiated from conventional endolysosomes as once thought. On the other hand, antigen processing compartments in DCs are particularly sensitive to functional and morphological changes in response to signals from receptors such as TLRs. For example, TLR stimulation of DCs leads to a reorganization of the MIICs to a tubular morphology, favoring the delivery of peptide:MHC-II complexes to the plasma membrane at the IS with T cells [274–277]. Tubular MIICs may coexist with bona fide unstimulated lysosomes in DCs [276,277], and thus might more strictly adhere to the LRO definition.

While MHC-II-dependent antigen presentation is impaired in CHS [278], it is not generally disrupted in HPS [279,280] (although see the following section for a unique exception in HPS2). However, AP-3 deficiency in HPS2 or its mouse models causes impaired presentation of microbial lipid antigens on human CD1b molecules or the mouse orthologue, CD1d [281–283]. CD1b binding to AP-3 is required for its localization to MIICs in B cells and for appropriate sorting to the plasma membrane [281–283]. This defect in antigen presentation, together with the defects described as follows in phagosomal TLR activation and antigen presentation, may in part account for the increased susceptibility to bacterial infections in HPS2.

12.5.7 Phagosomes and Autophagosomes as New Candidate LROs

Phagosomes are intracellular structures that surround large internalized particles following their uptake through the specialized mechanism of phagocytosis. Based on the definition of LROs as specialized cell-type specific organelles derived from the endosomal system, one could argue that maturing phagosomes can be considered to be LROs (Figure 12.3, left). We base this argument on the following criteria: (i) Phagosomes are only generated in specific phagocytic cell types such as macrophages, neutrophils, and DCs. (ii) Although they form by the unique process of phagocytosis and thus do not strictly derive from endosomes, they mature by continuously receiving input from the endosomal system [284,285]. (iii) Although they eventually fuse with lysosomes, they harbor unique cargoes such as NOX2 and perform unique functions within the cell that differentiate them from conventional lysosomes. (iv) Maturing phagosomes fuse with other phagocyte-specific LROs such as NOX2 positive LROs and MIICs, and coexist with conventional lysosomes. (v) They harbor MHC molecules and provide peptides for antigen presentation and thus behave functionally like autonomous MIICs [286]. (vi) At least in conventional DCs, TLR stimulation induces the extension of tubules from maturing phagosomes that facilitate MHC-II presentation much like the extension of tubules from endolysosomes. However, these tubules appear to function in a different way from endolysosomal tubules and do not seem to drive surface expression of MHC-II [287]. (vii) Finally, phagosome maturation is influenced by AP-3 and is altered in models of HPS2. In conventional DCs, AP-3 facilitates the recruitment of TLR4 and possibly other TLRs to maturing phagosomes, favoring the MHC-II-dependent presentation of phagosomal antigens and TLR-dependent proinflammatory cytokine stimulation [288]. Hence, in the mouse *pearl* model of HPS2, the impaired capacity for MHC-II presentation and reduced proinflammatory cytokine production from phagocytosed bacteria in DCs leads to reduced T-cell activation and IFN-γ secretion [288]. These defects in DC performance likely contribute to the recurrent bacterial infections observed in HPS2 patients even after restoration of normal neutrophil counts [41].

Another degradative process that also converges with lysosomes and LROs is macroautophagy. This process, generally referred to simply as autophagy, involves the enclosure of intracellular components – including organelles – into a double-membrane structure called the autophagosome. This structure ultimately fuses with lysosomes for degradation of its contents [289,290]. Autophagosomes and the machinery involved in their formation intersect LROs in a number of ways. Like phagosomes, autophagosomes represent a source of peptides for MHC-II presentation [291,292], and thus can feed into the MIIC pathway (Figure 12.3, left). As autophagosomes mature, they also receive input from the endosomal system; indeed, components involved in endosome maturation, such as the ESCRTs, are key components in autophagosome maturation and fusion with lysosomes [293–295]. More recently, the HOPS complex – which is required for the biogenesis of LROs such as melanosomes, platelet dense granules and α granules, eye pigment granules in *D. melanogaster* and gut granules in *C. elegans*, is required for autophagosome fusion with lysosomes through an interaction with the autophagosomal SNARE syntaxin 17 [296,297].

Components of the autophagy pathway have also been implicated in the biogenesis of LROs. Depletion of a number of autophagy components in a human melanoma cell line impairs pigmentation, and the lipidated form of LC3 might associate with melanosomes of distinct stages [298,299]. Similarly, TLR-dependent phagosome maturation in DCs involves recruitment of LC3 in a process requiring the key regulators of autophagosome formation, ATG5 and ATG7 [300,301]. ATG5 and ATG7 have also recently been implicated in agonist-stimulated vWF release from WPBs in epithelial cells, reflecting either storage of a cohort of vWF in autophagosomes or contribution of these components in WPB biogenesis [302]. Thus, the formation of autophagosomal intermediates might play a more general role in LRO biogenesis.

12.6 PERSPECTIVES

Over the last two to three decades, LROs have emerged as key effectors of metazoan physiology and important models for subcellular compartmentation and function. Analyses of LROs have provided insights into molecular mechanisms controlling organelle formation, nonredundant protein sorting pathways, novel modes of secretion, new physiological mechanisms, and the pathogenesis of a host of inherited diseases. In addition, the unusual structural components of some LROs – such as the fibrillar structures underlying melanosomes and WPBs – have given us insights into unexpected areas such as the role of oligomer size in cell physiology and the control of amyloid formation. Future studies will likely focus on integrating the "parts list" that human diseases have provided for us to understand how different components of biogenetic and secretory machinery function together to effect processes such as the formation of transport intermediates or the exocytosis and expulsion of large cargoes. Analyses of some of these processes within living organisms are likely to provide new insights into considerations that are not typically addressed under culture conditions, such as physical requirements of secretion of large cargoes against hemostatic pressure (e.g., see Ref. [303]). At the same time, the exponential rise in the ability to define genetic mutations in patients will provide a rich source of new information concerning components required for LRO biogenesis, secretion and function and on how these components might interact with each other. This information will in turn provide new avenues for inexpensive diagnostic tools and therapies for LRO-related diseases.

ACKNOWLEDGMENTS

The authors thank current and former members of the Marks laboratory, and particularly thank Graça Raposo and the members of her laboratory for many years of collaborative research, helpful discussions, and the original images shown in Figures 12.1 and 12.2. We also thank our many colleagues, particularly Dan Cutler, Miguel Seabra, Gillian Griffiths, Juan Bonifacino, Esteban Dell'Angelica, Scottie Robinson, John Hammer, Dick Swank, Wei Li, Greg Hermann, Sebastian Amigorena, Akiko Iwasaki, and Sergio Grinstein, for thoughtful discussions that contributed to this chapter. We apologize to those whose work we omitted.

REFERENCES

[1] Doyle EL, Ridger V, Ferraro F, Turmaine M, Saftig P, et al. CD63 is an essential cofactor to leukocyte recruitment by endothelial P-selectin. Blood 2011;118:4265–4273.

[2] van Niel G, Charrin S, Simoes S, Romao M, Rochin L, et al. The Tetraspanin CD63 integrates ESCRT-independent and dependent sorting at a common endosome. Dev Cell 2011;21:708–721.

[3] Kraft S, Jouvin MH, Kulkarni N, Kissing S, Morgan ES, et al. The tetraspanin CD63 is required for efficient IgE-mediated mast cell degranulation and anaphylaxis. J Immunol 2013;191:2871–2878.

[4] Hannah MJ, Williams R, Kaur J, Hewlett LJ, Cutler DF. Biogenesis of Weibel–Palade bodies. Semin Cell Dev Biol 2002;13:313–324.

[5] Stinchcombe J, Bossi G, Griffiths GM. Linking albinism and immunity: the secrets of secretory lysosomes. Science 2004;305:55–59.

[6] Andrews NW. Regulated secretion of conventional lysosomes. Trends Cell Biol 2000; 10:316–321.

[7] McNeil PL, Steinhardt RA. Plasma membrane disruption: repair, prevention, adaptation. Annu Rev Cell Dev Biol 2003;19:697–731.

[8] Ellis K, Bagwell J, Bagnat M. Notochord vacuoles are lysosome-related organelles that function in axis and spine morphogenesis. J Cell Biol 2013;200:667–679.

[9] Nordenfelt P, Tapper H. Phagosome dynamics during phagocytosis by neutrophils. J Leukocyte Biol 2011;90:271–284.

[10] Savina A, Jancic C, Hugues S, Guermonprez P, Vargas P, et al. NOX2 controls phagosomal pH to regulate antigen processing during crosspresentation by dendritic cells. Cell 2006;126:205–218.

[11] Wu XS, Masedunskas A, Weigert R, Copeland NG, Jenkins NA, et al. Melanoregulin regulates a shedding mechanism that drives melanosome transfer from melanocytes to keratinocytes. Proc Natl Acad Sci U S A 2012;109:E2101–E2109.

[12] Tarafder AK, Bolasco G, Correia MS, Pereira FJ, Iannone L, et al. Rab11b mediates melanin transfer between donor melanocytes and acceptor keratinocytes via coupled exo/endocytosis. J Invest Dermatol 2014;134:1056–1066.

[13] Raposo G, Marks MS, Cutler DF. Lysosome-related organelles: driving post-Golgi compartments into specialisation. Curr Opin Cell Biol 2007;19(4):394–401.

[14] Stinchcombe JC, Griffiths GM. Secretory mechanisms in cell-mediated cytotoxicity. Annu Rev Cell Dev Biol 2007;23:495–517.

[15] Segal AW. How neutrophils kill microbes. Annu Rev Immunol 2005;23:197–223.

[16] Mantegazza AR, Magalhaes JG, Amigorena S, Marks MS. Presentation of phagocytosed antigens by MHC class I and II. Traffic 2013;14:135–152.

[17] Machlus KR, Italiano JE Jr. The incredible journey: from megakaryocyte development to platelet formation. J Cell Biol 2013;201:785–796.

[18] Ren Q, Ye S, Whiteheart SW. The platelet release reaction: just when you thought platelet secretion was simple. Curr Opin Hematol 2008;15:537–541.

[19] Peters PJ, Raposo G, Neefjes JJ, Oorschot V, Leijendekker RL, et al. Major histocompatibility complex class II compartments in human B lymphoblastoid cells are distinct from early endosomes. J Exp Med 1995;182:325–334.

[20] Sasai M, Linehan MM, Iwasaki A. Bifurcation of Toll-like receptor 9 signaling by adaptor protein 3. Science 2010;329:1530–1534.

[21] Marks MS, Seabra MC. The melanosome: membrane dynamics in black and white. Nature Rev Mol Cell Biol 2001;2:738–748.

[22] Schmitz G, Muller G. Structure and function of lamellar bodies, lipid-protein complexes involved in storage and secretion of cellular lipids. J Lipid Res 1991;32:1539–1570.

[23] Wang H, Wan H, Li X, Liu W, Chen Q, et al. Atg7 is required for acrosome biogenesis during spermatogenesis in mice. Cell Res 2014;24:852–869.

[24] Marks MS, Heijnen HFG, Raposo G. Lysosome-related organelles: unusual compartments become mainstream. Curr Opin Cell Biol 2013;25:495–505.

[25] de Saint BG, Ménasché G, Fischer A. Molecular mechanisms of biogenesis and exocytosis of cytotoxic granules. Nat Rev Immunol 2010;10:568–579.

[26] Shiflett SL, Kaplan J, Ward DM. Chediak–Higashi syndrome: a rare disorder of lysosomes and lysosome related organelles. Pigment Cell Res 2002;15:251–257.

[27] Introne W, Boissy RE, Gahl WA. Clinical, molecular, and cell biological aspects of Chediak–Higashi Syndrome. Mol Genet Metabol 1999;68:283–303.

[28] Kaplan J, De Domenico I, Ward DM. Chediak–Higashi syndrome. Curr Opin Hematol 2008;15:22–29.

[29] Sepulveda FE, Burgess A, Heiligenstein X, Goudin N, Ménager MM, et al. LYST controls the biogenesis of the endosomal compartment required for secretory lysosome function. Traffic 2015. DOI: 10.1111/tra.12244.

[30] Holland P, Torgersen ML, Sandvig K, Simonsen A. LYST affects lysosome size and quantity, but not trafficking or degradation through autophagy or endocytosis. Traffic 2014;15:1390–1405.

[31] Barbosa MD, Nguyen QA, Tchernev VT, Ashley JA, Detter JC, et al. Identification of the homologous beige and Chediak–Higashi syndrome genes. Nature 1996;382: 262–265.

[32] Perou CM, Moore KJ, Nagle DL, Misumi DJ, Woolf EA, et al. Identification of the murine beige gene by YAC complementation and positional cloning. Nat Genet 1996;13:303–308.

[33] Durchfort N, Verhoef S, Vaughn MB, Shrestha R, Adam D, et al. The enlarged lysosomes in beige j cells result from decreased lysosome fission and not increased lysosome fusion. Traffic 2012;13:108–119.

[34] Falkenstein K, De Lozanne A. Dictyostelium LvsB has a regulatory role in endosomal vesicle fusion. J Cell Sci 2014;127:4356–4367.

[35] Ward DM, Shiflett SL, Huynh D, Vaughn MB, Prestwich G, et al. Use of expression constructs to dissect the functional domains of the CHS/beige protein: identification of multiple phenotypes. Traffic 2003;4:403–415.

[36] Cullinane AR, Schäffer AA, Huizing M. The BEACH is hot: a LYST of emerging roles for BEACH-domain containing proteins in human disease. Traffic 2013;14:749–766.

[37] Albers CA, Cvejic A, Favier R, Bouwmans EE, Alessi MC, et al. Exome sequencing identifies NBEAL2 as the causative gene for gray platelet syndrome. Nat Genet 2011;43:735–737.

[38] Gunay-Aygun M, Falik-Zaccai TC, Vilboux T, Zivony-Elboum Y, Gumruk F, et al. NBEAL2 is mutated in gray platelet syndrome and is required for biogenesis of platelet α-granules. Nat Genet 2011;43:732–734.

[39] Kahr WH, Hinckley J, Li L, Schwertz H, Christensen H, et al. Mutations in NBEAL2, encoding a BEACH protein, cause gray platelet syndrome. Nat Genet 2011;43: 738–740.

[40] Kahr WH, Lo RW, Li L, Pluthero FG, Christensen H, et al. Abnormal megakaryocyte development and platelet function in Nbeal2(-/-) mice. Blood 2013;122:3349–3358.

[41] Huizing M, Helip-Wooley A, Westbroek W, Gunay-Aygun M, Gahl WA. Disorders of lysosome-related organelle biogenesis: clinical and molecular genetics. Annu Rev Genomics Hum Genet 2008;9:359–386.

[42] Wei ML. Hermansky–Pudlak syndrome: a disease of protein trafficking and organelle function. Pigment Cell Res 2006;19:19–42.

[43] Wei A-H, Li W. Hermansky–Pudlak syndrome: pigmentary and non-pigmentary defects and their pathogenesis. Pigment Cell Melanoma Res 2013;26:176–192.

[44] Dell'Angelica EC. The building BLOC(k)s of lysosomes and related organelles. Curr Opin Cell Biol 2004;16:458–464.

[45] Di Pietro SM, Dell'Angelica EC. The cell biology of Hermansky–Pudlak syndrome: recent advances. Traffic 2005;6:525–533.

[46] Sitaram A, Marks MS. Mechanisms of protein delivery to melanosomes in pigment cells. Physiology 2012;27:85–99.

[47] Hermann GJ, Schroeder LK, Hieb CA, Kershner AM, Rabbitts BM, et al. Genetic analysis of lysosomal trafficking in *Caenorhabditis elegans*. Mol Biol Cell 2005; 16:3273–3288.

[48] Hermann GJ, Scavarda E, Weis AM, Saxton DS, Thomas LL, et al. *C. elegans* BLOC-1 functions in trafficking to lysosome-related gut granules. PLoS One 2012;7:e43043.

[49] Delahaye JL, Foster OK, Vine A, Saxton DS, Curtin TP, et al. *Caenorhabditis elegans* HOPS and CCZ-1 mediate trafficking to lysosome-related organelles independently of RAB-7 and SAND-1. Mol Biol Cell 2014;25:1073–1096.

[50] Cheli VT, Daniels RW, Godoy R, Hoyle DJ, Kandachar V, et al. Genetic modifiers of abnormal organelle biogenesis in a *Drosophila* model of BLOC-1 deficiency. Hum Mol Genet 2010;19:861–878.

[51] Ooi CE, Moreira JE, Dell'Angelica EC, Poy G, Wassarman DA, et al. Altered expression of a novel adaptin leads to defective pigment granule biogenesis in the *Drosophila* eye color mutant garnet. EMBO J 1997;16:4508–4518.

[52] Simpson F, Peden AA, Christopoulou L, Robinson MS. Characterization of the adaptor-related protein complex, AP-3. J Cell Biol 1997;137:835–845.

[53] Mullins C, Hartnell LM, Wassarman DA, Bonifacino JS. Defective expression of the mu3 subunit of the AP-3 adaptor complex in the *Drosophila* pigmentation mutant carmine. Mol Gen Genet 1999;262:401–412.

[54] Mullins C, Hartnell LM, Bonifacino JS. Distinct requirements for the AP-3 adaptor complex in pigment granule and synaptic vesicle biogenesis in *Drosophila melanogaster*. Mol Gen Genet 2000;263:1003–1014.

[55] Kretzschmar D, Poeck B, Roth H, Ernst R, Keller A, et al. Defective pigment granule biogenesis and aberrant behavior caused by mutations in the *Drosophila* AP-3beta adaptin gene ruby. Genetics 2000;155:213–223.

[56] Falcón-Pérez JM, Romero-Calderón R, Brooks ES, Krantz DE, Dell'Angelica EC. The *Drosophila* pigmentation gene *pink* (p) encodes a homologue of human Hermansky–Pudlak Syndrome 5 (HPS5). Traffic 2007;8:154–168.

[57] Syrzycka M, McEachern LA, Kinneard J, Prabhu K, Fitzpatrick K, et al. The pink gene encodes the *Drosophila* orthologue of the human Hermansky–Pudlak syndrome 5 (HPS5) gene. Genome 2007;50:548–556.

[58] Cheli VT, Dell'angelica EC. Early origin of genes encoding subunits of biogenesis of lysosome-related organelles complex-1, -2 and -3. Traffic 2010;11:579–586.

[59] Suzuki T, Oiso N, Gautam R, Novak EK, Panthier JJ, et al. The mouse organellar biogenesis mutant buff results from a mutation in Vps33a, a homologue of yeast vps33 and *Drosophila* carnation. Proc Natl Acad Sci U S A 2003;100:1146–1150.

[60] Brooks BP, Larson DM, Chan CC, Kjellstrom S, Smith RS, et al. Analysis of ocular hypopigmentation in Rab38cht/cht mice. Invest Ophthalmol Vis Sci 2007;48: 3905–3913.

[61] Lopes VS, Wasmeier C, Seabra MC, Futter CE. Melanosome maturation defect in Rab38-deficient retinal pigment epithelium results in instability of immature melanosomes during transient melanogenesis. Mol Biol Cell 2007;18:3914–3927.

[62] Oiso N, Riddle SR, Serikawa T, Kuramoto T, Spritz RA. The rat Ruby (R) locus is Rab38: identical mutations in Fawn-hooded and Tester-Moriyama rats derived from an ancestral Long Evans rat sub-strain. Mamm Genome 2004;15:307–314.

[63] Wasmeier C, Romao M, Plowright L, Bennett DC, Raposo G, et al. Rab38 an Rab32 control early post-Golgi trafficking of melanogenic enzymes. J Cell Biol 2006;175: 271–281.

[64] Warner TS, Sinclair DA, Fitzpatrick KA, Singh M, Devlin RH, et al. The light gene of *Drosophila melanogaster* encodes a homologue of VPS41, a yeast gene involved in cellular-protein trafficking. Genome 1998;41:236–243.

[65] Ma J, Plesken H, Treisman JE, Edelman-Novemsky I, Ren M. Lightoid and Claret: a rab GTPase and its putative guanine nucleotide exchange factor in biogenesis of *Drosophila* eye pigment granules. Proc Natl Acad Sci U S A 2004;101:11652–11657.

[66] Pulipparacharuvil S, Akbar MA, Ray S, Sevrioukov EA, Haberman AS, et al. *Drosophila* Vps16A is required for trafficking to lysosomes and biogenesis of pigment granules. J Cell Sci 2005;118:3663–3673.

[67] Sevrioukov EA, He JP, Moghrabi N, Sunio A, Kramer H. A role for the deep orange and carnation eye color genes in lysosomal delivery in *Drosophila*. Mol Cell 1999;4: 479–486.

[68] Hirst J, Irving C, Borner GH. Adaptor protein complexes AP-4 and AP-5: new players in endosomal trafficking and progressive spastic paraplegia. Traffic 2013;14: 153–164.

[69] Dell'Angelica EC. AP-3-dependent trafficking and disease: the first decade. Curr Opin Cell Biol 2009;21:552–559.

[70] Balderhaar HJ, Ungermann C. CORVET and HOPS tethering complexes – coordinators of endosome and lysosome fusion. J Cell Sci 2013;126:1307–1316.

[71] Gerondopoulos A, Langemeyer L, Liang J-R, Linford A, Barr FA. BLOC-3 mutated in Hermansky–Pudlak syndrome is a Rab32/38 guanine nucleotide exchange factor. Curr Biol 2012;22:2135–2139.

[72] Raposo G, Marks MS. Melanosomes – dark organelles enlighten endosomal membrane transport. Nat Rev Mol Cell Biol 2007;8:786–798.

[73] Seiji M, Fitzpatrick TM, Simpson RT, Birbeck MSC. Chemical composition and terminology of specialized organelles (melanosomes and melanin granules) in mammalian melanocytes. Nature 1963;197:1082–1084.

[74] Raposo G, Tenza D, Murphy DM, Berson JF, Marks MS. Distinct protein sorting and localization to premelanosomes, melanosomes, and lysosomes in pigmented melanocytic cells. J Cell Biol 2001;152:809–823.

[75] Berson JF, Harper D, Tenza D, Raposo G, Marks MS. Pmel17 initiates premelanosome morphogenesis within multivesicular bodies. Mol Biol Cell 2001;12:3451–3464.

[76] Hurbain I, Geerts WJC, Boudier T, Marco S, Verkleij A, et al. Electron tomography of early melanosomes: implications for melanogenesis and the generation of fibrillar amyloid sheets. Proc Natl Acad Sci U S A 2008;105:19726–19731.

[77] Watt B, van Niel G, Raposo G, Marks MS. PMEL: a pigment cell-specific model for functional amyloid formation. Pigment Cell Melanoma Res 2013;26:300–315.

[78] Edgar JR, Eden ER, Futter CE. Hrs- and CD63-dependent competing mechanisms make different sized endosomal intraluminal vesicles. Traffic 2014;15:197–211.

[79] Berson JF, Theos AC, Harper DC, Tenza D, Raposo G, et al. Proprotein convertase cleavage liberates a fibrillogenic fragment of a resident glycoprotein to initiate melanosome biogenesis. J Cell Biol 2003;161:521–533.

[80] Kummer MP, Maruyama H, Huelsmann C, Baches S, Weggen S, et al. Formation of Pmel17 amyloid is regulated by juxtamembrane metalloproteinase cleavage, and the resulting C-terminal fragment is a substrate for gamma-secretase. J Biol Chem 2009;284:2296–2306.

[81] Leonhardt RM, Vigneron N, Rahner C, Cresswell P. Proprotein convertases process Pmel17 during secretion. J Biol Chem 2011;286:9321–9337.

[82] Rochin L, Hurbain I, Serneels L, Fort C, Watt B, et al. BACE2 processes PMEL to form the melanosome amyloid matrix in pigment cells. Proc Natl Acad Sci U S A 2013;110:10658–10663.

[83] Fowler DM, Koulov AV, Alory-Jost C, Marks MS, Balch WE, et al. Functional amyloid formation within mammalian tissue. PLoS Biol 2006;4:e6.

[84] Fowler DM, Koulov AV, Balch WE, Kelly JW. Functional amyloid – from bacteria to humans. Trends Biochem Sci 2007;32:217–224.

[85] Watt B, van Niel G, Fowler DM, Hurbain I, Luk KC, et al. N-terminal domains elicit formation of functional Pmel17 amyloid fibrils. J Biol Chem 2009;284:35543–35555.

[86] Hu KN, McGlinchey RP, Wickner RB, Tycko R. Segmental polymorphism in a functional amyloid. Biophys J 2011;101:2242–2450.

[87] McGlinchey RP, Shewmaker F, McPhie P, Monterroso B, Thurber K, et al. The repeat domain of the melanosome fibril protein Pmel17 forms the amyloid core promoting melanin synthesis. Proc Natl Acad Sci U S A 2009;106:13731–13736.

[88] McGlinchey RP, Shewmaker F, Hu KN, McPhie P, Tycko R, et al. The repeat domains of melanosome matrix protein Pmel17 orthologs form amyloid fibrils at the acidic melanosomal pH. J Biol Chem 2011;286:8385–8393.

[89] Pfefferkorn CM, McGlinchey RP, Lee JC. Effects of pH on aggregation kinetics of the repeat domain of a functional amyloid, Pmel17. Proc Natl Acad Sci U S A 2010;107:21447–21452.

[90] Leonhardt RM, Vigneron N, Hee JS, Graham M, Cresswell P. Critical residues in the PMEL/Pmel17 N-terminus direct the hierarchical assembly of melanosomal fibrils. Mol Biol Cell 2013;24:964–981.

[91] Watt B, Tenza D, Lemmon MA, Kerje S, Raposo G, et al. Mutations in or near the transmembrane domain alter PMEL amyloid formation from functional to pathogenic. PLoS Genet 2011;7:e1002286.

[92] Brunberg E, Andersson L, Cothran G, Sandberg K, Mikko S, et al. A missense mutation in PMEL17 is associated with the Silver coat color in the horse. BMC Genet 2006;7:46.

[93] Kerje S, Sharma P, Gunnarsson U, Kim H, Bagchi S, et al. The Dominant white, Dun and Smoky color variants in chicken are associated with insertion deletion polymorphisms in the PMEL17 gene. Genetics 2004;168:1507–1518.

[94] Hamilton H. A study of the physiological properties of melanophores with special reference to their role in feather coloration. Anat Rec 1940;78:525–548.

[95] Hellström AR, Watt B, Fard SS, Tenza D, Mannström P, et al. Inactivation of the Pmel gene alters melanosome shape but has only a subtle effect on visible pigmentation. PLoS Genet 2011;7:e1002285.

[96] Cullinane AR, Curry JA, Carmona-Rivera C, Summers CG, Ciccone C, et al. A BLOC-1 mutations screen reveals that *PLDN* is mutated in Hermansky–Pudlak syndrome type 9. Am J Human Genet 2011;88:778–787.

[97] Setty SRG, Tenza D, Truschel ST, Chou E, Sviderskaya EV, et al. BLOC-1 is required for cargo-specific sorting from vacuolar early endosomes toward lysosome-related organelles. Mol Biol Cell 2007;18:768–780.

[98] Setty SRG, Tenza D, Sviderskaya EV, Bennett DC, Raposo G, et al. Cell-specific ATP7A transport sustains copper-dependent tyrosinase activity in melanosomes. Nature 2008;454:1142–1146.

[99] Sitaram A, Dennis MK, Chaudhuri R, De Jesus-Rojas W, Tenza D, et al. Differential recognition of a dileucine-based sorting signal by AP-1 and AP-3 reveals a requirement for both BLOC-1 and AP-3 in delivery of OCA2 to melanosomes. Mol Biol Cell 2012;23:3178–3192.

[100] Delevoye C, Hurbain I, Tenza D, Sibarita J-B, Uzan-Gafsou S, et al. AP-1 and KIF13A coordinate endosomal sorting and positioning during melanosome biogenesis. J Cell Biol 2009;187:247–264.

[101] Delevoye C, Miserey-Lenkei S, Montagnac G, Gilles-Marsens F, Paul-Gilloteaux P, et al. Recycling endosome tubule morphogenesis from sorting endosomes requires the kinesin motor KIF13A. Cell Rep 2014;6:445–454.

[102] Di Pietro SM, Falcon-Perez JM, Tenza D, Setty SRG, Marks MS, et al. BLOC-1 interacts with BLOC-2 and the AP-3 complex to facilitate protein trafficking on endosomes. Mol Biol Cell 2006;17:4027–4038.

[103] Newell-Litwa K, Salazar G, Smith Y, Faundez V. Roles of BLOC-1 and adaptor protein-3 complexes in cargo sorting to synaptic vesicles. Mol Biol Cell 2009;20: 1441–1453.

[104] Newell-Litwa K, Chintala S, Jenkins S, Pare J, McGaha L, et al. Hermansky–Pudlak protein complexes, AP-3 and BLOC-1, differentially regulate presynaptic composition in the striatum and hippocampus. J Neurosci 2010;30:820–831.

[105] Salazar G, Craige B, Styers ML, Newell-Litwa KA, Doucette MM, et al. BLOC-1 complex deficiency alters the targeting of Adaptor Protein complex-3 cargoes. Mol Biol Cell 2006;17:4014–4026.

[106] Theos AC, Tenza D, Martina JA, Hurbain I, Peden AA, et al. Functions of AP-3 and AP-1 in tyrosinase sorting from endosomes to melanosomes. Mol Biol Cell 2005;16: 5356–5372.

[107] Ghiani CA, Starcevic M, Rodriguez-Fernandez IA, Nazarian R, Cheli VT, et al. The dysbindin-containing complex (BLOC-1) in brain: developmental regulation, interaction with SNARE proteins and role in neurite outgrowth. Mol Psychiatry 2010;15:204–215.

[108] Huang L, Kuo YM, Gitschier J. The pallid gene encodes a novel, syntaxin 13-interacting protein involved in platelet storage pool deficiency. Nature Genet 1999;23:329–332.

[109] Moriyama K, Bonifacino JS. Pallidin is a component of a multi-protein complex involved in the biogenesis of lysosome-related organelles. Traffic 2002;3:666–677.

[110] Ilardi JM, Mochida S, Sheng ZH. Snapin: a SNARE-associated protein implicated in synaptic transmission. Nature Neurosci 1999;2:119–124.

[111] Ryder PV, Vistein R, Gokhale A, Seaman MN, Puthenveedu M, et al. The WASH complex, and endosomal Arp2/3 activator, interacts with the Hermansky–Pudlak syndrome

complex BLOC-1 and its cargo phosphatidylinositol-4 kinase type II alpha. Mol Biol Cell 2013;24:2269–2284.

[112] Salazar G, Zlatic S, Craige B, Peden AA, Pohl J, et al. Hermansky–Pudlak syndrome protein complexes associate with phosphatidylinositol-4-kinase type II alpha in neuronal and non-neuronal cells. J Biol Chem 2009;284:1790–1802.

[113] Gokhale A, Larimore J, Werner E, So L, Moreno-De-Luca A, et al. Quantitative proteomic and genetic analyses of the schizophrenia susceptibility factor dysbindin identify novel roles of the biogenesis of lysosome-related organelles complex 1. J Neurosci 2012;32:3697–3711.

[114] John Peter AT, Lachmann J, Rana M, Bunge M, Cabrera M, et al. The BLOC-1 complex promotes endosomal maturation by recruiting the Rab5 GTPase-activating protein Msb3. J Cell Biol 2013;201:97–111.

[115] Lee HH, Nemecek D, Schindler C, Smith WJ, Ghirlando R, et al. Assembly and architecture of Biogenesis of Lysosome-related Organelles Complex-1 (BLOC-1). J Biol Chem 2012;287:5882–5890.

[116] Suetsugu S, Kurisu S, Takenawa T. Dynamic shaping of cellular membranes by phospholipids and membrane-deforming proteins. Physiol Rev 2014;94:1219–1248.

[117] Helip-Wooley A, Westbroek W, Dorward HM, Koshoffer A, Huizing M, et al. Improper trafficking of melanocyte-specific proteins in Hermansky–Pudlak syndrome type-5. J Invest Dermatol 2007;127:1471–1478.

[118] Richmond B, Huizing M, Knapp J, Koshoffer A, Zhao Y, et al. Melanocytes derived from patients with Hermansky–Pudlak Syndrome types 1, 2, and 3 have distinct defects in cargo trafficking. J Invest Dermatol 2005;124:420–427.

[119] Boissy RE, Richmond B, Huizing M, Helip-Wooley A, Zhao Y, et al. Melanocyte-specific proteins are aberrantly trafficked in melanocytes of Hermansky–Pudlak syndrome-type 3. Am J Pathol 2005;166:231–240.

[120] Bultema JJ, Ambrosio AL, Burek CL, Di Pietro SM. BLOC-2, AP-3, and AP-1 function in concert with Rab38 and Rab32 to mediate protein trafficking to lysosome-related organelles. J Biol Chem 2012;287:19550–19563.

[121] Zhang Q, Zhao B, Li W, Oiso N, Novak EK, et al. Ru2 and Ru encode mouse orthologs of the genes mutated in human Hermansky–Pudlak syndrome types 5 and 6. Nature Genet 2003;33:145–153.

[122] Helip-Wooley A, Westbroek W, Dorward H, Mommaas M, Boissy RE, et al. Association of the Hermansky–Pudlak syndrome type-3 protein with clathrin. BMC Cell Biol 2005;6:33.

[123] Li K, Yang L, Zhang C, Niu Y, Li W, et al. HPS6 interacts with dynactin p150Glued to mediate retrograde trafficking and maturation of lysosomes. J Cell Sci 2014;127(Pt 21): 4574–4588.

[124] Zhang L, Yu K, Robert KW, DeBolt KM, Hong N, et al. Rab38 targets to lamellar bodies and normalizes their sizes in lung alveolar type II epithelial cells. Am J Physiol Lung Cell Mol Physiol 2011;301:461–477.

[125] Ambrosio AL, Boyle JA, Di Pietro SM. Mechanism of platelet dense granule biogenesis: study of cargo transport and function of Rab32 and Rab38 in a model system. Blood 2012;120:4072–4081.

[126] Nguyen T, Wei ML. Hermansky–Pudlak HPS1/pale ear gene regulates epidermal and dermal melanocyte development. J Invest Dermatol 2007;127:421–428.

[127] Nguyen T, Novak EK, Kermani M, Fluhr J, Peters LL, et al. Melanosome morphologies in murine models of Hermansky–Pudlak syndrome reflect blocks in organelle development. J Invest Dermatol 2002;119:1156–1164.

[128] Bultema JJ, Boyle JA, Malenke PB, Martin FE, Dell'Angelica EC, et al. Myosin vc interacts with rab32 and rab38 proteins and works in the biogenesis and secretion of melanosomes. J Biol Chem 2014;289:33513–33528.

[129] Hesketh GG, Pérez-Dorado I, Jackson LP, Wartosch L, Schäfer IB, et al. VARP is recruited on to endosomes by direct interaction with retromer, where together they function in export to the cell surface. Dev Cell 2014;29:591–606.

[130] Zhang X, He X, Fu XY, Chang Z. Varp is a Rab21 guanine nucleotide exchange factor and regulates endosome dynamics. J Cell Sci 2006;119:1053–1062.

[131] Wang F, Zhang H, Zhang X, Wang Y, Ren F, et al. Varp interacts with Rab38 and functions as its potential effector. Biochem Biophys Res Commun 2008;372:162–167.

[132] Burgo A, Sotirakis E, Simmler MC, Verraes A, Chamot C, et al. Role of Varp, a Rab21 exchange factor and TI-VAMP/VAMP7 partner, in neurite growth. EMBO Rep 2009;10:1117–1124.

[133] Tamura K, Ohbayashi N, Maruta Y, Kanno E, Itoh T, et al. Varp is a novel Rab32/38-binding protein that regulates Tyrp1 trafficking in melanocytes. Mol Biol Cell 2009;20:2900–2908.

[134] Schäfer IB, Hesketh GG, Bright NA, Gray SR, Pryor PR, et al. The binding of Varp to VAMP7 traps VAMP7 in a closed, fusogenically inactive conformation. Nat Struct Mol Biol 2012;19:1300–1309.

[135] McGough IJ, Steinberg F, Gallon M, Yatsu A, Ohbayashi N, et al. Identification of molecular heterogeneity in SNX27-retromer-mediated endosome-to-plasma-membrane recycling. J Cell Sci 2014;127:4940–4953.

[136] Tamura K, Ohbayashi N, Ishibashi K, Fukuda M. Structure-function analysis of VPS9-ankyrin-repeat protein (Varp) in the trafficking of tyrosinase-related protein 1 in melanocytes. J Biol Chem 2011;286:7507–7521.

[137] Gerritsen SM, Akkerman JW, Nijmeijer B, Sixma JJ, Witkop CJ, et al. The Hermansky–Pudlak syndrome. Evidence for a lowered 5-hydroxytryptamine content in platelets of heterozygotes. Scand J Haematol 1977;18:249–256.

[138] Novak EK, Hui SW, Swank RT. Platelet storage pool deficiency in mouse pigment mutations associated with seven distinct genetic loci. Blood 1984;63:536–544.

[139] Novak EK, Sweet HO, Prochazka M, Parentis M, Soble R, et al. Cocoa: a new mouse model for platelet storage pool deficiency. Br J Haematol 1988;69:371–378.

[140] Rao GH, White JG. An improved method for measuring endogenous serotonin in platelets of patients with Hermansky–Pudlak syndrome. Thromb Res 1988;51:225–227.

[141] White JG. Serotonin storage organelles in human megakaryocytes. Am J Pathol 1971; 63:403–410.

[142] Witkop CJ, Krumwiede M, Sedano H, White JG. Reliability of absent platelet dense bodies as a diagnostic criterion for Hermansky–Pudlak syndrome. Am J Hematol 1987;26:305–311.

[143] Meng R, Wu J, Harper DC, Wang Y, Kowalska MA, et al. Defective release of α granule and lysosome contents from platelets in mouse Hermansky–Pudlak syndrome models. Blood 2015;125:1623–1632.

[144] Rendu F, Maclouf J, Launay JM, Boinot C, Levy-Toledano S, et al. Hermansky–Pudlak platelets: further studies on release reaction and protein phosphorylations. Am J Hematol 1987;25:165–174.

[145] Sharda A, Kim SH, Jasuja R, Gopal S, Flaumenhaft R, et al. Defective PDI release from platelets and endothelial cells impairs thrombus formation in Hermansky–Pudlak syndrome. Blood 2015;125:1633–1642.

[146] Graham GJ, Ren Q, Dilks JR, Blair P, Whiteheart SW, et al. Endobrevin/VAMP-8-dependent dense granule release mediates thrombus formation *in vivo*. Blood 2009;114: 1083–1090.

[147] Seward SL Jr, Gahl WA. Hermansky–Pudlak syndrome: health care throughout life. Pediatrics 2013;132:153–160.

[148] Ninkovic I, White JG, Rangel-Filho A, Datta YH. The role of Rab38 in platelet dense granule defects. J Thromb Haemost 2008;6:2143–2151.

[149] Jedlitschky G, Tirschmann K, Lubenow LE, Nieuwenhuis HK, Akkerman JW, et al. The nucleotide transporter MRP4 (ABCC4) is highly expressed in human platelets and present in dense granules, indicating a role in mediator storage. Blood 2004; 104:3603–3610.

[150] Meng R, Wang Y, Yao Y, Zhang Z, Harper DC, et al. SLC35D3 delivery from megakaryocyte early endosomes is required for platelet dense granule biogenesis and differentially defective in Hermansky–Pudlak syndrome models. Blood 2012;120:404–414.

[151] Chintala S, Tan J, Gautam R, Rusiniak ME, Guo X, et al. The Slc35d3 gene, encoding an orphan nucleotide sugar transporter, regulates platelet dense granules. Blood 2007;109:1533–1540.

[152] McGarry MP, Borchers M, Novak EK, Lee NA, Ohtake PJ, et al. Pulmonary pathologies in pallid mice result from nonhematopoietic defects. Exp Mol Pathol 2002;72:213–220.

[153] McGarry MP, Reddington M, Novak EK, Swank RT. Survival and lung pathology of mouse models of Hermansky–Pudlak syndrome and Chediak–Higashi syndrome. Proc Soc Exp Biol Med 1999;220:162–168.

[154] Young LR, Borchers MT, Allen HL, Gibbons RS, McCormack FX. Lung-restricted macrophage activation in the pearl mouse model of Hermansky–Pudlak syndrome. J Immunol 2006;176:4361–4368.

[155] Young LR, Pasula R, Gulleman PM, Deutsch GH, McCormack FX. Susceptibility of Hermansky–Pudlak mice to bleomycin-induced type II cell apoptosis and fibrosis. Am J Respir Cell Mol Biol 2007;37:67–74.

[156] Guttentag SH, Akhtar A, Tao JQ, Atochina E, Rusiniak ME, et al. Defective surfactant secretion in a mouse model of Hermansky–Pudlak syndrome. Am J Respir Cell Mol Biol 2005;33:14–21.

[157] Gautam R, Novak EK, Tan J, Wakamatsu K, Ito S, et al. Interaction of Hermansky–Pudlak syndrome genes in the regulation of lysosome-related organelles. Traffic 2006; 7:779–792.

[158] Lyerla TA, Rusiniak ME, Borchers M, Jahreis G, Tan J, et al. Aberrant lung structure, composition, and function in a murine model of Hermansky–Pudlak syndrome. Am J Physiol Lung Cell Mol Physiol 2003;285:L643–653.

[159] Young LR, Gulleman PM, Bridges JP, Weaver TE, Deutsch GH, et al. The alveolar epithelium determines susceptibility to lung fibrosis in Hermansky–Pudlak syndrome. Am J Respir Crit Care Med 2012;186:1014–1024.

[160] Osanai K, Iguchi M, Takahashi K, Nambu Y, Sakuma T, et al. Expression and localization of a novel Rab small G protein (Rab38) in the rat lung. Am J Pathol 2001; 158:1665–1675.

[161] Osanai K, Higuchi J, Oikawa R, Kobayashi M, Tsuchihara K, et al. Altered lung surfactant system in a Rab38-deficient rat model of Hermansky–Pudlak syndrome. Am J Lung Cell Mol Physiol 2010;298:L243–L251.

[162] Faúndez V, Horng J-T, Kelly RB. A function for the AP3 coat complex in synaptic vesicle formation from endosomes. Cell 1998;93:423–432.

[163] Blumstein J, Faundez V, Nakatsu F, Saito T, Ohno H, et al. The neuronal form of adaptor protein-3 is required for synaptic vesicle formation from endosomes. J Neurosci 2001;21:8034–8042.

[164] Nakatsu F, Okada M, Mori F, Kumazawa N, Iwasa H, et al. Defective function of GABA-containing synaptic vesicles in mice lacking the AP-3B clathrin adaptor. J Cell Biol 2004;167:293–302.

[165] Kantheti P, Qiao X, Diaz ME, Peden AA, Meyer GE, et al. Mutation in AP-3d in the mocha mouse links endosomal transport to storage deficiency in platelets, melanosomes, and synaptic vesicles. Neuron 1998;21:111–122.

[166] Salazar G, Love R, Werner E, Doucette MM, Cheng S, et al. The zinc transporter ZnT3 interacts with AP-3 and it is preferentially targeted to a distinct synaptic vesicle subpopulation. Mol Biol Cell 2004;15:575–587.

[167] Salazar G, Craige B, Wainer BH, Guo J, De Camilli P, et al. Phosphatidylinositol-4-kinase type II a is a component of AP-3-derived vesicles. Mol Biol Cell 2005;16: 3692–3704.

[168] Salazar G, Love R, Styers ML, Werner E, Peden A, et al. AP-3-dependent mechanisms control the targeting of a chloride channel (ClC-3) in neuronal and non-neuronal cells. J Biol Chem 2004;279:25430–25439.

[169] Larimore J, Tornieri K, Ryder PV, Gokhale A, Zlatic SA, et al. The schizophrenia susceptibility factor dysbindin and its associated complex sort cargoes from cell bodies to the synapse. Mol Biol Cell 2011;22:4854–4867.

[170] Chen XW, Feng YQ, Hao CJ, Guo XL, He X, et al. DTNBP1, a schizophrenia susceptibility gene, affects kinetics of transmitter release. J Cell Biol 2008;181:791–801.

[171] Erway LC, Fraser AS, Hurley LS. Prevention of congenital otolith defect in pallid mutant mice by manganese supplementation. Genetics 1971;67:97–108.

[172] Trune DR, Lim DJ. The behavior and vestibular nuclear morphology of otoconia-deficient pallid mutant mice. J Neurogenet 1983;1:53–69.

[173] Ryder PV, Faundez V. Schizophrenia: the "BLOC" may be in the endosomes. Sci Signal 2009;2:pe66.

[174] Weibel ER, Palade GE. New cytoplasmic components in arterial endothelia. J Cell Biol 1964;23:101–112.

[175] Michaux G, Cutler DF. How to roll an endothelial cigar: the biogenesis of Weibel–Palade bodies. Traffic 2004;5:69–78.

[176] Nightingale T, Cutler D. The secretion of von Willebrand factor from endothelial cells; an increasingly complicated story. J Thromb Haemost 2013;11:192–201.

[177] Wagner DD. The Weibel–Palade body: the storage granule for von Willebrand factor and P-selectin. Thromb Haemost 1993;70:105–110.

[178] Journet AM, Saffaripour S, Cramer EM, Tenza D, Wagner DD. von Willebrand factor storage requires intact prosequence cleavage site. Eur J Cell Biol 1993;60:31–41.

[179] Wagner DD, Marder VJ. Biosynthesis of von Willebrand protein by human endothelial cells: processing steps and their intracellular localization. J Cell Biol 1984; 99:2123–2130.

[180] Berriman JA, Li S, Hewlett LJ, Wasilewski S, Kiskin FN, et al. Structural organization of Weibel–Palade bodies revealed by cryo-EM of vitrified endothelial cells. Proc Natl Acad Sci U S A 2009;106:17407–17412.

[181] Valentijn KM, Valentijn JA, Jansen KA, Koster AJ. A new look at Weibel–Palade body structure in endothelial cells using electron tomography. J Struct Biol 2008; 161:447–458.

[182] Zenner HL, Collinson LM, Michaux G, Cutler DF. High-pressure freezing provides insights into Weibel–Palade body biogenesis. J Cell Sci 2007;120:2117–2125.

[183] Ferraro F, Kriston-Vizi J, Metcalf DJ, Martin-Martin B, Freeman J, et al. A two-tier Golgi-based control of organelle size underpins the functional plasticity of endothelial cells. Dev Cell 2014;29:292–304.

[184] Michaux G, Abbitt KB, Collinson LM, Haberichter SL, Norman KE, et al. The physiological function of von Willebrand's factor depends on its tubular storage in endothelial Weibel–Palade bodies. Dev Cell 2006;10:223–232.

[185] Lui-Roberts WWY, Collinson LM, Hewlett LJ, Michaux G, Cutler DF. An AP-1/clathrin coat plays a novel and essential role in forming the Weibel–Palade bodies of endothelial cells. J Cell Biol 2005;170:627–636.

[186] Harrison-Lavoie KJ, Michaux G, Hewlett L, Kaur J, Hannah MJ, et al. P-selectin and CD63 use different mechanisms for delivery to Weibel–Palade bodies. Traffic 2006;7: 647–662.

[187] Michaux G, Pullen TJ, Haberichter SL, Cutler DF. P-selectin binds to the D′-D3 domains of von Willebrand factor in Weibel–Palade bodies. Blood 2006;107: 3922–3924.

[188] Hannah MJ, Hume AN, Arribas M, Williams R, Hewlett LJ, et al. Weibel–Palade bodies recruit Rab27 by a content-driven, maturation-dependent mechanism that is independent of cell type. J Cell Sci 2003;116:3939–3948.

[189] Nightingale TD, Pattni K, Hume AN, Seabra MC, Cutler DF. Rab27a and MyRIP regulate the amount and multimeric state of VWF released from endothelial cells. Blood 2009;113:5010–5018.

[190] Rojo Pulido I, Nightingale TD, Darchen F, Seabra MC, Cutler DF, et al. Myosin Va acts in concert with Rab27a and MyRIP to regulate acute von-Willebrand factor release from endothelial cells. Traffic 2011;12:1371–1382.

[191] Bierings R, Hellen N, Kiskin N, Knipe L, Fonseca AV, et al. The interplay between the Rab27A effectors Slp4-a and MyRIP controls hormone-evoked Weibel–Palade body exocytosis. Blood 2012;120:2757–2767.

[192] van Nispen tot Pannerden H, de Haas F, Geerts W, Posthuma G, van Dijk S, et al. The platelet interior revisited: electron tomography reveals tubular alpha-granule subtypes. Blood 2010;116:1147–1156.

[193] Van Gele M, Dynoodt P, Lambert J. Griscelli syndrome: a model system to study vesicular trafficking. Pigment Cell Melanoma Res 2009;22:268–282.

[194] Hume AN, Collinson LM, Rapak A, Gomes AQ, Hopkins CR, et al. Rab27a regulates the peripheral distribution of melanosomes in melanocytes. J Cell Biol 2001;152:795–808.

[195] Hume AN, Collinson LM, Hopkins CR, Strom M, Barral DC, et al. The leaden gene product is required with rab27a to recruit myosin Va to melanosomes in melanocytes. Traffic 2002;3:193–202.

[196] Wu X, Rao K, Bowers MB, Copeland NG, Jenkins NA, et al. Rab27a enables myosin Va-dependent melanosome capture by recruiting the myosin to the organelle. J Cell Sci 2001;114:1091–1100.

[197] Wu XS, Rao K, Zhang H, Wang F, Sellers JR, et al. Identification of an organelle receptor for myosin-Va. Nature Cell Biol 2002;4:271–278.

[198] Wu X, Wang F, Rao K, Sellers JR, Hammer JA III. Rab27a is an essential component of melanosome receptor for myosin Va. Mol Biol Cell 2002;13:1735–1749.

[199] Ménasché G, Pastural E, Feldmann J, Certain S, Ersoy F, et al. Mutations in RAB27A cause Griscelli syndrome associated with haemophagocytic syndrome. Nature Genet 2000;25:173–176.

[200] Matesic LE, Yip R, Reuss AE, Swing DA, O'Sullivan TN, et al. Mutations in Mlph, encoding a member of the Rab effector family, cause the melanosome transport defects observed in leaden mice. Proc Natl Acad Sci U S A 2001;98:10238–10243.

[201] Bahadoran P, Aberdam E, Mantoux F, Busca R, Bille K, et al. Rab27a: a key to melanosome transport in human melanocytes. J Cell Biol 2001;152:843–850.

[202] Wu X, Bowers B, Wei Q, Kocher B, Hammer JA 3rd.. Myosin V associates with melanosomes in mouse melanocytes: evidence that myosin V is an organelle motor. J Cell Sci 1997;110:847–859.

[203] Wu X, Bowers B, Rao K, Wei Q, Hammer JA III. Visualization of melanosome dynamics within wild-type and dilute melanocytes suggests a paradigm for myosin V function *in vivo*. J Cell Biol 1998;143:1899–1918.

[204] Evans RD, Robinson C, Briggs DA, Tooth DJ, Ramalho JS, et al. Myosin-Va and dynamic actin oppose microtubules to drive long-range organelle transport. Curr Biol 2014;24:1743–1750.

[205] Futter CE, Ramalho JS, Jaissle GB, Seeliger MW, Seabra MC. The role of Rab27a in the regulation of melanosome distribution within retinal pigment epithelial cells. Mol Biol Cell 2004;15:2264–2275.

[206] Lopes VS, Ramalho JS, Owen DM, Karl MO, Strauss O, et al. The ternary Rab27a-Myrip-myosin VIIa complex regulates melanosome motility in the retinal pigment epithelium. Traffic 2007;8:486–499.

[207] Hammer JA 3rd., Wagner W. Functions of class V myosins in neurons. J Biol Chem 2013;288:28428–28434.

[208] Wagner W, Brenowitz SD, Hammer JA 3rd.. Myosin-Va transports the endoplasmic reticulum into the dendritic spines of Purkinje neurons. Nat Cell Biol 2011;13:40–48.

[209] Ménasché G, Ho CH, Sanal O, Feldmann J, Tezcan I, et al. Griscelli syndrome restricted to hypopigmentation results from a melanophilin defect (GS3) or a MYO5A F-exon deletion (GS1). J Clin Invest 2003;112:450–456.

[210] Huang JD, Mermall V, Strobel MC, Russell LB, Mooseker MS, et al. Molecular genetic dissection of mouse unconventional myosin-VA: tail region mutations. Genetics 1998;148:1963–1972.

[211] Haddad EK, Wu X, Hammer JA, Henkart PA. Defective granule exocytosis in Rab27a-deficient lymphocytes from Ashen mice. J Cell Biol 2001;152:835–842.

[212] Stinchcombe JC, Barral DC, Mules EH, Booth S, Hume AN, et al. Rab27a is required for regulated secretion in cytotoxic T lymphocytes. J Cell Biol 2001;152:825–834.

[213] Fukuda M. Rab27 effectors, pleiotropic regulators in secretory pathways. Traffic 2013;14:949–963.

[214] Krzewski K, Cullinane AR. Evidence for defective Rab GTPase-dependent cargo traffic in immune disorders. Exp Cell Res 2013;319:2360–2367.

[215] Singh RK, Mizuno K, Wasmeier C, Wavre-Shapton ST, Recchi C, et al. Distinct and opposing roles for Rab27a/Mlph/MyoVa and Rab27b/Munc13-4 in mast cell secretion. FEBS J 2013;280:892–903.

[216] Kim JD, Willetts L, Ochkur S, Srivastava N, Hamburg R, et al. An essential role for Rab27a GTPase in eosinophil exocytosis. J Leukocyte Biol 2013;94:1265–1274.

[217] Munafó DB, Johnson JL, Ellis BA, Rutschmann S, Beutler B, et al. Rab27a is a key component of the secretory machinery of azurophilic granules in granulocytes. Biochem J 2007;402:229–239.

[218] Herrero-Turrión MJ, Calafat J, Janssen H, Fukuda M, Mollinedo F. Rab27a regulates exocytosis of tertiary and specific granules in human neutrophils. J Immunol 2008;181:3793–3803.

[219] Brzezinska AA, Johnson JL, Munafo DB, Crozat K, Beutler B, et al. The Rab27a effectors JFC1/Slp1 and Munc13-4 regulate exocytosis of neutrophil granules. Traffic 2008;9:2151–2164.

[220] Yi Z, Yokota H, Torii S, Aoki T, Hosaka M, et al. The Rab27a/granuphilin complex regulates the exocytosis of insulin-containing dense-core granules. Mol Cell Biol 2002;22:1858–1867.

[221] Ostrowski M, Carmo NB, Krumeich S, Fanget I, Raposo G, et al. Rab27a and Rab27b control different steps of the exosome secretion pathway. Nat Cell Biol 2010;12:19–30.

[222] Tolmachova T, Abrink M, Futter CE, Authi KS, Seabra MC. Rab27b regulates number and secretion of platelet dense granules. Proc Natl Acad Sci U S A 2007;104:5872–5877.

[223] Aricò M, Janka G, Fischer A, Henter JI, Blanche S, et al. Hemophagocytic lymphohistiocytosis. Report of 122 children from the International Registry. FHL Study Group of the Histiocyte Society. Leukemia 1996;10:197–203.

[224] Janka GE. Familial and acquired hemophagocytic lymphohistiocytosis. Annu Rev Med 2012;63:233–246.

[225] Ménasché G, Feldmann J, Fischer A, de Saint BG. Primary hemophagocytic syndromes point to a direct link between lymphocyte cytotoxicity and homeostasis. Immunol Rev 2005;203:165–179.

[226] Jessen B, Maul-Pavicic A, Ufheil H, Vraetz T, Enders A, et al. Subtle differences in CTL cytotoxicity determine susceptibility to hemophagocytic lymphohistiocytosis in mice and humans with Chediak–Higashi syndrome. Blood 2011;118:4620–4629.

[227] Jessen B, Bode SF, Ammann S, Chakravorty S, Davies G, et al. The risk of hemophagocytic lymphohistiocytosis in Hermansky–Pudlak syndrome type 2. Blood 2013. DOI: 10.1182/blood-2012-1110-463166.

[228] Clark RH, Stinchcombe JC, Day A, Blott E, Booth S, et al. Adaptor protein 3-dependent microtubule-mediated movement of lytic granules to the immunological synapse. Nature Immunol 2003;4:1111–1120.

[229] Fontana S, Parolini S, Vermi W, Booth S, Gallo F, et al. Innate immunity defects in Hermansky–Pudlak type 2 syndrome. Blood 2006;107:4857–4864.

[230] Stinchcombe JC, Majorovits E, Bossi G, Fuller S, Griffiths GM. Centrosome polarization delivers secretory granules to the immunological synapse. Nature 2006;443: 462–465.

[231] Stinchcombe JC, Salio M, Cerundolo V, Pende D, Arico M, et al. Centriole polarisation to the immunological synapse directs secretion from cytolytic cells of both the innate and adaptive immune systems. BMC Biol 2011;9:45.

[232] Kurowska M, Goudin N, Nehme NT, Court M, Garin J, et al. Terminal transport of lytic granules to the immune synapse is mediated by the kinesin-1/Slp3/Rab27a complex. Blood 2012;119:3879–3889.

[233] Ménager MM, Ménasché G, Romao M, Knapnougel P, Ho CH, et al. Secretory cytotoxic granule maturation and exocytosis require the effector protein hMunc13-4. Nat Immunol 2007;8:257–267.

[234] Tsun A, Qureshi I, Stinchcombe JC, Jenkins MR, de la Roche M, et al. Centrosome docking at the immunological synapse is controlled by Lck signaling. J Cell Biol 2011;192:663–674.

[235] Mentlik AN, Sanborn KB, Holzbaur EL, Orange JS. Rapid lytic granule convergence to the MTOC in natural killer cells is dependent on dynein but not cytolytic commitment. Mol Biol Cell 2010;21:2241–2256.

[236] Sieni E, Cetica V, Mastrodicasa E, Pende D, Moretta L, et al. Familial hemophagocytic lymphohistiocytosis: a model for understanding the human machinery of cellular cytotoxicity. Cell Mol Life Sci 2012;69:29–40.

[237] Al Hawas R, Ren Q, Ye S, Karim ZA, Filipovich AH, et al. Munc18b/STXBP2 is required for platelet secretion. Blood 2012;120:2493–2500.

[238] Ren Q, Wimmer C, Chicka MC, Ye S, Ren Y, et al. Munc13-4 is a limiting factor in the pathway required for platelet granule release and hemostasis. Blood 2010;116:869–877.

[239] Ye S, Karim ZA, Al Hawas R, Pessin JE, Filipovich AH, et al. Syntaxin-11, but not syntaxin-2 or syntaxin-4, is required for platelet secretion. Blood 2012;120:2484–2492.

[240] Nakamura L, Bertling A, Brodde MF, Zur SU, Schulz AS, et al. First characterization of platelet secretion defect in patients with familial hemophagocytic lymphohistiocytosis type 3 (FHL-3). Blood 2015;125:412–414.

[241] Neeft M, Wieffer M, de Jong AS, Negroiu G, Metz CHG, et al. Munc13-4 Is an effector of Rab27a and controls secretion of lysosomes in haematopoietic cells. Mol Biol Cell 2005;16:731–741.

[242] Elstak ED, Neeft M, Nehme NT, Voortman J, Cheung M, et al. The munc13-4-rab27 complex is specifically required for tethering secretory lysosomes at the plasma membrane. Blood 2011;118:1570–1578.

[243] Rizo J, Südhof TC. The membrane fusion enigma: SNAREs, Sec1/Munc18 proteins, and their accomplices – guilty as charged? Annu Rev Cell Dev Biol 2012;28:279–308.

[244] Tang BL, Low DY, Hong W. Syntaxin 11: a member of the syntaxin family without a carboxyl terminal transmembrane domain. Biochem Biophys Res Commun 1998;245:627–632.

[245] Valdez AC, Cabaniols JP, Brown MJ, Roche PA. Syntaxin 11 is associated with SNAP-23 on late endosomes and the trans-Golgi network. J Cell Sci 1999;112:845–854.

[246] Arneson LN, Brickshawana A, Segovis CM, Schoon RA, Dick CJ, et al. Cutting edge: syntaxin 11 regulates lymphocyte-mediated secretion and cytotoxicity. J Immunol 2007;179:3397–3401.

[247] Halimani M, Pattu V, Marshall MR, Chang HF, Matti U, et al. Syntaxin11 serves as a t-SNARE for the fusion of lytic granules in human cytotoxic T lymphocytes. Eur J Immunol 2014;44:573–584.

[248] Dabrazhynetskaya A, Ma J, Guerreiro-Cacais AO, Arany Z, Rudd E, et al. Syntaxin 11 marks a distinct intracellular compartment recruited to the immunological synapse of NK cells to colocalize with cytotoxic granules. J Cell Mol Med 2012;16:129–141.

[249] Côte M, Ménager MM, Burgess A, Mahlaoui N, Picard C, et al. Munc18-2 deficiency causes familial hemophagocytic lymphohistiocytosis type 5 and impairs cytotoxic granule exocytosis in patient NK cells. J Clin Invest 2009;119:3765–3773.

[250] Spessott WA, Sanmillan ML, McCormick ME, Patel N, Villanueva J, et al. Hemophagocytic lymphohistiocytosis caused by dominant negative mutations in STXBP2 that inhibit SNARE-mediated membrane fusion. Blood 2015;125:1566–1577.

[251] Dotta L, Parolini S, Prandini A, Tabellini G, Antolini M, et al. Clinical, laboratory and molecular signs of immunodeficiency in patients with partial oculo-cutaneous albinism. Orphanet J Rare Dis 2013;8:168.

[252] Meng R, Bridgman R, Toivio-Kinnucan M, Niemeyer GP, Vernau W, et al. Neutrophil elastase processing defect in cyclic hematopoietic dogs. Exp Hematol 2010;38:104–115.

[253] Horwitz MS, Duan Z, Korkmaz B, Lee HH, Mealiffe ME, et al. Neutrophil elastase in cyclic and severe congenital neutropenia. Blood 2007;109:1817–1824.

[254] Aprikyan AA, Dale DC. Mutations in the neutrophil elastase gene in cyclic and congenital neutropenia. Curr Opin Immunol 2001;13:535–538.

[255] Chiang P-W, Spector E, Thomas M, Frei-Jones M. Novel mutation causing Hermansky–Pudlak Syndrome Type 2. Pediatr Blood Cancer 2010;55:1438.

[256] Shotelersuk V, Dell'Angelica EC, Hartnell L, Bonifacino JS, Gahl WA. A new variant of Hermansky–Pudlak syndrome due to mutations in a gene responsible for vesicle formation. Am J Med 2000;108:423–427.

[257] Wenham M, Grieve S, Cummins M, Jones ML, Booth S, et al. Two patients with Hermansky–Pudlak syndrome type 2 and novel mutations in AP3B1. Haematologica 2010;95:333–337.

[258] White JG, Clawson CC. The Chédiak-Higashi syndrome; the nature of the giant neutrophil granules and their interactions with cytoplasm and foreign particulates. I. Progressive enlargement of the massive inclusions in mature neutrophils. II. Manifestations of cytoplasmic injury and sequestration. III. Interactions between giant organelles and foreign particulates. Am J Pathol 1980;98:151–196.

[259] Nunes P, Demaurex N, Dinauer MC. Regulation of the NADPH oxidase and associated ion fluxes during phagocytosis. Traffic 2013;14:1118–1131.

[260] Mantegazza AR, Savina A, Vermeulen M, Perez L, Geffner J, et al. NADPH oxidase controls phagosomal pH and antigen cross-presentation in human dendritic cells. Blood 2008;112:4712–4722.

[261] Jancic C, Savina A, Wasmeier C, Tolmachova T, El-Benna J, et al. Rab27a regulates phagosomal pH and NADPH oxidase recruitment to dendritic cell phagosomes. Nature Cell Biol 2007;9:367–378.

[262] Iwasaki A. A virological view of innate immune recognition. Annu Rev Microbiol 2012;66:177–196.

[263] Kawai T, Akira S. The role of pattern-recognition receptors in innate immunity: update on Toll-like receptors. Nat Immunol 2010;11:373–384.

[264] Blasius AL, Arnold CN, Georgel P, Rutschmann S, Xia Y, et al. Slc15a4, AP-3, and Hermansky–Pudlak syndrome proteins are required for Toll-like receptor signaling in plasmacytoid dendritic cells. Proc Natl Acad Sci U S A 2010;107:19973–19978.

[265] Nakamura N, Lill JR, Phung Q, Jiang Z, Bakalarski C, et al. Endosomes are specialized platforms for bacterial sensing and NOD2 signalling. Nature 2014;509:240–244.

[266] Cresswell P. Assembly, transport, and function of MHC class II molecules. Annu Rev Immunol 1994;12:259–293.

[267] Peters PJ, Neefjes JJ, Oorschot V, Ploegh HL, Geuze HJ. Segregation of MHC class II molecules from MHC class I molecules in the Golgi complex for transport to lysosomal compartments. Nature 1991;349:669–676.

[268] Kleijmeer MJ, Ossevoort MA, van Veen CJ, van Hellemond JJ, Neefjes JJ, et al. MHC class II compartments and the kinetics of antigen presentation in activated mouse spleen dendritic cells. J Immunol 1995;154:5715–5724.

[269] Nijman HW, Kleijmeer MJ, Ossevoort MA, Oorschot VM, Vierboom MP, et al. Antigen capture and major histocompatibility class II compartments of freshly isolated and cultured human blood dendritic cells. J Exp Med 1995;182:163–174.

[270] Sanderson F, Kleijmeer MJ, Kelly A, Verwoerd D, Tulp A, et al. Accumulation of HLA-DM, a regulator of antigen presentation, in MHC class-II compartments. Science 1994;266:1566–1569.

[271] Liljedahl M, Kuwana T, Fung-Leung W-P, Jackson MR, Peterson PA, et al. HLA-DO is a lysosomal resident which requires association with HLA-DM for efficient intracellular transport. EMBO J 1996;15:4817–4824.

[272] Calafat J, Nijenhuis M, Janssen H, Tulp A, Dusseljee S, et al. Major histocompatibility complex class II molecules induce the formation of endocytic MIIC-like structures. J Cell Biol 1994;126:967–977.

[273] Vyas JM, Van der Veen AG, Ploegh HL. The known unknowns of antigen processing and presentation. Nat Rev Immunol 2008;8:607–618.

[274] Chow A, Toomre D, Garrett W, Mellman I. Dendritic cell maturation triggers retrograde MHC class II transport from lysosomes to the plasma membrane. Nature 2002;418:988–994.

[275] Boes M, Cerny J, Massol R, Op den Brouw M, Kirchhausen T, et al. T-cell engagement of dendritic cells rapidly rearranges MHC class II transport. Nature 2002;418:983–988.

[276] Kleijmeer M, Ramm G, Schuurhuis D, Griffith J, Rescigno M, et al. Reorganization of multivesicular bodies regulates MHC class II antigen presentation by dendritic cells. J Cell Biol 2001;155:53–64.

[277] Turley SJ, Inaba K, Garrett WS, Ebersold M, Unternaehrer J, et al. Transport of peptide-MHC class II complexes in developing dendritic cells. Science 2000;288:522–527.

[278] Faigle W, Raposo G, Tenza D, Pinet V, Vogt AB, et al. Deficient peptide loading and MHC class II endosomal sorting in a human genetic immunodeficiency disease: the Chediak–Higashi syndrome. J Cell Biol 1998;141:1121–1134.

[279] Caplan S, Dell'Angelica EC, Gahl WA, Bonifacino JS. Trafficking of major histocompatibility complex class II molecules in human B-lymphoblasts deficient in the AP-3 adaptor complex. Immunol Lett 2000;72:113–117.

[280] Sevilla LM, Richter SS, Miller J. Intracellular transport of MHC class II and associated invariant chain in antigen presenting cells from AP-3-deficient mocha mice. Cell Immunol 2001;210:143–153.

[281] Sugita M, Cao X, Watts GFM, Rogers RA, Bonifacino JS, et al. Failure of trafficking and antigen presentation by CD1 in AP-3-deficient cells. Immunity 2002;16:697–706.

[282] Briken V, Jackman RM, Dasgupta S, Höning S, Porcelli SA. Intracellular trafficking pathway of newly synthesized CD1b molecules. EMBO J 2002;21:825–834.

[283] Elewaut D, Lawton AP, Nagarajan NA, Maverakis E, Khurana A, et al. The adaptor protein AP-3 is required for CD1d-mediated antigen presentation of glycosphingolipids and development of Vα14i NKT cells. J Exp Med 2003;198:1133–1146.

[284] Fairn GD, Grinstein S. How nascent phagosomes mature to become phagolysosomes. Trends Cell Biol 2012;33:397–405.

[285] Kinchen JM, Ravichandran KS. Phagosome maturation: going through the acid test. Nat Rev Mol Cell Biol 2008;9:781–795.

[286] Blander JM, Medzhitov R. On regulation of phagosome maturation and antigen presentation. Nat Immunol 2007;7:1029–1035.

[287] Mantegazza AR, Zajac AL, Twelvetrees A, Holzbaur ELF, Amigorena S, et al. TLR-dependent phagosome tubulation in dendritic cells promotes phagosome cross-talk to optimize MHC-II antigen presentation. Proc Natl Acad Sci U S A 2014;111: 15508–15513.

[288] Mantegazza AR, Guttentag SH, El-Benna J, Sasai M, Iwasaki A, et al. Adaptor Protein-3 in dendritic cells facilitates phagosomal Toll-like receptor signaling and antigen presentation to CD4+ T cells. Immunity 2012;36:782–794.

[289] Klionsky DJ, Codogno P. The mechanism and physiological function of macroautophagy. J Innate Immun 2013;5:427–433.

[290] Yang Z, Klionsky DJ. Eaten alive: a history of macroautophagy. Nat Cell Biol 2010; 12:814–822.

[291] Nimmerjahn F, Milosevic S, Behrends U, Jaffee EM, Pardoll DM, et al. Major histocompatibility complex class II-restricted presentation of a cytosolic antigen by autophagy. Eur J Immunol 2003;33:1250–1259.

[292] Paludan C, Schmid D, Landthaler M, Vockerodt M, Kube D, et al. Endogenous MHC class II processing of a viral nuclear antigen after autophagy. Science 2005;307: 593–596.

[293] Lee J-A, Beigneux A, Ahmad ST, Young SG, Gao F-B. ESCRT-III dysfunction causes autophagosome accumulation and neurodegeneration. Curr Biol 2007;17:1561–1567.

[294] Rusten TE, Vaccari T, Lindmo K, Rodahl LMW, Nezis IP, et al. ESCRTs and Fab1 regulate distinct steps of autophagy. Curr Biol 2007;17:1817–1825.

[295] Rusten TE, Simonsen A. ESCRT functions in autophagy and associated disease. Cell Cycle 2008;7:1166–1172.

[296] Jiang P, Nishimura T, Sakamaki Y, Itakura E, Hatta T, et al. The HOPS complex mediates autophagosome-lysosome fusion through interaction with syntaxin 17. Mol Biol Cell 2014;25:1327–1337.

[297] Liang C, Lee JS, Inn KS, Gack MU, Li Q, et al. Beclin1-binding UVRAG targets the class C Vps complex to coordinate autophagosome maturation and endocytic trafficking. Nat Cell Biol 2008;10:776–787.

[298] Ganesan AK, Ho H, Bodemann B, Petersen S, Aruri J, et al. Genome-wide siRNA-based functional genomics of pigmentation identifies novel genes and pathways that impact melanogenesis in human cells. PLoS Genet 2008;4:e1000298.

[299] Ho H, Kapadia R, Al-Tahan S, Ahmad S, Ganesan AK. WIPI1 coordinates melanogenic gene transcription and melanosome formation via TORC1 inhibition. J Biol Chem 2011;286:12509–12523.

[300] Sanjuan MA, Dillon CP, Tait SW, Moshiach S, Dorsey F, et al. Toll-like receptor signalling in macrophages links the autophagy pathway to phagocytosis. Nature 2007; 450:1253–1257.

[301] Lee HK, Mattei LM, Steinberg BE, Alberts P, Lee YH, et al. *In vivo* requirement for Atg5 in antigen presentation by dendritic cells. Immunity 2010;32:227–239.

[302] Torisu T, Torisu K, Lee IH, Liu J, Malide D, et al. Autophagy regulates endothelial cell processing, maturation and secretion of von Willebrand factor. Nat Med 2013;19:1281–1287.

[303] Masedunskas A, Sramkova M, Parente L, Sales KU, Amornphimoltham P, et al. Role for the actomyosin complex in regulated exocytosis revealed by intravital microscopy. Proc Natl Acad Sci U S A 2011;108:13552–13557.

13

AUTOPHAGY INHIBITION AS A STRATEGY FOR CANCER THERAPY

XIAOHONG MA, SHENGFU PIAO, QUENTIN MCAFEE, AND RAVI K. AMARAVADI

Department of Medicine, University of Pennsylvania, Philadelphia, PA, USA

Within the cell, protein degradation is carried out by at least two mechanisms, the proteasome and lysosome. For the degradation of larger molecules and organelles, lysosome-dependent autophagy is the only mechanism available. Our understanding of the role autophagy plays in disease pathology is reaching the point where therapeutic modulation of autophagy in a number of diseases may be possible. This chapter focuses on targeting lysosomal function to block autophagy in hopes of augmenting the effectiveness of cancer therapy.

Autophagy can be subdivided into at least three forms: macroautophagy, microautophagy, and chaperone-mediated autophagy. Each defined on the basis by the way of the degradative material delivered to the lysosome [1–4]. Microautophagy is a nonselective process that involves direct engulfment of cytoplasmic cargo by the lysosomal membrane [5,6], while chaperone-mediated autophagy is selective for specific KFERQ-containing proteins through their interaction with heat shock cognate 70 (HSC70) and lysosome-associated membrane glycoprotein 2A (LAMP2A) [7,8]. Both microautophagy and chaperone-mediated autophagy directly deliver cytoplasmic components to the lysosomal compartment. In contrast, macroautophagy is the most multifunctional and well-studied form of autophagy and consists of the generation of double-membraned autophagic vesicles known as autophagosomes, which

Lysosomes: Biology, Diseases, and Therapeutics, First Edition.
Edited by Frederick R. Maxfield, James M. Willard and Shuyan Lu.

sequester cytoplasmic material before delivering it to the lysosome for degradation. For this reason, we refer to macroautophagy hereafter as autophagy.

Autophagy can also be described as canonical autophagy and noncanonical autophagy. The molecular machinery of canonical autophagy is directed by a number of autophagy-related (*ATG*) genes (see Table 13.1) as well as non-ATG proteins such as phosphatidylinositol 3-kinase (PtdIns3KC3), p150 and activating molecule in Beclin 1-regulated autophagy (AMBRA1). Although far less well understood, noncanonical autophagy can occur in an *ATG7*- or *ATG5*-deficient background [32] through the unconventional biogenesis of autophagosomes [33]. The observation of noncanonical autophagy complicates that view of autophagy as a druggable target for therapy in human disease, since specific modulation of a canonical autophagy protein may be compensated for by an increase in the activity of noncanonical autophagy. One common denominator for all of the autophagy programs is that ultimately they rely on lysosomal degradation.

Autophagy plays a critical role in a number of disease states, so it might make sense to either induce or inhibit autophagy to treat disease depending on the cellular context. While there is some evidence that autophagic cell death plays a role in specific human disease contexts, more studies have provided strong evidence indicating autophagy can increase the survival of both normal and malignant cells under stressful conditions both *in vitro* and *in vivo* [34–36]. The critical role of autophagy in the survival of cancer cells is further underscored by its cell survival role in other biological contexts. For instance, during organismal development, autophagy is needed for differentiation, development, and survival [37]. The autophagy gene *BECN1* (Beclin 1) on chromosome 17q21, a human homologue of *ATG6/VPS30* in yeast, is essential for early embryonic development. Homozygous deletion of *Becn1* in mice results in early embryonic lethality, whereas monoallelic deletion (*Becn1* +/−) reproduces a commonly observed genotype in human cancers and results in the development of spontaneous malignancy in mice [38]. *ATG5* and *ATG7* are also essential autophagy genes; complete knockout of either *Atg5* or *Atg7* in mice have indicated that these genes are important for mammalian embryogenesis and survival during the early neonatal starvation period [39–41] (see Chapter 5). Defective autophagy has been implicated in many pathological conditions including neurodegeneration, cancer, heart disease, and infectious disease [42–45].

Based on this potential role of autophagy as a survival pathway in many cell contexts including advanced cancer cells, our group and others have focused on inhibiting autophagy in the context of cancer therapy. This chapter first provides a detailed description of the steps and stages of autophagy and the cellular regulators of autophagy. Next, the evidence that autophagy is a tumor suppressor and tumor promoting pathway in cancer cells depending on the cellular context is reviewed. Finally, chemical and pharmacological inhibitors of autophagy, and their use in laboratory experiments and early-phase clinical trials in cancer patients are reviewed. We focus especially on lysosomal autophagy inhibitors including antimalarial aminoquinolines.

TABLE 13.1 Yeast Autophagy Genes and Mammalian Homologs

Yeast	Mammal	Major Function	References
Atg1	Ulk1 and Ulk 2	Serine/threonine protein kinase	[9,10]
Atg13	Atg13	ULK substrate	[9,10]
	Atg101	Atg13-binding protein	[11]
Atg17	FIP200 (RB1CC1)	Modulates response of autophagy	[9]
Atg24 (Snx4)		PtdIns(3)P-binding protein	[12]
Atg6 (Vps30)	Beclin 1	Component of PI3K complex	[13]
Atg14	Atg14L	Component of PI3K complex	[14,15]
Vps34	PIK3C3	Kinase	[13]
Vps15	PIK3R4	Component of PI3K complex	[13]
Vps38	UVRAG	Involved in activation and maturation	[15]
Atg18	WIPI proteins	Binds PtdIns(3)P and PtdIns(3,5)P_2	[16]
Atg21	WIPI proteins	Binds PtdIns(3)P and PtdIns(3,5)P_2	[17]
Atg8	MAPLC3, GABARAP, GATE-16, mATG8L	Ubiquitin-like conjugation system	[18,19]
	ATG3	Ubiquitin-conjugating enzyme (E2) analogue	[20]
	ATG4A ATG4B	Cysteine protease that processes Atg8/LC3	[21,22]
	ATG5	Component of the Atg12–Atg5–Atg16 complex	[23]
	ATG7	Ubiquitin-activating (E1) enzyme	[24]
	ATG10	Similar to the E2 ubiquitin-conjugating enzyme	[25]
	ATG12	Ubiquitin-like protein	[23]
Atg16	ATG16L1 ATG16L2	Component of the Atg12–Atg5–Atg16 complex	[26]
Atg9	mAtg9A and ATG9B	Transmembrane protein, serves as a lipid carrier	[27,28]
Atg12		Ubiquitin-like protein	[23]
Ypt1	Rab1	Required for localization of Atg8 to thc PAS	[29]
Ypt7	Rab7	Small GTP-binding protein	[30]
Sec18	NSF	ATPase responsible for SNARE disassembly	[31]

Abbreviations: Atg, autophagy related; ULK, Unc51-like kinase; FIP200, focal adhesion kinase (FAK) family interacting protein of 200 kDa; RB1CC1, retinoblastoma 1-inducible coiled coil-1; Beclin 1. Bcl-2 interacting myosin/moesin-like coiled-coil protein 1; Vps, vacuolar protein sorting; UVRAG, UV irradiation resistance-associated gene; WIPI, WD repeat protein interacting with phosphoinositides; GATE-16, Golgi-associated ATPase enhancer of 16 kDa/GABARAPL2; GABARAP, gamma-aminobutyric acid receptor-associated protein; LC3, microtubule-associated protein 1 light chain 3; TSC1/2, tuberous sclerosis complex 1/2; Ypt, yeast protein; PE, phosphatidylethanolamine.

13.1 STAGES AND STEPS OF AUTOPHAGY

Autophagy consists of two stages. The proximal stage is characterized by the formation of autophagosome (Steps 1–3 below), and the distal stage is defined by the formation of autolysosome and degradation of vesicle contents (Steps 4–5 below). At present, the proximal stage of autophagy is relatively well studied, whereas the molecular mechanisms controlling the distal stage of autophagy remain poorly understood. The whole process of autophagy proceeds through the following sequential steps (Figure 13.1):

Step 1, Initiation: The initiation of autophagy begins by activation of the Atg1 complex, also known as the ULK complex, which is composed of the mammalian Atg1 homologue Unc-51-like kinases 1 or 2 (mammalian Atg1 homologue Unc-51-like kinases 1 or 2, ULK1 and ULK2), the mammalian autophagy-related 13 homologue (ATG13). 5′ adenosine monophosphate-activated protein kinase (AMPK), and mammalian target of rapamycin (mTOR) regulate autophagy through direct phosphorylation of ULK1 [9,10]. The tight regulation of ULK activity by intracellular energy and nutrient levels is critical for autophagy initiation, and an activated ULK complex induces autophagy by phosphorylating Beclin 1 and activating Vps34, a class III PtdIns3KC3 [9–11,46].

Step 2, Nucleation: Nucleation of membrane that eventually becomes the autophagic vesicle depends on the Beclin 1–Vps34–X complex, where X can be ATG14L, Rubicon, Ambra, or others depending on the cellular and subcellular context. Stimulation of this complex generates phosphatidylinositol-3-phosphate (PI3P), which promotes autophagosomal membrane nucleation arising from various sources in the cell, including plasma membrane [47], mitochondria [48], endoplasmic reticulum (ER) [49], or the Golgi apparatus [32]. ATG14L can directly interact with Beclin 1 and enhance VPS34 activity to induce autophagosome membrane nucleation [50].

Step 3, Elongation and maturation: As the autophagic vesicle forms, elongation and maturation require two ubiquitin-like conjugation systems. ATG12 is conjugated to ATG5 by ATG7 (an E1-like ligase) and ATG10 (an E2-like ligase) enzymes. ATG5-ATG12 forms a complex with ATG16L, which participates in maturation of the autophagosome. The second ubiquitylation-like reaction involves the conjugation of the ubiquitin-like protein microtubule-associated protein1 light chain 3 (LC3-I) to the lipid phosphatidylethanolamine (PE) to form membrane-associated LC3-II through ATG7 and ATG3 (an E2-like ligase). LC3-1 is generated from the cleavage of the precursor forms of LC3 by the cysteine protease ATG4. ATG5-ATG12 conjugation depends on Vps34 function, and is positively regulated by the small GTPase Rab5 [51]. ATG4 has a critical function by both processing LC3 precursors and deconjugating LC3-PE. Once LC3 is integrated into the bilayer, it recruits cargo adaptor proteins (also known as autophagy receptors), such as p62, Nbr1, or NIX. These proteins, in turn, recruit cargo from the cytoplasm (i.e., ubiquitinated protein aggregates in the case of p62 and damaged organelles in the case of NIX) to promote AV closure and maturation.

Step 4, Fusion and degradation: Once autophagic vesicles are formed and cargo is engulfed the next step is fusion of the autophagosome with the lysosome. This fusion event involves the trafficking of autophagosomes along microtubules and fusion with the lysosome, forming the autolysosome. This process is not well understood. What is known is that it is regulated by lysosomal membrane proteins such as LAMP2, small GTPases such as Rab7, adaptor proteins such as endosomal sorting complex required for transport (ESCRT) [11,52] and other vesicular structures such as multivesicular bodies (MVBs) [52,53]. Within the acidic environment of the autophagolysosome, the cytosolic cargo is degraded via pH-dependent hydrolases.

Step 5, Recycling and autophagic lysosome reformation: The breakdown products of degradation within the autolysosome that include sugars, amino acids, and nucleic acids are likely released back into the cytosol through permeases such as Spinster [54]. AV components not exposed to lysosomal hydrolases are recycled via a system involving multiple components of the outer membrane such as ATG9, ATG2, ATG18, and ATG21. Alternatively, autophagosomes may also fuse with the plasma membrane and release their contents [55]. The fusion of lysosomes and autophagosomes rapidly leads to depletion of free lysosomes and thus a cellular mechanism is required for maintaining lysosome homeostasis. Recent evidence indicates the presence of a lysosomal recycling mechanism after cargo degradation within the autolysosome [56].

13.2 INDUCTION OF AUTOPHAGY

Autophagy can be induced by many upstream signaling cascades. For instance, as mentioned, mTORC1 [57] is a major controller of autophagy, and the intimate connection between the lysosome and mTORC1 is being more appreciated. The mTOR protein resides in mTORC1, which contains the regulatory-associated protein of mTOR (Raptor), G protein beta subunit-like protein (GβL), and proline-rich AKT/PKB substrate 40 kDa (PRAS40), which can regulate the mTOR substrate, the ULK1/2 complex. In response to abundant nutrient supply and/or growth factor activation, mTORC1 inhibits autophagy through inhibiting ULK1. Class I PI3K activates mTOR, and therefore inhibits autophagy [58]. Rapamycin inhibits mTORC1 and is therefore a powerful activator of autophagy. The ULK1 kinase complex receives signals from the mTORC1 and controls the initiation of autophagy. Once mTORC1 kinase activity is inhibited, the ULK1 complex is activated. MTORC1 is bound to lysosomal membranes through the Ragulator complex [59]. MTORC1 has associated proteins that act as sensors of amino acid concentrations in the cytoplasm

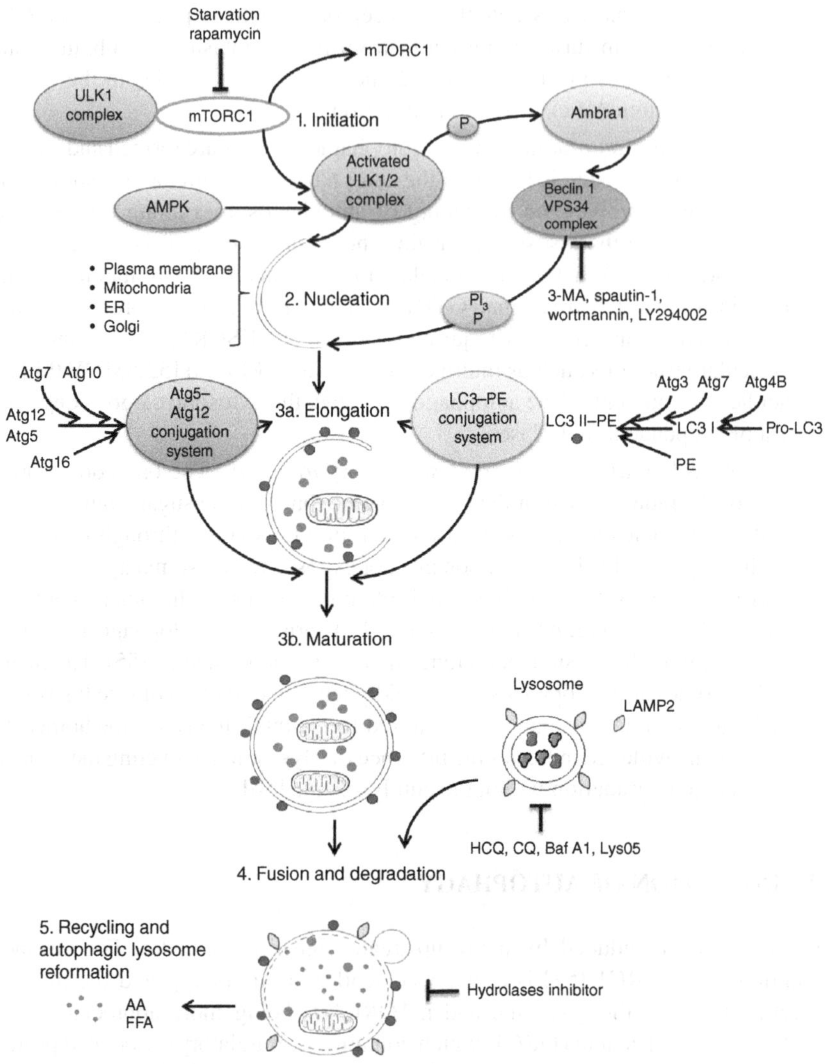

Figure 13.1 Molecular signaling of macroautophagy.

as well as within the lysosomal lumen. This sensing mechanism is dependent on the lysosomal H+-vacuolar ATPase [60]. Finally, mTOR signaling and MAPK signaling coordinately regulate lysosomal biogenesis through control of key transcription factors, such as transcription factor EB (TFEB) [61]. AMPK is another key regulator of autophagy, and can do so by either downregulating mTORC1 signaling or through direct phosphorylation and disinhibition of the ULK1 complex (see Section 13.6). Other pathways that regulate autophagy include the unfolded protein response (UPR)/endoplasmic reticulum stress (ER stress) response [62,63] and the regulation of autophagy through DNA damage and p53 at the transcriptional level. For instance, the ER stress response gene activating transcription factor 4 (ATF4) induces LC3 gene expression [64], and p53 induces transcription of multiple ATG genes [65]. Throughout the process of autophagy, the lysosomal positioning tightly regulates mTORC1 signaling and in turn regulates other signaling [66,67], which profoundly influences autophagosome formation and autophagosome–lysosome fusion, as well as autophagic flux by acting both at the proximal stage and the distal stages of autophagy [68]. The translational significance of these molecular connections is that therapeutic modulation in cancer with almost any drug will affect one or more of these pathways upstream of autophagy. Therefore, the majority of cancer therapeutics tend to activate autophagy [69–73].

13.3 STUDIES IN MOUSE MODELS UNRAVEL THE DUAL ROLES OF AUTOPHAGY IN TUMOR BIOLOGY

Autophagy plays a complex and controversial role in oncogenesis and tumor progression. The dual function of autophagy in cancer, as both a tumor suppressor mechanism and mechanism of cell survival, has been widely recognized. Autophagy levels within tumor cells and within cells within the tumor microenvironment may have a different influence on early tumorigenesis, progression of established cancer, and the impact of therapeutic interventions.

On the one hand, autophagy is a degradative process that was originally designated as type II programmed cell death [74]. Autophagy, if elevated or persistent can lead to autophagic, apoptotic, or necrotic cell death due to depletion of cellular components [75]. Autophagy can limit tumorigenesis through the elimination of damaged proteins and dysfunctional organelles thus mitigating oxidative stress and genomic instability that is characteristic of cancer cells [76]. These are, in general, arguments that support the notion that autophagy should be promoted and not inhibited in cancer cells.

In contrast, it is becoming clear that in later stages of tumor progression, autophagy's role as a garbage disposal and recycling plant enhance the survival of cancer cells and engender therapeutic resistance to cancer therapies. In established tumors, genetic suppression of autophagy significantly diminishes the ability of cancer cells to withstand starvation and increases cell death in response to metabolic stress [35].

In mice, homozygous deletion of Beclin 1 results in embryonic lethality, whereas monoallelic loss of *Becn1* (*Becn1* +/−) leads to spontaneous tumorigenesis, identifying Beclin 1 as a haploinsufficient tumor suppressor protein [38,77]. *BECN1* is monoallelically deleted in 40–75% of sporadic human breast cancers and ovarian cancers [78,79]. To date, *Becn1* heterozygote mice are the only "autophagy-deficient" mice that develop spontaneous malignant tumors. It is important to understand that *Becn1* heterozygote mice also have concurrently low levels of the tumor suppressor p53, due to their common regulation by specific deubiquitinases [80]. It is also important to note that the retained allele of *Becn1* in *Becn1* +/− cells is always wild type, and the protein expression of Beclin in *Becn1* +/− cells can be equal to *Becn1* +/+ cells if cells are stressed. Therefore, the relative "autophagy deficiency of the *Becn1* +/− model needs to be evaluated further. It is likely for this reason and others that many studies have instead used genetic inhibition of *Atg7* or *Atg5* to render cells "autophagy deficient." Given the possibility of a noncanonical autophagy program that is independent of Beclin, ATG5 or ATG7, there is concern that no model system involving proliferating cells can be truly and completely autophagy deficient.

Despite its role in preventing early tumor development, once tumors are established, tumor cell autophagy-related survival function can lead to tumor dormancy, progression, and therapeutic resistance [81]. Cancer cells have high metabolic demands due to increased cellular proliferation and a challenging microenvironment. One of the first demonstrations of autophagy's potential role as a cell survival mechanism in cancer cells came from studying the effects of growth factor withdrawal in growth factor–dependent cancer cells defective in the apoptotic proteins *Bax* and *Bak*. Growth factor withdrawal in *Bax/Bak*-deficient cells resulted in growth arrest and cell death, but even after many weeks in culture, a subpopulation of cells survived that were highly autophagic. If the growth factor was added back to the medium, these cells would resume exponential growth. If the lysosomal inhibitor chloroquine was added to cells undergoing autophagy, the cells would die providing clear evidence that autophagy was promoting survival in these deprived cells [34].

These findings have been explored further in animal models. It is becoming clear that cancers driven by mutations in the MAPK pathway are often "addicted" to autophagy. Cancer cells need high levels of autophagy to maintain mitochondrial function for tumorigenesis and tumor growth in *Kras*-driven [82] and *Braf*V600E-induced [83,84] non-small-cell lung cancer mouse models. Lack of *Atg5* or *Atg7* in *Kras*-driven pancreatic cancer cells causes accumulation of dysfunctional mitochondria upon nutrient starvation, impairing tumor growth. Autophagy deficiency also impairs the tumorigenicity of *KRAS*-transformed cells and human cancer cell lines with KRAS activation [82,85,86]. Similarly, autophagy can oppose p53-mediated tumor suppression function and promote breast cancer induced by partner and localizer of Brca2 (*Palb2*) loss [87]. Increased basal levels of autophagy were detected in human pancreatic cancer cell lines and tumor specimens, and autophagy can promote tumor cell growth by maintaining cellular energy production. Inhibition of autophagy resulted in tumor regression and extended the life span of nude mice harboring p53 mutant pancreatic xenografts [88]. In most of these mouse model studies across tumor types, lack of *Atg7* or *Atg5* accelerated early growth of premalignant cells but prevented or delayed the growth of aggressive and metastatic late stage tumors (Table 13.2). The recurring theme in these studies is that autophagy can be exploited by cancer cells to generate nutrients and energy during various stresses, including the stress of maintaining a rapidly proliferative cancer.

13.4 CLINICAL STUDIES ON AUTOPHAGY'S DUAL ROLE IN TUMORIGENESIS

Measurements of the expression of autophagy genes in cancer tissues and correlation to clinical outcomes have started to yield some important but conflicting results

TABLE 13.2 Mouse Model of Cancers with Genetically Manipulated Autophagy

Cancer Type	Genetic Background	Autophagy Gene Manipulated	Impact on Early Tumorigenesis	Impact on Established Tumor Growth	References
Lung	*Kras-mutant, p53* +/+	*Atg5* deletion	Accelerated	Delayed	[89]
Lung	*Kras-mutant, p53* −/−	*Atg5* deletion	Not examined	Accelerated	[89]
Lung	*Kras-mutant, p53* +/+	*Atg7* deletion		Delayed	[90]
Lung	*BRAFV600E*	*Atg7* deletion	Accelerated	Delayed	[83]
Pancreas	*Kras-mutant, p53* +/+	*Atg7* deletion *Atg5* deletion	Accelerated	Delayed	[91]
Pancreas	*Kras-mutant, p53* −/−	*Atg7* deletion *Atg5* deletion	Accelerated	Accelerated	[31]
Breast	*Palb2* −/−, *P53* +/+	*Becn1* monoallelic loss	Delayed	Delayed	[87]
Breast	*Palb2* −/−, *P53* −/−	*Becn1* monoallelic loss	No effect	Accelerated	[87]

on the role of autophagy in cancer. Numerous studies have looked at Beclin 1 expression and have come to conflicting views on the prognostic significance of Beclin 1 expression in cancer tissues [92–96]. One study identified elevated levels of LC3 puncta in the vast majority of advanced solid tumors [97]. High levels of autophagy in melanoma tumors predicted poor response to chemotherapy and shortened survival in melanoma [98]. On the other hand, one study has found that promoter methylation and silencing of *ATG5* was associated with more aggressive features of primary melanoma [99], supporting the role of autophagy as a tumor suppressor mechanism in early stages of malignancy. Additional work understanding the significance of autophagy levels in normal, premalignant, and tumor tissue as potential prognostic and predictive markers are sorely needed, but few methods that can be used to measure autophagy are available for clinical specimens [100].

13.5 MOUSE MODELS PROVIDE THE RATIONALE FOR AUTOPHAGY MODULATION IN THE CONTEXT OF CANCER THERAPY

The impact of autophagy inhibition, through RNAi targeting of essential autophagy genes, or pharmacological inhibitors, on cancer therapy has been evaluated in different tumor models, resulting in contrasting outcomes: prosurvival or prodeath, depending on the specific context. Broadly speaking, silencing of some essential autophagy genes, such as *ATG3*, *ATG4B*, *ATG5*, *Atg6/BECN1*, *ATG7*, *ATG10*, and *ATG12* has been found to increase cell death to a wide spectrum of therapeutic stresses (detailed in Table 13.3). These data indicate that autophagy is generally a survival mechanism for cancer cells especially for advanced tumor cells in response to inherent stress or treatment-associated genotoxic and metabolic stress [137,138], and considered a mechanism for acquired resistance of tumor cells to chemotherapeutic agents. These studies have largely been conducted *in vitro* and immunodeficient mouse models (xenograft tumors in nude mice). While most of these studies have found that genetic or pharmacological autophagy inhibition augments anticancer properties of a wide range of anticancer agents, more work is needed in immunocompetent mouse models. At least one study has found that genetic autophagy inhibition with *Atg7* knockdown blunted antitumor immunity elicited by chemotherapy in a syngeneic mouse flank tumor model [139]. In this paper, the investigators found a role for autophagy in releasing ATP from damaged cells, which serves as a key chemoattractant for cytotoxic T cells to enter the tumor microenvironment. If this is true, then the therapeutic strategy that should be pursued to augment existing cancer therapies is concurrent autophagy induction rather than autophagy inhibition. Other groups have found the opposite role for autophagy in tumor immunity. One group found that hypoxia-induced autophagy can increase the granzyme B degradation thus alleviating natural killer cell-mediated cell killing in breast cancer cell both *in vitro* and *in vivo*, while autophagy inhibition by suppressing Beclin 1 can restore granzyme B level *in vitro* and induce tumor regression by facilitating NK-mediated tumor cell killing [140].

TABLE 13.3 Preclinical Studies Supporting Autophagy Inhibition in Cancer Treatment

Treatment	Mechanism	Cancer Type	Agents Used to Block Autophagy	References
Irradiation	Genotoxic stress	Breast and colorectal	RNAi (Beclin 1, Atg3, Atg4B, Atg5)	[101]
Irradiation		Glioma	Bafilomycin A1(Baf A), RNAi(Beclin 1, Atg5)	[102]
Cisplatin	Genotoxic stress	Skin	3-MA, RNAi (Atg5)	[103]
Photodynamic therapy (photosensitizer:hypericin)	ER-directed oxidative stress	Fibrosarcoma	3-MA, RNAi (Atg5)	[104]
		Cervical	3-MA, RNAi (Atg5)	[105]
		Bladder carcinoma	RNAi (Atg5)	[106]
		Melanoma	RNAi (Atg5)	[106]
Tamoxifen	Hormone therapy	Breast	3-MA, RNAi (Beclin 1)	[107]
Androgen deprivation	Hormone therapy	Prostate	3-MA, RNAi (Beclin 1)	[108]
Imatinib	Tyrosine kinase inhibitor (TKI)	Chronic myeloid leukemia (CML)	Chloroquine (CQ), Baf A, RNAi (Atg5, Atg7)	[109]
		CML	3-MA, RNAi (Beclin 1, Atg7)	[110]
		Gastrointestinal stromal tumors (GISTs)	CQ, Quinacrine, RNAi (Atg7, Atg12, Lamp-2)	[111]
PI-103	Tyrosine kinase inhibitor	Glioma	3-MA, Baf A, monensin, RNAi (Lamp-2)	[112]
Saracatinib	Tyrosine kinase inhibitor	Prostate	3-MA, CQ, RNAi (Atg7)	[113]
Quercetin	Tyrosine kinase inhibitor	Gastric	CQ, RNAi (Beclin 1, Atg5)	[114]
Bortezomib	Proteasome inhibitor	Breast	RNAi [LC3B, ATF4, histone deacetylases (HDAC)]	[115]
Trastuzumab	Anti-HER2 monoclonal antibody	Breast	3-MA, Baf A, RNAi (LC3)	[116]
Suberoylanilidehydroxamic acid (SAHA)	HDAC inhibitor	CML	3-MA, CQ	[117]
5-Fluorouracil (5-FU)	Thymidylate synthase inhibitor	Colorectal	3-MA, RNAi (Atg7)	[118]

TABLE 13.3 ***(Continued)***

Treatment	Mechanism	Cancer Type	Agents Used to Block Autophagy	References
N-(4-hydroxyphenyl)retinamide (4-HPR)	Angiogenesis inhibitor	Glioma	3-MA, Baf A	[119]
Sulforaphane	NF-κB inhibitor	Breast	Baf A	[120]
		Colorectal	3-MA	[121]
Pegylated arginine deiminase (ADI-PEG20)	Arginine deprivation	Prostate	CQ, RNAi (Beclin 1)	[122]
Temozolomide	Genotoxic stress	Melanoma	HCQ, RNAi (Atg5)	[98]
PLX4720	BRAF inhibitor	Melanoma	Lys05	[123]
Camptothecin	DNA topoisomerase II inhibitor	Breast	3-MA, Baf A, RNAi (Beclin 1, Atg7)	[124]
Vorinostat	HDAC inhibitor	Colorectal	CQ , RNAi (Atg7)	[125]
5-FU	Thymidylate synthase inhibitor	Colorectal	3-MA	[126]
Cisplatin	Genotoxic stress	Esophageal	3-MA	[127]
Temozolomide	Genotoxic stress	Malignant glioma	3-MA, Baf A	[128]
Tamoxifen	Hormone therapy	Breast	3-MA, RNAi (Beclin 1, Atg5, Atg7)	[129]
Faslodex	Hormone therapy	Breast	RNAi (Beclin 1)	[130]
INNO-406	Tyrosine kinase inhibitor	CML	CQ	[131]
Imatinib	Tyrosine kinase inhibitor	CML	CQ	[132]
TPA	Megakaryocytic differentiation	CML	CQ	[132]
TRAIL	TNF receptor	Colorectal	3-MA, RNAi (Beclin 1, Atg5, Atg7, UVRAG)	[133]
TFT, 5-FU	Thymidylate synthase inhibitor	Colorectal	3-MA	[134]
Imatinib	Tyrosine kinase inhibitor	Malignant glioma	Bafilomycin A	[135]
FK228	HDAC inhibitor	Malignant rhabdoid tumor	CQ	[136]

Abbreviations: 3MA, 3-methyladenine; CQ, chloroquine; HCQ, hydroxychloroquine.

13.6 MULTIPLE DRUGGABLE TARGETS IN THE AUTOPHAGY PATHWAY

There are multiple druggable targets in the autophagy pathways [141]. Autophagy initiation is controlled by ULK1 complex, consisting of ULK1 or ULK2, FIP200, ATG101, and ATG13. Activated mTORC1 inhibits autophagy by causing hyperphosphorylation of ATG13, reducing its interaction with ATG1/ULK1, and by controlling phosphorylation of autophagy effectors such as the Beclin 1–Vps34 complex. The kinase domain of ULK1 and ULK2 could be attractive targets for future drug development. Proteomic studies investigating how inhibition of the mTORC1 pathway affects the global features of autophagy control showed no large-scale changes in core conjugation, lipid kinase, and recycling complexes. This finding implies that post-translational modifications may be involved in AV accumulation when the autophagy pathway is activated and may be a potential means to control autophagy [142].

AV nucleation represents a second major autophagy control point, involving Vps34 and interacting partners Beclin 1 and p150. Beclin has become a candidate for drug development for both autophagy inhibitors and autophagy promoters. As described below, the specific and potent autophagy inhibitor (Spautin, described below) was found to impact Beclin levels [80]. A novel fusion peptide called tat-Beclin 1 was found to specifically activate Beclin [143,144] and could eventually serve as a new class of specific autophagy inducers. Vps34 is a class III lipid kinase, and existing chemical platforms have already identified numerous candidate inhibitors of Vps34. For example, drugs that interfere with recruitment of Vps34 to membranes, including wortmannin and 3-methyladenine, are powerful proximal inhibitors of autophagy.

ATG8 family proteins are central coordinators of proper AV membrane development [145] and maturation and represent a third autophagy control point. LC3, the most widely studied ATG8 family member, is cleaved by ATG4 and conjugated to PE by an ATG7- and ATG3-dependent activation and transfer cascade. ATG4 is a cysteine protease, and efforts are underway to develop peptidomimetic inhibitors of its protease activity [146,147]. ATG7 is an E2-like ligase, and at least one pharmaceutical company is developing an ATG7 inhibitor [147]. LC3-PE conjugation is very important for AV elongation and maturation including the cargo recruitment by the cargo adaptor proteins, such as p62, Nbr1, and Nix, and the transfer of AV cargo to lysosomes.

The main concern with inhibiting the targets stated earlier (ULK1, ATG7, and ATG4) is that components of autophagy exhibit a high degree of redundancy. There are five ULK homologues and four ATG4 homologues [148]. While *ATG7* has no genetic redundancy, the description of a noncanonical autophagy program [32] in *Atg7* –/– and *Atg5* –/– cells suggests that autophagy is such a core function within the cell, that in the absence of canonical autophagy signaling, the cell may recruit other components of the endovesicular system to serve as the cargo collectors and delivery system needed to maintain a clean intracellular environment.

Delivery and degradation of AV contents represents a fourth autophagy control point. Because AVs and lysosomes move along microtubules, drugs that disrupt microtubules, such as nocodazole, colchicines, taxanes, and vinca alkaloids, inhibit AV fusion with lysosomes, resulting in AV accumulation. Rab GTPases likely play a role in vesicle maturation and fusion with lysosomes, but currently there are no known chemical approaches that have provided clinical leads to targeting Rab GTPases effectively. Lysosomes are acidic organelles, with their digesting hydrolases dependent on low pH. Consequently, agents such as bafilomycin A1 or chloroquine derivatives, which disrupt the vacuolar H-ATPase responsible for acidifying lysosomes, block autophagy in its final step, resulting in the accumulation of AVs. In the following section, the available tool compounds to inhibit autophagy in the laboratory are reviewed.

13.7 OVERVIEW OF PRECLINICAL AUTOPHAGY INHIBITORS AND EVIDENCE SUPPORTING COMBINATION WITH EXISTING AND NEW ANTICANCER AGENTS

In laboratory studies, the inhibition of autophagy can be accomplished with a number of tool compounds including 3-methyladenine (3-MA), selective and potent autophagy inhibitor 1 (Spautin-1), bafilomycin A (Baf A) and CQ/HCQ. Table 13.3 provides a partial list of preclinical studies that demonstrate that chemical autophagy inhibition enhances the efficacy of anticancer therapy. The literature has exploded with these reports, so it is not possible to review every study in detail. What is remarkable is that the synergy between autophagy inhibitors and anticancer agents spans a wide variety of histotypes and holds true with a wide variety of anticancer agents including alkylating agents, tyrosine kinase inhibitors, protease inhibitors, HDAC inhibitors, angiogenesis inhibitors, proteasome inhibitor, and hormonal therapy. One area of active research is to understand the mechanistic underpinnings of how some of the synergistic combinations manifest. For instance, melanomas that contains a point mutation in the *BRAF* proto-oncogene that encodes for BRAF V600E is well known for resistance to chemotherapies and immunomodulating therapies. PLX4032 is a selective BRAF inhibitor and, in clinical trials, has consistently induced tumor responses in the majority of patients with *BRAF* V600E mutant melanoma. However, duration of response has been limited due to the development of acquired resistance. From our recent study, we found that after PLX4032 treatment in melanoma cell lines, mutant BRAF bound the ER stress gatekeeper GRP78, and the disassociation of GRP78 from the PKR-like ER-kinase (PERK) promoted a PERK-dependent ER stress response. The ER stress subsequently activated autophagy, which is responsible for *survival* and drug resistance. Combined BRAF and autophagy inhibition promoted tumor regression in BRAFi-resistant xenografts. Our data provide a rationale for combinational approaches targeting the resistance pathway [123].

Understanding the mechanistic underpinnings of how cancer therapies initiate autophagy provides a roadmap to multiple possible cancer drug-autophagy inhibitor combinations. Much like our study in BRAF mutant melanoma, most of the autophagy inhibition studies carried out in Table 13.3 use a lysosomotropic chloroquine derivative to block autophagy distally as proof of principle, mainly because this approach is the most immediately translatable. One key point that needs further research is needed to determine if proximal autophagy inhibition would be more or less effective than distal autophagy inhibition. Toward the end, we next review the characteristics and proximal autophagy inhibitors briefly and provide a more extensive historical and mechanistic perspective on quinolines.

13.8 PROXIMAL AUTOPHAGY INHIBITORS

3-MA is a proximal autophagy inhibitor, which can prevent autophagic vesicle from forming by inhibiting multiple forms of PI3 kinases includingVps34, a key enzyme of the Beclin 1–Vps34 complex involved in the initiation of autophagosome formation [149]. Preclinical studies have shown that 3-MA when given at millimolar concentrations can enhance killing of cancer cells when combined with a variety of therapeutic stresses, including alkylating agents, tyrosine kinase inhibitors, hormone therapy, TNF-related apoptosis-inducing ligand (TRAIL), and photodynamic therapy (Table 13.3). Given the high concentrations needed for 3-MA to block autophagy and the promiscuity of the compound at these concentrations it is unlikely that a 3-MA derivative will ever be suitable for *in vivo* or clinical testing.

Spautin-1 is a newly developed small-molecule inhibitor of autophagy [80], which can promote the degradation of Beclin 1–Vps34 complexes through inhibiting the ubiquitin-specific peptidases (USP)10 and USP13 that are responsible for preventing Beclin 1 from undergoing proteasomal degradation. By inhibiting these USPs, Spautin, therefore, induces degradation of beclin and thus inhibits autophagy. Spautin-1 did not affect cell growth and survival of human breast cancer Bcap-37 cells under normal culture conditions but enhanced the cell death upon starvation. Similarly, spautin-1 could induce the death of the ovarian cancer ES-2 and OVCAR-3 cells in glucose-free condition [80]. Another study shows that in confluent conditions, in subclones of colon cancer cell line HCT16 engineered to harbor all 17 different cancer-associated mutant p53, spautin-1 treatment increased the degradation of the mutant p53 protein through increased chaperon-mediated autophagy (CMA) activity that was produced by effective inhibition of macroautophagy. In addition, depletion of mutant p53 expression due to macroautophagy inhibition sensitizes the death of dormant cancer cells under nonproliferating conditions [150]. Spautin-1 selectively induced cell death of mutant p53-expressing cancer cell lines under confluency but not in dispersed growth conditions. In complete medium with unstressed cells, no or little effect of spautin-1 on cell survival was detected in wild-type p53 or p53-null cancer cells. Recently, one study has shown that Spautin can enhance tyrosine kinase inhibitor imatinib-induced cell apoptosis in chronic myeloid leukemia [151].

13.9 QUINOLINES: FROM ANTIMALARIALS TO PROTOTYPICAL DISTAL AUTOPHAGY INHIBITORS

Chloroquine and related quinolines, the prototypical lysosomotropic autophagy inhibitors, were developed primarily as antimalarial drugs. Significant development of quinolines has taken place and continues to take place in the context of treating malaria. Observations that many quinoline antimalarials inhibit autophagy in mammalian cells have prompted development of the use of quinolines for treatment of cancer and investigation of the role of autophagy in other processes.

Development of quinoline antimalarials began in earnest with the synthesis of chloroquine in 1934. After the First World War, chloroquine, which was inexpensive and highly effective, became the dominant monotherapy used against malaria [152]. However, resistance to chloroquine was first reported in 1961 (reviewed in Ref. [153]) and quinoline resistance in general has become pervasive in *Plasmodium falciparum* [154,155].

The most cited mechanism of action of quinolines in regard to their antimalarial properties is that quinolines prevent the plasmodium species from successfully sequestering hemin, a toxic byproduct of the organism's digestion of hemoglobin. In the erythrocytic stage of its life cycle, plasmodium species phagocytize hemoglobin, which is delivered intracellularly to its acidic lysosome-like digestive vacuole, where hemoglobin undergoes proteolytical degradation liberating amino acids. Amino acids produced from the breakdown of hemoglobin are used as food, while the iron-containing free hemin (protoporphyrin IX) resulting from this breakdown poses a threat to the organism. Free hemin is prone to catalyzing Fenton reactions that generate reactive oxygen species, which in turn causes lipid peroxidation and bursting of the food vacuole [156]. Damage to the organism is prevented by the biomineralization of the abundant hemin molecules into a harmless insoluble crystal, hemozoin. Chloroquine and other quinolines are thought to interfere with this biomineralization process by interacting directly with free hemin, preventing sequestration of hemin in this crystal, leaving toxic quantities of hemin in solution [157]. While interference with hemin sequestration is considered to be the canonical mechanism of action of aminoquinolines, the precise detailed mechanisms by which quinolines inhibit hemin polymerization and promote the formation of reactive oxygen species remains unsettled [158–160].

For decades, quinolines have been clinically useful for the treatment of malaria, but since the 1960s [154], single agent quinolines for the treatment of malaria has fallen out of favor due to emergence of plasmodium strains that have mutations in the ABC-like drug efflux pump pfCRT (*P. falciparum* chloroquine resistance transporter) and pfMDR (*P. falciparum* multidrug resistance transporter) [155,161]. Both of these transporters remove quinolines from the organism's digestive vacuole, the main site of quinoline drug action. No other mutations in the parasite genome have been identified definitely to explain the resistance to quinolines.

One byproduct of the extensive investigation of the use of quinolines in malaria has been the investigation of the action of chloroquine on the vacuolar system in mammalian cells as it was beginning to be described in the 1960s through the 1970s [162,163]. Chloroquine has been used in mammalian cell culture experiments to improve DNA transfection efficiency [164], enhance antigen presentation for immunology research [165], and increase the expression of specific proteins that may be degraded through autophagy. For each of these experimental purposes, chloroquine's effects on autophagy and its intersection with each of the intended consequences was likely only partially appreciated, though the connection between chloroquine and mammalian autophagy was made by Fedorko all the way back in 1967 [166]. While an autophagy-like pathway with multiple conserved autophagy proteins (ATGs and others) exists in malaria [159], the common mechanism between chloroquine's activities in malaria and chloroquine's ability to inhibit autophagy in mammalian cells has not been established.

A defining mechanism of aminoquinolines is their ability to be concentrated within the lysosome by a process known as lysosomotropism or proton trapping [163,167]. This process is general, and any weak base that is hydrophobic at cytosolic pH and capable of diffusing into the lysosome will become concentrated there according to its permeability, pK_a, and the pHs of the two compartments. Chloroquine can be concentrated several hundred- to several thousand-fold in the lysosome or digestive vacuole [168–170] and the lysosomal compartment can represent the bulk of the cell's drug burden [171]. This partitioning can sequester compounds away from their site of action, as with daunorubicin [172], and more generally the effect is exploited by multidrug-resistant cancer cells to sequester toxic chemotherapeutics [173]. Alternatively, if a compound not normally sequestered in the lysosome is appended with basic amine groups, the compound will then be trapped within the lysosome [167,174]. Together, these effects confer the ability to target a compound to or away from the lysosome, or render a compound unable to reach its intended target. Because of the pivotal role of autophagy in multiple broadly diverse disorders and processes, especially cancer, and the measured successes of current upstream autophagy inhibitors, the development of inhibitors that specifically target the final and indispensable stage of autophagy – the lysosome – development of more effective lysosomal autophagy inhibitors is critical.

Though quinolines have been thoroughly investigated in malaria, less has been done in the context of understanding their mechanism of action as an autophagy inhibitor in cancer cells. Due to their availability and favorable pharmacology, numerous preclinical trials have established chloroquine and hydroxychloroquine as the first potential autophagy inhibitors that should be tested in clinical trials. Hydroxychloroquine, because of its safety and history of use in humans for treating maladies ranging from malaria to rheumatoid arthritis, has been the focus of many ongoing oncology clinical trials (reviewed in Section 13.10). However, despite the fact that micromolar concentrations of chloroquine are the most reproducible means of modulating autophagy *in vitro*, concerns have been raised that clinically achieving micromolar concentrations with currently available quinolines may prove to be difficult [175,176]. For this reason, more potent and specific inhibitors of autophagy that target the lysosome are sorely needed.

13.10 SUMMARY FOR THE CLINICAL TRIALS FOR CQ/HCQ

Currently there are six trials involving HCQ combinations, five in human cancer patients and one on dog cancer patients, that have completed enrollment and have reported outcomes (Table 13.4). In all mature clinical trials except one there have

TABLE 13.4 Completed Clinical Trials for HCQ

Trial	Therapeutic Agents	Pathway	Disease	Sites/Funding	DLT/MTD	N	Unique Features
1	Phase I/II temozolomide/ radiation + HCQ	DNA damage	Glioma	ABTC (15 centers), NCI	Myelo-suppression at 800 mg 600 mg MTD	92	Survival outcome. National trial
2	Phase I bortezomib + HCQ	Proteasome	Myeloma	Penn, Millenium	None MAD 1200 mg	30	Comparison of purified tumor cells with PBMC
3	Phase I temozolomide + HCQ	DNA damage	Solid tumors, melanoma	Penn, Merck	None MAD 1200 mg	40	High dose Chemo and HCQ well tolerated Clinical activity
4	Phase I temsirolimus + HCQ	mTOR	Solid tumors, melanoma	Penn, Pfizer	None MAD 1200 mg	35	Serial PET scans, tumor biopsies; clinical activity
5	Phase I vorinostat + HCQ	HDAC	Solid tumors, renal, colon	San Antonio, Merck, NCI	Grade 3 fatigue at 800 mg 600 mg MTD	24	Tumor biopsies; clinical activity
6	Phase I doxorubicin + HCQ	DNA damage	Dog lymphoma	Colorado State	Sepsis, death, 12.5 mg/kg MTD	27	100% clinical benefit

Abbreviations: HCQ, hydroxychloroquine; ABTC, Adult Brain Tumor Consortium; NCI, National Cancer Institute; DLT, dose-limiting toxicity; MTD, maximum tolerated dose; MAD, maximum administered dose.

been no recurrent dose-limiting toxicities identified and the maximal administered dose of 1200 mg HCQ per day in combination with different anticancer agents was easily achieved. In most trials, grade 2 anorexia, nausea, fatigue, and diarrhea were the most common side effects attributable to high-dose HCQ. Some of these clinical trials have produced encouraging signs of activity [177–180]. The most encouraging HCQ trial as far as antitumor activity is a phase I trial of temsirolimus and HCQ in advanced melanoma patients. We have observed a 74% stable disease rate in contrast to a 0% stable disease rate observed with temsirolimus alone in previous phase II trial [180]. Some of the conclusions that can be drawn from this series of papers are outlined as follows.

Autophagy modulation: These six clinical trials are the first deliberate attempt to modulate autophagy therapeutically on cancer or any disease. Each trial included a pharmacodynamic assay to determine if autophagy was indeed modulated by an HCQ-containing therapeutic regimen. The conclusion was that HCQ exposure in patients dictated whether or not autophagy inhibition could be detected in patient tissues using mainly an electron microscopy-based assay. Only patients with high HCQ exposure had detectable accumulation of autophagic vesicles. Even in these patients, the degree of therapy-associated autophagic vesicle accumulation was one- to twofold higher than the baseline sample. These studies indicate that HCQ can modulate autophagy in patients, and it is likely that more potent inhibitors are needed to block autophagy more profoundly.

Dosing and concentration achieved: The highest concentration of HCQ achieved in the blood was roughly 5000 ng/mL, which translates into 11.5 μM HCQ. Concentrations in tumor may be 10- to 100-fold much higher. In all clinical trials, HCQ concentration range is from 200 to 1200 mg per day. In the dog lymphoma trial [181], concentrations of HCQ were100-fold higher in the tumor than in the plasma. This sort of tissue concentration may be critical if HCQ is to have any future role in anticancer therapy.

Safety: The safety of adding HCQ to cancer regimens was established in multiple combinations and multiple malignancies. Further development of HCQ combinations is now possible, and novel autophagy inhibitors can enter clinical trials without an a priori assumption that blocking autophagy is too toxic. Some HCQ combinations produced dose-limiting toxicities while others did not. For instance, low-dose temozolomide without interruption [182] produced more toxicity than high-dose temozolomide with scheduled drug holidays [180] when combined with HCQ.

Efficacy: While clinical responses were not a primary endpoint in some the studies, there were a number of remarkable clinical responses and prolonged stable disease in patients with refractory cancers including myeloma, melanoma, renal cell carcinoma, and colon cancer. These results suggest that autophagy inhibition could be an active therapeutic approach in a number of cancers. Future preclinical studies should focus on identifying biomarkers that can predict sensitivity to autophagy inhibitors.

In a phase III randomized study of carmustine, radiation and low-dose chloroquine, (NCT00224978), CQ at 150 mg once daily combined with chemoradiation improved the median survival after surgery from 11 months (SOC alone) to 24 months [183]. These results were not statistically significant but were nevertheless provocative. However, in the adult brain tumor consortium phase I/II trial of temozolomide, radiation, and hydroxychloroquine, no significant improvement in survival was observed compared to historical controls. The lack of a survival benefit in our trial may be related to lack of effective and consistent autophagy inhibition in this population, as the dose of HCQ was lower, and the pharmacokinetic variability was high.

Biomarkers of autophagy modulation: Markers of autophagy inhibition and vesicle accumulation were consistently more pronounced and observed in tumor tissue than in PBMC. Therefore, EM-based measurement of autophagy in patient samples is feasible in a clinical research setting. However, better markers of tumor cell autophagy that can be clinically translated are sorely needed because following autophagy in easily accessible surrogate tissue does not fully reflect tumor autophagy. A significant pharmacokinetic–pharmacodynamic interaction was discovered in multiple trials that correlated HCQ exposure with autophagic vesicle accumulation in peripheral blood mononuclear cells. The degree of autophagy inhibition (AV accumulation by EM) achieved in most patients was usually one- to twofold over baseline. Therefore, we are intrigued to speculate that more potent autophagy inhibitors may produce more substantial clinical benefit. The pharmacokinetics of HCQ in cancer patients taking other cancer drugs demonstrated a higher degree of variability across studies than in previous studies conducted for single agent hydroxychloroquine in rheumatoid arthritis patients. The pharmacology of HCQ could hamper its utility as an effective anticancer agent and improved pharmacology could be a focus for future autophagy inhibitors.

Selection of patients: As is typical for a first-in-class inhibitor, the patients that participated in these first autophagy modulation studies were not preselected based on a biomarker of sensitivity. If an enriched patient population within a specific malignancy or based on a specific biomarker could be identified, and a significant incremental benefit of adding HCQ to existing cancer therapies is found in this subpopulation, incorporation of this cheap well-studied drug into the cancer pharmacopeia could have a remarkable impact on the burden of cancer care. This is especially true in the developing world, where cancer rates are exploding, and the expense of cancer therapy is poised to bankrupt economies.

13.11 DEVELOPING MORE POTENT ANTICANCER AUTOPHAGY INHIBITORS

While other groups are developing potent and specific inhibitors of upstream components of autophagy (as described earlier), further efforts to develop clinically useful drugs that more effectively impair lysosomal function and therefore autophagy is a focus of intense research. An underexploited avenue of elaboration on the quinoline structure is the use of polyvalent quinolines that include two or more quinoline rings with varying linkage, orientation, and substituents. The use of doubling or tripling structural motifs offers multiple advantages [184]. In the case of a target with multiple binding sites, multimerism may allow multiple sites to be bound with a single drug molecule, promoting a cooperative like binding action, but within the same molecule, which by virtue of multimerism is already present in close proximity. In the case of a target that is undefined, the use of a flexible linker can allow the two active moieties to conform the target without specific knowledge of how they should be positioned. In the malarial literature, examples of bisquinolines already exist and have showed some improvement over the activities of chloroquine in malaria [185–188]; however, no dimeric chloroquines have been developed for clinical use.

Recently, our group synthesized a series of dimeric chloroquine analogues to evaluate the basic requirements for inhibiting autophagy in mammalian cancer cells with a dimeric chloroquine molecule (Figure 13.2)[189]. Compounds synthesized were chosen to evaluate the activity of dimeric versus monomeric chloroquine compounds, the importance of the 4-chloro substituent, and the effect of the linker joining the quinoline rings, with HCQ serving as a reference compound to be improved upon. Lys01 was chosen as the first prototype to be investigated. Lys02 was designed to evaluate the linker present in Lys02 but not be dimeric. Lys03 was chosen to confirm the necessity of the 4-chloro group to the bisquinoline scaffold. Lys04 was chosen to evaluate the aminoalkyl linkage compared to a polyether linkage.

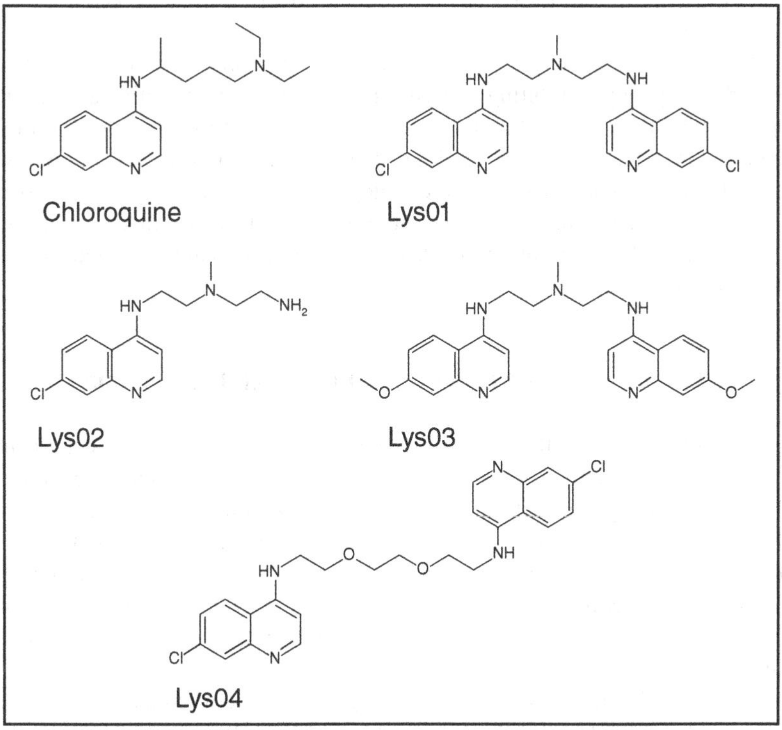

Figure 13.2 Chloroquine and Lys-series autophagy inhibitors.

When initially evaluated for ability to inhibit autophagy vial LC3-I/II immunoblotting and proliferation via MTT, all of the compounds (including hydroxychloroquine) were active at some concentration. However, Lys01, which combined dimeric 4-chloroquinoline groups linked by an aminoalkyl group, was significantly more active than all of the other compounds at inhibiting autophagy and proliferation in cancer cells. Lys01 showed an approximately 10-fold improvement in the ability to induce GFP-LC3 puncta formation than hydroxychloroquine. That the compound was a distal autophagy inhibitor through the lysosome was confirmed using electron microscopy and the bafilomycin clamp assay, and its ability to deacidify the lysosome and cause lysosomal membrane permeabilization was confirmed using LysoTracker red and acridine orange. Treatment of mouse xenografts with the soluble Lys01 Tris-HCl salt, Lys05 produced significantly greater antitumor activity compared to hydroxychloroquine and significant reduction in tumor growth at doses as low as 10 mg/kg over 14 days. In both treated tumor and cell fractions, Lys05 showed a significantly greater ability to concentrate in the lysosome. Higher doses of 80 mg/kg induced toxicity in the form of apparent colonic pseudo-obstruction and Paneth cell granule dysmorphia, a symptom set characteristic of Crohn's disease patients with inactivating polymorphisms in the essential upstream autophagy gene *ATG16L1*, as well as mice hypomorphic for *Atg16L1*mice [190,191]. That the phenotype of Lys05 treatment resembles the phenotype of a deficiency of an essential autophagy protein [192] shows that Lyso5 has a profound effect on the autophagic system at and beyond the lysosome.

Lys05 has shown that dimeric chloroquine compounds can have significantly improved autophagy inhibiting activity in cancer than the clinically available autophagy inhibitors and that there is significant room for the development of these compounds. Mechanistic understanding of quinoline action across kingdoms will continue facilitate understanding of the biology quinolines and their interaction with the endovesicular system. Further exploitation of the structural diversity of dimeric quinolines and allow for rationally designed and specifically targeted anticancer autophagy inhibitors.

13.12 SUMMARY, CONCLUSION, AND FUTURE DIRECTIONS

Drug resistance has been the major barrier for the treatment of metastatic cancers, and autophagy has emerged as a drug-resistance mechanism in clinic cancer therapies [123,141]. Autophagy inhibition through genetic modification or pharmacological inhibitors has been shown to sensitize cancer cells to different treatment modalities in both preclinical animal models and clinic trials. However, the role autophagy plays in cancer is complex and context-specific, our understanding of autophagy in tumorigenesis and in cancer therapy is still incomplete and some key questions remained unsolved.

Most cancer therapeutics induce autophagy, but the consequence of autophagy induction is not very clear, and the sensitivity of tumor cells in response to autophagy inhibition may depend on the genetic background and epigenetic modification.

A better understanding of the mechanisms of autophagy in tumorigenesis and how autophagy is regulated in tumor cells and inducted in cancer therapy can help identify which tumor types are more "addicted" to autophagy, and more accurate and informative biomarkers are wanted to properly monitor the tumor autophagy status and identify patients who can benefit from the autophagy inhibition treatment, and determine which drugs to combine with autophagy inhibition. More efforts are needed to understand the following aspects to better guide the autophagy inhibition in clinical cancer treatment:

1. The complex cross talk between autophagy and apoptosis, and other signals or pathways. The interplay between apoptosis and autophagy is quite complicated [193] and has not been well elucidated. Many signaling pathways induced by cell-intrinsic stress regulate both autophagy and apoptosis. For example, p53 can regulate both apoptosis and autophagy, and autophagy inhibition can have both proapoptotic and antiapoptotic effect, and Bcl-2 family members are dual regulators of apoptosis and autophagy, Bcl-2, Bcl-XL, Bcl-w, and Mcl-1 can not only inhibit apoptosis but also inhibit autophagy, while proapoptotic BH3-only proteins such as BNIP3, Bad, Noxa, and Puma can induce autophagy [194]. Autophagy primarily acts as a protective mechanism that may prevent cell death. There are complicated interaction between regulatory elements of autophagy and apoptosis, for example, the inhibitory interaction between BCL-2 and Beclin 1 and the interaction between LC3B and Fas. Similarly, better understanding of the cross talk between autophagy and ER stress, and the autophagy and DNA repair is also very helpful to target autophagy inhibition in cancer treatment.
2. The interaction of autophagy with immune response to cancer. A large amount of evidence has shown autophagy can profoundly influence the immune systems [195]. It is clear that autophagy plays a pivotal role in both the innate immune responses and adaptive immune responses such as antigen presentation and lymphocyte development [43,196–199]. Currently, most experiments have generally been performed in xenograft models, thereby eliminating the involvement of the immune system, while immunosurveillance plays very critical role in cancer prevention and therapy, and it may also play a critical role in determining effectiveness of autophagy inhibition in chemosensitization or radiosensitization. Autophagy can either increase anticancer immune responses induced by chemotherapeutic agents in mice [139] or decrease tumor cell susceptibility to NK cell-mediated lysis under hypoxia [140]. Moreover, chloroquine treatment improves CD8 immunity during vaccination [165]. Many unanswered questions remain to be tackled in the field of autophagy and immunity.
3. To develop specific and potent autophagy inhibitors. Currently, there are no clinically available specific autophagy inhibitors. All preclinical reports and clinical trials investigating pharmacological inhibition of autophagy have used chloroquine or hydroxychloroquine, which interfere with lysosomal function and block autophagy at a later stage. The capacity for sensitization by chloroquine appears to be quite wide-ranging, with dramatic effects for

some drugs/tumor models and modest or minimal effects in others. The pharmacology of HCQ could hamper its utility as an effective anticancer agent and improved pharmacology could be a focus for future autophagy inhibitors. More specific and potent autophagy inhibitors are critical for the clinical application of autophagy inhibition. However, autophagy inhibitors may be toxic to normal tissue because of the autophagy's cytoprotective role. Therefore, the side-effect profile of these compounds needs to be closely monitored. Besides inhibiting autophagy, CQ and HCQ may affect the tumor/cancer therapy through other mechanisms. It will be important to establish whether any anticancer activity of HCQ is due to autophagy impairment, as it may act by other mechanisms [200].

4. Identify better markers of tumor autophagy and biomarkers for patients' selection. Accurate and informative cancer biomarkers hold significant promise for improvements in the selection of the most effective therapeutic strategies. Better markers of tumor cell autophagy that can be clinically translated are sorely needed, because currently although Beclin 1 by immunohistochemistry as a measure of autophagy competence, and the measurement of AV number directly by electron microscopy, LC3, and p62 levels as markers of autophagy modulation, no autophagy marker fully reflects tumor autophagy. We need better ways to assess and monitor autophagy in human patient samples to determine which tumors are addicted to autophagy, and which tumors will respond favorably to autophagy inhibitors. Future preclinical studies should focus on identifying markers that can predict sensitivity to autophagy inhibitors and best drugs to combine with autophagy inhibition. Proper biomarkers are needed to identify the best subsets of patients can most benefit from a specific treatment.

13.13 IN SUMMARY

Autophagy can both suppress tumorigenesis and promote the tumor cell growth in established cancers. More studies are needed to better understand how autophagy is regulated in tumor cells, the cross talk between autophagy and apoptosis, autophagy and ER stress, autophagy and immune systems, and the specific mechanism by which autophagy confers treatment resistance. An increased understanding of autophagy in cancer is important for identifying better autophagy biomarker to detect which types of tumor to be treated combined with autophagy intervention, which group of patients can be most benefited from autophagy inhibition and developing most specific and potent autophagy inhibitors, and eventually benefit the patients with the treatment.

REFERENCES

[1] Yang Z, Klionsky DJ. Eaten alive: a history of macroautophagy. Nat Cell Biol 2010; 12:814–822.

[2] Mizushima N, Yoshimori T, Ohsumi Y. The role of Atg proteins in autophagosome formation. Annu Rev Cell Dev Biol 2011;27:107–132.

[3] Bandyopadhyay U, Kaushik S, Varticovski L, Cuervo AM. The chaperone-mediated autophagy receptor organizes in dynamic protein complexes at the lysosomal membrane. Mol Cell Biol 2008;28:5747–5763.

[4] Cuervo AM. Autophagy: many paths to the same end. Mol Cell Biochem 2004;263: 55–72.

[5] Ahlberg J, Glaumann H. Uptake – microautophagy – and degradation of exogenous proteins by isolated rat liver lysosomes. Effects of pH, ATP, and inhibitors of proteolysis. Exp Mol Pathol 1985;42:78–88.

[6] Ahlberg J, Marzella L, Glaumann H. Uptake and degradation of proteins by isolated rat liver lysosomes. Suggestion of a microautophagic pathway of proteolysis. Lab Invest 1982;47:523–532.

[7] Orenstein SJ, Cuervo AM. Chaperone-mediated autophagy: molecular mechanisms and physiological relevance. Semin Cell Dev Biol 2010;21:719–726.

[8] Kaushik S, Bandyopadhyay U, Sridhar S, Kiffin R, Martinez-Vicente M, et al. Chaperone-mediated autophagy at a glance. J Cell Sci 2011;124:495–499.

[9] Jung CH, Jun CB, Ro SH, Kim YM, Otto NM, et al. ULK-Atg13-FIP200 complexes mediate mTOR signaling to the autophagy machinery. Mol Biol Cell 2009;20: 1992–2003.

[10] Mizushima N. The role of the Atg1/ULK1 complex in autophagy regulation. Curr Opin Cell Biol 2010;22:132–139.

[11] Kim J, Kundu M, Viollet B, Guan KL. AMPK and mTOR regulate autophagy through direct phosphorylation of Ulk1. Nat Cell Biol 2011;13:132–141.

[12] Nice DC, Sato TK, Stromhaug PE, Emr SD, Klionsky DJ. Cooperative binding of the cytoplasm to vacuole targeting pathway proteins, Cvt13 and Cvt20, to phosphatidylinositol 3-phosphate at the pre-autophagosomal structure is required for selective autophagy. J Biol Chem 2002;277:30198–30207.

[13] Funderburk SF, Wang QJ, Yue Z. The Beclin 1-VPS34 complex – at the crossroads of autophagy and beyond. Trends Cell Biol 2010;20:355–362.

[14] Matsunaga K, Saitoh T, Tabata K, Omori H, Satoh T, et al. Two Beclin 1-binding proteins, Atg14L and Rubicon, reciprocally regulate autophagy at different stages. Nat Cell Biol 2009;11:385–396.

[15] Itakura E, Kishi C, Inoue K, Mizushima N. Beclin 1 forms two distinct phosphatidylinositol 3-kinase complexes with mammalian Atg14 and UVRAG. Mol Biol Cell 2008; 19:5360–5372.

[16] Guan J, Stromhaug PE, George MD, Habibzadegah-Tari P, Bevan A, et al. Cvt18/Gsa12 is required for cytoplasm-to-vacuole transport, pexophagy, and autophagy in *Saccharomyces cerevisiae* and *Pichia pastoris*. Mol Biol Cell 2001;12:3821–3838.

[17] Stromhaug PE, Reggiori F, Guan J, Wang CW, Klionsky DJ. Atg21 is a phosphoinositide binding protein required for efficient lipidation and localization of Atg8 during uptake of aminopeptidase I by selective autophagy. Mol Biol Cell 2004;15:3553–3566.

[18] Kabeya Y, Mizushima N, Ueno T, Yamamoto A, Kirisako T, et al. LC3, a mammalian homologue of yeast Apg8p, is localized in autophagosome membranes after processing. EMBO J 2000;19:5720–5728.

[19] Kabeya Y, Mizushima N, Yamamoto A, Oshitani-Okamoto S, Ohsumi Y, et al. LC3, GABARAP and GATE16 localize to autophagosomal membrane depending on form-II formation. J Cell Sci 2004;117:2805–2812.

[20] Tanida I, Tanida-Miyake E, Komatsu M, Ueno T, Kominami E. Human Apg3p/Aut1p homologue is an authentic E2 enzyme for multiple substrates, GATE-16, GABARAP, and MAP-LC3, and facilitates the conjugation of hApg12p to hApg5p. J Biol Chem 2002;277:13739–13744.

[21] Tanida I, Sou YS, Minematsu-Ikeguchi N, Ueno T, Kominami E. Atg8L/Apg8L is the fourth mammalian modifier of mammalian Atg8 conjugation mediated by human Atg4B, Atg7 and Atg3. FEBS J 2006;273:2553–2562.

[22] Li M, Hou Y, Wang J, Chen X, Shao ZM, et al. Kinetics comparisons of mammalian Atg4 homologues indicate selective preferences toward diverse Atg8 substrates. J Biol Chem 2011;286:7327–7338.

[23] Mizushima N, Noda T, Yoshimori T, Tanaka Y, Ishii T, et al. A protein conjugation system essential for autophagy. Nature 1998;395:395–398.

[24] Tanida I, Mizushima N, Kiyooka M, Ohsumi M, Ueno T, et al. Apg7p/Cvt2p: a novel protein-activating enzyme essential for autophagy. Mol Biol Cell 1999;10:1367–1379.

[25] Mizushima N, Sugita H, Yoshimori T, Ohsumi Y. A new protein conjugation system in human. The counterpart of the yeast Apg12p conjugation system essential for autophagy. J Biol Chem 1998;273:33889–33892.

[26] Mizushima N, Noda T, Ohsumi Y. Apg16p is required for the function of the Apg12p-Apg5p conjugate in the yeast autophagy pathway. EMBO J 1999;18:3888–3896.

[27] Webber JL, Tooze SA. New insights into the function of Atg9. FEBS Lett 2010; 584:1319–1326.

[28] Young AR, Chan EY, Hu XW, Kochl R, Crawshaw SG, et al. Starvation and ULK1-dependent cycling of mammalian Atg9 between the TGN and endosomes. J Cell Sci 2006;119:3888–3900.

[29] Lynch-Day MA, Bhandari D, Menon S, Huang J, Cai H, et al. Trs85 directs a Ypt1 GEF, TRAPPIII, to the phagophore to promote autophagy. Proc Natl Acad Sci U S A 2010;107:7811–7816.

[30] Balderhaar HJ, Arlt H, Ostrowicz C, Brocker C, Sundermann F, et al. The Rab GTPase Ypt7 is linked to retromer-mediated receptor recycling and fusion at the yeast late endosome. J Cell Sci 2010;123:4085–4094.

[31] Nair U, Jotwani A, Geng J, Gammoh N, Richerson D, et al. SNARE proteins are required for macroautophagy. Cell 2011;146:290–302.

[32] Nishida Y, Arakawa S, Fujitani K, Yamaguchi H, Mizuta T, et al. Discovery of Atg5/Atg7-independent alternative macroautophagy. Nature 2009;461:654–658.

[33] Codogno P, Mehrpour M, Proikas-Cezanne T. Canonical and non-canonical autophagy: variations on a common theme of self-eating? Nat Rev Mol Cell Biol 2012;13:7–12.

[34] Lum JJ, Bauer DE, Kong M, Harris MH, Li C, et al. Growth factor regulation of autophagy and cell survival in the absence of apoptosis. Cell 2005;120:237–248.

[35] Mathew R, Karantza-Wadsworth V, White E. Role of autophagy in cancer. Nat Rev Cancer 2007;7:961–967.

[36] Degenhardt K, Mathew R, Beaudoin B, Bray K, Anderson D, et al. Autophagy promotes tumor cell survival and restricts necrosis, inflammation, and tumorigenesis. Cancer Cell 2006;10:51–64.

[37] Mizushima N, Levine B. Autophagy in mammalian development and differentiation. Nat Cell Biol 2010;12:823–830.

[38] Yue Z, Jin S, Yang C, Levine AJ, Heintz N. Beclin 1, an autophagy gene essential for early embryonic development, is a haploinsufficient tumor suppressor. Proc Natl Acad Sci U S A 2003;100:15077–15082.

[39] Kuma A, Hatano M, Matsui M, Yamamoto A, Nakaya H, et al. The role of autophagy during the early neonatal starvation period. Nature 2004;432:1032–1036.

[40] Komatsu M, Waguri S, Ueno T, Iwata J, Murata S, et al. Impairment of starvation-induced and constitutive autophagy in Atg7-deficient mice. J Cell Biol 2005;169: 425–434.

[41] Tsukamoto S, Kuma A, Murakami M, Kishi C, Yamamoto A, et al. Autophagy is essential for preimplantation development of mouse embryos. Science 2008;321:117–120.

[42] Komatsu M, Waguri S, Chiba T, Murata S, Iwata J, et al. Loss of autophagy in the central nervous system causes neurodegeneration in mice. Nature 2006;441:880–884.

[43] Choi AM, Ryter SW, Levine B. Autophagy in human health and disease. N Engl J Med 2013;368:1845–1846.

[44] Levine B, Kroemer G. Autophagy in the pathogenesis of disease. Cell 2008;132: 27–42.

[45] Fujita N, Yoshimori T. Ubiquitination-mediated autophagy against invading bacteria. Curr Opin Cell Biol 2011;23:492–497.

[46] Russell RC, Tian Y, Yuan H, Park HW, Chang YY, et al. ULK1 induces autophagy by phosphorylating Beclin-1 and activating VPS34 lipid kinase. Nat Cell Biol 2013; 15:741–750.

[47] Ravikumar B, Moreau K, Jahreiss L, Puri C, Rubinsztein DC. Plasma membrane contributes to the formation of pre-autophagosomal structures. Nat Cell Biol 2010; 12:747–757.

[48] Hailey DW, Rambold AS, Satpute-Krishnan P, Mitra K, Sougrat R, et al. Mitochondria supply membranes for autophagosome biogenesis during starvation. Cell 2010;141:656–667.

[49] Hayashi-Nishino M, Fujita N, Noda T, Yamaguchi A, Yoshimori T, et al. A subdomain of the endoplasmic reticulum forms a cradle for autophagosome formation. Nat Cell Biol 2009;11:1433–1437.

[50] Kim HJ, Zhong Q, Sheng ZH, Yoshimori T, Liang C, et al. Beclin-1-interacting autophagy protein Atg14L targets the SNARE-associated protein Snapin to coordinate endocytic trafficking. J Cell Sci 2012;125:4740–4750.

[51] Ravikumar B, Imarisio S, Sarkar S, O'Kane CJ, Rubinsztein DC. Rab5 modulates aggregation and toxicity of mutant huntingtin through macroautophagy in cell and fly models of Huntington disease. J Cell Sci 2008;121:1649–1660.

[52] Djeddi A, Michelet X, Culetto E, Alberti A, Barois N, et al. Induction of autophagy in ESCRT mutants is an adaptive response for cell survival in C. elegans. J Cell Sci 2012;125:685–694.

[53] Fader CM, Colombo MI. Autophagy and multivesicular bodies: two closely related partners. Cell Death Differ 2009;16:70–78.

[54] Rong Y, McPhee CK, Deng S, Huang L, Chen L, et al. Spinster is required for autophagic lysosome reformation and mTOR reactivation following starvation. Proc Natl Acad Sci U S A 2011;108:7826–7831.

[55] Manjithaya R, Anjard C, Loomis WF, Subramani S. Unconventional secretion of *Pichia pastoris* Acb1 is dependent on GRASP protein, peroxisomal functions, and autophagosome formation. J Cell Biol 2010;188:537–546.

[56] Chen Y, Yu L. Autophagic lysosome reformation. Exp Cell Res 2013;319:142–146.

[57] Kim DH, Sarbassov DD, Ali SM, King JE, Latek RR, et al. mTOR interacts with raptor to form a nutrient-sensitive complex that signals to the cell growth machinery. Cell 2002;110:163–175.

[58] Kroemer G, Marino G, Levine B. Autophagy and the integrated stress response. Mol Cell 2010;40:280–293.

[59] Sancak Y, Bar-Peled L, Zoncu R, Markhard AL, Nada S, et al. Ragulator-Rag complex targets mTORC1 to the lysosomal surface and is necessary for its activation by amino acids. Cell 2010;141:290–303.

[60] Zoncu R, Bar-Peled L, Efeyan A, Wang S, Sancak Y, et al. mTORC1 senses lysosomal amino acids through an inside-out mechanism that requires the vacuolar H(+)-ATPase. Science 2011;334:678–683.

[61] David R. Autophagy: TFEB perfects multitasking. Nat Rev Mol Cell Biol 2011;12:404.

[62] Shen J, Chen X, Hendershot L, Prywes R. ER stress regulation of ATF6 localization by dissociation of BiP/GRP78 binding and unmasking of Golgi localization signals. Dev Cell 2002;3:99–111.

[63] Rouschop KM, van den Beucken T, Dubois L, Niessen H, Bussink J, et al. The unfolded protein response protects human tumor cells during hypoxia through regulation of the autophagy genes MAP1LC3B and ATG5. J Clin Invest 2010;120:127–141.

[64] Rzymski T, Milani M, Pike L, Buffa F, Mellor HR, et al. Regulation of autophagy by ATF4 in response to severe hypoxia. Oncogene 2010;29:4424–4435.

[65] Kenzelmann Broz D, Spano Mello S, Bieging KT, Jiang D, Dusek RL, et al. Global genomic profiling reveals an extensive p53-regulated autophagy program contributing to key p53 responses. Genes Dev 2013;27:1016–1031.

[66] Korolchuk VI, Rubinsztein DC. Regulation of autophagy by lysosomal positioning. Autophagy 2011;7:927–928.

[67] Pous C, Codogno P. Lysosome positioning coordinates mTORC1 activity and autophagy. Nat Cell Biol 2011;13:342–344.

[68] Laplante M, Sabatini DM. mTOR signaling in growth control and disease. Cell 2012; 149:274–293.

[69] Degtyarev M, De Maziere A, Orr C, Lin J, Lee BB, et al. Akt inhibition promotes autophagy and sensitizes PTEN-null tumors to lysosomotropic agents. J Cell Biol 2008;183:101–116.

[70] Carew JS, Kelly KR, Nawrocki ST. Autophagy as a target for cancer therapy: new developments. Cancer Manag Res 2012;4:357–365.

[71] Chen N, Karantza V. Autophagy as a therapeutic target in cancer. Cancer Biol Ther 2011;11:157–168.

[72] Maes H, Rubio N, Garg AD, Agostinis P. Autophagy: shaping the tumor microenvironment and therapeutic response. Trends Mol Med 2013;19:428–446.

[73] Li X, Xu HL, Liu YX, An N, Zhao S, et al. Autophagy modulation as a target for anticancer drug discovery. Acta Pharmacol Sin 2013;34:612–624.

[74] Kroemer G, Galluzzi L, Vandenabeele P, Abrams J, Alnemri ES, et al. Classification of cell death: recommendations of the Nomenclature Committee on Cell Death 2009. Cell Death Differ 2009;16:3–11.

[75] Amaravadi RK, Thompson CB. The roles of therapy-induced autophagy and necrosis in cancer treatment. Clin Cancer Res 2007;13:7271–7279.

[76] Mathew R, Kongara S, Beaudoin B, Karp CM, Bray K, et al. Autophagy suppresses tumor progression by limiting chromosomal instability. Genes Dev 2007;21:1367–1381.

[77] Qu X, Yu J, Bhagat G, Furuya N, Hibshoosh H, et al. Promotion of tumorigenesis by heterozygous disruption of the beclin 1 autophagy gene. J Clin Invest 2003;112:1809–1820.

[78] Liang XH, Jackson S, Seaman M, Brown K, Kempkes B, et al. Induction of autophagy and inhibition of tumorigenesis by beclin 1. Nature 1999;402:672–676.

[79] Saito H, Inazawa J, Saito S, Kasumi F, Koi S, et al. Detailed deletion mapping of chromosome 17q in ovarian and breast cancers: 2-cM region on 17q21.3 often and commonly deleted in tumors. Cancer Res 1993;53:3382–3385.

[80] Liu J, Xia H, Kim M, Xu L, Li Y, et al. Beclin1 controls the levels of p53 by regulating the deubiquitination activity of USP10 and USP13. Cell 2011;147:223–234.

[81] Yang ZJ, Chee CE, Huang S, Sinicrope F. Autophagy modulation for cancer therapy. Cancer Biol Ther 2011;11:169–176.

[82] Guo JY, Chen HY, Mathew R, Fan J, Strohecker AM, et al. Activated Ras requires autophagy to maintain oxidative metabolism and tumorigenesis. Genes Dev 2011;25: 460–470.

[83] Strohecker AM, Guo JY, Karsli-Uzunbas G, Price SM, Chen GJ, et al. Autophagy sustains mitochondrial glutamine metabolism and growth of BrafV600E-driven lung tumors. Cancer Discov 2013;3:1272–1285.

[84] Strohecker AM, White E. Autophagy promotes driven lung tumorigenesis by preserving mitochondrial metabolism. Autophagy 2014;10:384–385.

[85] Guo JY, Karsli-Uzunbas G, Mathew R, Aisner SC, Kamphorst JJ, et al. Autophagy suppresses progression of K-ras-induced lung tumors to oncocytomas and maintains lipid homeostasis. Genes Dev 2013;27:1447–1461.

[86] Guo JY, Xia B, White E. Autophagy-mediated tumor promotion. Cell 2013;155: 1216–1219.

[87] Huo Y, Cai H, Teplova I, Bowman-Colin C, Chen G, et al. Autophagy opposes p53-mediated tumor barrier to facilitate tumorigenesis in a model of PALB2-associated hereditary breast cancer. Cancer Discov 2013;3:894–907.

[88] Yang S, Wang X, Contino G, Liesa M, Sahin E, et al. Pancreatic cancers require autophagy for tumor growth. Genes Dev 2011;25:717–729.

[89] Rao S, Tortola L, Perlot T, Wirnsberger G, Novatchkova M, et al. A dual role for autophagy in a murine model of lung cancer. Nat Commun 2014;5:3056.

[90] Guo JY, White E. Autophagy is required for mitochondrial function, lipid metabolism, growth and fate of KRAS (G12D)-driven lung tumors. Autophagy 2013;9:1636–1638.

[91] Rosenfeldt MT, O'Prey J, Morton JP, Nixon C, MacKay G, et al. p53 status determines the role of autophagy in pancreatic tumour development. Nature 2013;504:296–300.

[92] Wang J, Pan XL, Ding LJ, Liu DY, Da-Peng L, et al. Aberrant expression of Beclin-1 and LC3 correlates with poor prognosis of human hypopharyngeal squamous cell carcinoma. PLoS One 2013;8:e69038.

[93] Lin HX, Qiu HJ, Zeng F, Rao HL, Yang GF, et al. Decreased expression of Beclin 1 correlates closely with Bcl-xL expression and poor prognosis of ovarian carcinoma. PLoS One 2013;8:e60516.

[94] Geng QR, Xu DZ, He LJ, Lu JB, Zhou ZW, et al. Beclin-1 expression is a significant predictor of survival in patients with lymph node-positive gastric cancer. PLoS One 2012;7:e45968.

[95] Spowart JE, Townsend KN, Huwait H, Eshragh S, West NR, et al. The autophagy protein LC3A correlates with hypoxia and is a prognostic marker of patient survival in clear cell ovarian cancer. J Pathol 2012;228:437–447.

[96] Morani F, Titone R, Pagano L, Galetto A, Alabiso O, et al. Autophagy and thyroid carcinogenesis: genetic and epigenetic links. Endocr Relat Cancer 2014;21:R13–R29.

[97] Lazova R, Camp RL, Klump V, Siddiqui SF, Amaravadi RK, et al. Punctate LC3B expression is a common feature of solid tumors and associated with proliferation, metastasis, and poor outcome. Clin Cancer Res 2012;18:370–379.

[98] Ma XH, Piao S, Wang D, McAfee QW, Nathanson KL, et al. Measurements of tumor cell autophagy predict invasiveness, resistance to chemotherapy, and survival in melanoma. Clin Cancer Res 2011;17:3478–3489.

[99] Liu H, He Z, von Rutte T, Yousefi S, Hunger RE, et al. Down-regulation of autophagy-related protein 5 (ATG5) contributes to the pathogenesis of early-stage cutaneous melanoma. Sci Transl Med 2013;5:202ra123.

[100] Galluzzi L, Aaronson SA, Abrams J, Alnemri ES, Andrews DW, et al. Guidelines for the use and interpretation of assays for monitoring cell death in higher eukaryotes. Cell Death Differ 2009;16:1093–1107.

[101] Apel A, Herr I, Schwarz H, Rodemann HP, Mayer A. Blocked autophagy sensitizes resistant carcinoma cells to radiation therapy. Cancer Res 2008;68:1485–1494.

[102] Lomonaco SL, Finniss S, Xiang C, Decarvalho A, Umansky F, et al. The induction of autophagy by gamma-radiation contributes to the radioresistance of glioma stem cells. Int J Cancer 2009;125:717–722.

[103] Claerhout S, Verschooten L, Van Kelst S, De Vos R, Proby C, et al. Concomitant inhibition of AKT and autophagy is required for efficient cisplatin-induced apoptosis of metastatic skin carcinoma. Int J Cancer 2010;127:2790–2803.

[104] Rubio N, Coupienne I, Di Valentin E, Heirman I, Grooten J, et al. Spatiotemporal autophagic degradation of oxidatively damaged organelles after photodynamic stress is amplified by mitochondrial reactive oxygen species. Autophagy 2012;8:1312–1324.

[105] Dewaele M, Martinet W, Rubio N, Verfaillie T, de Witte PA, et al. Autophagy pathways activated in response to PDT contribute to cell resistance against ROS damage. J Cell Mol Med 2011;15:1402–1414.

[106] Garg AD, Dudek AM, Ferreira GB, Verfaillie T, Vandenabeele P, et al. ROS-induced autophagy in cancer cells assists in evasion from determinants of immunogenic cell death. Autophagy 2013;9:1292–1307.

[107] Samaddar JS, Gaddy VT, Duplantier J, Thandavan SP, Shah M, et al. A role for macroautophagy in protection against 4-hydroxytamoxifen-induced cell death and the development of antiestrogen resistance. Mol Cancer Ther 2008;7:2977–2987.

[108] Li M, Jiang X, Liu D, Na Y, Gao GF, et al. Autophagy protects LNCaP cells under androgen deprivation conditions. Autophagy 2008;4:54–60.

[109] Bellodi C, Lidonnici MR, Hamilton A, Helgason GV, Soliera AR, et al. Targeting autophagy potentiates tyrosine kinase inhibitor-induced cell death in Philadelphia chromosome-positive cells, including primary CML stem cells. J Clin Invest 2009; 119:1109–1123.

[110] Crowley LC, Elzinga BM, O'Sullivan GC, McKenna SL. Autophagy induction by Bcr-Abl-expressing cells facilitates their recovery from a targeted or nontargeted treatment. Am J Hematol 2011;86:38–47.

[111] Gupta A, Roy S, Lazar AJ, Wang WL, McAuliffe JC, et al. Autophagy inhibition and antimalarials promote cell death in gastrointestinal stromal tumor (GIST). Proc Natl Acad Sci U S A 2010;107:14333–14338.

[112] Li X, Fan Z. The epidermal growth factor receptor antibody cetuximab induces autophagy in cancer cells by downregulating HIF-1alpha and Bcl-2 and activating the beclin 1/hVps34 complex. Cancer Res 2010;70:5942–5952.

[113] Wu Z, Chang PC, Yang JC, Chu CY, Wang LY, et al. Autophagy blockade sensitizes prostate cancer cells towards Src family kinase inhibitors. Genes Cancer 2010;1:40–49.

[114] Wang K, Liu R, Li J, Mao J, Lei Y, et al. Quercetin induces protective autophagy in gastric cancer cells: involvement of Akt-mTOR- and hypoxia-induced factor 1alpha-mediated signaling. Autophagy 2011;7:966–978.

[115] Milani M, Rzymski T, Mellor HR, Pike L, Bottini A, et al. The role of ATF4 stabilization and autophagy in resistance of breast cancer cells treated with Bortezomib. Cancer Res 2009;69:4415–4423.

[116] Vazquez-Martin A, Oliveras-Ferraros C, Menendez JA. Autophagy facilitates the development of breast cancer resistance to the anti-HER2 monoclonal antibody trastuzumab. PLoS One 2009;4:e6251.

[117] Carew JS, Nawrocki ST, Kahue CN, Zhang H, Yang C, et al. Targeting autophagy augments the anticancer activity of the histone deacetylase inhibitor SAHA to overcome Bcr-Abl-mediated drug resistance. Blood 2007;110:313–322.

[118] Li J, Hou N, Faried A, Tsutsumi S, Kuwano H. Inhibition of autophagy augments 5-fluorouracil chemotherapy in human colon cancer *in vitro* and *in vivo* model. Eur J Cancer 2010;46:1900–1909.

[119] Tiwari M, Bajpai VK, Sahasrabuddhe AA, Kumar A, Sinha RA, et al. Inhibition of N-(4-hydroxyphenyl)retinamide-induced autophagy at a lower dose enhances cell death in malignant glioma cells. Carcinogenesis 2008;29:600–609.

[120] Kanematsu S, Uehara N, Miki H, Yoshizawa K, Kawanaka A, et al. Autophagy inhibition enhances sulforaphane-induced apoptosis in human breast cancer cells. Anticancer Res 2010;30:3381–3390.

[121] Nishikawa T, Tsuno NH, Okaji Y, Shuno Y, Sasaki K, et al. Inhibition of autophagy potentiates sulforaphane-induced apoptosis in human colon cancer cells. Ann Surg Oncol 2010;17:592–602.

[122] Kim RH, Coates JM, Bowles TL, McNerney GP, Sutcliffe J, et al. Arginine deiminase as a novel therapy for prostate cancer induces autophagy and caspase-independent apoptosis. Cancer Res 2009;69:700–708.

[123] Ma XH, Piao SF, Dey S, McAfee Q, Karakousis G, et al. Targeting ER stress-induced autophagy overcomes BRAF inhibitor resistance in melanoma. J Clin Invest 2014; 124:1406–1417.

[124] Abedin MJ, Wang D, McDonnell MA, Lehmann U, Kelekar A. Autophagy delays apoptotic death in breast cancer cells following DNA damage. Cell Death Differ 2007;14:500–510.

[125] Carew JS, Medina EC, Esquivel JA 2nd, Mahalingam D, Swords R, et al. Autophagy inhibition enhances vorinostat-induced apoptosis via ubiquitinated protein accumulation. J Cell Mol Med 2010;14:2448–2459.

[126] Li J, Hou N, Faried A, Tsutsumi S, Takeuchi T, et al. Inhibition of autophagy by 3-MA enhances the effect of 5-FU-induced apoptosis in colon cancer cells. Ann Surg Oncol 2009;16:761–771.

[127] Liu D, Yang Y, Liu Q, Wang J. Inhibition of autophagy by 3-MA potentiates cisplatin-induced apoptosis in esophageal squamous cell carcinoma cells. Med Oncol 2011;28:105–111.

[128] Kanzawa T, Germano IM, Komata T, Ito H, Kondo Y, et al. Role of autophagy in temozolomide-induced cytotoxicity for malignant glioma cells. Cell Death Differ 2004;11:448–457.

[129] Schoenlein PV, Periyasamy-Thandavan S, Samaddar JS, Jackson WH, Barrett JT. Autophagy facilitates the progression of ERalpha-positive breast cancer cells to antiestrogen resistance. Autophagy 2009;5:400–403.

[130] Clarke R, Shajahan AN, Riggins RB, Cho Y, Crawford A, et al. Gene network signaling in hormone responsiveness modifies apoptosis and autophagy in breast cancer cells. J Steroid Biochem Mol Biol 2009;114:8–20.

[131] Kamitsuji Y, Kuroda J, Kimura S, Toyokuni S, Watanabe K, et al. The Bcr-Abl kinase inhibitor INNO-406 induces autophagy and different modes of cell death execution in Bcr-Abl-positive leukemias. Cell Death Differ 2008;15:1712–1722.

[132] Mishima Y, Terui Y, Mishima Y, Taniyama A, Kuniyoshi R, et al. Autophagy and autophagic cell death are next targets for elimination of the resistance to tyrosine kinase inhibitors. Cancer Sci 2008;99:2200–2208.

[133] Han J, Hou W, Goldstein LA, Lu C, Stolz DB, et al. Involvement of protective autophagy in TRAIL resistance of apoptosis-defective tumor cells. J Biol Chem 2008; 283:19665–19677.

[134] Bijnsdorp IV, Peters GJ, Temmink OH, Fukushima M, Kruyt FA. Differential activation of cell death and autophagy results in an increased cytotoxic potential for trifluorothymidine compared to 5-fluorouracil in colon cancer cells. Int J Cancer 2010;126:2457–2468.

[135] Shingu T, Fujiwara K, Bogler O, Akiyama Y, Moritake K, et al. Inhibition of autophagy at a late stage enhances imatinib-induced cytotoxicity in human malignant glioma cells. Int J Cancer 2009;124:1060–1071.

[136] Watanabe M, Adachi S, Matsubara H, Imai T, Yui Y, et al. Induction of autophagy in malignant rhabdoid tumor cells by the histone deacetylase inhibitor FK228 through AIF translocation. Int J Cancer 2009;124:55–67.

[137] Kondo Y, Kanzawa T, Sawaya R, Kondo S. The role of autophagy in cancer development and response to therapy. Nat Rev Cancer 2005;5:726–734.

[138] Gewirtz DA. Autophagy, senescence and tumor dormancy in cancer therapy. Autophagy 2009;5:1232–1234.

[139] Michaud M, Martins I, Sukkurwala AQ, Adjemian S, Ma Y, et al. Autophagy-dependent anticancer immune responses induced by chemotherapeutic agents in mice. Science 2011;334:1573–1577.

[140] Baginska J, Viry E, Berchem G, Poli A, Noman MZ, et al. Granzyme B degradation by autophagy decreases tumor cell susceptibility to natural killer-mediated lysis under hypoxia. Proc Natl Acad Sci U S A 2013;110:17450–17455.

[141] Amaravadi RK, Lippincott-Schwartz J, Yin XM, Weiss WA, Takebe N, et al. Principles and current strategies for targeting autophagy for cancer treatment. Clin Cancer Res 2011;17:654–666.

[142] Behrends C, Sowa ME, Gygi SP, Harper JW. Network organization of the human autophagy system. Nature 2010;466:68–76.

[143] Shoji-Kawata S, Sumpter R, Leveno M, Campbell GR, Zou Z, et al. Identification of a candidate therapeutic autophagy-inducing peptide. Nature 2013;494:201–206.

[144] Liu Y, Shoji-Kawata S, Sumpter RM Jr, Wei Y, Ginet V, et al. Autosis is a Na+,K+-ATPase-regulated form of cell death triggered by autophagy-inducing peptides, starvation, and hypoxia-ischemia. Proc Natl Acad Sci U S A 2013;110:20364–20371.

[145] Sou YS, Waguri S, Iwata J, Ueno T, Fujimura T, et al. The Atg8 conjugation system is indispensable for proper development of autophagic isolation membranes in mice. Mol Biol Cell 2008;19:4762–4775.

[146] Shu CW, Madiraju C, Zhai D, Welsh K, Diaz P, et al. High-throughput fluorescence assay for small-molecule inhibitors of autophagins/Atg4. J Biomol Screen 2011; 16:174–182.

[147] Gorski SM, Ries J, Lum JJ. Targeting autophagy: the Achilles' heel of cancer. Autophagy 2012;8:1279–1280.

[148] Marino G, Uria JA, Puente XS, Quesada V, Bordallo J, et al. Human autophagins, a family of cysteine proteinases potentially implicated in cell degradation by autophagy. J Biol Chem 2003;278:3671–3678.

[149] Seglen PO, Gordon PB. 3-Methyladenine: specific inhibitor of autophagic/lysosomal protein degradation in isolated rat hepatocytes. Proc Natl Acad Sci U S A 1982; 79:1889–1892.

[150] Vakifahmetoglu-Norberg H, Kim M, Xia HG, Iwanicki MP, Ofengeim D, et al. Chaperone-mediated autophagy degrades mutant p53. Genes Dev 2013;27:1718–1730.

[151] Shao S, Li S, Qin Y, Wang X, Yang Y, et al. Spautin-1, a novel autophagy inhibitor, enhances imatinib-induced apoptosis in chronic myeloid leukemia. Int J Oncol 2014; 44:1661–1668.

[152] Krafts K, Hempelmann E, Skorska-Stania A. From methylene blue to chloroquine: a brief review of the development of an antimalarial therapy. Parasitol Res 2012;111:1–6.

[153] Payne D. Spread of chloroquine resistance in *Plasmodium falciparum*. Parasitol Today 1987;3:241–246.

[154] Le Bras J, Durand R. The mechanisms of resistance to antimalarial drugs in *Plasmodium falciparum*. Fundam Clin Pharmacol 2003;17:147–153.

[155] Ecker A, Lehane AM, Clain J, Fidock DA. PfCRT and its role in antimalarial drug resistance. Trends Parasitol 2012;28:504–514.

[156] Sugioka Y, Suzuki M, Sugioka K, Nakano M. A ferriprotoporphyrin IX-chloroquine complex promotes membrane phospholipid peroxidation. A possible mechanism for antimalarial action. FEBS Lett 1987;223:251–254.

[157] Yayon A, Cabantchik ZI, Ginsburg H. Identification of the acidic compartment of *Plasmodium falciparum*-infected human erythrocytes as the target of the antimalarial drug chloroquine. EMBO J 1984;3:2695–2700.

[158] Roepe PD. Molecular and physiologic basis of quinoline drug resistance in *Plasmodium falciparum* malaria. Future Microbiol 2009;4:441–455.

[159] Gaviria D, Paguio MF, Turnbull LB, Tan A, Siriwardana A, et al. A process similar to autophagy is associated with cytocidal chloroquine resistance in *Plasmodium falciparum*. PLoS One 2013;8:e79059.

[160] Haynes RK, Cheu KW, Chan HW, Wong HN, Li KY, et al. Interactions between artemisinins and other antimalarial drugs in relation to the cofactor model – a unifying proposal for drug action. ChemMedChem 2012;7:2204–2226.

[161] Pleeter P, Lekostaj JK, Roepe PD. Purified *Plasmodium falciparum* multi-drug resistance protein (PfMDR 1) binds a high affinity chloroquine analogue. Mol Biochem Parasitol 2010;173:158–161.

[162] Weissmann G. Labilization and stabilization of lysosomes. Fed Proc 1964;23:1038–1044.

[163] de Duve C, de Barsy T, Poole B, Trouet A, Tulkens P, et al. Commentary. Lysosomotropic agents. Biochem Pharmacol 1974;23:2495–2531.

[164] Luthman H, Magnusson G. High efficiency polyoma DNA transfection of chloroquine treated cells. Nucleic Acids Res 1983;11:1295–1308.

[165] Accapezzato D, Visco V, Francavilla V, Molette C, Donato T, et al. Chloroquine enhances human CD8+ T cell responses against soluble antigens *in vivo*. J Exp Med 2005;202:817–828.

[166] Fedorko M. Effect of chloroquine on morphology of cytoplasmic granules in maturing human leukocytes – an ultrastructural study. J Clin Invest 1967;46:1932–1942.

[167] Ndolo RA, Luan Y, Duan S, Forrest ML, Krise JP. Lysosomotropic properties of weakly basic anticancer agents promote cancer cell selectivity *in vitro*. PLoS One 2012;7:e49366.

[168] Trapp S, Rosania GR, Horobin RW, Kornhuber J. Quantitative modeling of selective lysosomal targeting for drug design. Eur Biophys J 2008;37:1317–1328.

[169] Ohkuma S, Poole B. Cytoplasmic vacuolation of mouse peritoneal macrophages and the uptake into lysosomes of weakly basic substances. J Cell Biol 1981;90:656–664.

[170] Wibo M, Poole B. Protein degradation in cultured cells. II. The uptake of chloroquine by rat fibroblasts and the inhibition of cellular protein degradation and cathepsin B1. J Cell Biol 1974;63:430–440.

[171] Bulychev A, Trouet A, Tulkens P. Uptake and intracellular distribution of neutral red in cultured fibroblasts. Exp Cell Res 1978;115:343–355.

[172] Binaschi M, Bigioni M, Cipollone A, Rossi C, Goso C, et al. Anthracyclines: selected new developments. Curr Med Chem Anticancer Agents 2001;1:113–130.

[173] Chen VY, Rosania GR. The great multidrug-resistance paradox. ACS Chem Biol 2006;1:271–273.

[174] Duvvuri M, Konkar S, Funk RS, Krise JM, Krise JP. A chemical strategy to manipulate the intracellular localization of drugs in resistant cancer cells. Biochemistry 2005;44:15743–15749.

[175] Fan QW, Cheng C, Hackett C, Feldman M, Houseman BT, et al. Akt and autophagy cooperate to promote survival of drug-resistant glioma. Sci Signal 2010;3:ra81.

[176] Saleem A, Dvorzhinski D, Santanam U, Mathew R, Bray K, et al. Effect of dual inhibition of apoptosis and autophagy in prostate cancer. Prostate 2012;72:1374–1381.

[177] Algazy KM SL, Demichele A, Vaughn DJ, Torigian DA, Chang CY, Redlinger M, Davis LE, O'Dwyer PJ, Amaravadi RK (2011) Combined mTOR inhibition and autophagy inhibition: phase I trial of temsirolimus and hydroxchloroquine in patients with advanced solid tumors. AACR annual meeting Abstract 4500.

[178] Fecher LASL, Evans T, Chang CY, Gallagher M, Kramer A, O'Dwyer PJ, Amaravadi RK (2011) Combining chemotherapy with autophagy inhibition: phase I trial of dose-intense temozolomide and hydroxchloroquine in patients with advanced solid tumors. AACR annual meeting 2011 abstract 2521.

[179] Vogl DTSE, Bradner JE, Davis L, Carroll M, Nichols CW, Shank D, Carberry M, Swider CR, Mangan P, Shelley B, Thompson CB, Amaravadi RK. Combined autophagy and proteasome inhibition for multiple myeloma: preliminary results of a phase 1/2 trial of hydroxychloroquine and standard dose bortezomib for relapsed or refractory myeloma. Blood 2008;112:3684(ASH abstract).

[180] Rangwala R, Chang YC, Hu J, Algazy KM, Evans TL, et al. Combined MTOR and autophagy inhibition: phase I trial of hydroxychloroquine and temsirolimus in patients with advanced solid tumors and melanoma. Autophagy 2014;10:1391–1402.

[181] Barnard RA, Wittenburg LA, Amaravadi RK, Gustafson DL, Thorburn A, et al. Phase I clinical trial and pharmacodynamic evaluation of combination hydroxychloroquine and doxorubicin treatment in pet dogs treated for spontaneously occurring lymphoma. Autophagy 2014;10:1415–1425.

[182] Rosenfeld MR, Ye X, Supko JG, Desideri S, Grossman SA, et al. A phase I/II trial of hydroxychloroquine in conjunction with radiation therapy and concurrent and adjuvant temozolomide in patients with newly diagnosed glioblastoma multiforme. Autophagy 2014;10:1359–1368.

[183] Sotelo J, Briceno E, Lopez-Gonzalez MA. Adding chloroquine to conventional treatment for glioblastoma multiforme: a randomized, double-blind, placebo-controlled trial. Ann Intern Med 2006;144:337–343.

[184] Vance D, Shah M, Joshi A, Kane RS. Polyvalency: a promising strategy for drug design. Biotechnol Bioeng 2008;101:429–434.

[185] Vennerstrom JL, Ellis WY, Ager AL Jr, Andersen SL, Gerena L, et al. Bisquinolines. 1. N,N-bis(7-chloroquinolin-4-yl)alkanediamines with potential against chloroquine-resistant malaria. J Med Chem 1992;35:2129–2134.

[186] Raynes K. Bisquinoline antimalarials: their role in malaria chemotherapy. Int J Parasitol 1999;29:367–379.

[187] Ridley RG, Matile H, Jaquet C, Dorn A, Hofheinz W, et al. Antimalarial activity of the bisquinoline trans-N1,N2-bis (7-chloroquinolin-4-yl)cyclohexane-1,2-diamine: comparison of two stereoisomers and detailed evaluation of the S,S enantiomer, Ro 47-7737. Antimicrob Agents Chemother 1997;41:677–686.

[188] Girault S, Grellier P, Berecibar A, Maes L, Lemiere P, et al. Antiplasmodial activity and cytotoxicity of bis-, tris-, and tetraquinolines with linear or cyclic amino linkers. J Med Chem 2001;44:1658–1665.

[189] McAfee Q, Zhang Z, Samanta A, Levi SM, Ma XH, et al. Autophagy inhibitor Lys05 has single-agent antitumor activity and reproduces the phenotype of a genetic autophagy deficiency. Proc Natl Acad Sci U S A 2012;109:8253–8258.

[190] Cadwell K, Liu JY, Brown SL, Miyoshi H, Loh J, et al. A key role for autophagy and the autophagy gene Atg16l1 in mouse and human intestinal Paneth cells. Nature 2008;456:259–263.

[191] Adolph TE, Tomczak MF, Niederreiter L, Ko HJ, Bock J, et al. Paneth cells as a site of origin for intestinal inflammation. Nature 2013;503:272–276.

[192] Fujita N, Itoh T, Omori H, Fukuda M, Noda T, et al. The Atg16L complex specifies the site of LC3 lipidation for membrane biogenesis in autophagy. Mol Biol Cell 2008;19:2092–2100.

[193] Marino G, Niso-Santano M, Baehrecke EH, Kroemer G. Self-consumption: the interplay of autophagy and apoptosis. Nat Rev Mol Cell Biol 2014;15:81–94.

[194] Levine B, Sinha S, Kroemer G. Bcl-2 family members: dual regulators of apoptosis and autophagy. Autophagy 2008;4:600–606.

[195] Amaravadi RK. Cancer. Autophagy in tumor immunity. Science 2011;334:1501–1502.

[196] Deretic V, Levine B. Autophagy, immunity, and microbial adaptations. Cell Host Microbe 2009;5:527–549.

[197] Levine B, Mizushima N, Virgin HW. Autophagy in immunity and inflammation. Nature 2011;469:323–335.

[198] Crotzer VL, Blum JS. Autophagy and adaptive immunity. Immunology 2010;131:9–17.

[199] Crotzer VL, Blum JS. Autophagy and its role in MHC-mediated antigen presentation. J Immunol 2009;182:3335–3341.

[200] Maycotte P, Aryal S, Cummings CT, Thorburn J, Morgan MJ, et al. Chloroquine sensitizes breast cancer cells to chemotherapy independent of autophagy. Autophagy 2012;8:200–212.

14

AUTOPHAGY ENHANCERS, ARE WE THERE YET?

SHUYAN LU

Drug Safety Research and Development, Pfizer Inc., San Diego, CA, USA

RALPH A. NIXON

Department of Cell Biology, New York University Langone Medical Center, New York, NY, USA

14.1 INTRODUCTION

Autophagy is a self-eating process by which long-lived proteins and organelles within cells are delivered to lysosomes, ultimately degraded and recycled. This evolutionarily conserved process plays a critical role in preserving cellular homeostasis and maintaining a balance between synthesis and degradation/recycling under conditions of changing nutrition or stress. At least three forms of autophagy have been identified, including macroautophagy, microautophagy, and chaperone-mediated autophagy (CMA). Among these subtypes, macroautophagy is the best characterized form and the process involves autophagosomal sequestration of targets, which are then delivered to lysosomes for degradation. Discovery of core autophagy-related genes (Atgs) initiated a molecular-genetic era for autophagy research and a rapid expansion of knowledge about autophagy. Many Atgs are involved in the multiple steps of autophagosome formation, which include the initiation of a phagophore, elongation, and closure of this structure, followed by the final maturation step, fusion of the autophagosome with lysosomes. Current data suggest that membranes from multiple organelle sources potentially contribute to autophagosome biogenesis, including

Lysosomes: Biology, Diseases, and Therapeutics, First Edition.
Edited by Frederick R. Maxfield, James M. Willard and Shuyan Lu.

mitochondria, endoplasmic reticulum, Golgi, plasma membrane, and nucleus. A modest basal level of autophagy occurs under normal physiological conditions, but its rate can be modulated by various stimuli. The best characterized mechanism of autophagy regulation is through the canonical mTOR signaling pathways. When mTOR1 activity is downregulated under conditions of nutrient depletion, autophagy is induced by phosphorylation and activation of the ULK complex that initiates phagophore formation. Additional pathways that act independently of mTOR may also play a critical role in regulating autophagy. For instance, intracytosolic calcium, cAMP, and 1,4,5-inositol trisphosphate (IP3) have been shown to modulate autophagy [1,2]. Mounting evidence has revealed that autophagy is involved in various physiological processes, including regulating nutrient supply essential for survival and maintaining quality control of intracellular proteins and organelles. Mice that lack autophagic activity due to a knockout of various Atgs die either in utero or within 24 h after birth [3–5]. The vital role of constitutive autophagy maintaining cellular/tissue homeostasis is supported by various conditional knockout studies. For example, mice with a liver-specific knockout of Atg7 exhibit various liver lesions, including hepatomegaly and hepatocyte hypertrophy [4], and mice lacking Atg5 in cardiac muscle display cardiac hypertrophy and left ventricular dilation [6].

14.2 AUTOPHAGY IMPAIRMENT AND DISEASES

The link between autophagy and human disease was first demonstrated in studies showing that decreased expression of beclin 1 (mammalian homolog of year Atg6) contributes to the development or progression of human malignancies [7]. Recent efforts have revealed the role of autophagy dysfunction in additional pathological conditions such as aging and other aging-related diseases including neurodegenerative disorders, metabolic diseases, cardiomyopathy, and age-related macular degeneration. Reduced degradation of proteins and organelles may contribute to accumulation of damaged proteins and organelles and to the aging process. Both macroautophagy and CMA become markedly less efficient during the aging process, which contributes to malfunction of many biological processes [8] and may accelerate cell aging itself [9]. Decreased hepatic autophagy has been observed in both genetic and dietary mouse models of obesity and insulin resistance [10]. Surprisingly, restoring autophagy by increasing Atg7 expression alone can counteract insulin resistance and improve hepatic fat metabolism [10]. Impaired mitophagy (a specialized form of macroautophagy for eliminating dysfunctional mitochondria) is believed to contribute to cardiac aging [11]. Recent data also suggested that autophagy deficiency is associated with age-related macular degeneration [12]. AMPK activator AICAR can accelerate removal of protein aggregates and improve survival of retinal pigment epithelial cells in this disorder. Because of their large expanses of dendritic and axonal cytoplasm, neurons rely heavily on autophagy as evidenced by observations that the central nervous system (CNS) is usually the most severely affected organ in various lysosomal storage disorders (LSDs) [13]. Indeed, mounting data indicate a key role of impaired autophagy in a range of neurodegenerative conditions, such as AD (Alzheimer's disease), PD (Parkinson's disease), HD (Huntington's disease), and ALS (amyotrophic lateral sclerosis). Behavioral defects, neuronal loss, and abundant accumulation of polyubiquitinylated proteins were observed in mice after conditional knockout of Atg5 or Atg7 [14,15] underscoring the role of basal autophagy in the CNS and supporting the notion that compromised autophagy can lead to neurodegeneration. A mechanism underlying disease pathogenesis of many neurodegenerative diseases is the adoption of an abnormal conformation by certain proteins, for example, Aβ (amyloid-β) and tau in AD, α-syn (α-synuclein) in PD, and Htt (huntingtin) in HD [16], all of which are potential autophagy substrates. Notably, neurons, cardiomyocytes, and retinal pigment epithelial cells are all terminally differentiated postmitotic cells and cell division is not an available mechanism to dilute accumulated waste. Efficient removal of damaged proteins or organelles by autophagy is, therefore, particularly vital in these cells to preserve homeostasis during aging and in disease. Not surprisingly, the brain, heart, and eyes are particularly vulnerable to damage related to autophagy impairment.

14.3 AUTOPHAGY ENHANCER SCREENING

Given its pivotal role in multiple diseases, and especially neurodegeneration, autophagy deficiency is an attractive therapeutic target. Effort has focused particularly on identifying small molecules that can enhance autophagy, lower the burden of toxic protein aggregates, and ultimately provide therapeutic benefit. Indeed, autophagy upregulation is associated with beneficial effects on diverse disease-associated phenotypes in cell, fly, worm, and mouse models [17]. For instance, autophagy enhancement using mTOR inhibitors (e.g., rapamycin and analogs) improves deficits in numerous neurodegeneration models, including AD, PD, HD, and ALS [18]. Several researchers have conducted extensive compound screenings to identify positive autophagy regulators, which has resulted in a range of interesting chemical leads [19–21].

14.3.1 Methods for Monitoring Autophagy

As shown in Table 14.1, most investigators have used multiple methods to demonstrate the steady-state accumulation of autophagosomes. Transmission electron microscopy (TEM) was the first technology used to detect autophagosomes, the morphological hallmark of autophagy. Despite limitations of this technology, TEM is extremely informative and remains one of the most used methodologies to monitor autophagy. Atg8/LC3 is currently the most widely used protein markers of autophagosomes. LC3 is initially synthesized in an unprocessed form, which is converted into an LC3-I form by removal of amino acids from the C terminus. Bound principally to microtubules, LC3-I is modified to LC3-II upon PE-conjugation, which becomes associated with the phagophore membrane. Changes in autophagosome formation are commonly assessed by measuring the abundance of LC3-II on Western blots or detecting the presence of LC3-II on vesicular organelles with LC3-specific

TABLE 14.1 List of Autophagy Enhancers Reported in the Literature

Compound	Basic pK_a[a]	clog P[a]	Mode of Action/ Pharmacological Class	Primary	Secondary	*In Vivo*	Other Findings[b]	References
2′,5′-Dideoxyadenosine			Adenylyl cyclase inhibitor	a	g			[1]
2-Deoxyglucose	0	−1.46	Inhibits the production of glucose-6-p	a, b			ER stress	[22]
5-Fluorouracil	−8	−0.66	Pyrimidine analog	a, b	i			[23]
A23187	8.27	6.01	Calcium ionophore	a, b, c			ER stress	[24]
Aicar	3.71	−4.12	Inhibits AMPK	a, c				[25]
Amiodarone	8.47	7.64	Class iii antiarrhythmic2	a, b, c, e	f, g		ROS	[1,19,21, 26,27]
Apicidin	0	3.31	Histone deacetylase (hdac) inhibitor	a, d, e			ER stress and ROS	[28,29]
Arsenic trioxide	0	−0.13		a, b, c, d			ER stress	[30,31]
Azd8055	6.48	4.79	ATP-competitive inhibitors of both mTORC1 and mTORC2	a, d, e				[32]
Beta-elemene	0	5.77	Traditional Chinese medicine	a, b, e				[33]
Bortezomib	−0.7	1.53	Proteasome inhibitor	a, b, c				[34]
Brefeldin A	0	2.7	Protein transport inhibitor	a, b, c			ER stress	[24]
Bufalin	0	2.78	Phytochemicals in herbal medicine	a, b, c	f		ROS	[35]
Calpastatin			Calpain inhibitors	a, b	f, g			[1]
Calpeptin			Calpain inhibitors	a, b	f, g			[1]

Camptothecin	3.07	1.22	Cytotoxic quinoline alkaloid	b, c, d				[36]
Carbamazepine	−3.8	2.77	Antiepileptics2 (MIP synthase inhibitor)	a	f, g	X		[37]
Carfilzomib	4.96	4.2	Proteasome inhibitor	a, b	f			[38]
Celecoxib	0	4.01	Cox-2 selective nonsteroidal anti-inflammatory	a, b, d	f		ER stress	[39]
Cepharanthine	7.61	6.29	Natural alkaloid small molecules	a, b	f			[40]
Cetuximab			Epidermal growth factor receptor inhibitor	a, b, c, d				[41]
Chlorpromazine	9.2	4.54	Antipsychotic	a				[42]
Cisplatin	5.06	0.04	Platinum-containing anticancer drugs	a, b, c			LMP	[43,44]
Clonidine	8.16	2.49	Imidazoline-1 receptor agonists	a, b	g			[1]
Coibamide A			*n*-Methyl-stabilized depsipeptide	a, b, c, d	f			[45]
Compound C	8.53	2.03	Inhibits AMPK activity	a, c, d, e	f, g			[46,47]
Curcumin	0	2.64		a, b	i		LMP	[47]
Dauricine	7.88	4.34	Natural alkaloid small molecules	a, b	f			[40]
Deforolimus	0	3.91	Inhibits mTORC1	a, b		X		[48–50]
Dexamethasone	−3.3	1.68	Glucocorticoid2	b, c	f			[51]
Eb 1089	0	6.5	Vitamin D receptor agonist	a, c				[52]

(*continued*)

TABLE 14.1 (*Continued*)

Compound	Basic pK_a [a]	clog P [a]	Mode of Action/ Pharmacological Class	Primary	Secondary	*In Vivo*	Other Findings[b]	References
Erlotinib	5.32	3	Epidermal growth factor receptor inhibitor	a, b, c, e				[53]
Everolimus	−2.7	7.4	Inhibits mTORC1	a, b		X		[48–50]
Fangchinoline	7.71	3.35	Isolate of *Stephania tetrandra*	a, b, d	f			[54]
Fasudil	9.73	1.72	Rho-kinase inhibitor	a, b				[55]
Flavopiridol	9.28	3.27	Flavonoid alkaloid cdk9 kinase inhibitor	a, b	g		ER stress	[56]
Fluoxetine	10.05	3.93	Antidepressant of the selective serotonin reuptake inhibitor (SSRI)	c, e			ROS	[57,58]
Fluspirilene	9.31	5.86	Dopamine antagonists2	a, b	g			[21,42]
Fty720	8.71	4.18	Sphingosine analog, is a novel immunosuppressant	a, b, c	f, h			[59]
Gefitinib	7	2.7	Epidermal growth factor receptor inhibitor	a, b, c, e				[53]
Gemcitabine	−1.3	−1.5	Nucleoside analog	a, c				[60]
Gf 109203X	9.61	6.57	Selective competitive inhibitor of protein kinase c and of glycogen synthase kinase-3	a	g			[61]

Ginsenoside F2	0	3.24	Steroid glycosides from ginseng	a, b, d, e				[62]
Imatinib	7.55	2.89	Tyrosine kinase inhibitor	b, c, e	f		LMP	[63,64]
Imipramine	9.49	4.36	Tricyclic antidepressant	a, b, c, d				[65]
Isoliensinine	7.94	4.06	Natural alkaloid small molecules	a, b	f			[40]
L-690,330	0	−0.06	IMPase inhibitor	a, b	g			[66]
Latrepirdine	9.05	3.49	Antihistamine drug	a, b	g, i			[67]
Liensinine	7.94	4.06	Natural alkaloid small molecules	a, b	f			[40]
Lithium	0	0	IMPase inhibitor	a, b	g	X		[66,68]
Loperamide	9.41	4.77	Ca^{2+} channel blockers	a, b	g			[1,21,42]
Maprotiline	10.62	4.36	Tetracyclic antidepressant	c, e				[57]
Mesoridazine	8.19	3.57	Antipsychotic	a				[42]
Metformin	12.33	−1.25	Antidiabetic drug	a, c				[25]
Minoxidil	4.34	1.3	K+ATP channel opener2	a	g			[1]
N(10)-substituted phenoxazine	9.2	4.52		a, b, c	f, g			[42]
Nelfinavir	4.72	8.18	Protease inhibitors	a, b				[69]
Nerve growth factor	NA	NA		a, c, d				[70]
Nf449			Gαs inhibitor	a	g			[1]
Nicardipine	8.18	3.82	Ca^{2+} channel blockers	a, b	g			[1,21]
Niclosamide	6.89	3.91	Antiparasitic2	a, b	h			[19]
Niguldipine	8.45	6.04	Ca^{2+} channel blockers	a, b	g			[1,21,42]
Nilotinib	5.64	4.42	Tyrosine kinase inhibitor	a, d	f			[71]
Nimodipine	5.41	3.05	Ca^{2+} channel blockers	a, b	g			[1,21]
Nitrendipine	5.43	2.88	Ca^{2+} channel blockers	a, b	g			[1,21]
Nortriptyline	10.47	4.43	Tricyclic antidepressant	a	g			[42]
Npi-0052	0	2.28	Proteasome inhibitor	a, b, c				[34]

(continued)

TABLE 14.1 (*Continued*)

Compound	Basic pK_a[a]	*c*log *P*[a]	Mode of Action/ Pharmacological Class	Primary	Secondary	*In Vivo*	Other Findings[b]	References
Nvp-Bez235	6.41	4.73	Dual pi3k-mTOR inhibitor	a, d				[72]
Obatoclax	7.91	6.31	Inhibitor of the bcl-2 family of proteins	a, b				[73]
Oprozomib	1.62	1.5	Proteasome inhibitor	a, b	f			[38]
Osu-Hdac42			Histone deacetylases inhibitor	a, b, c			ER stress	[74]
Penitrem A	0	6.36	Inhibits high conductance Ca^{2+}-activated K^+ channel2	a, b	g			[21]
Perhexiline	10.58	5.53	Ca^{2+} channel blocker	a, b	h			[19]
Phenethyl isothiocyanate (Peitc)	0	3.47	Unknown	a, b, c, d				[75]
Pi103	2.97	3.04	Dual pi3k and mTOR inhibitor	a, b	f, i			[76]
Pimozide	8.38	5.83	Antipsychotic	a, b	g			[1,21,42]
Promazine	9.2	3.93	Antipsychotic	a				[42]
Promethazine	9.05	4.29	Antihistamine	a				[42]
Pterostilbene	0	4.06		a, c				[77]
Quercetin	−4	2.16		a, d, e	h	X		[78]
Rapamycin	−3	7.45	Inhibits mTORC1	a, b	g, i	X		[48–50]

Resveratrol	−5.7	3.4	Activates sirtuin-1 (histone deacetylases), inhibits p70 s6 kinase	a, b, c, d			LMP	[79–81]
Rilmenidine	7.88	1.48	Imidazoline-1 receptor agonists	a, b	g	X		[1,82]
Rosiglitazone	6.5	3.02	Antidiabetic drug in the thiazolidinedione class	a, b, c				[83]
Rottlerin	0	4.8	Inhibitor of PKC delta and calmodulin kinase iii2	a, b, c	f, h			[84,85]
Safingol	8.57	5.33	Lyso-sphingolipid protein kinase c inhibitor	a, c, d, e			ROS	[86]
Sertraline	9.47	5.08	Antidepressant of the SSRI	c				[87]
Siramesine	9.45	4.36		a, b, c			LMP and ROS	[88–90]
Smer10	0	2.71	Unknown	a, b	g	X		[91]
Smer18	0	3.68	Unknown	a, b	g	X		[91]
Smer28	6.12	2.1	Unknown	a, b	g	X		[91]
Spermidine	10.9	−1.1	Inhibits histone acetyl transferases	a, b				[79]
Tamoxifen	8.76	6.35	Estrogen receptor antagonist	c			LMP and ROS	[92–94]
Temozolomide	−3.6	−0.28	Chemotherapy drug/alkylating agent	a, b, c, d	f			[95]
Temsirolimus	−3.9	7.13	Inhibits mTOR1	a, b		X		[48–50]
Tetrandrine	7.7	3.23	Calcium channel blocker	a, b, d, e			ROS	[96]
Thapsigargin	0	4.44	Sarco/endoplasmic reticulum Ca^{2+} ATPase inhibitor	a, b, c			ER stress	[24]

(continued)

TABLE 14.1 (*Continued*)

Compound	Basic pK_a [a]	*c*log P [a]	Mode of Action/ Pharmacological Class	Primary	Secondary	*In Vivo*	Other Findings[b]	References
Thioridazine	8.93	5.47	Antipsychotic	a, d				[42]
Torin1	4.19	4.04	ATP-competitive mTOR inhibitor	a, b				[97]
TRAIL				a, b, c	f		LMP	[98,99]
Trehalose	−3.6	−4.8	Chemical chaperone	a, b	f, g	X		[100,101]
Trifluoperazine	8.39	5.03	Antipsychotic	a, b	g		LMP and ROS	[21,42, 102]
Triflupromazine	9.2	4.81	Antipsychotic	a				[42]
Tunicamycin	8.41	−1.22	Mixture of homologous antibiotics	a, b, c			LMP and ER stress	[24,103]
Valproic acid	0	2.58	Antiepileptics (MIP synthase inhibitor)	a, b, c, d, e			LMP and ROS	[104,105]
Verapamil	9.68	5.04	Ca^{2+} channel blockers	a, b	g			[1,21]
Vorinostat	2	−3.5	Histone deacetylases inhibitor	a, b, c			ER stress	[74]
Y-27632	10.74	1.04	Rho-associated protein kinase inhibitor	a	g			[106]

List of *in vitro* assays that are used to demonstrate autophagy induction from literature includes the following: a, increase in LC3II protein level (western blot); b, LC3 puncta; c, autophagic vacuoles by electron microscopy; d, increased vesicular organelles using acridine orange or lysotracker; e, upregulation of autophagy or lysosomal genes; f, inhibitor blocking assay using bafilomycin A or chloroquine; g, clearance of autophagic substrate; h, free GFP assessment; i, p62 abundance assessment.

[a] Compound's physicochemical properties basic pK_a and *c*log P are obtained from ChEMBL [107].

[b] ER stress, endoplasmic reticulum stress; ROS, reactive oxygen species; LMP, lysosomal membrane permeabilization.

antibodies using immunofluorescence (IF) labeling. An additional advantage of an immunohistochemical approach is the capability to colocalize LC3 with other organelle markers. LC3 tagged with a fluorescent protein such as GFP may be introduced into cells to detect autophagosomes and has recently been used to assess autophagy changes in animals *in vivo* [108]. Both endpoints (LC3-GFP puncta and LC3 IF puncta) can be quantified with Image J or the high content imaging platform for automatic image capturing and quantification. So far there is no specific dye to stain autophagosomes although acridine orange is frequently used in the literature to demonstrate development of acidic vesicular organelles (AVOs) as a surrogate marker for autophagosome [78]. However, staining from acridine orange mainly reflects the change of lysosome since this is a dye that accumulates in acidic lysosomes [109].

One challenge in screening for autophagy modulators is to discern enhancers from inhibitors, both of which may increase the number of autophagosomes. All previously mentioned methods reflect autophagosomes only in their static state rather revealing information about the dynamic balance between autophagosome formation and clearance ("autophagy flux"). Because increased autophagosome number can indicate autophagy induction or failed autophagosome clearance (e.g., blocked fusion with lysosomes or degradation by lysosomes), secondary assays are typically required to further understand autophagy flux. Direct evidence of autophagy flux activity can be obtained by measuring the degradation of long-lived proteins; however, this method involves somewhat labor-intensive radiolabeling of the cells and subsequent separation of acid-soluble from acid-insoluble radioactivity. Its wide application is further limited by the need to use inhibitors to establish specificity of proteolytic systems involved, potential for released amino acids to be metabolized and low autophagic activity of cells under certain conditions. The most common assay used in the literature is the determination of autophagosome formation rates by measuring levels of LC3-II in the presence and absence of an inhibitor (e.g., chloroquine, bafilomycin A, or proteases inhibitors) that blocks LC3 degradation. A rise in LC3-II levels in the presence of inhibitors can therefore reflect autophagosome formation. While relatively easy to apply once proper conditions of inhibitor concentration and incubation time are established, this assay provides no information about the lysosomal steps of autophagy (fusion and degradation). Therefore, this endpoint should not be considered as a reliable endpoint to evaluate autophagic flux. The substrate targeting protein, p62 protein, also called sequestosome 1 (SQSTM1), binds to ubiquitinated proteins facilitating their sequestration and ultimate delivery to lysosomes for degradation, during which process p62 is also degraded. Since p62 accumulates when autophagy is inhibited and its levels decrease when autophagy is induced, p62 turnover has been used as a marker of autophagic flux. Special consideration needs to be taken when using p62 data for autophagy enhancers since reduction of p62 synthesis by poor cell health could confound the interpretation. A dose-dependent response can certainly increase the confidence of the data interpretation using this assay. In addition, free GFP abundance relative to complete fusion of GFP-LC3 has also been used as a marker to reflect autophagic flux [110]. GFP-LC3-containing autophagosomes will be degraded after fusion with lysosomes. However, the free GFP moiety generated from the GFP-LC3 degradation is relatively resistant to proteolysis; hence, the appearance of free GFP to some extent reflects cargo delivery (GFP-LC3) from the autophagosome to the lysosome and subsequent degradation. Caution must be used, however, when conducting the GFP-LC3 cleavage assay as a determinant of autophagic flux because GFP fragments can be further degraded and thus do not accumulate when lysosomal acidity is high, for example, during starvation [111]. Recently, tandem mRFP/mCherry-GFP fluorescence microscopy has been particularly designed to monitor flux. The principle of the assay is based on the differential pH dependence of green and red fluorescent protein tags: GFP signal is typically quenched in a strongly acidic environment while mRFP and mCherry remain fluorescent. This assay can be used in the format of single time point data, where an increase of total red+green+ and red+green− vesicular compartments indicates autophagosome formation (i.e., induction of autophagy) and an increase of red+green− compartments alone reflects successful autophagosome–lysosome fusion and lysosome acidification associated with lysosomal clearance of autophagy substrate. Since the autophagosome is a transient organelle existing for less than 10 min before fusing with the lysosome [112], evaluating the dynamic behavior of autophagosomes provides more critical information on autophagic flux than single time point data. Tandem technology can be used to capture this dynamic behavior in real time and reveal the rate of formation and clearance of autophagosomes over time. This technique requires a high-resolution microscope (e.g., confocal) with temperature and CO_2 control to facilitate the kinetic monitoring. In addition, upregulation of Atgs (e.g., Atg5 and Atg7) has also been used as supporting evidence for autophagy activation [26,53]. However, most of the Atgs do not always show significant changes in mRNA and protein levels when autophagy is induced and the extent of increase varies among different cell and tissue types. More importantly, only induction of autophagy can be concluded from the Atgs data and this assay still does not provide information about the lysosomal degradation step.

14.3.2 Autophagy Enhancers Identified from Early Literature

Because of the potential of therapeutic benefits of autophagy enhancement, large-scale screening for autophagy modulators has been carried out in various labs [19–21,42]. There are many additional reports of fortuitously identified autophagy enhancers/activators in the literature (Table 14.1). Compounds exhibiting autophagy modulation activity are structurally and therapeutically diverse. Categorization of these agents is mostly based on mechanism of action and inducers can be broadly grouped as mTOR-dependent and mTOR-independent mechanisms. Compounds work through the mTOR-dependent pathway include not only classic mTOR inhibitors, such as rapamycin and torin 1, but also compounds from different pharmacological classes, for example, calcium channel blockers (perhexiline and verapamil), antiarrhythmics (amiodarone), and antiparasitics (niclosamide) [19]. Nevertheless, for a large number of compounds, the mechanism(s) by which they modulate autophagy have yet to be identified.

Interestingly, many compounds (e.g., tamoxifen, amiodarone, and verapamil) that are believed to be autophagy enhancers are basic lipophilic, which allows the compound to accumulate in acidic organelles (i.e., lysosome) by pH partitioning [113]. Lysosomotropic compounds were recently shown to inhibit autophagy by perturbing lysosomal function [114]. Historically, vinblastine, a microtubule disruptor, was once believed to be an autophagy "enhancer" but later identified to actually inhibit autophagosome–lysosome fusion. It may therefore be prudent to reevaluate the classification of these basic lipophilic compounds as autophagy enhancers.

Lysosomes are the final destination of autophagosomal substrates for degradation. Lysosomes are membrane-enclosed compartments filled with acid hydrolytic enzymes (e.g., cathepsins) used to digest macromolecules and are found in the cytosol of nearly all mammalian cells. For optimal activity of acid hydrolases, lysosomes must maintain a low internal pH of about 4–5. The acidic pH in the lumen is achieved by the vacuolar H+ ATPase, which uses the energy of ATP hydrolysis to pump H+ into the lysosome. The pH gradient between the lysosomal lumen and the cytosol (pH ≈ 7.2) can drive overaccumulation of basic lipophilic compounds via pH partitioning. Generally, lipophilic free bases can pass through the plasma membrane and lysosome membrane; however, the majority will lose permeability and become trapped inside lysosomes due to their protonation by the acidic lysosomal pH. For instance, chloroquine, a well-known lysosomotropic compound, can easily reach concentrations in excess of 20 mM inside lysosomes equal to a several hundred-fold high level relative to that outside of the cells [113]. Two physicochemical properties, basic pK_a (acid dissociation constant for the conjugate acid for the weak base) and $c\log P$ (partition coefficient between octanol and water, representing membrane permeability) both affect the accumulation by influencing the extent of lysosomal trapping and regulating the kinetics of passive permeation. The compounds that accumulate in lysosomes are classified as lysosomotropic agents. It was demonstrated previously that compounds with certain physicochemical properties (basic $pK_a > 6.5$ and $c\log P > 2$) tend to be lysosomotropic [115]. To understand if physicochemical properties could contribute to autophagy modulation by the compounds identified so far, we evaluated relationships between physicochemical properties (i.e., basic pK_a and $c\log P$) and autophagy modulation. Surprisingly, half of autophagy "enhancers" fit into the profile of being basic with $pK_a > 6$, in which majority of them are lipophilic with $c\log P > 2$ (Table 14.1 and Figure 14.1). It is noteworthy that commonly known "autophagy enhancers" amiodarone and tamoxifen are located in the basic lipophilic region (Figure 14.1)

It has been well established that lysosomotropic compounds can impair lysosome function potentially through multiple mechanisms. Lysosomotropic compounds, chloroquine and methylamine, increase lysosomal pH drastically (0.5–2.0 pH units) after accumulation [116]. In addition, significant pH elevation by chloroquine is observed *in vivo* [117]. In a previous study [115], lysosomotropic compounds were shown to decrease staining of cells by Lysotracker, which requires low pH to fluoresce. An increase in pH is expected to suppress lysosomal degradation since most lysosome enzyme activities have acidic pH optima. An increase of pH may also decrease the fusion capability of lysosomes in some cells as seen for chloroquine [118] although pH dependence of lysosome fusion is not universally

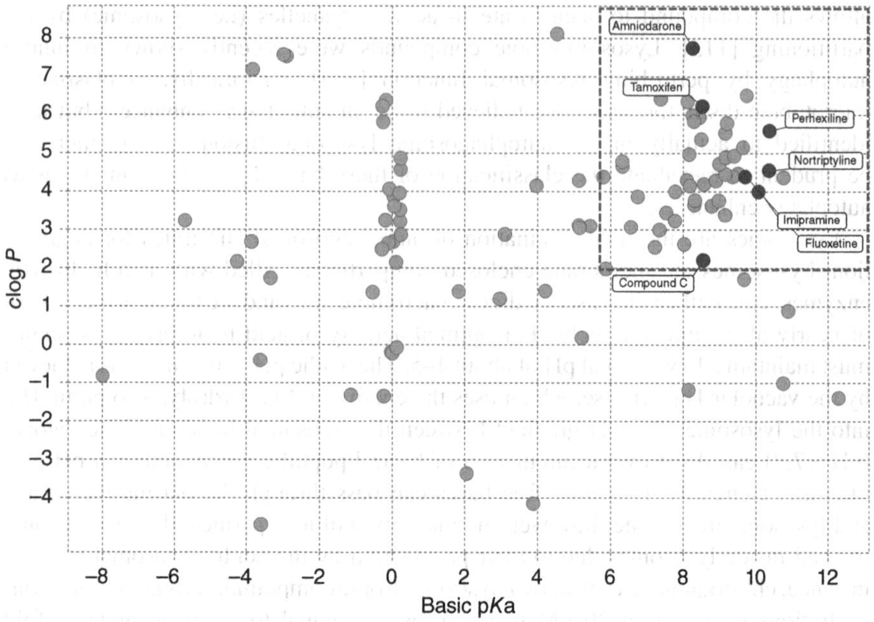

Figure 14.1 Correlation of physicochemical properties with autophagy screening: A scatter plot showing the distribution of autophagy enhancers within the clog*P*-basic pK_a physicochemical property space. A group of compounds clustered within the area (blue dotted square) where the clog*P* was >2 and basic pK_a was greater than 6. (*See color plate section for the color representation of this figure.*)

observed [119]. Lysosomotropic compounds have also been shown to decrease lysosomal enzyme activity. For instance, chlorpromazine and chloroquine inhibit the lysosomal phospholipase A1 *in vitro* [120,121] and desipramine, chlorpromazine, and chloroquine downregulate acid ceramidase [122]. Multiple lysosomotropic compounds can redistribute the mannose 6-phosphate receptor from the trans-Golgi network to endosomes and concomitantly increase the secretion of lysosomal enzymes causing intracellular lysosomal enzyme levels to decline [123] further exacerbating lysosomal dysfunction.

Given that lysosomal substrate degradation is an essential step of autophagy, impairment of lysosomal function, by definition, impairs autophagy activity. Not surprisingly, various lysosomotropic compounds have recently been shown to be autophagy inhibitors [114]. Interestingly in the same study, transcription of plentiful long-lived proteins, including cytoskeletal and extracellular matrix proteins, was suppressed further suggesting that autophagic turnover of these proteins is hindered. Notably, lysosomotropic compounds also impact other trafficking process (i.e., endocytosis and phagocytosis process) that delivers materials to lysosomes. The lysosomotropic compounds chloroquine, tamoxifen, and verapamil decrease phagocytosis activity [124,125] and chloroquine is recognized as a clathrin-dependent endocytosis inhibitor [126]. Amiodarone impairs trafficking through late endosomes and induces a Niemann–Pick C-like phenotype [127]. In addition, multiple cationic amphiphilic compounds (e.g., amiodarone, fluoxetine, nortriptyline, and imipramine) identified as "autophagy enhancers" are reported to induce phospholipidosis (PLD). Although the toxicological impact of PLD is still debatable, it is believed to be a drug-induced LSD characterized by the excess accumulation of phospholipids. Multiple mechanisms are proposed to contribute to PLD including lysosomal dysfunction via inhibition of lysosomal phospholipase enzyme activity or transport. Basic pK_a and $c\log P$ are important descriptors for *in silico* prediction of phospholipidosis [128]. Taken together, the evidence supports the conclusion that compounds with basic lipophilic properties have the propensity to impair lysosomal functions, consequently hampering autophagy activity rather than boosting it.

The foregoing evidence on current autophagy enhancer screening assays raises the critical question of whether they are sufficient to distinguish enhancers from inhibitors. The most commonly used autophagy flux assay is LC3 assessment with and without lysosome inhibitors. Compounds that increase LC3 abundance together with lysosomal proteolysis inhibitor treatment, as compared to single agent effect on LC3 in the absence of inhibitor, have often been interpreted as showing that the compound increased "autophagic flux." Autophagic flux in this context is a measure of autophagosome formation and provides no information about the lysosomal steps of autophagy. More appropriately, autophagic flux should reflect the activity of the entire autophagic process from substrate sequestration through complete substrate degradation. TFEB (transcription factor EB) stimulates the expression of genes for lysosome biogenesis as well as many involved in autophagosome formation and fusion of autophagosomes with lysosomes [129,130]. Recently, TFE3, another family member of the microphthalmia-associated transcription factor (MiTF) and TFE (miTF/TFE) families, was also shown to increase the expression of the proteins involved in autophagy and lysosomal biogenesis [131]. Interestingly, both transcriptional factors respond to lysosomal dysfunction, as shown by nuclear translocation of both TFEB and TFE3 after chloroquine treatment [130,131]. Possibly these transcriptional factors are activated as a compensatory feedback response to overcome the lysosomal stress triggered by this lysosomal inhibitor, which likely results in reduced signaling to the lysosomal amino acid sensing complex that regulates TFEB phosphorylation and translocation. Further experiments are required to confirm whether or not other basic lipophilic compounds trigger a similar nuclear translocation of TFEB and TFE3 to increase the gene expression of the protein involved in autophagosome formation and lysosomal biogenesis. In this regard, LAMP1 expression is increased by the basic lipophilic compound imatinib [63] and various autophagy genes (e.g., beclin 1, Atg5, or Atg7) are upregulated by amiodarone, gefitinib, and compound C [26,46,53]. In addition, as observed for chloroquine, multiple basic lipophilic compounds, including perhexiline, compound C, siramesine, and imipramine, decrease mTOR signaling, suggesting that nuclear translocation of TFEB and TFE3 may be occurring since mTOR activation contributes to the cytosolic localization of these transcription factors [130,131]. It is therefore plausible that basic lipophilic compounds generally act similarly to the mechanism proposed for chloroquine, namely by decreasing mTOR activity, triggering nuclear translocation of TFEB and TFE3, and ultimately inducing the gene expression of lysosomal and autophagy genes as a feedback response to the accumulation of substrates in autolysosomes, often referred to as "autophagic stress." Accordingly, activation of autophagy demonstrated by lysosomal inhibitor assay might not be sufficient to identify effective autophagy enhancers. Similar precautions need to be taken when interpreting the upregulation of autophagy or lysosome genes as evidence of autophagy enhancement since basic lipophilic autophagy inhibitors can activate those genes following lysosomal stress. Increased expression of autophagy or lysosome genes may indicate initiation of autophagosome or lysosome biogenesis but does not necessarily reflect increased autophagy activity, which should be a measure of both autophagosome formation and clearance.

An additional assay previously used to support autophagy enhancement is measurements of numbers of AVOs labeled by acridine orange. The development of AVOs, a characteristic of autophagy [41,78], can be monitored by lysosomotropic acridine orange, which selectively accumulates in acidic organelles, enabling quantitation of both number and size of lysosomes, although the lysosome expansion can imply higher or lower lysosomal activity. Multiple lysosomotropic compounds, for example, chloroquine that could perturb lysosomal functions have been shown to trigger marked expansion of lysosomal volume [132,133]. In addition, lysosomal volume expands in multiple lysosomal storage diseases, which are also associated with lysosomal dysfunction [134,135]. Reduction of lysosomal volume in these disease states has been used as a phenotypic screening to identify the compounds that could improve lysosomal function. [135,136]. Therefore, increase of AVO might not be the appropriate marker for identifying true autophagy enhancers due to the potential association with lysosomal dysfunction.

Assays based on p62 turnover and the tandem-tagged GFP-mCherry-LC3 probe can provide suitable appraisals of autophagy flux. Induction of autophagy by arsenic trioxide has been demonstrated by TEM, IF staining with EGFP-LC3, and increased LC3 in the presence of the lysosomal proton pump (vacuolar ATPase) inhibitor bafilomycin A [30,137,138]. However, arsenic trioxide inhibited autophagy flux as evidenced by the tandem-tagged GFP-mCherry-LC3 probe assay [139]. In this assay, starved cells exhibit both yellow and red puncta but cells treated with arsenic trioxide exhibited persistent yellow/orange puncta, similar to the effects of bafilomycin A, indicating a block of autophagy flux. Arsenic also induced accumulation of p62 puncta, further indicating inhibition of autophagy flux. Although arsenic is not basic lipophilic, lysosomal uptake of heavy metals is well documented [140,141]. Indeed, arsenic caused rapid destabilization of lysosomes [142], which might underlie the negative effect of arsenic on autophagy. Thapsigargin, an inhibitor of the ER SERCA calcium pump, has been frequently used as an autophagy inducer [24,143], as reflected by increased LC3 in the presence of bafilomycin A [143]; however, in the tandem GFP-mCherry-LC3 assay, loss of GFP signal is delayed after thapsigargin indicating impaired fusion of autophagosomes with lysosomes [144]. Fluoxetine, nortriptyline, and imipramine have all been classified as "autophagy enhancers" based on earlier studies, but they were recently shown to increase p62 protein abundance in a dose-dependent manner [114], supporting an inhibition of autophagic flux.

Collectively, these findings suggest that many autophagy enhancers identified so far might not represent true enhancers of autophagy flux. In particular, compounds with basic and lipophilic moieties need to be evaluated very carefully due to their high potential to disturb lysosomal function. Even compounds lacking a basic lipophilic moiety may also perturb lysosome function as illustrated by arsenic trioxide, which destabilized lysosomal membranes [142]. TRAIL triggers lysosomal membrane permeabilization (LMP) via death receptor-5 [98]. LMP and relocation of cathepsins contribute to the cell death by curcumin [47]. Pharmacological inducers of ER stress, such as tunicamycin, have also been reported to associate with LMP [103]. Notably, LMP has recently been shown to stimulate protein aggregate formation in neurons [145]. In addition, oxidative stress or ER stress was associated with multiple putative autophagy enhancers including amiodarone, apicidin, and fluoxetine (Table 14.1). Compounds with liability of LMP, oxidative stress, or ER stress are more likely to weaken their therapeutic benefit for conditions requiring autophagy enhancement.

Since all membrane trafficking including autophagy, phagocytosis, and endocytosis converge on lysosomes compounds that promote autophagy flux should not compromise lysosomal function. For instance, chloroquine, a lysosomotropic compounds that inhibit lysosomal function, is a well-known autophagy inhibitor and has also been shown as an inhibitor for endocytosis and phagocytosis [124,126]. As we discussed previously, many reported "autophagy enhancers" target lysosomes either due to their lysosomotropic nature (e.g., amiodarone and latrepirdine) or by compromising lysosomal membrane integrity (e.g., TRAIL). Analysis of other trafficking processes, such as endocytosis and phagocytosis, could provide further certainty on the lysosomal functions. In other words, if compounds shown to augment autophagy perturb endocytic, phagocytic, or lysosomal functions, extra consideration should be taken when classifying these compounds as "autophagy enhancers." Especially if "increased" autophagy is surmised on the basis of only elevated LC3 expression with proteolysis inhibitors, increased expression of Atgs, or increased AVOs, this "increase" may merely reflect a compensatory response to lysosomal dysfunction. Therefore, evaluation of the general function of lysosomes could provide additional supporting evidence for autophagy modulation.

14.3.3 mTOR Inhibitors

Mammalian target of rapamycin (mTOR) is a conserved serine/threonine protein kinase that plays a significant role in cell growth, proliferation, and protein synthesis by regulating the activities of factors such as 4E-BP1, S6K, and Akt. mTOR1 is also a master negative regulator of autophagy by blocking the ULK1-ATG13-FIP200 complex. Various mTOR inhibitors were developed to enhance autophagosome biogenesis and thereby promote clearance of aggregate-prone proteins. Notably, genetic or pharmacological inhibition of TOR activity multiple model systems is associated with life span extension [146–149], possibly through inhibitory effects on protein synthesis and stimulation of autophagy. Numerous studies show that rapamycin can alleviate toxicity exerted by protein aggregates in a range of neurodegeneration models (e.g., HD, PD, and AD) [150]. Surprisingly, rapamycin was shown to induce pulmonary PLD demonstrated by lamella body formation using EM (NDA 021083) indicating alteration of lysosomal function by mTOR inhibition. In order to have a thorough understanding of mTOR inhibitors, we compared the biological effects of TOR inhibitors (putatively to enhance autophagy induction), like rapamycin, and lysosomotropic compounds, such as chloroquine (putatively to block lysosomal degradation) (Table 14.2). Multiple endpoints such as vacuolation, PLD, endocytosis, and phagocytosis inhibition were employed and unexpectedly, mTOR inhibitors and lysosomotropic compounds, which supposedly have opposite effects, behaved indistinguishably in all the endpoints that we compared. Lysosomotropic compounds are well known to cause cytoplasmic vacuolization [151] after they accumulate in acidic organelles. The mTOR inhibitor, everolimus, also induces cytoplasmic vacuolization in macrophages [152]. Large, clear vacuoles by rapamycin were shown to be lysosomes that presumably had not returned to normal size after proteolysis when autophagy is induced heavily due to mTOR inhibition [153]. Although the biological meaning of vacuolization is not well understood, the appearance of vacuoles in the brain is used as a typical hallmark of neurodegeneration in *Drosophila* [154] and is observed in aged tissue [155]. Cytoplasmic vacuolization is also seen in various lysosomal storage diseases [156]. However, it is possible that vacuolation represents the enlarged lysosome that is associated with increase in lysosomal function. Drug-induced PLD, which is characterized by accumulation of phospholipid and drugs in lysosomes, are often regarded as an acquired LSD [157]. Lysosomotropic compounds, which typically contain a hydrophobic ring structure and a hydrophilic amine moiety, are best known to trigger PLD [158]. As discussed earlier, the "autophagy enhancers" amiodarone

TABLE 14.2 Comparable Biological Effects from mTOR Inhibitor (Autophagy Enhancer) and Lysosomotropic Compounds (Autophagy Inhibitor)

Biological Effect	TOR Inhibitor	Lysosomotropic Compounds
Vacuolation	+	+
Phospholipidosis	+	+
Endocytosis inhibition	+	+
Phagocytosis inhibition	+	+
mTOR inhibition	+	+
TFEB nuclear translocation	+	+
LAMP1 upregulation	+	+
Increase Lysotracker staining	+	+
Immune response inhibition	+	+
Cancer therapy	+	+

Effects presented are not necessarily related to autophagy modulation.

and tamoxifen are classic inducers of PLD [159] and mTOR inhibitors, including everolimus, sirolimus, and temsirolimus [160], are all reported to induce PLD, suggesting that they perturb lysosomal functions. Besides PLD induction, both mTOR inhibitors and lysosomotropic compounds may also impair other aspects of membrane trafficking, such as endocytosis and phagocytosis. Chloroquine is recognized as a clathrin-dependent endocytosis inhibitor [126] and the phagocytosis of rod outer segments is inhibited by chloroquine and tamoxifen in retinal pigment epithelial cell cultures [124]. Azithromycin, a lysosomotropic antibiotic, delays sequestration of receptor-bound transferrin and peroxidase–antiperoxidase immune complexes into cell-surface endocytic pits and vesicles [161]. Although the mechanism is not clear, mTOR may play a critical role in clathrin-mediated endocytosis [162], as suggested by the ability of rapamycin to reduce albumin uptake [163]. Rapamycin also impairs macropinocytosis and mannose receptor-mediated endocytosis of antigens by bone marrow-derived DCs [164] and decreases phagocytosis of bacteria and apoptotic cells [165].

Many lysosomotropic compounds, for example, niclosamide and imipramine have been shown to lower mTOR1 activity [65,166], possibly attributable to lysosomal dysfunction that reduces the generation of amino acids critical for the mTOR. TFEB is a master regulator of lysosomal biogenesis and autophagy. Interestingly, both mTOR inhibitor and lysosomotropic compounds can trigger TFEB nuclear translocation [130,167], leading to the upregulation of a broad range of lysosomal genes and increased lysosomal biogenesis [168,169]. Lysosomotropic compounds can robustly increase lysotracker staining [132,170]. Similarly, both allosteric mTOR inhibitor (rapamycin) [146] and catalytic inhibitors (PP242 and Torin1) [169] also markedly increase lysotracker fluorescence intensity. Increased lysotracker staining in these situations indicates increased numbers or size of lysosomal-related compartments, which can potentially reflect both an attempted adaptive response and a pathological phenomenon (lysosomal degradative impairment) as seen in lysosomal storage diseases and aging.

Autophagy and autophagy proteins have multifaceted roles in the innate and adaptive immune responses, for example, antigen presentation and lymphocyte development [171]. Surprisingly, with supposedly opposite actions on autophagy, both types of agents (mTOR inhibitors and chloroquine) are shown to inhibit immune response. Tor kinase has multiple effects on innate and adaptive immune responses. The mTOR inhibitors sirolimus and everolimus block the response of T- and B-cell activation by cytokines, which prevents cell-cycle progression and proliferation [172]. Rapamycin has also been shown to suppress the mature dendritic cell function and decrease the production of cytokine IFNγ produced by dendritic cells [173]. As noted earlier, rapamycin has an inhibitory effect on dendritic cells endocytosis [164], which could further decrease the immune response. The mTOR inhibitor (e.g., rapamycin) has been used as an immunosuppressant to prevent transplant rejection. The efficacy of the lysosomotropic compound hydroxychloroquine in the treatment of systemic lupus erythematosus and rheumatoid arthritis has been well demonstrated [174,175], and it has become a standard component of therapy for patients with these diseases. Anti-inflammatory response by hydroxychloroquine could be ascribed to lysosomal dysfunction, which diminishes the formation of peptide–MHC protein complexes required to stimulate CD4+ T cells and results in downregulation of the immune response against autoantigenic peptides. Additional lysosomotropic compounds, for example, fluoxetine and citalopram have also shown anti-inflammatory effects in arthritis animal models [176].

In addition, both mTOR inhibitors and lysosomotropic compounds, such as chloroquine, have been used in cancer therapy. Currently, rapamycin analog, temsirolimus, has been used for the treatment of renal cell carcinoma and everolimus for various cancer types [177]. Owing to their nature of autophagy inhibition, lysosomotropic compounds, for example, chloroquine or, more commonly, hydroxychloroquine in combination with other antitumor agents have been explored in numerous oncology clinical trials [178]. Interestingly, the combination of hydroxychloroquine with the mTOR inhibitor demonstrated an encouraging clinical outcome in a phase I trial with advanced melanoma patients [179].

The unanticipated similarity between mTOR inhibitors and lysosomotropic compounds further accentuate the challenge of autophagy enhancer screening, granting some of the effects could be attributed to pharmacological actions unrelated to autophagy. Data interpretation of endpoints for autophagy screening needs to be considered carefully. In addition, the comparable biological effects raised a critical question about the role of mTOR on lysosomal function. The ability of mTOR inhibitors and lysosomotropic compounds to induce PLD and inhibit endocytosis and phagocytosis suggests that both classes of compounds may have a negative impact on lysosomal functions. An increase of lysotracker staining from both types of agents further supports this notion since a similar phenotype is often observed when lysosomal functions is perturbed, for example, cells from LSD patients.

As discussed earlier, it is understandable as to why lysosomotropic compounds have global negative impact on lysosomes. A negative impact of mTOR inhibitors is seemingly more puzzling, given their ability to stimulate TFEB-mediated lysosomal biogenesis. Recent findings, however, reveal positive roles of mTOR activity in trafficking processes that lead to lysosomes and lysosomal function. Although the mechanisms have not been fully elucidated, mTOR activity may influence microtubule and actin dynamics through the actions of ROCKs, which play a critical role in regulating actin polymerization and associated cytoskeletal rearrangements. mTOR inhibition by rapamycin abolishes ROCK-1 synthesis in macrophages thereby inhibiting chemotaxis and phagocytosis [180]. Rapamycin slows microtubule polymerization and lowers binding of Bik1P to microtubules [181]. mTOR hyperactivation seems to disturb microtubule organization [182], suggesting that a proper mTOR activity balance is important in this process. Disruption of the cytoskeleton obviously hinders various trafficking processes, which prevents cargo delivery to lysosomes and substrate degradation. Beyond these effects, mTOR activity is required to generate protolysosomal tubules and facilitate their maturation into functional lysosomes under conditions of nutrition deprivation [183]. This process, described as autophagic lysosome reformation (ALR), is observed after prolonged starvation and blocked by rapamycin treatment. Similar phenomena such as ALR were also observed in phagosome maturation process where phagosomes and entotic vacuoles undergo a fission process, which redistributes vacuolar contents into lysosomal networks [184]. mTOR inhibition slows down the shrinkage of phagosomes indicating that membrane fission is regulated by mTOR. Recently, mTOR was also shown to positively regulate lysosomal ATP-sensitive two-pore Na^+ channels, which modulate membrane potential and may influence pH stability of lysosomes under certain cellular conditions [185]. This finding underscores the reciprocal relationship between mTOR and lysosomes in which amino acids generated during lysosomal proteolysis modulate mTOR activity and, in turn, potentially regulate additional endosomal and lysosomal functions via lysosomal ATP-sensitive two-pore Na^+ channels and additional changes in intralumenal pH and ion fluxes.

Although it has been demonstrated that phosphorylation of TFEB by mTORC1 contributes to the cytoplasmic location of TFEB [131,167,186], contradictory results have also been reported, in which TFEB nuclear translocation was promoted by mTORC1 activity [163]. The latter study employed an mTORC1 overactivation model by knocking out tuberous sclerosis complex 2(TSC2), which negatively regulates mTORC1. Statistically significant overrepresentation of lysosomal genes and especially genes encoding vacuolar H^+-ATPase (V-ATPase) was identified with higher expression when mTOR was activated in Tsc2−/− cells, but low expression in TSC2−/− with rapamycin treatment indicating V-ATPase genes were tightly regulated by mTORC1. Interestingly, TFEB depletion downregulates V-ATPase expression suggesting that this mTORC1-dependent V-ATPase upregulation requires TFEB. TSC inactivation indeed promoted nuclear translocation of TFEB and this was reverted by rapamycin treatment suggesting that TFEB nuclear localization is mTORC1 dependent. The multiple roles of mTORC1 on various biological processes (e.g., protein synthesis and cell growth) may well complicate the interpretation of the net outcomes of mTORC1 modulation and its ultimate impact on autophagy. It is plausible that mTORC1 might play varying roles in lysosomal function under different physiological/pathological conditions. It is worth noting that the substantial benefits of rapamycin in mouse models of different neurodegenerative diseases may also derive from autophagy-independent therapeutic effects such as rapamycin's antiaging properties, reduced protein translation [18], or change in immune responses. Further characterization of the role of mTORC1 especially on late stages of autophagy, phagocytosis, and lysosomal function under diverse conditions should provide a stronger mechanistic rationale for mTORC1 modulation as a therapeutic approach to neurodegeneration.

14.4 OTHER AGENTS THAT BOOST AUTOPHAGY AND LYSOSOMAL FUNCTIONS

As a promising therapeutic approach, autophagy enhancement has attracted extensive research interest. Given the foregoing discussion about the negative impact on lysosomal function by some of the current "autophagy enhancers" and the complex effects of mTOR inhibition on autophagy, further efforts are warranted for identifying positive autophagy regulators especially the ones that increase autophagic substrate clearance. Emerging evidence suggests lysosomal clearance defects as a mechanism for various adult-onset neurodegenerative diseases, supported by findings on neuronal cell death in LSDs [17]. When the lysosomal clearance process is impaired, autophagy induction might exacerbate the pathology. As a result, the success of any autophagy enhancer in a given condition may depend on relieving blockages in lysosomal clearance. Lysosomal function or general membrane trafficking as therapeutic targets in neurodegeneration is relatively unexplored. Nonetheless, numerous therapeutic targets have been suggested to have positive effect, including cathepsin activity enhancement, pH restoration, and increasing lysosomal stability [17].

14.4.1 HDAC Inhibition

Histone deacetylase inhibitors (HDAC inhibitors, HDACi) are a class of compounds that interfere with the function of histone deacetylase, an enzyme with diverse biological effects owing to broad actions on gene transcription. HDACi clear the excess accumulation of cholesterol and correct cholesterol storage defects in human NPC1 mutant cells [187]. HDACi have also been shown to correct the cellular phenotype of cystic fibrosis [188]. How HDACi restore the phenotype is not thoroughly understood and multiple mechanisms have been suggested in the literature. Proteostasis of various misfolded proteins could potentially contribute to the efficacy observed since HDACi increase expression and restore activity of NPC1, glucocerebrosidase, and α1-antitrypsin [187,189], thereby promoting lysosomal function. Protein acetylation may regulate autophagy and HDACi promote autophagy via acetylation of ULK1 and Atgs [190–192]. Moreover, increased tubulin acetylation by HDACi and subsequent recruitment of kinesin and dynein may also increase flux of vesicles during autophagy [193]. Interestingly, HDAC6 inhibition rescues both anterograde and retrograde transport [193,194]. Many studies have indeed demonstrated that HDACi can ameliorate deficits in mouse models of a wide range of neurologic disorders including HD, PD, AD, and ALS [195–198]. Many questions about HDAC inhibition, however, remain unanswered. For example, effects of HDACi on autophagy may be cell type-specific since HDACi suppress autophagy in cardiomyocytes [199] but induce autophagy in other cell types (e.g., skeletal muscle cell) [200]. Also, restoration of protein function by HDACi may be selective: increased expression of NPC1 protein with dramatic correction of NPC phenotype is observed with HDACi treatment using human fibroblast carrying NPC1 mutation; however, HDACi has no effect on HDAC in an NPC2 mutant human fibroblast line [187]. Moreover, acetylation may regulate autophagy in opposite directions: KAT2B regulates autophagy negatively, whereas KAT5 upregulates autophagy [201]. Underlying mechanisms that contribute to these different effects require additional studies. Functional substrate identification could improve the druggability of HDACi and more selective HDACi could potentially achieve greater efficacy with less toxicity.

14.4.2 pH Restoration

Lumen acidity of lysosomes is critical not only for optimal lysosomal enzyme activity during autophagy but also for efficient membrane trafficking. Abnormal lysosomal pH could be a key aspect in diseases of accumulation. For example, presenilin 1 mutations, which are the most common genetic form of early-onset familial AD, elevate pH and reduce activation of cathepsins [202,203]. Even a few tenths of a unit increase may be sufficient to reduce the activity of certain lysosomal proteases, such as cathepsin D [204]. Restoring lysosomal pH could have broad therapeutic benefits by enhancing degradation and vesicular trafficking [202]. Indeed, acidic nanoparticles that were targeted to lysosomes of retinal pigment epithelium cells in a model of macular degeneration produced a sustained lowering of lysosomal pH and improved degradative activity [205]. Multiple signaling molecules, which could potentially be therapeutic targets, play roles in regulating lysosomal pH. For example, cAMP elevation can lower lysosome pH and restore cathepsin D activity in fibroblasts from patients with familial AD due to a presenilin 1 mutation [203]. Phosphodiesterase PDE10A inhibition has also been shown to ameliorate striatal and cortical pathology in an HD animal model, further supporting cAMP regulation as a treatment approach for lysosomal accumulation diseases. Glycogen synthase kinase-3 (GSK 3) has also attracted considerable interest in neurodegeneration research. Among its multiple effects, inhibition of GSK3 restores lysosomal acidification by rescuing impaired glycosylation and lysosomal delivery of the v-ATPase subunit V0a1 in a presenilin 1/APP model of AD and consequently reducing β-amyloid pathology [206].

14.4.3 TRP Activator

Mucolipin transient receptor potential channel 1 (TRPML1) is believed to be a Ca^{2+} and Fe^{2+} dually permeable cation channel [207], which predominantly resides in late endosomes and lysosomes [208]. The critical role of TRPML1 in lysosomal function is supported by the lysosomal storage disease mucolipidosis type IV, which is caused by mutations in the gene MCOLN1 (codes for TRPML1). Increasing TRPML1 activity by its agonist is adequate to restore lysosomal degradative functions and prevent abnormal lipid accumulation, possibly by controlling Ca^{2+}-dependent lysosomal trafficking [209]. In addition, TRPML1 might also facilitate H^+ import and lysosomal pH acidification [210]. Both particle ingestion and lysosomal exocytosis are inhibited by synthetic TRPML1 blockers and are defective in macrophages isolated from TRPML1 knockout mice further supporting the positive role of TRPML1 in membrane trafficking [211]. Although these findings support the therapeutic benefit for LSDs, similar approaches might also be beneficial for other neurodegenerative diseases such as AD and PD, which involve defects in lysosomal hydrolytic activity [17,212].

14.4.4 TFEB Overexpression/Activation

Since the discovery of TFEB, overexpression or activation of TFEB has become an appealing therapeutic approach, especially targeting LSDs. TFEB overexpression has been shown to reduce glycosaminoglycans in mouse models of MSD and MPSIIA [213] and glycogen load and lysosomal size in a Pompe disease model [214]. In the hepatic disease associated with alpha-1-anti-trypsin (AAT) deficiency, gene transfer of TFEB increases clearance of the mutant AAT protein and corrects the pathological phenotype in liver [215]. Remarkably, in an MPTP cellular model of PD, TFEB overexpression reversed MPP^+-induced lysosomal depletion and attenuated MPP^+-induced cell death [216]. More recently, TFEB has been shown to partially block proteotoxicity by PGC1α in a mouse model of HD [217]. The therapeutic benefits of TFEB may be attributed to its multiple biological functions as a multitasking transcription factor. Increased lysosomal biogenesis, autophagy, and lysosomal exocytosis by TFEB can collectively contribute to amelioration of toxicity exerted by various pathogenic proteins. In addition, TFEB activation has been shown to enhance folding, trafficking, and lysosomal activity of a destabilized glucocerebrosidase and rescue the activity of a beta-hexosaminidase mutant [218], which could further contribute to the efficacy of TFEB in LSD and neurodegenerative disorders. Obviously, a small molecule approach to promote nuclear translocation could be a very powerful approach to upregulate lysosomal function; however, additional caution needs to be taken to discern the true TFEB enhancing agent from the molecules that induce TFEB nuclear translocation after lysosomal stress (e.g., CQ) [167]. One potential caveat for this approach is that pathogenic proteins released into the extracellular space by TFEB upregulation might not be cleared adequately [219]. Reuptake of pathogenic proteins such as α-synuclein and amyloid-β could seed the misfolding of their normal conformers and promote prion-like spreading [16].

14.4.5 Lysosomal Efficiency

The function of lysosomes is critically dependent on soluble lysosomal hydrolases as well as on lysosomal membrane integrity. Numerous promising therapeutic directions have been evaluated to improve lysosomal efficiency. Enzyme replacement therapies, which replace an enzyme in patients in whom that particular enzyme is deficient, are quite successful for multiple LSDs, including Gaucher disease, Fabry disease, and Pompe's disease. Genetic deletion of cystatin B, an endogenous lysosomal cysteine protease inhibitor, enhanced multiple cathepsin activities, reduced abnormal accumulations of amyloid-ß peptide, rescued autophagic-lysosomal pathology, and restored normal cognition in a mouse model of AD [220]. Recently, the same approach also promoted clearance of accumulated lipids in this disease model [221] Similarly, genetic ablation of cystatin C in a similar mouse AD model significantly decreased plaque load and restored normal synaptic plasticity [222]. Pharmacological chaperone therapy has also proven effective in several models of LSDs [223]. In principle, these reversible small molecule inhibitors bind to a target protein in the ER, stabilize its correct conformation, and thus enable a functional enzyme to traffic to lysosomes. This principle can potentially apply to neurodegenerative diseases, such as AD, to promote the folding and proper function of vATPase [202] or boost enzyme activity to clear lipid storage that can impair lysosomal function [224].

Promoting lysosomal membrane stability can also have therapeutic benefit. LMP impairs lysosomal function, accelerates protein aggregate formation, and contributes to neuronal cell death [88,145,225]. Heat shock protein 70 (Hsp70) acts as a molecular chaperone to stabilize lysosomal membranes [136] and in recombinant form, was able to enhance the enzymatic activity of endogenous mutated acid sphingomyelinase and reverse lysosomal pathology in fibroblasts derived from Niemann–Pick disease. Calpains are abnormally activated in AD [226] and may act as a mediator of LMP [227]. Calpain inhibitors have striking therapeutic effects in mouse models of AD and several other neurological disorders [228–231]. The effects of calpain inhibitors toward inducing autophagy and protecting against LMP, in addition to reducing cytotoxicity of calpain overactivation, have created considerable excitement around calpain inhibitor development.

14.4.6 MicroRNA

MicroRNAs (miRNAs), small noncoding RNAs ~22 nucleotides in length have been shown to regulate gene expression at the posttranscriptional and translational levels. miRNAs guide the binding of the RNA-induced silencing complex to the target mRNA via base-pair interactions with mRNA transcripts, resulting in target mRNA degradation and/or translational inhibition. In general, each individual miRNA has multiple target proteins. miRNAs play an important role in various biological process, such as proliferation, differentiation, and apoptosis [232]. Recently, miRNAs have been found to modulate mammalian autophagy. Proteins involved in multiple steps of autophagy, such as induction, nucleation, and elongation, have been shown to be targeted by various miRNAs [233,234]. For instance, miR181a attenuates autophagy through Atg5 [235]. Interestingly, an emerging role of miRNAs in neurodegeneration has also been revealed. An exclusive group of miRNAs, such as let-7 miRNAs, are expressed in brain [236]. Critical neuronal functions such as plasticity and memory may be regulated by miRNAs [237] and global disruption of miRNAs triggers neurodegeneration [238]. In *Drosophila*, loss of miR-34 triggers a late-onset brain degenerative state, whereas miR-34 upregulation extends life span and alleviates neurodegeneration by polyglutamine disease protein [239], providing the evidence for involvement of specific miRNAs in neuronal functions. Notably, of the multiple miRNAs that become dysregulated in neurodegenerative diseases [240], many are involved in autophagy regulation [235]. Some miRNAs, such as miR-7, have been shown to induce autophagy [241] and downregulate α-synuclein protein expression through the 3-untranslated region (UTR) of α-synuclein mRNA [242], both of which could contribute to the neuron protection. By contrast, miR-128 aggravates α-synuclein toxicity by downregulating TFEB [243]. While miRNA research is still in its infancy, the multitude of ways that miRNAs regulate autophagy and influence neurodegeneration suggest that miRNA modulation could be a viable target in the future treatment of neurodegeneration. A phase I clinical trial of miR-34 for oncology has recently started and could provide a proof of principle for using miRNAs in the clinic. A major challenge for the clinical use of miRNAs in neurodegenerative disorders is the efficiency of delivery into the brain. Possibly intranasal delivery could provide a noninvasive method of bypassing the blood–brain barrier to rapidly deliver therapeutic agents [244]. Although multiple target modulation by miRNAs provides therapeutic benefits, pathways, or targets, off-target adverse effects represent another potential limitation. Ideally, miRNAs that are only associated with disease are the optimal targets. Further understanding the role of miRNAs under pathological conditions could provide the rationale for miRNA target selection.

14.5 CONCLUDING REMARKS

Mounting evidence supports the pathologic role of autophagy impairment and lysosomal dysfunction in the development of neurodegeneration, which provides a strong rationale for developing therapeutics to modulate autophagy by targeting steps in autophagy substrate recognition and sequestration, and increasing the efficiency of autophagy substrate clearance by modulating lysosome trafficking and hydrolytic function. As promising as these therapeutic directions are, there is still a critical need to understand the molecule mechanism of pathogenesis of neurodegeneration and further decipher the complex regulation of membrane trafficking. For instance, the existence of an autophagy protein interactome including over 400 interacting proteins not only underscores the intricacy of autophagy regulation [245] but also reveals an encouraging breath of druggable candidate targets for autophagy modulation. A better understanding of this network could shed light on strategic interventions with more specific therapeutic outcomes and fewer off-target effects. As we discussed earlier, compounds of interest are not yet fully characterized, especially their dual effect on both early and late stages of membrane trafficking. Positive effects on upstream steps in autophagy (autophagy substrate recognition/sequestration) could be masked by blockage at downstream steps of autophagy (e.g., lysosomal hydrolysis of substrate) with an overall negative effect on autophagic substrate turnover as seen with niclosamide [166]. Current screening paradigms can be further enhanced to better understand the multiple biological effects of a compound. In particular, comprehensive efforts are required to distinguish autophagy enhancers from inhibitors.

Another major barrier for compound characterization is the scarcity of tools to measure their activity *in vivo* and establish drug efficacy. Most assay methods involve static endpoints, which do not provide much information on the kinetics of autophagic substrate turnover. Novel techniques are especially needed to ideally evaluate the various stages of autophagosome clearance and substrate degradation. These more dynamic assessments of autophagy turnover will better establish target engagement by a test compound and the development of surrogate biomarkers will enable new approaches for noninvasive monitoring. Although genetically engineered animals are very popular efficacy models for neurodegeneration study, the translation between efficacy of a therapy in preclinical models and success in clinical trials is still very poor [246]. Improvements in disease modeling, study design, and standardization across multiple models will be needed to enhance predictive power.

Additional important consideration for the application of autophagy augmentation strategies in a neurodegenerative disease setting is the stage of the disease. At an early stage prior to a general failure of lysosomal function, enhancement of autophagy could provide therapeutic benefit by reinstating normal autophagic rate to eliminate the formation of aggregates. Conversely, in the late stage of the disease where lysosomal function has become markedly impaired, simple autophagy stimulation might be not only ineffective but also detrimental. A better understanding of the molecular mechanisms of disease could obviously strengthen the foundation for therapeutic strategies. Autophagy enhancement can act as a double-edged sword in neurodegeneration diseases. On the one side are the cytoprotective roles of increased autophagy; however, on the other side, unchecked autophagy is a distinct mechanism for non-apoptotic cell death in some neuropathological states, such as ischemia and exposure to certain toxins [247].Similarly, LMP induces necrosis of neurons [227,248] and upregulation of lysosomal activity conceivably might facilitate cell death through cathepsin release. When boosting lysosomal activity, increasing lysosomal membrane stability might be necessary to minimize the potential for cellular damage. Balancing beneficial and potential detrimental roles of autophagy regulation will be an important consideration in developing effective autophagy modulators.

Beyond the requirements for optimizing activity and selectivity of candidate therapeutics, an additional challenge with drug development for neurodegeneration is achieving sufficient bioavailability to the CNS from systemic administration due to the blood–brain barrier. Interestingly, both basicity and lipophilicity can promote penetration through the blood–brain barrier [249] and the CNS drugs have an average $c\log P$ of 2.8 and an average basic pK_a of 8.4 [250]. Because of their lysosomotropic properties, potential lysosomal dysfunction by this type of autophagy modulator may lessen the efficacy, as discussed before. The use of liposome or polymeric nanoparticles could potentially facilitate drug entry in the brain [249]. Intranasal administration is also an emerging area of investigation due to the rapid uptake of the drug, ease of self-administration, and potential for frequent, chronic dosing [251].

REFERENCES

[1] Williams A, Sarkar S, Cuddon P, Ttofi EK, Saiki S, et al. Novel targets for Huntington's disease in an mTOR-independent autophagy pathway. Nat Chem Biol 2008;4:295–305.

[2] Criollo A, Maiuri MC, Tasdemir E, Vitale I, Fiebig AA, et al. Regulation of autophagy by the inositol trisphosphate receptor. Cell Death Differ 2007;14:1029–1039.

[3] Kuma A, Hatano M, Matsui M, Yamamoto A, Nakaya H, et al. The role of autophagy during the early neonatal starvation period. Nature 2004;432:1032–1036.

[4] Komatsu M, Waguri S, Ueno T, Iwata J, Murata S, et al. Impairment of starvation-induced and constitutive autophagy in Atg7-deficient mice. J Cell Biol 2005;169: 425–434.

[5] Sou YS, Waguri S, Iwata J, Ueno T, Fujimura T, et al. The Atg8 conjugation system is indispensable for proper development of autophagic isolation membranes in mice. Mol Biol Cell 2008;19:4762–4775.

[6] Nakai A, Yamaguchi O, Takeda T, Higuchi Y, Hikoso S, et al. The role of autophagy in cardiomyocytes in the basal state and in response to hemodynamic stress. Nat Med 2007;13:619–624.

[7] Liang XH, Jackson S, Seaman M, Brown K, Kempkes B, et al. Induction of autophagy and inhibition of tumorigenesis by beclin 1. Nature 1999;402:672–676.

[8] Martinez-Vicente M, Sovak G, Cuervo AM. Protein degradation and aging. Exp Gerontol 2005;40:622–633.

[9] Ma Q, Qiang J, Gu P, Wang Y, Geng Y, et al. Age-related autophagy alterations in the brain of senescence accelerated mouse prone 8 (SAMP8) mice. Exp Gerontol 2011;46: 533–541.

[10] Yang L, Li P, Fu S, Calay ES, Hotamisligil GS. Defective hepatic autophagy in obesity promotes ER stress and causes insulin resistance. Cell Metab 2010;11:467–478.

[11] Dutta D, Calvani R, Bernabei R, Leeuwenburgh C, Marzetti E. Contribution of impaired mitochondrial autophagy to cardiac aging: mechanisms and therapeutic opportunities. Circ Res 2012;110:1125–1138.

[12] Viiri J, Amadio M, Marchesi N, Hyttinen JM, Kivinen N, et al. Autophagy activation clears ELAVL1/HuR-mediated accumulation of SQSTM1/p62 during proteasomal inhibition in human retinal pigment epithelial cells. PLoS One 2013;8:e69563.

[13] Desnick RJ, Schuchman EH. Enzyme replacement and enhancement therapies: lessons from lysosomal disorders. Nat Rev Genet 2002;3:954–966.

[14] Komatsu M, Waguri S, Chiba T, Murata S, Iwata J, et al. Loss of autophagy in the central nervous system causes neurodegeneration in mice. Nature 2006;441:880–884.

[15] Hara T, Nakamura K, Matsui M, Yamamoto A, Nakahara Y, et al. Suppression of basal autophagy in neural cells causes neurodegenerative disease in mice. Nature 2006; 441:885–889.

[16] Costanzo M, Zurzolo C. The cell biology of prion-like spread of protein aggregates: mechanisms and implication in neurodegeneration. Biochem J 2013;452:1–17.

[17] Nixon RA. The role of autophagy in neurodegenerative disease. Nat Med 2013;19: 983–997.

[18] Bove J, Martinez-Vicente M, Vila M. Fighting neurodegeneration with rapamycin: mechanistic insights. Nat Rev Neurosci 2011;12:437–452.

[19] Balgi AD, Fonseca BD, Donohue E, Tsang TC, Lajoie P, et al. Screen for chemical modulators of autophagy reveals novel therapeutic inhibitors of mTORC1 signaling. PLoS One 2009;4:e7124.

[20] Sarkar S, Perlstein EO, Imarisio S, Pineau S, Cordenier A, et al. Small molecules enhance autophagy and reduce toxicity in Huntington's disease models. Nat Chem Biol 2007;3:331–338.

[21] Zhang L, Yu J, Pan H, Hu P, Hao Y, et al. Small molecule regulators of autophagy identified by an image-based high-throughput screen. Proc Natl Acad Sci U S A 2007; 104:19023–19028.

[22] Xi H, Kurtoglu M, Liu H, Wangpaichitr M, You M, et al. 2-deoxy-D-glucose activates autophagy via endoplasmic reticulum stress rather than ATP depletion. Cancer Chemother Pharmacol 2011;67:899–910.

[23] Li J, Hou N, Faried A, Tsutsumi S, Kuwano H. Inhibition of autophagy augments 5-fluorouracil chemotherapy in human colon cancer *in vitro* and *in vivo* model. Eur J Cancer 2010;46:1900–1909.

[24] Ding WX, Ni HM, Gao W, Hou YF, Melan MA, et al. Differential effects of endoplasmic reticulum stress-induced autophagy on cell survival. J Biol Chem 2007;282:4702–4710.

[25] Buzzai M, Jones RG, Amaravadi RK, Lum JJ, DeBerardinis RJ, et al. Systemic treatment with the antidiabetic drug metformin selectively impairs p53-deficient tumor cell growth. Cancer Res 2007;67:6745–6752.

[26] Lee KY, Oh S, Choi YJ, Oh SH, Yang YS, et al. Activation of autophagy rescues amiodarone-induced apoptosis of lung epithelial cells and pulmonary toxicity in rats. Toxicol Sci 2013;136:193–204.

[27] Nicolescu AC, Ji Y, Comeau JL, Hill BC, Takahashi T, et al. Direct mitochondrial dysfunction precedes reactive oxygen species production in amiodarone-induced toxicity in human peripheral lung epithelial HPL1A cells. Toxicol Appl Pharmacol 2008;227:370–379.

[28] Ahn MY, Ahn SG, Yoon JH. Apicidin, a histone deaceylase inhibitor, induces both apoptosis and autophagy in human oral squamous carcinoma cells. Oral Oncol 2011; 47:1032–1038.

[29] Choi JH, Lee JY, Choi AY, Hwang KY, Choe W, et al. Apicidin induces endoplasmic reticulum stress- and mitochondrial dysfunction-associated apoptosis via phospholipase Cgamma1- and Ca(2+)-dependent pathway in mouse Neuro-2a neuroblastoma cells. Apoptosis 2012;17:1340–1358.

[30] Kanzawa T, Zhang L, Xiao L, Germano IM, Kondo Y, et al. Arsenic trioxide induces autophagic cell death in malignant glioma cells by upregulation of mitochondrial cell death protein BNIP3. Oncogene 2005;24:980–991.

[31] Yang X, An L, Li X. Arsenic trioxide induced endoplasmic reticulum stress in laryngeal squamous cell line Hep-2 cells. Auris Nasus Larynx 2014;41:81–83.

[32] Hu M, Huang H, Zhao R, Li P, Li M, et al. AZD8055 induces cell death associated with autophagy and activation of AMPK in hepatocellular carcinoma. Oncol Rep 2014;31:649–656.

[33] Liu J, Hu XJ, Jin B, Qu XJ, Hou KZ, et al. Beta-elemene induces apoptosis as well as protective autophagy in human non-small-cell lung cancer A549 cells. J Pharm Pharmacol 2012;64:146–153.

[34] Zhu K, Dunner K Jr, McConkey DJ. Proteasome inhibitors activate autophagy as a cytoprotective response in human prostate cancer cells. Oncogene 2010;29:451–462.

[35] Xie CM, Chan WY, Yu S, Zhao J, Cheng CH. Bufalin induces autophagy-mediated cell death in human colon cancer cells through reactive oxygen species generation and JNK activation. Free Radic Biol Med 2011;51:1365–1375.

[36] Abedin MJ, Wang D, McDonnell MA, Lehmann U, Kelekar A. Autophagy delays apoptotic death in breast cancer cells following DNA damage. Cell Death Differ 2007;14:500–510.

[37] Hidvegi T, Ewing M, Hale P, Dippold C, Beckett C, et al. An autophagy-enhancing drug promotes degradation of mutant alpha1-antitrypsin Z and reduces hepatic fibrosis. Science 2010;329:229–232.

[38] Zang Y, Thomas SM, Chan ET, Kirk CJ, Freilino ML, et al. The next generation proteasome inhibitors carfilzomib and oprozomib activate prosurvival autophagy via induction of the unfolded protein response and ATF4. Autophagy 2012;8:1873–1874.

[39] Huang S, Sinicrope FA. Celecoxib-induced apoptosis is enhanced by ABT-737 and by inhibition of autophagy in human colorectal cancer cells. Autophagy 2010;6:256–269.

[40] Law BY, Chan WK, Xu SW, Wang JR, Bai LP, et al. Natural small-molecule enhancers of autophagy induce autophagic cell death in apoptosis-defective cells. Sci Rep 2014; 4:5510.

[41] Li X, Fan Z. The epidermal growth factor receptor antibody cetuximab induces autophagy in cancer cells by downregulating HIF-1alpha and Bcl-2 and activating the beclin 1/hVps34 complex. Cancer Res 2010;70:5942–5952.

[42] Tsvetkov AS, Miller J, Arrasate M, Wong JS, Pleiss MA, et al. A small-molecule scaffold induces autophagy in primary neurons and protects against toxicity in a Huntington disease model. Proc Natl Acad Sci U S A 2010;107:16982–16987.

[43] Guo XL, Li D, Hu F, Song JR, Zhang SS, et al. Targeting autophagy potentiates chemotherapy-induced apoptosis and proliferation inhibition in hepatocarcinoma cells. Cancer Lett 2012;320:171–179.

[44] Nilsson C, Roberg K, Grafstrom RC, Ollinger K. Intrinsic differences in cisplatin sensitivity of head and neck cancer cell lines: correlation to lysosomal pH. Head Neck 2010;32:1185–1194.

[45] Hau AM, Greenwood JA, Lohr CV, Serrill JD, Proteau PJ, et al. Coibamide A induces mTOR-independent autophagy and cell death in human glioblastoma cells. PLoS One 2013;8:e65250.

[46] Vucicevic L, Misirkic M, Janjetovic K, Vilimanovich U, Sudar E, et al. Compound C induces protective autophagy in cancer cells through AMPK inhibition-independent blockade of Akt/mTOR pathway. Autophagy 2011;7:40–50.

[47] Chen QY, Shi JG, Yao QH, Jiao DM, Wang YY, et al. Lysosomal membrane permeabilization is involved in curcumin-induced apoptosis of A549 lung carcinoma cells. Mol Cell Biochem 2012;359:389–398.

[48] Ravikumar B, Duden R, Rubinsztein DC. Aggregate-prone proteins with polyglutamine and polyalanine expansions are degraded by autophagy. Hum Mol Genet 2002;11:1107–1117.

[49] Ravikumar B, Vacher C, Berger Z, Davies JE, Luo S, et al. Inhibition of mTOR induces autophagy and reduces toxicity of polyglutamine expansions in fly and mouse models of Huntington disease. Nat Genet 2004;36:585–595.

[50] Bjorkoy G, Lamark T, Brech A, Outzen H, Perander M, et al. p62/SQSTM1 forms protein aggregates degraded by autophagy and has a protective effect on huntingtin-induced cell death. J Cell Biol 2005;171:603–614.

[51] Laane E, Tamm KP, Buentke E, Ito K, Kharaziha P, et al. Cell death induced by dexamethasone in lymphoid leukemia is mediated through initiation of autophagy. Cell Death Differ 2009;16:1018–1029.

[52] Hoyer-Hansen M, Bastholm L, Mathiasen IS, Elling F, Jaattela M. Vitamin D analog EB1089 triggers dramatic lysosomal changes and Beclin 1-mediated autophagic cell death. Cell Death Differ 2005;12:1297–1309.

[53] Han W, Pan H, Chen Y, Sun J, Wang Y, et al. EGFR tyrosine kinase inhibitors activate autophagy as a cytoprotective response in human lung cancer cells. PLoS One 2011;6:e18691.

[54] Wang N, Pan W, Zhu M, Zhang M, Hao X, et al. Fangchinoline induces autophagic cell death via p53/sestrin2/AMPK signalling in human hepatocellular carcinoma cells. Br J Pharmacol 2011;164:731–742.

[55] Iorio F, Bosotti R, Scacheri E, Belcastro V, Mithbaokar P, et al. Discovery of drug mode of action and drug repositioning from transcriptional responses. Proc Natl Acad Sci U S A 2010;107:14621–14626.

[56] Mahoney E, Lucas DM, Gupta SV, Wagner AJ, Herman SE, et al. ER stress and autophagy: new discoveries in the mechanism of action and drug resistance of the cyclin-dependent kinase inhibitor flavopiridol. Blood 2012;120:1262–1273.

[57] Cloonan SM, Williams DC. The antidepressants maprotiline and fluoxetine induce Type II autophagic cell death in drug-resistant Burkitt's lymphoma. Int J Cancer 2011; 128:1712–1723.

[58] Lee CS, Kim YJ, Jang ER, Kim W, Myung SC. Fluoxetine induces apoptosis in ovarian carcinoma cell line OVCAR-3 through reactive oxygen species-dependent activation of nuclear factor-kappaB. Basic Clin Pharmacol Toxicol 2010;106:446–453.

[59] Zhang N, Qi Y, Wadham C, Wang L, Warren A, et al. FTY720 induces necrotic cell death and autophagy in ovarian cancer cells: a protective role of autophagy. Autophagy 2010;6:1157–1167.

[60] Mukubou H, Tsujimura T, Sasaki R, Ku Y. The role of autophagy in the treatment of pancreatic cancer with gemcitabine and ionizing radiation. Int J Oncol 2010; 37:821–828.

[61] Jiang Q, Wang Y, Li T, Shi K, Li Z, et al. Heat shock protein 90-mediated inactivation of nuclear factor-kappaB switches autophagy to apoptosis through becn1 transcriptional inhibition in selenite-induced NB4 cells. Mol Biol Cell 2011;22:1167–1180.

[62] Mai TT, Moon J, Song Y, Viet PQ, Phuc PV, et al. Ginsenoside F2 induces apoptosis accompanied by protective autophagy in breast cancer stem cells. Cancer Lett 2012;21:144–153.

[63] Ertmer A, Huber V, Gilch S, Yoshimori T, Erfle V, et al. The anticancer drug imatinib induces cellular autophagy. Leukemia 2007;21:936–942.

[64] Puissant A, Colosetti P, Robert G, Cassuto JP, Raynaud S, et al. Cathepsin B release after imatinib-mediated lysosomal membrane permeabilization triggers BCR-ABL cleavage and elimination of chronic myelogenous leukemia cells. Leukemia 2010;24:115–124.

[65] Jeon SH, Kim SH, Kim Y, Kim YS, Lim Y, et al. The tricyclic antidepressant imipramine induces autophagic cell death in U-87MG glioma cells. Biochem Biophys Res Commun 2011;413:311–317.

[66] Sarkar S, Floto RA, Berger Z, Imarisio S, Cordenier A, et al. Lithium induces autophagy by inhibiting inositol monophosphatase. J Cell Biol 2005;170:1101–1111.

[67] Steele JW, Ju S, Lachenmayer ML, Liken J, Stock A, et al. Latrepirdine stimulates autophagy and reduces accumulation of alpha-synuclein in cells and in mouse brain. Mol Psychiatry 2013;18:882–888.

[68] Jia DD, Zhang L, Chen Z, Wang CR, Huang FZ, et al. Lithium chloride alleviates neurodegeneration partly by inhibiting activity of GSK3beta in a SCA3 *Drosophila* model. Cerebellum 2013;12:892–901.

[69] Gills JJ, Lopiccolo J, Tsurutani J, Shoemaker RH, Best CJ, et al. Nelfinavir, A lead HIV protease inhibitor, is a broad-spectrum, anticancer agent that induces endoplasmic reticulum stress, autophagy, and apoptosis *in vitro* and *in vivo*. Clin Cancer Res 2007;13:5183–5194.

[70] Hansen K, Wagner B, Hamel W, Schweizer M, Haag F, et al. Autophagic cell death induced by TrkA receptor activation in human glioblastoma cells. J Neurochem 2007;103:259–275.

[71] Yu HC, Lin CS, Tai WT, Liu CY, Shiau CW, et al. Nilotinib induces autophagy in hepatocellular carcinoma through AMPK activation. J Biol Chem 2013;288:18249–18259.

[72] Chresta CM, Davies BR, Hickson I, Harding T, Cosulich S, et al. AZD8055 is a potent, selective, and orally bioavailable ATP-competitive mammalian target of rapamycin kinase inhibitor with *in vitro* and *in vivo* antitumor activity. Cancer Res 2010;70:288–298.

[73] Heidari N, Hicks MA, Harada H. GX15-070 (obatoclax) overcomes glucocorticoid resistance in acute lymphoblastic leukemia through induction of apoptosis and autophagy. Cell Death Dis 2010;1:e76.

[74] Liu YL, Yang PM, Shun CT, Wu MS, Weng JR, et al. Autophagy potentiates the anti-cancer effects of the histone deacetylase inhibitors in hepatocellular carcinoma. Autophagy 2010;6:1057–1065.

[75] Bommareddy A, Hahm ER, Xiao D, Powolny AA, Fisher AL, et al. Atg5 regulates phenethyl isothiocyanate-induced autophagic and apoptotic cell death in human prostate cancer cells. Cancer Res 2009;69:3704–3712.

[76] Fan QW, Cheng C, Hackett C, Feldman M, Houseman BT, et al. Akt and autophagy cooperate to promote survival of drug-resistant glioma. Sci Signal 2010;3:ra81.

[77] Wang Y, Ding L, Wang X, Zhang J, Han W, et al. Pterostilbene simultaneously induces apoptosis, cell cycle arrest and cyto-protective autophagy in breast cancer cells. Am J Transl Res 2012;4:44–51.

[78] Wang K, Liu R, Li J, Mao J, Lei Y, et al. Quercetin induces protective autophagy in gastric cancer cells: involvement of Akt-mTOR- and hypoxia-induced factor 1alpha-mediated signaling. Autophagy 2011;7:966–978.

[79] Morselli E, Marino G, Bennetzen MV, Eisenberg T, Megalou E, et al. Spermidine and resveratrol induce autophagy by distinct pathways converging on the acetylproteome. J Cell Biol 2011;192:615–629.

[80] Opipari AW Jr, Tan L, Boitano AE, Sorenson DR, Aurora A, et al. Resveratrol-induced autophagocytosis in ovarian cancer cells. Cancer Res 2004;64:696–703.

[81] Mena S, Rodriguez ML, Ponsoda X, Estrela JM, Jaattela M, et al. Pterostilbene-induced tumor cytotoxicity: a lysosomal membrane permeabilization-dependent mechanism. PLoS One 2012;7:e44524.

[82] Rose C, Menzies FM, Renna M, Acevedo-Arozena A, Corrochano S, et al. Rilmenidine attenuates toxicity of polyglutamine expansions in a mouse model of Huntington's disease. Hum Mol Genet 2010;19:2144–2153.

[83] Wu J, Wu JJ, Yang LJ, Wei LX, Zou DJ. Rosiglitazone protects against palmitate-induced pancreatic beta-cell death by activation of autophagy via 5'-AMP-activated protein kinase modulation. Endocrine 2013;44:87–98.

[84] Kumar D, Shankar S, Srivastava RK. Rottlerin induces autophagy and apoptosis in prostate cancer stem cells via PI3K/Akt/mTOR signaling pathway. Cancer Lett 2013; 343:179–189.

[85] Singh BN, Kumar D, Shankar S, Srivastava RK. Rottlerin induces autophagy which leads to apoptotic cell death through inhibition of PI3K/Akt/mTOR pathway in human pancreatic cancer stem cells. Biochem Pharmacol 2012;84:1154–1163.

[86] Ling LU, Tan KB, Lin H, Chiu GN. The role of reactive oxygen species and autophagy in safingol-induced cell death. Cell Death Dis 2011;2:e129.

[87] Chen J, Korostyshevsky D, Lee S, Perlstein EO. Accumulation of an antidepressant in vesiculogenic membranes of yeast cells triggers autophagy. PLoS One 2012;7:e34024.

[88] Boya P, Kroemer G. Lysosomal membrane permeabilization in cell death. Oncogene 2008;27:6434–6451.

[89] Ostenfeld MS, Hoyer-Hansen M, Bastholm L, Fehrenbacher N, Olsen OD, et al. Anti-cancer agent siramesine is a lysosomotropic detergent that induces cytoprotective autophagosome accumulation. Autophagy 2008;4:487–499.

[90] Cesen MH, Repnik U, Turk V, Turk B. Siramesine triggers cell death through destabilisation of mitochondria, but not lysosomes. Cell Death Dis 2013;4:e818.

[91] Sarkar S, Rubinsztein DC. Small molecule enhancers of autophagy for neurodegenerative diseases. Mol Biosyst 2008;4:895–901.

[92] Bursch W, Ellinger A, Kienzl H, Torok L, Pandey S, et al. Active cell death induced by the anti-estrogens tamoxifen and ICI 164 384 in human mammary carcinoma cells (MCF-7) in culture: the role of autophagy. Carcinogenesis 1996;17:1595–1607.

[93] Cho KS, Yoon YH, Choi JA, Lee SJ, Koh JY. Induction of autophagy and cell death by tamoxifen in cultured retinal pigment epithelial and photoreceptor cells. Invest Ophthalmol Vis Sci 2012;53:5344–5353.

[94] Kallio A, Zheng A, Dahllund J, Heiskanen KM, Harkonen P. Role of mitochondria in tamoxifen-induced rapid death of MCF-7 breast cancer cells. Apoptosis 2005;10:1395–1410.

[95] Kanzawa T, Germano IM, Komata T, Ito H, Kondo Y, et al. Role of autophagy in temozolomide-induced cytotoxicity for malignant glioma cells. Cell Death Differ 2004;11:448–457.

[96] Gong K, Chen C, Zhan Y, Chen Y, Huang Z, et al. Autophagy-related gene 7 (ATG7) and reactive oxygen species/extracellular signal-regulated kinase regulate tetrandrine-induced autophagy in human hepatocellular carcinoma. J Biol Chem 2012;287:35576–35588.

[97] Thoreen CC, Kang SA, Chang JW, Liu Q, Zhang J, et al. An ATP-competitive mammalian target of rapamycin inhibitor reveals rapamycin-resistant functions of mTORC1. J Biol Chem 2009;284:8023–8032.

[98] Werneburg NW, Bronk SF, Guicciardi ME, Thomas L, Dikeakos JD, et al. Tumor necrosis factor-related apoptosis-inducing ligand (TRAIL) protein-induced lysosomal translocation of proapoptotic effectors is mediated by phosphofurin acidic cluster sorting protein-2 (PACS-2). J Biol Chem 2012;287:24427–24437.

[99] Han J, Hou W, Goldstein LA, Lu C, Stolz DB, et al. Involvement of protective autophagy in TRAIL resistance of apoptosis-defective tumor cells. J Biol Chem 2008;283:19665–19677.

[100] Sarkar S, Davies JE, Huang Z, Tunnacliffe A, Rubinsztein DC. Trehalose, a novel mTOR-independent autophagy enhancer, accelerates the clearance of mutant huntingtin and alpha-synuclein. J Biol Chem 2007;282:5641–5652.

[101] Rodriguez-Navarro JA, Rodriguez L, Casarejos MJ, Solano RM, Gomez A, et al. Trehalose ameliorates dopaminergic and tau pathology in parkin deleted/tau overexpressing mice through autophagy activation. Neurobiol Dis 2010;39:423–438.

[102] Zong D, Haag P, Yakymovych I, Lewensohn R, Viktorsson K. Chemosensitization by phenothiazines in human lung cancer cells: impaired resolution of gammaH2AX

and increased oxidative stress elicit apoptosis associated with lysosomal expansion and intense vacuolation. Cell Death Dis 2011;2:e181.

[103] Huang WC, Lin YS, Chen CL, Wang CY, Chiu WH, et al. Glycogen synthase kinase-3beta mediates endoplasmic reticulum stress-induced lysosomal apoptosis in leukemia. J Pharmacol Exp Ther 2009;329:524–531.

[104] Fu J, Shao CJ, Chen FR, Ng HK, Chen ZP. Autophagy induced by valproic acid is associated with oxidative stress in glioma cell lines. Neuro Oncol 2010;12:328–340.

[105] Pourahmad J, Eskandari MR, Kaghazi A, Shaki F, Shahraki J, et al. A new approach on valproic acid induced hepatotoxicity: involvement of lysosomal membrane leakiness and cellular proteolysis. Toxicol In Vitro 2012;26:545–551.

[106] Bauer PO, Wong HK, Oyama F, Goswami A, Okuno M, et al. Inhibition of Rho kinases enhances the degradation of mutant huntingtin. J Biol Chem 2009;284:13153–13164.

[107] Gaulton A, Bellis LJ, Bento AP, Chambers J, Davies M, et al. ChEMBL: a large-scale bioactivity database for drug discovery. Nucleic Acids Res 2012;40:D1100–D1107.

[108] Mizushima N, Kuma A. Autophagosomes in GFP-LC3 transgenic mice. Methods Mol Biol 2008;445:119–124.

[109] Moriyama Y, Takano T, Ohkuma S. Acridine orange as a fluorescent probe for lysosomal proton pump. J Biochem 1982;92:1333–1336.

[110] Eng CH, Yu K, Lucas J, White E, Abraham RT. Ammonia derived from glutaminolysis is a diffusible regulator of autophagy. Sci Signal 2010;3:ra31.

[111] Ni HM, Bockus A, Wozniak AL, Jones K, Weinman S, et al. Dissecting the dynamic turnover of GFP-LC3 in the autolysosome. Autophagy 2011;7:188–204.

[112] Ohsumi Y. Historical landmarks of autophagy research. Cell Res 2014;24:9–23.

[113] de Duve C, de Barsy T, Poole B, Trouet A, Tulkens P, et al. Commentary. Lysosomotropic agents. Biochem Pharmacol 1974;23:2495–2531.

[114] Ashoor R, Yafawi R, Jessen B, Lu S. The contribution of lysosomotropism to autophagy perturbation. PLoS One 2013;8:e82481.

[115] Nadanaciva S, Lu S, Gebhard DF, Jessen BA, Pennie WD, et al. A high content screening assay for identifying lysosomotropic compounds. Toxicol In Vitro 2011;25:715–723.

[116] Myers BM, Tietz PS, Tarara JE, LaRusso NF. Dynamic measurements of the acute and chronic effects of lysosomotropic agents on hepatocyte lysosomal pH using flow cytometry. Hepatology 1995;22:1519–1526.

[117] Ndolo RA, Forrest ML, Krise JP. The role of lysosomes in limiting drug toxicity in mice. J Pharmacol Exp Ther 2010;333:120–128.

[118] Kawai A, Uchiyama H, Takano S, Nakamura N, Ohkuma S. Autophagosome-lysosome fusion depends on the pH in acidic compartments in CHO cells. Autophagy 2007; 3:154–157.

[119] Klionsky DJ, Elazar Z, Seglen PO, Rubinsztein DC. Does bafilomycin A1 block the fusion of autophagosomes with lysosomes? Autophagy 2008;4:849–850.

[120] Grabner R. Influence of cationic amphiphilic drugs on the phosphatidylcholine hydrolysis by phospholipase A2. Biochem Pharmacol 1987;36:1063–1067.

[121] Kubo M, Hostetler KY. Mechanism of cationic amphiphilic drug inhibition of purified lysosomal phospholipase A1. Biochemistry 1985;24:6515–6520.

[122] Elojeimy S, Holman DH, Liu X, El-Zawahry A, Villani M, et al. New insights on the use of desipramine as an inhibitor for acid ceramidase. FEBS Lett 2006;580:4751–4756.

[123] Ikeda K, Hirayama M, Hirota Y, Asa E, Seki J, et al. Drug-induced phospholipidosis is caused by blockade of mannose 6-phosphate receptor-mediated targeting of lysosomal enzymes. Biochem Biophys Res Commun 2008;377:268–274.

[124] Mannerstrom M, Maenpaa H, Toimela T, Salminen L, Tahti H. The phagocytosis of rod outer segments is inhibited by selected drugs in retinal pigment epithelial cell cultures. Pharmacol Toxicol 2001;88:27–33.

[125] Kogut MH, Genovese KJ, Lowry VK. Differential activation of signal transduction pathways mediating phagocytosis, oxidative burst, and degranulation by chicken heterophils in response to stimulation with opsonized Salmonella enteritidis. Inflammation 2001;25:7–15.

[126] Chen CL, Hou WH, Liu IH, Hsiao G, Huang SS, et al. Inhibitors of clathrin-dependent endocytosis enhance TGFbeta signaling and responses. J Cell Sci 2009;122:1863–1871.

[127] Piccoli E, Nadai M, Caretta CM, Bergonzini V, Del Vecchio C, et al. Amiodarone impairs trafficking through late endosomes inducing a Niemann–Pick C-like phenotype. Biochem Pharmacol 2011;82:1234–1249.

[128] Ploemen JP, Kelder J, Hafmans T, van de Sandt H, van Burgsteden JA, et al. Use of physicochemical calculation of pKa and CLogP to predict phospholipidosis-inducing potential: a case study with structurally related piperazines. Exp Toxicol Pathol 2004;55:347–355.

[129] Palmieri M, Impey S, Kang H, di Ronza A, Pelz C, et al. Characterization of the CLEAR network reveals an integrated control of cellular clearance pathways. Hum Mol Genet 2011;20:3852–3866.

[130] Settembre C, Di Malta C, Polito VA, Garcia Arencibia M, Vetrini F, et al. TFEB links autophagy to lysosomal biogenesis. Science 2011;332:1429–1433.

[131] Martina JA, Diab HI, Lishu L, Jeong AL, Patange S, et al. The nutrient-responsive transcription factor TFE3 promotes autophagy, lysosomal biogenesis, and clearance of cellular debris. Sci Signal 2014;7:ra9.

[132] Lu S, Jessen B, Strock C, Will Y. The contribution of physicochemical properties to multiple *in vitro* cytotoxicity endpoints. Toxicol In Vitro 2012;26:613–620.

[133] Funk RS, Krise JP. Cationic amphiphilic drugs cause a marked expansion of apparent lysosomal volume: implications for an intracellular distribution-based drug interaction. Mol Pharm 2012;9:1384–1395.

[134] te Vruchte D, Speak AO, Wallom KL, Al Eisa N, Smith DA, et al. Relative acidic compartment volume as a lysosomal storage disorder-associated biomarker. J Clin Invest 2014;124:1320–1328.

[135] Xu M, Liu K, Swaroop M, Sun W, Dehdashti SJ, et al. A phenotypic compound screening assay for lysosomal storage diseases. J Biomol Screen 2014;19:168–175.

[136] Kirkegaard T, Roth AG, Petersen NH, Mahalka AK, Olsen OD, et al. Hsp70 stabilizes lysosomes and reverts Niemann–Pick disease-associated lysosomal pathology. Nature 2010;463:549–553.

[137] Smith DM, Patel S, Raffoul F, Haller E, Mills GB, et al. Arsenic trioxide induces a beclin-1-independent autophagic pathway via modulation of SnoN/SkiL expression in ovarian carcinoma cells. Cell Death Differ 2010;17:1867–1881.

[138] Bolt AM, Zhao F, Pacheco S, Klimecki WT. Arsenite-induced autophagy is associated with proteotoxicity in human lymphoblastoid cells. Toxicol Appl Pharmacol 2012;264:255–261.

[139] Lau A, Zheng Y, Tao S, Wang H, Whitman SA, et al. Arsenic inhibits autophagic flux, activating the Nrf2-Keap1 pathway in a p62-dependent manner. Mol Cell Biol 2013;33:2436–2446.

[140] Fowler BA. General subcellular effects of lead, mercury, cadmium, and arsenic. Environ Health Perspect 1978;22:37–41.

[141] Havelaar AC, de Gast IL, Snijders S, Beerens CE, Mancini GM, et al. Characterization of a heavy metal ion transporter in the lysosomal membrane. FEBS Lett 1998;436:223–227.

[142] Kitareewan S, Roebuck BD, Demidenko E, Sloboda RD, Dmitrovsky E. Lysosomes and trivalent arsenic treatment in acute promyelocytic leukemia. J Natl Cancer Inst 2007;99:41–52.

[143] Grotemeier A, Alers S, Pfisterer SG, Paasch F, Daubrawa M, et al. AMPK-independent induction of autophagy by cytosolic Ca^{2+} increase. Cell Signal 2010;22:914–925.

[144] Ganley IG, Wong PM, Gammoh N, Jiang X. Distinct autophagosomal-lysosomal fusion mechanism revealed by thapsigargin-induced autophagy arrest. Mol Cell 2011; 42:731–743.

[145] Micsenyi MC, Sikora J, Stephney G, Dobrenis K, Walkley SU. Lysosomal membrane permeability stimulates protein aggregate formation in neurons of a lysosomal disease. J Neurosci 2013;33:10815–10827.

[146] Bjedov I, Toivonen JM, Kerr F, Slack C, Jacobson J, et al. Mechanisms of life span extension by rapamycin in the fruit fly *Drosophila melanogaster*. Cell Metab 2010;11:35–46.

[147] Kaeberlein M, Powers RW 3rd, Steffen KK, Westman EA, Hu D, et al. Regulation of yeast replicative life span by TOR and Sch9 in response to nutrients. Science 2005;310:1193–1196.

[148] Harrison DE, Strong R, Sharp ZD, Nelson JF, Astle CM, et al. Rapamycin fed late in life extends lifespan in genetically heterogeneous mice. Nature 2009;460:392–395.

[149] Miller RA, Harrison DE, Astle CM, Baur JA, Boyd AR, et al. Rapamycin, but not resveratrol or simvastatin, extends life span of genetically heterogeneous mice. J Gerontol A Biol Sci Med Sci 2011;66:191–201.

[150] Wong M. Mammalian target of rapamycin (mTOR) pathways in neurological diseases. Biomed J 2013;36:40–50.

[151] Aki T, Nara A, Uemura K. Cytoplasmic vacuolization during exposure to drugs and other substances. Cell Biol Toxicol 2012;28:125–131.

[152] Martinet W, Verheye S, De Meyer GR. Everolimus-induced mTOR inhibition selectively depletes macrophages in atherosclerotic plaques by autophagy. Autophagy 2007; 3:241–244.

[153] Boland B, Kumar A, Lee S, Platt FM, Wegiel J, et al. Autophagy induction and autophagosome clearance in neurons: relationship to autophagic pathology in Alzheimer's disease. J Neurosci 2008;28:6926–6937.

[154] Kretzschmar D, Hasan G, Sharma S, Heisenberg M, Benzer S. The swiss cheese mutant causes glial hyperwrapping and brain degeneration in *Drosophila*. J Neurosci 1997;17:7425–7432.

[155] Haines DC, Chattopadhyay S, Ward JM. Pathology of aging B6;129 mice. Toxicol Pathol 2001;29:653–661.

[156] Di Malta C, Fryer JD, Settembre C, Ballabio A. Astrocyte dysfunction triggers neurodegeneration in a lysosomal storage disorder. Proc Natl Acad Sci U S A 2012;109:E2334–2342.

[157] Shayman JA, Abe A. Drug induced phospholipidosis: an acquired lysosomal storage disorder. Biochim Biophys Acta 2013;1831:602–611.

[158] Sun H, Shahane S, Xia M, Austin CP, Huang R. Structure based model for the prediction of phospholipidosis induction potential of small molecules. J Chem Inf Model 2012;52:1798–1805.

[159] Nioi P, Perry BK, Wang EJ, Gu YZ, Snyder RD. *In vitro* detection of drug-induced phospholipidosis using gene expression and fluorescent phospholipid based methodologies. Toxicol Sci 2007;99:162–173.

[160] Orogo AM, Choi SS, Minnier BL, Kruhlak NL. Construction and consensus performance of (Q)SAR models for predicting phospholipidosis using a dataset of 743 compounds. Mol Inform 2012;31:725–739.

[161] Tyteca D, Van Der Smissen P, Mettlen M, Van Bambeke F, Tulkens PM, et al. Azithromycin, a lysosomotropic antibiotic, has distinct effects on fluid-phase and receptor-mediated endocytosis, but does not impair phagocytosis in J774 macrophages. Exp Cell Res 2002;281:86–100.

[162] Pelkmans L, Fava E, Grabner H, Hannus M, Habermann B, et al. Genome-wide analysis of human kinases in clathrin- and caveolae/raft-mediated endocytosis. Nature 2005;436:78–86.

[163] Pena-Llopis S, Vega-Rubin-de-Celis S, Schwartz JC, Wolff NC, Tran TA, et al. Regulation of TFEB and V-ATPases by mTORC1. EMBO J 2011;30:3242–3258.

[164] Hackstein H, Taner T, Logar AJ, Thomson AW. Rapamycin inhibits macropinocytosis and mannose receptor-mediated endocytosis by bone marrow-derived dendritic cells. Blood 2002;100:1084–1087.

[165] Monti P, Mercalli A, Leone BE, Valerio DC, Allavena P, et al. Rapamycin impairs antigen uptake of human dendritic cells. Transplantation 2003;75:137–145.

[166] Li M, Khambu B, Zhang H, Kang JH, Chen X, et al. Suppression of lysosome function induces autophagy via a feedback down-regulation of MTOR complex 1 (MTORC1) activity. J Biol Chem 2013;288:35769–35780.

[167] Settembre C, Zoncu R, Medina DL, Vetrini F, Erdin S, et al. A lysosome-to-nucleus signalling mechanism senses and regulates the lysosome via mTOR and TFEB. EMBO J 2012;31:1095–1108.

[168] Enzenmuller S, Gonzalez P, Debatin KM, Fulda S. Chloroquine overcomes resistance of lung carcinoma cells to the dual PI3K/mTOR inhibitor PI103 by lysosome-mediated apoptosis. Anticancer Drugs 2013;24:14–19.

[169] Zhou J, Tan SH, Nicolas V, Bauvy C, Yang ND, et al. Activation of lysosomal function in the course of autophagy via mTORC1 suppression and autophagosome-lysosome fusion. Cell Res 2013;23:508–523.

[170] Logan R, Kong AC, Axcell E, Krise JP. Amine-containing molecules and the induction of an expanded lysosomal volume phenotype: a structure–activity relationship study. J Pharm Sci 2014;103:1572–1580.

[171] Kuballa P, Nolte WM, Castoreno AB, Xavier RJ. Autophagy and the immune system. Annu Rev Immunol 2012;30:611–646.

[172] Hardinger KL, Koch MJ, Brennan DC. Current and future immunosuppressive strategies in renal transplantation. Pharmacotherapy 2004;24:1159–1176.

[173] Haidinger M, Poglitsch M, Geyeregger R, Kasturi S, Zeyda M, et al. A versatile role of mammalian target of rapamycin in human dendritic cell function and differentiation. J Immunol 2010;185:3919–3931.

[174] Costedoat-Chalumeau N, Dunogue B, Morel N, Le Guern V, Guettrot-Imbert G. *Hydroxychloroquine: A Multifaceted Treatment in Lupus*. Presse Med; 2014.

[175] Morris SJ, Wasko MC, Antohe JL, Sartorius JA, Kirchner HL, et al. Hydroxychloroquine use associated with improvement in lipid profiles in rheumatoid arthritis patients. Arthritis Care Res (Hoboken) 2011;63:530–534.

[176] Sacre S, Medghalchi M, Gregory B, Brennan F, Williams R. Fluoxetine and citalopram exhibit potent antiinflammatory activity in human and murine models of rheumatoid arthritis and inhibit toll-like receptors. Arthritis Rheum 2010;62:683–693.

[177] Benjamin D, Colombi M, Moroni C, Hall MN. Rapamycin passes the torch: a new generation of mTOR inhibitors. Nat Rev Drug Discov 2011;10:868–880.

[178] Thorburn A, Thamm DH, Gustafson DL. Autophagy and cancer therapy. Mol Pharmacol 2014;85:830–838.

[179] Rangwala R, Chang YC, Hu J, Algazy KM, Evans TL, et al. Combined MTOR and autophagy inhibition: phase I trial of hydroxychloroquine and temsirolimus in patients with advanced solid tumors and melanoma. Autophagy 2014;10:1391–1402.

[180] Fox R, Nhan TQ, Law GL, Morris DR, Liles WC, et al. PSGL-1 and mTOR regulate translation of ROCK-1 and physiological functions of macrophages. EMBO J 2007;26:505–515.

[181] Choi JH, Adames NR, Chan TF, Zeng C, Cooper JA, et al. TOR signaling regulates microtubule structure and function. Curr Biol 2000;10:861–864.

[182] Jiang X, Yeung RS. Regulation of microtubule-dependent protein transport by the TSC2/mammalian target of rapamycin pathway. Cancer Res 2006;66:5258–5269.

[183] Yu L, McPhee CK, Zheng L, Mardones GA, Rong Y, et al. Termination of autophagy and reformation of lysosomes regulated by mTOR. Nature 2010;465:942–946.

[184] Krajcovic M, Krishna S, Akkari L, Joyce JA, Overholtzer M. mTOR regulates phagosome and entotic vacuole fission. Mol Biol Cell 2013;24:3736–3745.

[185] Cang C, Zhou Y, Navarro B, Seo YJ, Aranda K, et al. mTOR regulates lysosomal ATP-sensitive two-pore Na(+) channels to adapt to metabolic state. Cell 2013;152:778–790.

[186] Roczniak-Ferguson A, Petit CS, Froehlich F, Qian S, Ky J, et al. The transcription factor TFEB links mTORC1 signaling to transcriptional control of lysosome homeostasis. Sci Signal 2012;5:ra42.

[187] Pipalia NH, Cosner CC, Huang A, Chatterjee A, Bourbon P, et al. Histone deacetylase inhibitor treatment dramatically reduces cholesterol accumulation in Niemann–Pick type C1 mutant human fibroblasts. Proc Natl Acad Sci U S A 2011;108:5620–5625.

[188] Hutt DM, Herman D, Rodrigues AP, Noel S, Pilewski JM, et al. Reduced histone deacetylase 7 activity restores function to misfolded CFTR in cystic fibrosis. Nat Chem Biol 2010;6:25–33.

[189] Lu J, Yang C, Chen M, Ye DY, Lonser RR, et al. Histone deacetylase inhibitors prevent the degradation and restore the activity of glucocerebrosidase in Gaucher disease. Proc Natl Acad Sci U S A 2011;108:21200–21205.

[190] Lee IH, Finkel T. Regulation of autophagy by the p300 acetyltransferase. J Biol Chem 2009;284:6322–6328.

[191] Yi C, Yu L. How does acetylation regulate autophagy? Autophagy 2012;8:1529–1530.

[192] Lin SY, Li TY, Liu Q, Zhang C, Li X, et al. GSK3-TIP60-ULK1 signaling pathway links growth factor deprivation to autophagy. Science 2012;336:477–481.

[193] Dompierre JP, Godin JD, Charrin BC, Cordelieres FP, King SJ, et al. Histone deacetylase 6 inhibition compensates for the transport deficit in Huntington's disease by increasing tubulin acetylation. J Neurosci 2007;27:3571–3583.

[194] Badding MA, Dean DA. Highly acetylated tubulin permits enhanced interactions with and trafficking of plasmids along microtubules. Gene Ther 2013;20:616–624.

[195] Taes I, Timmers M, Hersmus N, Bento-Abreu A, Van Den Bosch L, et al. Hdac6 deletion delays disease progression in the SOD1G93A mouse model of ALS. Hum Mol Genet 2013;22:1783–1790.

[196] Ferrante RJ, Kubilus JK, Lee J, Ryu H, Beesen A, et al. Histone deacetylase inhibition by sodium butyrate chemotherapy ameliorates the neurodegenerative phenotype in Huntington's disease mice. J Neurosci 2003;23:9418–9427.

[197] Govindarajan N, Rao P, Burkhardt S, Sananbenesi F, Schluter OM, et al. Reducing HDAC6 ameliorates cognitive deficits in a mouse model for Alzheimer's disease. EMBO Mol Med 2013;5:52–63.

[198] Kidd SK, Schneider JS. Protective effects of valproic acid on the nigrostriatal dopamine system in a 1-methyl-4-phenyl-1,2,3,6-tetrahydropyridine mouse model of Parkinson's disease. Neuroscience 2011;194:189–194.

[199] Cao DJ, Wang ZV, Battiprolu PK, Jiang N, Morales CR, et al. Histone deacetylase (HDAC) inhibitors attenuate cardiac hypertrophy by suppressing autophagy. Proc Natl Acad Sci U S A 2011;108:4123–4128.

[200] Moresi V, Carrer M, Grueter CE, Rifki OF, Shelton JM, et al. Histone deacetylases 1 and 2 regulate autophagy flux and skeletal muscle homeostasis in mice. Proc Natl Acad Sci U S A 2012;109:1649–1654.

[201] Yi C, Ma M, Ran L, Zheng J, Tong J, et al. Function and molecular mechanism of acetylation in autophagy regulation. Science 2012;336:474–477.

[202] Lee JH, Yu WH, Kumar A, Lee S, Mohan PS, et al. Lysosomal proteolysis and autophagy require presenilin 1 and are disrupted by Alzheimer-related PS1 mutations. Cell 2010;141:1146–1158.

[203] Coffey EE, Beckel JM, Laties AM, Mitchell CH. Lysosomal alkalization and dysfunction in human fibroblasts with the Alzheimer's disease-linked presenilin 1 A246E mutation can be reversed with cAMP. Neuroscience 2014;263C:111–124.

[204] Schwartz W, Bird JW. Degradation of myofibrillar proteins by cathepsins B and D. Biochem J 1977;167:811–820.

[205] Baltazar GC, Guha S, Lu W, Lim J, Boesze-Battaglia K, et al. Acidic nanoparticles are trafficked to lysosomes and restore an acidic lysosomal pH and degradative function to compromised ARPE-19 cells. PLoS One 2012;7:e49635.

[206] Avrahami L, Farfara D, Shaham-Kol M, Vassar R, Frenkel D, et al. Inhibition of glycogen synthase kinase-3 ameliorates beta-amyloid pathology and restores lysosomal acidification and mammalian target of rapamycin activity in the Alzheimer disease mouse model: *in vivo* and *in vitro* studies. J Biol Chem 2013;288:1295–1306.

[207] Dong XP, Wang X, Shen D, Chen S, Liu M, et al. Activating mutations of the TRPML1 channel revealed by proline-scanning mutagenesis. J Biol Chem 2009;284: 32040–32052.

[208] Bach G. Mucolipin 1: endocytosis and cation channel – a review. Pflugers Arch 2005; 451:313–317.

[209] Shen D, Wang X, Li X, Zhang X, Yao Z, et al. Lipid storage disorders block lysosomal trafficking by inhibiting a TRP channel and lysosomal calcium release. Nat Commun 2012;3:731.

[210] Weiss N. Cross-talk between TRPML1 channel, lipids and lysosomal storage diseases. Commun Integr Biol 2012;5:111–113.

[211] Samie M, Wang X, Zhang X, Goschka A, Li X, et al. A TRP channel in the lysosome regulates large particle phagocytosis via focal exocytosis. Dev Cell 2013;26:511–524.

[212] Settembre C, Fraldi A, Medina DL, Ballabio A. Signals from the lysosome: a control centre for cellular clearance and energy metabolism. Nat Rev Mol Cell Biol 2013;14:283–296.

[213] Medina DL, Fraldi A, Bouche V, Annunziata F, Mansueto G, et al. Transcriptional activation of lysosomal exocytosis promotes cellular clearance. Dev Cell 2011;21:421–430.

[214] Spampanato C, Feeney E, Li L, Cardone M, Lim JA, et al. Transcription factor EB (TFEB) is a new therapeutic target for Pompe disease. EMBO Mol Med 2013;5:691–706.

[215] Pastore N, Blomenkamp K, Annunziata F, Piccolo P, Mithbaokar P, et al. Gene transfer of master autophagy regulator TFEB results in clearance of toxic protein and correction of hepatic disease in alpha-1-anti-trypsin deficiency. EMBO Mol Med 2013;5:397–412.

[216] Dehay B, Bove J, Rodriguez-Muela N, Perier C, Recasens A, et al. Pathogenic lysosomal depletion in Parkinson's disease. J Neurosci 2010;30:12535–12544.

[217] Tsunemi T, Ashe TD, Morrison BE, Soriano KR, Au J, et al. PGC-1alpha rescues Huntington's disease proteotoxicity by preventing oxidative stress and promoting TFEB function. Sci Transl Med 2012;4:142ra197.

[218] Song W, Wang F, Savini M, Ake A, di Ronza A, et al. TFEB regulates lysosomal proteostasis. Hum Mol Genet 2013;22:1994–2009.

[219] Schultz ML, Tecedor L, Chang M, Davidson BL. Clarifying lysosomal storage diseases. Trends Neurosci 2011;34:401–410.

[220] Yang DS, Stavrides P, Mohan PS, Kaushik S, Kumar A, et al. Reversal of autophagy dysfunction in the TgCRND8 mouse model of Alzheimer's disease ameliorates amyloid pathologies and memory deficits. Brain 2011;134:258–277.

[221] Yang DS, Stavrides P, Saito M, Kumar A, Rodriguez-Navarro JA, Pawlik M, Huo C, Walkley SU, Saito M, Cuervo AM, Nixon RA. Defective macroautophagic turnover of brain lipids in the TgCRND8 Alzheimer mouse model: prevention by correcting lysosomal proteolytic deficits. Brain. 2014;137(Pt 12):3300–3318.

[222] Sun B, Zhou Y, Halabisky B, Lo I, Cho SH, et al. Cystatin C-cathepsin B axis regulates amyloid beta levels and associated neuronal deficits in an animal model of Alzheimer's disease. Neuron 2008;60:247–257.

[223] Boyd RE, Lee G, Rybczynski P, Benjamin ER, Khanna R, et al. Pharmacological chaperones as therapeutics for lysosomal storage diseases. J Med Chem 2013;56:2705–2725.

[224] Tamboli IY, Hampel H, Tien NT, Tolksdorf K, Breiden B, et al. Sphingolipid storage affects autophagic metabolism of the amyloid precursor protein and promotes Abeta generation. J Neurosci 2011;31:1837–1849.

[225] Ditaranto K, Tekirian TL, Yang AJ. Lysosomal membrane damage in soluble Abeta-mediated cell death in Alzheimer's disease. Neurobiol Dis 2001;8:19–31.

[226] Rao MV, Mohan PS, Peterhoff CM, Yang DS, Schmidt SD, et al. Marked calpastatin (CAST) depletion in Alzheimer's disease accelerates cytoskeleton disruption

and neurodegeneration: neuroprotection by CAST overexpression. J Neurosci 2008;28:12241–12254.

[227] Yamashima T, Tonchev AB, Tsukada T, Saido TC, Imajoh-Ohmi S, et al. Sustained calpain activation associated with lysosomal rupture executes necrosis of the postischemic CA1 neurons in primates. Hippocampus 2003;13:791–800.

[228] Trinchese F, Fa M, Liu S, Zhang H, Hidalgo A, et al. Inhibition of calpains improves memory and synaptic transmission in a mouse model of Alzheimer disease. J Clin Invest 2008;118:2796–2807.

[229] Nikkel AL, Martino B, Markosyan S, Brederson JD, Medeiros R, et al. The novel calpain inhibitor A-705253 prevents stress-induced tau hyperphosphorylation *in vitro* and *in vivo*. Neuropharmacology 2012;63:606–612.

[230] Rao MV, McBrayer MK, Campbell J, Kumar A, Hashim A, et al. Specific calpain inhibition by calpastatin prevents tauopathy and neurodegeneration and restores normal lifespan in tau P301L mice. J Neurosci 2014;34:9222–9234.

[231] Diepenbroek M, Casadei N, Esmer H, Saido TC, Takano J, et al. Overexpression of the calpain-specific inhibitor calpastatin reduces human alpha-Synuclein processing, aggregation and synaptic impairment in [A30P]alphaSyn transgenic mice. Hum Mol Genet 2014;23:3975–3989.

[232] Ha TY. MicroRNAs in human diseases: from cancer to cardiovascular disease. Immune Netw 2011;11:135–154.

[233] Fullgrabe J, Klionsky DJ, Joseph B. The return of the nucleus: transcriptional and epigenetic control of autophagy. Nat Rev Mol Cell Biol 2014;15:65–74.

[234] Frankel LB, Lund AH. MicroRNA regulation of autophagy. Carcinogenesis 2012;33: 2018–2025.

[235] Tekirdag KA, Korkmaz G, Ozturk DG, Agami R, Gozuacik D. MIR181A regulates starvation- and rapamycin-induced autophagy through targeting of ATG5. Autophagy 2013;9:374–385.

[236] Coolen M, Bally-Cuif L. MicroRNAs in brain development and physiology. Curr Opin Neurobiol 2009;19:461–470.

[237] Bredy TW, Lin Q, Wei W, Baker-Andresen D, Mattick JS. MicroRNA regulation of neural plasticity and memory. Neurobiol Learn Mem 2011;96:89–94.

[238] Hebert SS, Papadopoulou AS, Smith P, Galas MC, Planel E, et al. Genetic ablation of Dicer in adult forebrain neurons results in abnormal tau hyperphosphorylation and neurodegeneration. Hum Mol Genet 2010;19:3959–3969.

[239] Liu N, Landreh M, Cao K, Abe M, Hendriks GJ, et al. The microRNA miR-34 modulates ageing and neurodegeneration in *Drosophila*. Nature 2012;482:519–523.

[240] Goodall EF, Heath PR, Bandmann O, Kirby J, Shaw PJ. Neuronal dark matter: the emerging role of microRNAs in neurodegeneration. Front Cell Neurosci 2013;7:178.

[241] Tazawa H, Yano S, Yoshida R, Yamasaki Y, Sasaki T, et al. Genetically engineered oncolytic adenovirus induces autophagic cell death through an E2F1-microRNA-7-epidermal growth factor receptor axis. Int J Cancer 2012;131:2939–2950.

[242] Junn E, Lee KW, Jeong BS, Chan TW, Im JY, et al. Repression of alpha-synuclein expression and toxicity by microRNA-7. Proc Natl Acad Sci U S A 2009;106: 13052–13057.

[243] Decressac M, Mattsson B, Weikop P, Lundblad M, Jakobsson J, et al. TFEB-mediated autophagy rescues midbrain dopamine neurons from alpha-synuclein toxicity. Proc Natl Acad Sci U S A 2013;110:E1817–E1826.

[244] Musumeci T, Pellitteri R, Spatuzza M, Puglisi G. Nose-to-brain delivery: evaluation of polymeric nanoparticles on olfactory ensheathing cells uptake. J Pharm Sci 2014;103:628–635.

[245] Behrends C, Sowa ME, Gygi SP, Harper JW. Network organization of the human autophagy system. Nature 2010;466:68–76.

[246] Jucker M. The benefits and limitations of animal models for translational research in neurodegenerative diseases. Nat Med 2010;16:1210–1214.

[247] Rami A. Review: autophagy in neurodegeneration: firefighter and/or incendiarist? Neuropathol Appl Neurobiol 2009;35:449–461.

[248] Yamashima T, Oikawa S. The role of lysosomal rupture in neuronal death. Prog Neurobiol 2009;89:343–358.

[249] Kasinathan N, Jagani HV, Alex AT, Volety SM, Rao JV. Strategies for drug delivery to the central nervous system by systemic route. Drug Deliv 2014;22:243–257.

[250] Wager TT, Chandrasekaran RY, Hou X, Troutman MD, Verhoest PR, et al. Defining desirable central nervous system drug space through the alignment of molecular properties, *in vitro* ADME, and safety attributes. ACS Chem Neurosci 2010;1:420–434.

[251] Pardridge WM. Drug transport across the blood–brain barrier. J Cereb Blood Flow Metab 2012;32:1959–1972.

15

PHARMACOLOGICAL CHAPERONES AS POTENTIAL THERAPEUTICS FOR LYSOSOMAL STORAGE DISORDERS: PRECLINICAL RESEARCH TO CLINICAL STUDIES

ROBERT E. BOYD, ELFRIDA R. BENJAMIN, SU XU, RICHIE KHANNA, AND KENNETH J. VALENZANO

Amicus Therapeutics, Cranbury, NJ, USA

15.1 INTRODUCTION

Lysosomes are cellular organelles in which a variety of glycosphingolipids (GSLs), glycosaminoglycans, glycoproteins, and oligosaccharides are degraded into simpler substances that can be recycled or excreted from the cell. To metabolize these complex biomolecules, a variety of hydrolases has evolved to function optimally at the mildly acidic pH (~5.5) of the lysosome. Disruption of any of these metabolic pathways can lead to accumulation of one or more substrates and emergence of the associated disease pathology. Collectively, these diseases are known as lysosomal storage disorders (LSDs), with currently more than 50 described [1]. The pathology associated with LSDs varies significantly depending on the deficient enzyme and accumulated substrate(s). Interestingly, disease severity as well as organ involvement also can be markedly different between individuals with the same LSD [2].

The current standard of care for several LSDs is enzyme replacement therapy (ERT) in which a manufactured enzyme is infused on a weekly or biweekly basis.

Lysosomes: Biology, Diseases, and Therapeutics, First Edition.
Edited by Frederick R. Maxfield, James M. Willard and Shuyan Lu.

While ERT does improve clinical outcomes for many patients, immunological responses, sometimes severe, and insufficient efficacy in key target tissues and organs leave a significant unmet medical need for many patients. This is most evident for those LSDs that involve CNS impairment since currently available ERTs are unable to cross the blood–brain barrier (BBB), though brain-targeted ERTs are an active area of preclinical research [3]. An alternative therapeutic approach for LSDs is substrate reduction therapy (SRT). This treatment option relies on inhibition of a key enzyme in the biosynthetic pathway of the accumulated substrate. A potential advantage of this approach is the fact that the small-molecule enzyme inhibitor has the potential to cross the BBB, and therefore addresses the CNS pathology associated with the LSD that it is intended to treat. In the case of Gaucher disease (GD), the side effect profile associated with the only approved SRT molecule, *N*-butyl-deoxynojirimycin (NB-DNJ, miglustat, Zavesca™; Actelion, Allschwil, Switzerland), has precluded its use in the vast majority of patients [4], though this same molecule was recently approved to treat the CNS manifestations of another LSD, Niemann–Pick type C disease (NPC) [5]. Second-generation SRT molecules for GD are currently in clinical development [6].

Over the last half century, it has been recognized that accumulation of a particular substrate(s) results from reduced activity of a specific lysosomal enzyme. To this end, Brady et al. [7] recognized that for GD, the reduced enzymatic activity results from mutations in the gene (*GBA1*) that encodes the lysosomal enzyme acid β-glucosidase (GCase). Since that time, the connection between genetic mutations and virtually all LSDs has been firmly established. While these mutations may be large deletions, insertions, or premature stop codons, they are more often missense mutations that involve a single amino acid change in the enzyme's primary amino acid sequence. These mutations are often outside the enzyme's active site and in many cases lead to the expression of a catalytically competent form of the enzyme. However, missense mutations often result in an enzyme that is thermodynamically less stable than its wild-type form; consequently, a substantially smaller quantity of enzyme is able to fold properly and pass the quality control mechanisms of the endoplasmic reticulum (ER) immediately after protein synthesis [8]. As a consequence, the majority of these proteins are tagged for proteasomal degradation via the ER-associated degradation (ERAD) pathway, which ultimately results in insufficient lysosomal activity [9].

Given the fact that the mutated enzymes associated with many LSDs are often catalytically competent but unable to traffic to lysosomes, an alternative therapeutic approach for LSDs involves the use of pharmacological chaperones (PCs). PCs are small molecules that bind with high affinity and selectivity to increase the physical stability of their otherwise unstable enzyme target, thereby improving cellular trafficking from the ER to the lysosome [10,11]. In principle, there are many sites that such small molecules could bind on their respective target; in practice, however, an overwhelming majority of PCs bind to the active site of their target enzyme, generally with high specificity [10,11]. The idea of using a small molecule that binds to an enzyme's active site (by definition, an inhibitor) to increase total cellular enzymatic activity is counterintuitive; however, there are several reasons that account for the viability of this approach [10]. First, PCs are intended to bind and stabilize their target enzyme in the neutral pH environment of the ER, and to dissociate in the acidic environment of the lysosome. Consequently, reduced binding affinity at acidic pH is an important property of many PCs. In addition, high concentrations of accumulated substrate in the lysosome can compete with the PC for binding to the enzyme's active site, helping to drive kinetics toward substrate turnover rather than PC-mediated inhibition. Finally, understanding the lysosomal half-life of the target enzyme relative to the half-life of the small-molecule PC can aid in the development of administration regimens and doses that allow for maximized substrate turnover [10,12,13]. While nonactive site PCs have the potential to greatly simplify the need to optimize administration regimens as inhibition likely would not be present, the identification of such compounds has proved challenging. However, Marugan and coworkers [14] have recently made some progress with a series of molecules that seem to increase the trafficking of GCase to the lysosome without inhibiting its catalytic activity. While it is unclear if this series of compounds can also increase substrate turnover, it does represent an important step in the identification of nonactive site PCs.

Chemical chaperones, such as dimethylsulfoxide or glycerol, are another class of small molecules that also can bind and stabilize a variety of proteins [15]. However, these molecules are nonselective and have low potency; as such, they are yet to reach clinical practice. Hence, chemical chaperones should not be confused with PCs, which bind with high affinity to a specific target protein to stabilize its three-dimensional conformation and permit proper trafficking through the secretory pathway. Comprehensive reviews on the discovery/characterization, mechanism of action, and medicinal chemistry aspects of PCs have been previously published [10,11]. The remainder of this chapter therefore focuses on the select LSDs for which specific PCs have been identified and moved through preclinical evaluation and into clinical development.

15.2 FABRY DISEASE

Fabry disease (FD) is an X-linked LSD caused by mutations in the gene (*GLA*) that encodes the lysosomal hydrolase α-galactosidase A (α-Gal A; EC 3.2.1.22) [16]. Deficiency of α-Gal A results in accumulation of neutral GSLs with terminal α-galactose residues, primarily globotriaosylceramide (GL-3), in plasma and cells of blood vessels, skin, heart, kidney, brain, and other tissues [16–19]. A deacylated analog of GL-3, globotriaosylsphingosine (known as lyso-Gb_3), is also elevated in the plasma of Fabry patients by more than 10-fold compared to GL-3, and is an important new indicator of FD [20–22]. High levels of plasma lyso-Gb_3 correlate with increased risk for cerebrovascular disease and left ventricular hypertrophy in males and females with FD, respectively, and greater lifetime exposure to lyso-Gb_3 has been positively correlated with disease severity [21].

The clinical manifestations of FD span a broad spectrum of severity and can include progressive renal failure, cardiac disease, cerebrovascular disease, small-fiber peripheral neuropathy, and skin lesions, among others [16,23]. Males with FD who have little or no detectable α-Gal A activity are commonly referred to as "classic"

Fabry patients and are most severely affected. If not treated, the life expectancy of these patients is reduced and death usually occurs in the fourth or fifth decade of life from renal failure, cardiac dysfunction, or stroke. Female Fabry patients may be mildly symptomatic or as severely affected as classic males [24]. In addition, many individuals with FD present with a later-onset form, and generally have higher residual α-Gal A activity than classic males [25]. Furthermore, recent screening studies suggest that there could be a large undiagnosed population of later-onset Fabry patients [26–32].

The currently available treatment for FD is ERT, with two approved products: Fabrazyme® (agalsidase beta; Genzyme, A Sanofi Company, Cambridge, MA), which is available worldwide, and Replagal® (agalsidase alfa; Shire, Cambridge, MA), available outside the United States only. Each is administered as a biweekly infusion. These ERTs are generally well tolerated and in many patients lead to lower levels of GL-3 in plasma, urine, and microvascular endothelium, stabilize kidney function, and improve FD-related symptoms [33–38]. The reduction of plasma lyso-Gb_3 levels in response to ERT has also been demonstrated [20,39,40], although Fabry males who developed neutralizing antibodies toward the infused enzyme had significantly less reduction compared to males who did not [41].

PCs may serve as a new approach to the treatment of FD for some patients [10,11,42–46]. The first PC investigated as a potential treatment for FD was galactose, which is also the terminal residue of GL-3 and lyso-Gb_3 that is removed by α-Gal A. Incubation of male FD patient-derived lymphoblast cell lines with galactose led to significantly greater cellular α-Gal A protein levels and enzyme activity for 7 of the 11 different mutant forms tested [43]. In a subsequent clinical case study, intravenous infusion of galactose (1 g/kg every other day) led to a significant improvement in cardiac function in a male FD patient with severe myocardial disease [47]. After a 2-year treatment period, it was determined that the patient no longer required cardiac transplantation, offering the first proof of concept for PC therapy in an LSD.

A few years later, Fan et al. [44,46] described the use of a natural substrate mimetic, the iminosugar 1-deoxygalactonojirimycin (DGJ, AT1001, migalastat), as a PC for α-Gal A. DGJ binds reversibly and selectively to the active site of α-Gal A with high affinity [48]. Furthermore, DGJ increased the cellular levels and activity of 49 of 75 different missense mutant forms of α-Gal A in cultured lymphoblasts or fibroblasts derived from male FD patients [42,49–51]. Importantly, elevated GL-3 levels were also reduced in Fabry fibroblasts that showed increased α-Gal A levels in response to DGJ incubation [42,49,51].

Initial preclinical *in vivo* studies with DGJ used a transgenic mouse model (TgM/KO) that lacked the endogenous α-Gal A gene but expressed a human mutant R301Q transgene driven by a β-actin promoter [52]. The R301Q mutant form of α-Gal A has been identified in patients with both classic and late-onset FD, and has shown increased levels and activity in FD patient-derived cell lines in response to DGJ [49,53–56]. Following 1-week oral administration of DGJ, TgM/KO mice showed significant increases in α-Gal A activity in the heart, kidney, spleen, and liver at doses of 3 and 30 mg/kg [52]. Subsequent studies using a different transgenic mouse that also expresses the human R301Q transgene, but driven by the human α-Gal A promoter, also showed significant increases in α-Gal A activity in various disease-relevant tissues [13]. Most importantly, a concomitant reduction of GL-3 levels in the skin, heart, and kidney following daily administration of DGJ for 4 weeks was shown; 24-week administration resulted in even greater reductions [13]. These *in vivo* studies also demonstrated that less-frequent administration (e.g., repeated cycles of 4 days with drug followed by 3 days without drug, or every other day with drug (QOD) resulted in even greater GL-3 reduction compared to daily administration, with reductions comparable to those obtained with once weekly 1 mg/kg agalsidase beta [13]. These data indicate that oral administration of DGJ can increase α-Gal A activity and decrease GL-3 levels in disease-relevant tissues of FD mice. Furthermore, they highlight the importance of identifying an optimal administration regimen that allows sufficient time for DGJ clearance from the lysosome in order to restore α-Gal A activity and enable maximum substrate turnover. Taken together, the results of these preclinical studies indicate a potentially positive effect on FD patients that express mutant forms of α-Gal A that can be elevated by DGJ, and have supported moving DGJ into clinical investigations.

To this end, DGJ is the active component of an investigational new drug, migalastat hydrochloride (hereafter, migalastat; Amicus Therapeutics, Cranbury, NJ), that is in development as a potential treatment for FD. In support of migalastat clinical development, four Phase 1 studies in healthy volunteers were conducted to determine the pharmacokinetics, pharmacodynamics, safety, and tolerability of migalastat HCl [57]. The pharmacokinetic profile of migalastat was dose-proportional between 25 and 2000 mg, with a half-life of 3–4 h. No abnormal cardiac effects were observed. Importantly, in a multiple ascending dose study with repeat administration of migalastat (50 or 150 mg twice a day) over 7 days, increases in α-Gal A activity were observed in both dose groups in white blood cell lysates on Days 5 and 7 [57]. Also noteworthy, the increase over baseline persisted for an additional 7 days after the final administration of DGJ. Hence, orally administered DGJ can bind and stabilize endogenous wild-type α-Gal A in white blood cells of healthy volunteers, thus leading to greater cellular α-Gal A levels. These data highlighted the key pharmacodynamic outcome that supported the advancement of DGJ into Phase 2 clinical studies.

In two open-label Phase 2 studies, a total of nine male FD subjects was administered 150 mg of migalastat orally every other day (QOD) for up to 48 weeks (clinicaltrials.gov: NCT00283959 and NCT00283933) [58]. Increases in α-Gal A activity of at least 50% were seen in white blood cells, skin, and kidney in six of nine subjects. Decreases in GL-3 levels were seen in the skin, urine, and/or kidney in these six subjects. These six subjects all expressed mutant forms of α-Gal A that were elevated following incubation with 10 μM migalastat in a HEK-293 cell-based assay; the three subjects who did not show a consistent response had mutant forms of α-Gal A that did not respond to migalastat in the HEK-293 cell-based assay [56,58].

The selection of female FD subjects for PC therapy has been more challenging than for male subjects. Changes in the biomarkers that have been used historically to evaluate the effect of new treatments, such as GL-3 levels in plasma, urine, skin, or kidney, are difficult to interpret in females, relative to males, since they generally have lower GL-3 burden, and have different patterns of GL-3 accumulation due to

X-linked mosaicism [24,59–61]. Thus, an open-label Phase 2 study in females ($n = 9$) with symptomatic FD was conducted to evaluate the safety, tolerability, and pharmacodynamic effects of migalastat (clinicaltrials.gov: NCT003045120) [62]. Migalastat (50, 150, or 250 mg QOD) was administered for 12 weeks, with extension to 48 weeks. The *GLA* mutation of each subject was retrospectively assessed for response to migalastat based on the magnitude of the mutant α-Gal A response to 10 μM migalastat in the HEK-293 cell-based assay described above [56,62]. Female subjects with responsive mutant forms of α-Gal A tended to demonstrate a greater pharmacodynamic response compared to female subjects with nonresponsive mutant forms. The greatest declines in urine GL-3 were observed in three subjects with responsive mutant forms who were administered 150 or 250 mg migalastat; these three also demonstrated reduced GL-3 inclusions in kidney peritubular capillaries. These results suggested that migalastat potentially may be a novel genotype-specific therapy for FD in patients with responsive mutant forms of α-Gal A, including heterozygous females.

Preliminary data from ongoing extension studies of four open-label Phase 2 migalastat clinical trials (the three studies described above, plus another that is unpublished) are available (clinicaltrials.gov: NCT00526071) [63,64]. To date, 22 subjects received migalastat for over 3 years, of which 7 received migalastat for over 4 years. Migalastat was generally safe and well tolerated, with encouraging preliminary renal function data. In addition, a trend toward reduced urine protein was observed in subjects with responsive mutant forms. Taken together, these results suggested that further clinical investigation was warranted.

As such, migalastat has been investigated in two Phase 3 clinical studies. The first was a double-blind, randomized, placebo-controlled study to evaluate the efficacy, safety, and pharmacodynamics of 150 mg QOD oral migalastat administration in male and female FD subjects with the following characteristics: (i) treatment-naïve or off ERT for at least 6 months, (ii) elevated urine GL-3 at screening, (iii) estimated glomerular filtration rate (eGFR) $\geq$30 mL/min/1.73 m^2, and (iv) express a mutant form of α-Gal A that can be elevated in the presence of migalastat as determined using a research-purpose HEK-293 cell-based assay (referred to as the "clinical trial HEK assay") (ClinicalTrials.gov: NCT00925301) [65]. Subjects were randomized 1:1 to receive either placebo or migalastat. The first 6 months of this study were double-blind and placebo-controlled; thereafter, all remaining subjects received migalastat for 6 months (open-label), followed by an optional 12-month extension. The primary endpoint was the proportion of subjects ($n = 67$) that showed a $\geq$50% reduction in kidney interstitial capillary (IC) GL-3 levels following migalastat treatment using a novel quantitative histological scoring method [66]. The 6-month results indicated that the primary endpoint was not met ($p = 0.3$, based on the exact Cochran–Mantel–Haenszel test stratified by gender) [67]. Review of the data revealed that many subjects had relatively low IC GL-3 levels, and that larger decreases in IC GL-3 were observed with increasingly higher baseline IC GL-3 values in migalastat-treated subjects [68]. Furthermore, some subjects' *GLA* mutations were recategorized as "amenable" or "nonamenable" based on a GLP-validated form of the HEK-293 cell-based assay ("GLP HEK assay") that was completed before the 12-month data were unblinded. When evaluating the mean change from baseline in the number of inclusions per capillary (IPC) for subjects with amenable mutations only during the 6-month double-blind phase, a statistically significant decrease in the number of IPCs was seen in the migalastat group ($n = 25$) relative to the placebo group ($n = 21$), which showed an increase in the number of IPCs. This effect was durable, with similar findings at the 12-month time point. Kidney function, measured by eGFR, also remained stable for up to 24 months of treatment. These results show the ability of migalastat to reduce substrate (i.e., kidney IC GL-3), and to stabilize renal function in subjects with *GLA* mutations that were determined to be amenable in the GLP HEK assay.

The second Phase 3 trial is an ongoing randomized, open-label study to compare the efficacy and safety of migalastat to that of ERT in subjects with FD and amenable mutations (ClinicalTrials.gov: NCT01218659) [69]. The study enrolled 60 male and female FD subjects with the following characteristics: (i) on ERT for ≥12 months, with ≥3 months at the labeled dose and regimen, (ii) *GLA* mutation corresponding to an amenable mutant form of α-Gal A as determined by the clinical trial HEK assay, and (iii) eGFR ≥30 mL/min/1.73 m^2. The study was designed to randomize subjects 1.5:1 to treatment with migalastat or continued ERT for 18 months (i.e., $n = 36$ migalastat, $n = 24$ ERT), with those finishing the study offered an optional 12-month migalastat extension. The primary outcome measures were renal function as assessed by measured (iohexol) GFR and eGFR, with secondary outcome measures including (i) renal function assessed by eGFR (based on a different methodology than used for the primary endpoint) and 24-h urine protein, (ii) composite clinical outcome, as assessed by time to occurrence of renal, cardiac, or cerebrovascular events, or death, (iii) cardiac function, as assessed by echocardiography, and (iv) patient-reported outcomes including pain and quality of life. Following randomization, 34 of 36 subjects who switched to migalastat and 18 of 24 subjects who continued with ERT completed the primary 18-month treatment period. Among subjects completing the 18-month primary treatment period, 32 out of 34 in the migalastat group, and 16 out of 18 in the ERT group had GLP HEK amenable mutations. In the migalastat group, 31 out of 32 subjects with GLP HEK amenable mutations elected to continue to receive migalastat in the 12-month treatment extension, and in the ERT group, 15 out of 16 subjects with GLP HEK amenable mutations elected to switch from ERT to migalastat for the 12-month treatment extension, which is ongoing.

15.3 GAUCHER DISEASE

GD is generally thought to be the most prevalent LSD and is caused by inherited mutations in the gene (*GBA1*) that encodes acid β-glucosidase (GCase, glucocerebrosidase; EC 3.2.1.45), the lysosomal enzyme responsible for removal of the terminal β-linked glucose from glucosylceramide (GlcCer) to yield ceramide [70]. Mutations in GCase result in reduced cellular enzyme activity and progressive accumulation of GlcCer primarily within macrophages (Gaucher cells) of the liver, bone marrow, and spleen. The typical clinical presentation of GD includes anemia, thrombocytopenia, hepatosplenomegaly, bone lesions, and in some cases, central nervous system

(CNS) impairment [70]. GD without CNS involvement is classified as type I, while neuronopathic GD is classified as type II (infantile acute) or type III (subacute). Over 200 mutations in *GBA1* have been identified, though the two most prevalent missense mutant forms are N370S and L444P GCase [71]. The N370S mutant form has approximately 30% of the cellular enzymatic activity relative to cells that express wild-type GCase [72]. Patients homozygous or heterozygous for N370S GCase typically present with nonneuronopathic GD. In contrast, the L444P mutant form has significantly less residual activity (10–12% of wild type); consequently, individuals homozygous for the L444P mutation often present with CNS impairment that is associated with type III disease. More than 70% of Gaucher patients within the Ashkenazi Jewish population carry at least one N370S allele, while 38% of non-Jewish Gaucher patients carry the L444P allele [71,73,74].

Currently, ERT and SRT are the only two approved treatment options for patients with GD [75–79]. GD was the first LSD for which an ERT was available, with the placental-derived Ceredase® (alglucerase) reaching clinical practice in 1988. Ceredase was replaced with a recombinant form, Cerezyme® (imiglucerase), in 2001; both ERTs are marketed by Genzyme. Several other ERTs have more recently been approved for GD in various geographies, including velaglucerase alfa (VPRIV®, Shire), and taliglucerase alfa (ELELYSO™, Pfizer, New York, NY and Protalix, Carmiel, Israel). Treatment of GD with ERT is most effective against type I disease, as well as the visceral manifestations of type II and III diseases, with significant reductions in spleen and liver weights, as well as increases in platelet counts and hemoglobin levels, typically observed [80–83]. The CNS manifestations of type II and III GD do not respond well to ERT due to the inability of the exogenous enzyme to cross the BBB [84].

As discussed, SRT drugs that are used to treat GD act as inhibitors of glucosyltransferase, the enzyme responsible for the synthesis of GlcCer, thereby leading to reduced cellular levels of this substrate. Zavesca® (*N*-butyl-1-deoxynojirimycin (NB-DNJ), miglustat) is currently the only approved SRT for mild-to-moderate type I GD patients [78,79,85]. Unfortunately, many patients treated with Zavesca experience side effects that include diarrhea, weight loss, tremor, and peripheral neuropathy [4]. Recently, another small molecule, orally available SRT drug, eliglustat tartrate (Genzyme), has shown promise in Phase 2 and 3 studies for treatment of type I GD [6]. As the therapeutic agents of SRT are small molecules, these drugs have the potential for CNS access and treatment of the neurological aspects of GD. To this end, Zavesca has been evaluated as a treatment for neuronopathic GD, though no significant neurological benefit was seen in type III patients [86]. Unfortunately, eliglustat tartrate does access the CNS, but is also a PgP substrate, which transports eliglustat out of the brain, preventing significant drug levels from being achieved in the CNS [87]. It is the hope that future drugs in this class may provide benefit to patients with neuronopathic forms of GD, though the long-term effects of inhibiting this natural biosynthetic pathway are not known.

PCs also have been proposed as a potential therapy for GD [88–90]. Since none of the currently approved therapies address the CNS manifestations of GD, there is also a hope that small-molecule PCs may provide treatment of the neuronopathic forms of the disease. A wide variety of compounds that increase the cellular activity of various mutant forms of GCase in cell lines derived from GD patients have been evaluated as PCs. These molecules can be categorized as carbohydrate mimetics (iminosugars, azasugars, or carbasugars), or noncarbohydrate compounds identified by high-throughput screening initiatives [10,11]. From this collection, two have advanced through preclinical animal studies and into early-stage clinical development, namely isofagomine (IFG; AT2101; afegostat tartrate) and Ambroxol.

IFG is an azasugar that binds and inhibits both wild-type and mutant forms of GCase, resulting in stabilization and increased cellular and lysosomal levels [91,92], particularly for the N370S variant [92,93]. Similarly, the cellular levels of a number of other missense mutant forms of GCase, including L444P, are also elevated in response to incubation with IFG [92,94–96]. These encouraging *in vitro* data supported the advancement of IFG into multiple preclinical animal studies. In the first study, 4-week administration of IFG (3–30 mg/kg) led to statistically significant increases in GCase activity in liver, spleen, lung, bone, and brain, as well as in liver macrophages, in a mouse model homozygous for L444P GCase [96]. Furthermore, increased lysosomal levels of L444P GCase were demonstrated in mouse liver homogenates. More importantly, 6-month administration of approximately 8 mg/kg IFG to these mice modestly reduced liver and spleen weights, providing some evidence of *in vivo* efficacy. In a second study using a GD mouse model homozygous for V394L GCase that to some extent mimics neuronopathic forms of the disease, increases in GCase protein levels and activity were demonstrated in peripheral tissues and brain following daily administration of IFG [97]. Importantly, reduced neuroinflammation and delayed onset of neurological disease were also noted, with life spans increased by up to 18 days (~35%). In a third study, the effects of daily oral IFG administration were evaluated in three different mouse models, each homozygous for a different variant of GCase (V394L, D409H, or D409V) [92]. In all models, increased GCase activity was observed in visceral tissues and brain following administration of 30 mg/kg IFG for up to 8 weeks. Finally, using transgenic mice that express either the human N370S or L444P mutant forms of GCase, daily subcutaneous administration of 20 mg/kg IFG for 14 days resulted in consistently elevated GCase levels in both spleen and liver, with modest increases in cerebrum also noted in the L444P line [98]. These experimental results, as well as a clean safety profile, led to the advancement of IFG into clinical studies.

In support of IFG clinical development, two randomized, double-blind Phase 1 studies in healthy volunteers were conducted by Amicus Therapeutics to determine the pharmacokinetics, pharmacodynamics, safety, and tolerability of IFG [99]. In the single ascending dose study, 8, 25, 75, 150, and 300 mg IFG all showed good systemic and dose-proportional plasma exposures. Mean plasma levels peaked at 3.4 h, and the plasma elimination half-life was 14 h. In the multiple ascending dose study, 25, 75, or 225 mg IFG was administered daily for 7 days. On Day 7, the pharmacokinetic behavior remained linear with dose, with no accumulation of IFG. In both studies, IFG was generally well tolerated and treatment-emergent adverse events were mild; no serious adverse events occurred.

In the multiple-dose healthy volunteer study, GCase activity in white blood cells was measured on Days 1, 3, 5, and 7 during IFG administration and on Days 9, 14, and 21 during the posttreatment washout period [10]. In all subjects receiving IFG, there was a marked increase in GCase levels during the treatment period, followed by a decrease upon withdrawal of the drug, and a return to near baseline levels by Day 21. The increase in enzyme level was dose related, reaching approximately 3.5-fold above baseline levels at the highest dose level. These results for the safety, pharmacokinetics, and preliminary pharmacodynamic effects in healthy volunteers supported the further evaluation of IFG in patients for the treatment of GD.

As such, two Phase 2 clinical studies with IFG were initiated by Amicus Therapeutics. The first study evaluated the safety and pharmacodynamic effects of IFG in type I GD subjects who discontinued ERT for the 4-week duration of the study (ClinicalTrials.gov: NCT00433147). Twenty-two women and eight men between the ages of 18 and 63, representing 12 different mutations (including N370S and L444P) and a 9-year average duration of ERT treatment, were enrolled. IFG (25 and 150 mg daily, 150 mg 3 days on/4 days off, or 7 days on/7 days off) was generally well tolerated at all dose levels evaluated, with no serious adverse events reported. GCase activity as measured in white blood cells was increased in 20 of the 26 subjects. Five of the six subjects without a clear increase were in the lowest dose cohort or the cohort that was administered IFG least frequently. As expected for a short-term study, the levels of relevant markers of GD, including platelet counts, hemoglobin levels, GlcCer levels, chitotriosidase activity, and pulmonary activation-related chemokine were unchanged. The second Phase 2 study was a 6-month study designed to evaluate the safety of IFG and its effect on parameters that are commonly abnormal in GD (ClinicalTrials.gov: NCT00446550). This study was also conducted in type I GD subjects who had never received ERT. Two dose regimens of IFG (225 mg 3 days on/4 days off or 7 days on/7 days off) were studied. While all subjects experienced an increase in the levels of GCase as measured in white blood cells, clinically meaningful improvements in key measures of disease were observed in just 1 of the 18 subjects who completed the study. Importantly, the white blood cell GCase levels attained in this one subject following IFG treatment were close to those of healthy volunteers, which was substantially greater than the magnitude of increase in all other subjects. While the results suggested that treatment with IFG was generally well tolerated, with no serious adverse events reported, Phase 3 studies were not pursued.

Ambroxol, an approved expectorant, was identified from a screen of 1040 FDA-approved drugs by its ability to stabilize wild-type GCase against thermal denaturation [95]. Ambroxol also showed pH-dependent affinity for GCase, with decreasing inhibition at lysosomal pH values. In cell-based assays using patient-derived cell lines homozygous for N370S, Ambroxol increased cellular GCase activity up to twofold at a concentration of 60 μM, with evidence of increased lysosomal levels; similar cellular effects were seen in a heterozygous cell line expressing F213I and L444P GCase. Evidence was also provided for modest reductions in GlcCer levels in patient-derived cells. In other *in vitro* studies, Ambroxol (at concentrations up to 100 μM) elevated the cellular and lysosomal levels of multiple GCase mutants in patient-derived cells expressing R120W, R131C, N188S, G193W, F213I, N370S, L444P, and/or P415R [100,101]. In wild-type mice, Ambroxol was generally well tolerated, with detrimental effects on water intake and body weight occurring in the highest dose groups only [100]. Finally, N370S or L444P transgenic mice were subcutaneously administered 100 mg/kg Ambroxol daily for 14 days, resulting in consistently elevated GCase levels in both spleen and liver [98]. These results led to the advancement of Ambroxol into a physician-sponsored study in GD patients.

This pilot study assessed the tolerability and efficacy of 150 mg Ambroxol daily in 12 subjects with symptomatic type I GD who were not on ERT, 11 of whom were homozygous for the N370S allele [102]. Of the nine subjects that completed the 6-month study, none showed deterioration in the measured GD-related parameters (e.g., body weight, hemoglobin levels, platelet counts, liver and spleen volumes, and chitotriosidase activity), and three continued on therapy for an additional 6 months. Over the 12-month course, these three subjects showed reductions in spleen volumes (mean reduction approximately 15%) and chitotriosidase activity (mean reduction approximately 40%), with stabilized hemoglobin levels and liver volumes. Importantly, in one subject platelet counts also increased over 50%. Given these promising results, follow-up placebo-controlled studies that include larger numbers of GD subjects with more diverse genotypes, and potentially with CNS manifestations, are warranted.

15.4 GM2 GANGLIOSIDOSES (TAY–SACHS/SANDHOFF DISEASES)

The terminal sugar *N*-acetylgalactosamine, which is present on several GSLs as well as various other oligosaccharides and glycoproteins, can be enzymatically removed in the lysosome by β-hexosaminidase (β-Hex; EC 3.2.1.52). β-Hex is composed of two subunits and can exist as a heterodimer comprised of α- and β-subunits (β-Hex A), or as one of two homodimers, one containing two α-subunits (β-Hex S, which is exceptionally rare and whose function is not clearly understood) and the other containing two β-subunits (β-Hex B). While most *N*-acetylgalactosamine-containing substrates can be processed by either the β-Hex A or β-Hex B isoform, the ganglioside GM2 and several oligosaccharides can only be processed by β-Hex A. Mutations in either gene result in GM2 gangliosidosis [103]. More specifically, mutations in the gene that encodes the β-subunit primarily lead to accumulation of GM2 and result in an LSD known as Tay–Sachs disease (TSD). In contrast, α-subunit mutations result in decreased activity of both β-Hex A and β-Hex B, and the accumulation of multiple GSLs. The resultant LSD is an equally or sometimes more severe form of GM2 gangliosidosis known as Sandhoff disease (SD). Progressive neurodegeneration is the primary pathology associated with both forms of GM2 gangliosidosis. As with many LSDs, the infantile or early-onset form of the disease has the most severe pathology with rapid neurodegeneration typically occurring at approximately 6 months of age, with an overall lifespan of 4 years or less. The late-onset form typically tends to present clinically in the second decade of life, again with progressive neurodegeneration over the next two to three decades. An intermediate juvenile form has also been characterized, with life expectancies typically not beyond two decades. Currently, there are no effective treatment options for GM2 gangliosidosis patients.

Pyrimethamine (PYR), an approved antimalarial drug, was identified as a potent inhibitor of β-Hex from a screen of 1040 FDA-approved drugs [104]. In follow-up cell-based assays using 10 TSD and 7 SD patient-derived cell lines, PYR led to significant increases in cellular and lysosomal β-HEX levels and activity in one cell line with a β-subunit mutation (TSD) as well as all seven cell lines with α-subunit mutations (SD). Consequently, escalating doses of PYR (maximum: 100 mg/day) administered orally in a single daily dose over a 16-week period were evaluated in a Phase 1/2 open-label clinical study to determine tolerability and efficacy in the treatment of late-onset GM2 gangliosidosis subjects [105]. A total of 11 subjects were enrolled, 8 males and 3 females, aged 23–50 years. For the eight subjects that completed the study, up to fourfold increases in β-Hex activity were observed in white blood cell lysates at doses of 50 mg per day or less, despite marked individual variations in PYR's pharmacokinetic profile. Unfortunately, significant side effects (increased ataxia, lack of coordination, and seizures) were observed in most subjects at doses of 75 mg per day or higher and the trial was discontinued. These data suggest that PYR can enhance β-Hex activity in peripheral cells of subjects with late-onset GM2 gangliosidosis at doses lower than those that are associated with adverse effects. Future studies will be necessary to determine the clinical efficacy of PYR in TSD and SD patients.

15.5 POMPE DISEASE

Pompe disease (acid maltase deficiency, glycogen storage disease type II) is an LSD caused by mutations in the gene (*GAA*) that encodes the lysosomal enzyme acid α-glucosidase (GAA; EC 3.2.1.20) [106,107], the enzyme that hydrolyzes the α1,4- and α1,6-glycosidic bonds of glycogen [108]. Deficiency in GAA activity results in glycogen accumulation and deposition in the lysosomes of heart, skeletal muscle, and other tissues [109]. GAA is synthesized as a 110 kDa immature glycoprotein precursor in the ER and undergoes a series of proteolytic and *N*-glycan processing events to yield the 95 kDa intermediate and the 76 and 70 kDa mature isoforms [110–113]. This processing into the intermediate and mature forms occurs in late endosomes and lysosomes, respectively [110], with the final mature isoforms showing significantly increased affinity and activity toward glycogen compared to the precursor forms [111,114]. Increasing the amount of mature, active GAA in lysosomes is an important step toward increasing glycogen hydrolysis and reducing substrate accumulation in this disease.

Pompe disease shows a broad phenotypic spectrum that ranges from the severe infantile-onset form to more slowly progressing, later-onset forms. Infantile-onset Pompe disease patients have little or no GAA activity, present with hypotonia, cardiomegaly, and cardiorespiratory distress, and have a life expectancy of about 2 years [115]. Late-onset forms of the disease typically show some detectable GAA activity, present in childhood or adulthood, and progress more slowly, with musculoskeletal and pulmonary involvement leading to progressive weakness and respiratory insufficiency [116]. Compared to an age- and gender-matched healthy population, untreated late-onset Pompe patients have higher mortality rates [117], while those treated with ERT show substantially reduced mortality [118].

Currently, ERT is the only approved treatment for Pompe disease, administered as a biweekly intravenous infusion of recombinant human GAA (rhGAA; alglucosidase alfa, marketed as Myozyme® and Lumizyme®, Genzyme-Sanofi). Treatment with rhGAA improves cardiac function, motor skills, and life span in infantile-onset patients [119–124] and leads to some improvements in motor and respiratory function in late-onset patients [125,126]. Despite the clinical benefits of ERT, the efficacy of rhGAA may be limited by insufficient targeting and uptake into key disease-relevant tissues, as well as poor tolerability due to sometimes severe immunogenic and anaphylactic reactions to the exogenous enzyme [124,126–132]. Thus, a clear unmet medical need still exists for many Pompe patients.

Small-molecule PCs have been proposed as a potential alternative to ERT for the treatment of Pompe disease [133–136]. Approximately 150 *GAA* mutations are predicted to produce a full-length protein containing a single amino acid substitution. However, these GAA mutants likely do not fold as efficiently or are less stable compared to wild-type GAA and, therefore, are susceptible to premature degradation via ERAD. It has been demonstrated that some mutant forms of GAA are responsive to the PCs 1-deoxynojirimycin hydrochloride (DNJ, duvoglustat hydrochloride, AT2220) and NB-DNJ, as evidenced by increased trafficking, maturation, and activity [133,135]. To this end, Flanagan et al. [136] characterized the effects of DNJ on 76 different mutant forms of GAA. A homology model of GAA was also constructed using the crystal structure of a related eukaryotic α-glucosidase, maltase-glucoamylase, to gain conformational insights into GAA and to map the locations of the residues of the mutant forms of GAA responsive to DNJ. Given the large number of pathogenic GAA mutants identified to date and the need to determine which forms are amenable to PC therapy, this study provided additional evidence and support for the continued evaluation of DNJ as a potential therapeutic for Pompe disease. Mechanistically, DNJ has multiple modes of action during the synthesis and maturation of mutant GAA, including increased specific activity prior to proteolytic processing in lysosomes, facilitated export from the ER with subsequent trafficking and processing through the secretory pathway to lysosomes, and stabilization of the mature isoforms in lysosomes [12].

Furthermore, the *in vivo* effects of DNJ were tested in a new mouse model of Pompe disease [12]. This mouse model expresses the human mutant GAA transgene P545L on a *Gaa* knockout background (hP545L GAA Tg/KO) and shows low GAA activity and elevated glycogen levels in disease-relevant tissues including heart, diaphragm, multiple skeletal muscles, and brain. Daily oral administration of DNJ to hP545L GAA Tg/KO mice resulted in significant and dose-dependent increases in GAA activity with concomitant reduction in tissue glycogen levels; less-frequent DNJ administration regimens (such as 4 days on drug followed by 3 days off, or 5 days on drug followed by 2 days off) resulted in even greater glycogen reduction compared to daily administration. Collectively, these results provided support for the evaluation of DNJ as a potential treatment for Pompe disease.

To date, five Phase 1 clinical studies have been completed in healthy adult volunteers by Amicus Therapeutics, with single ascending DNJ doses ranging from 50 to 2500 mg, and repeat-administration (up to 14 days) ascending doses ranging from

50 to 2000 mg. Across these studies, DNJ was generally safe and well tolerated at all doses and was orally available with a plasma half-life of 4–5 h. There were no drug-related serious adverse events in any of these studies, and no adverse events were considered to be definitely or probably related to study treatment. In the multiple ascending dose study, all possibly related adverse events were mild in severity and resolved spontaneously.

Following these Phase 1 studies, a Phase 2 study (ClinicalTrials.gov: NCT00688 597) was conducted in adult Pompe subjects. The subjects were to be enrolled sequentially to one of three dosing cohorts: 2.5 g for 3 days followed by no drug for 4 days, 5 g for 3 days followed by no drug for 4 days, and 5 g for 7 days followed by no drug for 7 days. This study was terminated after enrollment of three subjects at the 2.5 g dose level due to the occurrence of severe adverse events (muscle weakness) in two subjects, which were judged to be drug related. Fortunately, these effects were reversible over the 4–11 months following DNJ withdrawal. At this high dose of DNJ, these adverse events were deemed to be due to sustained DNJ-mediated inhibition of the low levels of endogenous GAA activity present in these subjects. A follow-up Phase 1 study in healthy volunteers [137] was conducted to evaluate the pharmacokinetics of DNJ in muscle of healthy volunteers after a single oral dose of 100 or 1000 mg. Following administration of 1000 mg DNJ, the drug rapidly appeared and remained in muscle for over 1 week at concentrations that were in excess of its IC_{50} value for inhibition of GAA (approximately 20 nM at pH 5.2). Muscle exposure to DNJ, as assessed by peak DNJ concentration, increased in an approximately dose-proportional manner. DNJ levels in the muscle rapidly declined to 1% of peak levels within 24 h after administration; the remainder was slowly cleared with a terminal half-life of approximately 90 h [137]. Similar to other PCs, DNJ is a reversible, competitive inhibitor that noncovalently binds to the active site of GAA; hence, it is necessary to balance DNJ-mediated increases in GAA stability and lysosomal levels with the potential for *in situ* GAA inhibition. The pharmacokinetic profile of DNJ in human muscles suggests that the oral dose of DNJ used in the Phase 2 study (2.5 g) was too high. Furthermore, the endogenous GAA activity was quite low in these subjects (less than 1% of wild type), suggesting that an appropriate balance between chaperoning and inhibition was not achieved, resulting in prolonged inhibition of endogenous GAA activity, which may have led to the clinically significant adverse, but reversible, muscle-related events in these Pompe subjects. As such, reducing the dose and/or administration frequency of DNJ in future clinical studies may avoid such adverse events.

15.6 PC-ERT COMBINATION THERAPY

As discussed earlier, ERT is used to treat several LSDs, often representing the only approved therapy (with SRT for GD and NPC being the only exception). While providing improved clinical outcomes, ERT in general has many limitations. The infused enzymes tend to have short circulating half-lives, insufficient biodistribution to some key cell types, tissues, and organs, and often elicit immunologic responses that adversely affect tolerability and efficacy. For instance, a recent study with Fabry patients revealed that ERT provided only minimal improvement on cardiovascular parameters, again underscoring the significant unmet medical need in LSDs [138].

PCs offer another approach to the treatment of these diseases, with the potential benefits of being noninvasive, well tolerated, and broadly biodistributed, including the CNS. At the same time, PC therapy also has some limitations, especially as PC monotherapy is only applicable for patients that express responsive mutant forms of the enzymes. However, it was recently hypothesized and demonstrated that PCs can improve the biochemical and pharmacological properties of the exogenous enzymes used in ERTs through binding and stabilization, potentially opening up a new approach for LSD treatment: combination therapy using PCs and ERTs together for improved therapeutic activity.

One of the first examples of PC-ERT combination therapy was described in the context of GD [139]. Incubation of recombinant human GCase (Cerezyme) with IFG significantly increased the stability of the recombinant enzyme and increased cellular uptake into GD patient-derived cells to levels greater than those obtained with enzyme alone. Parenti and coworkers [140] extended the findings on combination therapy to the treatment of Pompe disease using the *Gaa* knockout mouse model. Animals receiving an oral dose of NB-DNJ (4.3 mg/kg) for 2 days prior to an intravenous (IV) infusion of rhGAA showed a significant increase in enzyme activity in multiple disease-relevant organs as compared to those receiving ERT only, including muscles that usually show little-to-no response to ERT alone. Furthermore, DNJ prevented rhGAA denaturation and loss of activity at neutral pH and 37 °C *in vitro*. In rats, oral coadministration of DNJ (3 or 30 mg/kg), 60 min prior to intravenous infusion of rhGAA, increased the circulating half-life and overall exposure of the enzyme by up to twofold. In *Gaa* knockout mice, four biweekly IV administrations of rhGAA with oral coadministration of DNJ (30 mg/kg) prior to ERT resulted in a significant increase of up to 2.5-fold in GAA activity in disease-relevant tissues, including skeletal muscles, compared to ERT alone. More importantly, PC coadministration led to robust glycogen reduction in these same tissues, up to 2.6-fold greater compared to ERT alone [141]. Similarly, increased enzyme stability and cellular uptake and improved substrate reduction in an FD mouse model were observed with DGJ in combination with recombinant human α-Gal A [142].

These promising preclinical results led to the initiation of several Phase 2 PC/ERT coadministration clinical studies. In a Pompe Phase 2 study conducted by Amicus Therapeutics (ClinicalTrials.gov: NCT01380743), subjects who had been on ERT as their standard of care were orally coadministered a single dose of DNJ at 50, 100, 250, or 600 mg 1 h prior to ERT infusion; GAA activities in plasma and muscle were evaluated in comparison to administration of ERT alone. Similar to the findings of the preclinical studies, the PC coadministration led to increased GAA activity in plasma as well as in muscle. Dose-dependent increases in plasma GAA exposure were observed for all subjects, attaining 1.5- to 2.8-fold greater exposures compared to ERT alone. In muscle biopsy samples taken on Day 3 or 7 after administration, increases in total GAA activity were observed in 16 of 24 (67%) subjects with evaluable data. While all DNJ doses evaluated showed some increase in plasma GAA activity, the lowest dose (50 mg) was the least robust, while the highest dose (600 mg) showed the greatest effect on both plasma and muscle GAA levels [143]. In addition, another clinical study based on the combination of NB-DNJ and rhGAA was initiated in Italy (Telethon foundation, trial GUP09017), and has similarly shown greater circulating GAA levels in plasma with PC coadministration compared to ERT alone [144].

Finally, a Fabry Phase 2 study (ClinicalTrials.gov: NCT01196871) initiated by Amicus Therapeutics explored a single oral coadministered dose of 150 or 450 mg DGJ 2 h prior to the infusion of ERT. Again, increased plasma exposures of α-Gal A activity of 1.2- to 5.0-fold compared to ERT alone were seen in 22 of 23 subjects. Both tested doses of DGJ led to greater total α-Gal A activity in Day 2 skin biopsies in 19 of 23 (82.6%) subjects relative to ERT alone, with modest improvements still noted in some Day 7 skin biopsies. Of the two DGJ doses tested, 450 mg resulted in greater increases of active α-Gal A enzyme levels in skin tissue [145].

These clinical studies extend preclinical observations and support the notion that oral administration of a PC shortly before ERT infusion can stabilize the recombinant enzyme in the circulation, protecting the infused proteins from denaturing and leading to greater exposure and improved cellular uptake. These proof-of-concept studies lay the foundation to further explore whether PC/ERT combination therapy can alter ERT biodistribution and therapeutic efficacy in target tissues that respond poorly to ERT alone. Furthermore, the stabilization by the PC also may reduce the immunogenicity of the infused enzyme and thereby ameliorate the immune response elicited by many ERTs. This could improve tolerability and safety, and further enhance therapeutic outcomes in some patients. The synergism of an ERT combined with a PC offers a new therapeutic option with promises for patients suffering with LSDs, especially those who are unable to use PC monotherapy, or who do not get sufficient benefit from ERT treatment alone.

REFERENCES

[1] Parkinson-Lawrence EJ, Shandala T, Prodoehl M, Plew R, Borlace GN, Brooks DA. Lysosomal storage disease: revealing lysosomal function and physiology. Physiology 2010;25:102–115.

[2] Ron I, Horowitz M. ER retention and degradation as the molecular basis underlying Gaucher disease heterogeneity. Hum Mol Genet 2005;14:2387–2398.

[3] Wang D, El-Amouri SS, Dai M, Kuan C-Y, Hui DY, Brady RO, Pan D. Engineering a lysosomal enzyme with a derivative of receptor-binding domain of apoE enables delivery across the blood–brain barrier. Proc Natl Acad Sci U S A 2013;110:2999–3004.

[4] Zimran A, Elstein D. Gaucher disease and the clinical experience with substrate reduction therapy. Philos Trans Roy Soc Lond Ser B Biol Sci 2003;358:961–966.

[5] Wraith JE, Vecchio D, Jacklin E, Abel L, Chadha-Boreham H, Luzy C, Giorgino R, Patterson MC. Miglustat in adult and juvenile patients with Niemann–Pick disease Type C: long-term data from a clinical trial. Mol Genet Metab 2010;99:351–357.

[6] Cox TM. Eliglustat tartrate, an orally active glucocerebroside synthase inhibitor for the potential treatment of Gaucher disease and other lysosomal storage diseases. Curr Opin Investig Drugs 2010;11:1169–1181.

[7] Brady RO, Kanfer JN, Bradley RM, Shapiro D. Demonstration of a deficiency of glucocerebroside-cleaving enzyme in Gaucher's disease. J Clin Invest 1966;45: 1112–1115.

[8] Germain DP, Fan JQ. Pharmacological chaperone therapy by active-site-specific chaperones in Fabry disease: *in vitro* and preclinical studies. Int J Clin Pharmacol Ther 2009;47:S111–S117.

[9] Bonifacino JS, Weissman AM. Ubiquitin and the control of protein fate in the secretory and endocytic pathways. Annu Rev Cell Dev Biol 1998;14:19–57.

[10] Valenzano KJ, Khanna R, Powe AC, Boyd R, Lee G, Flanagan JJ, Benjamin ER. Identification and characterization of pharmacological chaperones to correct enzyme deficiencies in lysosomal storage disorders. Assay Drug Develop Technol 2011;9:213–235.

[11] Boyd RE, Lee G, Rybczynski P, Benjamin ER, Khanna R, Wustman BA, Valenzano KJ. Pharmacological chaperones as therapeutics for lysosomal storage diseases. J Med Chem 2013;56:2705–2725.

[12] Khanna R, Powe AC, Lun Y, Soska R, Feng J, Dhulipala R, Frascella M, Garcia A, Pellegrino LJ, Xu S, Brignol N, Toth MJ, Do HV, Lockhart DJ, Wustman BA, Valenzano KJ. The pharmacological chaperone AT2220 increases the specific activity and lysosomal delivery of mutant acid α-glucosidase, and promotes glycogen reduction in a transgenic mouse model of Pompe disease. PLoS ONE 2014;9:e102092.

[13] Khanna R, Soska R, Lun Y, Feng J, Frascella M, Young B, Brignol N, Pellegrino L, Sitaraman SA, Desnick RJ, Benjamin ER, Lockhart DJ, Valenzano KJ. The pharmacological chaperone 1-deoxygalactonojirimycin reduces tissue globotriaosylceramide levels in a mouse model of Fabry disease. Mol Ther 2010;18:23–33.

[14] Patnaik S, Zheng W, Choi JH, Motabar O, Southall N, Westbroek W, Lea WA, Velayat A, Goldin E, Sidransky E, Leister W, Marugan JJ. Discovery, SAR, and biological evaluation of noninhibitory small molecule chaperones of glucocerebrosidase. J Med Chem 2012;55:5734–5748.

[15] Perlmutter DH. Chemical chaperones: a pharmacological strategy for disorders of protein folding and trafficking. Pediatr Res 2002;52:832–836.

[16] Desnick R, Ioannou Y, Eng C. a-Galactosidase A deficiency: Fabry disease. In: Sriver CR, Beaudet AL, Sly WS, Valle D, editors. *The Metabolic and Molecular Bases of Inherited Disease*. New York: McGraw-Hill; 2001. p 3733–3774.

[17] Sweeley CC, Klionsky B. Fabry's disease: classification as a sphingolipidosis and partial characterization of a novel glycolipid. J Biol Chem 1963;238:PC3148–PC3150.

[18] Brady OR, Gal AE, Bradley RM, Martensson E, Warshaw AL, Laster L. Enzymatic defect in Fabry's disease: ceramidetrihexosidase deficiency. N Engl J Med 1967;276: 1163–1167.

[19] Askari H, Kaneski CR, Semino-Mora C, Desai P, Ang A, Kleiner DE, Perlee LT, Quezado M, Spollen LE, Wustman BA, Schiffmann R. Cellular and tissue localization of globotriaosylceramide in Fabry disease. Virchows Arch 2007;451:823–834.

[20] Aerts JM, Groener JE, Kuiper S, Donker-Koopman WE, Strijland A, Ottenhoff R, van Roomen C, Mirzaian M, Wijburg FA, Linthorst GE, Vedder AC, Rombach SM, Cox-Brinkman J, Somerharju P, Boot RG, Hollak CE, Brady RO, Poorthuis BJ. Elevated globotriaosylsphingosine is a hallmark of Fabry disease. Proc Natl Acad Sci U S A 2008;105:2812–2817.

[21] Rombach SM, Dekker N, Bouwman MG, Linthorst GE, Zwinderman AH, Wijburg FA, Kuiper S, vd Bergh Weerman MA, Groener JEM, Poorthuis BJ, Hollak CEM, Aerts

JMFG. Plasma globotriaosylsphingosine: diagnostic value and relation to clinical manifestations of Fabry disease. Biochim Biophys Acta 2010;1802:741–748.

[22] Togawa T, Kodama T, Suzuki T, Sugawara K, Tsukimura T, Ohashi T, Ishige N, Suzuki K, Kitagawa T, Sakuraba H. Plasma globotriaosylsphingosine as a biomarker of Fabry disease. Mol Genet Metab 2010;100:257–261.

[23] Germain D. Fabry disease. Orphanet J Rare Dis 2010;5:30.

[24] Wilcox WR, Oliveira JP, Hopkin RJ, Ortiz A, Banikazemi M, Feldt-Rasmussen U, Sims K, Waldek S, Pastores GM, Lee P, Eng CM, Marodi L, Stanford KE, Breunig F, Wanner C, Warnock DG, Lemay RM, Germain DP. Females with Fabry disease frequently have major organ involvement: lessons from the Fabry registry. Mol Genet Metab 2008;93:112–128.

[25] von Scheidt W, Eng CM, Fitzmaurice TF, Erdmann E, Hubner G, Olsen EG, Christomanou H, Kandolf R, Bishop DF, Desnick RJ. An atypical variant of Fabry's disease with manifestations confined to the myocardium. N Engl J Med 1991;324:395–399.

[26] Bekri S, Enica A, Ghafari T, Plaza G, Champenois I, Choukroun G, Unwin R, Jaeger P. Fabry disease in patients with end-stage renal failure: the potential benefits of screening. Nephron Clin Pract 2005;101:c33–38.

[27] Chimenti C, Pieroni M, Morgante E, Antuzzi D, Russo A, Russo MA, Maseri A, Frustaci A. Prevalence of Fabry disease in female patients with late-onset hypertrophic cardiomyopathy. Circulation 2004;110:1047–1053.

[28] Nakao S, Takenaka T, Maeda M, Kodama C, Tanaka A, Tahara M, Yoshida A, Kuriyama M, Hayashibe H, Sakuraba H, Tanaka H. An atypical variant of Fabry's disease in men with left ventricular hypertrophy. N Engl J Med 1995;333:288–293.

[29] Nakao S, Kodama C, Takenaka T, Tanaka A, Yasumoto Y, Yoshida A, Kanzaki T, Enriquez AL, Eng CM, Tanaka H, Tei C, Desnick RJ. Fabry disease: detection of undiagnosed hemodialysis patients and identification of a "renal variant" phenotype. Kidney Int 2003;64:801–807.

[30] Rolfs A, Bottcher T, Zschiesche M, Morris P, Winchester B, Bauer P, Walter U, Mix E, Lohr M, Harzer K, Strauss U, Pahnke J, Grossmann A, Benecke R. Prevalence of Fabry disease in patients with cryptogenic stroke: a prospective study. Lancet 2005;366:1794–1796.

[31] Sachdev B, Takenaka T, Teraguchi H, Tei C, Lee P, McKenna WJ, Elliott PM. Prevalence of Anderson–Fabry disease in male patients with late onset hypertrophic cardiomyopathy. Circulation 2002;105:1407–1411.

[32] Spada M, Pagliardini S, Yasuda M, Tukel T, Thiagarajan G, Sakuraba H, Ponzone A, Desnick RJ. High incidence of later-onset Fabry disease revealed by newborn screening. Am J Hum Genet 2006;79:31–40.

[33] Eng CM, Guffon N, Wilcox WR, Germain DP, Lee P, Waldek S, Caplan L, Linthorst GE, Desnick RJ. Safety and efficacy of recombinant human α-galactosidase A replacement therapy in Fabry's disease. N Engl J Med 2001;345:9–16.

[34] Hughes DA, Elliott PM, Shah J, Zuckerman J, Coghlan G, Brookes J, Mehta AB. Effects of enzyme replacement therapy on the cardiomyopathy of Anderson–Fabry disease: a randomised, double-blind, placebo-controlled clinical trial of agalsidase alfa. Heart 2008;94:153–158.

[35] Schiffmann R, Floeter MK, Dambrosia JM, Gupta S, Moore DF, Sharabi Y, Khurana RK, Brady RO. Enzyme replacement therapy improves peripheral nerve and sweat function in Fabry disease. Muscle Nerve 2003;28:703–710.

[36] Schiffmann R, Kopp JB, Austin Iii HA, Sabnis S, Moore DF, Weibel T, Balow JE, Brady RO. Enzyme replacement therapy in Fabry disease: a randomized controlled trial. JAMA 2001;285:2743–2749.

[37] Warnock DG, Ortiz A, Mauer M, Linthorst GE, Oliveira JP, Serra AL, Maródi L, Mignani R, Vujkovac B, Beitner-Johnson D, Lemay R, Cole JA, Svarstad E, Waldek S, Germain DP, Wanner C, Registry obotF. Renal outcomes of agalsidase beta treatment for Fabry disease: role of proteinuria and timing of treatment initiation. Nephrol Dial Transplant 2012;27:1042–1049.

[38] West M, Nicholls K, Mehta A, Clarke JTR, Steiner R, Beck M, Barshop BA, Rhead W, Mensah R, Ries M, Schiffmann R. Agalsidase alfa and kidney dysfunction in Fabry disease. J Am Soc Nephrol 2009;20:1132–1139.

[39] Togawa T, Kawashima I, Kodama T, Tsukimura T, Suzuki T, Fukushige T, Kanekura T, Sakuraba H. Tissue and plasma globotriaosylsphingosine could be a biomarker for assessing enzyme replacement therapy for Fabry disease. Biochem Biophys Res Commun 2010;399:716–720.

[40] van Breemen MJ, Rombach SM, Dekker N, Poorthuis BJ, Linthorst GE, Zwinderman AH, Breunig F, Wanner C, Aerts JM, Hollak CE. Reduction of elevated plasma globotriaosylsphingosine in patients with classic Fabry disease following enzyme replacement therapy. Biochim Biophys Acta 2011;1812:70–76.

[41] Rombach SM, Aerts JMFG, Poorthuis BJHM, Groener JEM, Donker-Koopman W, Hendriks E, Mirzaian M, Kuiper S, Wijburg FA, Hollak CEM, Linthorst GE. Long-term effect of antibodies against infused alpha-galactosidase A in Fabry disease on plasma and urinary (Lyso)Gb3 reduction and treatment outcome. PLoS One 2012;7:e47805.

[42] Yam GH, Bosshard N, Zuber C, Steinmann B, Roth J. Pharmacological chaperone corrects lysosomal storage in Fabry disease caused by trafficking-incompetent variants. Am J Physiol Cell Physiol 2006;290:C1076–1082.

[43] Okumiya T, Ishii S, Takenaka T, Kase R, Kamei S, Sakuraba H, Suzuki Y. Galactose stabilizes various missense mutants of α-galactosidase in Fabry disease. Biochem Biophys Res Commun 1995;214:1219–1224.

[44] Fan J-Q, Nakao S, Kaneski CR, Brady RO, Desnick RJ, Suzuki S, Ishii S. Intracellular enhancement of α-galactosidase A activity in 31 Fabry lymphoblasts and fibroblasts by 1-deoxy-galactonojirimycin. Am J Hum Genet 1999;65:A308.

[45] Fan J-Q. A contradictory treatment for lysosomal storage disorders: inhibitors enhance mutant enzyme activity. Trends Pharmacol Sci 2003;24:355–360.

[46] Fan J-Q, Ishii S, Asano N, Suzuki Y. Accelerated transport and maturation of lysosomal α-galactosidase A in Fabry lymphoblasts by an enzyme inhibitor. Nat Med 1999; 5:112–115.

[47] Frustaci A, Chimenti C, Ricci R, Natale L, Russo MA, Pieroni M, Eng CM, Desnick RJ. Improvement in cardiac function in the cardiac variant of Fabry's disease with galactose-infusion therapy. N Engl J Med 2001;345:25–32.

[48] Guce AI, Clark NE, Rogich JJ, Garman SC. The molecular basis of pharmacological chaperoning in human α-galactosidase. Chem Biol 2011;18:1521–1526.

[49] Benjamin E, Flanagan J, Schilling A, Chang H, Agarwal L, Katz E, Wu X, Pine C, Wustman B, Desnick R, Lockhart D, Valenzano K. The pharmacological chaperone 1-deoxygalactonojirimycin increases α-galactosidase A levels in Fabry patient cell lines. J Inherit Metab Dis 2009;32:424–440.

[50] Ishii S, Chang H-H, Yoshioka H, Shimada T, Mannen K, Higuchi Y, Taguchi A, Fan J-Q. Preclinical efficacy and safety of 1-deoxygalactonojirimycin in mice for Fabry disease. J Pharmacol Exp Therap 2009;328:723–731.

[51] Yam GH, Zuber C, Roth J. A synthetic chaperone corrects the trafficking defect and disease phenotype in a protein misfolding disorder. FASEB J 2005;19:12–18.

[52] Ishii S, Yoshioka H, Mannen K, Kulkarni AB, Fan JQ. Transgenic mouse expressing human mutant α-galactosidase A in an endogenous enzyme deficient background: a biochemical animal model for studying active-site specific chaperone therapy for Fabry disease. Biochim Biophys Acta 2004;1690:250–257.

[53] Germain DP, Poenaru L. Fabry disease: identification of novel α-galactosidase A mutations and molecular carrier detection by use of fluorescent chemical cleavage of mismatches. Biochem Biophys Res Commun 1999;257:708–713.

[54] Ashton-Prolla P, Tong B, Shabbeer J, Astrin KH, Eng CM, Desnick RJ. Fabry disease: twenty-two novel mutations in the α-galactosidase A gene and genotype/phenotype correlations in severely and mildly affected hemizygotes and heterozygotes. J Invest Med 2000;48:227–235.

[55] Ishii S, Sakuraba H, Suzuki Y. Point mutations in the upstream region of the α-galactosidase A gene exon 6 in an atypical variant of Fabry disease. Hum Genet 1992;89:29–32.

[56] Wu X, Katz E, Valle CD, Mascioli K, Flanagan J, Castelli JP, Schiffmann R, Boudes P, Lockhart DJ, Valenzano KJ, Benjamin ER. A pharmacogenetic approach to identify mutant forms of α-galactosidase A that respond to a pharmacological chaperone for Fabry disease. Hum Mutat 2011;32:965–977.

[57] Johnson FK, Mudd PN, Bragat A, Adera M, Boudes P. Pharmacokinetics and safety of migalastat HCl and effects on agalsidase activity in healthy volunteers. Clin Pharmacol Drug Develop 2013;2:120–132.

[58] Germain D, Giugliani R, Hughes D, Mehta A, Nicholls K, Barisoni L, Jennette C, Bragat A, Castelli J, Sitaraman S, Lockhart D, Boudes P. Safety and pharmacodynamic effects of a pharmacological chaperone on α-galactosidase A activity and globotriaosylceramide clearance in Fabry disease: report from two phase 2 clinical studies. Orphanet J Rare Dis 2012;7:91.

[59] Bekri S, Lidove O, Jaussaud R, Knebelmann B, Barbey F. The role of ceramide trihexoside (globotriaosylceramide) in the diagnosis and follow-up of the efficacy of treatment of Fabry disease: a review of the literature. Cardiovasc Hematol Agents Med Chem 2006;4:289–297.

[60] Aerts J, Kallemeijn W, Wegdam W, Joao Ferraz M, van Breemen M, Dekker N, Kramer G, Poorthuis B, Groener J, Cox-Brinkman J, Rombach S, Hollak C, Linthorst G, Witte M, Gold H, van der Marel G, Overkleeft H, Boot R. Biomarkers in the diagnosis of lysosomal storage disorders: proteins, lipids, and inhibodies. J Inherit Metab Dis 2011;34:605–619.

[61] Young E, Mills K, Morris P, Vellodi A, Lee P, Waldek S, Winchester B. Is globotriaosylceramide a useful biomarker in Fabry disease? Acta Paediatr Suppl 2005;94:51–54.

[62] Giugliani R, Waldek S, Germain DP, Nicholls K, Bichet DG, Simosky JK, Bragat AC, Castelli JP, Benjamin ER, Boudes PF. A phase 2 study of migalastat hydrochloride in females with Fabry disease: selection of population, safety and pharmacodynamic effects. Mol Genet Metab 2013;109:86–92.

[63] Schiffmann R, Fernhoff P, Germain D, Hughes D, Mehta A, Nicholls K, Pastores G, Giugliani R, Waldek S, Wilcox W, Adera M.. Long-term safety, tolerability, and assessments of renal function in adult Fabry patients receiving treatment with AT1001 (migalastat hydrochloride), a pharmacological chaperone. Presented at the 7th Annual WORLD Symposium. Poster 113, Las Vegas, NV; 2011.

[64] Hughes D, Adera M, Castelli J, Bragat A, Marsden D, Boudes P. Preliminary long-term safety, tolerability, and assessments of renal function of adult Fabry patients receiving treatment with AT1001 (migalastat hydrochloride), a pharmacological chaperone, for up to 3 years. J Inherit Metab Dis 2010;33:S148.

[65] Adera M, Overton C, Boudes P. A double-blind, randomized, placebo-controlled study to evaluate the efficacy, safety and pharmacodynamics of AT1001 in patients with Fabry disease and AT1001-responsive GLA mutations. Mol Genet Metab 2011;102:S4–S5.

[66] Barisoni L, Jennette JC, Colvin R, Sitaraman S, Bragat A, Castelli J, Walker D, Boudes P. Novel quantitative method to evaluate globotriaosylceramide inclusions in renal peritubular capillaries by virtual microscopy in patients with Fabry disease. Arch Pathol Lab Med 2012;136:816–824.

[67] Nicholls K, Germain DP, Feliciani C, Shankar S, Ezgu F, Janmohamed SG, Laing SM, Schroyer RO, Bragat AC, Sitaraman S, Boudes PF. Phase 3 study of migalastat HCl for Fabry disease: stage 1 results. Mol Genet Metab 2013;108:S70.

[68] Barlow C, Castelli J, Benjamin E, Yu J, France N, Ludington E, Lockhart DJ. Phase 3 facets study of migalastat HCl for Fabry disease: *post hoc GLA* mutation-based identification of subjects likely to show a drug effect. Mol Genet Metab 2014;111:S24.

[69] Adera M, Gladden A, Boudes P. A randomized, open-label study to compare the efficacy and safety of AT1001 and enzyme replacement therapy (ERT) in patients with Fabry disease and AT1001-responsive *GLA* mutations, who were previously treated with ERT. Mol Genet Metab 2011;102:S4.

[70] Beutler E, Grabowski G. Gaucher disease. In: Scriver C, Beaudet A, Sly W, Valle D, editors. *The Metabolic and Molecular Bases of Inherited Disease*. New York: McGraw-Hill; 2001. p 3635–3668.

[71] Grabowski GA. Gaucher disease: gene frequencies and genotype/phenotype correlations. Genet Test 1997;1:5–12.

[72] Sawkar AR, Adamski-Werner SL, Cheng WC, Wong CH, Beutler E, Zimmer KP, Kelly JW. Gaucher disease-associated glucocerebrosidases show mutation-dependent chemical chaperoning profiles. Chem Biol 2005;12:1235–1244.

[73] Horowitz M, Pasmanik-Chor M, Borochowitz Z, Falik-Zaccai T, Heldmann K, Carmi R, Parvari R, Beit-Or H, Goldman B, Peleg L, Levy-Lahad E, Renbaum P, Legum S, Shomrat R, Yeger H, Benbenisti D, Navon R, Dror V, Shohat M, Magal N, Navot N, Eyal N. Prevalence of glucocerebrosidase mutations in the Israeli Ashkenazi Jewish population. Hum Mutat 1998;12:240–244.

[74] Stenson PD, Ball EV, Mort M, Phillips AD, Shiel JA, Thomas NST, Abeysinghe S, Krawczak M, Cooper DN. Human gene mutation database (HGMD®): 2003 update. Hum Mutat 2003;21:577–581.

[75] Barton NW, Brady RO, Dambrosia JM, Di Bisceglie AM, Doppelt SH, Hill SC, Mankin HJ, Murray GJ, Parker RI, Argoff CE, et al. Replacement therapy for inherited enzyme deficiency – macrophage-targeted glucocerebrosidase for Gaucher's disease. N Engl J Med 1991;324:1464–1470.

[76] Cox T. Gaucher disease: understanding the molecular pathogenesis of sphingolipidoses. J Inherit Metab Dis 2001;24:107–123.

[77] Figueroa ML, Rosenbloom BE, Kay AC, Garver P, Thurston DW, Koziol JA, Gelbart T, Beutler E. A less costly regimen of alglucerase to treat Gaucher's disease. N Engl J Med 1992;327:1632–1636.

[78] Platt FM, Jeyakumar M, Andersson U, Priestman DA, Dwek RA, Butters TD, Cox TM, Lachmann RH, Hollak C, Aerts JM, Van Weely S, Hrebicek M, Moyses C, Gow I, Elstein D, Zimran A. Inhibition of substrate synthesis as a strategy for glycolipid lysosomal storage disease therapy. J Inherit Metab Dis 2001;24:275–290.

[79] Tifft CJ, Proia RL. Stemming the tide: glycosphingolipid synthesis inhibitors as therapy for storage diseases. Glycobiology 2000;10:1249–1258.

[80] Barton NW, Furbish FS, Murray GJ, Garfield M, Brady RO. Therapeutic response to intravenous infusions of glucocerebrosidase in a patient with Gaucher disease. Proc Natl Acad Sci U S A 1990;87:1913–1916.

[81] Grabowski GA, Leslie N, Wenstrup R. Enzyme therapy for Gaucher disease: the first 5 years. Blood Rev 1998;12:115–133.

[82] Lebel E, Dweck A, Foldes AJ, Golowa Y, Itzchaki M, Zimran A, Elstein D. Bone density changes with enzyme therapy for Gaucher disease. J Bone Mineral Metabol 2004;22:597–601.

[83] Schiffmann R, Mankin H, Dambrosia JM, Xavier RJ, Kreps C, Hill SC, Barton NW, Rosenthal DI. Decreased bone density in splenectomized Gaucher patients receiving enzyme replacement therapy. Blood Cells Mol Dis 2002;28:288–296.

[84] Futerman AH, Sussman JL, Horowitz M, Silman I, Zimran A. New directions in the treatment of Gaucher disease. Trends Pharmacol Sci 2004;25:147–151.

[85] Cox T, Lachmann R, Hollak C, Aerts J, van Weely S, Hrebicek M, Platt F, Butters T, Dwek R, Moyses C, Gow I, Elstein D, Zimran A. Novel oral treatment of Gaucher's disease with *N*-butyldeoxynojirimycin (OGT 918) to decrease substrate biosynthesis. Lancet 2000;355:1481–1485.

[86] Schiffmann R, FitzGibbon EJ, Harris C, DeVile C, Davies EH, Abel L, Schaik INV, Benko WS, Timmons M, Ries M, Vellodi A. Randomized, controlled trial of miglustat in Gaucher's disease type 3. Ann Neurol 2008;64:514–522.

[87] Shayman JA. Eliglustat tartrate: glucosylceramide synthase inhibitor treatment of type 1 Gaucher disease. Drugs Future 2010;35:613–620.

[88] Beck M. New therapeutic options for lysosomal storage disorders: enzyme replacement, small molecules and gene therapy. Hum Genet 2007;121:1–22.

[89] Sawkar AR, Cheng WC, Beutler E, Wong CH, Balch WE, Kelly JW. Chemical chaperones increase the cellular activity of N370S β-glucosidase: a therapeutic strategy for Gaucher disease. Proc Natl Acad Sci U S Am 2002;99:15428–15433.

[90] Sawkar AR, D'Haeze W, Kelly JW. Therapeutic strategies to ameliorate lysosomal storage disorders – a focus on Gaucher disease. Cell Mol Life Sci 2006;63:1179–1192.

[91] Lieberman RL, Wustman BA, Huertas P, Powe AC Jr, Pine CW, Khanna R, Schlossmacher MG, Ringe D, Petsko GA. Structure of acid β-glucosidase with pharmacological chaperone provides insight into Gaucher disease. Nat Chem Biol 2007;3:101–107.

[92] Sun Y, Liou B, Xu Y-H, Quinn B, Zhang W, Hamler R, Setchell KDR, Grabowski GA. *Ex vivo* and *in vivo* effects of isofagomine on acid b-glucosidase variants and substrate levels in Gaucher disease. J Biol Chem 2012;287:4275–4287.

[93] Steet RA, Chung S, Wustman B, Powe A, Do H, Kornfeld SA. The iminosugar isofagomine increases the activity of N370S mutant acid β-glucosidase in Gaucher fibroblasts by several mechanisms. Proc Natl Acad Sci U S A 2006;109:13813–13818.

[94] Yu Z, Sawkar AR, Whalen LJ, Wong CH, Kelly JW. Isofagomine- and 2,5-anhydro-2,5-imino-D-glucitol-based glucocerebrosidase pharmacological chaperones for Gaucher disease intervention. J Med Chem 2007;50:94–100.

[95] Maegawa GHB, Tropak MB, Buttner JD, Rigat BA, Fuller M, Pandit D, Tang L, Kornhaber GJ, Hamuro Y, Clarke JTR, Mahuran DJ. Identification and characterization of ambroxol as an enzyme enhancement agent for Gaucher disease. J Biol Chem 2009;284:23502–23516.

[96] Khanna R, Benjamin ER, Pellegrino L, Schilling A, Rigat BA, Soska R, Nafar H, Ranes BE, Feng J, Lun Y, Powe AC, Palling DJ, Wustman BA, Schiffmann R, Mahuran DJ, Lockhart DJ, Valenzano KJ. The pharmacological chaperone isofagomine increases the activity of the Gaucher disease L444P mutant form of b-glucosidase. FEBS J 2010;277:1618–1638.

[97] Sun Y, Ran H, Liou B, Quinn B, Zamzow M, Zhang W, Bielawski J, Kitatani K, Setchell KDR, Hannun YA, Grabowski GA. Isofagomine *in vivo* effects in a neuronopathic Gaucher disease mouse. PLoS One 2011;6:e19037.

[98] Sanders A, Hemmelgarn H, Melrose HL, Hein L, Fuller M, Clarke LA. Transgenic mice expressing human glucocerebrosidase variants: utility for the study of Gaucher disease. Blood Cells Mol Dis 2013;51:109–115.

[99] Palling DJ, Ludwig K, Casar M, Sitaraman S, Duke C, Marion AS, Stiles M, Duey S, Tsikos T, Goblot D, Jurima-Romet M., Roy M, DiMarco M, Demnati R. Phase 1 studies of the safety, pharmacokinetics and pharmacodynamics of AT2101, a pharmacological chaperone for the treatment of Gaucher disease. Presented at the American College of Medical Genetics Annual Meeting, Nashville, TN; 2007.

[100] Luan Z, Li L, Higaki K, Nanba E, Suzuki Y, Ohno K. The chaperone activity and toxicity of ambroxol on Gaucher cells and normal mice. Brain Develop 2013;35:317–322.

[101] Bendikov-Bar I, Maor G, Filocamo M, Horowitz M. Ambroxol as a pharmacological chaperone for mutant glucocerebrosidase. Blood Cells Mol Dis 2013;50:141–145.

[102] Zimran A, Altarescu G, Elstein D. Pilot study using ambroxol as a pharmacological chaperone in Type 1 Gaucher disease. Blood Cells Mol Dis 2013;50:134–137.

[103] Gravel R, Kaback M, Proia RL, Sandhoff K, Suzuki K, Suzuki K. The GM2 gangliosidoses. In: Scriver C, Beaudet A, Sly W, Valle D, editors. *The Metabolic and Molecular Bases of Inherited Disease Online*. New York: McGraw-Hill; 2001. p 3827–3876.

[104] Maegawa GHB, Tropak M, Buttner J, Stockley T, Kok F, Clarke JTR, Mahuran DJ. Pyrimethamine as a potential pharmacological chaperone for late-onset forms of GM2 gangliosidosis. J BiolChem 2007;282:9150–9161.

[105] Clarke JTR, Mahuran DJ, Sathe S, Kolodny EH, Rigat BA, Raiman JA, Tropak MB. An open-label phase I/II clinical trial of pyrimethamine for the treatment of patients affected with chronic GM2 gangliosidosis (Tay-Sachs or Sandhoff variants). Mol Genet Metab 2011;102:6–12.

[106] Hirschhorn R, Reuser AJJ. Glycogen storage disease type II: acid a-glucosidase (acid maltase) deficiency. In: Scriver C, Beaudet A, Sly W, Valle D, editors. *The Metabolic and Molecular Bases of Inherited Disease*. New York: McGraw-Hill; 2001. p 3389–3420.

[107] van der Ploeg AT, Reuser AJJ. Pompe's disease. Lancet 2008;372:1342–1353.

[108] Mancini R, Aebi M, Helenius A. Multiple endoplasmic reticulum-associated pathways degrade mutant yeast carboxypeptidase Y in mammalian cells. J Biol Chem 2003; 278:46895–46905.

[109] Raben N, Plotz P, Byrne BJ. Acid α-glucosidase deficiency (glycogenosis type II, Pompe disease). Curr Mol Med 2002;2:145–166.

[110] Moreland RJ, Jin X, Zhang XK, Decker RW, Albee KL, Lee KL, Cauthron RD, Brewer K, Edmunds T, Canfield WM. Lysosomal acid α-glucosidase consists of four different peptides processed from a single chain precursor. J Biol Chem 2005;280:6780–6791.

[111] Wisselaar HA, Kroos MA, Hermans MM, van Beeumen J, Reuser AJ. Structural and functional changes of lysosomal acid α-glucosidase during intracellular transport and maturation. J Biol Chem 1993;268:2223–2231.

[112] Hasilik A, Klein U, Waheed A, Strecker G, von Figura K. Phosphorylated oligosaccharides in lysosomal enzymes: identification of α-*N*-acetylglucosamine (1)phospho(6)mannose diester groups. Proc Natl Acad Sci U S A 1980;77:7074–7078.

[113] Hasilik A, Neufeld EF. Biosynthesis of lysosomal enzymes in fibroblasts. Synthesis as precursors of higher molecular weight. J Biol Chem 1980;255:4937–4945.

[114] Bijvoet AG, Kroos MA, Pieper FR, Van der Vliet M, De Boer HA, Van der Ploeg AT, Verbeet MP, Reuser AJ. Recombinant human acid α-glucosidase: high level production in mouse milk, biochemical characteristics, correction of enzyme deficiency in GSDII KO mice. Hum Mol Genet 1998;7:1815–1824.

[115] Kishnani PS, Howell RR. Pompe disease in infants and children. Journal of Pediatrics 2004;144:S35–S43.

[116] Hagemans ML, Winkel LP, Hop WC, Reuser AJ, Van Doorn PA, Van der Ploeg AT. Disease severity in children and adults with Pompe disease related to age and disease duration. Neurology 2005;64:2139–2141.

[117] Gungor D, de Vries J, Hop W, Reuser A, van Doorn P, van der Ploeg A, Hagemans M. Survival and associated factors in 268 adults with Pompe disease prior to treatment with enzyme replacement therapy. Orphanet J Rare Dis 2011;6:34.

[118] Gungor D, Kruijshaar M, Plug I, D'Agostino R, Hagemans M, van Doorn P, Reuser A, van der Ploeg A. Impact of enzyme replacement therapy on survival in adults with Pompe disease: results from a prospective international observational study. Orphanet J Rare Dis 2013;8:49.

[119] Van den Hout JM, Kamphoven JH, Winkel LP, Arts WF, De Klerk JB, Loonen MC, Vulto AG, Cromme-Dijkhuis A, Weisglas-Kuperus N, Hop W, Van Hirtum H, Van Diggelen OP, Boer M, Kroos MA, Van Doorn PA, Van der Voort E, Sibbles B, Van Corven EJ, Brakenhoff JP, Van Hove J, Smeitink JA, de Jong G, Reuser AJ, Van der Ploeg AT. Long-term intravenous treatment of Pompe disease with recombinant human α-glucosidase from milk. Pediatrics 2004;113:e448–e457.

[120] Nicolino M, Byrne B, Wraith JE, Leslie N, Mandel H, Freyer DR, Arnold GL, Pivnick EK, Ottinger CJ, Robinson PH, Loo J-CA, Smitka M, Jardine P, Tatò L, Chabrol B, McCandless S, Kimura S, Mehta L, Bali D, Skrinar A, Morgan C, Rangachari L, Corzo D, Kishnani PS. Clinical outcomes after long-term treatment with alglucosidase alfa in infants and children with advanced Pompe disease. Genet Med 2009;11:210–219.

[121] Chen L-R, Chen C-A, Chiu S-N, Chien Y-H, Lee N-C, Lin M-T, Hwu W-L, Wang J-K, Wu M-H. Reversal of cardiac dysfunction after enzyme replacement in patients with infantile-onset Pompe disease. J Pediatr 2009;155:271–275.

[122] Kishnani PS, Hwu WL, Mandel H, Nicolino M, Yong F, Corzo D. A retrospective, multinational, multicenter study on the natural history of infantile-onset Pompe disease. J Pediatr 2006;148:671–676.

[123] Kishnani PS, Nicolino M, Voit T, Rogers RC, Tsai AC, Waterson J, Herman GE, Amalfitano A, Thurberg BL, Richards S, Davison M, Corzo D, Chen YT. Chinese hamster ovary cell-derived recombinant human acid α-glucosidase in infantile-onset Pompe disease. J Pediatr 2006;149:89–97.

[124] Kishnani PS, Steiner RD, Bali D, Berger K, Byrne BJ, Case LE, Crowley JF, Downs S, Howell RR, Kravitz RM, Mackey J, Marsden D, Martins AM, Millington DS, Nicolino M, O'Grady G, Patterson MC, Rapoport DM, Slonim A, Spencer CT, Tifft CJ, Watson MS. Pompe disease diagnosis and management guideline. Genet Med 2006;8:267–288.

[125] Strothotte S, Strigl-Pill N, Grunert B, Kornblum C, Eger K, Wessig C, Deschauer M, Breunig F, Glocker F, Vielhaber S, Brejova A, Hilz M, Reiners K, Müller-Felber W, Mengel E, Spranger M, Schoser B. Enzyme replacement therapy with alglucosidase alfa in 44 patients with late-onset glycogen storage disease type 2: 12-month results of an observational clinical trial. J Neurol 2010;257:91–97.

[126] van der Ploeg AT, Clemens PR, Corzo D, Escolar DM, Florence J, Groeneveld GJ, Herson S, Kishnani PS, Laforet P, Lake SL, Lange DJ, Leshner RT, Mayhew JE, Morgan C, Nozaki K, Park DJ, Pestronk A, Rosenbloom B, Skrinar A, van Capelle CI, van der Beek NA, Wasserstein M, Zivkovic SA. A randomized study of alglucosidase alfa in late-onset Pompe's disease. N Engl J Med 2010;362:1396–1406.

[127] Raben N, Danon M, Gilbert AL, Dwivedi S, Collins B, Thurberg BL, Mattaliano RJ, Nagaraju K, Plotz PH. Enzyme replacement therapy in the mouse model of Pompe disease. Mol Genet Metab 2003;80:159–169.

[128] Raben N, Nagaraju K, Lee A, Lu N, Rivera Y, Jatkar T, Hopwood JJ, Plotz PH. Induction of tolerance to a recombinant human enzyme, acid α-glucosidase, in enzyme deficient knockout mice. Transgenic Res 2003;12:171–178.

[129] Kishnani PS, Goldenberg PC, DeArmey SL, Heller J, Benjamin D, Young S, Bali D, Smith SA, Li JS, Mandel H, Koeberl D, Rosenberg A, Chen YT. Cross-reactive immunologic material status affects treatment outcomes in Pompe disease infants. Mol Genet Metab 2010;99:26–33.

[130] Cardone M, Porto C, Tarallo A, Vicinanza M, Rossi B, Polishchuk E, Donaudy F, Andria G, De Matteis MA, Parenti G. Abnormal mannose-6-phosphate receptor trafficking impairs recombinant α-glucosidase uptake in Pompe disease fibroblasts. Pathogenetics 2008;1:6.

[131] de Vries JM, van der Beek NAME, Kroos MA, Özkan L, van Doorn PA, Richards SM, Sung CCC, Brugma J-DC, Zandbergen AAM, van der Ploeg AT, Reuser AJJ. High antibody titer in an adult with Pompe disease affects treatment with alglucosidase alfa. Mol Genet Metab 2010;101:338–345.

[132] Fukuda T, Ahearn M, Roberts A, Mattaliano RJ, Zaal K, Ralston E, Plotz PH, Raben N. Autophagy and mistargeting of therapeutic enzyme in skeletal muscle in Pompe disease. Mol Ther 2006;14:831–839.

[133] Parenti G, Zuppaldi A, Gabriela Pittis M, Rosaria Tuzzi M, Annunziata I, Meroni G, Porto C, Donaudy F, Rossi B, Rossi M, Filocamo M, Donati A, Bembi B, Ballabio A, Andria G. Pharmacological enhancement of mutated α-glucosidase activity in fibroblasts from patients with Pompe disease. Mol Ther 2007;15:508–514.

[134] Fan J-Q. A counterintuitive approach to treat enzyme deficiencies: use of enzyme inhibitors for restoring mutant enzyme activity. Biol Chem 2008;389:1–11.

[135] Okumiya T, Kroos MA, Vliet LV, Takeuchi H, Van der Ploeg AT, Reuser AJ. Chemical chaperones improve transport and enhance stability of mutant α-glucosidases in glycogen storage disease type II. Mol Genet Metab 2007;90:49–57.

[136] Flanagan JJ, Rossi B, Tang K, Wu X, Mascioli K, Donaudy F, Tuzzi MR, Fontana F, Cubellis MV, Porto C, Benjamin E, Lockhart DJ, Valenzano KJ, Andria G, Parenti G, Do HV. The pharmacological chaperone 1-deoxynojirimycin increases the activity and lysosomal trafficking of multiple mutant forms of acid α-glucosidase. Hum Mutat 2009;30:1683–1692.

[137] Adera M, Boudes P, Bragat A, Sitaraman S, Lazauskas R, Valenzano K, Lankas G, Stoltz R, Royalty J, Greene D. Pharmacokinetics and muscle distribution of AT2220, a pharmacological chaperone of acid-glucosidase, in healthy volunteers. Mol Genet Metab 2011;102:S4.

[138] Koskenvuo J, Hartiala J, Nuutila P, Kalliokoski R, Viikari J, Engblom E, Penttinen M, Knuuti J, Mononen I, Kantola I. Twenty-four-month α-galactosidase A replacement therapy in Fabry disease has only minimal effects on symptoms and cardiovascular parameters. J Inherit Metab Dis 2008;31:432–441.

[139] Shen J-S, Edwards NJ, Hong YB, Murray GJ. Isofagomine increases lysosomal delivery of exogenous glucocerebrosidase. Biochem Biophys Res Commun 2008; 369:1071–1075.

[140] Porto C, Cardone M, Fontana F, Rossi B, Tuzzi MR, Tarallo A, Barone MV, Andria G, Parenti G. The pharmacological chaperone *N*-butyldeoxynojirimycin enhances enzyme replacement therapy in Pompe disease fibroblasts. Mol Ther 2009;17:964–971.

[141] Khanna R, Flanagan JJ, Feng J, Soska R, Frascella M, Pellegrino LJ, Lun Y, Guillen D, Lockhart DJ, Valenzano KJ. The pharmacological chaperone AT2220 increases recombinant human acid α-glucosidase uptake and glycogen reduction in a mouse model of Pompe disease. PLoS One 2012;7:e40776.

[142] Benjamin ER, Khanna R, Schilling A, Flanagan JJ, Pellegrino LJ, Brignol N, Lun Y, Guillen D, Ranes BE, Frascella M, Soska R, Feng J, Dungan L, Young B, Lockhart DJ, Valenzano KJ. Co-administration with the pharmacological chaperone AT1001 increases recombinant human α-galactosidase A tissue uptake and improves substrate reduction in Fabry mice. Mol Ther 2012;20:717–726.

[143] Kishnani P, Tarnopolsky M, Sivakumar K, Roberts M, Byrne B, Goker-Alpan O, Guter K, Pervaiz MA, Dasouki M, Mozaffar T, Finanger E, Johnson F, Boudes P. A phase 2A study to investigate drug–drug interactions between escalating doses of AT2220 (Duvoglustat hydrochloride) and acid alfa-glucosidase in subjects with Pompe disease. Mol Genet Metab 2013;108:S54.

[144] Parenti G, Fecarotta S, Marca Gl, Rossi B, Ascione S, Donati MA, Morandi LO, Ravaglia S, Pichiecchio A, Ombrone D, Sacchini M, Pasanisi MB, Filippi PD, Danesino C, Casa RD, Romano A, Mollica C, Rosa M, Agovino T, Nusco E, Porto C, Andria G. A chaperone enhances blood α-glucosidase activity in Pompe disease patients treated with enzyme replacement therapy. Mol Ther 2014;22:2004–2012.

[145] Warnock D, Bichet D, Holida M, Goker-Alpan O, Nicholls K, Thomas M, Eyskens F, Shankar S, Dasouki M, Johnson F, Boudes P. A phase 2A study to investigate the effect of a single dose of migalastat HCl, a pharmacological chaperone, on agalsidase activity in subjects with Fabry disease. Mol Genet Metab 2013;108:S96.

16

ENDOSOMAL ESCAPE PATHWAYS FOR DELIVERY OF BIOLOGICS

PHILIP L. LEOPOLD

Department of Chemistry, Chemical Biology, and Biomedical Engineering, Stevens Institute of Technology, Hoboken, NJ, USA

16.1 INTRODUCTION

The term "biologics" refers to a group of clinical therapies that employ biological macromolecules, either as purified components, or in the context of biological preparations, cells, and/or tissues that are intended to modify the health and well-being of patients based on the inherent biological structure and function of the therapeutic. In the United States, the responsibility for approval and use of biologics lies with the United States Food and Drug Administration (FDA). The FDA provides a list of classifications of biologics that includes Allergenics, Blood and Blood Products, Cellular and Gene Therapy Products, Tissue and Tissue Products, Vaccines, and Xenotransplantation (use of nonhuman biologics in humans). Among these major classes, only biologics related to Cellular and Gene Therapy Products are likely to require endosomal escape for full activity. Of interest, as of January 2015, 9 of the 10 approved products in this category included live human cells while the 10th is a live mycobacterium for use as a vaccine, none of which require endosomal escape (US Food and Drug Administration Vaccines, Blood & Biologics: Cellular and Gene Therapy Products; http://www.fda.gov/biologicsbloodvaccines/cellulargenetherapyproducts/ApprovedProducts/default.htm; viewed on 01/30/15). Therefore, while a tremendous amount of research has gone into the development of intracellular therapies that will require endosomal escape, the benefits of that

Lysosomes: Biology, Diseases, and Therapeutics, First Edition.
Edited by Frederick R. Maxfield, James M. Willard and Shuyan Lu.

research remain to be realized, and the difficulties surrounding intracellular delivery of large molecules can certainly be identified as one of the factors delaying the development of these therapies. Among the therapies in the pipeline, gene therapy holds tremendous promise for correction of genetic defects as well as *in vivo* production of therapeutic proteins. Despite the fact that the first approved human gene therapy trial was conducted in the United States in 1990 [1], there remain no FDA-approved gene therapies as of the writing of this chapter (US Food and Drug Administration Vaccines, Blood & Biologics: Cellular and Gene Therapy Products; http://www.fda.gov/biologicsbloodvaccines/cellulargenetherapyproducts/default.htm; viewed on 01/30/15). Nonetheless, the role of intracellular therapeutics, involving nucleic acid-based drugs as well as other macromolecules, is expected to grow rapidly in coming years. In particular, the recent focus on the development of small interfering RNA (siRNA) technology has pushed the development of new solutions to passing the plasma membrane [2–4].

Biologics, as a group of therapeutic agents, have unique challenges in terms of preparation and delivery. Compared with small molecules, biologics require specific formulations that preserve their complex molecular structure, are limited in terms of routes of administration due to tissue barriers, and have unique pharmacokinetic and pharmacodynamics profiles based on their interactions with other biologic molecules in the patient [5]. Thus, it is not surprising that delivery of biologics to the cytosol of a target cell raises a new set of issues. While small molecules typically rely on organic phase partitioning or use of pumps or channels to enter the cytosol, the larger, more complex biologic drugs must find other ways to cross the membrane and gain access to the cytosol.

This chapter addresses common strategies for delivering biologics past the endosomal membrane and into the cytosol. The major routes for biologic delivery from the endosome are summarized in Table 16.1. The majority, but not all, mechanisms for leaving the endosome involve disruption of the endosomal membrane. Major mechanisms for disrupting the endosomal are summarized in Figure 16.1. This review focuses on biologics that must escape from endosomes to be active. For some biologics, the site of action is within the endosomal–lysosomal system and does not require passage across a membrane. For example, exogenous delivery of lysosomal enzymes to treat lysosomal storage diseases does not require crossing a membrane and is not addressed here. As noted here, the vast majority of biologics are either in cellular form or constitute vaccines, and thus not relevant to this topic, vaccines that are meant to enter the endosomal processing pathway of antigen-presenting cells but that do not escape, or molecules that act primarily in the blood or on the cell surface such as therapeutic antibodies (e.g., Herceptin or Avastin) or clotting factors that act extracellularly and, thus, do not require endocytosis to reach their site of action. These classes of biologics are not addressed in this chapter.

16.2 ENDOSOME CHARACTERISTICS

The first step in developing strategies for endosomal escape during delivery of biologics is to understand the environment within organelles of the endosomal

TABLE 16.1 Examples of Natural or Engineered Endosomal Escape for Delivery of Biologics to the Cytosol

Mode of Endosomal Escape	Examples of Delivery of Natural Biologics	Examples of Delivery of Therapeutic Biologics
Endosomal membrane rupture	**Amphipathic helix**	
	Influenza hemagglutinin HA2 protein [6–8]	Liposomes/lipoplexes with peptides derived from HA2 [9–12]
	HIV gp41 [13]	A peptide derived from HIV gp41 retains endosomolytic activity and enhances siRNA and plasmid DNA delivery [14,15]
	Flock house virus capsid protein autocleavage product [16]	
	Adenovirus protein VI [17]	
	Bluetongue virus VP5 [18]	
	Rotavirus VP7 [19]	
	Rotavirus NSP4 [20]	
	Poliovirus VP1[a] [21,22]	
	Hepatitis A virus VP4 [23]	
	Human rhinovirus VP1 [24]	A peptide from the human rhinovirus VP1 protein was used to enhance liposome-mediated gene delivery [24]
	Listeria listeriolysin O[b] [25–27]	Liposomes/lipoplexes with peptides derived from listeriolysin O [28–30]
	Staphylococcus aureus PSMα toxins [31,32]	
	Insertion of hydrophobic domains	
	Reovirus μN1 [33–35]	
	Murine polyoma virus VP2 [36]	
	Human rhinovirus VP4 [37]	
	Poliovirus VP4[a] [38]	
	Acid-catalyzed phospholipase	
	Adeno-associated virus VP1 [39]	
	Canine parvovirus VP1 [40,41]	
	B19 VP1 [42]	
	Porcine parvovirus VP1 [42]	

(*continued*)

TABLE 16.1 (*Continued*)

Mode of Endosomal Escape	Examples of Delivery of Natural Biologics	Examples of Delivery of Therapeutic Biologics
		Phospholipase A2 from honey bee venom was combined with a plasmid in a nanohydrogel-enhanced plasmid delivery [43]
Proton sponge effect	**Production of small pores that could admit K^+ ions**	
	Shigella IpaB and IpaC pore-forming proteins[b] [44] *Francisella* VgrG pore-forming protein[b] [45]	
	Secondary amine- and imidazole-containing polymers	
		Polyethylenimine (PEI) [46] Polyamidoamine (PAA) [47] Branched polylysine [48] Polyhistidine [49–51]
Membrane fusion	**Peptide-induced membrane fusion**	
	Dengue virus glycoprotein E [52] Ebola virus glycoprotein [53,54] SARS coronavirus [55] [56]	
	Fusion induced by high local positive charge	
		Cationic liposomes/ lipoplexes with conical zwitterionic colipids [57–59] Conversion of zwitterionic head groups to positive head groups by protonation of carboxyl groups [3]
Alternative endocytic trafficking	**Retrograde secretory pathway trafficking**	
	Pseudomonas exotoxin [60]	Toxin conjugates for anticancer drug delivery [61]
	Diphtheria toxin [60] *Cholera* toxin [62] Ricin [63] *Shiga* toxin [64]	

[a]Poliovirus appears to use more than one strategy for endosomal escape.

[b]Since bacterial escape from phagosomes is complex, mechanistic contributions of bacterial proteins listed here are proposed based on structure and comparison to other endosomal escape mechanisms.

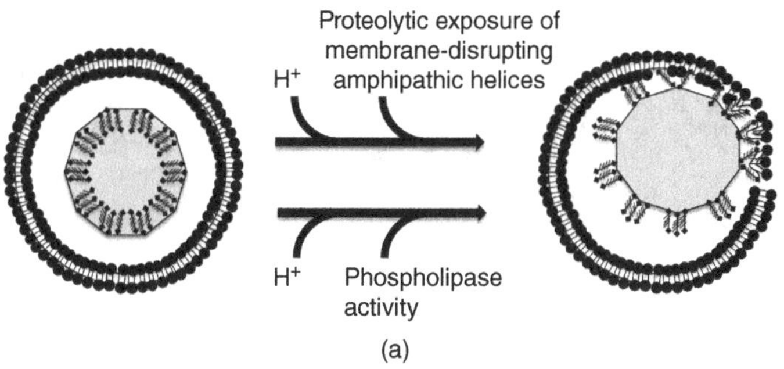

Figure 16.1 Endosomal escape pathways. Delivery of macromolecular biologic drugs to the cytosol requires passage across at least one membrane barrier. The endosome membrane represents a convenient point of transit to the cytosol due to the fact that the characteristic, rapid acidification of the endosome provides a useful environmental trigger to deploy molecules that might otherwise be cytotoxic. At the same time, the endosome is a confined space where the environment can be manipulated and close apposition to the membrane is guaranteed. (a) Rupture of the endosomal membrane can be accomplished by membrane-lytic molecules such as amphipathic helices that display a series of positively charged R groups on one side of the helix facilitating interaction with negatively charged lipid head groups. The resulting disruption to the lipid bilayer can result in pore formation or larger fissures, sufficient to release endosomal contents to the cytosol including viral particles with a diameter up to 80 nm and bacteria with short-axis diameters up to 0.5 μm. The amphipathic helices are potentially cytotoxic and, thus, are stored in an inactive form as part of either the capsid (e.g., adenovirus), in the membrane of an enveloped virus (e.g., influenza virus), or within a bacterium (e.g., *Listeria*). Typically, the deployment of helices is linked to acidification of the endosome and activation of either a viral protease, such as the L3/p23 protease of adenovirus, or an endosomal protease as in the case of influenza virus HA2 or listeriolysin O. When an enveloped virus utilizes a pH-triggered insertion of an amphipathic helix into the endosomal membrane, the insertion can be coupled with a conformational change in the helix-bearing membrane protein to enhance the fusion of the viral membrane and the endosomal membrane. (b) Endosomal membrane rupture can also be accomplished by osmotic lysis via the proton sponge effect. The formation of a particle that includes a high concentration of functional groups that convert from neutral to proton-accepting moieties at slightly acidic pH, for example, secondary or tertiary amino groups, results in fixation of protons in association with weak bases inside of the endosome. As a result, far more protons are pumped into the endosome to achieve a drop in pH. The massive influx of protons is accompanied by chloride counterions resulting in a locally high osmolarity. An influx of water by osmosis then elevates the interior pressure in the endosome resulting in lysis of the membrane. The high transfection efficiency of polyethylenimine (PEI) has been attributed to the proton sponge effect. Other weak bases such as histidine residues can also generate osmotic lysis of endosomes. (c) Membrane fusion can also contribute to the delivery of biologics to the cytosol across the endosomal membrane. Membrane fusion occurs when using cationic liposomes or lipoplexes due to the locally high number of positive charges in close apposition to the inner leaflet of the endosomal membrane. While acidification is not required for fusion to occur, the close association of membranes that occurs within the endosome is likely to favor fusion.

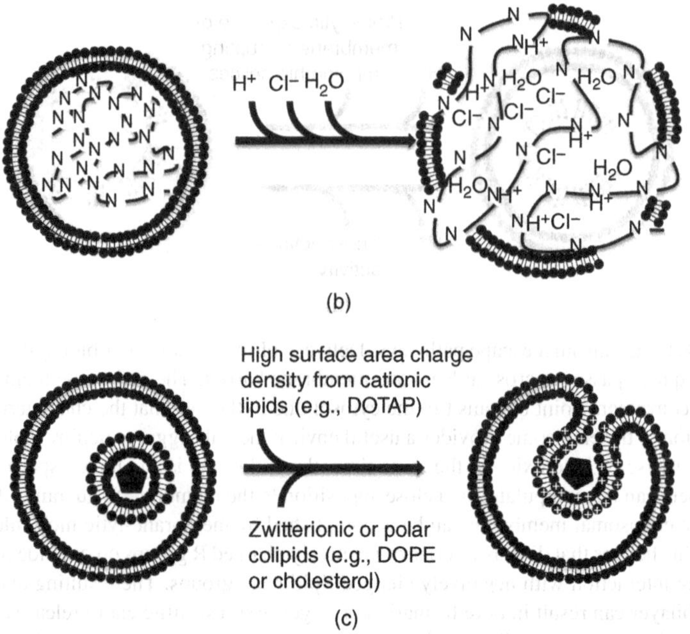

Figure 16.1 (*Continued*)

and phagocytic trafficking pathways to lysosomes. Relevant parameters include compartment size and pH as well as the kinetics of formation and maturation of endosomal compartments, the dichotomy between the flow of membrane versus fluid contents within the endosomal pathway, and the complement of lytic enzymes found in these compartments. Several detailed reviews of these topics are available [65–67].

Several distinct pathways exist for entry into the endocytic or phagocytic pathways [65]. Endocytic pathways include both clathrin-mediated endocytosis and clathrin-independent endocytosis, both of which involve the invagination of membrane from the plasma membrane creating vesicles of approximately 100 nm in diameter. There do not appear to be any markers of nonclathrin-mediated endocytosis that are distinct from the clathrin pathway, suggesting that clathrin-independent endocytosis may be simply a consequence of a particular set of conditions rather than a distinct pathway with unique molecular control. Both classes of endocytic vesicles rapidly (<3 min) acidify to a pH of approximately 6.0 and then fuse with a larger organelle, the sorting endosome. Sorting endosomes are larger than endocytic vesicles and feature a tightly controlled pH in the range from 5.9 to 6.0. Residence time for components of the sorting endosome depends on whether the cargo adheres to the membrane or is present in the fluid phase. Ligands for cell surface receptors either remain bound to their receptor or detach from their receptor as pH decreases. Membranous components including receptors and bound ligands return to the cell surface while fluid-phase components including released ligands and other materials that may have entered endosomes by diffusion during endocytosis remain associated with the sorting endosome until the sorting endosome matures to become a late endosome. Due to this dichotomy in routing at the sorting endosome, residence time depends on fractionation. Membrane-bound components reside in sorting endosomes for approximately 2.5 min, while fluid-phase components leave the sorting endosome with a half-life of 8 min. Membrane-bound components, in general, are trafficked to recycling endosomes before returning the plasma membrane while late endosomes ultimately fuse with primary lysosomes where the internal pH decreases to 5.0–5.5 and a host of hydrolytic enzymes are activated. A subset of the hydrolytic enzymes are encountered earlier in the endocytic pathway, with cathepsins, type II transmembrane serine proteases, furin, and other preprotein convertases having been reported in sorting endosomes [68–70].

Caveolae, smaller invaginations of the plasma membrane having a characteristic "flask" shape, represent an important plasma membrane specialization that plays a critical role in cell signaling and appear to serve as an alternative site of internalization for a subset of ligands and receptors [71]. Caveolae have a distinct protein coat, often including caveolin, and serve as sites for accumulation of aggregated, lipid-anchored proteins on the cell surface. Ligands that are internalized via caveolae enter the endocytic sorting system and are likely to encounter similar conditions as those internalized by endocytosis.

While phagocytosis accomplishes the same mission as endocytosis, that is, bringing a cargo into the cell with a surrounding envelope comprised of plasma membrane, the process is structurally and functionally distinct from endocytosis [72]. Phagocytosis is normally restricted to a set of cells in the immune system including macrophages, dendritic cells, and neutrophils. These cells play key roles in internalization of bacteria and other pathogens, hydrolytic degradation of those pathogens, and antigen presentation [73]. The process of phagocytosis in macrophages and neutrophils is not directly relevant to the general topic of endosomal or lysosomal escape, and thus is not discussed in detail here. However, a related process occurs in a larger variety of cells that normally do not exhibit endocytosis where a certain group of pathogens induce phagocytic entry of the pathogen to the cytosol by escape from phagocytic compartments to establish colonization of the cytosol [74]. These species (include the trypanosomes, which are eukaryotic protists, as well as several genuses of bacteria, including *Shigella*, *Listeria*, *Rickettsia*, and *Francisella*) are able to gain access to the cytosol by breaching a phagosome membrane and, thus, are of interest to a discussion of delivery of biologics to the cytosol.

In summary, the design of mechanisms to enable escape of biologics from endosomes after internalization must be cognizant of the rate of acidification, size of the compartment, presence of potentially catalytic and/or hydrolytic enzymes, and residence time in that compartment prior to encountering a new set of conditions.

16.3 DELIVERY OF NATURE'S BIOLOGICS: LESSONS ON ENDOSOMAL ESCAPE FROM PATHOGENS

Nature has been a consistent contributor as a starting point for engineering. In seeking solutions to problems, a careful consideration of nature often points toward a reasonable, well-tested solution. In the case of endosomal escape, nature does not disappoint. Viruses, bacteria, and toxins have all been observed to escape from endosomes.

16.3.1 Viruses

Viruses provide, perhaps, the richest set of examples for molecular solutions to the problem of endosomal escape. While some viruses are capable of fusing with the plasma membrane, many viruses depend on endosomal escape mechanisms to deliver their genetic payload to the cytoplasm, with the ultimate target being either the cytoplasm, for some RNA genome-containing viruses, or the nucleus, for DNA genome-containing viruses or for those viruses that utilize reverse transcriptase to create a DNA genome from an RNA genome. In both cases, the endosomal membrane must be breached as an initial step in infection.

Viruses, in general, fall into two categories: enveloped and nonenveloped viruses. Enveloped viruses are formed by budding from an infected cell. Budding is an exocytic process during which viral proteins, host cellular proteins, and a small portion of host plasma membrane are collected around the viral core (nucleocapsid). In contrast, nonenveloped viruses simply contain a protein shell, or capsid, that encases their genetic material. The encounter of enveloped viruses and nonenveloped viruses with the endosomal membrane, while fundamentally different, can share similar molecular mechanisms for escape.

The most prevalent shared mechanism is the requirement for a pH trigger. Many viruses that escape from endosomes require acidification of the endosome as part of the triggering mechanism for escape. Influenza virus hemagglutinin protein was the first protein for which the lytic activity was unambiguously assigned [6,7]. This requirement has been experimentally documented by demonstrating that infection of target cells is inhibited in the presence of molecules that inhibit endosome acidification. Endosome acidification can be prevented through the use of weak bases such as ammonium chloride or chloroquine that enter acidic compartments as neutral compounds before acting as weak bases and associating with protons at low pH where the charged molecule becomes trapped [75]. Alternatively, ionophores such as monensin A1 can be used to prevent the acidification of endosomes by exchanging protons with other monovalent cations across the endosome membrane [76]. Finally, acidification of endosomes can be blocked through the use of inhibitors of the vacuolar-type H^+ ATPase, such as bafilomycin A1 [77]. In the presence of these agents, the infection of many viruses is reduced or delayed.

Further investigation into the mechanism of the pH trigger has revealed several different models. The hemagglutinin 2 (HA2) protein from the influenza virus, an enveloped virus, was shown to undergo proteolytic cleavage mediated by ubiquitous endosomal proteases, such as furin [78] or preprotein convertase-6 [79]. Proteolytic modification reveals an amphipathic helix that is capable of insertion into a biological membrane. Upon association of a sufficient number of helices with the membrane, the endosomal membrane is effectively destabilized and the virus is able to escape to the cytosol. Several models of membrane destabilization by amphipathic helices have been proposed [80]. Membrane destabilization results from the interaction of positive amino acid side chains (largely lysine and histidine) with negatively charged head groups of lipids. Two models, the barrel-stave model and the carpet model, describe insertion of the amphipathic helices into membranes following binding to the lipid head groups and can result in disruptions ranging from pore formation to membrane solubilization. Relevant to the role of amphipathic helices in endosomal escape, low pH enhances the ability of amphipathic helices to disrupt membranes [81]. In the case of enveloped viruses, membrane destabilization can lead to membrane fusion. Three models for membrane fusion exist: Class I, Class II, and Class III, which generally involve peptide–peptide interactions leading to the formation of hairpins that drive the fusion process [82].

Similar amphipathic peptides have also been implicated in endosomal escape mediated by other viral proteins [83]. In addition to influenza virus, proteins from other enveloped viruses including HIV gp41 [13] and the γ fragment of Flock house virus capsid protein [16], as well as nonenveloped viruses such as adenovirus protein VI [17], bluetongue virus VP5 [18], rotavirus VP7 [84], rotavirus NSP4 [20], poliovirus VP1 [21,22], and human rhinovirus VP1 [24] have also been implicated in endosomal escape via an amphipathic alpha helix.

While the involvement of an amphipathic alpha helix is a common mechanism found among viruses, other lytic systems are also employed. The reovirus μN1 [33,34], murine polyomavirus VP2 [36], hepatitis A virus VP4 protein [23], and human rhinovirus VP4 protein [37] use the insertion of a myristoylated proteins into the endosomal membrane to create pores that could permit translocation of viral RNA to the cytosol. Of interest, poliovirus also contains a myristoylated protein, VP4, suggesting that the poliovirus escape route might combine more than one endosomal exit strategy [38].

In contrast, four parvoviruses including adeno-associated virus [39], canine parvovirus [40,41], and B19 and porcine parvovirus [42] use a very different strategy. The VP1 proteins of these viruses contain a phospholipase A2 (PLA2) enzyme activity. After entry into the endosome, the PLA2 activity is thought to destabilize the inner leaflet of the endosomal membrane leading to membrane rupture and capsid escape. Finally, flaviviruses such as dengue virus use the acidic environment of the endosome to induce a conformational change in an envelope protein (glycoprotein E in dengue virus) exposing a series of hydrophobic peptide loops that can insert into the membrane inducing fusion of the viral envelope and organelle [85].

Viruses may harbor additional mechanisms and/or control points for endosomal escape beyond those discussed above. For example, in the case of adenovirus, the acid dependence of the escape is mediated by the L3/p23 protease. Upon exposure to an acidic environment, L3/p23 cleaves a second protein, protein VI [17]. Protein VI is present in hundreds of copies just under the facets of the viral capsid [86]. A proteolytic cleavage of protein VI reveals the amphipathic helix that is ultimately responsible for destabilizing the endosomal membrane [17,87]. Similar to adenovirus, the mammalian reovirus is the beneficiary of a proteolytic trigger wherein the μ1 protein is cleaved in the endosome to reveal the myristoylated μ1N fragment [33–35]. In the case of dengue virus, the fusion loops of glycoprotein E are not active in the early endosome where they are revealed. Instead, fusion is delayed until the virus encounters the late endosome where the loops interact with anionic lipids as cofactors for fusion [52]. Although the precise molecular events of endosomal escape by Ebola virus, a flavivirus, are not known, they bear some resemblance to that of dengue virus. Both dengue virus and Ebola virus are activated by proteases during entry. For Ebola virus, members of the cathepsin cysteine protease family prime the glycoprotein for interaction with cellular membranes [88,89]. Although Ebola virus infection does not require low pH, cathepsin activity is enhanced at low pH, but the fusion event, similar to that of dengue virus, does not occur in early endosomes. Instead, the cleaved Ebola virus glycoprotein binds to the Niemann–Pick C protein leading to membrane fusion and late endosomal escape [53,54]. Finally, the SARS coronavirus shares a similar strategy to Ebola in that it employs a proteolytically activated fusion protein with multiple, bundled alpha helices, but, like reovirus, a posttranslationally added lipid molecule helps to guide the fusion reaction [55,56].

16.3.2 Bacteria, Protozoa, and Fungi

Some bacteria and lower eukaryotic cells have developed mechanisms to escape from the endolysosomal system to avoid hydrolytic enzymes and to take safe harbor in the cytoplasm of host cells. The escape problem for these pathogens is literally magnified relative to viruses since viruses are typically less than 500 nm in their largest dimension, with all nonenveloped viruses being less than 100 nm in diameter [90], a property that enables uptake by endocytosis. In contrast, bacteria are hundreds to thousands of nanometers in diameter while protozoa and fungal cells are larger. For most cells, engulfment by a eukaryotic phagosome leads to degradation and death, but a select group of pathogens live within eukaryotic host cells where they have developed extraordinary means of avoiding their own demise.

Phagosomes are large, often on the order of microns in diameter, and require actin-dependent extension of the plasma membrane, intracellular signaling, and interaction with endosomes and lysosomes within the cell to engulf, internalize, acidify, and hydrolyze a cargo [91]. The maturation of the phagosome includes a period during which membrane can be exchanged with early endosomes and recycling endosomes, although the contents of the phagosome remain in a single compartment. The phagosome first acquires the GTP-binding protein, rab5, which is known for its roles in directing membrane trafficking events in endocytic systems, but the regulation and, perhaps, the function of phagosomal rab5 appears to be distinct from that of endosomal rab5, suggesting common functions but a clear delineation between the trafficking pathways [92]. Eventually, rab5 is exchanged for rab7 and maturation concludes with accumulation of lysosomal markers such as lysosomal-associated membrane protein 1 (LAMP-1) and LAMP-2 in the compartment due to fusion with primary lysosomes [92].

While some bacteria, protozoans, or fungi modify the phagosome to create a viable environment, others escape to the cytosol [74], and these methods of escape are relevant to this discussion. The mechanisms of phagosomal escape are as varied as observed earlier for viral escape from endosomes. Species of *Shigella*, *Listeria*, *Rickettsia*, *Francisella*, *Burkholderia*, and *Cryptococcus* genuses have developed unique solutions to this problem. In each case, the effectors of escape have been investigated. In some cases, the mechanisms are quite complex and involve multiple proteins or even multiple protein complexes. Among these, the bacterial secretion systems are often involved. Gram-negative bacteria are characterized by seven distinct secretion systems that enable bacteria to facilitate delivery of large molecules across membranes [93]. In more than one case, elements of these systems do double duty enabling escape of the entire organism from the phagosome.

Recent data on the escape mechanism for *Francisella tularensis* illustrate the complexity with which bacteria escape from phagosomes. The ability of *Francisella* to cause disease is dependent upon escape from the endosome. Escape from the phagosome has been genetically mapped to a group of genes known as the *Francisella* pathogenicity island, which itself is a conserved part of type 6 secretion systems. Among the genes encoded in this domain, some gene products create a pore through which other pathogenicity island proteins can be secreted into host cells. To give a sense of the complexity in this system, some proteins, such as IglI, have been implicated in pathogenicity [94]. IglI is a protein that is secreted by *Francisella* into the cytoplasm of the host cell and requires the presence of VgrG to accomplish secretion. In contrast, VgrG, another protein that is secreted, does not require any other protein in the pathogenicity island to accomplish secretion [94]. VgrG appears to be a key protein since oligomers of VgrG can create holes in membranes [45]. Other members of the pathogenicity island are also required for phagosome escape, despite the fact that their roles appear to be in secretion. DotU mutants do not escape from phagosomes [45], and IglE, a palmitylated outer membrane protein, is also required for both secretion and escape [95]. It is not clear where the line between secretion and endosomal escape is drawn.

Shigella flexneri is another Gram-negative human pathogen that begins its colonization by entry into macrophages or epithelial cells by phagocytosis followed by escape from the phagosome. Similar to several of the other pathogenic intracellular bacteria, *Shigella* enters cells and then spreads from cell to cell by a nonphagocytic mechanism. This discussion focuses on the initial escape from the phagosome. *Shigella* relies on two proteins, IpaB and IpaC, to form a 2.5 nm pore, far too small to permit escape of an intact bacterium [44]. These pores likely admit potassium ions that disrupt the phagosome membrane, since a separate model using IpaB alone is able to cause endosomal disruption [96]. While IpaB and IpaC are integral membrane proteins, a soluble protein, IpaD facilitates their role in pore formation [97,98]. The same proteins are involved in a type 2 secretion system that enables cell-to-cell spread of the bacteria. Separating the mechanical role of the Ipa proteins in phagosome escape compared with intercellular spread is complex.

Listeria monocytogenes, a Gram-positive bacteria, is able to escape from phagolysosomes in macrophages with expression of bacterial phospholipases and listeriolysin O, a pore-forming protein of the cytolysin family that initially opens holes of 20–30 nm that appear to expand over time, eventually allowing release of the entire bacterium into the cytosol of the host cell. The initial activity requires proximity triggers in a similar manner to many of the viruses described earlier. Activation of host calpain proteases by high Ca^{2+} ion concentrations, the presence of a pH-activated reductase activity, and host cell cholesterol are keys to initiation of *Listeria* escape [25,26,99]. The involvement of a host protease to initiate escape is reminiscent of the pH trigger that initiates adenovirus or influenza virus escape. Recall that low pH triggers both endogenous furin to activate influenza virus hemagglutinin protein and the adenovirus L3/p23 protease. *Listeria* escape also resembles adenovirus and influenza virus in that the main actor, listeriolysin O, contains an amphipathic helix [27]. Of interest, *Listeria* fails to infect neutrophils, despite the phagocytic potential of that cell type. Arnett et al. [100] showed that neutrophils avoid infection due to the fact that neutrophil metalloproteinase-8 lyses listeriolysin O.

Staphylococcus aureus is the organism responsible for the cytolytic, methicillin-resistant *Staphylococcus aureus* (MRSA) infections of skin. This same organism is also capable of escape from endosomes. In this case, the pathogen synthesizes a series of toxins that either individually or cooperatively have the ability to lyse cholesterol-containing membranes [101,102].

The escape of Gram-negative *Rickettsia* species from phagosomes was initially thought to be due to Rickettsial PL2, an enzyme that is required for infection of target cells [103]. However, Walker et al. [104] later showed that under conditions that inhibited the bacterial enzyme, there was no difference noted in the proportion of bacteria that escaped from phagosomes by electron microscopic observation. Therefore, while it is tempting to propose that phospholipase enzyme plays a role in escape from endosomes as it does in some parvoviruses, it appears that the enzyme plays a role in a different aspect of infection such as internalization from the cell surface.

Perhaps the most unusual escape route is shared by *Cryptococcus neoformans*, a fungal pathogen, and two species of yeast, *Candida albicans* and *Candida krusei*. *C. neoformans* use a nonlytic method of escape from the endosome, termed vomocytosis [105]. Vomocytosis involves secretion of phospholipase B_1, [106], and while it at first appears to use a comparable mechanism to the parvoviruses-PLA2 activity and *Rickettsia* secretion of PLA2 discussed previously, the end result of vomocytosis permits phagosome escape without causing lysis of the organelle. Vomocytosis also exhibits a unique, inverted, two-stage pH dependence wherein acidification must occur, perhaps to trigger the process to begin, while inhibition of acidification ultimately enhances the end result [74].

16.3.3 Toxins

Organisms in every kingdom have evolved variety of toxic molecules that provide offensive capability for overcoming prey or defensive properties that reduce the likelihood of predation. Toxins can take the form of a number of different biological molecules from small molecule metabolites to genetically encoded peptides. In the latter case, the large size and relative hydrophilicity of the molecules require that the toxin carry a mechanism for translocating across the plasma membrane in order to access the vulnerable cellular machinery in the cytoplasm. *Pseudomonas* exotoxin A, for example, is produced by the *Pseudomonas* bacterium and, upon entry to the cytoplasm, poisons cells by interrupting the process of protein translation by catalyzing ADP ribosylation of an elongation factor. In order to enter the cytoplasm, the exotoxin requires three functions: cell surface binding, translocation from the endosome to the cytosol, and enzymatic activity. The three functions are compartmentalized in the structure of the exotoxin, with domain II enabling escape from endosomes [107]. Domain II contains a series of positively charged arginine residues, and site-directed mutagenesis of either one of two critical Arg residues can knock out the translocation activity of *Pseudomonas* exotoxin [108]. The presence of the Arg residues might lead to the conclusion that the exotoxin utilizes a membrane disruption strategy similar to that described for cationic liposomes. However, replacement of the Arg residues with other basic amino acids did not restore activity, and translocation of the exotoxin exhibited two other distinctions from a general membrane disruption strategy: exotoxin translocation was demonstrated to be a saturable process and was dependent on the presence of Ca^{2+} ions suggesting that a defined molecular interaction rather than a general chemical reaction was occurring in the endosome [108,109]. Subsequent characterization revealed that *Pseudomonas* exotoxin exemplified a class of molecules that escaped from endosomes without disrupting membranes. Instead, the exotoxin exhibited the ability to stay within membrane-bound organelles as it adopted a rather unorthodox intracellular trafficking pathway that takes the toxin from endosomes to the trans-Golgi apparatus to the Golgi apparatus and finally to the endoplasmic reticulum – a retrograde trip through the secretory pathway [60]. During this translocation, the exotoxin interacts with both protein disulfide isomerases (PDIs) and the endosomal protease furin, both of which change the structure of the toxin by releasing the C-terminal catalytic domain from the N-terminal targeting domain. Finally, by virtue of an endogenous RDEL peptide, the toxin rides from the Golgi to the endoplasmic reticulum where it escapes to the cytosol via the Sec61p peptide transporter, a protein complex designed to rid the endoplasmic reticulum of misfolded proteins [110].

The example of *Pseudomonas* exotoxin A is but one of many naturally occurring cytotoxic molecules that have been commandeered due to their potential for inducing death of tumor cells in clinical settings. Gilabert-Oriol et al. [61] recently reviewed the topic and include an exhaustive list of toxins that have been targeted for anticancer purposes. Notably, diphtheria toxin [60], cholera toxin [62], ricin [111], and Shiga toxin [64] all use substantially the same route to escape the endosomes and gain access to the cytosol.

16.4 ENDOSOMAL ESCAPE USING ENGINEERED SYSTEMS

Much of what is known about engineered endosomal escape has been derived from the field of gene therapy in which the cargo is a large, hydrophilic macromolecule such as DNA or RNA, and the goal is delivery to the cytosol or even to the nucleus of the target cell. As a result, methods for breaching the barrier presented by the plasma membrane are in demand. In the case of nucleic acid delivery, designs for gene delivery vectors have often borrowed ideas from nature's gene delivery vectors, namely, viruses. A description of engineered endosome escape systems will, therefore, intermittently introduce new chemical and material science applications interspersed with virus-inspired designs. In retrospect, the two are sometimes difficult to separate, but it should suffice to say that one of the principal tenets of gene delivery was the requirement to form small condensed particles containing the polyanionic nucleic acid. Naturally, polycations were an early solution to this problem and were introduced in the form of polymeric polycations as well as cationic liposomes that simulated polycations by virtue of their sequestration into a lipid bilayer [112]. In both cases, the clustered positive charges and clustered negative charges happily formed the requisite small, <100 nm, condensed particles, and with assistance from targeting molecules that conferred high-affinity binding to the cell surface, could lead to uptake of those particles via endocytosis. Having gained entry into cells, the next stage of design was to exit the endosome before the contents of the endosome encountered the hydrolytic environment of the lysosome. As it turned out, the very same polymers and lipids used for condensing nucleic acids were also amenable to modification to enhance endosomal escape.

16.4.1 Peptides and Polymers

Among the first polymers to be used for condensation of DNA was polyethylenamine or PEI, a very simple polymer containing a repeating structure that has secondary amino groups separated by ethyl group [113]. PEI's efficiency as a gene delivery vehicle has been attributed to its ability to lyse endosomes. The proposed mechanism by which PEI lyses endosomes is known as the proton sponge effect [46]. This effect is predicated on the presence of functional groups that can accept protons only when the pH drops well below neutral. Secondary and tertiary amines are ideal for this role since they do not become proton acceptors until the pH falls to approximately 6, the pH found in endosomes. In contrast, primary amines would be fully protonated at a neutral pH as they entered the endosome, and thus, would have no further capacity to absorb protons. Following this logic, the multiple secondary amino groups in PEI are available to bind protons that are pumped into the endosome by the vacuolar ATPase. However, far more protons are required to acidify the endosome due to the effective buffering provided by the PEI, so a second phenomenon begins to occur. Chloride ions, passively following the protons into the endosome down a charge gradient, begin to raise the local osmotic pressure inside of the endosome, finally resulting in rupture of the endosome and release of the contents. Some reports have questioned the mechanism proposed by the proton sponge effect based on an absence of discernable effect on lysosomes in PEI-transfected cells [114], but the efficiency of PEI along with the relatively long time required for endocytic cargo to access lysosomes suggests that PEI complexes may accomplish their escape from endosomes by the proton sponge effect well before the encounter with lysosomes becomes an issue. Now, 20 years after it was introduced for the delivery of macromolecular biologics to the cytosol of target cells, PEI remains an active area of research with well over 100 PubMed-listed articles on PEI- or modified-PEI-mediated drug delivery having been published in 2014 alone.

Using the same proton sponge logic, other polymers incorporating weak bases with the potential to protonate at endosomal pH values have been incorporated into cationic polymers to accomplish DNA packaging and delivery. The proton sponge effect helps to explain why a dendrimer made of branched polylysine is far more effective in gene delivery than a linear polylysine polymer [48]. Polyamidoamine (PAA), an early polymer that achieved success in gene transfer with low toxicity, is also thought to utilize the proton sponge effect [47]. Another prominent entity that has been incorporated into gene-transfer carriers to take advantage of the proton sponge effect has been the amino acid, histidine, which contains a secondary amine in its imidazole ring. As in the case of secondary and tertiary amines in the other polymers, histidine remains uncharged at neutral pH but can become protonated below pH 6. The well-known biochemistry of amino acids and their polymerization into peptides has facilitated the incorporation of histidine residues into a wide variety of molecules ranging from histidine-containing peptides [49] to histidylated polymers [115] to histidylated lipids [116]. A more complete description of histidine-containing peptides has been provided by Martin and Rice [50].

Other peptides (amino acid polymers) have been utilized in endosomal escape systems, but rather than relying on osmotic lysis of the endosome, some peptide-based strategies have utilized membrane fusion or membrane-lytic strategies, reminiscent of the solutions developed by viruses or bacteria using similar tools. Recall that several viruses made use of amphipathic alpha helices to induce membrane disruption. Both the HA2 protein from influenza virus and protein VI from adenovirus exhibited this property [6,17]. Not surprisingly, the amphipathic helix from the HA2 protein has been incorporated into engineered drug delivery vectors for accessing the cytosol. Both Subbarao et al. [9] and Wagner et al. [10] used the HA2 protein as a starting point for developing a set of potential fusogenic peptides to enhance the delivery of DNA into cells. While the native HA2 peptide showed the ability to lyse membranes with a gradient of increasing lytic activity as the pH decreased from 7 to 4.5 [10], the engineered peptides exhibited no lytic activity at pH 7 and maximal activity below pH 5.5 [9,10]. By focusing the lytic activity only at acidic pH, the engineered peptides were less likely to cause damage to target cells. This example is an elegant demonstration of the principle that while nature has provided a wide array of biologically active molecules for us to discover, we should continue to use our own knowledge to attempt to bend those molecules to address our needs.

Before leaving the topic of peptides, a note should be made about cell-penetrating peptides. These peptides belong to a unique class of peptide that has a seemingly magical ability to translocate across phospholipid bilayers, breaking many of the rules that we learn in introductory biology class [117,118]. Many of these peptides were discovered in natural settings such as the Tat protein of human immunodeficiency virus I [119] or the Antennapedia protein that was identified in studies of fruit fly development [120]. Not only can these peptides translocate themselves across membranes, but, when conjugated to other large, hydrophilic molecules, these small adducts can drag much larger hydrophilic molecules across membranes. Translocation of biologics including siRNA and full biologically active proteins has been accomplished by conjugation of the bioactive molecule with cell-penetrating peptides [121,122]. However, since there is no requirement for entry into endosomes, the topic is beyond the scope of this review.

16.4.2 Lipids

Just as cationic polymers provide a very useful strategy for cytosolic delivery of polyanionic nucleic acids, cationic lipids have also been employed in nucleic acid delivery. Lipids, however, have a feature that sets them apart from polymers, and that is the natural formation of a container through the arrangement of the hydrophobic tails into a lipid bilayer and, ultimately, into the bag-like liposome that has a natural ability to sequester a payload. As a result, liposomes can also be employed to deliver other biologics such as antibodies or enzymes [123,124]. But for any of these payloads, the critical problem remains delivery to the cytosol, so the same set of barriers and similar solutions are employed.

Liposomes have the added advantage that the investigator controls the components of the liposome since liposomes are typically synthesized from stock solutions of lipids in organic solvents that are mixed to create the overall lipid formula, dried to a thin film under vacuum, and then hydrated to create multilamellar liposomes. In order to create unilamellar liposomes, a shear force is typically applied [124]. When hydrated in the presence of large molecular solutes, such as biologics, the solutes are often trapped within the layers of the liposome. Alternatively, the cargo can be combined with the liposome after the lipid bilayers have formed. The latter strategy is commonly used to combine cationic liposomes with DNA, and the term for the resulting complex structure is a "lipoplex" [125]. The structure and activity of lipoplexes has been studied at great length, such that sophisticated structure–activity relationships can now be identified and associated with the commonly observed lamellar or hexagonal packing of DNA with cationic lipids [125]. The cationic liposomes include a mixture of cationic lipids and the so-called "colipids," which are lipids with zwitterionic head groups that enhance the ability of liposome bilayer to fuse with endosomal or plasma membranes during delivery of cargo to the cytoplasm [57]. Recent evidence points to the geometry of the colipid (conical vs cylindrical) as well as the neutral net charge as being determining factors in driving fusion in endosomes [58]. In fact, the composition of cationic lipids and colipids can be chosen to enhance fusion events in the endosome [59], just as peptide chemistry can be designed to favor membrane lysis in a specified pH range as we saw earlier.

Toita et al. [43] employed a unique strategy for changing membrane lipid geometry after internalization. By packaging plasmid DNA with PLA2 in a carbohydrate-based cationic hydrogel, these authors were able to demonstrate enhanced delivery of DNA to target cells. The use of PLA2 in this context recalls the use of PLA2 by viruses to achieve endosomal disruption, as described earlier.

Liposomes, like polymers, are amenable to modification. Following nature's lead, several groups have incorporated fusogenic or lysogenic molecules from pathogens into lipid-based drug delivery systems. Provoda [28] incorporated variations on the listeriolysin O protein to enhance liposome-mediated delivery of a tumor-killing toxin to cancer cells while Lorenzi and Lee [29] used listeriolysin O to enhance plasmid delivery. In the same manner that the delivery properties of the HA2 proteins have been improved by amino acid substitutions, Walls et al. [30] employed site-directed mutagenesis to develop a highly efficient version of listeriolysin O that is effective in far smaller doses.

16.4.3 Other Chemical and Physical Strategies

The issue of delivery of biologics to the cytosol is such an important problem that it has sparked the imagination of many scientists. While the major endosomal escape strategies have been reviewed earlier, there remains an array of interesting strategies that may yet take center stage. Some of these strategies include ultrasound-induced delivery of drugs to cells via induction of microbubbles, also known as sonoporation [126], photochemical internalization [127], photothermal endosome disruption [128], and lipid emulsions [129], all of which have been used to deliver macromolecules to the cytoplasm through an endocytic route. Perhaps the most exciting new technology involves exosomes, which can either be naturally occurring plasma membrane-bound vesicles released from cells that can carry bioactive molecules, such as micro-RNAs, to other cells and are being examined for potential roles in understanding and perhaps treating conditions related to oncology, infectious disease, regenerative medicine, and a host of other settings [130–132]. One key to utilizing exosomes for drug delivery may lie in virally engineered exosomes with a specific drug delivery goal in mind [133]. Evidence is mounting that exosomes are taken into cells by endocytosis before they fuse with the endosomal membrane to escape [134].

16.5 CONCLUSION

Based on the discussion of natural and engineered endosomal escape systems above, it is clear that a variety of successful strategies exist for delivering large macromolecules to the cytoplasm of target cells. More importantly, the fact that replication-deficient forms of these pathogens routinely employ a number of these systems during successful delivery of genes to cells without appreciable cytotoxicity implies that the delivery can be accomplished without irreparable harm to the target cell. Therefore, we can look forward to a time when a range of biologics including siRNA molecules, full-length genes, or even therapeutic proteins can be developed for deployment to the cytosol of cells. However, a very practical consideration should not be overlooked: many of the natural and/or engineered systems for delivery of biologics to the cytosol discussed above involve full-length foreign proteins, foreign peptides, or foreign polymers. Use of foreign proteins is obviously a problem for drug delivery since the administration of a foreign protein will almost certainly initiate an acquired immune response leading to humoral immunity against any therapeutic agent bearing the foreign protein. As a result, readministration of the therapy will be difficult if any foreign proteins are exposed on the surface of a therapeutic particle. While peptides, in general, are not as immunogenic as larger proteins, it should also be kept in mind that the use of peptides derived from foreign proteins as discussed above does not really constitute delivery of peptides *per se*.

Rather, as envisioned in engineered systems, the peptides will be part of a larger particle that combines a therapeutic cargo as well as the packaging for that cargo. Therefore, the peptides will not escape detection of the immune system but will be seen by the immune system as part of a pathogen-sized particle. Even polymers used in engineered systems run the risk of evoking an immune response. Recall that the immune system can respond to antigens as small as a dinitrophenol hapten when conjugated with an appropriate carrier. In order to fool the immune system, it is likely that use of self-peptides will be preferred to the use of foreign peptides and that a shielding system or transient immunosuppression will be required to prevent exposure of engineered particles during transit to the site of action. These are not simple problems to solve. The immune system has successfully thwarted the widespread implementation of gene therapy strategies for more than two decades.

REFERENCES

[1] Culver K, Cornetta K, Morgan R, Morecki S, Aebersold P, et al. Lymphocytes as cellular vehicles for gene therapy in mouse and man. Proc Natl Acad Sci 1991;88:3155–3159.

[2] Miller AD. Delivery of RNAi therapeutics: work in progress. Expert Rev Med Devices 2013;10:781–811.

[3] Li T, Wu M, Zhu YY, Chen J, Chen L. Development of RNA interference-based therapeutics and application of multi-target small interfering RNAs. Nucleic Acid Therap 2014;24:302–312.

[4] Deng Y, Wang CC, Choy KW, Du Q, Chen J, et al. Therapeutic potentials of gene silencing by RNA interference: principles, challenges, and new strategies. Gene 2014; 538:217–227.

[5] Skalko-Basnet N. Biologics: the role of delivery systems in improved therapy. Biologics 2014;8:107–114.

[6] Huang RTC, Wahn K, Klenk HD, Rott R. Fusion between cell membrane and liposomes containing the glycoproteins of influenza virus. Virology 1980;104:294–302.

[7] Gething MJ. Studies on the mechanism of membrane fusion: site-specific mutagenesis of the hemagglutinin of influenza virus. J Cell Biol 1986;102:11–23.

[8] Cross KJ, Langley WA, Russell RJ, Skehel JJ, Steinhauer DA. Composition and functions of the influenza fusion peptide. Protein Pept Lett 2009;16:766–778.

[9] Subbarao NK, Parente RA, Szoka FC, Nadasdi L, Pongracz K. The pH-dependent bilayer destabilization by an amphipathic peptide. Biochemistry 1987;26:2964–2972.

[10] Wagner E, Plank C, Zatloukal K, Cotten M, Birnstiel ML. Influenza virus hemagglutinin HA-2 N-terminal fusogenic peptides augment gene transfer by transferrin-polylysine-DNA complexes: toward a synthetic virus-like gene-transfer vehicle. Proc Natl Acad Sci 1992;89:7934–7938.

[11] Liou JS, Liu BR, Martin AL, Huang YW, Chiang HJ, et al. Protein transduction in human cells is enhanced by cell-penetrating peptides fused with an endosomolytic HA2 sequence. Peptides 2012;37:273–284.

[12] El-Sayed A, Masuda T, Khalil I, Akita H, Harashima H. Enhanced gene expression by a novel stearylated INF7 peptide derivative through fusion independent endosomal escape. J Control Release 2009;138:160–167.

[13] Chernomordik L, Chanturiya AN, Suss-Toby E, Nora E, Zimmerberg J. An amphipathic peptide from the C-terminal region of the human immunodeficiency virus envelope glycoprotein causes pore formation in membranes. J Virol 1994;68:7115–7123.

[14] Kwon EJ, Bergen JM, Pun SH. Application of an HIV gp41-derived peptide for enhanced intracellular trafficking of synthetic gene and siRNA delivery vehicles. Bioconjug Chem 2008;19:920–927.

[15] Kwon EJ, Liong S, Pun SH. A truncated HGP peptide sequence that retains endosomolytic activity and improves gene delivery efficiencies. Mol Pharm 2010;7:1260–1265.

[16] Bong DT, Steinern C, Janshoff A, Johnson JE, Reza Ghadiri M. A highly membrane-active peptide in Flock House virus: implications for the mechanism of nodavirus infection. Chem Biol 1999;6:473–481.

[17] Wiethoff CM, Wodrich H, Gerace L, Nemerow GR. Adenovirus protein VI mediates membrane disruption following capsid disassembly. J Virol 2005;79:1992–2000.

[18] Hassan SH, Wirblich C, Forzan M, Roy P. Expression and functional characterization of bluetongue virus VP5 protein: role in cellular permeabilization. J Virol 2001; 75:8356–8367.

[19] Elaid S, Libersou S, Ouldali M, Morellet N, Desbat B, et al. A peptide derived from the rotavirus outer capsid protein VP7 permeabilizes artificial membranes. Biochim Biophys Acta 2014;1838:2026–2035.

[20] Browne EP, Bellamy AR, Taylor JA. Membrane-destabilizing activity of rotavirus NSP4 is mediated by a membrane-proximal amphipathic domain. J Gen Virol 2000; 81:1955–1959.

[21] Fricks CE, Hogle JM. Cell-induced conformational change in poliovirus: externalization of the amino terminus of VP1 is responsible for liposome binding. J Virol 1990; 64:1934–1945.

[22] Roberts JA, Kuiper MJ, Thorley BR, Smooker PM, Hung A. Investigation of a predicted N-terminal amphipathic α-helix using atomistic molecular dynamics simulation of a complete prototype poliovirus virion. J Mol Graph Model 2012;38:165–173.

[23] Shukla A, Padhi AK, Gomes J, Banerjee M. The VP4 peptide of hepatitis a virus ruptures membranes through formation of discrete pores. J Virol 2014;88:12409–12421.

[24] Zauner W, Blaas D, Kuechler E, Wagner E. Rhinovirus-mediated endosomal release of transfection complexes. J Virol 1995;69:1085–1092.

[25] Radtke AL, Anderson KL, Davis MJ, DiMagno MJ, Swanson JA, et al. Listeria monocytogenes exploits cystic fibrosis transmembrane conductance regulator (CFTR) to escape the phagosome. Proc Natl Acad Sci 2011;108:1633 1638.

[26] Lopez-Castejon G, Corbett D, Goldrick M, Roberts IS, Brough D. Inhibition of calpain blocks the phagosomal escape of Listeria monocytogenes. PLoS One 2012;7:e35936.

[27] Safley SA, Cluff CW, Marshall NE, Ziegler HK. Role of listeriolysin-O (LLO) in the T lymphocyte response to infection with Listeria monocytogenes. Identification of T cell epitopes of LLO. J Immunol 1991;146:3604–3616.

[28] Provoda CJ. Tumor cell killing enabled by listeriolysin O-liposome-mediated delivery of the protein toxin gelonin. J Biol Chem 2003;278:35102–35108.

[29] Lorenzi GL, Lee K-D. Enhanced plasmid DNA delivery using anionic LPDII by listeriolysin O incorporation. J Gene Med 2005;7:1077–1085.

[30] Walls ZF, Goodell S, Andrews CD, Mathis J, Lee K-D. Mutants of listeriolysin O for enhanced liposomal delivery of macromolecules. J Biotechnol 2013;164:500–502.

[31] Grosz M, Kolter J, Paprotka K, Winkler AC, Schafer D, et al. Cytoplasmic replication of Staphylococcus aureus upon phagosomal escape triggered by phenol-soluble modulin alpha. Cell Microbiol 2014;16:451–465.

[32] Laabei M, Jamieson WD, Yang Y, van den Elsen J, Jenkins AT. Investigating the lytic activity and structural properties of Staphylococcus aureus phenol soluble modulin (PSM) peptide toxins. Biochim Biophys Acta 2014;1838:3153–3161.

[33] Corcoran JA, Syvitski R, Top D, Epand RM, Epand RF, et al. Myristoylation, a protruding loop, and structural plasticity are essential features of a nonenveloped virus fusion peptide motif. J Biol Chem 2004;279:51386–51394.

[34] Odegard AL, Chandran K, Zhang X, Parker JSL, Baker TS, et al. Putative autocleavage of outer capsid protein 1, allowing release of myristoylated peptide 1N during particle uncoating, is critical for cell entry by reovirus. J Virol 2004;78:8732–8745.

[35] Nibert ML, Odegard AL, Agosto MA, Chandran K, Schiff LA. Putative autocleavage of reovirus μ1 protein in concert with outer-capsid disassembly and activation for membrane permeabilization. J Mol Biol 2005;345:461–474.

[36] Sahli R, Freund R, Dubensky T, Garcea R, Bronson R, et al. Defect in entry and altered pathogenicity of a polyoma virus mutant blocked in VP2 myristylation. Virology 1993;192:142–153.

[37] Panjwani A, Strauss M, Gold S, Wenham H, Jackson T, et al. Capsid protein VP4 of human rhinovirus induces membrane permeability by the formation of a size-selective multimeric pore. PLoS Pathog 2014;10:e1004294.

[38] Tuthill TJ, Groppelli E, Hogle JM, Rowlands DJ. *Picornaviruses. Cell Entry by Non-Enveloped Viruses*. Springer Science + Business Media; 2010. p 43–89.

[39] Stahnke S, Lux K, Uhrig S, Kreppel F, Hösel M, et al. Intrinsic phospholipase A2 activity of adeno-associated virus is involved in endosomal escape of incoming particles. Virology 2011;409:77–83.

[40] Zádori Z, Szelei J, Lacoste M-C, Li Y, Gariépy S, et al. A viral phospholipase A2 is required for parvovirus infectivity. Dev Cell 2001;1:291–302.

[41] Vihinen-Ranta M, Wang D, Weichert WS, Parrish CR. The VP1 N-terminal sequence of canine parvovirus affects nuclear transport of capsids and efficient cell infection. J Virol 2002;76:1884–1891.

[42] Canaan S, Zadori Z, Ghomashchi F, Bollinger J, Sadilek M, et al. Interfacial enzymology of parvovirus phospholipases A2. J Biol Chem 2004;279:14502–14508.

[43] Toita S, S-i S, Akiyoshi K. Polysaccharide nanogel gene delivery system with endosome-escaping function: co-delivery of plasmid DNA and phospholipase A2. J Control Release 2011;155:54–59.

[44] Blocker A. The tripartite type III secreton of *Shigella flexneri* inserts IpaB and IpaC into host membranes. J Cell Biol 1999;147:683–693.

[45] Bröms JE, Meyer L, Lavander M, Larsson P, Sjöstedt A. DotU and VgrG, core components of Type VI secretion systems, are essential for francisella LVS pathogenicity. PLoS One 2012;7:e34639.

[46] Boussif O, Lezoualc'h F, Zanta MA, Mergny MD, Scherman D, et al. A versatile vector for gene and oligonucleotide transfer into cells in culture and *in vivo*: polyethylenimine. Proc Natl Acad Sci 1995;92:7297–7301.

[47] Haensler J, Szoka FC. Polyamidoamine cascade polymers mediate efficient transfection of cells in culture. Bioconjugate Chem 1993;4:372–379.

[48] Yamagata M, Kawano T, Shiba K, Mori T, Katayama Y, et al. Structural advantage of dendritic poly(L-lysine) for gene delivery into cells. Bioorg Med Chem 2007;15:526–532.

[49] Kichler A, Leborgne C, Danos O, Bechinger B. Characterization of the gene transfer process mediated by histidine-rich peptides. J Mol Med 2006;85:191–201.

[50] Martin ME, Rice KG. Peptide-guided gene delivery. AAPS J 2007;9:E18–E29.

[51] Kichler A, Leborgne C, Danos O, Bechinger B. Characterization of the gene transfer process mediated by histidine-rich peptides. J Mol Med (Berl) 2007;85:191–201.

[52] Zaitseva E, Yang S-T, Melikov K, Pourmal S, Chernomordik LV. Dengue virus ensures its fusion in late endosomes using compartment-specific lipids. PLoS Pathog 2010;6:e1001131.

[53] Carette JE, Raaben M, Wong AC, Herbert AS, Obernosterer G, et al. Ebola virus entry requires the cholesterol transporter Niemann–Pick C1. Nature 2011;477:340–343.

[54] Miller EH, Obernosterer G, Raaben M, Herbert AS, Deffieu MS, et al. Ebola virus entry requires the host-programmed recognition of an intracellular receptor. EMBO J 2012;31:1947–1960.

[55] Heald-Sargent T, Gallagher T. Ready, set, fuse! The coronavirus spike protein and acquisition of fusion competence. Viruses 2012;4:557–580.

[56] Simmons G, Zmora P, Gierer S, Heurich A, Pöhlmann S. Proteolytic activation of the SARS-coronavirus spike protein: cutting enzymes at the cutting edge of antiviral research. Antiviral Res 2013;100:605–614.

[57] Hafez IM, Cullis PR. Roles of lipid polymorphism in intracellular delivery. Adv Drug Deliv Rev 2001;47:139–148.

[58] Lonez C, Lensink MF, Kleiren E, Vanderwinden J-M, Ruysschaert J-M, et al. Fusogenic activity of cationic lipids and lipid shape distribution. Cell Mol Life Sci 2009;67:483–494.

[59] Hafez IM, Ansell S, Cullis PR. Tunable pH-sensitive liposomes composed of mixtures of cationic and anionic lipids. Biophys J 2000;79:1438–1446.

[60] Weldon JE, Pastan I. A guide to taming a toxin – recombinant immunotoxins constructed from *Pseudomonas* exotoxin A for the treatment of cancer. FEBS J 2011; 278:4683–4700.

[61] Gilabert-Oriol R, Weng A, Mallinckrodt B, Melzig M, Fuchs H, et al. Immunotoxins constructed with ribosome-inactivating proteins and their enhancers: a lethal cocktail with tumor specific efficacy. Curr Pharm Des 2014;20:6584–6643.

[62] Lencer WI. Microbes and microbial Toxins: paradigms for microbial-mucosal toxins. V. Cholera: invasion of the intestinal epithelial barrier by a stably folded protein toxin. Am J Physiol Gastrointest Liver Physiol 2001;280:G781–786.

[63] Hartley MR, Lord JM. Cytotoxic ribosome-inactivating lectins from plants. Biochim Biophys Acta 2004;1701:1–14.

[64] Sandvig K, Bergan J, Dyve A-B, Skotland T, Torgersen ML. Endocytosis and retrograde transport of Shiga toxin. Toxicon 2010;56:1181–1185.

[65] Maxfield FR, McGraw TE. Endocytic recycling. Nat Rev Mol Cell Biol 2004; 5:121–132.

[66] Mukherjee S, Ghosh RN, Maxfield FR. Endocytosis. Physiol Rev 1997;77:759–803.

[67] Duncan R, Richardson SCW. Endocytosis and intracellular trafficking as gateways for nanomedicine delivery: opportunities and challenges. Mol Pharm 2012;9:2380–2402.

[68] Pillay CS, Elliott E, Dennison C. Endolysosomal proteolysis and its regulation. Biochem J 2002;363:417.

[69] Bugge TH, Antalis TM, Wu Q. Type II transmembrane serine proteases. J Biol Chem 2009;284:23177–23181.

[70] Seidah NG. What lies ahead for the proprotein convertases? Ann N Y Acad Sci 2011;1220:149–161.

[71] Parton RG, del Pozo MA. Caveolae as plasma membrane sensors, protectors and organizers. Nat Rev Mol Cell Biol 2013;14:98–112.

[72] Huynh KK, Kay JG, Stow JL, Grinstein S. Fusion, fission, and secretion during phagocytosis. Physiology 2007;22:366–372.

[73] Ramachandra L, Chu RS, Askew D, Noss EH, Canaday DH, et al. Phagocytic antigen processing and effects of microbial products on antigen processing and T-cell responses. Immunol Rev 1999;168:217–239.

[74] Smith Leanne M, May Robin C. Mechanisms of microbial escape from phagocyte killing. Biochem Soc Trans 2013;41:475–490.

[75] Titus EO. Recent developments in the understanding of the pharmacokinetics and mechanism of action of chloroquine. Ther Drug Monit 1989;11:369–379.

[76] Tartakoff AM. Perturbation of vesicular traffic with the carboxylic ionophore monensin. Cell 1983;32:1026–1028.

[77] Crider BP, Xie XS, Stone DK. Bafilomycin inhibits proton flow through the H+ channel of vacuolar proton pumps. J Biol Chem 1994;269:17379–17381.

[78] Stieneke-Grober A, Vey M, Angliker H, Shaw E, Thomas G, et al. Influenza virus hemagglutinin with multibasic cleavage site is activated by furin, a subtilisin-like endoprotease. EMBO J 1992;11:2407–2414.

[79] Horimoto T, Kawaoka Y. The hemagglutinin cleavability of a virulent avian influenza virus by subtilisin-like endoproteases is influenced by the amino acid immediately downstream of the cleavage Site. Virology 1995;210:466–470.

[80] Shai Y. Mechanism of the binding, insertion and destabilization of phospholipid bilayer membranes by α-helical antimicrobial and cell non-selective membrane-lytic peptides. Biochim Biophys Acta Biomembr 1999;1462:55–70.

[81] Vogt TCB, Bechinger B. The interactions of histidine-containing amphipathic helical peptide antibiotics with Lipid Bilayers: the effects of charges and pH. J Biol Chem 1999;274:29115–29121.

[82] Weissenhorn W, Hinz A, Gaudin Y. Virus membrane fusion. FEBS Lett 2007;581: 2150–2155.

[83] Moyer CL, Wiethoff CM, Maier O, Smith JG, Nemerow GR. Functional genetic and biophysical analyses of membrane disruption by human adenovirus. J Virol 2011; 85:2631–2641.

[84] Elaid S, Libersou S, Ouldali M, Morellet N, Desbat B, et al. A peptide derived from the rotavirus outer capsid protein VP7 permeabilizes artificial membranes. Biochim Biophys Acta Biomembr 2014;1838:2026–2035.

[85] Fuzo CA, Degrève L. The pH dependence of flavivirus envelope protein structure: insights from molecular dynamics simulations. J Biomol Struct Dyn 2013;32: 1563–1574.

[86] Liu H, Jin L, Koh SBS, Atanasov I, Schein S, et al. Atomic structure of human adenovirus by cryo-EM reveals interactions among protein networks. Science 2010; 329:1038–1043.

[87] Moyer CL, Nemerow GR. Viral weapons of membrane destruction: variable modes of membrane penetration by non-enveloped viruses. Curr Opin Virol 2011;1:44–49.

[88] Chandran K. Endosomal proteolysis of the ebola virus glycoprotein is necessary for infection. Science 2005;308:1643–1645.

[89] Schornberg K, Matsuyama S, Kabsch K, Delos S, Bouton A, et al. Role of endosomal cathepsins in entry mediated by the ebola virus glycoprotein. J Virol 2006; 80:4174–4178.

[90] Leopold P. *Microtubule-Dependent Motility during Intracellular Trafficking of Vector Genome to the Nucleus*. Nanotechnology in Biology and Medicine: Informa UK Limited; 2007.

[91] Swanson JA. Shaping cups into phagosomes and macropinosomes. Nat Rev Mol Cell Biol 2008;9:639–649.

[92] Kinchen JM, Ravichandran KS. Phagosome maturation: going through the acid test. Nat Rev Mol Cell Biol 2008;9:781–795.

[93] Saier MH. Protein secretion and membrane insertion systems in Gram-negative bacteria. J Membr Biol 2006;214:75–90.

[94] Barker JR, Chong A, Wehrly TD, Yu J-J, Rodriguez SA, et al. The Francisella tularensis pathogenicity island encodes a secretion system that is required for phagosome escape and virulence. Mol Microbiol 2009;74:1459–1470.

[95] Robertson GT, Child R, Ingle C, Celli J, Norgard MV. IglE is an outer membrane-associated lipoprotein essential for intracellular survival and murine virulence of Type A Francisella tularensis. Infect Immun 2013;81:4026–4040.

[96] Senerovic L, Tsunoda SP, Goosmann C, Brinkmann V, Zychlinsky A, et al. Spontaneous formation of IpaB ion channels in host cell membranes reveals how *Shigella* induces pyroptosis in macrophages. Cell Death Dis 2012;3:e384.

[97] Picking WL, Nishioka H, Hearn PD, Baxter MA, Harrington AT, et al. IpaD of *Shigella flexneri* is independently required for regulation of ipa protein secretion and efficient insertion of IpaB and IpaC into host membranes. Infect Immun 2005;73:1432–1440.

[98] Roehrich AD, Guillossou E, Blocker AJ, Martinez-Argudo I. *Shigella* IpaD has a dual role: signal transduction from the type III secretion system needle tip and intracellular secretion regulation. Mol Microbiol 2013;87:690–706.

[99] Parihar SP, Guler R, Lang DM, Suzuki H, Marais AD, et al. Simvastatin enhances protection against Listeria monocytogenes infection in mice by counteracting Listeria-induced phagosomal escape. PLoS One 2013;8:e75490.

[100] Arnett E, Vadia S, Nackerman CC, Oghumu S, Satoskar AR, et al. The pore-forming toxin listeriolysin O is degraded by neutrophil metalloproteinase-8 and fails to mediate Listeria monocytogenes intracellular survival in neutrophils. J Immun 2013; 192:234–244.

[101] Grosz M, Kolter J, Paprotka K, Winkler A-C, Schäfer D, et al. Cytoplasmic replication of *Staphylococcus aureus* upon phagosomal escape triggered by phenol-soluble modulin α. Cell Microbiol 2013;16:451–465.

[102] Laabei M, Jamieson WD, Yang Y, van den Elsen J, Jenkins ATA. Investigating the lytic activity and structural properties of *Staphylococcus aureus* phenol soluble modulin (PSM) peptide toxins. Biochim Biophys Acta Biomembr 2014;1838:3153–3161.

[103] Silverman DJ. *Oxidative Cell Injury and Spotted Fever Group Rickettsiae Infectious Agents and Pathogenesis*. Springer Science + Business Media; 2002. p 79–98.

[104] Walker DH, Feng HM, Popov VL. Rickettsial phospholipase A2 as a pathogenic mechanism in a model of cell injury by typhus and spotted fever group rickettsiae. Am J Trop Med Hyg 2001;65:936–942.

[105] Johnston SA, May RC. Cryptococcus interactions with macrophages: evasion and manipulation of the phagosome by a fungal pathogen. Cell Microbiol 2012;15:403–411.

[106] Chayakulkeeree M, Johnston SA, Oei JB, Lev S, Williamson PR, et al. SEC14 is a specific requirement for secretion of phospholipase B1 and pathogenicity of *Cryptococcus neoformans*. Mol Microbiol 2011;80:1088–1101.

[107] Pastan I, FitzGerald D. *Pseudomonas* exotoxin: chimeric toxins. J Biol Chem 1989; 264:15157–15160.

[108] Jinno Y, Ogata M, Chaudhary VK, Willingham MC, Adhya S, et al. Domain II mutants of *Pseudomonas* exotoxin deficient in translocation. J Biol Chem 1989;264: 15953–15959.

[109] Pirker R, FitzGerald DJ, Willingham MC, Pastan I. Enhancement of the activity of immunotoxins made with either ricin A chain or *Pseudomonas* exotoxin in human ovarian and epidermoid carcinoma cell lines. Cancer Res 1988;48:3919–3923.

[110] Plemper RK, Bohmler S, Bordallo J, Sommer T, Wolf DH. Mutant analysis links the translocon and BiP to retrograde protein transport for ER degradation. Nature 1997;388:891–895.

[111] Hartley MR, Lord JM. Cytotoxic ribosome-inactivating lectins from plants. Biochim Biophys Acta Prot Proteom 2004;1701:1–14.

[112] Mintzer MA, Simanek EE. Nonviral vectors for Gene Delivery. Chem Rev 2009;109: 259–302.

[113] Godbey WT, Wu KK, Mikos AG. Poly(ethylenimine) and its role in gene delivery. J Control Release 1999;60:149–160.

[114] Godbey WT, Barry MA, Saggau P, Wu KK, Mikos AG. Poly(ethylenimine)-mediated transfection: a new paradigm for gene delivery. J Biomed Mater Res 2000;51:321–328.

[115] Lächelt U, Kos P, Mickler FM, Herrmann A, Salcher EE, et al. Fine-tuning of proton sponges by precise diaminoethanes and histidines in pDNA polyplexes. Nanomed Nanotechnol Biol Med 2014;10:35–44.

[116] Tarwadi JJA, Prankerd RJ, Pouton CW. Preparation and *in vitro* evaluation of novel lipopeptide transfection agents for efficient gene delivery. Bioconjugate Chem 2008;19:940–950.

[117] Bechara C, Sagan S. Cell-penetrating peptides: 20years later, where do we stand? FEBS Lett 2013;587:1693–1702.

[118] Copolovici DM, Langel K, Eriste E, Langel Ü. Cell-penetrating peptides: design, synthesis, and applications. ACS Nano 2014;8:1972–1994.

[119] Fawell S, Seery J, Daikh Y, Moore C, Chen LL, et al. Tat-mediated delivery of heterologous proteins into cells. Proc Natl Acad Sci 1994;91:664–668.

[120] Derossi D, Chassaing G, Prochiantz A. Trojan peptides: the penetratin system for intracellular delivery. Trends Cell Biol 1998;8:84–87.

[121] van den Berg A, Dowdy SF. Protein transduction domain delivery of therapeutic macromolecules. Curr Opin Biotechnol 2011;22:888–893.

[122] Presente A, Dowdy SF. PTD/CPP peptide-mediated delivery of siRNAs. Curr Pharm Des 2013;19:2943–2947.

[123] Lim SB, Banerjee A, Önyüksel H. Improvement of drug safety by the use of lipid-based nanocarriers. J Control Release 2012;163:34–45.

[124] Akbarzadeh A, Rezaei-Sadabady R, Davaran S, Joo SW, Zarghami N, et al. Liposome: classification, preparation, and applications. Nanoscale Res Lett 2013;8:102.

[125] Ewert KK, Zidovska A, Ahmad A, Bouxsein NF, Evans HM, et al. Cationic liposome–nucleic acid complexes for gene delivery and silencing: pathways and mechanisms for plasmid DNA and siRNA. Nucleic acid transfection. Top Curr Chem 2010; 296:191–226.

[126] Lentacker I, De Cock I, Deckers R, De Smedt SC, Moonen CTW. Understanding ultrasound induced sonoporation: definitions and underlying mechanisms. Adv Drug Deliv Rev 2014;72:49–64.

[127] Adigbli DK, MacRobert AJ. Photochemical internalisation: the journey from basic scientific concept to the threshold of clinical application. Curr Opin Pharmacol 2012;12:434–438.

[128] Kim H, Kim WJ. Photothermally controlled gene delivery by reduced graphene oxide-polyethylenimine nanocomposite. Small 2013;10:117–126.

[129] Nam HY, Park JH, Kim K, Kwon IC, Jeong SY. Lipid-based emulsion system as non-viral gene carriers. Arch Pharm Res 2009;32:639–646.

[130] De Jong OG, Van Balkom BWM, Schiffelers RM, Bouten CVC, Verhaar MC. Extracellular vesicles: potential roles in regenerative medicine. Front Immunol 2014;5:608.

[131] Vella LJ. The emerging role of exosomes in epithelial mesenchymal-transition in cancer. Frontiers in Oncology 2014;4:361.

[132] Schorey JS, Cheng Y, Singh PP, Smith VL. Exosomes and other extracellular vesicles in host-pathogen interactions. EMBO Rep 2014;16:24–43.

[133] Koppers-Lalic D, Hogenboom MM, Middeldorp JM, Pegtel DM. Virus-modified exosomes for targeted RNA delivery: a new approach in nanomedicine. Adv Drug Deliv Rev 2013;65:348–356.

[134] Mulcahy LA, Pink RC, Carter DR. Routes and mechanisms of extracellular vesicle uptake. J Extracell Vesicles 2014;3:24641, http://dx.doi.org/10.3402/jev.v3.24641.

17

LYSOSOMES AND ANTIBODY–DRUG CONJUGATES

Michelle Mack, Jennifer Kahler, Boris Shor, Michael Ritchie, Maureen Dougher, Matthew Sung,and Puja Sapra

Oncology Research Unit, World Wide Research and Development, Pfizer Inc., Pearl River, NY, USA

17.1 INTRODUCTION

Antibody–drug conjugates (ADCs) are emerging therapeutic modalities that utilize the specificity of antibodies to deliver chemotherapeutic drugs directly to tumors with the promise of maintaining antitumor efficacy while minimizing nontumor toxicities. With improved tumor targeting, ADCs are strategically designed to increase the therapeutic index of a given chemotherapeutic drug. The lysosome is a key organelle central to the biology of ADCs. In this chapter, we discuss the cellular biology that impacts ADC efficacy and design a strategy as well as propose the lysosome as an antitumor target for novel ADC development.

With ever-increasing interest in precision medicine and targeted therapy, in the past decade, there has been an explosion in the development of therapeutic biologics, including naked antibodies and ADCs. Many of the targets for these therapeutic entities are located on the plasma membrane with the target epitopes exposed to the extracellular space. Therapeutic naked antibodies bind to their target antigen and presumptively inhibit their function (e.g., reduced ligand binding and partner interactions) [1–3]. For example, trastuzumab (Herceptin) is a therapeutic antibody targeting the receptor tyrosine kinase human epidermal growth factor receptor 2 (HER2) and is Food and Drug Administration (FDA)-approved to treat HER2+

Lysosomes: Biology, Diseases, and Therapeutics, First Edition.
Edited by Frederick R. Maxfield, James M. Willard and Shuyan Lu.

breast cancer. The antibody component of ADCs also binds extracellular epitopes of plasma-membrane-bound antigens. However, in order to achieve the therapeutic effect of ADCs, the target antigen must also internalize into the cell. First, we discuss the mechanisms that mediate internalization of plasma-bound antigens.

17.2 RECEPTOR INTERNALIZATION

Molecules can be internalized through multiple mechanisms, including clathrin-mediated endocytosis (CME) or clathrin-independent mechanisms such as phagocytosis, macropinocytosis, and caveolin-dependent endocytosis (Figure 17.1) [4,5]. The internalization of cell surface receptors provides ADCs with a target-specific entry point [4,5]. CME is a strikingly complex process involving the recruitment of a series of both adaptor and accessory proteins, along with a clathrin polymer lattice to both the proximal phospholipids of the membrane and the intracellular segment of the receptor [6]. Adaptor proteins serve to target receptors for endocytosis.

Sequence motifs in the cytoplasmic domains of transmembrane proteins are known to recruit different adaptor proteins; this allows segregation and packaging of their internalized cargo into coated vesicles. Similar motifs, including tyrosine-based and dileucine-based sorting signals, can mediate sorting of transmembrane proteins to endosomes and lysosomes and facilitate direct segregation of lysosomal proteins at the *trans*-Golgi network (TGN) [7,8]. In addition, small RAB GTPases play a key role in controlling the different sorting fates of cargoes in endosomal compartments [9]. The best characterized adaptor protein is adaptor complex 2 (AP2), which binds to short, linear tyrosine- and dileucine-based sequences on the cytoplasmic tails of receptors [10].

AP-2 plays a key role in clathrin-coated vesicle (CCV) formation and function, being responsible for the assembly of clathrin triskelia at the plasma membrane and selection of cargo receptors that are internalized by forming CCVs [11–13]. Upon receptor internalization, clathrin begins to polymerize, causing membrane rearrangement and the formation of intracellular vesicles [14]. This intracellular vesicle is released from the plasma membrane via GTP-dependent proteins, such as the large GTPase, dynamin (Dyn), which hydrolyzes GTP to remove the vesicle from the plasma membrane [15–17]. After the internalized vesicles are released from the plasma membrane, they continue to fuse with each other in the cytoplasm to form the intracellular organelle known as the early endosome (EE) [18].

Fusion of these vesicles is crucial to the physiology of the endosome and regulates the vesicle maturation from the EE into the late endosome (LE). The EE is surrounded by a complex and compartmentalized system of proteins within the cytoplasm that function to regulate the intracellular distribution of all internalized cargo. Within the EE, internalized receptors and their ligands are subjected to segregation into separate intracellular trafficking pathways, resulting in receptor recycling or degradation. In one scenario, internalized receptors discharge their ligands in EEs and recycle back to the cell surface. Examples of receptors that traverse this pathway are the transferrin receptor (TfR) [19] and HER2 [20]. Alternatively, cargo can be retained within

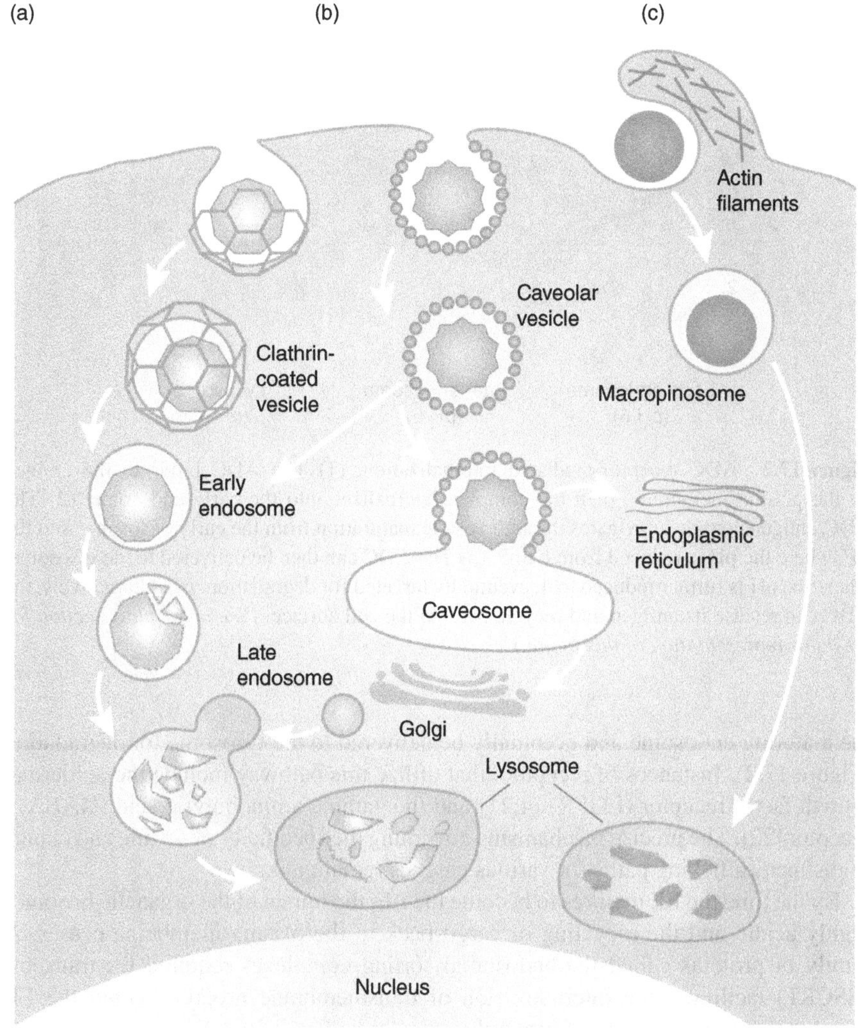

Figure 17.1 Internalization pathways mediating ADC uptake. (a) Clathrin-mediated endocytosis (CME) originates with adaptor proteins targeting receptors for internalization by forming clathrin-coated vesicles (CCVs), followed by membrane rearrangement and the formation of intracellular vesicles. These intracellular vesicles are released and fuse to form the early endosome. (b) Caveolae-mediated endocytosis is initiated by caveolae, a lipid raft containing sphingolipids, cholesterol, and caveolin proteins. Cargoes contained in caveolin-coated vesicles traffic to the caveosome, an intermediate compartment, *en route* to the early endosome. (c) Macropinocytosis mediates nonspecific uptake of soluble antigens. Intake is an actin-dependent process mediated by plasma membrane projections that give rise to macropinosomes, large endocytic vesicles (>1 μm). (*See color plate section for the color representation of this figure.*)

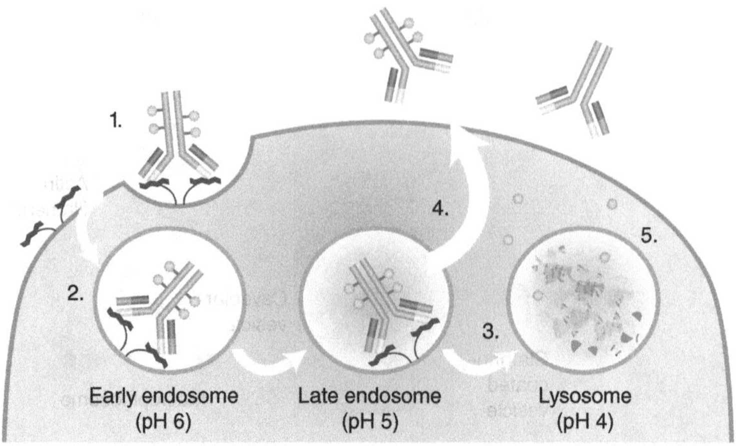

Figure 17.2 ADC receptor-mediated internalization. (1) The ADC binds to the antigen on the plasma membrane, then the complex internalizes into the early endosome. (2) The ADC/antigen complex navigates through vesicle maturation from the early endosome into the LE, where the pH is reduced from 6 to 5. (3) The ADC can then be delivered to the lysosome, where the pH is further reduced to 4, eventually targeted for degradation. (4) Alternatively, the ADC can release its antigen and recycle back to the cell surface. (*See color plate section for the color representation of this figure.*)

the maturing endosome and eventually be delivered to the lysosome for degradation (Figure 17.2). Instances of receptors that utilize this pathway include the epidermal growth factor receptor (EGFR) [4,21] and the gamma aminobutyric acid ($GABA_A$) receptor [22]. The precise mechanisms governing the specificity of sorting and choice of distinct trafficking paths for various cargoes remain unknown.

By the time the EE matures to become the LE, the lumen of the organelle becomes highly acidic and the recycling of cargo back to the plasma membrane ceases. A family of proteins called the endosomal sorting complexes required for transport (ESCRT) facilitates the internalization of transmembrane proteins within the LE membrane into the lumen to form multivesicular bodies (MVBs).

Mammalian ESCRTs have been implicated in diverse cellular processes including MVB biogenesis, cytokinesis, viral budding, and autophagy. The primary function of ESCRT counterparts is to downregulate signaling receptors among other membrane proteins. Protein sorting into the internal vesicles of MVBs is regulated by the four endosomal sorting complexes required for transport, ESCRT-0, -I, -II, and -III [23]. Notably, the main function of complexes 0–II is to perform sorting by recognizing ubiquitin-tagged membrane proteins in the endocytic pathway and mediate the membrane involution of those proteins into MVBs. The ESCRT-III and Vps4 complexes expedite the completion of cargo sorting, deubiquitination, and induction of membrane curvature for vesicle formations [23]. Specifically, ESCRT-III complexes are only transiently assembled on endosomes and facilitate recruitment of deubiquitination machinery. Protein sorting in the MVB pathway is initiated by ESCRT-0, which localizes to the endosomal membrane via phosphatidylinositol-3-phosphate (PI3P) binding. This complex contains ubiquitin- and clathrin-binding domains that aid in the recognition, binding, and clustering of ubiquitylated proteins tagged for degradation [24]. ESCRT-I and -II cooperate with neighboring functional oligomeric units to further facilitate cargo sorting and membrane involution [25]. Finally, Vps4 disassembles the complex to complete the formation of intraluminal vesicles (ILVs). ILVs fuse with lysosomes to deliver the processed contents and the remaining molecules are degraded in the lumen by acid hydrolysis [23].

Next, we outline the strategic design of ADCs and the mechanisms that drive receptor-ADC internalization.

17.3 ANTIBODY–DRUG CONJUGATES

There are four components to consider when designing successful ADCs, a monoclonal antibody (mAb) specifically targeting tumor-associated cell surface antigens, a cytotoxic chemotherapeutic payload, and a chemical linker that attaches and stabilizes the payload onto the mAb (Figure 17.3). Lysosomal biology has helped to delineate how mAbs exploit receptor internalization and act as transporters for cytotoxic payloads, by delivering them to the cancer cell microenvironment. Keeping in mind, each antigen may utilize different intracellular sorting pathways. In addition, effective ADCs are reliant on both linker release and payload delivery to the lysosomal compartment of the targeted cells [26].

ADC efficacy is dependent on its binding to the tumor-associated antigen (TAA) followed by receptor internalization to guarantee delivery of the cytotoxic payload inside the tumor cell. Often, targets that internalize have the capability of also trafficking to the lysosome. Therefore, selection of the appropriate ADC target is a critical component of any successful ADC development program. In general, several parameters should be considered, including overall target expression on the cell

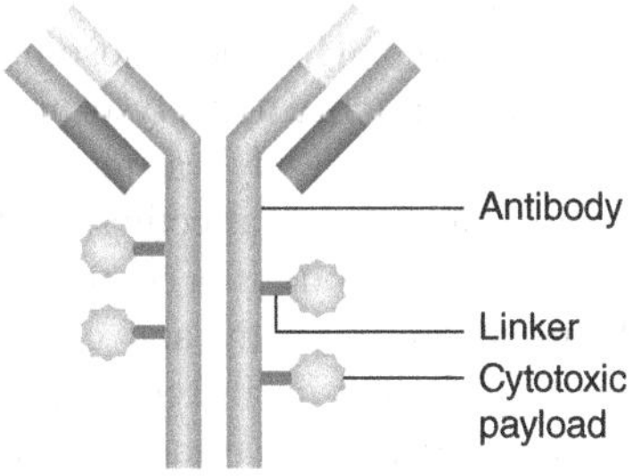

Figure 17.3 ADC structure. Antibody–monoclonal antibody (mAb) that specifically targets tumor-associated cell surface antigens. Linker – attaches and stabilizes payload to the mAb, releases cytotoxic payload within the target cell. Cytotoxic payload – highly potent cytotoxic agent. For example, those causing DNA damage or microtubule disruption. (*See color plate section for the color representation of this figure.*)

membrane, relative levels of overexpression in malignant tissues compared with normal counterparts, rates of antigen internalization, and the intracellular trafficking dynamics of internalized receptors. Interestingly, the level of TAA expression needed for an efficacious ADC is dependent on the ADC payload. Programs that utilize less potent payloads, such as microtubule inhibitors, should have TAA expressed at sufficiently high density to ensure delivery of cytotoxic payload to tumor cells. ADCs with more toxic DNA-damaging agents, for example, Mylotarg® (calicheamicin payload) can be delivered with a relatively low TAA copy number [27,28].

Advances in antibody engineering have exploited differences between the environments of the extracellular and endosomal space. Despite the route that the target antigen follows after internalization, antibody engineering can modify the intracellular trafficking dynamics. Engineered antibodies can improve lysosomal localization thereby ensuring the ADC/antigen complex is dissociated in the endosome, allowing the ADC to independently traverse endosomal sorting pathways regardless of its receptor. In two separate studies, scientists have altered the intracellular route of antibody trafficking by designing antibodies with high-affinity binding for the extracellular target at neutral pH and low-affinity binding for the receptor within the acidic EE [29,30]. Both studies utilized a histidine scanning approach, in which histidine residues were introduced in critical areas of the antibody. This screen selected for mutations that disrupt antibody binding at a low pH environment without affecting binding at neutral pH. Histidine residues were prioritized based on the fact that they show pK_a values of 6.0, and thus these histidine residues become protonated in pH environments below 6.0, causing disruption of antibody binding. In both cases, this approach was used to identify the antibody that can dissociate from its receptor within the EE and enter the recycling pathway while its target antigen underwent lysosomal degradation. This yielded an antibody capable of escaping lysosomal degradation, leading to an increase in the serum half-life and efficacy of the therapeutic [29,30]. In recent studies, a similar method was used to generate anti-IL6 neutralizing antibodies with a range of pH dependencies for binding [31]. Following uptake into the cell, anti-IL-6 antibodies bind antigens with higher affinity at near-neutral pH (relative to acidic endosomal pH). After ADC uptake, the complex is expected to release antigen within the endosome, with concomitant antibody recycling and exocytosis occurring in FcRn-expressing cells. In the acidic pH of the endosome, the cytokine dissociated from the engineered antibody and trafficked to the lysosome, thus decreasing its level in the circulation, while the anti-IL-6 antibody was recycled out of the cell [31].

Linker-payloads are additional factors affecting ADC efficacy whereby the intracellular drug release from the antibody is accomplished through the disruption of the chemical linker between the antibody and cytotoxic payload. Therefore, ADC efficacy can be directly impacted by linker choice. Early-generation ADCs relied on the use of nonproteolytically cleavable linkers such as hydrazone, or disulfide-based linker approaches [32]. Hydrazone linkers are stable at a neutral pH but are cleaved within the low pH environment of intracellular endosome and lysosome compartments. Proteolytically cleavable linkers, such as the peptide linker valine-citrulline-*p*-aminobenzyl carbamate (vc), are designed to be selectively cleaved by lysosomal proteases, allowing for specific cleavage after internalization [33]. Noncleavable linkers, such as thioether- or amide-based linkers, have been utilized more recently and are intended to retain stability throughout the plasma and most of the intracellular space. Current evidence suggests that the early and LEs provide a highly reductive environment and contain sufficient proteases to process both cleavable types of linkers [34,35]. Nonetheless, one study has disputed this notion and has shown that the endosomes are in fact highly oxidative in nature, thereby limiting the cleavage of disulfide-based linkers [36]. This finding suggested the possibility that using disulfide linkers may result in reduced antitumor activity in some situations. Interestingly, payloads attached through disulfide linkers can form lipophilic drug metabolites after cleavage in the lysosome [37]. These metabolites have the ability to cross membranes and contribute to the bystander killing effect when effluxed out of tumor cells. Such approaches may be beneficial when targeting tumors with heterogeneous antigen expression, but, at the same time, can result in higher levels of undesirable toxicity. Another study has shown that the specific sites on which the linker-payload is conjugated onto the antibody can influence stability of a proteolytically cleavable linker [38].

One of the most common linkers across the ADCs in clinical trials is the aforementioned vc-linker. Brentuximab vedotin (Adcetris®, SGN-35), a currently approved ADC, is an anti-CD30-auristatin conjugate that incorporates a peptide cleavable vc-linker conjugated to monomethyl auristatin E (MMAE) [39]. In Brentuximab vedotin, the vc-linker is conjugated via a cysteine sulfhydryl group on the antibody to the maleimide functionality on the linker. Upon internalization of the ADC into the endosomes and lysosomes, the amide bond between the citrulline residue and the *p*-aminobenzyl carbamate portion of the linker is cleaved by lysosomal proteases [33]. Self-elimination of the *p*-aminobenzyl carbamate group, which is important to spatially separate the drug from the site of enzymatic cleavage, releases free MMAE in the tumor cell. While ADCs with cleavable linkers tend to show good efficacy, the nature of a cleavable linker generates certain liability due to potential extracellular cleavage of the payload and the resultant off-target toxicity. As a consequence, some recent ADCs have been designed to possess linkers that are not cleavable by conventional mechanisms, such as thioether or amide bonds [40]. While this creates an ADC that is highly stable in the extracellular environment, the use of such a linker requires that the ADC be routed to the lysosomal compartment of the cell, where the antibody is completely degraded, liberating the linker-payload. The payload is then released from the lysosome into the cytosol, where it elicits cytotoxic activity. However, some targets are not delivered to the lysosomal compartment at a sufficient concentration and are instead recycled back to the plasma membrane, making it challenging to develop ADCs with a noncleavable linker against such targets.

The development of novel cytotoxic payloads with optimized potency and appropriate linker compatibility remains an important goal in ADC development. Early-generation ADCs contained payloads that provided little antitumor activity due to the lack of potency (e.g., methotrexate, doxorubicin, and vinblastine) [41–46]. As a result, newer ADCs have incorporated highly potent cytotoxic agents that interfere with microtubule dynamics or cause irreversible disruptions in DNA integrity. Usage of highly potent microtubule-disrupting agents has become widely adopted due to the efficacy observed against rapidly dividing cells, which provides an additional mechanism for tumor selectivity. Compared to conventional spindle poison cytotoxics, DNA-damaging agents are less dependent on cell cycle progression to induce apoptosis, a property that may be advantageous when targeting malignant cells that are not rapidly proliferating. Significant efforts are underway to develop new and more tumor-selective linker-payloads.

Lysosomal sequestration should be considered when developing novel payloads for ADC platforms. The concept of pH partitioning has been well described for payloads linked by amine residues [47]. When payloads exist in a pure basic form, they are capable of free diffusion across membranes. For example, when a small molecule exhibits a pK_a value of 7–8, it exists predominantly in the free base form in the cytosol. However, if this molecule crosses into organelles with acidic lumens, such as endosomes and lysosomes, the molecule exists primarily in an ionized or protonated form potentially producing a membrane impermeable molecule that subsequently traps the molecule inside the organelle. This is believed to be one of the reasons why tumors develop resistance to drug therapy [47]. Lysosomal sequestration has been described for molecules such as sunitinib [48]. In addition, it has been suggested that lysosomal sequestration of molecules can occur through the presence of multidrug transporters, such as P-glycoprotein (Pgp), in the lysosomal membrane [49]. In this case, cytotoxic agents are pumped out of the cytoplasm into the lysosomal compartment by Pgp, where they are subsequently ionized within the lysosome, rendering them ineffective agents. This mode of sequestration has been described for molecules such as doxorubicin, daunorubicin, and vinblastine [49]. To date, lysosomal sequestration has not been reported for ADC payloads. Novel payload development strategies should be considered to reduce potential issues of lysosomal sequestration. Therefore, developing payloads or agents that interfere with the integrity of the lysosomal compartment is an attractive option to consider in preclinical studies with future ADCs [50]. Disruption of lysosomes can lead to autophagic cell death, resulting in the release of detrimental lysosomal proteases into the cytoplasm. Bafilomycin A, salicylihalamide A, and concanamycin A are interesting examples of natural products that disrupt the lysosomes. These agents inhibit vacuolar ATPase (vATPase), which maintains the proton gradient across the lysosomal membrane [51–54]. Accordingly, inhibition of this vATPase impedes the ability of lysosomes to acidify, which can result in cell death.

17.4 MECHANISMS OF RESISTANCE TO ADCs

The efficacy and duration of any chemotherapeutic regimen is dictated by the innate resistance of cancer cells to a therapy or the acquisition of resistance following drug exposure and resistant cell population selection. Multiple ADCs have entered clinical development, yet little data is available that describes the molecular mechanisms of resistance that emerges within this drug class. Published clinical data for acute myeloid leukemia (AML) suggests patients that initially responded but eventually failed Mylotarg therapy had an enrichment of CD33+ cells with an upregulation of

Pgp or multidrug resistance (MDR1) [55,56]. As Mylotarg's payload is a known substrate for MDR1, the mechanism of resistance in Mylotarg refractory patients could be due to efflux of payload out of leukemic cells [57,58]. However, published data about clinical resistance to other FDA-approved ADCs (i.e., Adcetris or Kadcyla®) are lacking. Recently, several studies have reported the acquisition of resistance to trastuzumab-based ADCs in *in vitro* models. The first study showed that chronic treatment of HER2+ cell lines with Kadcyla resulted in the emergence of resistant cell populations with varying mechanisms of resistance [59]. Mechanisms of resistance observed in this study included the compensatory expression of neuregulin-1, MDR1 protein overexpression, HER2 protein downregulation, and changes in mitogen-activated protein kinase (MAPK) family signaling. A second study revealed that cyclical treatment of HER2+ cell lines with a trastuzumab-maytansinoid ADC, which is structurally similar to Kadcyla, yielded resistant cell lines. The observed resistance was mediated by payload drug efflux through overexpression of multidrug resistance protein 1 (MRP1) and HER2 protein downregulation [60]. Whether these mechanisms of resistance translate from *in vitro* models to the clinic in HER2+ breast cancer patients remain to be seen.

Interestingly, the aforementioned class of lysosomal-targeted payloads may be further attractive to develop given that some analogs do not appear to be substrates for drug efflux proteins. The classical models of drug resistance mediated by MDR1- or MRP1-induced drug efflux use HL60/Vinc and HL60/Adr cell lines [61]. Marquardt and Center revealed a regulatory role for vATPase in the inhibition of drug efflux in these cell models. Surprisingly, bafilomycin A, a vATPase inhibitor, appears not to be a substrate for MDR1 [61]. This sets the stage for proposing the appealing combination of vATPase inhibitors with the payload classes of current clinical ADCs.

17.5 SUMMARY

With the promise of targeting chemotherapeutic drugs specifically to tumors, ADCs effectively increase the therapeutic index of conventional chemotherapeutic compounds by increasing efficacy while decreasing toxicities in many cases. ADCs utilize TAAs, receptor internalization, and intracellular trafficking through the endolysosomal pathway to deliver their cytotoxic payloads to tumor cells while protecting nontumor cells from drug exposure. Using antibody targeting as a platform, we can reconsider compounds previously thought to be too toxic, such as the lysosomal inhibitors described herein, as exciting new strategies for ADC development and patient treatment regimens going forward.

REFERENCES

[1] Franklin MC, Carey KD, Vajdos FF, Leahy DJ, de Vos AM, et al. Insights into ErbB signaling from the structure of the ErbB2-pertuzumab complex. Cancer Cell 2004;5: 317–328.

[2] Modjtahedi H, Komurasaki T, Toyoda H, Dean C. Anti-EGFR monoclonal antibodies which act as EGF, TGF alpha, HB-EGF and BTC antagonists block the binding of epiregulin to EGFR-expressing tumours. Int J Cancer 1998;75:310–316.

[3] Teramoto T, Onda M, Tokunaga A, Asano G. Inhibitory effect of anti-epidermal growth factor receptor antibody on a human gastric cancer. Cancer 1996;77:1639–1645.

[4] Le Roy C, Wrana JL. Clathrin- and non-clathrin-mediated endocytic regulation of cell signalling. Nat Rev Mol Cell Biol 2005;6:112–126.

[5] Mayor S, Pagano RE. Pathways of clathrin-independent endocytosis. Nat Rev Mol Cell Biol 2007;8:603–612.

[6] Honing S, Ricotta D, Krauss M, Spate K, Spolaore B, et al. Phosphatidylinositol-(4,5)-bisphosphate regulates sorting signal recognition by the clathrin-associated adaptor complex AP2. Mol Cell 2005;18:519–531.

[7] Bonifacino JS, Traub LM. Signals for sorting of transmembrane proteins to endosomes and lysosomes. Annu Rev Biochem 2003;72:395–447.

[8] Trowbridge IS, Collawn JF, Hopkins CR. Signal-dependent membrane protein trafficking in the endocytic pathway. Annu Rev Cell Biol 1993;9:129–161.

[9] Stenmark H. Rab GTPases as coordinators of vesicle traffic. Nat Rev Mol Cell Biol 2009;10:513–525.

[10] Owen DJ, Evans PR. A structural explanation for the recognition of tyrosine-based endocytotic signals. Science 1998;282:1327–1332.

[11] Conner SD, Schmid SL. Identification of an adaptor-associated kinase, AAK1, as a regulator of clathrin-mediated endocytosis. J Cell Biol 2002;156:921–929.

[12] Fingerhut A, von Figura K, Honing S. Binding of AP2 to sorting signals is modulated by AP2 phosphorylation. J Biol Chem 2001;276:5476–5482.

[13] Semerdjieva S, Shortt B, Maxwell E, Singh S, Fonarev P, et al. Coordinated regulation of AP2 uncoating from clathrin-coated vesicles by rab5 and hRME-6. J Cell Biol 2008;183:499–511.

[14] McMahon HT, Boucrot E. Molecular mechanism and physiological functions of clathrin-mediated endocytosis. Nat Rev Mol Cell Biol 2011;12:517–533.

[15] Ferguson SM, Raimondi A, Paradise S, Shen H, Mesaki K, et al. Coordinated actions of actin and BAR proteins upstream of dynamin at endocytic clathrin-coated pits. Dev Cell 2009;17:811–822.

[16] Sundborger A, Soderblom C, Vorontsova O, Evergren E, Hinshaw JE, et al. An endophilin-dynamin complex promotes budding of clathrin-coated vesicles during synaptic vesicle recycling. J Cell Sci 2011;124:133–143.

[17] Wigge P, Kohler K, Vallis Y, Doyle CA, Owen D, et al. Amphiphysin heterodimers: potential role in clathrin-mediated endocytosis. Mol Biol Cell 1997;8:2003–2015.

[18] Helenius A, Marsh M. Endocytosis of enveloped animal viruses. Ciba Found Symp 1982; 92:59–76.

[19] van Dam EM, Stoorvogel W. Dynamin-dependent transferrin receptor recycling by endosome-derived clathrin-coated vesicles. Mol Biol Cell 2002;13:169–182.

[20] Rubin I, Yarden Y. The basic biology of HER2. Ann Oncol 2001;12(Suppl 1):S3–8.

[21] Sirisaengtaksin N, Gireud M, Yan Q, Kubota Y, Meza D, et al. UBE4B protein couples ubiquitination and sorting machineries to enable epidermal growth factor receptor (EGFR) degradation. J Biol Chem 2014;289:3026–3039.

[22] Arancibia-Carcamo IL, Yuen EY, Muir J, Lumb MJ, Michels G, et al. Ubiquitin-dependent lysosomal targeting of GABA(A) receptors regulates neuronal inhibition. Proc Natl Acad Sci U S A 2009;106:17552–17557.

[23] Schmidt O, Teis D. The ESCRT machinery. Curr Biol 2012;22:R116–120.

[24] Raiborg C, Stenmark H. The ESCRT machinery in endosomal sorting of ubiquitylated membrane proteins. Nature 2009;458:445–452.

[25] Henne WM, Buchkovich NJ, Emr SD. The ESCRT pathway. Dev Cell 2011;21:77–91.

[26] Beck A, Haeuw JF, Wurch T, Goetsch L, Bailly C, et al. The next generation of antibody–drug conjugates comes of age. Discov Med 2010;10:329–339.

[27] DiJoseph JF, Armellino DC, Boghaert ER, Khandke K, Dougher MM, et al. Antibody-targeted chemotherapy with CMC-544: a CD22-targeted immunoconjugate of calicheamicin for the treatment of B-lymphoid malignancies. Blood 2004;103:1807–1814.

[28] DiJoseph JF, Popplewell A, Tickle S, Ladyman H, Lawson A, et al. Antibody-targeted chemotherapy of B-cell lymphoma using calicheamicin conjugated to murine or humanized antibody against CD22. Cancer Immunol Immunother 2005;54:11–24.

[29] Chaparro-Riggers J, Liang H, DeVay RM, Bai L, Sutton JE, et al. Increasing serum half-life and extending cholesterol lowering *in vivo* by engineering antibody with pH-sensitive binding to PCSK9. J Biol Chem 2012;287:11090–11097.

[30] Igawa T, Ishii S, Tachibana T, Maeda A, Higuchi Y, et al. Antibody recycling by engineered pH-dependent antigen binding improves the duration of antigen neutralization. Nat Biotechnol 2010;28:1203–1207.

[31] Devanaboyina SC, Lynch SM, Ober RJ, Ram S, Kim D, et al. The effect of pH dependence of antibody–antigen interactions on subcellular trafficking dynamics. MAbs 2013;5:851–859.

[32] Ducry L, Stump B. Antibody–drug conjugates: linking cytotoxic payloads to monoclonal antibodies. Bioconjug Chem 2010;21:5–13.

[33] Doronina SO, Toki BE, Torgov MY, Mendelsohn BA, Cerveny CG, et al. Development of potent monoclonal antibody auristatin conjugates for cancer therapy. Nat Biotechnol 2003;21:778–784.

[34] Lewis Phillips GD, Li G, Dugger DL, Crocker LM, Parsons KL, et al. Targeting HER2-positive breast cancer with trastuzumab-DM1, an antibody–cytotoxic drug conjugate. Cancer Res 2008;68:9280–9290.

[35] Tanaka M, Kano Y, Akutsu M, Tsunoda S, Izumi T, et al. The cytotoxic effects of gemtuzumab ozogamicin (mylotarg) in combination with conventional antileukemic agents by isobologram analysis *in vitro*. Anticancer Res 2009;29:4589–4596.

[36] Austin CD, Wen X, Gazzard L, Nelson C, Scheller RH, et al. Oxidizing potential of endosomes and lysosomes limits intracellular cleavage of disulfide-based antibody–drug conjugates. Proc Natl Acad Sci U S A 2005;102:17987–17992.

[37] Erickson HK, Park PU, Widdison WC, Kovtun YV, Garrett LM, et al. Antibody–maytansinoid conjugates are activated in targeted cancer cells by lysosomal degradation and linker-dependent intracellular processing. Cancer Res 2006;66:4426–4433.

[38] Dorywalska M, Strop P, Melton-Witt JA, Hasa-Moreno A, Farias SE, et al. Effect of attachment site on stability of cleavable antibody drug conjugates. Bioconjug Chem 2015; 26:650–659.

[39] Gerber HP. Emerging immunotherapies targeting CD30 in Hodgkin's lymphoma. Biochem Pharmacol 2010;79:1544–1552.

[40] Kovtun YV, Audette CA, Ye Y, Xie H, Ruberti MF, et al. Antibody–drug conjugates designed to eradicate tumors with homogeneous and heterogeneous expression of the target antigen. Cancer Res 2006;66:3214–3221.

[41] Friedman PN, Chace DF, Trail PA, Siegall CB. Antitumor activity of the single-chain immunotoxin BR96 sFv-PE40 against established breast and lung tumor xenografts. J Immunol 1993;150:3054–3061.

[42] Ghose T, Ferrone S, Blair AH, Kralovec Y, Temponi M, et al. Regression of human melanoma xenografts in nude mice injected with methotrexate linked to monoclonal antibody 225.28 to human high molecular weight-melanoma associated antigen. Cancer Immunol Immunother 1991;34:90–96.

[43] Schrappe M, Bumol TF, Apelgren LD, Briggs SL, Koppel GA, et al. Long-term growth suppression of human glioma xenografts by chemoimmunoconjugates of 4-desacetylvinblastine-3-carboxyhydrazide and monoclonal antibody 9.2.27. Cancer Res 1992;52:3838–3844.

[44] Smyth MJ, Pietersz GA, McKenzie IF. The *in vitro* and *in vivo* anti-tumour activity of N-AcMEL-(Fab')2 conjugates. Br J Cancer 1987;55:7–11.

[45] Trail PA, King HD, Dubowchik GM. Monoclonal antibody drug immunoconjugates for targeted treatment of cancer. Cancer Immunol Immunother 2003;52:328–337.

[46] Yang HM, Reisfeld RA. Pharmacokinetics and mechanism of action of a doxorubicin-monoclonal antibody 9.2.27 conjugate directed to a human melanoma proteoglycan. J Natl Cancer Inst 1988;80:1154–1159.

[47] Raghunand N, Gillies RJ. pH and drug resistance in tumors. Drug Resist Updat 2000; 3:39–47.

[48] Gotink KJ, Broxterman HJ, Labots M, de Haas RR, Dekker H, et al. Lysosomal sequestration of sunitinib: a novel mechanism of drug resistance. Clin Cancer Res 2011; 17:7337–7346.

[49] Yamagishi T, Sahni S, Sharp DM, Arvind A, Jansson PJ, et al. P-glycoprotein mediates drug resistance via a novel mechanism involving lysosomal sequestration. J Biol Chem 2013;288:31761–31771.

[50] Gerber HP, Koehn FE, Abraham RT. The antibody–drug conjugate: an enabling modality for natural product-based cancer therapeutics. Nat Prod Rep 2013;30:625–639.

[51] Drose S, Altendorf K. Bafilomycins and concanamycins as inhibitors of V-ATPases and P-ATPases. J Exp Biol 1997;200:1–8.

[52] Forgac M. Vacuolar ATPases: rotary proton pumps in physiology and pathophysiology. Nat Rev Mol Cell Biol 2007;8:917–929.

[53] Xie XS, Padron D, Liao X, Wang J, Roth MG, et al. Salicylihalamide A inhibits the V0 sector of the V-ATPase through a mechanism distinct from bafilomycin A1. J Biol Chem 2004;279:19755–19763.

[54] Yoshimori T, Yamamoto A, Moriyama Y, Futai M, Tashiro Y. Bafilomycin A1, a specific inhibitor of vacuolar-type H(+)-ATPase, inhibits acidification and protein degradation in lysosomes of cultured cells. J Biol Chem 1991;266:17707–17712.

[55] Walter RB, Gooley TA, van der Velden VH, Loken MR, van Dongen JJ, et al. CD33 expression and P-glycoprotein-mediated drug efflux inversely correlate and predict clinical outcome in patients with acute myeloid leukemia treated with gemtuzumab ozogamicin monotherapy. Blood 2007;109:4168–4170.

[56] Linenberger ML, Hong T, Flowers D, Sievers EL, Gooley TA, et al. Multidrug-resistance phenotype and clinical responses to gemtuzumab ozogamicin. Blood 2001;98:988–994.

[57] Cianfriglia M. The biology of MDR1-P-glycoprotein (MDR1-Pgp) in designing functional antibody drug conjugates (ADCs): the experience of gemtuzumab ozogamicin. Ann Ist Super Sanita 2013;49:150–168.

[58] Kung Sutherland MS, Walter RB, Jeffrey SC, Burke PJ, Yu C, et al. SGN-CD33A: a novel CD33-targeting antibody–drug conjugate using a pyrrolobenzodiazepine dimer is active in models of drug-resistant AML. Blood 2013;122:1455–1463.

[59] Phillips GD, Fields CT, Li G, Dowbenko D, Schaefer G, et al. Dual targeting of HER2-positive cancer with trastuzumab emtansine and pertuzumab: critical role for neuregulin blockade in antitumor response to combination therapy. Clin Cancer Res 2014;20: 456–468.

[60] Loganzo F, Tan X, Sung M, Jin G, Myers JS, et al. Tumor cells chronically treated with a trastuzumab-maytansinoid antibody–drug conjugate develop varied resistance mechanisms but respond to alternate treatments. Mol Cancer Ther 2015;14(4):952–963.

[61] Marquardt D, Center MS. Involvement of vacuolar H(+)-adenosine triphosphatase activity in multidrug resistance in HL60 cells. J Natl Cancer Inst 1991;83:1098–1102.

18

THE MECHANISMS AND THERAPEUTIC CONSEQUENCES OF AMINE-CONTAINING DRUG SEQUESTRATION IN LYSOSOMES

NADIA HAMID AND JEFFREY P. KRISE

Department of Pharmaceutical Chemistry, The University of Kansas, Lawrence, KS, USA

18.1 INTRODUCTION

It is well known that certain low-molecular-weight, weakly basic drugs become highly concentrated in acidic intracellular organelles including lysosomes. Substrates for this accumulation are routinely referred to as being lysosomotropic, and the mechanism for accumulation is known as ion trapping or pH partitioning. Lysosomotropic drugs have been shown to reach concentrations within lysosomes that are more than 1000 times greater than in the cytosol or extracellular space at steady state. This chapter reviews the mechanism by which lysosomal trapping occurs and the drug- and cell-related factors that influence the process. The chapter also examines some of the techniques used to assess lysosomal trapping in cultured cells. Subsequent discussions focus on the therapeutic implications associated with lysosomal drug sequestration. As discussed, the lysosomal sequestration of drugs can impact the extent of interaction with intracellular targets. Toward this end, lysosomotropic behavior of drugs has been purposefully exploited in an effort to increase drug potency and selectivity. Moreover, the sequestration of drugs in lysosomes can have a profound impact on the pharmacokinetic properties of drugs. Specifically, lysosomotropic drugs often have large volumes of distribution and, therefore, long elimination half-lives.

Lysosomes: Biology, Diseases, and Therapeutics, First Edition.
Edited by Frederick R. Maxfield, James M. Willard and Shuyan Lu.

Finally, the chapter reviews instances where it has been shown that one drug is capable of influencing the lysosomal accumulation of a secondarily administered drug, establishing the basis for a new type of pharmacokinetic drug–drug interaction.

18.2 LYSOSOMAL TRAPPING OVERVIEW

For over a century, light microscopy studies have revealed that certain low-molecular-weight dyes known as vital stains are able to selectively stain cellular compartments of living cells. Vital stains that were thought to specifically stain lysosomes were first described in the literature in the mid-19th century and originally included weakly basic molecules such as neutral red and acridine orange (see Figure 18.1) [1,2].

In the mid-1970s, de Duve [3], a Nobel laureate, credited with the initial characterization of lysosomes, and his colleagues wrote an elegant theoretical commentary on the mechanism for weakly basic drug accumulation in lysosomes. In their commentary, they referred to the substrates for lysosomal accumulation as lysosomotropic agents. The authors argued that in order for ion trapping in lysosomes to occur, a weakly basic drug must be relatively membrane permeable when it is unionized and relatively membrane impermeable when ionized. Under these conditions, the driving force for lysosomal sequestration is based on the preferential ionization of weakly basic drug that occurs inside the acidic lysosomes due to the low lysosomal pH (see Figure 18.2). This relatively low pH (approximately pH 4.5) is maintained by the vacuolar proton ATPase [4]. The activity of vacuolar proton ATPase is therefore central to the ion trapping phenomenon, and inhibitors of this enzyme have been shown to abolish lysosomal ion trapping of drugs [5].

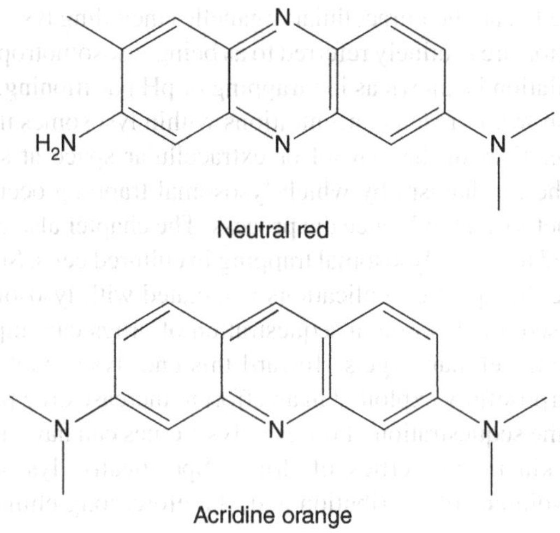

Figure 18.1 Structures of neutral red and acridine orange, which are early examples of lysosomal vital stains.

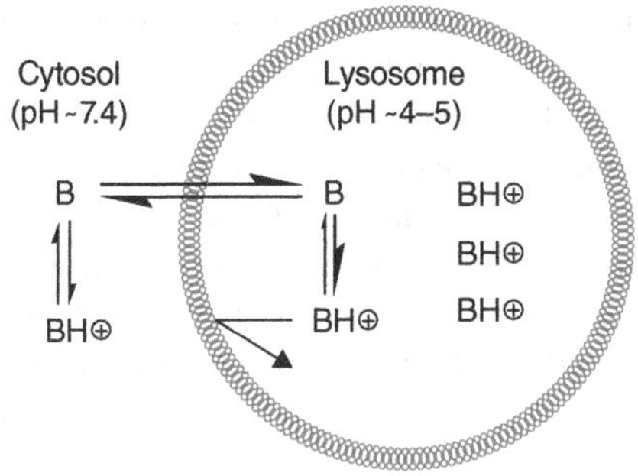

$$\frac{[\text{lysosome}]}{[\text{extracellular}]} = \frac{(\alpha[\text{H}^+]_\text{E} + K_\text{a})}{([\text{H}^+]_\text{E} + K_\text{a})} \; \frac{([\text{H}^+]_\text{L} + K_\text{a})}{(\alpha[\text{H}^+]_\text{L} + K_\text{a})}$$

Figure 18.2 Diagram illustrating the pH partitioning-based mechanism for accumulation of weakly basic drugs (B) in the acidic lysosomes. The equation at the bottom of the figure represents the lysosome to extracellular space steady-state concentration ratio for a base. The dissociation constant for the conjugate acid of the weak base is denoted as K_a and $[H^+]$ is the proton concentration (subscript E represents extracellular and L represents lysosomal). The ratio of permeabilities of the ionized base to that of the unionized base in the lysosomal lipid bilayer is denoted by the α term.

In their commentary, de Duve and his colleagues derive an equation relating the steady-state concentration ratio of a weakly basic drug inside lysosomes compared to the extracellular space (see Figure 18.2). According to this equation, the maximum possible concentration ratio for a weakly basic substance in lysosomes (relative to the extracellular space) is equal to the ratio of the hydrogen ion concentration in lysosomes relative to the fluid surrounding the cells. For example, if the extracellular pH was 7.4 and the pH of lysosomes was 4.4, the maximal lysosomal concentration ratio would be 1000. According to de Duve's derivations, there are two drug-related terms that can theoretically influence the steady-state accumulation of weakly basic drugs in lysosomes. The first variable is the weak base pK_a, and the second is termed alpha (α), which represents the ratio of permeabilities of the weakly basic drug across the lysosomal lipid bilayer in its ionized and unionized forms.

The pK_a of the weakly basic drug is predicted to impact the steady-state lysosomal accumulation ratio. Molecules with pK_a values of 6 or below were not considered to be lysosomotropic because they would not extensively ionize in lysosomes. Low pK_a molecules can never reach the maximal steady-state accumulation ratio of 1000. Molecules with pK_a values of 8 or greater can theoretically achieve 1000-fold higher levels in lysosomes relative to the extracellular space. However, molecules with very high pK_a values will start to become less lysosomotropic in a therapeutically relevant timescale because they are predicted to take exceedingly long times to reach steady state. Specifically, the permeation rate was predicted to decrease by a factor of 10 for each 1-unit increase in drug pK_a. Interestingly, de Duve's theoretical calculations predict that the time to reach one-half of steady state for molecules with a pK_a of 12 would be over 1 1/2 years. In contrast, bases with pK_a values near 7 are predicted to take 10 min to reach one-half steady-state levels.

Duvvuri et al. [6] later quantitatively tested the theoretical predictions of de Duve and colleagues in cultured cells. Using a series of weakly basic structural isomers that varied in pK_a from 4 to 9, the authors experimentally investigated the influence of pK_a on lysosomal accumulation. Consistent with theoretical predictions, the authors reported that molecules with pK_a values below 6 were not appreciably lysosomotropic. The molecule with a pK_a value of 9 had the greatest degree of lysosomal accumulation.

As a consequence of extensive entrapment of drugs, the lysosomes can take on a vacuolated appearance. Using vacuolization as a readout for lysosomotropic behavior, early work by Ohkuma and Poole [7] investigated how weak base structure influences lysosomal accumulation. Collectively, with few exceptions, both theoretical and experimental approaches support the notion that most lysosomotropic agents will have optimal pK_a values between 7 and 10.

As previously mentioned, the term alpha (α) refers to the ratio of the permeability coefficients across the lysosomal lipid bilayer for the ionized base divided by that of the unionized base [3]. This term can theoretically vary from zero to one. Molecules with an alpha value of zero, meaning the ionized base is completely impermeable, will be the most lysosomotropic and will reach a theoretical maximal steady-state ratio of accumulation dictated by the pH differential. However, as the alpha parameter increases in magnitude, the maximal steady-state accumulation ratio of a drug in the lysosomes significantly decreases. For example, a drug with an alpha value equal to 1 will have equal membrane permeability regardless of ionization. Under these circumstances, extensive protonation of the drug in lysosomes would not slow the rate of diffusion back to the cytosol. A theoretical relationship illustrating the combined influence of pK_a and alpha on lysosomal trapping is shown in Figure 18.3.

Duvvuri et al. [8] have experimentally estimated alpha values for a series of drugs and model compounds and correlated this value with the experimentally measured lysosome-to-cytosol concentration ratio obtained in cell culture experiments. As a proxy for lysosomal membrane permeability, which is difficult to measure, the authors measured the octanol/water partition coefficients for drugs as a function of pH. Estimates for the log D of the ionized and unionized species were obtained, and the ratio of these values was used to obtain alpha. As anticipated, it was found that compounds with lower alpha values accumulated to a greater extent in lysosomes relative to compounds with higher alpha values. The authors also examined how alpha correlated with mitochondrial versus lysosomal accumulation of weakly basic drugs. They found that molecules with low alpha values accumulated almost exclusively in lysosomes, whereas molecules with high alpha values accumulated within mitochondria. It was noted that the alpha parameter for a given drug appeared to correlate with the degree of charge delocalization present in the molecule. Weak bases with fixed localized positive charges tended to have very low alpha values, whereas those with delocalized charges tended to have higher alpha values near 1. The relationship between charge delocalization and alpha was attributed to

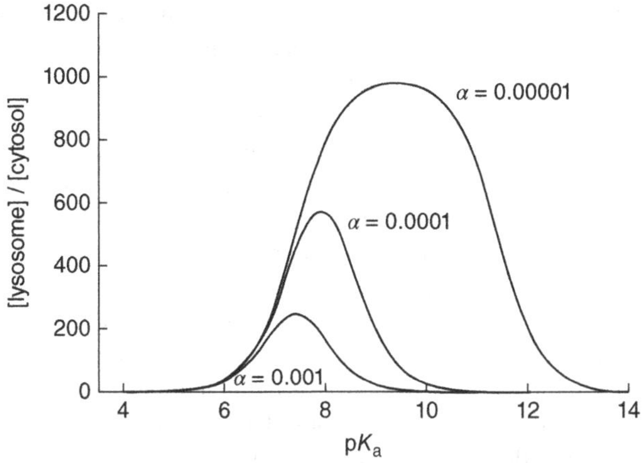

Figure 18.3 Theoretical relationship between weak base pK_a and the alpha permeability parameter. The relationships were derived using the equation shown in Figure 18.2, with indicated values for alpha.

differences in how tightly the molecules bind water, which subsequently relates to the energy required to pass into a nonpolar lipid-like environment.

The relative lipophilicity (i.e., log P) does not influence the predicted steady-state accumulation ratios for weakly basic drugs. However, it will most definitely influence the time to reach steady state and the propensity to exhibit lysosomal trapping in a therapeutically relevant timescale. As a consequence, most experimentally observed lysosomotropic substrates tend to be relatively lipophilic and have log P values greater than 1 [9].

18.3 TECHNIQUES TO ASSESS LYSOSOMAL TRAPPING

For therapeutic reasons that are discussed below, there is a keen interest in understanding how drugs localize and distribute within the cells and tissues of our body. Toward this end, a number of assays and approaches have been described over the years. These approaches have been more extensively reviewed elsewhere [10,11], and the reader should refer to these examples for more details. The purpose of this section is to give the reader a brief overview of the different techniques used and some of the associated strengths and weaknesses.

By far, the simplest approach is to utilize fluorescence microscopy on cultured cells that have been exposed to a drug that is intrinsically fluorescent. The anticancer drug daunorubicin is highly fluorescent, and, as a consequence, numerous publications have examined the intracellular distribution of this drug in cultured cells [12–14]. Unfortunately, most drugs are not sufficiently fluorescent to allow for this technique. However, extensive structure–localization relationship studies with regard to lysosomal accumulation have been conducted using weakly basic fluorescent compounds that are not drugs *per se*. For example, Horobin and colleagues [15,16] have compiled correlations between intracellular distribution and physical and structural properties of a variety of fluorophores. Similarly, Rosania and coworkers [17,18] have employed results from such studies to develop models and simulations that describe lysosomal accumulation of weak bases.

Microscopic detection methods that do not rely on fluorescence have also been investigated. Recently, Fu et al. [19] described the use of hyperspectral stimulated Raman scattering microscopy to examine the intracellular localization and concentrations of tyrosine kinase inhibitors and chloroquine in cultured cells. The authors were able to show that the weakly basic drug imatinib accumulated in lysosomes at concentrations that were approximately 1000-fold higher than extracellular levels. The authors provided evidence that the drugs accumulated in lysosomes to an extent greater than ion trapping theory would have predicted and that this was, for some drugs, due to precipitation within lysosomes. The authors also used this method to show that chloroquine administration decreases lysosomal accumulation of imatinib and that this could provide the basis for the positive therapeutic advantages of taking two drugs simultaneously. Although the current limit of detection for drugs in cells described in this manuscript was quite high (approximately 1–2 mM), it was commented that improvements in the instrumentation and techniques such as deuterium labeling of drugs can be expected to increase sensitivity.

It is also possible to isolate lysosomes from cultured cells or tissues in an effort to quantify lysosome and drug association. The purification can either occur before or after the cells have been exposed to drug. In the former scenario, the test drug is added to the isolated lysosomes, and the degree of accumulation is assessed either directly or indirectly. For example, Ishizaki et al. [20] have used highly purified rat liver lysosomes to study the mechanism for imipramine accumulation and have also studied the ability of other basic molecules to compete with imipramine uptake.

Andrew et al. [21] developed an indirect approach to assess the permeability of weakly basic drugs across the lysosomal lipid bilayer using an unpurified cell lysate. The assay was based on the measured ability of a test drug to protect lysosomes from osmotic pressure-induced rupturing, which is directly related to the propensity for a test drug to permeate across the lipid bilayer of the isolated lysosomes. The readout for lysosomal rupturing involved testing for release of free lysosomal enzymes into the cell lysate. The advantages of this approach are that lysosomes need not be purified, and the test drug needs no special properties or tags. Accordingly, this approach is amenable to relatively high-throughput analysis, and the authors have used this method to demonstrate a strong correlation between hydrogen bonding capacity and lysosomal membrane permeability.

A unique analytical approach for evaluating the accumulation of fluorescent compounds into acidic organelles, such as lysosomes, was described by Arriaga and coworkers [22]. The authors utilized capillary electrophoresis with laser-induced fluorescence with dual-channel detection to identify lysosomes containing fluorescent nanospheres that were preloaded into lysosomes using an endocytic uptake-based pulse-chase technique. Selecting organelles possessing both nanospheres and doxorubicin fluorescence allowed them to specifically determine the content of the drug in lysosomes without potential interference from other organelles in which the drug could associate. The authors were able to identify heterogeneity among lysosomes in their doxorubicin content, an observation not readily achievable using alternative techniques.

Duvvuri et al. [23] describe an approach whereby living cells are incubated, in a pulse-chase manner, with magnetic dextran particles to allow them to be specifically localized in lysosomes. The cells were subsequently incubated with a test drug and then lysed so that the lysosomes could be isolated using magnetic chromatography. The drug associated with the isolated lysosomes was then quantified using conventional strategies including high-performance liquid chromatography-mass spectrometry/mass spectrometry (HPLC-MS/MS). Similar to other studies, the authors show that model drugs including quinacrine accumulated in lysosomes to levels much greater than simple ion trapping theory would predict.

Lemieux et al. [9] have developed an approach for evaluating lysosomal sequestration that is amenable to high-throughput screening formats. The main principle of the method involves a competition between the test drug and the fluorescent basic amine probe LysoTracker Red (LTR) in cultured cells. The authors show that drugs that are cationic and amphiphilic decrease LTR fluorescence in a dose-dependent manner, presumably by increasing lysosomal pH. The authors use this assay to establish that lysosomotropism occurs more readily for drugs with a high $\log P$. This general assay format was further examined by Nadanaciva et al. [24], and they were able to establish that lysosomotropic drugs identified in the assay generally had $C\log P > 2$ and a basic pK_a between 6.5 and 11. This same assay format was more recently applied to an immortalized hepatocyte cell line [25].

18.4 INFLUENCE OF LYSOSOMOTROPISM ON DRUG ACTIVITY

Mammalian cells are highly compartmentalized, and drug targets are typically localized in discrete intracellular organelles as opposed to being evenly distributed across all intracellular compartments. How a drug distributes within a human cell can have a profound impact on its activity and side effect profile. Knowledge regarding the total amount of drug associated with a cell cannot by itself allow one to accurately predict the likelihood of experiencing a therapeutic effect. Instead, the most relevant piece of information to know would be the free concentration of the drug that exists in the local microenvironment that immediately surrounds the intended drug target.

Hypothetically speaking, if a drug distributed evenly across the entire cell and all of its compartments, one would anticipate activity at some dose, but the propensity for off-target effects would also be high. This is due to the drug having a high degree of interaction with unintended cellular components. However, if a scenario is considered where the drug has the propensity to specifically concentrate within the organelle containing the target, one would anticipate this causing an increase in the potency. In addition, the propensity for unintended off-target interactions would be minimized. The most unfavorable scenario would be realized if the drug were to exclusively concentrate in some discrete organelle that did not contain the target. In this scenario, the drug would have no activity, and off-target effects would be the only possible effects. Our understanding of these various scenarios is increasing and is spurring efforts in the design and development of drugs with known and optimized intracellular distribution in order to maximize potency and reduce side effects.

According to the previous discussion, lysosomotropic behavior of a drug can potentially increase or decrease therapeutic activity. In instances where the drug target is inside the lysosomes, the lysosomotropic behavior would be expected to increase potency; when the target is outside the lysosomes, the sequestration of drugs in lysosomes could decrease the availability of the drug to interact with its target and decrease activity.

There are examples in drug discovery where the lysosomotropic behavior of a drug was not recognized or exploited in the initial drug design, but later studies have revealed that the sequestration serendipitously contributed to the therapeutic activity. The most well-known example is the lysosomotropic antimalarial drug chloroquine (see Figure 18.4). This drug was originally synthesized in the 1930s and is recognized to have potent antimalarial properties [26]. It is now believed that the antimalarial effects are attributed to inhibition of hemozoin biocrystallization in the digestive vacuole of the malarial parasite that resides in the red blood cells. The digestive vacuole is lysosome-like in its function and the low pH facilitates drug accumulation in this space, thus maximizing the effectiveness of the drug. Other weakly basic antimalarials such as mefloquine and quinine (see Figure 18.4) are also thought to show increased therapeutic efficacy from the sequestration in the digestive vacuole but are not believed to be sequestered as extensively as chloroquine [27].

In the late 1970s, Firestone and colleagues, to our knowledge, were the first group to publish on the purposeful exploitation of lysosomotropic behavior in the design of low-molecular-weight drugs. In their work, the scientists synthesized and evaluated a series of long-chain alkyl amines as potential anticancer agents [28,29]. The authors selected amines with intermediate pK_a values in the range of 5–8, such as imidazole and morpholine, which had varied alkyl chain length (see Figure 18.5).

These agents were designed to accumulate in lysosomes by ion trapping and exist there predominantly in their ionized form. Upon reaching high concentrations,

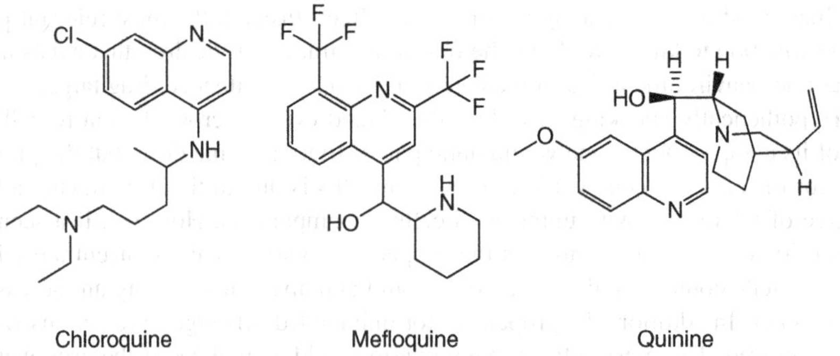

Figure 18.4 Structures of lysosomotropic antimalarial drugs chloroquine, mefloquine, and quinine.

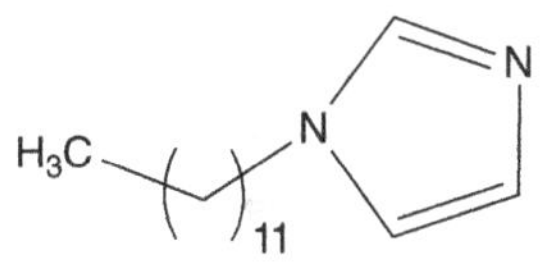

Figure 18.5 Example of lysosomotropic detergent investigated by Firestone and colleagues.

the agents would acquire surfactant properties that would lead to the selective permeabilization of the lysosomal membrane, which would precipitate the death of the cell. The authors provided several compelling pieces of evidence that the agents were indeed acting as lysosomotropic detergents. First, cell fractionation studies showed that the drugs were specifically accumulating in fractions rich in lysosomes. Cellular cytotoxicity studies revealed a sharp sigmoidal dose–response relationship, which is consistent with the requirement for the formation of detergent micelles. In addition, the drugs were toxic only to cells that had lysosomes and were not toxic to those that did not contain lysosomes (i.e., red blood cells). Finally, cell death was preceded by lysosomal vacuolization, led to an increase in lysosomal pH, and was associated with the release of lysosomal contents.

In a later work, Dubowchick and colleagues also rationally incorporated lysosomotropic features in a series of molecules that were designed to increase the luminal pH of endosomes and lysosomes. Such agents were hypothesized to have potential utility in the reversal of anticancer multidrug resistance that was mediated through increased lysosomal ion trapping of weakly basic anticancer drugs [30,31]. In addition, it was suggested that such agents could be effective against some viruses that exploit the low pH of lysosomes to facilitate entry in the cell cytosol to allow for subsequent replication [32]. The molecules were designed and optimized to be membrane active and capable of transporting protons across lipid bilayers resulting in the dissipation of the lysosomal pH gradient (see Figure 18.6 for a depiction of the strategy). The efficiency of proton transfer was optimized by designing hydrophobic nitrogenous bases that would have the tendency to remain embedded in the lysosomal lipid bilayer while mediating the proton transfer from the lumen of lysosomes to the cytosol. Toward this end, the authors identified a lipophilic imidazole-containing molecule that had the best ability to neutralize lysosomal pH in cells grown in culture and had physicochemical properties evaluated in *in vitro* tests that were consistent with the aforementioned mechanism of action in proton transfer (see Figure 18.6 for structure). This molecule also showed activity in reversing anticancer drug resistance and in the inhibition of influenza virus replication [33].

There are numerous examples whereby known lysosomotropic agents have been shown to interfere with endosomal/lysosomal trafficking and function. This is an important consideration from a potential toxicological point of view; however, this is the focus of a separate chapter and is not further discussed here. Interestingly, these otherwise potentially negative effects have been proposed to have potential relevance in the treatment of various diseases. For example, lysosomotropic agents have been shown to decrease the escape of viruses and other endocytosed membrane-impermeable toxins from late endosomes and lysosomes [34–38].

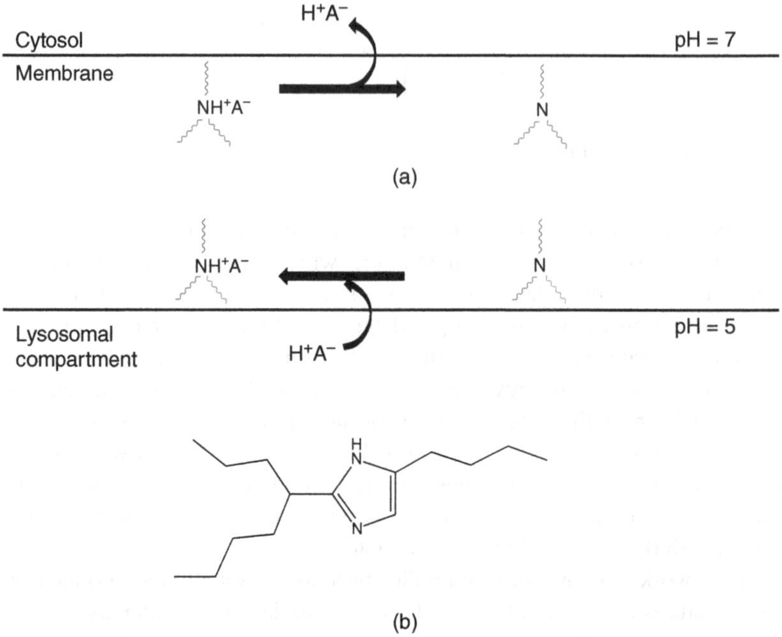

Figure 18.6 (a) Diagram of proposed lysosomal proton transfer mechanism. (b) Structure of lysosomotropic proton transfer reagent.

Similarly, there exists work that suggests that lysosomotropic amines can suppress major histocompatibility complex (MHC) class II antigen presentation [39] and decrease iron release from lysosomes [40]. Lysosomotropic amines have also been used to help decrease the exocytosis of anticancer drugs that are sequestered in lysosomes as a means to reverse the anticancer drug-resistance mechanism, as was previously described [41–44]. The mechanism by which lysosomotropic amines decrease lysosomal egress/function in the previous applications is not completely understood. One explanation is that the amines transiently increase the pH of lysosomes through a buffering effect. In this way, the drugs could decrease the activity of lysosomal hydrolases and this could lead to a buildup of undigested materials, which can interfere with lysosomal trafficking. Another possibility is that the lysosomotropic amines could directly modify lysosomal membranes such that fusion and fission events required for the vesicle-mediated trafficking could become impaired. Consistent with the latter notion, the weakly basic drug primaquine has been shown to impair budding events at the Golgi, which also has an acidic lumen relative to the cell cytosol [45].

It is well known that most anticancer agents can be classified as being weakly basic and therefore presumably lysosomotropic. However, the therapeutic advantage of lysosomal trapping, if any, in the treatment of cancer is not immediately obvious because most anticancer drug targets are not localized within the lysosomes, and trapping in this space would be anticipated to decrease interactions with the drug targets, as previously discussed. Recently, it has been proposed that the weakly basic nature of anticancer drugs contributes to the selective toxicity of the drugs against cancer cells. A number of studies have demonstrated that many cancer cells have defective acidification of lysosomes for reasons that have not been fully elucidated [46–49]. Based on this difference in lysosomal pH between normal and many cancer cells, it was hypothesized that lysosomotropic behavior of a cancer drug with extralysosomal targets could enhance the selective toxicity of the drug toward a cancer cell compared to a normal cell with normally acidified lysosomes. In other words, weakly basic anticancer agents that are exquisite substrates for lysosomal trapping would achieve very low concentrations in the cell cytosol and nucleus of normal cells and thus have reduced capacity to interact with targets and exert cytotoxic effects. Alternatively, when the lysosomotropic cancer drug is present in cancer cells with elevated lysosomal pH, the lysosomotropic drug would be relatively less concentrated in lysosomes and would exist to a greater degree in the cell cytosol and nucleus. According to this proposed mechanism, the lysosomotropic anticancer agent would possess increased selectivity against tumor cells compared to nonlysosomotropic anticancer agents (see Figure 18.7 for an illustration of this concept).

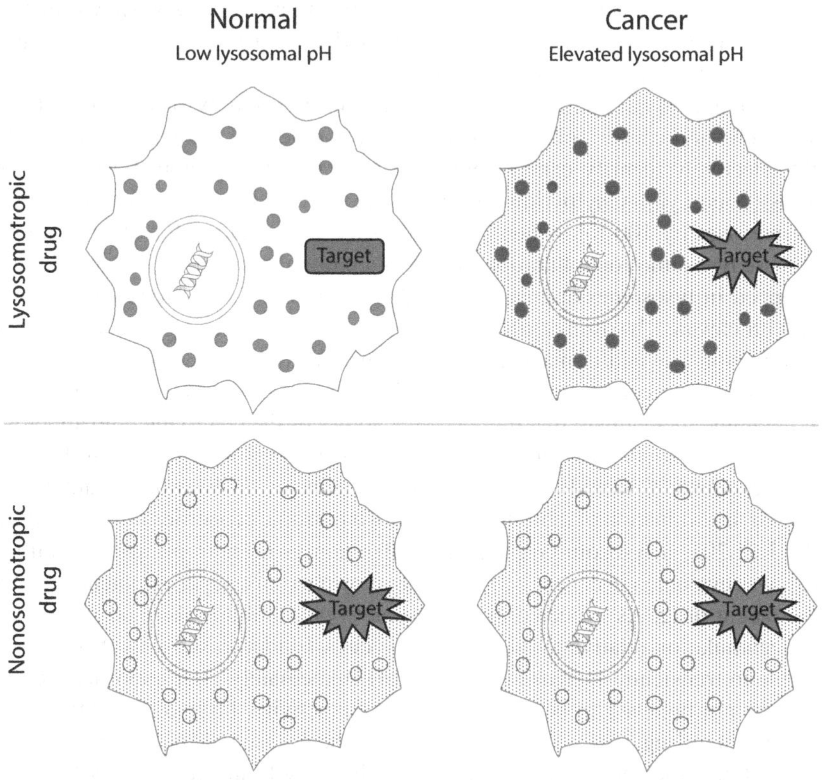

Figure 18.7 Overview of intracellular distribution-based anticancer drug targeting platform. (*See color plate section for the color representation of this figure.*)

The initial evaluations of these hypotheses were conducted in cultured cells that had variable lysosomal pH and with model anticancer agents that were derivatives of geldanamycin, a prototypic inhibitor of the molecular chaperone heat shock protein 90 (Hsp90). Geldanamycin is amenable to diverse chemical derivatization at the 17-position with little effect on binding to Hsp90 [50]. It was therefore possible to synthesize inhibitors that had variations at the 17-position that made them either weakly basic (i.e., lysosomotropic) or neutral (nonlysosomotropic). The selectivity of the drugs was measured by taking the ratio of the IC_{50} cytotoxicity value for a given drug in cells with low (normal) pH divided by the corresponding value in cells with elevated lysosomal pH (i.e., cancer cells). These studies reveal that weakly basic inhibitors had additional selective toxicity toward cells with elevated lysosomal pH, a characteristic that was not the case for nonlysosomotropic inhibitors [51]. Further investigation of this hypothesis was completed in non-tumor-bearing mice with and without elevated lysosomal pH, again using lysosomotropic and nonlysosomotropic Hsp90 inhibitors [52]. In this study, it was found that mice with elevated lysosomal pH were far more susceptible to the toxic effects of lysosomotropic Hsp90 inhibitors. As a control, nonlysosomotropic inhibitors had toxicities that did not vary with alterations in lysosomal pH. More recently, a more extensive series of Hsp90 derivatives were synthesized with variable pK_a in the 17-position, and it was shown that the degree of selectivity toward tumor cells increased as the pK_a was changed to values that were most conducive to lysosomal trapping [53].

Kang et al. [54] have similarly exploited this difference in lysosomal pH between normal cells and cancer cells to aid in the development of fluorescent probes that could selectively identify cancer cells. The authors have developed a fluorophore (3,6-bis-(1-methyl-4-vinylpyridinium)carbazole diiodide (BMVC)), whose quantum yield increases by 100-fold when bound to DNA. The authors show that in normal cells these lysosomotropic fluorophores are not particularly fluorescent because they are trapped in lysosomes and are unavailable to bind to DNA. However, in cancer cells with elevated lysosomal pH, the fluorophores escape the lysosomes and bind with mitochondrial and/or nuclear DNA and become highly fluorescent.

Prostate cancer cells, similar to many other cancer cell types, have alterations in their ceramide metabolism that influence the ability of the cell to undergo apoptosis. This has led investigators to target this pathway in the development of a novel class of anticancer agents. Ceramide metabolism occurs in the acidic lysosomes, and this has prompted the evaluation of inhibitors with lysosomotropic properties [55–57]. Specifically, LCL204, a lysosomotropic analog of a nonlysosomotropic parent inhibitor (B13), was evaluated in cancer cells and was shown to have superior potency relative to the parent drug. As expected, the lysosomotropic inhibitor also had greater cellular accumulation relative to the parent molecule. Evidence suggests that LCL204 was able to transiently increase lysosomal pH immediately after exposure, but this alone was not believed to be responsible for the activity. Instead, the authors proposed that the drug also needed to destabilize the lysosomal membrane such that lysosomal enzymes were released into the cell cytosol, an event that is known to induce apoptosis.

Lysosomotropic properties have also been optimized in the development of drugs used to treat osteoporosis [58]. The lysosomal protease cathepsin K is involved in the process of bone reabsorption, and efforts are underway to develop potent and specific inhibitors against this enzyme. Falgueyret and colleagues have synthesized and evaluated inhibitors of cathepsin K with varying degrees of lysosomotropic properties [58,59]. Interestingly, molecules with lysosomotropic properties had far greater activity in whole cell enzyme occupancy assays compared to activity evaluations conducted with isolated enzymes. Fortuitously, these compounds were inherently fluorescent, and their cellular distribution could be monitored using two-photon confocal fluorescence microscopy. Microscopic evaluations demonstrated that compounds localized in lysosomes, as was confirmed with colocalization studies with the lysosomal vital stain LTR. The lysosomotropic inhibitors of cathepsin K also had much improved pharmacokinetic properties in rats (see the following section for more details). The lysosomotropic inhibitors had high tissue-to-plasma concentration ratios in organs rich in lysosomes including liver, lungs, and kidneys. In addition, the half-life of lysosomotropic inhibitors was sevenfold longer than that of nonlysosomotropic inhibitors. One noteworthy obstacle with this approach was that basic cathepsin K inhibitors were also concentrated in the lysosomes of cells other than osteoclasts, and lysosomotropism tended to enhance their potential for off-target interactions with other members of the cathepsin family in these cells (i.e., cathepsin B, S, and L) [60].

The same principle in lysosomal targeting of cathepsin K inhibitors was additionally evaluated in a structure called the hemivacuole. The osteoclast resorbs bone by forming a ring in close contact with the bone surface within the confines of which it secretes protons and lysosomal enzymes, thus forming an extracellular digestive hemivacuole. In this hemivacuole, the mineral component of bone is dissolved by protons, and the organic component is digested by cathepsin K. Interestingly, lysosomotropism was shown to occur in the hemivacuole surrounding the osteoclast [61]. This space was shown to be capable of trapping lysosomotropic drugs, thereby increasing their potency and prolonging their duration of action relative to nonlysosomotropic inhibitors that had equal binding affinity to cathepsin K.

18.5 INFLUENCE OF LYSOSOMAL TRAPPING ON PHARMACOKINETICS

Shortly after lysosomotropic behavior of drugs grew to be appreciated in the literature, it became apparent that it could provide the driving force for large tissue accumulation and, subsequently, a large volume of distribution and long half-life [62,63]. In other words, lysosomotropic drugs had very large tissue-to-plasma concentration ratios, and this was particularly the case in tissues that had a high abundance of lysosomes such as the liver, lungs, and kidneys.

In an attempt to quantitatively describe the contribution of lysosomal trapping on volume of distribution, a hypothetical scenario will be employed (see Figure 18.8).

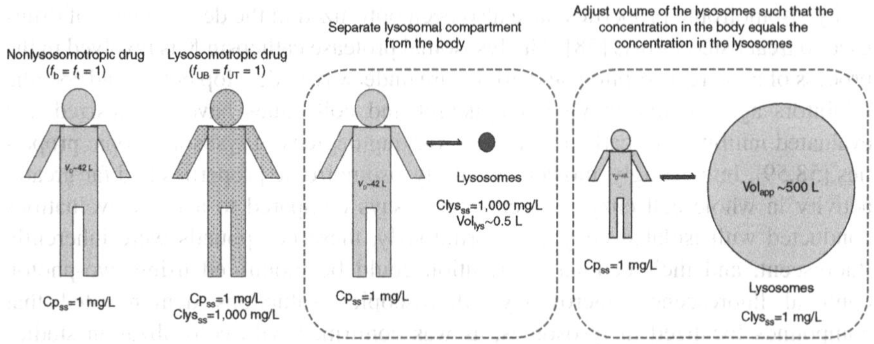

Figure 18.8 Hypothetical example illustrating the potential influence of lysosomotropism on the apparent volume of distribution of a drug. See text for details.

In this example, the volume of distribution will be comparatively approximated for a nonlysosomotropic drug and a lysosomotropic counterpart. For the lysosomotropic drug, it will be assumed that its steady-state concentration in the lysosomes is 1000 times greater than extracellular concentrations. We will assume that both drugs are dosed to achieve the same steady-state plasma concentration (i.e., 1 mg/L). The equation for the volume of distribution (V_D) is given by the following equation:

$$V_D = V_P + \frac{f_{UB}}{f_{UT}} V_T$$

This equation states the V_D is equal to the volume of the plasma (V_P) plus the volume of the tissues (V_T) multiplied by the ratio of the free fraction of the drug in the blood (f_{UB}) and the free fraction of the drug in the tissues (f_{UT}). For the sake of simplicity, we will assume that the drug does not physically bind to anything in the body (i.e., free fraction in the blood and tissues are both equal to one). Under these circumstances, the total volume of distribution for the nonlysosomotropic drug would be equal to the volume of plasma plus the volume of tissues, which is approximately 42 L for an average 70 kg person.

In this scenario, the nonlysosomotropic drug will distribute evenly throughout the body, while the lysosomotropic drug will be concentrated within the lysosomes. Accordingly, it is clear that the volume of distribution will be greater for the lysosomotropic drug relative to the nonlysosomotropic counterpart (i.e., 42 L). In order to quantitatively estimate the volume of distribution for the lysosomotropic drug, one needs estimates for both the volume of the lysosomes in the body and the concentration of drug in the lysosomal compartment.

To estimate the volume of the lysosomes in the body, it is necessary to know the number of cells in the body, the average volume of a cell, and the percent of the cell volume comprised by lysosomes. It will be assumed that the body has some 15 trillion cells with an average volume of 3.4×10^{-9} cm^3, a value which is the volume estimate of the average human liver cell. Using these values, we arrive at a total cell volume estimate of 51 L. Assuming that lysosomes take up 1% of the total volume of a cell, the volume of lysosomes in an average human body is approximately 0.51 L in total. This estimation can be rounded to 0.5 L.

In order for the physical volume of the lysosomes to be factored into the total body volume of distribution, the concentration of drug must be the same inside the lysosomes as it is in the plasma. To equalize concentrations, the apparent volume of the lysosomes must be increased by 1000-fold. Accordingly, the apparent lysosomal volume is now 500 L. Using these assumptions, one would anticipate that the total volume of distribution for the lysosomotropic drug would be 542 L, which is approximately 10 times greater than the value obtained with an identical nonlysosomotropic counterpart.

Such alterations in the volume of distribution of a drug will translate into changes in the elimination half-life for a drug. The half-life of a drug is referred to as a dependent pharmacokinetic parameter that is dictated by the independent pharmacokinetic parameters of clearance and volume of distribution, as is shown in the following equation:

$$t_{1/2} = \frac{0.693 \times V_{\mathrm{d}}}{\mathrm{Cl}}$$

According to this relationship, one would predict that a 10-fold increase in volume of distribution would translate into a 10-fold increase in half-life, assuming no differences in clearance. Collectively, the aforementioned scenario is only hypothetical and relies on a number of assumptions; nevertheless, it illustrates the impact that lysosomal trapping could have on a drug by influencing pharmacokinetic parameters including the volume of distribution and half-life.

In the preceding theoretical example, it was assumed that the concentration of drugs in lysosomes at steady state would be 1000 times greater than extracellular concentrations. Assuming that lysosomes constitute 1% of the total cell volume and that a drug does not bind with anything in the cell, one could predict that total cellular concentrations of a lysosomotropic drug might be approximately 10 times greater than extracellular or plasma concentrations. Interestingly, when quantified through experimental means, Duvvuri and Krise [23] have shown that lysosomal trapping occurs to a much larger extent than pure pH-partitioning theory would predict. This is further supported by the work of MacIntyre and Cutler [62]. Performing *in vitro* and *in vivo* studies in rats, and using typical therapeutic levels of chloroquine, the authors found tissue-to-plasma accumulation ratios of approximately 800. The work goes on to show that this sort of accumulation occurred as a result of lysosomal trapping. Furthermore, the accumulation ratio was constant at varying concentrations of chloroquine, but this accumulation ratio decreased greatly when the concentration of chloroquine was raised. This last observation likely occurs due to the tendency of high concentrations of chloroquine to raise lysosomal pH [62]. Cramb's [64] work shows similar results. In this work, the lysosomotropic drug propranolol, when administered at therapeutically relevant concentrations, exhibits a concentration in cultured cells that is 1000-fold greater than in extracellular media. Cramb posits that lysosomal trapping could allow for a continuation of drug activity even after therapy ceases abruptly [64]. A fourth study by Ishizaki et al. [20] takes a look at the accumulation of imipramine in isolated lysosomes and finds that accumulation is approximately 140 times greater than would be predicted through pH-partitioning theory. The authors attribute this finding to either the binding of the drug to lysosomal resident molecule, the aggregation of the drug, or a combination of both of these two scenarios. The scientists conclude that these basic drugs, upon sequestration and binding within the lysosome, shift the equilibrium to favor the entry of further drug molecules [20]. Taking these studies into consideration, it is plausible that lysosomal trapping might have even greater implications on volume of distribution and half-life of drugs than was contemplated in the previous theoretical example.

The lysosomal trapping phenomenon can occur unevenly in the various cells and tissues comprising the body. A study by Rodgers and colleagues examined the major factors governing the distribution of lysosomotropic β-blockers *in vivo*. These drugs are weakly basic, and the studies found that distribution into the lungs was more prevalent than distribution into other tissues, especially with lipophilic β-blockers. Lysosomal trapping is given as the cause of this observation, since lung tissues contain large numbers of lysosomes. In addition to lysosomal sequestration, the researchers suggest phospholipid binding and pulmonary phospholipidosis, occurring when the breakdown of phospholipids is impaired by the lysosomotropic drugs, as potential reasons for this phenomenon [65].

Hung et al. [66] have examined and modeled the disposition kinetics of several cationic drugs in isolated perfused rat liver preparations. The predictions of these scientists hold that the unbound concentrations of drugs in tissues are over seven times greater than the perfusate concentrations for lysosomotropic drugs such as propranolol. Conversely, the ratio is one-to-one for a nonlysosomotropic drug such as antipyrine.

A drug having a large volume of distribution can be viewed beneficially in that it may have a long half-life, thereby allowing for a prolonged dosage interval in a multiple dose setting. However, it is also possible that lysosomal trapping could be viewed negatively from a drug development standpoint, particularly if the tissue binding is strong enough to result in undetectable plasma concentrations. Along this line, Gong et al. [67] have demonstrated that lysosomotropic melanocortin receptor agonists were effectively trapped in the lysosomes of the liver after oral administration and that this limited their detection in the plasma. Counterparts with less affinity for lysosomes were more easily detectable in the plasma. Ultimately, this work suggests that there needs to be some sort of balance of lysosomotropic properties to achieve success in drug development.

18.6 PHARMACOKINETIC DRUG–DRUG INTERACTIONS INVOLVING LYSOSOMES

Given the aforementioned implications of lysosomal trapping on the pharmacokinetics of a drug, it is important to consider possible scenarios whereby the administration of one drug (i.e., perpetrator of the interaction) to a patient could influence the lysosomal accumulation of a second (i.e., victim of the interaction). Conceptually, such interactions could happen in one of three general ways. The first possibility is that a perpetrator drug could in some way modify the steady-state accumulation of a victim by either changing lysosomal pH or altering the binding of the victim with some component of lysosomes. In a second scenario, the steady-state concentration of a drug in the lysosomes remains constant, but a perpetrator drug, by some means, is able to modify the volume of the lysosomal compartment. Finally, it is possible that a perpetrator drug could modify the permeability characteristics of the lysosomal membrane, and this could impact retention of the victim drug.

Daniel et al. [68,69] have investigated drug–drug interactions involving lysosomes whereby one lysosomotropic drug could increase lysosomal pH and therefore decrease the accumulation of a second lysosomotropic drug that was coadministered or administered secondarily. Lysosomotropic antidepressant and antipsychotic drugs were the subject of the studies and were administered simultaneously either *in vivo* or to isolated tissues in relatively high concentrations. Moreover, the authors demonstrate that the perpetrator drug caused a shift in the *in vivo* disposition of victim drugs. Specifically, the victim drugs were shown to decrease in concentration in tissues high in lysosomes (i.e., lungs, liver, and kidneys) and increase in those that have lower abundance of lysosomes (i.e., heart, skeletal muscle, and brain) [69]. It is important to note that this type of interaction is unlikely to have profound clinical implications. Clinically, lysosomotropic drugs would not be administered at the doses required to cause an elevation in lysosomal pH. If a drug-induced elevation in lysosomal pH were to occur, it would likely be very short lived.

More recently, alternative mechanisms for drug-drug interactions involving lysosomes that may have more widespread therapeutic relevance have been considered. Such interactions are focused on the longer term effects that drugs can have on lysosome structure and function that would occur at therapeutically relevant drug concentrations. As previously discussed, under most therapeutic settings most drugs would not be expected to significantly raise lysosomal pH. Funk and coworkers have shown that numerous weakly basic drugs from a variety of therapeutic classes can cause an appreciable increase (two- to threefold) in the apparent volume of the lysosomes in a concentration- and time-dependent manner, without impacting lysosomal pH [70]. In these conditions, the presence of the first drug can dramatically increase the cellular accumulation of a secondarily administered lysosomotropic drug. The authors investigated the mechanism for this effect and have shown it to be due to changes in the vesicle-mediated trafficking associated with lysosomes [71]. Specifically, the amine-containing drugs that perpetrate the interaction do so by increasing the flux of membrane and volume into the lysosomes through the induction of autophagy. In addition, the perpetrating amines have also been shown to decrease the efficiency of vesicle-mediated flux out of the lysosomes. More recently, Logan et al. [72] have examined the structure–activity relationship of drugs that perpetrate this effect and have found that both the propensity to be lysosomotropic and having a capacity to intercalate within lipid bilayer membranes are important. So far, this drug interaction pathway has not been extensively studied in animals or humans. However, it is postulated that future studies pertaining to this drug–drug interaction pathway involving lysosomes could shed light on a novel basis for variability in drug pharmacokinetics.

Working with cultured cells, Kornhuber and colleagues have demonstrated that weakly basic, lysosomotropic drugs such as amitriptyline do not alter the lysosomal pH or lysosomal volume but instead work to increase the permeability of the lipid membrane of the lysosome. When lysosomotropic perpetrator drugs were administered concurrently with LTR, they caused a decrease in its fluorescence, suggestive of a displacement of LTR from the lysosomes. This finding fits in with these researchers' prediction that the lysosomotropic drug has increased the permeability of the lysosomal membrane, allowing for increased diffusion of LTR out of the lysosome [73].

REFERENCES

[1] Allison AC, Mallucci L. Histochemical studies of lysosomes and lysosomal enzymes in virus-infected cell cultures. J Exp Med 1965;121:463–476.

[2] Allison AC, Young MR. Uptake of dyes and drugs by living cells in culture. Life Sci 1964;3:1407–1414.

[3] de Duve C, de Barsy T, Poole B, Trouet A, Tulkens P, et al. Commentary. Lysosomotropic agents. Biochem Pharmacol 1974;23:2495–2531.

[4] Forgac M. Structure and function of vacuolar class of ATP-driven proton pumps. Physiol Rev 1989;69:765–796.

[5] Ouar Z, Bens M, Vignes C, Paulais M, Pringel C, et al. Inhibitors of vacuolar H+-ATPase impair the preferential accumulation of daunomycin in lysosomes and reverse the resistance to anthracyclines in drug-resistant renal epithelial cells. Biochem J 2003;370:185–193.

[6] Duvvuri M, Konkar S, Funk RS, Krise JM, Krise JP. A chemical strategy to manipulate the intracellular localization of drugs in resistant cancer cells. Biochemistry 2005;44:15743–15749.

[7] Ohkuma S, Poole B. Cytoplasmic vacuolation of mouse peritoneal macrophages and the uptake into lysosomes of weakly basic substances. J Cell Biol 1981;90:656–664.

[8] Duvvuri M, Gong Y, Chatterji D, Krise JP. Weak base permeability characteristics influence the intracellular sequestration site in the multidrug-resistant human leukemic cell line HL-60. J Biol Chem 2004;279:32367–32372.

[9] Lemieux B, Percival MD, Falgueyret JP. Quantitation of the lysosomotropic character of cationic amphiphilic drugs using the fluorescent basic amine Red DND-99. Anal Biochem 2004;327:247–251.

[10] Kaufmann AM, Krise JP. Lysosomal sequestration of amine-containing drugs: analysis and therapeutic implications. J Pharm Sci 2007;96:729–746.

[11] Goldman SD, Funk RS, Rajewski RA, Krise JP. Mechanisms of amine accumulation in, and egress from, lysosomes. Bioanalysis 2009;1:1445–1459.

[12] Seynhaeve AL, Dicheva BM, Hoving S, Koning GA, ten Hagen TL. Intact Doxil is taken up intracellularly and released doxorubicin sequesters in the lysosome: evaluated by *in vitro/in vivo* live cell imaging. J Control Release 2013;172:330–340.

[13] Noel G, Peterson C, Trouet A, Tulkens P. Uptake and subcellular localization of daunorubicin and adriamycin in cultured fibroblasts. Eur J Cancer 1978;14:363–368.

[14] Slapak CA, Lecerf JM, Daniel JC, Levy SB. Energy-dependent accumulation of daunorubicin into subcellular compartments of human leukemia cells and cytoplasts. J Biol Chem 1992;267:10638–10644.

[15] Horobin RW, Stockert JC, Rashid-Doubell F. Fluorescent cationic probes for nuclei of living cells: why are they selective? A quantitative structure-activity relations analysis. Histochem Cell Biol 2006;126:165–175.

[16] Rashid F, Horobin RW, Williams MA. Predicting the behaviour and selectivity of fluorescent probes for lysosomes and related structures by means of structure-activity models. Histochem J 1991;23:450–459.

[17] Zhang X, Zheng N, Rosania GR. Simulation-based cheminformatic analysis of organelle-targeted molecules: lysosomotropic monobasic amines. J Comput Aided Mol Des 2008; 22:629–645.

[18] Trapp S, Rosania GR, Horobin RW, Kornhuber J. Quantitative modeling of selective lysosomal targeting for drug design. Eur Biophys J 2008;37:1317–1328.

[19] Fu D, Zhou J, Zhu WS, Manley PW, Wang YK, et al. Imaging the intracellular distribution of tyrosine kinase inhibitors in living cells with quantitative hyperspectral stimulated Raman scattering. Nat Chem 2014;6:614–622.

[20] Ishizaki J, Yokogawa K, Ichimura F, Ohkuma S. Uptake of imipramine in rat liver lysosomes *in vitro* and its inhibition by basic drugs. J Pharmacol Exp Ther 2000;294: 1088–1098.

[21] Andrew CL, Klemm AR, Lloyd JB. Lysosome membrane permeability to amines. Biochim Biophys Acta 1997;1330:71–82.

[22] Chen Y, Walsh RJ, Arriaga EA. Selective determination of the doxorubicin content of individual acidic organelles in impure subcellular fractions. Anal Chem 2005; 77:2281–2287.

[23] Duvvuri M, Krise JP. A novel assay reveals that weakly basic model compounds concentrate in lysosomes to an extent greater than pH-partitioning theory would predict. Mol Pharm 2005;2:440–448.

[24] Nadanaciva S, Lu S, Gebhard DF, Jessen BA, Pennie WD, et al. A high content screening assay for identifying lysosomotropic compounds. Toxicol in Vitro 2011;25:715–723.

[25] Kazmi F, Hensley T, Pope C, Funk RS, Loewen GJ, et al. Lysosomal sequestration (trapping) of lipophilic amine (cationic amphiphilic) drugs in immortalized human hepatocytes (Fa2N-4 cells). Drug Metab Dispos 2013;41:897–905.

[26] Krafts K, Hempelmann E, Skorska-Stania A. From methylene blue to chloroquine: a brief review of the development of an antimalarial therapy. Parasitol Res 2012;111:1–6.

[27] Foley M, Tilley L. Quinoline antimalarials: mechanisms of action and resistance and prospects for new agents. Pharmacol Ther 1998;79:55–87.

[28] Firestone RA, Pisano JM, Bonney RJ. Lysosomotropic agents. 1. Synthesis and cytotoxic action of lysosomotropic detergents. J Med Chem 1979;22:1130–1133.

[29] Miller DK, Griffiths E, Lenard J, Firestone RA. Cell killing by lysosomotropic detergents. J Cell Biol 1983;97:1841–1851.

[30] Simon SM, Schindler M. Cell biological mechanisms of multidrug resistance in tumors. Proc Natl Acad Sci U S A 1994;91:3497–3504.

[31] Simon S, Roy D, Schindler M. Intracellular pH and the control of multidrug resistance. Proc Natl Acad Sci U S A 1994;91:1128–1132.

[32] Bullough PA, Hughson FM, Skehel JJ, Wiley DC. Structure of influenza haemagglutinin at the pH of membrane fusion. Nature 1994;371:37–43.

[33] Dubowchik GM, Padilla L, Edinger K, Firestone RA. Reversal of doxorubicin resistance and catalytic neutralization of lysosomes by a lipophilic imidazole. Biochim Biophys Acta 1994;1191:103–108.

[34] Helenius A, Marsh M, White J. Inhibition of Semliki forest virus penetration by lysosomotropic weak bases. J Gen Virol 1982;58(Pt 1):47–61.

[35] Marnell MH, Stookey M, Draper RK. Monensin blocks the transport of diphtheria toxin to the cell cytoplasm. J Cell Biol 1982;93:57–62.

[36] Gupta DK, Gieselmann V, Hasilik A, von Figura K. Tilorone acts as a lysosomotropic agent in fibroblasts. Hoppe Seylers Z Physiol Chem 1984;365:859–866.

[37] Schwartz AL, Hollingdale MR. Primaquine and lysosomotropic amines inhibit malaria sporozoite entry into human liver cells. Mol Biochem Parasitol 1985;14:305–311.

[38] Alcami A, Carrascosa AL, Vinuela E. The entry of African swine fever virus into Vero cells. Virology 1989;171:68–75.

[39] Schultz KR, Gilman AL. The lysosomotropic amines, chloroquine and hydroxychloroquine: a potentially novel therapy for graft-versus-host disease. Leuk Lymphoma 1997; 24:201–210.

[40] Komarov AM, Hall JM, Chmielinska JJ, Weglicki WB. Iron uptake and release by macrophages is sensitive to propranolol. Mol Cell Biochem 2006;288:213–217.

[41] Shiraishi N, Akiyama S, Kobayashi M, Kuwano M. Lysosomotropic agents reverse multiple drug resistance in human cancer cells. Cancer Lett 1986;30:251–259.

[42] Klohs WD, Steinkampf RW. The effect of lysosomotropic agents and secretory inhibitors on anthracycline retention and activity in multiple drug-resistant cells. Mol Pharmacol 1988;34:180–185.

[43] Zamora JM, Pearce HL, Beck WT. Physical-chemical properties shared by compounds that modulate multidrug resistance in human leukemic cells. Mol Pharmacol 1988;33:454–462.

[44] Klugmann FB, Decorti G, Crivellato E, Candussio L, Mallardi F, et al. Effect of lysosomotropic and membrane active substances on adriamycin uptake and histamine release. Anticancer Res 1990;10:1571–1577.

[45] Hiebsch RR, Raub TJ, Wattenberg BW. Primaquine blocks transport by inhibiting the formation of functional transport vesicles. Studies in a cell-free assay of protein transport through the Golgi apparatus. J Biol Chem 1991;266:20323–20328.

[46] Altan N, Chen Y, Schindler M, Simon SM. Defective acidification in human breast tumor cells and implications for chemotherapy. J Exp Med 1998;187:1583–1598.

[47] Kokkonen N, Rivinoja A, Kauppila A, Suokas M, Kellokumpu I, et al. Defective acidification of intracellular organelles results in aberrant secretion of cathepsin D in cancer cells. J Biol Chem 2004;279:39982–39988.

[48] Schindler M, Grabski S, Hoff E, Simon SM. Defective pH regulation of acidic compartments in human breast cancer cells (MCF-7) is normalized in adriamycin-resistant cells (MCF-7adr). Biochemistry 1996;35:2811–2817.

[49] Gong Y, Duvvuri M, Krise JP. Separate roles for the Golgi apparatus and lysosomes in the sequestration of drugs in the multidrug-resistant human leukemic cell line HL-60. J Biol Chem 2003;278:50234–50239.

[50] Tian ZQ, Liu Y, Zhang D, Wang Z, Dong SD, et al. Synthesis and biological activities of novel 17-aminogeldanamycin derivatives. Bioorg Med Chem 2004;12:5317–5329.

[51] Duvvuri M, Konkar S, Hong KH, Blagg BS, Krise JP. A new approach for enhancing differential selectivity of drugs to cancer cells. ACS Chem Biol 2006;1:309–315.

[52] Ndolo RA, Forrest ML, Krise JP. The role of lysosomes in limiting drug toxicity in mice. J Pharmacol Exp Ther 2010;333:120–128.

[53] Ndolo RA, Luan Y, Duan S, Forrest ML, Krise JP. Lysosomotropic properties of weakly basic anticancer agents promote cancer cell selectivity *in vitro*. PLoS One 2012;7:e49366.

[54] Kang CC, Huang WC, Kouh CW, Wang ZF, Cho CC, et al. Chemical principles for the design of a novel fluorescent probe with high cancer-targeting selectivity and sensitivity. Integrative Biol 2013;5:1217–1228.

[55] Szulc ZM, Mayroo N, Bai A, Bielawski J, Liu X, et al. Novel analogs of D-e-MAPP and B13. Part 1: synthesis and evaluation as potential anticancer agents. Bioorg Med Chem 2008;16:1015–1031.

[56] Bielawska A, Bielawski J, Szulc ZM, Mayroo N, Liu X, et al. Novel analogs of D-e-MAPP and B13. Part 2: signature effects on bioactive sphingolipids. Bioorg Med Chem 2008;16:1032–1045.

[57] Szulc ZM, Bielawski J, Gracz H, Gustilo M, Mayroo N, et al. Tailoring structure-function and targeting properties of ceramides by site-specific cationization. Bioorg Med Chem 2006;14:7083–7104.

[58] Black WC, Percival MD. The consequences of lysosomotropism on the design of selective cathepsin K inhibitors. Chembiochem 2006;7:1525–1535.

[59] Falgueyret JP, Desmarais S, Oballa R, Black WC, Cromlish W, et al. Lysosomotropism of basic cathepsin K inhibitors contributes to increased cellular potencies against off-target cathepsins and reduced functional selectivity. J Med Chem 2005;48:7535–7543.

[60] Desmarais S, Black WC, Oballa R, Lamontagne S, Riendeau D, et al. Effect of cathepsin k inhibitor basicity on *in vivo* off-target activities. Mol Pharmacol 2008;73:147–156.

[61] Fuller K, Lindstrom E, Edlund M, Henderson I, Grabowska U, et al. The resorptive apparatus of osteoclasts supports lysosomotropism and increases potency of basic versus non-basic inhibitors of cathepsin K. Bone 2010;46:1400–1407.

[62] MacIntyre AC, Cutler DJ. Role of lysosomes in hepatic accumulation of chloroquine. J Pharm Sci 1988;77:196–199.

[63] MacIntyre AC, Cutler DJ. The potential role of lysosomes in tissue distribution of weak bases. Biopharm Drug Dispos 1988;9:513–526.

[64] Cramb G. Selective lysosomal uptake and accumulation of the beta-adrenergic antagonist propranolol in cultured and isolated cell systems. Biochem Pharmacol 1986;35: 1365–1372.

[65] Rodgers T, Leahy D, Rowland M. Tissue distribution of basic drugs: accounting for enantiomeric, compound and regional differences amongst beta-blocking drugs in rat. J Pharm Sci 2005;94:1237–1248.

[66] Hung DY, Chang P, Weiss M, Roberts MS. Structure-hepatic disposition relationships for cationic drugs in isolated perfused rat livers: transmembrane exchange and cytoplasmic binding process. J Pharmacol Exp Ther 2001;297:780–789.

[67] Gong Y, Zhao Z, McConn DJ, Beaudet B, Tallman M, et al. Lysosomes contribute to anomalous pharmacokinetic behavior of melanocortin-4 receptor agonists. Pharm Res 2007;24:1138–1144.

[68] Daniel WA, Wojcikowski J. Interactions between promazine and antidepressants at the level of cellular distribution. Pharmacol Toxicol 1997;81:259–264.

[69] Daniel WA, Wojcikowski J, Palucha A. Intracellular distribution of psychotropic drugs in the grey and white matter of the brain: the role of lysosomal trapping. Br J Pharmacol 2001;134:807–814.

[70] Funk RS, Krise JP. Cationic amphiphilic drugs cause a marked expansion of apparent lysosomal volume: implications for an intracellular distribution-based drug interaction. Mol Pharm 2012;9:1384–1395.

[71] Logan R, Kong A, Krise JP. Evaluating the roles of autophagy and lysosomal trafficking defects in intracellular distribution-based drug–drug interactions involving lysosomes. J Pharm Sci 2013;102:4173–4180.

[72] Logan R, Kong AC, Axcell E, Krise JP. Amine-containing molecules and the induction of an expanded lysosomal volume phenotype: a structure-activity relationship study. Journal of Pharmaceutical Sciences 2014;103:1572–1580.

[73] Kornhuber J, Henkel AW, Groemer TW, Stadtler S, Welzel O, et al. Lipophilic cationic drugs increase the permeability of lysosomal membranes in a cell culture system. J Cell Physiol 2010;224:152–164.

19

LYSOSOME DYSFUNCTION: AN EMERGING MECHANISM OF XENOBIOTIC-INDUCED TOXICITY

SHUYAN LU AND BART JESSEN

Drug Safety Research and Development, Pfizer Inc., San Diego, CA, USA

YVONNE WILL

Drug Safety Research and Development, Pfizer Inc., Groton, CT, USA

GREG STEVENS

Drug Safety Research and Development, Pfizer Inc., San Diego, CA, USA

19.1 INTRODUCTION

All eukaryotic cells contain an elaborate array of membrane compartments including the endoplasmic reticulum (ER), Golgi cisternae, the trans-Golgi network (TGN), endosomes, lysosomes, mitochondria, various types of secretory vesicles, and the plasma membrane. The compartments are highly dynamic, and vesicular trafficking is an essential cellular process delivering cargo from one compartment to another. Lysosomes are single membrane-enclosed compartments filled with acid hydrolytic enzymes (e.g., cathepsins) that digest macromolecules and organelles from both external and internal origins via endocytosis, phagocytosis, and autophagy degradation pathways. For the optimal activity of the acid hydrolases, lysosomes require the maintenance of a low internal pH of about 4–5. Central to their role as metabolic regulators, recent compelling evidence has indicated that lysosomes play a critical

Lysosomes: Biology, Diseases, and Therapeutics, First Edition.
Edited by Frederick R. Maxfield, James M. Willard and Shuyan Lu.

role in nutrient sensing and signaling pathways that are involved in cell metabolism and growth. For instance, the kinase complex mammalian target of rapamycin complex 1 (mTORC1), a master controller of cell and organism growth, needs to be translocated onto the lysosomal surface to exerts its activity [1]. In this review, we discuss how perturbation of lysosomes by pharmaceuticals and other chemical substances can impact membrane trafficking and other biological processes that potentially can lead to cell injury and a variety of organ toxicities.

19.2 COMPOUNDS THAT IMPACT LYSOSOMAL FUNCTION

The list of compounds that can impact lysosomal function is steadily growing and can be classified into two main categories. One is associated with lysosomal accumulation by either passive permeation or active transport, while the other involves pharmacological inhibition of certain targets required for lysosomal function and/or the trafficking process. Collectively, these compounds are referred to as lysosomal toxicants.

19.2.1 Lysosomotropic Compounds

For many decades, it has been known that weakly basic lipophilic (also known as cationic amphiphilic) compounds can accumulate in acidic organelles, including lysosomes. Nobel laureate and discoverer of lysosomes, de Duve et al. [2] wrote an elegant commentary discussing the concept of lysosomotropism, a mechanism for accumulation, and the potential link to toxic side effects of pharmaceuticals. The acidic pH (≈4.5) in the lysosomal lumen is achieved by the vacuolar H^+ ATPase, which uses the energy of ATP hydrolysis to pump H^+ into the lysosome. The pH gradient between the lysosomal lumen and the cytosol (≈7.4) drives hyperaccumulation of basic lipophilic compounds via pH partitioning. Generally, the lipophilic free bases are believed to easily traverse lipid bilayers becoming trapped in the acidic environment of the lysosome due to ionization, which decreases the permeability of the compound. Large amounts of basic lipophilic compounds can accumulate in lysosomes. For instance, chloroquine, a well-known lysosomotropic compound, can easily reach a concentration in excess of 20 mM inside lysosomes yielding a ratio several 100-fold higher than outside of the cells [2]. Similarly, after 2 h incubation with amiodarone, the intracellular drug concentration can reach millimolar levels, which is 500-fold higher than the initial extracellular concentrations [3].

Two physicochemical properties, basic pK_a (acid dissociation constant for the conjugated acid of the weak base) and $c\log P$ (partition coefficient between octanol and water, representing membrane permeability) affect the drug accumulation by influencing the extent of lysosomal trapping and regulating the kinetics of passive permeation, respectively [2] (see Chapter 18). Since lysosomotropism is driven by physicochemical properties, its occurrence will not be restricted to a specific chemical or drug class. The physicochemical property distribution of close to 1500 marketed drugs is shown in Figure 19.1 (compounds and property information were downloaded from DrugBank [4]). A significant portion of marketed drugs (651) carry a basic moiety (basic $pK_a > 6$) and close to 400 of them are basic and lipophilic ($c\log P > 1.5$). Hypothetically, all these basic lipophilic drugs could accumulate in lysosomes by pH partitioning, potentially impacting lysosomal function. As illustrated in Table 19.1, a variety of drugs from multiple indications (analgesic, antiarrhythmic, antibiotics from different classes, antihypertensive, and antineoplastic) contain a basic moiety and most of them are lipophilic with $c\log P > 2$. It is noteworthy that numerous central nervous system (CNS) drug classes (antipsychotics, selective serotonin reuptake inhibitor, and tricyclic antidepressants) are basic and lipophilic, fitting the lysosomotropic profile. Indeed, CNS drugs have been reported to have an average $c\log P$ of 2.8 and an average basic pK_a of 8.4, both of which contribute to the blood–brain barrier (BBB) penetration [5].

Interestingly, compounds that are either not basic or lipophilic can also accumulate inside lysosomes. For instance, basic but polar aminoglycoside antibiotics (Table 19.1) has low cell membrane penetration by passive permeation. However,

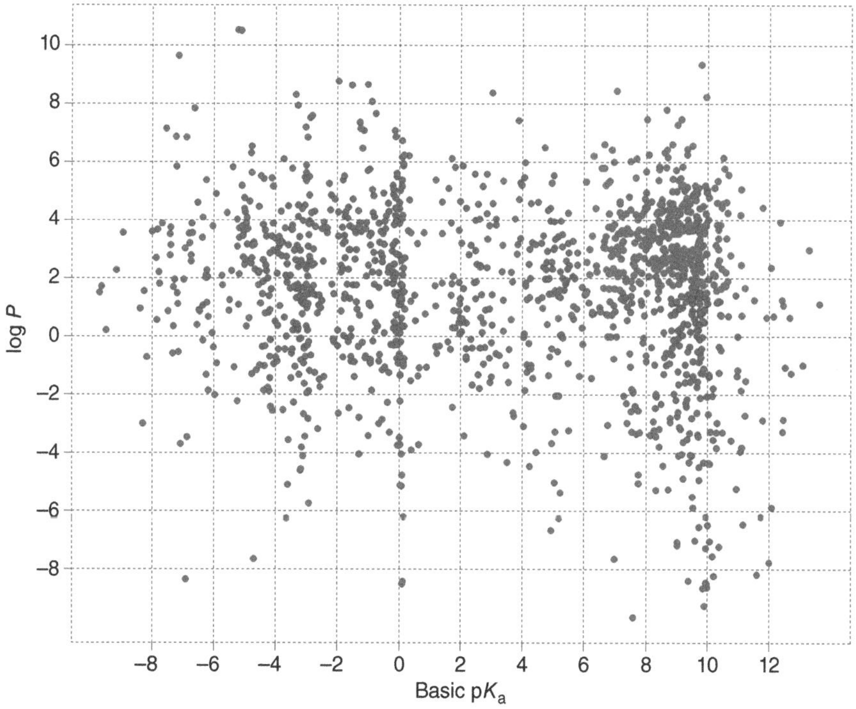

Figure 19.1 Property distribution of approved drugs. A scatter plot showing the distribution of approved (including withdrawn ones) small molecules drugs within the $c\log P$-basic pK_a physicochemical property space. Basic (basic $pK_a > 6$) and lipophilic ($c\log P > 1.5$) drugs are in red circles.

TABLE 19.1 Basic Compounds in Multiple Drug Classes

Indication	Class	Drugs	*c*log *P*	Basic pK_a
Analgesic	Narcotic	Levorphanol	2.9	9.66
		Levomethadyl acetate	4.88	9.87
	Opioid	Tapentadol	2.96	9.6
		Dezocine	3.23	9.67
		Dextromethorphan	3.49	9.85
Anesthetic		Phenazopyridine	2.69	6.86
		Mepivacaine	3.19	7.25
		Lidocaine	2.84	7.75
		Bupivacaine	4.52	8
		Levobupivacaine	4.52	8
Antiallergic	Antihistimine	Chlorpheniramine	3.58	9.47
		Dexbrompheniramine	3.75	9.48
		Brompheniramine	3.75	9.48
		Phenindamine	3.62	9
		Promethazine	4.29	9.05
		Antazoline	2.88	9.24
		Loperamide	4.77	9.41
		Pheniramine	2.98	9.48
Antiarrhythmic	Noncompetitive a- and b-adrenergic inhibitor	Amiodarone	7.64	8.47
	Sodium channel blockers	Encainide	4.49	9.48
		Mexiletine	2.46	9.52
		Flecainide	3.19	9.62
		Terfenadine	6.48	9.02
Antibiotic	Aminoglycoside	Kanamycin	−7.1	9.75
		Amikacin	−8.4	9.79
		Tobramycin	−6.5	9.83
		Gentamicin	−3.1	10.18
		Streptomycin	−7.7	10.88
		Netilmicin	−3.5	9.97
	Fluoroquinolones	Ciprofloxacin	8.68	−0.81
		Levofloxacin	6.2	
		Moxifloxacin	9.42	−0.5
		Norfloxacin	8.68	−1
		Ofloxacin	6.2	−0.39
		Gemifloxacin	9.53	−0.92

TABLE 19.1 (*Continued*)

Indication	Class	Drugs	*c*log *P*	Basic pK_a
	Ketolid	Telithromycin	5.37	7.65
	Macrolide	Troleandomycin	4.3	7.87
		Josamycin	3.22	7.9
		Erythromycin	2.6	8.38
		Clarithromycin	3.24	8.38
		Azithromycin	2.44	9.57
		Rifapentine	3.56	7.98
		Quinupristin	2.18	8.28
		Rifabutin	4.19	8.62
Antifungal		Ketoconazole	4.19	6.75
		Tioconazole	5.3	6.77
		Miconazole	5.96	6.77
		Econazole	5.35	6.77
		Sertaconazole	6.23	6.77
		Butoconazole	6.55	6.78
Anti-HIV	Chemokine receptor antagonist	Maraviroc	3.63	9.38
	Protease inhibitor	Indinavir	2.81	7.37
		Nelfinavir	4.72	8.18
		Saquinavir	3.16	8.47
Antihypertensive	Angiotensin II receptor antagonist	Eprosartan	3.8	6.93
	Beta-1 adrenoceptor antagonist	Bevantolol	3.03	9.31
	Calcium blocker	Nicardipine	3.56	8.18
		Bepridil	5.49	9.16
		Lercanidipine	6.41	9.36
		Amlodipine	1.64	9.45
		Verapamil	5.04	9.68
		Perhexiline	5.53	10.58
	Ganglionic blocker	Mecamylamine	2.37	10.88
Antimalarials		Quinine	2.51	9.05
		Hydroxychloroquine	2.89	9.76
		Lumefantrine	9.19	9.78
		Halofantrine	8.06	10.05
		Proguanil	1.89	10.12
		Primaquine	1.64	10.2
		Amodiaquine	3.76	10.23

(*continued*)

TABLE 19.1 (*Continued*)

Indication	Class	Drugs	$c\log P$	Basic pK_a
Antineoplastic	Nonsteroidal selective estrogen receptor modulator	Toremifene	6.27	8.76
		Tamoxifen	6.35	8.76
	Protein synthesis inhibitor	Homoharringtonine	1.88	9.42
	Tyrosine kinase inhibitor	Gefitinib	3.75	6.85
		Vandetanib	4.54	9.13
		Nilotinib	4.41	6.3
		Lapatinib	4.64	7.2
		Dasatinib	3.82	7.22
		Ponatinib	4.97	8.03
		Imatinib	4.38	8.27
		Bosutinib	4.09	8.43
		Afatinib	3.76	8.81
		Sunitinib	2.93	9.04
		Daunorubicin	1.73	8.94
CNS	Antipsychotics	Promazine	3.93	9.2
		Quetiapine	2.81	7.06
		Chlorpromazine	4.54	9.2
		Chlorprothixene	5.07	9.76
	Serotonin–norepinephrine reuptake inhibitor	Desvenlafaxine	2.29	8.87
		Venlafaxine	2.74	8.91
	Selective serotonin reuptake inhibitor	Fluvoxamine	2.8	9.16
		Fluoxetine	4.17	9.8
		Sertraline	5.15	9.85
	Tricyclic antidepressant	Trimipramine	4.76	9.42
		Amitriptyline	4.81	9.76
		Amoxapine	3.08	8.83
		Imipramine	4.28	9.2
		Clomipramine	4.88	9.2
		Doxepin	3.84	9.76
		Desipramine	3.9	10.02
		Nortriptyline	4.43	10.47
		Protriptyline	4.5	10.54
		Maprotiline	4.37	10.54

lysosomal accumulation has been demonstrated for multiple aminoglycoside antibiotics [6], specifically in the kidney proximal tubule cells [7] and sensory hair cells of the inner ear [8]. Rather than passive permeation, specialized uptake transporters and endocytosis is utilized by those polar molecules to transport themselves into lysosomes [9]. The major transport of aminoglycosides into proximal tubule cells involves interaction with acidic, negatively charged phospholipid-binding sites at the level of the brush border membrane. The megalin receptor, a scavenger receptor with broad tissue distribution, including the cochlea of the inner ear [10] and apical surface of polarized epithelial cell membranes facing the proximal tubule [11], has been shown to contribute to the aminoglycoside accumulation in these tissues [12].

Another chemical class that is not associated with the basic lipophilic properties but is lysosomotropic involves heavy metals such as arsenic, cadmium [13,14], and platinum-containing compounds (e.g., cisplatin) [15]. Cadmium and arsenic are environmental toxicants, and cisplatin is one of the most widely used antitumor drugs. Similar to aminoglycoside antibiotics, various membrane transporters have been shown to be involved in heavy metal uptake. Cellular uptake of cisplatin is mediated, at least in part, by transport proteins including the copper transporters 1 and 2 (Ctr1 and Ctr2), the P-type copper-transporting ATPases ATP7A and ATP7B, the organic cation transporter 2 (OCT2), and the multidrug extrusion transporter 1 (MATE1) [16]. Similarly, arsenic is transported into mammalian cells via multiple pathways including the aquaglyceroporins (AQP3, AQP7, AQP9, AQP10), the glucose permeases (GLUT1, GLUT2, GLUT5), and the organic anion transporting polypeptides (OATPB, OATPC) [17]. The major pathway for cadmium uptake by kidney proximal tubule cells involves the endocytosis of cadmium complexes followed by binding to the high-affinity metal-binding protein metallothionein [18]. Cadmium may also be transported into cells via zinc and/or copper transporters [19]. Interestingly, various metal transporters are located on the lysosomes and play a role in metal import/export from lysosomes [20]. The proton-coupled divalent metal transporter DMT1, for example, is expressed exclusively in endosomes/lysosomes and transports cadmium into lysosomes [21]. The exact mechanism of lysosomal sequestration of heavy metals and what lysosomal transporters are involved needs to be further investigated.

19.2.2 Nonlysosomotropic Compounds

Lysosomal accumulation is not the only mechanism by which xenobiotics can impact the lysosomal pathway. Lysosomal toxicants can also affect lysosomal function via pharmacological inhibition of molecular targets involved in lysosomal regulation. Bafilomycin A (Baf A) and concanamycin, for example, can block lysosomal acidification through selective inhibition of the V-type ATPase [22]. The ionophores monensin and nigericin, which mediate the exchange of H^+ and Na^+, can also increase the pH of lysosomes [23]. Evidently, inhibition of various cathepsins can disrupt lysosomal proteolysis. Microtubule disruptors, such as paclitaxel and vincristine, can impair lysosomes by disrupting microtubule-dependent trafficking. Bacterial proteins such as staphylococcal a-toxin and anthrax toxin have been shown to impact lysosomal function by disrupting lysosomal membrane integrity [24,25]. Recently, it was shown that botulinum neurotoxins (BoNTs, serotypes A–G) and tetanus neurotoxin (TeNT) induce perturbations of the fusogenic SNARE (soluble *N*-ethylmaleimide-sensitive factor attachment protein receptor) complex dynamics, which could impact the fusion of lysosomes with other vesicles. In addition, tumor necrosis factor-related apoptosis-inducing ligand (TRAIL) was observed to trigger lysosomal change after binding to death receptor 5 via recruitment of the proapoptotic proteins Bim and Bax to the lysosomes [26].

19.3 CELLULAR CONSEQUENCES

Lysosomes are kinetic organelles and have an intimate link with the endocytic apparatus, phagocytosis, and autophagy, which deliver materials to lysosomes for degradation and recycling. Lysosomes are not the only acidic organelles; early endosomes, late endosomes, phagosomes, and autolysosomes all have a differing degrees of luminal acidity and thus are targets for accumulation of basic lipophilic xenobiotics. For simplicity, our discussions primarily focus on lysosomal changes and vesicular trafficking.

19.3.1 Effect of Drugs on pH and Lysosomal Volume

Accumulation of high concentrations of basic compounds in lysosomes can increase the luminal pH. The lysosomotropic compounds, chloroquine and methylamine, were shown to increase lysosomal pH drastically (0.5–2.0 pH units) after accumulation *in vitro* [27], and in the case of chloroquine, *in vivo* [28]. The LysoTracker probes consist of a fluorophore linked to a weak base that is only partially protonated at neutral pH. This allows LysoTracker probes to freely permeate cell membranes enabling them access to the luminal regions of acidic organelles in live cells. Similar to other lysosomotropic compounds, low pH is required for the lysosomal accumulation and subsequent detection of LysoTracker® probes. In a previous study using H9C2 cells, basic lipophilic compounds were shown to decrease the staining of LysoTracker Red DND-99 (LTR) after 4 h of treatment, indicating an increase in lysosomal pH [29]. The drugs used in this study were from multiple drug classes and pharmacological actions including antidepressants, antipsychotics, and anticancer, indicating the increase in pH is related to their physicochemical properties rather than specific pharmacological target inhibition. Indeed, clustering analysis demonstrated that compounds possessing a $c\log P > 2$ and a basic pK_a between 6.5 and 11 were associated with the increase in pH [29]. These data are consistent with the observation from a different publication using Fa2N-4 cells in which the decrease in LTR fluorescence was associated with basic and lipophilic compounds [30]. An increase in pH would be expected to decrease the lysosomal degradation capability since acidic pH is optimal for lysosomal enzyme activity. In addition, an increase in pH can decrease the fusion capability of lysosomes [31], and in fact, chloroquine has been shown to decrease the fusion between autophagosome and lysosome [32]. Besides basic lipophilic compounds, vacuole ATPase inhibitors, such as Baf A, can also inhibit the acidification of lysosomes, resulting in decreased LTR staining [33].

Interestingly, pH increase by basic lipophilic compounds is only a short-term effect. No significant change in lysosomal pH was observed after 24 h of imipramine exposure [34]. Consistent with a reversal of the pH change, prolonged exposure of numerous basic lipophilic drugs significantly increases the cellular accumulation of LTR [35,36], indicating an increase in lysosomal volume and normalization of luminal pH. A fourfold increase in lysosomal volume was estimated based on both the lysosomal pH and amount of LTR accumulated after 24 h exposure of imipramine [34]. One plausible reason for the increase in lysosomal volume could be due to an adaptive change by increasing lysosomal biogenesis with basic lipophilic compounds. For example, increases in lysosome-associated membrane protein 2 (LAMP-2) staining was observed *in vivo* with multiple basic lipophilic drugs [37] and marked lysosome-associated membrane protein 1 (LAMP-1) increase was observed *in vitro* with chloroquine treatment [38]. Transcription factor EB (TFEB) is a master transcription factor for lysosomal biogenesis [39] and TFEB nuclear translocation was triggered by chloroquine treatment, which could explain in part the increase in lysosomal volume [40]. TFEB also potentially plays a role in reversing the pH change via upregulation of vATPase [39].

A second reason for an increase in lysosomal volume is accumulation of undigested substrate in lysosomes. Lysosomal volume expansion by imipramine can be reversed by exposing cells to hydroxypropyl-β-cyclodextrin, which reduces lysosomal cholesterol burden, suggesting elevation in lysosomal cholesterol content contributes to the lysosomal volume increase [34]. Similar lysosomal changes were also observed for compounds that impact lysosomes via nonaccumulative process. For instance, increases in the acidic compartment and LAMP-2-positive vesicles were observed following vincristine treatment [41]. Interestingly, cells derived from patients with lysosomal storage diseases (LSDs) are often associated with an increase in LTR staining [34,42], implying the increase in LTR by compound treatment is indicative of lysosomal dysfunction. Hence, decreased LTR staining is a common approach as a phenotypic screen for LSDs drug discovery [42], which further supports this notion.

19.3.2 Effects on Lysosomal Enzymes

Besides the pH change, lysosomotropic compounds have demonstrated effects on lysosomal enzyme activity, either directly or indirectly. Chlorpromazine and chloroquine have demonstrated inhibition of lysosomal phospholipase A1 *in vitro* [43,44], as well as downregulation of acid ceramidase that was not caused by decreased transcription [45]. Furthermore, multiple lysosomotropic compounds have demonstrated the ability to redistribute the mannose 6-phosphate (M6P) receptor from the TGN to endosomes and concomitantly increase the secretion of lysosomal enzymes resulting in a decline of mature intracellular lysosomal enzyme levels [46]. Since the majority of enzymes delivered to lysosomes require the M6P receptor, the redistribution of the receptor by lysosomotropic compounds could dysregulate a range of lysosomal enzymes. Reduction of lysosomal enzyme activity (either due to alterations of pH, or by direct or indirect inhibition) could contribute to lysosomal dysfunction. Paradoxically lysosomotropic cathepsin K inhibitors were shown to increase other cathepsin activity [47]. In addition, upregulation of cathepsin D was also demonstrated in

Alzheimer's disease (AD) presumably due to a compensatory protective response [48,49]. It is therefore conceivable that lysosomal toxicants can also trigger upregulation of lysosomal cathepsins to offset the lysosomal dysfunction. Long-term impact of lysosomal toxicants on lysosomal enzymes needs to be evaluated further.

19.3.3 Lysosomal Substrate Accumulation

One of the critical functions of lysosomes is to break down all types of biological polymers including proteins, nucleic acids, carbohydrates, and lipids. Changes in pH or enzymatic activity by basic lipophilic compounds may decrease lysosomal degradation capacity, triggering accumulation of various substrates. One of the best characterized examples of this is the excessive accumulation of phospholipid, for example, phospholipidosis (PLD), induced by lysosomotropic compounds. The hallmark feature of PLD is the characteristic lamellar bodies that are visible by transmission electron microscopy [50] (see Chapter 20). An *in silico* model for PLD prediction proposed by [51] describes a simple method based on both pK_a of the most basic center and $c\log P$ value of the molecules. Interestingly, besides basic lipophilic compounds, polar aminoglycoside antibiotics, which carry a basic moiety, are also associated with PLD [52]. Lysosomal hydrolysis contributes to the mobilization of lipid droplet-associated cholesterol [53] and cholesterol accumulation is associated with lysosomal dysfunction. For instance, a mutation in NPC1 resulting in the accumulation of cholesterol in late endosomes or lysosomes leads to the autosomal recessive disease Niemann–Pick disease type C (NPC). The basic lipophilic compounds U18666A, clomiphene, and terconazole have been used to produce the NPC phenotype (cholesterol accumulation) experimentally [54,55]. Moreover, accumulation of lipofuscin, which is an intralysosomal waste material consisting of various components such as protein, lipid, carbohydrates, metals, and autofluorescent pigment [56] has been observed in postmitotic cells associated with aging, lysosomal protease inhibitors, and chloroquine treatment [57,58].

Excessive accumulation of various substrates triggers severe lysosomal dysfunctions exemplified by LSDs. While the majority of LSDs are a consequence of the deficiency of a single enzyme required for the metabolism of lipids or glycoproteins, lysosomotropic compounds may affect multiple lysosomal activities. In addition, lysosomotropic agents including ammonium chloride, chloroquine, methylamine, doxorubicin, and the lysosomal enzyme inhibitor E64D, cause ferritin accumulation [59], which prevents the redistribution of iron to vital cellular processes and triggers cellular damage as a result.

19.3.4 Lysosomal Membrane Permeabilization (LMP) and Cell Death

Lysosomal membrane integrity, as measured by acridine orange leakage, can be compromised by excessive compound accumulation [60,61]. Subsequent leakage of large amounts of lysosomal acid hydrolases following lysosomal membrane disruption leads to cellular toxicity through apoptotic mechanisms [62–65]. In general, the magnitude of lysosomal rupture and, consequently, the amount of hydrolytic enzymes released into the cytosol, is dose dependent and may induce sublethal damage, apoptosis, or necrosis depending on the exposure level and duration [66]. This dose-related differential outcome has been demonstrated by the lysosomotropic agent *O*-methyl-serine dodecylamide hydrochloride (MSDH), in which apoptosis accompanied by caspase-3 activation occurred at $<50\,\mu M$ while at $>75\,\mu M$ necrosis occurred and was associated with more extensive lysosomal rupture [67]. Tamoxifen also exhibited a similar toxicity profile in which high concentration ($10\,\mu M$) caused extensive necrosis while apoptotic-like cell death was observed at lower concentrations [68]. The close connection between lysosomal accumulation and cell death is further demonstrated by the observation that Baf A pretreatment of cells can rescue the toxicity caused by certain basic lipophilic compounds [69].

The intimate relationship between lysosomes and mitochondria in the execution of apoptosis is emphasized in the lysosomal–mitochondria axis theory of apoptosis, where released lysosomal enzymes trigger mitochondria permeability transition and apoptosis via the Bcl-2 family (e.g., bax and BAD) [70] (see Chapter 8). Lysosomal membrane permeabilization has also been reported to initiate a caspase-independent cell death pathway [71] as well as the release of iron that can catalyze the formation of highly reactive hydroxyl radicals leading to oxidative stress-induced apoptosis or necrosis [72].

19.3.5 Membrane Trafficking Changes

Lysosomes are the central hub for membrane trafficking including endocytosis, autophagy, and phagocytosis. Lysosomal dysfunction from xenobiotic effects on pH, enzymatic activity, substrate accumulation, and/or impairment of membrane integrity could lead to disruption of various trafficking pathways associated with lysosomes.

19.3.5.1 Autophagy Perturbation Autophagy refers to the catabolic process in eukaryotic cells that delivers cytoplasmic material to lysosomes for degradation. This highly conserved process is involved in the clearance of long-lived proteins and damaged organelles. Vinblastine (a microtubule inhibitor), leupeptin (a lysosomal protease inhibitor), and the lysosomotropic amines propylamine and chloroquine are all capable of inhibiting autophagy in various model systems [73,74]. Microtubule-associated protein light chain 3 (LC3) is an ubiquitin-like protein required for the formation of autophagosomal membranes. During autophagy, a cytosolic form of LC3 (LC3-I) is conjugated to phosphatidylethanolamine to form an LC3-phosphatidylethanolamine conjugate (LC3-II), which is recruited to autophagosomal membranes. Detection of LC3 by immunoblotting or immunofluorescence has become a reliable method for monitoring autophagy. Several lysosomotropic compounds have been shown to increase LC3 staining, indicating the modulation of autophagy [75]. The concentration-dependent increase in the abundance of the autophagy substrate, sequestosome 1 (SQSTM1/p62, hereafter referred to as p62) by these compounds supports the hypothesis that the increase in LC3 was due to autophagy inhibition rather than upstream enhancement. In addition, the downregulation of extracellular matrix and cytoskeletal gene transcription by these compounds also indicate ineffective turnover of long-lived proteins, supporting the notion of autophagy inhibition. Further evidence of this hypothesis is demonstrated by the finding that the lysosomotropic detergent siramesine decreases the colocalization of LC3-positive vesicles with LAMP-2-positive vesicles, indicating the impairment of autophagosomal fusion with lysosomes [76]. Conversely, many basic lipophilic compounds such as amiodarone [77], tamoxifen [78], amitriptyline [79], verapamil [80], and dimebon [81] have been proposed as autophagy stimulators based on their ability to increase LC3. Because of their basic lipophilic properties, these compounds have the propensity to accumulate in lysosomes [29].

One common assay format employed to study autophagy flux is to evaluate the effect on LC3-II abundance with test compound in the presence of protease inhibitors (e.g., E64d) or Baf A. An increase of LC3-II abundance with the combination treatment relative to the test compound alone is indicative of increased flux through the autophagy process. However, this assay only measures the initial stage of autophagosome formation without the evaluation of the degradation step. Recently, it was shown that lysosomotropic compounds (e.g., chloroquine) induce lysosomal stress and, consequently, provokes TFEB nuclear translocation [40], which could drive expression of autophagy genes leading to the induction of autophagy. It is possible that the increase of LC3-II caused by these basic lipophilic compounds is due to the adaptive increase in autophagy by TFEB nuclear translocation. Ultimately, downstream lysosomal dysfunction inflicted by lysosomotropic compounds would tend to block autophagy [75]. Other types of membrane trafficking perturbation discussed later further supports the notion of global lysosomal dysfunction.

19.3.5.2 Impairment of Endocytosis/Phagocytosis Endocytosis is an energy-consuming process by which cells take up extracellular molecules (such as proteins) by engulfing them. There are various mechanisms of endocytosis such as clathrin-mediated, caveolae, and macropinocytosis. Phagocytosis is a special form of endocytosis in which large particles are taken up via the phagosome. The endocytic pathway consists of distinct membrane compartments, which internalize molecules from the plasma membrane and then either recycle them back to the surface, as in early endosomes and recycling endosomes, or designate them for degradation, as in late endosomes and lysosomes. Lysosomes are the last compartment of the endocytic pathway and their dysfunction by lysosomal toxicants can negatively impact the endocytic process and various endocytic compartments.

Indeed, numerous lysosomal toxicants are known to impact various stages and compartments of endocytosis and are routinely used to study the endocytic process. For example, direct evidence for recycling came from the observation that the process can be inhibited by basic compounds, including chloroquine, NH_4Cl, methylamine, and ionophores, such as monensin [82]. Many receptors are recycled and return to the plasma membrane after ligand dissociation. Lysosomal toxicants have been shown to prevent recycling of receptors for LDL [83], transferrin [84], epidermal growth factor, and macroglobulin [85]. Instead of returning to the plasma membrane they became trapped in endosomes. The mechanism could be related to the pH elevation of acidic compartments and inhibition of the dissociating of the ligand from the receptor [86] or perturbation of the clathrin lattice assembly. Indeed, chloroquine is recognized as a clathrin-dependent endocytosis inhibitor [87]. In addition, using an *in vitro* macrophage model, monensin and chloroquine were shown to inhibit transfer to lysosomes of endocytosed macromolecules [88]. Azithromycin, a lysosomotropic antibiotic, delayed sequestration of receptor-bound transferrin and peroxidase–antiperoxidase immune complexes into cell-surface endocytic pits and vesicles [89].

Virus entry relies on either membrane fusion or endocytosis and, intriguingly, the viral entry has been shown to be impacted by lysosomotropic compounds. Ebola viruses enter the cell by a macropinocytic-like process after attachment to the cell surfaces [90], and multiple lysosomotropic compounds, including U18666A and clomiphene, were shown to inhibit Ebola virus entry and infection. Chloroquine and NH_4Cl also demonstrated more than a 50% reduction of HCV infectivity. Recently, another lysosomotropic compound, amiodarone, has been shown to inhibit the entry of the hepatitis C virus [91]. Together these data strongly support the inhibition of endocytosis by lysosomal toxicants.

19.3.5.3 Phagocytosis Basic lipophilic local anesthetics, dibucaine, tetracaine, and procaine, have been shown to exert a reversible inhibition of phagocytosis of opsonized sheep red blood cells using mouse peritoneal macrophages [92]. Interestingly, the potencies of these local anesthetics for inhibition of phagocytic uptake were proportional to their $\log P$, indicating the critical role of this physicochemical property in inhibiting phagocytosis. Retinal pigment epithelium (RPE) is one of the most phagocytotically active cell types. Chloroquine has been shown to decrease the uptake of rhodamine-labeled dextran in retinal pigment epithelial cells [38], suggesting disruption of the phagocytic pathway. The basic lipophilic compounds tamoxifen, toremifene, and chloroquine were also shown to decrease the uptake of FITC-labeled rod outer segments [93] in retinal pigment epithelial cells *in vitro*, further supporting the notion of inhibiting phagocytosis by lysosomotropic xenobiotics.

19.3.5.4 Exocytosis Exocytosis involves the transport of intracellular vesicles to the plasma membrane of the cell where vesicular fusion results in the delivery of membrane and protein to the cell surface, as well as secretion of the vesicular contents. It is well established that increases in lysosomal pH result in lysosomal exocytosis, with enhanced secretion of preformed hydrolases [46,94–96]. Chloroquine not only stimulated the release of porphyrins from yeast [97] but also enhanced the exocytosis of MTT formazan, which accumulates in the endosomal/lysosomal compartment [98]. Similarly, exocytosis of lysosomal contents was also observed from skeletal muscle *in vivo* following chloroquine treatment [99]. The association of increased exocytosis with lysosomal dysfunction is further supported by the findings in lysosomal storage and aging-related diseases. Lysosomal storage materials have been found in extracellular fluids, blood, and urine in some LSDs patients, providing direct evidence of lysosomal exocytosis [100,101]. The pathogenesis of different neurodegenerative diseases, including AD, Parkinson's disease (PD), and Huntington's disease (HD), share one common feature: abnormal accumulation and aggregation of disease-specific proteins that form either intracellular inclusions or extracellular aggregates, presumably by excessive lysosomal exocytosis. For instance the truncated, misfolded, and oxidatively modified α-synuclein forms appear to be enriched in the vesicular fraction and to be preferentially secreted [102]. Chloroquine has been shown to increase nuclear translocation of TFEB during lysosomal stress [40], promoting lysosomal exocytosis by increasing the pool of lysosomes in the proximity of, and promoting fusion with, the plasma membrane [103]. Besides the role of TFEB, Ca^{2+} release also contributes to the regulation of exocytosis. The lysosomotropic agent glycyl-L-phenylalanine-naphthylamide (GPN) evoked robust increases in the intracellular Ca^{2+} concentration and triggered release of β-hexosaminidase within 5 min of exposure, demonstrating the characteristic ruffling of the plasma membrane associated with exocytosis [104]. Ca^{2+} release from the lysosome in this case is believed to play a role in the immediate increase in exocytosis.

19.3.6 Other Cellular Impacts

Due to the central role of lysosomes in membrane trafficking, organelle homeostasis critically relies on effective lysosomal function. Membrane trafficking perturbation due to lysosomal dysfunction subsequently impacts other organelles. In addition, the intriguing discovery that activation of mTORC1 requires lysosomal recruitment in response to amino acid concentrations [105] positions these degradation organelles as a new hub for signaling transduction. However, how xenobiotics affect cell signaling via lysosomes is yet to be investigated.

19.3.6.1 Mitochondria The intimate relationship between lysosomes and mitochondria is twofold. First, as discussed earlier, released cathepsins can destabilize mitochondria, promote cytochrome *c* release, and ultimately induce apoptosis. Second, autophagic delivery to lysosomes is the major degradative pathway in mitochondrial turnover, which is critical for mitochondrial quality control [106]. This special form of mitochondrial degradation by autophagy is referred to as mitophagy. Mitochondrial abnormalities would be an expected outcome of drug-induced lysosomal dysfunction and the persistence of dysfunctional mitochondria has indeed been observed with multiple LSDs [107]. Abnormal mitochondria were also observed in multiple autophagy knockout studies further strengthening this critical role of autophagy [108]. The number of mitochondria and GFP-LC3 colocalized puncta increased in the presence of lysosomal inhibitors such as chloroquine or E64 D plus pepstatin A [109]. Accumulation of damaged mitochondria by lysosomotropic compounds could certainly lead to cellular perturbations.

19.3.6.2 Endoplasmic Reticulum The ER has a role in folding and modifying proteins, and secreting proteins via the secretory pathway. Elevated levels of misfolded proteins causes ER stress, which is commonly observed in cells derived from

LSD patients, indicating a role for lysosomes in maintaining of ER homeostasis [110]. The autophagy process has been shown to degrade sequestered ER within autophagosomes [111], for instance, autophagy-deficient T lymphocytes demonstrate abnormal expansion of ER and impaired calcium influx [112]. Induction of autophagy during ER stress [113,114] is believed to serve as a protective mechanism against cell death. Presumably, an autophagy defect due to lysosomal dysfunction induced by lysosomal toxicants could exacerbate the ER stress and enhance cytotoxicity. The lysosomotropic kinase inhibitors, imatinib and flavopiridol, have been associated with ER stress resulting from perturbation of autophagy and lysosomal membrane permeabilization [61] [115].

19.3.6.3 Oxidative Stress Several reports have associated reactive oxidative species (ROS) with lysosomotropic agents [116–119]. Interestingly, *in situ* imaging showed ROS production was present within 1 h of exposure to the lysosomotropic antibiotic gentamicin [120]. The links between autophagy dysregulation and increases in oxidative stress are evident by mouse lysosomal protein knockout studies and autophagy inhibitors [108]. These studies demonstrate the accumulation of oxidized ubiquitinated proteins and dysfunctional mitochondria. Mitochondria are a major site of ROS generation from multiple sites along the respiratory chain [121], and accumulation of damaged mitochondria due to lysosomal function impairment likely contributes to the increase of oxidative stress. Interestingly, lysosomes are susceptible to oxidative stress and an increase in oxidative stress can act as an amplifying loop to further destabilize the lysosomal membrane [122,123]. Antioxidant treatments have demonstrated that prevention of ROS production at least partially rescued cell death caused by lysosomal permeabilization [120]. Lysosomes are rich in redox-active iron due to degradation of iron-containing macromolecules and the lysosomal iron has been shown to play a role in exacerbating the effects of ROS via the Fenton reaction, in which an extremely reactive hydroxyl radical is formed during hydrogen peroxide oxidization of iron [124].

19.3.6.4 Signaling Modulation Besides membrane trafficking and degradation, lysosomes have a critical emerging role as a signaling hub, particularly for nutrient sensing and cellular metabolism. mTORC1 needs to be recruited to the lysosomal surface to exert its kinase activity [105]. Interestingly, it is the level of amino acids inside the lysosomal lumen that controls mTORC1 docking on the lysosomal surface and subsequent activation. Multiple protein complexes, including Rag GTPase, raptor and regulator, are involved in the mTORC1 translocation. p62, a substrate for autophagy, has been shown to bind to raptor, becoming an integral part of the mTORC1 complex and facilitating mTORC1 translocation to the lysosomal surface upon amino acid stimulation [125]. Furthermore, activation of mTORC1 inhibits both autophagy and lysosome biogenesis [126,127].

Lysosomal toxicants may affect signaling molecules via lysosomes in several ways. First of all, inadequate lysosomal degradation may decrease free amino acid generation, impeding mTORC1 activation. Indeed, a decrease in mTORC1 activity was not only observed in LSD models, such as in fibroblasts from mucopolysaccharidosis type I, Fabry disease, and aspartylglucosaminuria [128], but also observed after treatment with several lysosomotropic compounds (e.g., chloroquine [40], amiodarone [129], and imipramine [130]. In addition, p62 has been shown to regulate various other signal transduction pathways (Nrf2, NF-κB, caspase 8, and ERK1) through specific protein interactions [131]. As was discussed before, increases in p62 are typically associated with lysosomal dysfunction and autophagy deficiency, and this increase could modulate multiple signaling molecules. A noncanonical mechanism of Nrf2 activation by autophagy deficiency due to direct interaction between Keap1 and p62 has recently been proposed [132,133]. Essentially, p62 interacts with the Nrf2-binding site on Keap1, a component of Cullin-3-type ubiquitin ligase for Nrf2. When p62 accumulates due to autophagy deficiency, it competes with the interaction between Nrf2 and Keap1, resulting in stabilization of Nrf2 and transcriptional activation of Nrf2 target genes. Lysosomotropic compounds were clearly shown to induce multiple Nrf2-regulated genes such as glutamate–cysteine ligase and glutathione-S-transferase, indicating Nrf2 activation [75]. Remarkably, arsenic inhibits autophagic flux and activates the Nrf2-Keap1 pathway in a p62-dependent manner since Nrf2 activation by arsenic is diminished when p62 is knocked down [134]. In addition, activation of NF-κB has been observed in an autophagy-suppressed Atg7 knockout model accompanied by accumulation of p62 [135], whereas knockout of p62 suppressed NF-κB signaling, indicating a critical role of p62 in the regulation of NF-κB. The lysosomotropic compounds chloroquine and imatinib mesylate were indeed shown to induce NF-κB in astroglial cells and pancreatic islet cells, respectively [136,137]. However, whether the increase in NF-κB is generic for all lysosomotropic compounds and whether the effect is dependent on p62 needs to be further investigated. p62 also promotes the full activation of caspase 8 with subsequent triggering of apoptosis [138], and suppresses ERK1 signal during adipogenesis [139]. More recent data showed autolysosomal sequestration of active RhoA was via p62 and negative regulation of RhoA signaling activity by constitutive autophagy [140]. Overall, lysosomal changes and p62 accumulation due to autophagy deficiency from lysosomal toxicant treatment could set off a chain of altered signaling molecules, which in turn can further impact a variety of biological processes.

In conclusion, we propose a multifaceted model (Figure 19.2) in which lysosomal toxicants, either by accumulation or disturbance of lysosome function, not only induce cell death due to lysosomal membrane permeabilization but also impair lysosomal function. These functional changes can result in the perturbation of multiple membrane trafficking pathways including autophagy, endocytosis, phagocytosis, and possibly exocytosis enhancement. Besides the lysosomal change, other organelles such as mitochondria and ER could be damaged due to deficiency of autophagy, which subsequently leads to oxidative stress. Furthermore, multiple signal transduction processes including mTORC1, ERK1, Nrf2, and NFκB are modulated by lysosomal dysfunction due to the increase of p62 abundance resulting from autophagy deficiency. The pathophysiological consequences of modulating those signaling molecules by lysosomal toxicants needs to be further deciphered. Thus far, the focus has been on the impact of small molecules on lysosomes. However, nanotechnology is currently an area of intense scientific interest due to a

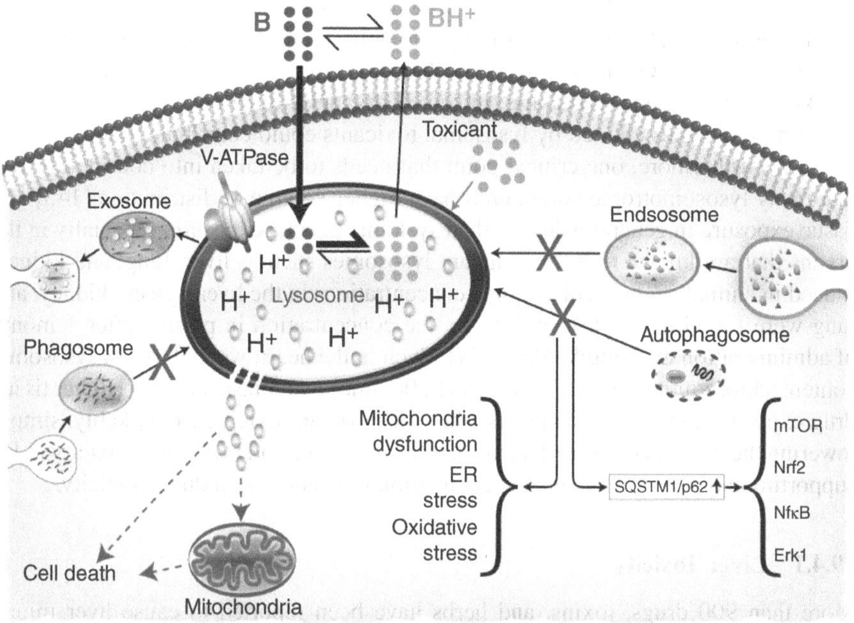

Figure 19.2 Biological impacts of the lysosomotropic compounds. Compounds can perturb lysosome functions either by accumulation inside the lysosomes or direct inhibition of various lysosomal functions (e.g., decrease of lysosomal enzyme activity). Lysosomal membrane permeabilization can trigger cell death. In addition, lysosomal dysfunction can tamper membrane trafficking pathways including autophagy, endocytosis, phagocytosis, and exocytosis. Multiple signal transduction processes including MTORC, ERK1, Nrf2, and NFκB can also be modulated by lysosomal dysfunction due to the increase of p62 abundance resulting from autophagy deficiency. (*See color plate section for the color representation of this figure.*)

wide variety of potential applications in biomedical, optical, and electronic fields, and various nanoparticles have been shown to be trapped inside the lysosomes and disrupt lysosome function including LMP and inhibition of autophagy [141,142]. Thus, lysosomal dysfunction could be an emerging toxicological mechanism for both small molecules and nanomaterials.

19.4 IMPAIRED LYSOSOMAL FUNCTION AS A MECHANISM FOR ORGAN TOXICITY

A number of cellular processes can be affected by lysosomal dysfunction that can ultimately lead to organ toxicity as illustrated in LSDs. Pathological manifestations in LSDs as a result of lysosomal defects strongly indicate that a similar mechanism could contribute to organ toxicity resulting from xenobiotics that perturb lysosomal function. In addition to the CNS, multiple peripheral organ systems including cardiac, kidney, skeletal muscle, and hepatocyte are impacted by LSDs [107]. As will be discussed later, a number of organ toxicities have been associated with lysosomal dysfunction. Obviously, many mechanisms discussed previously such as LMP, blocking of membrane trafficking, impacted organelle function, and modulation of signaling molecules elicited by lysosomal toxicants could contribute to target organ toxicity. Furthermore, one critical point that needs to be taken into consideration is that many lysosomotropic compounds have higher volume of distribution [143] and tissue exposure in general is higher than systemic plasma exposure, especially in the organs that are known to have abundant lysosomes such as liver, lung, and kidney. Indeed in animal studies, chloroquine concentrations in the liver, spleen, kidney, and lung were found to be 700–1600 times the concentration in plasma after 1 month of administration at 40 mg/kg/day [144]. Even in the heart, which has low lysosomal content, chloroquine concentrations were 200 times over plasma levels. Higher tissue drug exposure can certainly increase the risk of organ toxicity. Remarkably, simply lowering the pK_a can result in decreased tissue exposure and lower liver toxicity [145] supporting the role of lysosomal accumulation in xenobiotic-induced toxicity.

19.4.1 Liver Toxicity

More than 900 drugs, toxins, and herbs have been reported to cause liver injury. Different mechanisms such as reactive metabolite formation, oxidative stress, and mitochondrial impairment have been proposed to contribute to the pathogenesis of drug-induced liver injury. The involvement of the lysosomal pathway in liver toxicity has been suggested by numerous studies. Oral drugs are absorbed by the gastrointestinal system and enter the liver via the hepatic portal vein prior to systemic circulation. Thus, the liver typically receives the highest maximal drug exposures relative to other tissues. These high maximal exposures coupled with the lysosomal trapping of basic lipophilic drugs often leads to drug accumulation in the liver. For example, the lysosomotropic compounds amiodarone and chloroquine have been shown to have significantly higher levels in liver tissue compared to the plasma [146,147]. Many basic lipophilic compounds, such as amiodarone, chloroquine, and perhexiline, are indeed associated with hepatic morphologic effects, which include marked lysosomal changes described as lamellar lysosomal inclusions [148].

As discussed previously, nonbasic xenobiotics can affect the lysosomal pathway via a nonaccumulation mechanism. The cathepsin B–mitochondria apoptotic pathway has been demonstrated to play a critical role in hepatocyte apoptosis and liver injury induced by TNFα [149,150]. Either genetic knockout or pharmacological inhibitors of cathepsin B can mitigate the liver damage. Interestingly, acetaminophen, overdose of which is the most frequent cause of drug-induced liver failure in the United States, can also increase the lysosomal membrane instability as demonstrated by the release of cathepsin B from lysosomes [151]. In addition, lysosomal iron mobilization into mitochondria has been shown as a contributing factor to acetaminophen-induced hepatotoxicity [152,153]. Hepatotoxicity is also one of the adverse reactions caused by diclofenac, an arylacetic nonsteroidal anti-inflammatory drug (NSAID). Involvement of LMP and lysosomal/mitochondrial cross talk induced by reactive oxygen species (ROS) formation contributes to diclofenac hepatotoxicity [154]. It is noteworthy that the lysosomal pathway also contributes to liver injury in pathological conditions such as cholestasis and hepatic lipotoxicity [155,156]. Using a bile duct-ligated mouse model, both genetic and pharmacologic inactivation of cathepsin B reduces liver injury, inflammation, and hepatic fibrogenesis during cholestasis, indicating a role for lysosomes in liver damage [155]. Similarly, in a dietary murine model of nonalcoholic fatty liver disease and hepatic steatosis, liver injury was mitigated by the genetic and pharmacologic inactivation of cathepsin B [156]. Moreover, the loss of *Atg7* in the adult liver leads to hepatomegaly and accumulation of abnormal organelles in hepatic cells [157]. Histological analysis revealed disorganized hepatic lobules and cell swelling in the ATG7-deficient liver. These findings support the physiological role of the autophagy/lysosome pathway in the maintenance of liver homeostasis and lend credence to the involvement of xenobiotic disruption of this pathway in liver injury.

19.4.1.1 Heart The heart is low in lysosomal mass and in general does not accumulate lysosomotropic xenobiotics to levels measured in the liver [144]. However, cardiomyocytes can be very sensitive to tissue homeostasis perturbation due to membrane trafficking impairment. Cardiac muscle is characterized by a predominantly aerobic metabolism and consequently ROS are continuously generated [158]. Efficient removal of damaged cellular structures by the lysosomal pathway, particularly autophagy, is critical to the maintenance of the cardiac myocyte homeostasis. Indeed, controlled cardiac-specific deficiency of autophagy-related homolog 5 (Atg5) in adult mice led to cardiac hypertrophy, left ventricular dilation, and contractile dysfunction [159]. The findings could be due to the long-lived postmitotic nature of the cardiac myocytes, in which autophagy plays an important role in the quality control of proteins and organelles. Diminished lysosomal function has been associated with heart disease and aging as evidenced by accumulation of the age pigment lipofuscin in the lysosomal compartment and extralysosomal waste accumulation [160]. Danon disease in humans is a well-characterized example of how lysosomal dysfunction by LAMP-2 deficiency can lead to heart failure [161].

The classical lysosomal inhibitors chloroquine and hydroxychloroquine have been associated with cardiac adverse effects (more commonly cardiac conduction block) with long-term treatment. Electron microscopy analysis of myocardial samples from patients revealed large lysosomes and myelin figures [162], which is consistent with perturbed autophagy. Another lysosomotropic compound, isoproterenol, is a β-adrenergic agonist that induces myocardial infarction and triggers decreased stability of lysosomal membranes. This is reflected by lowered activities of β-glucuronidase and cathepsin D in the lysosomal fraction [163], linking lysosomes in the cardiac toxicity.

Imatinib mesylate (Gleevec) inhibits the catalytic activity of Bcr-abl and is the first targeted therapy approved for the treatment of Ph+CML. Despite its breakthrough success in the treatment of CML, imatinib treatment has been implicated in the development of congestive heart failure [164]. There have been controversial reports regarding the association of the inhibition of wild-type Abelson oncogene 1 (c-Abl) kinase and the observed cardiac findings. Through the use of RNA interference technology and kinase-inactive chemical analogs of imatinib, a lack of association between c-Abl inhibition and *in vitro* cardiomyocyte toxicity was demonstrated [69]. Instead, the toxic effects of imatinib were related to the compound's physicochemical properties. Imatinib and its kinase-inactive analogs are basic lipophilic compounds that accumulate in lysosomes and interrupt autophagy. The vacuolar-type H(+)-ATPase inhibitor Baf A inhibits the lysosomal acidification [165] and consequently prevents accumulation of basic compounds in lysosomes. The *in vitro* cytotoxicity of both imatinib and kinase-inactive analogs was shown to be rescued by Baf A, further supporting the contribution of the lysosomal accumulation in imatinib-induced cardiac toxicity. The *in vivo* findings from imatinib-treated spontaneous hypertensive rats further support this hypothesis. In the imatinib-treated rat, aggregated lysosomes and nonfusion vacuoles were detected in the cytoplasm of severely atrophic myocytes, indicative of autophagy disruption [166]. Interestingly, myocyte lesions were more severe in spontaneous hypertensive rats than in normal Sprague Dawley rat. Similarly, when cardiac-specific deficiency of Atg5 occurs early in cardiogenesis between embryonic day 7.5 and 8, no cardiac phenotype was observed under baseline conditions. However, cardiac dysfunction and left ventricular dilatation developed 1 week after treatment with pressure overload [159], suggesting the role of autophagy in response to this type of cardiac stress. As drug treatment gradually triggers lysosomal dysfunction and autophagy interruption, adaptive changes take place to maintain the basal function. This adaptive response can be compromised by the stress of pressure overload leading to greater sensitivity to autophagy disruption. Alternatively, autophagy disruption may limit the adaptive response to the pressure overload, resulting in the same net effect on cardiac function. As a result, utilizing a stress model might increase the sensitivity for identification of cardiac toxicants, especially those that trigger autophagy disruption.

The clinical use of the antineoplastic drug doxorubicin (DOX) is limited by its cumulative dose-dependent propensity to cause irreversible degenerative cardiomyopathy and congestive heart failure [167,168]. The mechanism of cardiotoxicity by DOX involves oxidative stress, mitochondrial dysfunction, and apoptosis. Interestingly, DOX is a basic lipophilic compound and has been shown to become trapped insides lysosomes [169] affecting lysosomal morphology and enzyme activity [170]. Recently, dysregulation of autophagy has been shown to play a contributing role in DOX-induced cardiotoxicity [171]. Using GFP-LC3 transgenic mice, acute DOX cardiotoxicity is associated with impaired autophagic functions demonstrated by an increase of GFP-LC3 puncta, and accumulation of LC3-II and p62 [171]. Interestingly, starvation, which is the most extensively studied condition that induces autophagy, can at least partially restore autophagosome formation and is cardioprotective against DOX-induced cardiotoxicity.

19.4.2 Kidney Toxicity

Cisplatin, aminoglycoside antibiotics, and cadmium have been extensively studied for their nephrotoxic effects on proximal tubule epithelial cells [14,172,173].

Lysosomal acid phosphatase staining has shown that proximal tubules are rich in lysosomes, while in distal tubules and glomerular cells, lysosomes are relatively sparse [174]. The local abundance of lysosomes in the proximal tubule could explain in part drug-induced kidney injury as a result of xenobiotic accumulation in the lysosomes. Functionally, proximal tubular epithelial cells have an extensive apical endocytotic apparatus that is critical for the reabsorption and degradation of proteins that traverse the glomerular filtration barriers [175] and endosomal acidification has been shown to be critical for this physiological function of proximal tubule cells [176]. Therefore, lysosomal disruption could adversely affect proximal tubule function. It has been demonstrated that cisplatin [177], aminoglycosides [178], and cadmium [14] can accumulate in proximal tubule cells, localizing in lysosomal vacuoles. Ultrastructural alterations in proximal tubular cells following aminoglycoside treatment span from enlargement of lysosomes and progressive deposition of polar lipids at low doses to lysosomal rupture and extensive mitochondrial swelling at higher doses [179]. Apoptosis induced by gentamicin in LLC-PK1 cells (pig kidney epithelial cells) involves the permeabilization of lysosomes followed by subsequent activation of the mitochondrial apoptosis pathway [180], indicating the critical role of LMP in the mitochondrial change and cell death. In addition, gentamicin has been shown to decrease the activity of cathepsins B and L in proximal tubules after *in vivo* treatment, which could reduce renal protein catabolism [181]. Cisplatin has also been shown to increase lysosome fragility [182] and accumulation of autophagosome-associated LC3 [173]. In addition, cisplatin also inhibits uptake of fluorescein-isothiocyanate (FITC)-albumin, a receptor-mediated endocytosis marker [183]. Ulinastatin can protect the kidney from gentamicin and cisplatin-induced toxicity, possibly through its lysosomal membrane-stabilizing effect [182].

Remarkably, both gentamicin and cisplatin are highly polar compounds with $c\log P$ values of −1.7 and −2.19, respectively. Compounds with this range of polarity typically do not cross lipid membranes by passive diffusion. Electron microscopic analysis demonstrated that aminoglycosides are localized in claritin-coated pits, endocytic compartments, and lysosomes in the proximal tubular cells, indicating the involvement of an endocytic pathway of accumulation [184]. Megalin, a multiligand, endocytic receptor abundantly expressed in the renal proximal tubule, has been associated with kidney tubular cell uptake of aminoglycosides [185], and renal accumulation of gentamicin can be decreased by competition with another megalin ligand [184]. This suggests an alternative route for basic compound accumulation in lysosomes besides permeation. It is noteworthy to mention that megalin is also expressed in the marginal cells of the stria vascularis of cochlear duct of the ear. Both aminoglycosides and cisplatin are known for their nephro- and ototoxicity, further highlighting the role of megalin in the uptake of these polar molecules. In fact, nonsynonymous single-nucleotide polymorphisms of the megalin gene have been shown to impact the individual susceptibility toward cisplatin-induced ototoxicity [186]. Clinically, the susceptibility of the kidney to these two compounds may be enhanced when administered by the IV route as this route bypasses absorption barriers and first-pass metabolism that an oral route would normally see. A relatively large volume of blood flow to the kidney increases the accessibility of these drugs to the kidney as well.

19.4.3 Retinal

The retina is a thin layer of light-sensitive neural tissue that lines the back of the eye. The retina is highly structured with multiple intricate layers responsible for converting of visible light into the electrochemical signals interpreted by the brain as vision. The RPE is a specialized monolayer epithelium that forms the outermost layer of the retina and is positioned between the neuroretina and choroid. The RPE performs multiple functions including nutritional transport, light absorption, and retinoid metabolism [187,188]. RPE cells are differentiated with numerous long microvilli on the apical side facing the photoreceptor outer segment. One RPE cell supports 30–50 photoreceptors, which shed ~5% of their outer segment mass daily [189]. The shredded discs are phagocytosed into the RPE and digested within phagolysosomes. RPE cells are considered one of the most active phagocytic cell types, making them particularly susceptible to membrane trafficking perturbation due to lysosomal dysfunction. Not surprisingly, cathepsin D-deficient mice have membrane-bound compartments containing granular osmiophilic deposits in the RPE, which leads to progressive retinal layer thinning and loss of cone and rod cells [190]. The critical role of the lysosome in retinal homeostasis is also corroborated by the LSDs, such as mucopolysaccharidoses (MPS). MPS is due to iduronate-2-sulfatase deficiency and characterized by the defective breakdown of glycosaminoglycans (GAGs). GAG deposition within the retinal pigment epithelial cells and in the photoreceptor matrix leads to progressive photoreceptor loss, retinal degeneration, and dysfunction of the remaining photoreceptors [191]. In addition, in the mouse model of Npc1 (Niemann–Pick type C) deficiency, accumulation of lipofuscin in the RPE layer and degeneration of photoreceptors was observed [192]. Many lysosomotropic compounds including amiodarone, imipramine, clomipramine, chlorphentermine, and chloroquine have been shown to induce retinal lipidosis [193]. In long-term treatments with chloroquine (10 months) in rats, complete destruction of the outer segments of photoreceptors is indeed observed [194]. Recently, a correlation among the physicochemical properties, lysosomal dysfunction *in vitro*, and retinal lesion *in vivo* was established, demonstrating the critical role of physicochemical properties in the xenobiotics-induced retinal toxicity [195].

19.4.4 Peripheral Neuropathy

Neurons are especially dependent on the membrane trafficking due to their complex, highly polarized, elongated structure, and active trafficking for synaptic activity. Anterograde axonal transport delivers cargo from the cell body toward distal terminals to maintain synaptic function, whereas retrograde axonal transport helps clear toxic components from nerve terminals and deliver distal trophic signals to the soma. LC3-positive autophagosomes have been shown to undergo exclusive retrograde movement and rapidly obtain the endolysosomal markers Rab7 and

LAMP1 [196]. In addition, signaling endosomes containing neurotrophic factor receptor signaling complexes and their retrograde transport is critical to deliver the signals from distal points to the cell body [197]. Axonal transport deficits have emerged as a contributing factor in multiple neurodegenerative diseases such as AD and PD [198,199]. Recent data has suggested that impairment of membrane trafficking could be, in part, the cause for drug-induced peripheral neuropathy. Taxanes, including paclitaxel, docetaxel, and cabazitaxel, stabilize GDP-bound tubulin in the microtubule, whereas plant alkaloids, such as vinblastine, vincristine, and vinorelbine, prevent formation of microtubules. All of these drugs that perturb microtubule dynamics are associated with peripheral neuropathy [200], predominately of sensory neurons. Microtubule-targeting drugs have demonstrated inhibition of axonal transport [201], which translates into length-dependent neuropathies after cumulative exposure. In addition, peripheral neuropathy also develops during platinum drug treatment including cisplatin, carboplatin, and oxaliplatin [200] due to perturbed lysosomal function demonstrated with the disrupted aggresome-autophagy pathway [202].

Lysosomotropic aminoglycosides have long been reported to cause peripheral neuropathy, and nerve biopsies have revealed an association with lysosomal abnormalities [203,204]. Recently, a new peripheral neuropathy warning has been issued by Food and Drug Administration (FDA) for the fluoroquinolone class of antibiotics including ciprofloxacin, levofloxacin, moxifloxacin, norfloxacin, ofloxacin, and gemifloxacin (http://www.fda.gov/Drugs/ DrugSafety/ucm365050.htm). Fluoroquinolones are basic drugs for which lysosomal accumulation has been demonstrated. It is fairly sensible to assume that lysosomal dysfunction and membrane trafficking perturbation play a significant role in those peripheral neuropathies. However, additional studies are needed to further strengthen the link. Remarkably, platinum drugs, aminoglycosides, and fluoroquinolones antibiotics are more hydrophilic than typical lysosomotropic compounds [205]. The potential effects of hydrophilicity on their mechanism of toxicity warrant further investigation.

19.4.5 Muscle Toxicity

Similar to the other postmitotic tissues, skeletal muscle is also vulnerable to dysfunctional organelles and accumulation of aggregation-prone proteins. Therefore, these degradation systems need to be carefully regulated in muscle cells. Indeed, excessive or defective activity of autophagic lysosomes or ubiquitin-proteasomes leads to detrimental effects on muscle homeostasis [206]. A growing number of studies have linked abnormalities in the regulation of these two pathways to myofiber degeneration and muscle weakness. The concept linking lysosomal failure or membrane trafficking perturbation to skeletal muscle toxicity is further supported by evidence from genetic diseases and a range of mouse knockout studies. Danon disease, an X-linked dominant disorder resulting from a mutation in the lysosomal protein LAMP-2 is characterized by cardiomyopathy and myopathy [207]. Extensive accumulation of autophagosomes in muscle was observed in LAMP-2-deficient mice and patients with Danon disease [208]. Conditional knockout of the Atg7 gene in skeletal muscle of mice induces myofiber degeneration and muscle loss [209]. In addition, autophagy-deficient muscles are extremely weak compared to muscles from control animals.

Multiple lysosomotropic drugs, such as chloroquine and hydroxychloroquine, are associated with lysosomal storage myopathy, where storage of curvilinear bodies within the lysosomes and autophagic vacuolation with phospholipid inclusion has been demonstrated [210]. Amiodarone has been associated with myopathy in humans and mouse models, demonstrating proximal muscle weakness with diffuse myalgia [211] and autophagic vacuolation and phospholipid inclusions, respectively [212]. Antimicrotubule drugs such as colchicines and vincristine have also been reported to cause myopathy [213,214], which, in the case of colchicine, causes a vacuolar myopathy with accumulation of lysosomes and autophagic vacuoles without necrosis [215].

19.4.6 Tumorigenesis

Autophagy has paradoxical roles in the initiation and progression of cancer. In established tumors, autophagy might provide a survival advantage for tumor cells under metabolic stress. The emerging hypothesis is that autophagy can be a safe guardian of the genome and provide an antitumorgenesis function, at least during initiation. Hypothetically, lysosomal toxicants could block the autophagy process, subsequently triggering mitochondrial dysfunction leading to an increase in oxidative stress that consequently elicits DNA damage and genetic instability. Thus, lysosomal perturbation and autophagy inhibition could play a role in drug-induced tumorigenesis. This notion was supported by a variety of autophagy gene deletion studies. The first evidence to show a link between autophagy and tumorigenesis was established in 1999, when the ATG gene beclin 1 was discovered as a candidate tumor suppressor [216]. A single copy loss of beclin-1 in the heterozygous knockout mice increases the frequency of spontaneous malignancies in the lung and liver and accelerates the development of hepatitis B virus-induced premalignant lesions [217]. Besides beclin-1, other autophagy gene knockouts (e.g., Bif-1, atg5, and atg7) have also been reported to lead to tumors [135,218]. Nrf2, a transcription factor that controls the expression of antioxidant and cytoprotective genes, was recently suggested to have an oncogenic function. In liver-specific autophagy-deficient mice, Nrf2 activation via p62 contributes to the development of hepatocellular carcinoma [219].

High incidences of cancer in various organs such as skin, lung, bladder, liver, and kidney have been associated with chronic exposure to inorganic arsenic from contaminated drinking water [220]. Multiple mechanisms including changes in DNA methylation and generation of ROS [221,222] have been proposed for the mechanism of arsenic-related carcinogenicity. In a notable study, arsenic was shown to block autophagy, resulting in accumulation of p62, sequestration of Keap1 in autophagosomes, and prolonged Nrf2 activation [134], similarly to autophagy-deficient mice. Interestingly, Nrf2 overexpression has been associated with many cancer types [223,224]. As a major regulator of cytoprotective responses, the Nrf2 pathway has also been demonstrated to be hijacked by cancer cells, resulting in intrinsic or acquired chemoresistance [225,226]. Although further studies are required, the noncanonical activation of Nrf2 due to autophagy impairment could be part of a novel mechanism for arsenic carcinogenicity.

Besides arsenic, other heavy metals can potentially employ the same pathway for carcinogenicity. Cadmium is classified as a human carcinogen [227] and can trigger a lysosomal membrane permeabilization-dependent necrosis pathway [228]. It is possible that autophagy deficiency induced by cadmium can also activate Nrf2, leading to cancer. Zebrafish larvae exposed to cadmium indeed showed an increase in glutathione-S-transferase glutamate–cysteine ligase catalytic subunit, heme oxygenase 1, and peroxiredoxin 1 mRNA levels indicative of Nrf2 activation [229].

19.4.7 General Considerations for Organ Toxicity

In this review, we only touch upon a limited set of organ toxicities but other organs such as thyroid and the immune system can also be impacted by lysosomal toxicants. There are a number of factors that need to be considered when dealing with toxicity related to lysosomal dysfunction.

1. Although the lysosome was discovered more than 50 years ago and the concept of lysosomotropism has been well established, the mechanistic link to toxicity has been overshadowed by the vast research effort focusing on PLD. The link between lysosomal dysfunction and organ toxicity discussed in this review is largely based on the information from LSDs, genetic modulation models, and lysosomal changes inflicted by xenobiotics. There are many gaps in the knowledge of the molecular pathways tying lysosomal perturbation to organ toxicity that requires further investigation. It is plausible that certain cells or organs are more susceptible to the particular cellular change due to lysosome dysfunction. For instance, phagocytosis inhibition could be more detrimental for retinal epithelial cells than liver cells, while cardiomyocytes may be more susceptible to autophagy inhibition based on their respective specific physiological functions.
2. Clearly not all lysosomal toxicants have the same *in vivo* toxicity profile. Dose and duration certainly can dictate the degree of lysosomal dysfunction thus resulting in different toxicity manifestations. More importantly, the tissue distribution could significantly contribute to the organs impacted. Although various analyses have been conducted to determine the correlation between physicochemical properties with lysosomal accumulation *in vitro*, further investigations are required to understand the translation to tissue accumulation *in vivo*. Asymmetrical tissue accumulation are often observed with xenobiotics and it is reasonable to assume that orally administered compounds with higher lipophilicity will result in higher relative liver concentrations due to the first-pass effects and fast permeation. This scenario may lead to decreased exposure in the other organs, while less lipophilic xenobiotics may have a better chance of accumulating in the other organs. Development of sophisticated mathematic modeling with incorporation of pharmacokinetic parameters could

certainly help uncover the relationship between physicochemical properties and tissue distribution, in hopes of predicting organ toxicity. However, it is more complex and challenging when considering the contribution of active transport in the model. The other convoluting factor is that adaptive responses will occur when cells or tissues are challenged with lysosomal toxicants. As was discussed earlier, lysosomal and autophagy genes can be upregulated by lysosomal toxicants. The ultimate physiological outcome will be determined by the net effects of the negative impact by the xenobiotic and adaptive response within the cells. In other words, an adequate adaptive response may spare some tissues from detrimental consequences, while inadequate adaptation might result in the toxicity in others. The possibility that different tissues are equipped with differing capabilities to respond to lysosomal insults warrants further investigation.

3. The occurrence of organ toxicities are a function of duration and dose, and some may only manifest following long-term exposure due to the protective adaptive response and dependence on interactions with other stressors, as was observed with cardiotoxicity from chloroquine. To some extent, the occurrence of organ toxicity mimics the progressive aging process. This presents a challenge for early detection of toxicity in drug development since longer term studies are not conducted until the later stages of development. The intriguing discovery that atg5 knockout mice only showed rapid and dramatic declines in cardiac function with pressure overload [159] suggests the value of using stress models in toxicity screening. It is conceivable that, at basal conditions, cells or tissues can maintain normal function with a certain degree of lysosomal dysfunction, particularly with some level of adaptive response, while additional stress can sway the delicate balance toward a pathophysiological perturbation. This concept can also offer some explanation for the occasional discrepancy between preclinical studies and clinical outcome. In general, young healthy animals are used in the preclinical studies, while patients in the clinical setting may carry various levels of lysosomal dysfunction (disease- or age-related) prior to drug exposure. As a result, a greater incidence or severity of toxicity, or a different toxicity profile altogether may be revealed in the clinical situation. Alternative animal models, like those under particular stress or with preexisting lysosomal dysfunction, could offer additional value in predicting clinical toxicity.
4. Toxicities triggered by impairment of lysosomal functional are potentially reversible. For instance, PLD is reversible after cessation of fluoxetine treatment and reversibility of fulminant amiodarone-induced hepatitis has also been shown for some cases [230]. While the degree of damage can determine the potential for reversibility, it is possible that certain dosing regimens could prevent toxicity from occurring. Further investigation on the tissue pharmacokinetics during the dosing and recovery phases may shed light on an innovative dosing regimen design that avoids toxicity. Growing evidence has demonstrated the benefit of various therapeutic approaches, such as antioxidants or compounds that increase the lysosomal membrane stability. The antioxidant vitamin E was shown to play a role in ameliorating

amiodarone toxicity [231,232]. In addition, gallic acid and lycopene have been shown to ameliorate isoproterenol-induced cardiotoxicity by preserving the integrity of the lysosomal membrane, presumably due to their free radical scavenging and antioxidant properties [163,233]. The concept of increasing stability of the lysosomal membrane to ameliorate toxicity is further supported by the finding that HSP70 can revert Niemann–Pick disease-associated lysosomal pathology by stabilizing lysosomes through binding to anionic phospholipid bis(monoacylglycero)phosphate [234]. Recently, curcumin has been shown to promote exosomes/microvesicles secretion that attenuates lysosomal cholesterol traffic impairment [235], indicating the potential of curcumin as a remedy for lysosomal dysfunction. However, curcumin was also reported to trigger LMP and apoptosis [236]. Further research in this area is needed to identify and characterize a genuine antidote. Appropriate combination of drugs and antidotes may well attenuate or diminish certain toxicity associated with lysosomal dysfunction.

5. Mechanisms of toxicity can be multifactorial. In this review, we focused on how a compound's impact on lysosomal function can lead to toxicity. However, other organelles (e.g., mitochondria) or biological processes can also be simultaneously disturbed by these same compounds further contributing to the manifestation of toxicity. Nefazodone, which was withdrawn from the US market in 2004 due to hepatotoxicity, has been studied extensively. As expected with a basic pK_a of 7.65 and $c\log P$ of 4.09, nefazodone has been shown to increase lysosomal content staining [35]. It was also demonstrated that nefazodone profoundly inhibited mitochondrial respiration in isolated rat liver mitochondria and in intact HepG2 cells [237]. Potentially, both mechanisms can contribute to the hepatotoxicity observed for nefazodone given the potential for interaction between the two pathways. Lysosomal dysfunction can aggravate the mitochondria function by perturbing efficient mitochondria turnover. This can lead to mitochondrial dysfunction-elicited ROS production that could further induce lysosomal membrane permeabilization. Compounds with multiple impacts on organelles or biological processes may carry a higher risk of toxicity.

19.5 CONCLUDING REMARKS

Growing evidence suggested that lysosomal dysfunction and membrane trafficking perturbation not only contribute to aging and aging-related diseases but also serve as unique emerging mechanisms for xenobiotic-induced toxicities. Lysosomal toxicants can disrupt lysosomal function either via accumulation, direct damage to the lysosome, or indirectly by vesicular trafficking perturbation. Lysosomal dysfunction not only impacts all membrane trafficking pathways that converge on lysosomes but also modulates various signal transduction molecules, both of which can impair various cellular biological functions, consequently contributing to the development of a variety of organ toxicities. Although membrane trafficking, especially autophagy research, has progressed noticeably, research efforts aimed at understanding the link between the lysosomal pathway and toxicity are still in imperative need. A molecular comparison among the toxicities arising from lysosomal toxicants, aging-related diseases, including neurodegeneration, and genetic lysosomal dysfunction could serve as a valuable model in understanding disease pathology related to lysosomal dysfunction.

REFERENCES

[1] Sancak Y, Bar-Peled L, Zoncu R, Markhard AL, Nada S, et al. Ragulator-Rag complex targets mTORC1 to the lysosomal surface and is necessary for its activation by amino acids. Cell 2010;141:290–303.

[2] de Duve C, de Barsy T, Poole B, Trouet A, Tulkens P, et al. Commentary. Lysosomotropic agents. Biochem Pharmacol 1974;23:2495–2531.

[3] Honegger UE, Zuehlke RD, Scuntaro I, Schaefer MH, Toplak H, et al. Cellular accumulation of amiodarone and desethylamiodarone in cultured human cells. Consequences of drug accumulation on cellular lipid metabolism and plasma membrane properties of chronically exposed cells. Biochem Pharmacol 1993;45:349–356.

[4] Law V, Knox C, Djoumbou Y, Jewison T, Guo AC, et al. DrugBank 4.0: shedding new light on drug metabolism. Nucleic Acids Res 2014;42:D1091–1097.

[5] Wager TT, Chandrasekaran RY, Hou X, Troutman MD, Verhoest PR, et al. Defining desirable central nervous system drug space through the alignment of molecular properties, *in vitro* ADME, and safety attributes. ACS Chem Neurosci 2010;1: 420–434.

[6] Tulkens P, Trouet A. The uptake and intracellular accumulation of aminoglycoside antibiotics in lysosomes of cultured rat fibroblasts. Biochem Pharmacol 1978;27:415–424.

[7] Beauchamp D, Gourde P, Bergeron MG. Subcellular distribution of gentamicin in proximal tubular cells, determined by immunogold labeling. Antimicrob Agents Chemother 1991;35:2173–2179.

[8] Hashino E, Shero M, Salvi RJ. Lysosomal targeting and accumulation of aminoglycoside antibiotics in sensory hair cells. Brain Res 1997;777:75–85.

[9] Nagai J, Takano M. Entry of aminoglycosides into renal tubular epithelial cells via endocytosis-dependent and endocytosis-independent pathways. Biochem Pharmacol 2014;90:331–337.

[10] Mizuta K, Saito A, Watanabe T, Nagura M, Arakawa M, et al. Ultrastructural localization of megalin in the rat cochlear duct. Hear Res 1999;129:83–91.

[11] Chatelet F, Brianti E, Ronco P, Roland J, Verroust P. Ultrastructural localization by monoclonal antibodies of brush border antigens expressed by glomeruli. I. Renal distribution. Am J Pathol 1986;122:500–511.

[12] Dagil R, O'Shea C, Nykjaer A, Bonvin AM, Kragelund BB. Gentamicin binds to the megalin receptor as a competitive inhibitor using the common ligand binding motif of complement type repeats: insight from the nmr structure of the 10th complement type repeat domain alone and in complex with gentamicin. J Biol Chem 2013;288:4424–4435.

[13] Sokolova IM, Ringwood AH, Johnson C. Tissue-specific accumulation of cadmium in subcellular compartments of eastern oysters *Crassostrea virginica* Gmelin (Bivalvia: Ostreidae). Aquat Toxicol 2005;74:218–228.

[14] Chargui A, Zekri S, Jacquillet G, Rubera I, Ilie M, et al. Cadmium-induced autophagy in rat kidney: an early biomarker of subtoxic exposure. Toxicol Sci 2011;121:31–42.

[15] Safaei R, Larson BJ, Cheng TC, Gibson MA, Otani S, et al. Abnormal lysosomal trafficking and enhanced exosomal export of cisplatin in drug-resistant human ovarian carcinoma cells. Mol Cancer Ther 2005;4:1595–1604.

[16] Ciarimboli G. Membrane transporters as mediators of Cisplatin effects and side effects. Scientifica (Cairo) 2012;2012:473829.

[17] Maciaszczyk-Dziubinska E, Wawrzycka D, Wysocki R. Arsenic and antimony transporters in eukaryotes. Int J Mol Sci 2012;13:3527–3548.

[18] Thevenod F. Nephrotoxicity and the proximal tubule. Insights from cadmium. Nephron Physiol 2003;93:87–93.

[19] Gachot B, Poujeol P. Effects of cadmium and copper on zinc transport kinetics by isolated renal proximal cells. Biol Trace Elem Res 1992;35:93–103.

[20] Lockwood TD. Lysosomal metal, redox and proton cycles influencing the CysHis cathepsin reaction. Metallomics 2013;5:110–124.

[21] Abouhamed M, Wolff NA, Lee WK, Smith CP, Thevenod F. Knockdown of endosomal/lysosomal divalent metal transporter 1 by RNA interference prevents cadmium-metallothionein-1 cytotoxicity in renal proximal tubule cells. Am J Physiol Renal Physiol 2007;293:F705–712.

[22] Kane MD, Schwarz RD, St Pierre L, Watson MD, Emmerling MR, et al. Inhibitors of V-type ATPases, bafilomycin A1 and concanamycin A, protect against beta-amyloid-mediated effects on 3-(4,5-dimethylthiazol-2-yl)-2,5-diphenyltetrazolium bromide (MTT) reduction. J Neurochem 1999;72:1939–1947.

[23] Reijngoud DJ, Oud PS, Tager JM. Effect of ionophores on intralysosomal pH. Biochim Biophys Acta 1976;448:303–313.

[24] Bernheimer AW, Schwartz LL. Lysosomal disruption by bacterial toxins. J Bacteriol 1964;87:1100–1104.

[25] Averette KM, Pratt MR, Yang Y, Bassilian S, Whitelegge JP, et al. Anthrax lethal toxin induced lysosomal membrane permeabilization and cytosolic cathepsin release is Nlrp1b/Nalp1b-dependent. PLoS One 2009;4:e7913.

[26] Werneburg NW, Bronk SF, Guicciardi ME, Thomas L, Dikeakos JD, et al. Tumor necrosis factor related apoptosis-inducing ligand (TRAIL) protein-induced lysosomal translocation of proapoptotic effectors is mediated by phosphofurin acidic cluster sorting protein-2 (PACS-2). J Biol Chem 2012;287:24427–24437.

[27] Myers BM, Tietz PS, Tarara JE, LaRusso NF. Dynamic measurements of the acute and chronic effects of lysosomotropic agents on hepatocyte lysosomal pH using flow cytometry. Hepatology 1995;22:1519–1526.

[28] Ndolo RA, Forrest ML, Krise JP. The role of lysosomes in limiting drug toxicity in mice. J Pharmacol Exp Ther 2010;333:120–128.

[29] Nadanaciva S, Lu S, Gebhard DF, Jessen BA, Pennie WD, et al. A high content screening assay for identifying lysosomotropic compounds. Toxicol In Vitro 2011;25:715–723.

[30] Kazmi F, Hensley T, Pope C, Funk RS, Loewen GJ, et al. Lysosomal sequestration (trapping) of lipophilic amine (cationic amphiphilic) drugs in immortalized human hepatocytes (Fa2N-4 cells). Drug Metab Dispos 2013;41:897–905.

[31] Kawai A, Uchiyama H, Takano S, Nakamura N, Ohkuma S. Autophagosome-lysosome fusion depends on the pH in acidic compartments in CHO cells. Autophagy 2007;3:154–157.

[32] Iwai-Kanai E, Yuan H, Huang C, Sayen MR, Perry-Garza CN, et al. A method to measure cardiac autophagic flux *in vivo*. Autophagy 2008;4:322–329.

[33] Hamacher-Brady A, Brady NR, Gottlieb RA. Enhancing macroautophagy protects against ischemia/reperfusion injury in cardiac myocytes. J Biol Chem 2006; 281:29776–29787.

[34] Funk RS, Krise JP. Cationic amphiphilic drugs cause a marked expansion of apparent lysosomal volume: implications for an intracellular distribution-based drug interaction. Mol Pharm 2012;9:1384–1395.

[35] Lu S, Jessen B, Strock C, Will Y. The contribution of physicochemical properties to multiple *in vitro* cytotoxicity endpoints. Toxicol In Vitro 2012;26:613–620.

[36] Logan R, Kong AC, Axcell E, Krise JP. Amine-containing molecules and the induction of an expanded lysosomal volume phenotype: a structure–activity relationship study. J Pharm Sci 2014;103:1572–1580.

[37] Obert LA, Sobocinski GP, Bobrowski WF, Metz AL, Rolsma MD, et al. An immunohistochemical approach to differentiate hepatic lipidosis from hepatic phospholipidosis in rats. Toxicol Pathol 2007;35:728–734.

[38] Chen PM, Gombart ZJ, Chen JW. Chloroquine treatment of ARPE-19 cells leads to lysosome dilation and intracellular lipid accumulation: possible implications of lysosomal dysfunction in macular degeneration. Cell Biosci 2011;1:10.

[39] Sardiello M, Palmieri M, di Ronza A, Medina DL, Valenza M, et al. A gene network regulating lysosomal biogenesis and function. Science 2009;325:473–477.

[40] Settembre C, Zoncu R, Medina DL, Vetrini F, Erdin S, et al. A lysosome-to-nucleus signalling mechanism senses and regulates the lysosome via mTOR and TFEB. EMBO J 2012;31:1095–1108.

[41] Groth-Pedersen L, Ostenfeld MS, Hoyer-Hansen M, Nylandsted J, Jaattela M. Vincristine induces dramatic lysosomal changes and sensitizes cancer cells to lysosome-destabilizing siramesine. Cancer Res 2007;67:2217–2225.

[42] Xu M, Liu K, Swaroop M, Sun W, Dehdashti SJ, et al. A phenotypic compound screening assay for lysosomal storage diseases. J Biomol Screen 2014;19:168–175.

[43] Grabner R. Influence of cationic amphiphilic drugs on the phosphatidylcholine hydrolysis by phospholipase A2. Biochem Pharmacol 1987;36:1063–1067.

[44] Kubo M, Hostetler KY. Mechanism of cationic amphiphilic drug inhibition of purified lysosomal phospholipase A1. Biochemistry 1985;24:6515–6520.

[45] Elojeimy S, Holman DH, Liu X, El-Zawahry A, Villani M, et al. New insights on the use of desipramine as an inhibitor for acid ceramidase. FEBS Lett 2006;580:4751–4756.

[46] Ikeda K, Hirayama M, Hirota Y, Asa E, Seki J, et al. Drug-induced phospholipidosis is caused by blockade of mannose 6-phosphate receptor-mediated targeting of lysosomal enzymes. Biochem Biophys Res Commun 2008;377:268–274.

[47] Falgueyret JP, Desmarais S, Oballa R, Black WC, Cromlish W, et al. Lysosomotropism of basic cathepsin K inhibitors contributes to increased cellular potencies against off-target cathepsins and reduced functional selectivity. J Med Chem 2005;48:7535–7543.

[48] Nixon RA, Cataldo AM. Lysosomal system pathways: genes to neurodegeneration in Alzheimer's disease. J Alzheimers Dis 2006;9:277–289.

[49] Cataldo AM, Nixon RA. Enzymatically active lysosomal proteases are associated with amyloid deposits in Alzheimer brain. Proc Natl Acad Sci U S A 1990;87:3861–3865.

[50] Reasor MJ, Hastings KL, Ulrich RG. Drug-induced phospholipidosis: issues and future directions. Expert Opin Drug Saf 2006;5:567–583.

[51] Ploemen JP, Kelder J, Hafmans T, van de Sandt H, van Burgsteden JA, Saleminki PJ, van Esch E. Use of physicochemical calculation of pKa and CLogP to predict phospholipidosis-inducing potential: a case study with structurally related piperazines. Exp Toxicol Pathol. 2004;55(5):347–355.

[52] Carlier MB, Laurent G, Claes PJ, Vanderhaeghe HJ, Tulkens PM. Inhibition of lysosomal phospholipases by aminoglycoside antibiotics: *in vitro* comparative studies. Antimicrob Agents Chemother 1983;23:440–449.

[53] Ouimet M, Franklin V, Mak E, Liao X, Tabas I, et al. Autophagy regulates cholesterol efflux from macrophage foam cells via lysosomal acid lipase. Cell Metab 2011;13:655–667.

[54] Ishibashi S, Yamazaki T, Okamoto K. Association of autophagy with cholesterol-accumulated compartments in Niemann–Pick disease type C cells. J Clin Neurosci 2009;16:954–959.

[55] Shoemaker CJ, Schornberg KL, Delos SE, Scully C, Pajouhesh H, et al. Multiple cationic amphiphiles induce a Niemann–Pick C phenotype and inhibit Ebola virus entry and infection. PLoS One 2013;8:e56265.

[56] Brunk UT, Terman A. Lipofuscin: mechanisms of age-related accumulation and influence on cell function. Free Radic Biol Med 2002;33:611–619.

[57] Ivy GO, Kanai S, Ohta M, Smith G, Sato Y, et al. Lipofuscin-like substances accumulate rapidly in brain, retina and internal organs with cysteine protease inhibition. Adv Exp Med Biol 1989;266:31–45; discussion 45–37.

[58] Ivy GO, Schottler F, Wenzel J, Baudry M, Lynch G. Inhibitors of lysosomal enzymes: accumulation of lipofuscin-like dense bodies in the brain. Science 1984;226:985–987.

[59] Kwok JC, Richardson DR. Examination of the mechanism(s) involved in doxorubicin-mediated iron accumulation in ferritin: studies using metabolic inhibitors, protein synthesis inhibitors, and lysosomotropic agents. Mol Pharmacol 2004;65:181–195.

[60] Olsson GM, Brunmark A, Brunk UT. Acridine orange-mediated photodamage of microsomal- and lysosomal fractions. Virchows Arch B Cell Pathol Incl Mol Pathol 1989;56:247–257.

[61] Erdal H, Berndtsson M, Castro J, Brunk U, Shoshan MC, et al. Induction of lysosomal membrane permeabilization by compounds that activate p53-independent apoptosis. Proc Natl Acad Sci U S A 2005;102:192–197.

[62] Ostenfeld MS, Hoyer-Hansen M, Bastholm L, Fehrenbacher N, Olsen OD, et al. Anti-cancer agent siramesine is a lysosomotropic detergent that induces cytoprotective autophagosome accumulation. Autophagy 2008;4:487–499.

[63] Uchimoto T, Nohara H, Kamehara R, Iwamura M, Watanabe N, et al. Mechanism of apoptosis induced by a lysosomotropic agent, L-Leucyl-L-Leucine methyl ester. Apoptosis 1999;4:357–362.

[64] Holman DH, Turner LS, El-Zawahry A, Elojeimy S, Liu X, et al. Lysosomotropic acid ceramidase inhibitor induces apoptosis in prostate cancer cells. Cancer Chemother Pharmacol 2008;61:231–242.

[65] van Nierop K, Muller FJ, Stap J, Van Noorden CJ, van Eijk M, et al. Lysosomal destabilization contributes to apoptosis of germinal center B-lymphocytes. J Histochem Cytochem 2006;54:1425–1435.

[66] Bursch W. The autophagosomal-lysosomal compartment in programmed cell death. Cell Death Differ 2001;8:569–581.

[67] Li W, Yuan X, Nordgren G, Dalen H, Dubowchik GM, et al. Induction of cell death by the lysosomotropic detergent MSDH. FEBS Lett 2000;470:35–39.

[68] Bursch W, Ellinger A, Kienzl H, Torok L, Pandey S, et al. Active cell death induced by the anti-estrogens tamoxifen and ICI 164 384 in human mammary carcinoma cells (MCF-7) in culture: the role of autophagy. Carcinogenesis 1996;17:1595–1607.

[69] Hu W, Lu S, McAlpine I, Jamieson JD, Lee DU, et al. Mechanistic investigation of imatinib-induced cardiac toxicity and the involvement of c-Abl kinase. Toxicol Sci 2012;129:188–199.

[70] Terman A, Kurz T, Gustafsson B, Brunk UT. Lysosomal labilization. IUBMB Life 2006;58:531–539.

[71] Boya P, Kroemer G. Lysosomal membrane permeabilization in cell death. Oncogene 2008;27:6434–6451.

[72] Terman A, Kurz T. Lysosomal iron, iron chelation, and cell death. Antioxid Redox Signal 2013;18:888–898.

[73] Amaravadi RK, Yu D, Lum JJ, Bui T, Christophorou MA, et al. Autophagy inhibition enhances therapy-induced apoptosis in a Myc-induced model of lymphoma. J Clin Invest 2007;117:326–336.

[74] Amaravadi RK, Lippincott-Schwartz J, Yin XM, Weiss WA, Takebe N, et al. Principles and current strategies for targeting autophagy for cancer treatment. Clin Cancer Res 2011;17:654–666.

[75] Ashoor R, Yafawi R, Jessen B, Lu S. The contribution of lysosomotropism to autophagy perturbation. PLoS One 2013;8:e82481.

[76] Cesen MH, Repnik U, Turk V, Turk B. Siramesine triggers cell death through destabilisation of mitochondria, but not lysosomes. Cell Death Dis 2013;4:e818.

[77] Zhang L, Yu J, Pan H, Hu P, Hao Y, et al. Small molecule regulators of autophagy identified by an image-based high-throughput screen. Proc Natl Acad Sci U S A 2007;104:19023–19028.

[78] de Medina P, Silvente-Poirot S, Poirot M. Tamoxifen and AEBS ligands induced apoptosis and autophagy in breast cancer cells through the stimulation of sterol accumulation. Autophagy 2009;5:1066–1067.

[79] Zschocke J, Zimmermann N, Berning B, Ganal V, Holsboer F, et al. Antidepressant drugs diversely affect autophagy pathways in astrocytes and neurons – dissociation from cholesterol homeostasis. Neuropsychopharmacology 2011;36:1754–1768.

[80] Renna M, Jimenez-Sanchez M, Sarkar S, Rubinsztein DC. Chemical inducers of autophagy that enhance the clearance of mutant proteins in neurodegenerative diseases. J Biol Chem 2010;285:11061–11067.

[81] Bharadwaj PR, Verdile G, Barr RK, Gupta V, Steele JW, et al. Latrepirdine (Dimebon) enhances autophagy and reduces intracellular GFP-Abeta42 levels in yeast. J Alzheimers Dis 2012;32:949–967.

[82] Brown MS, Anderson RG, Goldstein JL. Recycling receptors: the round-trip itinerary of migrant membrane proteins. Cell 1983;32:663–667.

[83] Wang LH, Rothberg KG, Anderson RG. Mis-assembly of clathrin lattices on endosomes reveals a regulatory switch for coated pit formation. J Cell Biol 1993;123:1107–1117.

[84] Hunt RC, Marshall-Carlson L. Internalization and recycling of transferrin and its receptor. Effect of trifluoperazine on recycling in human erythroleukemic cells. J Biol Chem 1986;261:3681–3686.

[85] DiPaola M, Keith CH, Feldman D, Tycko B, Maxfield FR. Loss of alpha 2-macroglobulin and epidermal growth factor surface binding induced by phenothiazines and naphthalene sulfonamides. J Cell Physiol 1984;118:193–202.

[86] Basu SK, Goldstein JL, Anderson RG, Brown MS. Monensin interrupts the recycling of low density lipoprotein receptors in human fibroblasts. Cell 1981;24:493–502.

[87] Chen CL, Hou WH, Liu IH, Hsiao G, Huang SS, et al. Inhibitors of clathrin-dependent endocytosis enhance TGFbeta signaling and responses. J Cell Sci 2009;122:1863–1871.

[88] Stenseth K, Thyberg J. Monensin and chloroquine inhibit transfer to lysosomes of endocytosed macromolecules in cultured mouse peritoneal macrophages. Eur J Cell Biol 1989;49:326–333.

[89] Tyteca D, Van Der Smissen P, Mettlen M, Van Bambeke F, Tulkens PM, et al. Azithromycin, a lysosomotropic antibiotic, has distinct effects on fluid-phase and receptor-mediated endocytosis, but does not impair phagocytosis in J774 macrophages. Exp Cell Res 2002;281:86–100.

[90] Saeed MF, Kolokoltsov AA, Albrecht T, Davey RA. Cellular entry of ebola virus involves uptake by a macropinocytosis-like mechanism and subsequent trafficking through early and late endosomes. PLoS Pathog 2010;6:e1001110.

[91] Cheng YL, Lan KH, Lee WP, Tseng SH, Hung LR, et al. Amiodarone inhibits the entry and assembly steps of hepatitis C virus life cycle. Clin Sci (Lond) 2013;125: 439–448.

[92] Fogler WE, Gersten DM, Fidler IJ. Inhibition by tertiary amine local anesthetics of phagocytosis in cultured mouse peritoneal macrophages. Biochem Pharmacol 1978;27:2447–2453.

[93] Mannerstrom M, Maenpaa H, Toimela T, Salminen L, Tahti H. The phagocytosis of rod outer segments is inhibited by selected drugs in retinal pigment epithelial cell cultures. Pharmacol Toxicol 2001;88:27–33.

[94] Brown JA, Novak EK, Swank RT. Effects of ammonia on processing and secretion of precursor and mature lysosomal enzyme from macrophages of normal and pale ear mice: evidence for two distinct pathways. J Cell Biol 1985;100:1894–1904.

[95] Tapper H, Sundler R. Role of lysosomal and cytosolic pH in the regulation of macrophage lysosomal enzyme secretion. Biochem J 1990;272:407–414.

[96] Tapper H, Sundler R. Bafilomycin A1 inhibits lysosomal, phagosomal, and plasma membrane H(+)-ATPase and induces lysosomal enzyme secretion in macrophages. J Cell Physiol 1995;163:137–144.

[97] Kotal P, Kotyk A. Exocytosis as a possible mechanism for the secretion of coproporphyrin from a chloroquine-treated yeast. Eur J Pharmacol 1988;155:301–303.

[98] Isobe I, Michikawa M, Yanagisawa K. Enhancement of MTT, a tetrazolium salt, exocytosis by amyloid beta-protein and chloroquine in cultured rat astrocytes. Neurosci Lett 1999;266:129–132.

[99] Stauber WT, Trout JJ, Schottelius BA. Exocytosis of intact lysosomes from skeletal muscle after chloroquine treatment. Exp Mol Pathol 1981;34:87–93.

[100] Meikle PJ, Fietz MJ, Hopwood JJ. Diagnosis of lysosomal storage disorders: current techniques and future directions. Expert Rev Mol Diagn 2004;4:677–691.

[101] Aerts JM, Kallemeijn WW, Wegdam W, Joao Ferraz M, van Breemen MJ, et al. Biomarkers in the diagnosis of lysosomal storage disorders: proteins, lipids, and inhibodies. J Inherit Metab Dis 2011;34:605–619.

[102] Jang A, Lee HJ, Suk JE, Jung JW, Kim KP, et al. Non-classical exocytosis of alpha-synuclein is sensitive to folding states and promoted under stress conditions. J Neurochem 2010;113:1263–1274.

[103] Medina DL, Fraldi A, Bouche V, Annunziata F, Mansueto G, et al. Transcriptional activation of lysosomal exocytosis promotes cellular clearance. Dev Cell 2011;21: 421–430.

[104] Sivaramakrishnan V, Bidula S, Campwala H, Katikaneni D, Fountain SJ. Constitutive lysosome exocytosis releases ATP and engages P2Y receptors in human monocytes. J Cell Sci 2012;125:4567–4575.

[105] Zoncu R, Bar-Peled L, Efeyan A, Wang S, Sancak Y, et al. mTORC1 senses lysosomal amino acids through an inside-out mechanism that requires the vacuolar H(+)-ATPase. Science 2011;334:678–683.

[106] Ashrafi G, Schwarz TL. The pathways of mitophagy for quality control and clearance of mitochondria. Cell Death Differ 2013;20:31–42.

[107] Platt FM, Boland B, van der Spoel AC. The cell biology of disease: lysosomal storage disorders: the cellular impact of lysosomal dysfunction. J Cell Biol 2012;199:723–734.

[108] Lee J, Giordano S, Zhang J. Autophagy, mitochondria and oxidative stress: cross-talk and redox signalling. Biochem J 2012;441:523–540.

[109] Ding WX, Yin XM. Mitophagy: mechanisms, pathophysiological roles, and analysis. Biol Chem 2012;393:547–564.

[110] Wei H, Kim SJ, Zhang Z, Tsai PC, Wisniewski KE, et al. ER and oxidative stresses are common mediators of apoptosis in both neurodegenerative and non-neurodegenerative lysosomal storage disorders and are alleviated by chemical chaperones. Hum Mol Genet 2008;17:469–477.

[111] Hamasaki M, Noda T, Baba M, Ohsumi Y. Starvation triggers the delivery of the endoplasmic reticulum to the vacuole via autophagy in yeast. Traffic 2005;6:56–65.

[112] Jia W, Pua HH, Li QJ, He YW. Autophagy regulates endoplasmic reticulum homeostasis and calcium mobilization in T lymphocytes. J Immunol 2011;186:1564–1574.

[113] Yorimitsu T, Nair U, Yang Z, Klionsky DJ. Endoplasmic reticulum stress triggers autophagy. J Biol Chem 2006;281:30299–30304.

[114] Ciechomska IA, Gabrusiewicz K, Szczepankiewicz AA, Kaminska B. Endoplasmic reticulum stress triggers autophagy in malignant glioma cells undergoing cyclosporine a-induced cell death. Oncogene 2013;32:1518–1529.

[115] Mahoney E, Lucas DM, Gupta SV, Wagner AJ, Herman SE, et al. ER stress and autophagy: new discoveries in the mechanism of action and drug resistance of the cyclin-dependent kinase inhibitor flavopiridol. Blood 2012;120:1262–1273.

[116] Pivtoraiko VN, Stone SL, Roth KA, Shacka JJ. Oxidative stress and autophagy in the regulation of lysosome-dependent neuron death. Antioxid Redox Signal 2009;11:481–496.

[117] Yu Z, Li W, Brunk UT. 3-Aminopropanal is a lysosomotropic aldehyde that causes oxidative stress and apoptosis by rupturing lysosomes. APMIS 2003;111:643–652.

[118] Ghosh M, Carlsson F, Laskar A, Yuan XM, Li W. Lysosomal membrane permeabilization causes oxidative stress and ferritin induction in macrophages. FEBS Lett 2011;585:623–629.

[119] Rouschop KM, Ramaekers CH, Schaaf MB, Keulers TG, Savelkouls KG, et al. Autophagy is required during cycling hypoxia to lower production of reactive oxygen species. Radiother Oncol 2009;92:411–416.

[120] Denamur S, Tyteca D, Marchand-Brynaert J, Van Bambeke F, Tulkens PM, et al. Role of oxidative stress in lysosomal membrane permeabilization and apoptosis induced by gentamicin, an aminoglycoside antibiotic. Free Radic Biol Med 2011;51:1656–1665.

[121] Murphy MP. How mitochondria produce reactive oxygen species. Biochem J 2009;417:1–13.

[122] Choi JH, Kim DH, Yun IJ, Chang JH, Chun BG, et al. Zaprinast inhibits hydrogen peroxide-induced lysosomal destabilization and cell death in astrocytes. Eur J Pharmacol 2007;571:106–115.

[123] Antunes F, Cadenas E, Brunk UT. Apoptosis induced by exposure to a low steady-state concentration of H2O2 is a consequence of lysosomal rupture. Biochem J 2001;356:549–555.

[124] Kurz T, Terman A, Gustafsson B, Brunk UT. Lysosomes and oxidative stress in aging and apoptosis. Biochim Biophys Acta 2008;1780:1291–1303.

[125] Duran A, Amanchy R, Linares JF, Joshi J, Abu-Baker S, et al. p62 is a key regulator of nutrient sensing in the mTORC1 pathway. Mol Cell 2011;44:134–146.

[126] Alers S, Loffler AS, Wesselborg S, Stork B. Role of AMPK-mTOR-Ulk1/2 in the regulation of autophagy: cross talk, shortcuts, and feedbacks. Mol Cell Biol 2012;32:2–11.

[127] Settembre C, Di Malta C, Polito VA, Garcia Arencibia M, Vetrini F, et al. TFEB links autophagy to lysosomal biogenesis. Science 2011;332:1429–1433.

[128] Yu L, McPhee CK, Zheng L, Mardones GA, Rong Y, et al. Termination of autophagy and reformation of lysosomes regulated by mTOR. Nature 2010;465:942–946.

[129] Balgi AD, Fonseca BD, Donohue E, Tsang TC, Lajoie P, et al. Screen for chemical modulators of autophagy reveals novel therapeutic inhibitors of mTORC1 signaling. PLoS One 2009;4:e7124.

[130] Jeon SH, Kim SH, Kim Y, Kim YS, Lim Y, et al. The tricyclic antidepressant imipramine induces autophagic cell death in U-87MG glioma cells. Biochem Biophys Res Commun 2011;413:311–317.

[131] Komatsu M, Kageyama S, Ichimura Y. p62/SQSTM1/A170: physiology and pathology. Pharmacol Res 2012;66:457–462.

[132] Lau A, Wang XJ, Zhao F, Villeneuve NF, Wu T, et al. A noncanonical mechanism of Nrf2 activation by autophagy deficiency: direct interaction between Keap1 and p62. Mol Cell Biol 2010;30:3275–3285.

[133] Komatsu M, Kurokawa H, Waguri S, Taguchi K, Kobayashi A, et al. The selective autophagy substrate p62 activates the stress responsive transcription factor Nrf2 through inactivation of Keap1. Nat Cell Biol 2010;12:213–223.

[134] Lau A, Zheng Y, Tao S, Wang H, Whitman SA, et al. Arsenic inhibits autophagic flux, activating the Nrf2-Keap1 pathway in a p62-dependent manner. Mol Cell Biol 2013;33:2436–2446.

[135] Takamura A, Komatsu M, Hara T, Sakamoto A, Kishi C, et al. Autophagy-deficient mice develop multiple liver tumors. Genes Dev 2011;25:795–800.

[136] Park J, Kwon D, Choi C, Oh JW, Benveniste EN. Chloroquine induces activation of nuclear factor-kappaB and subsequent expression of pro-inflammatory cytokines by human astroglial cells. J Neurochem 2003;84:1266–1274.

[137] Mokhtari D, Li T, Lu T, Welsh N. Effects of Imatinib Mesylate (Gleevec) on human islet NF-kappaB activation and chemokine production *in vitro*. PLoS One 2011;6:e24831.

[138] Jin Z, Li Y, Pitti R, Lawrence D, Pham VC, et al. Cullin3-based polyubiquitination and p62-dependent aggregation of caspase-8 mediate extrinsic apoptosis signaling. Cell 2009;137:721–735.

[139] Rodriguez A, Duran A, Selloum M, Champy MF, Diez-Guerra FJ, et al. Mature-onset obesity and insulin resistance in mice deficient in the signaling adapter p62. Cell Metab 2006;3:211–222.

[140] Belaid A, Ndiaye PD, Cerezo M, Cailleteau L, Brest P, et al. Autophagy and SQSTM1 on the RHOA(d) again: emerging roles of autophagy in the degradation of signaling proteins. Autophagy 2014;10:201–208.

[141] Ma X, Wu Y, Jin S, Tian Y, Zhang X, et al. Gold nanoparticles induce autophagosome accumulation through size-dependent nanoparticle uptake and lysosome impairment. ACS Nano 2011;5:8629–8639.

[142] Stern ST, Adiseshaiah PP, Crist RM. Autophagy and lysosomal dysfunction as emerging mechanisms of nanomaterial toxicity. Part Fibre Toxicol 2012;9:20.

[143] Bickel MH, Graber BE, Moor M. Distribution of chlorpromazine and imipramine in adipose and other tissues of rats. Life Sci 1983;33:2025–2031.

[144] McChesney EW, Banks WF Jr, Fabian RJ. Tissue distribution of chloroquine, hydroxychloroquine, and desethylchloroquine in the rat. Toxicol Appl Pharmacol 1967;10:501–513.

[145] Diaz D, Ford KA, Hartley DP, Harstad EB, Cain GR, et al. Pharmacokinetic drivers of toxicity for basic molecules: strategy to lower pK_a results in decreased tissue exposure and toxicity for a small molecule Met inhibitor. Toxicol Appl Pharmacol 2013;266:86–94.

[146] MacIntyre AC, Cutler DJ. Role of lysosomes in hepatic accumulation of chloroquine. J Pharm Sci 1988;77:196–199.

[147] Pirovino M, Honegger U, Muller O, Zysset T, Kupfer A, et al. Differences in hepatic drug accumulation and enzyme induction after chronic amiodarone feeding of two rat strains: role of the hydroxylator phenotype? Br J Pharmacol 1990;99:35–40.

[148] Schneider P, Korolenko TA, Busch U. A review of drug-induced lysosomal disorders of the liver in man and laboratory animals. Microsc Res Tech 1997;36:253–275.

[149] Guicciardi ME, Miyoshi H, Bronk SF, Gores GJ. Cathepsin B knockout mice are resistant to tumor necrosis factor-alpha-mediated hepatocyte apoptosis and liver injury: implications for therapeutic applications. Am J Pathol 2001;159: 2045–2054.

[150] Gezginci S, Bolkent S. The effect of Z-FA.FMK on D-galactosamine/TNF-alpha-induced liver injury in mice. Cell Biochem Funct 2007;25:277–286.

[151] Woolbright BL, Ramachandran A, McGill MR, Yan HM, Bajt ML, et al. Lysosomal instability and cathepsin B release during acetaminophen hepatotoxicity. Basic Clin Pharmacol Toxicol 2012;111:417–425.

[152] Uchiyama A, Kim JS, Kon K, Jaeschke H, Ikejima K, et al. Translocation of iron from lysosomes into mitochondria is a key event during oxidative stress-induced hepatocellular injury. Hepatology 2008;48:1644–1654.

[153] Kon K, Kim JS, Uchiyama A, Jaeschke H, Lemasters JJ. Lysosomal iron mobilization and induction of the mitochondrial permeability transition in acetaminophen-induced toxicity to mouse hepatocytes. Toxicol Sci 2010;117:101–108.

[154] Pourahmad J, Mortada Y, Eskandari MR, Shahraki J. Involvement of Lysosomal Labilisation and Lysosomal/mitochondrial Cross-Talk in Diclofenac Induced Hepatotoxicity. Iran J Pharm Res 2011;10:877–887.

[155] Canbay A, Guicciardi ME, Higuchi H, Feldstein A, Bronk SF, et al. Cathepsin B inactivation attenuates hepatic injury and fibrosis during cholestasis. J Clin Invest 2003;112:152–159.

[156] Feldstein AE, Werneburg NW, Canbay A, Guicciardi ME, Bronk SF, et al. Free fatty acids promote hepatic lipotoxicity by stimulating TNF-alpha expression via a lysosomal pathway. Hepatology 2004;40:185–194.

[157] Komatsu M, Waguri S, Ueno T, Iwata J, Murata S, et al. Impairment of starvation-induced and constitutive autophagy in Atg7-deficient mice. J Cell Biol 2005;169:425–434.

[158] Murray TV, Ahmad A, Brewer AC. Reactive oxygen at the heart of metabolism. Trends Cardiovasc Med 2014;24:113–120.

[159] Nakai A, Yamaguchi O, Takeda T, Higuchi Y, Hikoso S, et al. The role of autophagy in cardiomyocytes in the basal state and in response to hemodynamic stress. Nat Med 2007;13:619–624.

[160] Terman A, Kurz T, Gustafsson B, Brunk UT. The involvement of lysosomes in myocardial aging and disease. Curr Cardiol Rev 2008;4:107–115.

[161] Nishino I, Fu J, Tanji K, Yamada T, Shimojo S, et al. Primary LAMP-2 deficiency causes X-linked vacuolar cardiomyopathy and myopathy (Danon disease). Nature 2000;406:906–910.

[162] Soong TR, Barouch LA, Champion HC, Wigley FM, Halushka MK. New clinical and ultrastructural findings in hydroxychloroquine-induced cardiomyopathy – a report of 2 cases. Hum Pathol 2007;38:1858–1863.

[163] Stanely Mainzen Prince P, Priscilla H, Devika PT. Gallic acid prevents lysosomal damage in isoproterenol induced cardiotoxicity in Wistar rats. Eur J Pharmacol 2009;615:139–143.

[164] Kerkela R, Grazette L, Yacobi R, Iliescu C, Patten R, et al. Cardiotoxicity of the cancer therapeutic agent imatinib mesylate. Nat Med 2006;12:908–916.

[165] Yoshimori T, Yamamoto A, Moriyama Y, Futai M, Tashiro Y. Bafilomycin A1, a specific inhibitor of vacuolar-type H(+)-ATPase, inhibits acidification and protein degradation in lysosomes of cultured cells. J Biol Chem 1991;266:17707–17712.

[166] Herman EH, Knapton A, Rosen E, Thompson K, Rosenzweig B, et al. A multifaceted evaluation of imatinib-induced cardiotoxicity in the rat. Toxicol Pathol 2011;39:1091–1106.

[167] Bristow MR, Billingham ME, Mason JW, Daniels JR. Clinical spectrum of anthracycline antibiotic cardiotoxicity. Cancer Treat Rep 1978;62:873–879.

[168] Takemura G, Fujiwara H. Doxorubicin-induced cardiomyopathy from the cardiotoxic mechanisms to management. Prog Cardiovasc Dis 2007;49:330–352.

[169] Hurwitz SJ, Terashima M, Mizunuma N, Slapak CA. Vesicular anthracycline accumulation in doxorubicin-selected U-937 cells: participation of lysosomes. Blood 1997;89:3745–3754.

[170] Singal PK, Segstro RJ, Singh RP, Kutryk MJ. Changes in lysosomal morphology and enzyme activities during the development of adriamycin-induced cardiomyopathy. Can J Cardiol 1985;1:139–147.

[171] Kawaguchi T, Takemura G, Kanamori H, Takeyama T, Watanabe T, et al. Prior starvation mitigates acute doxorubicin cardiotoxicity through restoration of autophagy in affected cardiomyocytes. Cardiovasc Res 2012;96:456–465.

[172] Schmitz C, Hilpert J, Jacobsen C, Boensch C, Christensen EI, et al. Megalin deficiency offers protection from renal aminoglycoside accumulation. J Biol Chem 2002;277:618–622.

[173] Kaushal GP, Kaushal V, Herzog C, Yang C. Autophagy delays apoptosis in renal tubular epithelial cells in cisplatin cytotoxicity. Autophagy 2008;4:710–712.

[174] Saftig P, editor. *Lysosomes*. Landes Biosciences. 193 p; 2005.

[175] Marshansky V, Bourgoin S, Londono I, Bendayan M, Maranda B, et al. Receptor-mediated endocytosis in kidney proximal tubules: recent advances and hypothesis. Electrophoresis 1997;18:2661–2676.

[176] Marshansky V, Ausiello DA, Brown D. Physiological importance of endosomal acidification: potential role in proximal tubulopathies. Curr Opin Nephrol Hypertens 2002;11:527–537.

[177] Arany I, Safirstein RL. Cisplatin nephrotoxicity. Semin Nephrol 2003;23:460–464.

[178] Silverblatt FJ, Kuehn C. Autoradiography of gentamicin uptake by the rat proximal tubule cell. Kidney Int 1979;15:335–345.

[179] Mingeot-Leclercq MP, Tulkens PM. Aminoglycosides: nephrotoxicity. Antimicrob Agents Chemother 1999;43:1003–1012.

[180] Servais H, Van Der Smissen P, Thirion G, Van der Essen G, Van Bambeke F, et al. Gentamicin-induced apoptosis in LLC-PK1 cells: involvement of lysosomes and mitochondria. Toxicol Appl Pharmacol 2005;206:321–333.

[181] Olbricht CJ, Fink M, Gutjahr E. Alterations in lysosomal enzymes of the proximal tubule in gentamicin nephrotoxicity. Kidney Int 1991;39:639–646.

[182] Yamasaki F, Ishibashi M, Nakakuki M, Watanabe M, Shinkawa T, et al. Protective action of ulinastatin against cisplatin nephrotoxicity in mice and its effect on the lysosomal fragility. Nephron 1996;74:158–167.

[183] Takano M, Nakanishi N, Kitahara Y, Sasaki Y, Murakami T, et al. Cisplatin-induced inhibition of receptor-mediated endocytosis of protein in the kidney. Kidney Int 2002;62:1707–1717.

[184] Nagai J, Saito M, Adachi Y, Yumoto R, Takano M. Inhibition of gentamicin binding to rat renal brush-border membrane by megalin ligands and basic peptides. J Control Release 2006;112:43–50.

[185] Moestrup SK, Cui S, Vorum H, Bregengard C, Bjorn SE, et al. Evidence that epithelial glycoprotein 330/megalin mediates uptake of polybasic drugs. J Clin Invest 1995;96:1404–1413.

[186] Riedemann L, Lanvers C, Deuster D, Peters U, Boos J, et al. Megalin genetic polymorphisms and individual sensitivity to the ototoxic effect of cisplatin. Pharmacogenomics J 2008;8:23–28.

[187] Bok D. The retinal pigment epithelium: a versatile partner in vision. J Cell Sci Suppl 1993;17:189–195.

[188] Strauss O. The retinal pigment epithelium in visual function. Physiol Rev 2005;85:845–881.

[189] Bonilha VL. Age and disease-related structural changes in the retinal pigment epithelium. Clin Ophthalmol 2008;2:413–424.

[190] Koike M, Nakanishi H, Saftig P, Ezaki J, Isahara K, et al. Cathepsin D deficiency induces lysosomal storage with ceroid lipofuscin in mouse CNS neurons. J Neurosci 2000;20:6898–6906.

[191] Summers CG, Ashworth JL. Ocular manifestations as key features for diagnosing mucopolysaccharidoses. Rheumatology (Oxford) 2011;50(Suppl 5):v34–v40.

[192] Claudepierre T, Paques M, Simonutti M, Buard I, Sahel J, et al. Lack of Niemann–Pick type C1 induces age-related degeneration in the mouse retina. Mol Cell Neurosci 2010;43:164–176.

[193] Mecklenburg L, Schraermeyer U. An overview on the toxic morphological changes in the retinal pigment epithelium after systemic compound administration. Toxicol Pathol 2007;35:252–267.

[194] Ivanina TA, Zueva MV, Lebedeva MN, Bogoslovsky AI, Bunin AJ. Ultrastructural alterations in rat and cat retina and pigment epithelium induced by chloroquine. Graefes Arch Clin Exp Ophthalmol 1983;220:32–38.

[195] Khoh-Reiter S, Sokolowski SA, Jessen B, Evans M, Dalvie D, et al. Contribution of membrane trafficking perturbation to retinal toxicity. Toxicol Sci 2015;145:383–395.

[196] Lee S, Sato Y, Nixon RA. Lysosomal proteolysis inhibition selectively disrupts axonal transport of degradative organelles and causes an Alzheimer's-like axonal dystrophy. J Neurosci 2011;31:7817–7830.

[197] Ibanez CF. Message in a bottle: long-range retrograde signaling in the nervous system. Trends Cell Biol 2007;17:519–528.

[198] Millecamps S, Julien JP. Axonal transport deficits and neurodegenerative diseases. Nat Rev Neurosci 2013;14:161–176.

[199] Neefjes J, van der Kant R. Stuck in traffic: an emerging theme in diseases of the nervous system. Trends Neurosci 2014;37:66–76.

[200] Grisold W, Cavaletti G, Windebank AJ. Peripheral neuropathies from chemotherapeutics and targeted agents: diagnosis, treatment, and prevention. Neuro Oncol 2012;14(Suppl 4):iv45–iv54.

[201] LaPointe NE, Morfini G, Brady ST, Feinstein SC, Wilson L, et al. Effects of eribulin, vincristine, paclitaxel and ixabepilone on fast axonal transport and kinesin-1 driven microtubule gliding: implications for chemotherapy-induced peripheral neuropathy. Neurotoxicology 2013;37:231–239.

[202] Watanabe T, Nagase K, Chosa M, Tobinai K. Schwann cell autophagy induced by SAHA, 17-AAG, or clonazepam can reduce bortezomib-induced peripheral neuropathy. Br J Cancer 2010;103:1580–1587.

[203] Bischoff A, Meier C, Roth F. Gentamicin neurotoxicity (polyneuropathy–encephalopathy). Schweiz Med Wochenschr 1977;107:3–8.

[204] Grill MF, Maganti RK. Neurotoxic effects associated with antibiotic use: management considerations. Br J Clin Pharmacol 2011;72:381–393.

[205] Martorell C, Calpena AC, Escribano E, Poblet JM, Freixas J. Influence of the chromatographic capacity factor (log k′) as an index of lipophilicity in the antibacterial activity of a series of 6-fluoroquinolones. Relationship between physico-chemical and structural properties and their hydrophobicity. J Chromatogr A 1993;655:177–184.

[206] Sandri M, Coletto L, Grumati P, Bonaldo P. Misregulation of autophagy and protein degradation systems in myopathies and muscular dystrophies. J Cell Sci 2013;126:5325–5333.

[207] Yang Z, Vatta M. Danon disease as a cause of autophagic vacuolar myopathy. Congenit Heart Dis 2007;2:404–409.

[208] Tanaka Y, Guhde G, Suter A, Eskelinen EL, Hartmann D, et al. Accumulation of autophagic vacuoles and cardiomyopathy in LAMP-2-deficient mice. Nature 2000;406:902–906.

[209] Masiero E, Agatea L, Mammucari C, Blaauw B, Loro E, et al. Autophagy is required to maintain muscle mass. Cell Metab 2009;10:507–515.

[210] Guis S, Mattei JP, Liote F. Drug-induced and toxic myopathies. Best Pract Res Clin Rheumatol 2003;17:877–907.

[211] Clouston PD, Donnelly PE. Acute necrotising myopathy associated with amiodarone therapy. Aust N Z J Med 1989;19:483–485.

[212] Costa-Jussa FR, Guevara A, Brook GA, Duchen LW, Jacobs JM. Changes in denervated skeletal muscle of amiodarone-fed mice. Muscle Nerve 1988;11:627–637.

[213] Choi SS, Chan KF, Ng HK, Mak WP. Colchicine-induced myopathy and neuropathy. Hong Kong Med J 1999;5:204–207.

[214] Bradley WG, Lassman LP, Pearce GW, Walton JN. The neuromyopathy of vincristine in man. Clinical, electrophysiological and pathological studies. J Neurol Sci 1970;10:107–131.

[215] Kuncl RW, Duncan G, Watson D, Alderson K, Rogawski MA, et al. Colchicine myopathy and neuropathy. N Engl J Med 1987;316:1562–1568.

[216] Liang XH, Jackson S, Seaman M, Brown K, Kempkes B, et al. Induction of autophagy and inhibition of tumorigenesis by beclin 1. Nature 1999;402:672–676.

[217] Qu X, Yu J, Bhagat G, Furuya N, Hibshoosh H, et al. Promotion of tumorigenesis by heterozygous disruption of the beclin 1 autophagy gene. J Clin Invest 2003;112: 1809–1820.

[218] Takahashi Y, Coppola D, Matsushita N, Cualing HD, Sun M, et al. Bif-1 interacts with Beclin 1 through UVRAG and regulates autophagy and tumorigenesis. Nat Cell Biol 2007;9:1142–1151.

[219] Inami Y, Waguri S, Sakamoto A, Kouno T, Nakada K, et al. Persistent activation of Nrf2 through p62 in hepatocellular carcinoma cells. J Cell Biol 2011;193:275–284.

[220] Platanias LC. Biological responses to arsenic compounds. J Biol Chem 2009;284:18583–18587.

[221] Flora SJ. Arsenic-induced oxidative stress and its reversibility. Free Radic Biol Med 2011;51:257–281.

[222] Hughes MF, Beck BD, Chen Y, Lewis AS, Thomas DJ. Arsenic exposure and toxicology: a historical perspective. Toxicol Sci 2011;123:305–332.

[223] Muscarella LA, Parrella P, D'Alessandro V, la Torre A, Barbano R, et al. Frequent epigenetics inactivation of KEAP1 gene in non-small cell lung cancer. Epigenetics 2011;6:710–719.

[224] Jiang T, Chen N, Zhao F, Wang XJ, Kong B, et al. High levels of Nrf2 determine chemoresistance in type II endometrial cancer. Cancer Res 2010;70:5486–5496.

[225] Slocum SL, Kensler TW. Nrf2: control of sensitivity to carcinogens. Arch Toxicol 2011;85:273–284.

[226] Shim GS, Manandhar S, Shin DH, Kim TH, Kwak MK. Acquisition of doxorubicin resistance in ovarian carcinoma cells accompanies activation of the NRF2 pathway. Free Radic Biol Med 2009;47:1619–1631.

[227] Waalkes MP. Cadmium carcinogenesis. Mutat Res 2003;533:107–120.

[228] Messner B, Ploner C, Laufer G, Bernhard D. Cadmium activates a programmed, lysosomal membrane permeabilization-dependent necrosis pathway. Toxicol Lett 2012;212:268–275.

[229] Wang L, Gallagher EP. Role of Nrf2 antioxidant defense in mitigating cadmium-induced oxidative stress in the olfactory system of zebrafish. Toxicol Appl Pharmacol 2013;266:177–186.

[230] Iliopoulou A, Giannakopoulos G, Mayrikakis M, Zafiris E, Stamatelopoulos S. Reversible fulminant hepatitis following intravenous amiodarone loading. Amiodarone hepatotoxicity. Int J Clin Pharmacol Ther 1999;37:312–313.

[231] Agoston M, Orsi F, Feher E, Hagymasi K, Orosz Z, et al. Silymarin and vitamin E reduce amiodarone-induced lysosomal phospholipidosis in rats. Toxicology 2003;190:231–241.

[232] Zaki MS, Eid RA. Role of vitamin-E on rat liver-amiodarone: an ultrastructural study. Saudi J Gastroenterol 2009;15:104–110.

[233] Mohamadin AM, Elberry AA, Mariee AD, Morsy GM, Al-Abbasi FA. Lycopene attenuates oxidative stress and heart lysosomal damage in isoproterenol induced cardiotoxicity in rats: a biochemical study. Pathophysiology 2012;19:121–130.

[234] Kirkegaard T, Roth AG, Petersen NH, Mahalka AK, Olsen OD, et al. Hsp70 stabilizes lysosomes and reverts Niemann–Pick disease-associated lysosomal pathology. Nature 2010;463:549–553.

[235] Canfran-Duque A, Pastor O, Quintana-Portillo R, Lerma M, de la Pena G, et al. Curcumin promotes exosomes/microvesicles secretion that attenuates lysosomal cholesterol traffic impairment. Mol Nutr Food Res 2014;58:687–697.

[236] Chen QY, Shi JG, Yao QH, Jiao DM, Wang YY, et al. Lysosomal membrane permeabilization is involved in curcumin-induced apoptosis of A549 lung carcinoma cells. Mol Cell Biochem 2012;359:389–398.

[237] Dykens JA, Jamieson JD, Marroquin LD, Nadanaciva S, Xu JJ, et al. *In vitro* assessment of mitochondrial dysfunction and cytotoxicity of nefazodone, trazodone, and buspirone. Toxicol Sci 2008;103:335–345.

20

LYSOSOMES AND PHOSPHOLIPIDOSIS IN DRUG DEVELOPMENT AND REGULATION*

James M. Willard and Albert De Felice

Division of Cardiovascular and Renal Products, Office of Drug Evaluation I, Center for Drug Evaluation and Research, United States Food and Drug Administration, Silver Spring, MD, USA

20.1 INTRODUCTION

Drug development is inherently difficult, since most, if not all drugs carry some risk of an adverse event (AE). AEs vary in clinical importance, incidence, pathogenesis, and predictability. The AEs may result from pharmacodynamic activity, operation of collateral physiologic homeostatic systems, off-target direct cytotoxicity, or be due to other biochemical lesions. Drug-induced phospholipidosis (DIPL) is essentially a lysosomal storage disorder provoked by a variety of pharmacodynamically diverse drugs, usually cationic amphiphilic drugs (CADs). The lesion is characterized by the intracellular accumulation of tissue phospholipid–drug complexes. Not always disruptive, the location and pathological significance of the deposits vary, reminiscent of the hereditary lysosomal storage diseases. These complexes form concentric lamellar membranous vesicles that are diagnostic of DIPL [1–3]. The amphiphilic region of the molecule can enter the phospholipid bilayer of the plasma membrane, and the complex deposits in the lysosome during phagocytosis, pinocytosis, or plasma membrane

*This chapter reflects the views of the authors and should not be construed to represent FDA's views or policies.

Lysosomes: Biology, Diseases, and Therapeutics, First Edition.
Edited by Frederick R. Maxfield, James M. Willard and Shuyan Lu.

recycling. The subsequent acidification of the lysosome traps the phospholipid–drug complex. Although the inclusions typically dissipate upon drug withdrawal, DIPL represents a concern in risk assessment, and biomarkers are being sought for routine phospholipidosis assessment.

20.2 FDA INVOLVEMENT

FDA deliberation of the safety implications of positive nonclinical toxicity finding of an investigational new drug (IND) escalated after 1962 when the Food and Drug Act was passed by the US Congress. It required the Food and Drug Administration (FDA) to evaluate and approve pharmaceuticals based on evidence of safety as well as efficacy [4]. Since then, an increasing understanding of mechanisms of toxicologic pathology and organ dysfunction has afforded more reliable forecasting of risk of an AE in target patient populations. This can be attributed in part to the evolution and continued refinement of tests for toxicologic histopathology, safety pharmacology, and organ dysfunction that the FDA relies on to identify the safety margins and minimize the risk of an AE in a clinical trial [5], enabling increased safety of first-in-man trials [6]. Nevertheless, the safety implications of some toxicologic lesions that are encountered in preclinical trials remain contentious. This includes DIPL, especially when there is important collateral histopathology and organ dysfunction. However, DIPL, even when isolated, should be monitored, especially in drugs that will be prescribed to patients to long-term therapeutic use.

Toxicologic phospholipidosis was identified in the late 1960s as the cause of the lethal hepatotoxicity associated with use of coralgil and perhexiline [1], and it remains problematic. DIPL can be a regulatory issue even when encountered in IND trial at suprapharmacodynamic dosages due to the absence of validated biomarkers and the uncertainty of the toxicological significance. In 2004, the Center for Drug Evaluation and Research (CDER) of the FDA established the Phospholipidosis Working Group (PLWG) [7] after contentious findings of DIPL in nonclinical IND programs, and the absence of clinical information on its consequences, prevalence, and time course. There also had been no uniform regulatory review policy given the lack of consensus as to whether DIPL was an adaptive or toxic response [8] and, its safety implications, especially absent any other collateral histopathology. Clearly, this phenomenon required further study to enable informed and uniform regulatory responses because it is not infrequently observed during IND development [9,10]. The PLWG has considered the importance of the pathogenesis of DIPL primarily from a regulatory perspective.

Until recently, DIPL has been poorly understood. A big step in clarifying the collateral effects of DIPL occurred when Sawada et al. [11] reported that 12 phospholipidotic compounds altered the expression of genes controlling lysosomal phospholipase, lysosomal enzyme transport, and phospholipid and cholesterol biosynthesis. Real-time polymerase chain reaction analysis showed that expression of the 12 gene markers covaried with lysosomal lamellar myelin body formation. The affected lysosomes and/or endosomes accumulated phospholipids in a complex with the inducing drug or metabolite [1,12–16], leading to the signature multilamellar vesicles and leading to the cellular effects of DIPL.

20.3 AUTOPHAGY AND DIPL

Several drugs involved with DIPL have also been shown to be involved in autophagy, including rapamycin, for which the mammalian target of Rapamycin (mTOR) received its name. Conceivably, some consequences of DIPL likely depend on the extent of disruption, if any, of pathways that control autophagy. Studies of the activity of the lysosome, the primary target of DIPL, have elucidated autophagy as a homeostatic catabolic process in which the cell clears old organelles and dysfunctional proteins synthesized in the cell, thereby preserving cardiac, neuronal, and other cell function [17–23]. These studies have revealed signaling pathways cardinal to cell cycling that help determine whether cells are quiescent and differentiated or apoptotic. The process of autophagy is additionally regulated by cell conditions such as starvation or stress. Multiple pathways control autophagy, and the processes of internalization of exogenous or plasma membrane-bound targets, and protein traffic within the cell [20] could be perturbed by lipidosis.

Many drugs that cause DIPL are useful for investigating the regulation of autophagy and apoptosis. Compounds that regulate autophagy and apoptosis such as rapamycin, tamoxifen, and gentamicin also induce DIPL. The mTOR is an important protein in the autophagy cascade [18,24]. Binding of rapamycin, a CAD, to the mTOR complex inhibits autophagy [24]. Rapamycin can block neurodegeneration through its actions on mTOR [25]. Other inducers of autophagy cause clearance of mutant proteins such as β-amyloid and prevent neurodegeneration [18]. Tamoxifen induces autophagy and causes cell death in cultured retinal pigment epithelial and photoreceptor cells [26], while chloroquine inhibits autophagy in a cell line with retinal pigment epithelial properties and causes cell death [27]. The pathways controlling autophagy should be studied further, and CADs are strong candidates for manipulating such pathways.

20.4 EARLY EXPERIENCE WITH LETHAL DIPL

In the late 1960s, multiple hepatic deaths and hospitalizations of patients in Japan treated with the coronary vasodilator coralgil (4,4′-diethylaminoethoxyhexestrol) brought DIPL to the attention of regulatory agencies [28–30]. Electron microscopy of liver tissue revealed the signature lamellar myeloid bodies in lysosomes [8,30]. Bis(monoacylglycerol)phosphate, that is, lysobisphosphatidic acid (BMP) and acyl phosphatidylglycerol were recovered from coralgil-treated patients. The lesions and symptoms were noted to resemble those of patients with Niemann–Pick type C (NP-C), a heritable disorder of phospholipid metabolism [30]. Subsequently, use of perhexiline maleate (a vasodilator prescribed for angina pectoris) and telithromycin (Ketek) were restricted because of the risk of liver toxicity and phospholipidosis [16,31]. DIPL may also underlie the toxicity of several other drugs. For example, the quinolone-type antimalarial drugs carry a risk of important neurotoxicity [32] and myotoxicity [33]. Amiodarone labeling has a boxed warning of the risk of lethal pulmonary toxicity.

20.5 CLINICAL AND NONCLINICAL EXPRESSIONS OF DIPL

20.5.1 Clinical

The clinical expression and context of DIPL may vary markedly. For example, although foamy alveolar macrophages and cytoplasmic lamellar bodies are prominent features of amiodarone-induced pneumonitis, their presence alone may not distinguish the toxic from the asymptomatic patients [34,35], and their role in the pathogenesis of the pulmonary lesions in these cardiac patients remains uncertain. Some phospholipidotic compounds such as the antihistamine loratadine have few toxic manifestations other than the lipidosis, as in the case of the asymptomatic cardiac patients with amiodarone-induced DIPL. Obviously, such variations may confound the evaluation of phospholipidosis as a risk factor and cause difficulty in assessing the importance of DIPL. The genetic lysosomal storage diseases, for example, NP-C disease also varies incidence and clinical importance of the liver and spleen lesions.

Response to a particular CAD is qualitatively and quantitatively unpredictable, and forecasting clinical or veterinary risk of organ dysfunction, or asymptomatic tissue involvement, remains tenuous. Currently, it is not possible to predict the major target organ for DIPL. In the importantly involved organs, compromised function can reflect biochemical reactions beyond any concurrently altered phospholipid traffic. For example, a phospholipidotic molecule such as gentamicin may inhibit both protein synthesis and phospholipid degradation in a target tissue. Gentamicin is a nephrotoxic aminoglycoside that accumulates in lysosomes, the Golgi complex, and mitochondria in the renal cortex of the rat and disrupts function of these organelles early in the time course of the nephrotoxicity [36]. As in the rat, gentamicin-associated DIPL manifests in both the inner ear and kidney in humans, and deafness and kidney failure can be caused by exposure to this antibiotic. In some cases, phospholipidosis does not obviously underlie clinical disease. In NP-C disease, a hereditary lysosomal lipid storage disorder, respiratory distress is associated with depressed protein expression in alveolar macrophages. Although foamy concentrically laminated macrophages can be present in the lung, spleen, and liver of the NP-C patient, overt collateral pathology (alveolar proteinosis) can be confined to the lung at the time of biopsy [37]. Accordingly, for both DIPL and hereditary phospholipidosis, associated collateral pathology may be unpredictable, or not expressed concurrently with the lipidotic tissue lesions.

Drug accumulation and associated organ dysfunction in DIPL generally are reversible when drug treatment is discontinued, as with gentamicin nephrotoxicity.

20.5.2 Nonclinical

In nonclinical trials, DIPL is often multiorgan and not isolated, that is, it is associated with excess nonphospholipidotic histopathology. In an analysis of 419 internal toxicology reports of 46 proprietary phospholipogenic and 62 nonphospholipogenic compounds tested in 1- to 4-week studies in the rat and dog, AstraZeneca (London, UK) reported that half of the phospholipogenic compounds targeted three or more organs [35]. Furthermore, the phospholipogenic drugs were more likely to be associated with other histopathology than the nonphospholipogenic compounds in all organs, with significantly higher frequencies of liver necrosis, alveolitis, and pneumonitis, and lymphocytolysis in the thymus, lymph nodes, and spleen [35]. Analysis of nonclinical studies submitted to the FDA CDER showed that the lowest observed adverse effect level (LOAEL) for phospholipidosis was highly correlated with the LOAEL for the appearance of nonlipidotic histopathological lesions [38]. Therefore, the appearance of phospholipidosis may be associated with other histopathological findings in nonclinical studies.

20.6 PHYSICAL CHEMISTRY

Phospholipidotic activity is predictable based on physicochemical criteria [39], though the target organ is not. Most compounds that cause DIPL are CADs that have a positively charged primary, secondary, or tertiary amine group associated with an amphiphilic region that typically is incorporated into a ring structure [40]. DIPL may be predicted from the acid dissociation constant (pK_a) and logarithm of the partition coefficient between *n*-octanol and water ($c\log P$), an index of hydrophobicity [40]. Molecular modeling programs are used to calculate physicochemical properties from structural motifs [39]. The likelihood that a drug with $pK_a > 8$ and $c\log P > 1$ will concentrate in the lysosomes by pH partitioning and induce DIPL *in vivo* is increased when $([pK_a]^2 + [c\log P]^2) \geq 90$ [39]. Sensitivity and specificity are 80% in several models [39]. Modifications to the formula may increase predictive potential, and incorporation of Bayesian methods increases sensitivity to 93% [41] and specificity to 80%. Applying quantitative structure–activity relationship (QSAR) to the FDA database of 385 compounds that provoked DIPL and the 358 compounds that did not provoke DIPL yielded 5 false positives and 38 false negatives [42]. However, 6 of the 38 false negative compounds were amphiphilic; therefore, in that data set, 32 non-CADs provoked DIPL [42]. Based on this database and distribution of amines, more than 90% of the compounds that induced DIPL were CADs. Not surprisingly, metabolic transformation that appreciably affects cationic amphiphilic properties also affects phospholipidotic activity [43].

CADs have physicochemical properties that promote access to the cell cytoplasm [44], enabling access to the cell membrane, and promoting intracellular sequestering. The ionization pK_a would enable these molecules to be charged, and thus to be trapped, in the acidic interior of the lysosome [45]. One practical application of this property is in the design of drugs for central nervous system (CNS) activity. Cationic amphiphilic compounds cross the blood–brain barrier, and are not extruded by transporters. CADs that are pharmacologically active could be candidates for neurologic and psychiatric therapies. Indeed, lysosomal sequestering of CNS drugs is a well-known phenomenon, and not always accompanied by central lipidosis [46]. Nevertheless, DIPL in the CNS is a major concern, especially when irreversible, because DIPL occurring in nerves resembles the pathology of some severe genetic sphingomyelin storage disorders such as NP-C [46]. Neurotoxicity can be a problem with antimalarial drugs such as mefloquine that can achieve CNS concentrations that are effective against resident parasites, but that may also cause lipidosis and neurologic AEs [46].

20.7 QUANTITATIVE STRUCTURE–ACTIVITY RELATIONSHIP (QSAR)

QSAR screening is used to forecast toxicity as well as targeted pharmacodynamic activity in drug development [47]. The QSAR programs for predicting DIPL complement those for carcinogenicity, reproductive toxicity, genetic toxicity, and QT prolongation, and, together, are a valuable recourse ancillary to FDA safety assessment [40]. QSAR involves algorithms that deconstruct the primary structure of compounds and identify compounds that fit a common pharmacophore. Phospholipidotic compounds are especially amenable to QSAR modeling because they share relatively well-defined and discrete structural and physicochemical features.

Model building for QSAR-based predictions of DIPL has been evolving at the FDA [9,42]. The FDA PLWG has created a large database of compounds that vary in phospholipidotic activity by using the FDA and literature databases, and by soliciting FDA reviewers for candidate compounds [35,42]. This provided a composite database of 750 compounds, including 385 that provoked DIPL in at least one species. From this robust data set, QSAR models were developed, using different commercial algorithms that performed with 80% sensitivity and specificity. The early predictive models had high concordance with published or experimental results, comparing favorably to commercially available predictive tools. Modifications have since resulted in an increase in sensitivity and specificity to 84% and 80% [41].

Pharmaceutical companies have used QSAR models for DIPL activity and human ether-à-go-go-related gene (hERG) channel interaction to complement *in vitro* assay systems in forecasting potential IND safety issues [47]. The utility of such modeling depends on accounting for factors that affect activity *in vivo*. For example, a QSAR analysis of CNS drugs that had cationic amphiphilic properties generated disproportionately more false positives, in part because the expected therapeutic plasma concentrations may not have been attained [41]. It has been suggested that perhexiline, a phospholipidotic antianginal drug, be returned to use if patients are screened for deficiency in metabolic conversion [43,48,49]. Perhexiline can be problematic in patients with such a deficiency. However, loratadine and ketoconazole are more likely to be phospholipidotic because metabolism leads to the usual oxidation to the phospholipidotic metabolite [43], making it less hydrophobic.

The continuing refinement of QSAR platforms for both targeted pharmacodynamic activity and lipidotic activity should help further both the synthesis and selection of promising drug candidates devoid of lipidotic activity. This is aided by the fact that many physicochemical features of a drug that predispose to DIPL are independent of the pharmacophore that mediates pharmacodynamic activity, and capacity for lipidosis may be independent of pharmacologic activity. Lysosomotropic drugs, for example, chloroquine, can be viewed as converging in the lysosome [3], but diverging, markedly, in pharmacologic activity. Exceptions would include nanoparticles and nonorganic compounds without potent or specific pharmacodynamic activity that also accumulate in the lysosome [44].

Many of the hERG channel blockers have a positive charge on one end of the molecule and hydrophobic regions on the other end [50,51], recalling CADs and their association with DIPL. The QSAR for predicting hERG channel interference is sophisticated and includes van der Waals surface forces and hydrogen bonding. However, hERG channel blockers and CADs have similar behavior. To block the hERG channel, a compound must readily cross the plasma membrane [3] to access the interior of the cell, similar to CAD-like compounds. Therefore, the correspondence in physicochemical properties between compounds affecting lysosome behavior and hERG channel traffic, and the overlap between the hERG and DIPL pharmacophores [38,51] forecast the possibility of concurrent phospholipidosis and QT prolongation. This association is further probed below (see Section 20.12).

20.8 TOXICOGENOMICS

Changes in gene expression associated with lipid accumulation are being evaluated as markers of DIPL. Gene chip microarray technologies have been used to examine such changes in response to a phospholipidotic challenge [11,52]. In a hepatic cell line (HepG2) exposed to 12 compounds provoking DIPL, alterations in gene expression most correlated with lamellar myelin-like body formation changes were found in those genes regulating phospholipid and cholesterol metabolism and lysosomal enzyme transport [11]. When combined with histopathology and measures of cell and organ function, such a DNA microarray technique could identify early sensitive markers of DIPL and determine whether DIPL is a primary pathogenic factor or an adaptive response. From these studies, a profile of similarities between NP-C and phospholipidosis has been observed in regard to cholesterol transport that may define common features of disorders of lysosomal storage function. The early studies on DIPL in Japan noted a link between the increase in BMP levels in the liver and a concomitant increase in cholesterol levels [28]. These studies have begun to clarify the processes and systems underlying the toxic lysosomal disruption induced by phospholipidosis.

20.9 FLUORESCENCE, DYE, AND IMMUNOHISTOCHEMICAL METHODS FOR SCREENING

Fluorescent compounds and dyes such as Nile red and nitrobenzoxadiazole (NBD) have been used *in vitro* for the rapid examination of cell cultures as indicators of phospholipidosis, and these fluorophores may enable the rapid detection and quantification of DIPL [15,53–55]. The dye methods may enable the quantification of signal level but may be confounded by signal artifacts. In contrast, immunostaining methods, for example, for lysosomal proteins, provide a specific signal, depending on the choice of antibody target. Lysosomes have several specific surface markers that enable identification and characterization of the organelle. The lysosome-associated membrane proteins 1 and 2 (LAMP-1 and LAMP-2) are specific to the lysosome and comprise 50% of the surface membrane proteins of lysosomes [56]. There is functional overlap between LAMP-1 and LAMP-2, but knockouts of LAMP-1 have little effect. LAMP-2 knockouts cause accumulation of lysosomes and Danon disease, and the double knockout is lethal to the embryo [56]. Use of LAMP-2 antibody immunostaining and costaining with anti-adipophilin antibodies may enable the differentiation of phospholipidosis from lipidosis with nonphosphorylated lipids [57,58]. Therefore, immunostaining is a simple and effective method to determine the presence of phospholipidosis without recourse to electron microscopy.

20.10 FDA DATABASE AND QSAR MODELING

As mentioned, FDA pharmacologists have revisited the issue of DIPL, and its effect on drug development due to at least 50 INDs had caused the lesions [8]. A data set was created from information in NDA and IND submissions, using published methods [35]. At present, 384 compounds that caused DIPL in animals have afforded an examination of the physicochemical characteristics of CADs and QSAR analyses of such [41,42]. QSAR modeling is further described in the following sections.

20.11 LINKING PHOSPHOLIPIDOSIS AND OVERT TOXICITY

DIPL may be linked to symptomatic organ dysfunction such as arrhythmias. However, a causal relation is difficult to confirm [37], in part because of the difficulty of diagnosing DIPL in humans. Absent validated biomarkers, electron microscopy is the only recourse [8]. Therefore, evidence of DIPL linked with damage is typically observed only in catastrophic occurrences, for example, hepatotoxicity of coralgil. Its use in Japan led to over 100 deaths and many more hospitalizations. This provided an early link between a drug and DIPL [13,29,30,37] and also raised the possibility of BMP as a biomarker of liver damage [37]. Another coronary vasodilator indicated for angina pectoris, perhexiline, caused nonalcoholic steatohepatitis (NASH), similar to coralgil [59], along with neuropathy. These occurrences raised much concern. That caution continues to this day with the recent limitations placed on the use of telithromycin (Ketek) after the observation of DIPL and NASH [31].

Preclinical studies can be more forthcoming than clinical trials in probing for dose-related DIPL and its role in the pathogenesis of any histopathology beyond lipidosis, and any organ dysfunction. In IND toxicology studies, all animals dying spontaneously or killed in extremis or at the study end are subjected to gross and microscopic examination of tissues. If evidence of DIPL such as foamy macrophages is observed, further examination with electron microscopy can be done to confirm DIPL. An association has been observed between the appearance of DIPL in tissues and histopathological findings such as necrosis, apoptosis, and/or hypertrophy [35]. Tissues vary in susceptibility to DIPL, but there is enough evidence of colocalization of such conditions to suggest that DIPL may have a pathogenic role in the expression of other concurrent histopathology.

The lung, liver, and lymphoid tissue are the most susceptible to DIPL; based on tissue distribution of phospholipidosis in the 134 rat and 42 dog studies [35] that comprise the FDA database (Figure 20.1). These organs may be sentinel organs where the drug–lipid complex collects, but the lesions can present anywhere in the body, and predicting the target tissue in animals or man is not possible. In dogs, the most frequent target organ is the liver followed by the lung and lymphoid tissue. In rats and dogs, coexpression of DIPL and nonlipidotic histopathology has been investigated systematically [35]. Reevaluation of the behavior of phospholipogenic and nonphospholipogenic compounds (AstraZeneca) showed significantly higher frequencies of liver necrosis, alveolitis/pneumonitis, and lymphocytolysis in lymphoid tissue associated with the typically multiorgan systemic lipidosis, at a lower

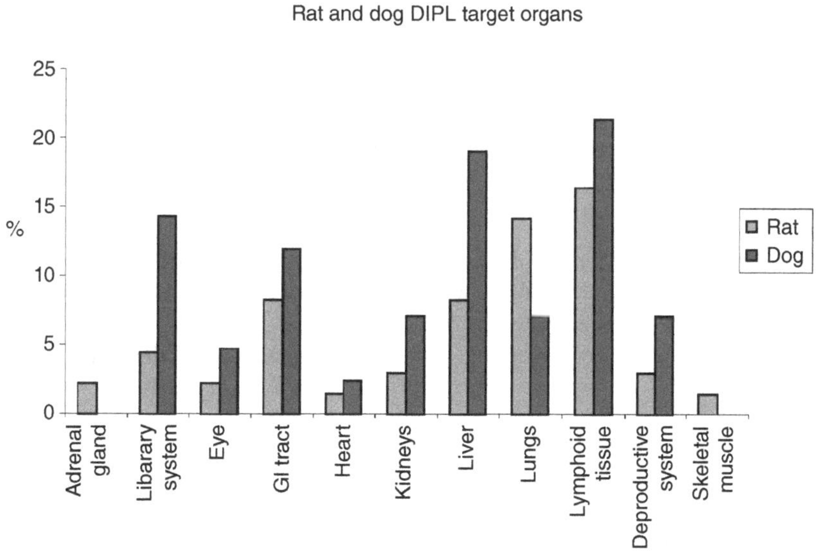

Figure 20.1 Comparison of dog (42 studies) and rat (134 studies) tissue distributions of drug-induced phospholipidosis from investigational new drug (IND) and new drug application (NDA) submissions to the USFDA from pharmaceutical companies [38]. GI, gastrointestinal; PLD, phospholipidosis. (*See color plate section for the color representation of this figure.*)

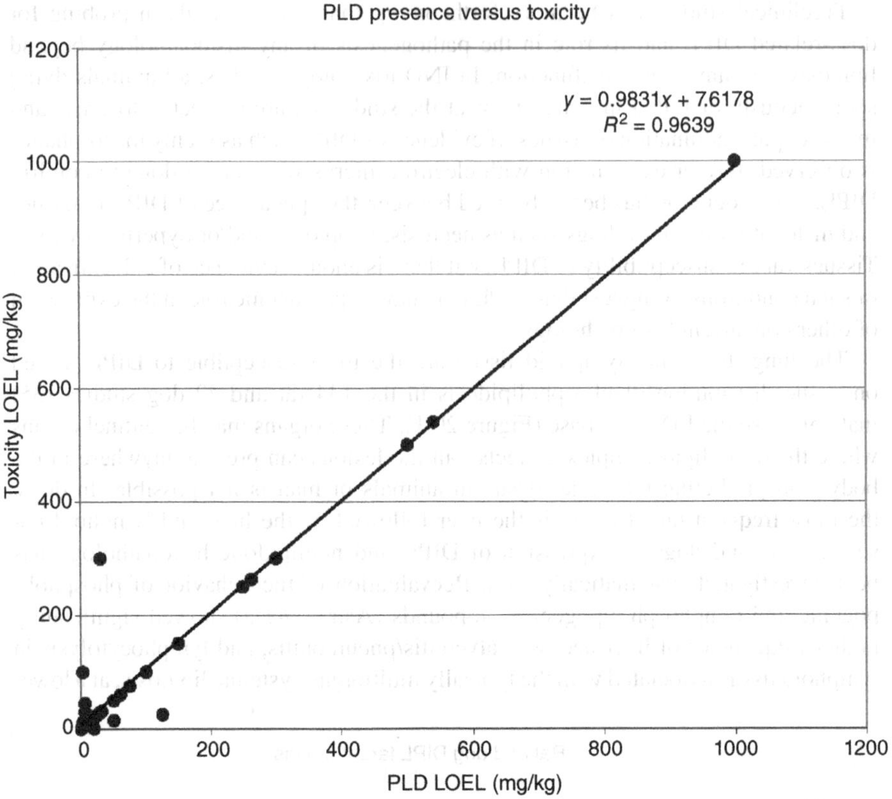

Figure 20.2 Relation between lowest observed drug-induced phospholipidosis (DIPL) dose and the lowest observed adverse effect level (LOAEL) toxicity dose in nonclinical studies submitted to the USFDA by pharmaceutical companies. A total 53 different studies were used to generate the graph [38]. LOEL, lowest observed effect level; PLD, phospholipidosis.

plasma exposure than in nonlipidotic animals. From the FDA DIPL database, there was good correlation ($R^2 = 0.96$) between the dosage that caused DIPL and the overt LOAEL for the drugs tested (Figure 20.2) [38]. These two results are evidence of a link between the appearance of DIPL and other histopathological damage.

20.12 PHOSPHOLIPIDOSIS AND QT INTERVAL PROLONGATION

Forecasting cardiovascular safety issues, including an important delay in cardiac ventricular repolarization, is an important goal in safety pharmacology testing of an IND. Prolonging the corrected QT (QTc) interval by altering hERG and/or other ion channel trafficking may cause Torsades de pointes (TdP) and sudden cardiac death. Therefore, early testing for QTc interval prolongation is required because hERG channel blockade may be proarrhythmic in some patients [5]. Proarrhythmia has led to the postmarketing discontinuation of potentially beneficial drugs because cardiac risk outweighed potential benefit.

The pharmacophore similarities in QSAR models for phospholipidosis and QT prolongation suggest that CADs that provoke DIPL may also carry risk of TdP although many drugs that caused phospholipidosis in preclinical animal toxicology studies are not potent inhibitors of the hERG potassium current, at least *in vitro*. A review of the Arizona Center for Education and Research on Therapeutics database [60] showed that 44% of 28 drugs that increased QT interval and thus the risk of TP were also associated with DIPL in animal studies (Table 20.1). This was not unexpected because 69% of more than 100 approved drugs in our database that provoked DIPL also prolonged QT (Table 20.2) usually by blocking the hERG

TABLE 20.1 Selection from the Combined List of Drugs that Prolong QT and/or cause Torsades de Pointes (TDP) [54]

Generic Name	Brand Name	Phospholipidosis
Amiodarone	Cordarone®, Pacerone®	Yes
Arsenic trioxide	Trisenox®	Yes
Astemizole	Hismanal®	Yes
Bepridil	Vascor®	No
Chloroquine	Aralen®	Yes
Chlorpromazine	Thorazine®	Yes
Cisapride	Propulsid®	No
Clarithromycin	Biaxin®	No
Disopyramide	Norpace®	No
Dofetilide	Tikosyn®	No
Domperidone	Motilium®	No
Droperidol	Inapsine	No
Erythromycin	Erythromycin, erythromycin ethylsuccinate	Yes
Halofantrine	Halfan®	No
Ibutilide	Corvert®	No
Levomethadyl	Orlaam®	No
Mesoridazine	Serentil®	Yes
Methadone	Dolophine®, Methadose®	Yes
Pentamidine	Pentam®, Nebupent®	Yes
Pimozide	Orap®	No
Probucol	Lorelco®	No
Procainamide	Pronestyl®, Procan®	No
Quinidine	Quinaglute®, Cardioquin®, Quinidex®	Yes
Sotalol	Betapace®	No
Sparfloxacin	Zagam®	Yes
Terfenadine	Seldane®	No
Thioridazine	Mellaril®	Yes

TABLE 20.2 From the USFDA database of 95 Approved New Drug Applications (NDAs) that have Drug-Induced Phospholipidosis (DIPL)

Generic Name	Drug Class	QT Positive
1-Chloroamitriptyline	Antidepressant	Yes
Amantadine	Antiviral	Yes
Amikacin	Antibacterial	Yes
Amiodarone	Antiarrhythmic	Yes
Amitriptyline	Antidepressant	Yes
Aripiprazole	Antipsychotic	Yes
Astemizole	Antihistamine	Yes
AY-9944		Yes
Azimilide	Antiarrhythmic	Yes
Azithromycin	Antibacterial	Yes
Bifeprunox	Antipsychotic	Yes
Bisoprolol fumarate with hydrochlorothiazide	Antihypertensive	Yes
Blonanserin	Antipsychotic	No
Boxidine	Anticholesteremic	No
Cethromycin	Anti-infective	Yes
Chlorcyclizine		Yes
Chloroquine	Antimalarial	Yes
Chloroquine mustard		Yes
Chlorpheniramine	Antihistamine	Yes
Chlorphentermine	Anorectic	Yes
Chlorpromazine	Antipsychotic	Yes
Citalopram	Antidepressant	Yes
Clindamycin	Antibiotic	Yes
Cloforex	Anorectic	No
Clomipramine	Antidepressant	Yes
Clozapine	Antipsychotic	Yes
Cyclizine	Antihistamine	No
Dapoxetine	Antidepressant	No
Darapladib	Phospholipase A2 inhibitor	Yes
Deramciclane fumarate	Anxiolytic	No
Desloratadine	Antihistamine	Yes
Dexchlorpheniramine	Antihistamine	Yes
Dibekacin	Antibiotic	No
Dibucaine	Anesthetic	No
Disobutamide		Yes
Dronedarone	Antiarrhythmic	Yes
Duloxetine	Antidepressant	No
Dutasteride	5-α-reductase inhibitor	No
Ebastine	Antihistamine	Yes
Erythromycin	Antibiotic	Yes
Escitalopram oxalate	Antidepressant	Yes
Fenfluramine	Anorectic	No

TABLE 20.2 (*Continued*)

Generic Name	Drug Class	QT Positive
Fluoxetine	Antidepressant	Yes
Gentamicin	Antibiotic	Yes
Haloperidol	Antipsychotic	Yes
Homochlorcyclizine	Antidepressant	Yes
Hydroxyzine	Antihistamine	No
Imipramine	Antidepressant	Yes
Indoramin	Antiarrhythmic	Yes
Iprindole	Antidepressant	Yes
Ketoconazole	Antifungal	Yes
Lansoprazole	Proton pump inhibitor	No
Lapatinib ditosylate	Oncology	No
Mefloquine	Antimalarial	No
Loratadine	Antihistamine	Yes
Mianserin	Antihypertensive	Yes
Mibefradil dihydrochloride	Antihypertensive	Yes
Nebivolol	Antihypertensive	No
Nilotinib	Oncology	Yes
Norchlorcyclizine	Antihistamine	Yes
Noxiptiline	Antidepressant	Yes
Orvepitant	Neurokinin 1 antagonist	No
Perhexiline	Antianginal	Yes
Phentermine	Antihistamine	Yes
Posaconazole	Antifungal	Yes
Promazine	Antipsychotic	Yes
Propranolol	β-blocker	Yes
Quinacrine	Antibiotic	No
Quinidine	Antiarrhythmic	No
Remeron	Sleep	No
Rilapladib	Phospholipase A2 inhibitor	No
Rimantadine	Antiviral	Yes
Ritonavir	Antiviral	Yes
Rotigotine		No
Satavaptan	Diuretic	Yes
Sertraline	Antidepressant	Yes
Sirolimus	Immunosuppressant	Yes
Spectinomycin	Antibiotic	No
TAK-032	Human immunodeficiency virus treatment	Yes
Tamoxifen	Antiestrogen receptor	Yes

(*continued*)

TABLE 20.2 (*Continued*)

Generic Name	Drug Class	QT Positive
Tecastemizole/norastemizole	Antihistamine	Yes
Telithromycin	Antibiotic	Yes
Tetracaine	Anesthetic	No
Thioridazine	Antipsychotic	Yes
Thiothixene	Antiviral	No
Tiotropium bromide	Muscarinic antagonist	Yes
Tobramycin	Antibiotic	Yes
Tocamide	Antiarrhythmic	Yes
Triparanol		No
Tripelennamine	Antihistamine	No
Valdecoxib	Nonsteroidal anti-inflammatory drug	No
Varenicline tartrate	Antismoking	Yes
Verapamil	Antihypertensive	No

channel [38,51]. Some of these compounds may affect the hERG channel by other mechanisms, for example, inhibiting protein trafficking. For example, pentamidine does not block the hERG channel directly, but at 10 μM, a level achieved in antiparasitic therapy, it inhibits hERG channel trafficking [61,62]. In contrast, the DIPL-inducing and QTc-prolonging drug aclarubicin, an anthracycline antineoplastic drug, is an inhibitor of the vacuolar-type adenosine triphosphatase (V-ATPase) that is part of the proton pump mechanism in lysosomes [63]. Such activities may facilitate accumulation of the anthracycline and inhibit phospholipase activity in the lysosome by inhibiting lysosomal acidification. Accordingly, there are multiple ways to interfere with endosomal/lysosomal functions, which are important not only for clearing endocytosed materials coming from the cell surface but also for transporting hERG channels and other materials from the Golgi apparatus to the cell surface or other locations in the cell. Culprit drugs also might disrupt lysosomal recycling of misfolded or incorrectly made proteins along with blocking cellular organelle recycling [23].

20.13 DIPL MECHANISMS

Multiple pathogenic mechanisms have been proposed for DIPL. Early studies on DIPL examined the role of lysosomal phospholipases in the pathogenesis of DIPL. It was hypothesized that the causative drug interfered with the activity of the phospholipases (sphingomyelin phosphodiesterase, lysosomal phospholipase A2 [LYPLA2], and lysosomal phospholipase A1) or that the CAD had increased cholesterol biosynthesis via the lanosterol synthase pathway, leading to accumulation of phospholipids in the lysosomes [64,65]. Mice that were LYPLA2-deficient presented with foam cells, splenomegaly, and phospholipidosis [65]. However, it has been argued that phospholipase inhibition is not a factor in DIPL as BMP is still excreted from cells and is subsequently observed in the serum and urine [37]. Since BMP is produced by phospholipase A2 action, increased synthesis of the former would not be expected if inhibition of phospholipase was a constant or important factor in phospholipidosis [37]. Evidence has been accumulating for the direct binding of the causative drug to a phospholipid such as the phospholipids present in l-α-dipalmitoyl phosphatidylcholine vesicles. Such binding distinguishes several CADs that cause DIPL from quinidine, which does not bind to phospholipids [66]. The phospholipidotic and antimalarial activities of chloroquine may occur from blocking the V-ATPase and associated increase in pH in the parasitic and host cell organelle, with associated indirect inhibition of acidic phospholipase and other activities requiring low pH [67]. Procainamide also blocks V-ATPase and raise lysosomal pH [24]. Pentamidine inhibits hERG protein trafficking in cells [61] but does not inhibit phospholipases.

20.14 TREATMENT

There are no approved treatments for DIPL, likely because it is usually reversible upon drug withdrawal, and the need may not arise if DIPL is detected early. It is not clear how a clinical trial could be acceptably designed to test for reversal of DIPL, rather prevention might be more feasible. Gentamicin DIPL causes apoptosis of renal tubules [68] that can be blocked by polyaspartic acid [69]. Vitamin E can also inhibit apoptosis induced by 7-ketocholesterol, an inducer of phospholipidosis [70]. DIPL caused by amiodarone is also reduced by vitamin E treatment [71], an antioxidant also reported to inhibit the formation of DIPL and multilamellar vesicles associated with desipramine and other CADs [72]. These isolated reports are encouraging, but a systematic study of the effects of these compounds on DIPL is needed to determine whether polyaspartic acid and vitamin E may be therapeutic or prophylactic in DIPL.

While it is not clear how DIPL would be treated other than recourse to prompt drug withdrawal, it is clear that CADs possess useful properties for developing drugs for treating CNS or other indications. Some CADs are valuable pharmacotherapeutics that can readily cross plasma membranes to reach targets in the cell interior. This property also enables such compounds to cross the blood–brain barrier, which is otherwise a major obstacle to the development of new centrally acting drugs [46,73–76]. The use of CADs for selectively promoting autophagy may afford new therapies for aging, neurodegeneration, and cancer [18–21,23,26,27,70,77].

20.15 DISCUSSION

Although the pathogenesis of DIPL is increasingly understood and the propensity for causing these lesions is now relatively predictable [64], the forecasting of important sequelae is not [9]. The safety implications remain obscure, and DIPL is not easily addressed in the development and regulation of an IND. At the beginning of the modern era of drug development in the late 1960s and early 1970s, regulators were confronted with deaths and severe hepatic injury associated with DIPL and elevated levels of circulating BMP [28–30,37]. There is also a concern that lipidotic drugs might be associated with cardiac arrhythmias due to QTc prolongation, hERG protein trafficking, or other effects related to endosome/lysosome complexes. At present, DIPL, when encountered in a toxicology study, cannot be dismissed without further investigation, especially if associated with cardiac arrhythmias, neuronal toxicity, or liver, lung, kidney, or cardiac parenchymal injury.

There is concern in the pharmaceutical industry when DIPL is encountered, or likely to be encountered, given the problems in early detection and the potential clinical importance. Ongoing research addressing multiple aspects of DIPL, and an earlier and easier diagnosis, reflects the importance of DIPL to industry and to regulators [1,11,15,35,37,39,45,47,54,55,70,76,78–82]. The issue has driven QSAR and other machine learning technologies for *in silico* prediction of phospholipidosis [35,42,45]. Such forecasting is important considering, for example, the overlap between hERG channel blockade and phospholipidosis [38,51]. Research has prompted development of rapid *in vitro* methods for cell culture and assays for phospholipid binding [15,53–55,66,74,78,81] and validation of their ability to predict DIPL. Toxicogenomic studies to evaluate responding genes [11], and immunohistochemical methods of identifying DIPL definitively [28,57,58], have helped to identify the lipidosis without recourse to TEM. A reliable urine or blood test would provide an important alternative to the definitive, but less convenient, current diagnostic procedure and would resolve the present difficulty in recognizing DIPL in the clinic.

DIPL and hereditary lysosomal storage disease both vary widely in clinical importance. Concern over DIPL is, of course, greater when there is additional nonphospholipidotic histopathology and important overt sequelae. The FDA phospholipidosis database (Figure 20.2) and the AstraZeneca database, with its spectrum of histologic lesions from toxicology studies, shows a close association between DIPL and excess nonphospholipidotic lesions [35]. However, DIPL provoked in animals by some compounds such as loratadine may occur absent other concurrent histopathology. However, even the risk of isolated DIPL with loratadine depends on the extent of metabolic conversion to the less lipidotic desloratadine metabolite [43].

The lysosomal storage disease NP-C, and DIPL, not only share the hallmark multilamellar vesicles and disrupted cholesterol transport, but both can be devastating in their extreme expressions. Even patients with nonlethal NP-C and DIPL may present with important hepatotoxicity, neuropathy, and myopathy. Elevated serum and urine levels of BMP are a feature of both NP-C and amiodarone-induced DIPL [37]. Patients with DIPL present with variable organ damage, for example, chloroquine myopathy, perhexiline neurogenic muscle atrophy, 4,4′-diethylaminoethoxyhexestrol hepatotoxicity, and gentamicin nephrotoxicity, evidence that DIPL and parenchymal injury involve those organ(s) where there is a critical level of drug accumulation and lysosomal disruption.

There are several practical approaches to balancing the potential benefits versus risks of CADs. Regulatory agencies, including the FDA, approve drug doses that favorably balance the risks of no-treatment (factoring in disease severity and natural history) versus adverse effects [83]. The initial doses for human phase 1 trials in healthy subjects are based primarily on safety margins for positive findings in preclinical toxicity studies; the incidence and severity of AEs in clinical trials would limit the maximum recommended human dose. Drugs occasionally must be administered based on milligram per kilogram or milligram per square meter when there is a relatively small therapeutic index. As with any condition that does not have a convenient validated biomarker, it may be difficult to establish the relation between DIPL and the study drug and whether organ dysfunction precedes or follows the appearance of DIPL.

Covert off-target toxicities can be difficult to detect and evaluate unless toxicity is an expected extension of pharmacodynamic activity, for example, systemic toxicity of corticosteroids [73,75]. Although QSAR and physicochemical analysis of lipophilicity and amphiphilicity may be reassuring when they predict a low risk of DIPL, when the latter is observed in animal safety studies, it may be an unexpected off-target event and imprudent to dismiss. The important lipophilicity and amphiphilicity components of the determinants of *in vivo* toxicity may be receptor-mediated, a consequence of nonspecific accumulation, or a detergent effect [75]. Drug discovery projects are heavily populated with cationic amphiphilic amines, which are common pharmacophores for histamine, muscarinic, and adrenal receptors. As exposure-toxicity data sets and other lipidotic pharmacophore domains are accumulated, the risks of developing DIPL may become clarified, and DIPL may be avoided. Extensive DIPL and other histopathological lesions may not be colocated or obviously associated; therefore, these additional data sets and focused mechanistic studies may identify the collateral pharmacodynamic activity that may cause collateral injury.

With the availability of QSAR and *in vitro* screening methods [80], the potential risk of developing DIPL may be assessed early in drug development. However, safety margins cannot be determined until after the start of clinical studies and the establishment of clinical dose ranges. When development of DIPL is predicted from QSAR or *in vitro* testing, the risk should be verified in preclinical toxicology studies that should determine propensity, safety margin, severity, and reversibility. Tissues that show DIPL should be monitored for signs of collateral histopathology, and the clinical relevance of histologic lesions such as liver or lung dysfunction or electrocardiographic changes adjudicated. There may be a link between severity of DIPL, AEs, and QTc interval prolongation by direct or indirect (protein trafficking) mechanisms.

Although BMP level may not be monitored in animal toxicity trials, except for cause, elevated levels would suggest its use in monitoring clinical studies. Early indications of DIPL from *in silico* or *in vitro* assays would indicate whether BMP monitoring may be useful in animal or clinical trials. A preclinical assay of protein trafficking (hERG-Lite assay, ChanTest, Cleveland, OH) may be advisable for DIPL-positive compounds. The possibility should be considered that metabolites may be more lipidotic than the parent compound, even though this may be uncommon [43]. It is important to evaluate lung, liver, brain, and lymph node parenchymal cells, resident tissue macrophages, and bile duct cells, and this evaluation may be a basis for classifying and monitoring DIPL according to organ systems most commonly involved. As with any toxicologic pathology, targets may be specific to species, organ, and CAD, and a reliable biomarker is needed that would be species- and organ-independent. Monitoring in the clinic with BMP would be advisable as an early sentinel for the collateral tissue damage, especially when there is important tissue damage in nonclinical toxicology studies and elevated BMP level. This may be important especially when a small safety margin exists for phospholipidotic versus targeted pharmacodynamic activity. Clinicians who oversee trials of CADs may benefit from having a biomarker, knowing the affected organ systems in animal studies and focusing on those systems. Recommendations are available for assessment of potential toxicity of DIPL and QT prolongation and protein trafficking [80] (Figure 20.3).

Although it may be difficult to develop a drug that may induce DIPL, there is much potential for therapeutic benefit. CADs are increasingly recognized as inhibitors and inducers of autophagy [27] and processes linked to aging [19], cancer [77], Alzheimer's disease [84], and neurodegenerative disease [21,22,25]. The physicochemical properties of CADs enable these drugs to access the interior of the cell and facilitate passage across the blood–brain barrier [3], which may not be achievable with other drugs. The multimodal study by Mesens et al. [85] provides an example of additional ways to characterize a CAD compound. With the potential for this group of compounds, Figure 20.4 offers a flowchart of steps and considerations in developing a compound that causes phospholipidosis (see also Ref. [80]).

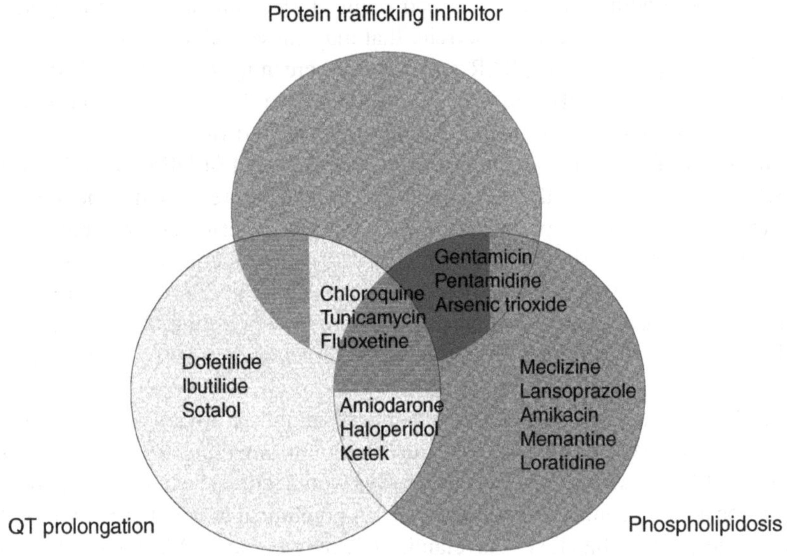

Figure 20.3 Relation between phospholipidosis, QT prolongation, protein trafficking effects, and adverse events. The QT-only group had adverse events that were significant but on target (i.e., related to the mechanism of action of drugs). The phospholipidosis-only group had mild adverse events. The overlap groups had significant, off-target adverse events associated with their use [38]. (*See color plate section for the color representation of this figure.*)

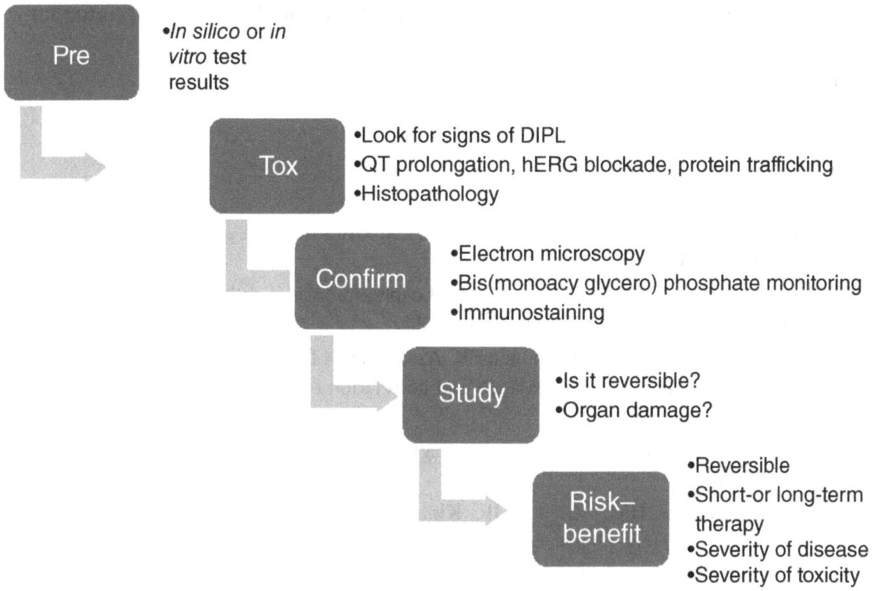

Figure 20.4 Flow chart for pharmaceutical development. When drug-induced phospholipidosis (DIPL) is predicted from *in silico* studies or *in vitro* assays, animal studies are used to evaluate DIPL. When DIPL is confirmed, it may be monitored with BMP and associated with tissue toxicity that may or may not be reversible. The BMP may be monitored clinically to assess potential drug toxicities.

All drugs have AEs that may or may not be dose related. Regulatory agencies understand the potential for AEs and use risk/benefit profiles in decision-making. Together with sponsors, regulatory agencies implement a Risk Evaluation and Mitigation Strategy (REMS) when it is necessary to ensure that the benefit of a drug or biological product will justify the risk. Multiple reports suggest that vitamin E and polyamino acids may decrease adverse symptoms of amiodarone [69–72], a phospholipidotic antiarrhythmic drug with major toxicity that restricts its usage. It would be important to know whether vitamin E and polyamino acids improved the safety of amiodarone without reducing its utility in atrial fibrillation.

20.16 FUTURE DIRECTIONS AND RECOMMENDATIONS

- Increase use of BMP as a biomarker in nonclinical and clinical trials of drugs that induce phospholipidosis.
- Further develop strategies to treat or prevent phospholipidosis, for example, vitamin E and polyamino acids versus amiodarone lung toxicity.
- Increase awareness of the importance of metabolism, and genetic variations of patients, in the expression of phospholipidosis.

- Further investigate the association between phospholipidosis and autophagy/ apoptosis.
- Develop new therapeutic uses for CADs.

REFERENCES

[1] Shikata T, Kanetaka T, Endo Y, Nagashima K. Drug-induced generalized phospholipidosis. Acta Pathol Jpn 1972;22(3):517–531.

[2] Nonoyama T, Fukuda R. Drug-induced phospholipidosis – pathological aspects and its prediction. J Toxicol Pathol 2008;21(1):9–24.

[3] Vitovic P, Alakoskela JM, Kinnunen PK. Assessment of drug-lipid complex formation by a high-throughput Langmuir-balance and correlation to phospholipidosis. J Med Chem 2008;51(6):1842–1848.

[4] Temple R. Policy developments in regulatory approval. Stat Med 2002;21(19): 2939–2948.

[5] Bass AS, Vargas HM, Valentin JP, Kinter LB, Hammond T, Wallis R, Siegl PK, Yamamoto K. Safety pharmacology in 2010 and beyond: survey of significant events of the past 10 years and a roadmap to the immediate-, intermediate- and long-term future in recognition of the tenth anniversary of the Safety Pharmacology Society. J Pharmacol Toxicol Methods 2011;64(1):7–15.

[6] Karsh LI. A clinical trial primer: historical perspective and modern implementation. Urol Oncol 2012;30(4 suppl):S28–S32.

[7] Berridge BR, Chatman LA, Odin M, Schultze AE, Losco PE, Meehan JT, Peters T, Vonderfecht SL, Society of Scientific and Regulatory Policy Committee Working Group. Phospholipidosis in nonclinical toxicity studies. Toxicol Pathol 2007;35(2):325.

[8] Reasor MJ, Hastings KL, Ulrich RG. Drug-induced phospholipidosis: issues and future directions. Expert Opin Drug Saf 2006;5(4):567–583.

[9] Reasor MJ, Kacew S. Drug-induced phospholipidosis: are there functional consequences? Exp Biol Med 2001;226(9):825–830.

[10] Shayman JA, Abe A. Drug induced phospholipidosis: an acquired lysosomal storage disorder. Biochim Biophys Acta 2013;1831(3):602–611.

[11] Sawada H, Takami K, Asahi S. A toxicogenomic approach to drug-induced phospholipidosis: analysis of its induction mechanism and establishment of a novel *in vitro* screening system. Toxicol Sci 2005;83(2):282–292.

[12] Kodavanti UP, Mehendale HM. Cationic amphiphilic drugs and phospholipid storage disorder. Pharmacol Rev 1990;42(4):327–354.

[13] Blohm TR. Drug-induced lysosomal lipidosis: biochemical interpretations. Pharmacol Rev 1978;30(4):593–603.

[14] Lüllmann H, Lüllmann-Rauch R, Wassermann O. Lipidosis induced by amphiphilic cationic drugs. Biochem Pharmacol 1978;27(8):1103–1108.

[15] Monteith DK, Morgan RE, Halstead B. *In vitro* assays and biomarkers for drug-induced phospholipidosis. Expert Opin Drug Metab Toxicol 2006;2(5):687–696.

[16] Farrell GC. Drugs and steatohepatitis. Semin Liver Dis 2002;22(2):185–194.

[17] Levine B, Kroemer G. Autophagy in the pathogenesis of disease. Cell 2008;132(1): 27–42.

[18] Gustafsson AB, Gottlieb RA. Recycle or die: the role of autophagy in cardioprotection. J Mol Cell Cardiol 2008;44(4):654–661.

[19] Rajawat YS, Hilioti Z, Bossis I. Aging: central role for autophagy and the lysosomal degradative system. Ageing Res Rev 2009;8(3):199–213.

[20] Mehrpour M, Esclatine A, Beau I, Codogno P. Overview of macroautophagy regulation in mammalian cells. Cell Res 2010;20(7):748–762.

[21] Renna M, Jimenez-Sanchez M, Sarkar S, Rubinsztein DC. Chemical inducers of autophagy that enhance the clearance of mutant proteins in neurodegenerative diseases. J Biol Chem 2010;285(15):11061–11067.

[22] Eskelinen EL, Illert AL, Tanaka Y, Schwarzmann G, Blanz J, Von Figura K, Saftig P. Role of LAMP-2 in lysosome biogenesis and autophagy. Mol Biol Cell 2002;13(9): 3355–3368.

[23] Shanware NP, Bray K, Abraham RT. The PI3K, metabolic, and autophagy networks: interactive partners in cellular health and disease. Annu Rev Pharmacol Toxicol 2013; 53:89–106.

[24] Pastor MD, García-Yébenes I, Fradejas N, Pérez-Ortiz JM, Mora-Lee S, Tranque P, Moro MA, Pende M, Calvo S. mTOR/S6 kinase pathway contributes to astrocyte survival during ischemia. J Biol Chem 2009;284(33):22067–22078.

[25] Bové J, Martinez-Vicente M, Vila M. Fighting neurodegeneration with rapamycin: mechanistic insights. Nature Reviews. Neuroscience 2011;12(8):437–452.

[26] Cho KS, Yoon YH, Choi JA, Lee SJ, Koh JY. Induction of autophagy and cell death by tamoxifen in cultured retinal pigment epithelial and photoreceptor cells. Invest Ophthalmol Vis Sci 2012;53(9):5344–5353.

[27] Yoon YH, Cho KS, Hwang JJ, Lee SJ, Choi JA, Koh JY. Induction of lysosomal dilatation, arrested autophagy, and cell death by chloroquine in cultured ARPE-19 cells. Invest Ophthalmol Vis Sci 2010;51(11):6030–6037.

[28] Matsuzawa Y, Yamamoto A, Adachi S, Nishikawa M. Studies on drug-induced lipidosis. VIII Correlation between drug accumulation and acidic phospholipids. J Biochem 1977; 82(5):1369–1377.

[29] Akeda S. A study on the lipidosis induced by a coronary vasodilator, 4,4'-diethylaminoethoxyhexestrol dihydrochoride. Mie Med J 1972–1973;22:65–96.

[30] Kasama K, Yoshida K, Takeda S, Akeda S, Kawai K. Bis-(monoacylglyceryl)phosphate and acyl phosphatidylglycerol isolated from human livers of lipidosis induced by 4,4'-diethylaminoethoxyhexestrol. Lipids 1974;9(4):235–243.

[31] Bolesta S, Roslund BP. Elevated hepatic transaminases associated with telithromycin therapy: a case report and literature review. Am J Health Syst Pharm 2008;65(1):37–41.

[32] Casado E, Gratacós J, Tolosa C, Martínez JM, Ojanguren I, Ariza A, Real J, Sanjuán A, Larrosa M. Antimalarial myopathy: an underdiagnosed complication? Prospective longitudinal study of 119 patients. Ann Rheum Dis 2006;65(3):385–390.

[33] Myers JL, Kennedy JI, Plumb VJ. Amiodarone lung: pathologic findings in clinically toxic patients. Hum Pathol 1987;18(4):349–354.

[34] Kennedy JI, Myers JL, Plumb VJ, Fulmer JD. Amiodarone pulmonary toxicity. Clinical, radiologic, and pathologic correlations. Arch Intern Med 1987;147(1):50–55.

[35] Barone LR, Boyer S, Damewood JR Jr, Fikes J, Ciaccio PJ. Phospholipogenic pharmaceuticals are associated with a higher incidence of histological findings than nonphospholipogenic pharmaceuticals in preclinical toxicology studies. J Toxicol 2012;2012:308594. DOI: 10.1155/2012/308594.

[36] Sundin DP, Sandoval R, Molitoris BA. Gentamicin inhibits renal protein and phospholipid metabolism in rats: implications involving intracellular trafficking. J Am Soc Nephrol 2001;12(1):114–123.

[37] Tengstrand EA, Miwa GT, Hsieh FY. Bis(monoacylglycerol)phosphate as a non-invasive biomarker to monitor the onset and time-course of phospholipidosis with drug-induced toxicities. Expert Opin Drug Metab Toxicol 2010;6(5):555–570.

[38] Willard JM. 2010. Phospholipidosis: data mining, modeling and laboratory research. United States Food and Drug Administration Web site. Available at http://www.fda.gov/downloads/AdvisoryCommittees/CommitteesMeetingMaterials/Drugs/AdvisoryCommitteeforPharmaceuticalScienceandClinicalPharmacology/UCM210798.pdf. Accessed 2015 June 29.

[39] Pelletier DJ, Gehlhaar D, Tilloy-Ellul A, Johnson TO, Greene N. Evaluation of a published *in silico* model and construction of a novel Bayesian model for predicting phospholipidosis inducing potential. J Chem Inf Model 2007;47(3):1196–1205.

[40] Kruhlak NL, Choi SS, Contrera JF, Weaver JL, Willard JM, Hastings KL, Sancilio LF. Development of a phospholipidosis database and predictive quantitative structure–activity relationship (QSAR) models. Toxicol Mech Methods 2008;18(2–3):217–227.

[41] Choi SS, Kim JS, Valerio LG Jr, Sadrieh N. *In silico* modeling to predict drug-induced phospholipidosis. Toxicol Appl Pharmacol 2013;269(2):195–204.

[42] Orogo AM, Choi SS, Minnier BL, Kruhlak NL. Construction and consensus performance of (Q)SAR models for predicting phospholipidosis using a dataset of 743 compounds. Mol Inform 2012;31(10):725–739.

[43] Goracci L, Ceccarelli M, Bonelli D, Cruciani G. Modeling phospholipidosis induction: reliability and warnings. J Chem Inf Model 2013;53(6):1436–1446.

[44] Kaufmann AM, Krise JP. Lysosomal sequestration of amine-containing drugs: analysis and therapeutic implications. J Pharm Sci 2007;96(4):729–746.

[45] Zhang X, Zheng N, Rosania GR. Simulation-based cheminformatic analysis of organelle-targeted molecules: lysosomotropic monobasic amines. J Comput Aided Mol Des 2008; 22(9):629–645.

[46] Mesens N, Steemans M, Hansen E, Verheyen GR, Van Goethem F, Van Gompel J. Screening for phospholipidosis induced by central nervous drugs: comparing the predictivity of an *in vitro* assay to high throughput *in silico* assays. Toxicol In Vitro 2010; 24(5):1417–1425.

[47] Valerio LG Jr. *In silico* toxicology models and databases as FDA Critical Path Initiative toolkits. Hum Genomics 2011;5(3):200–207.

[48] Gardiner SJ, Begg EJ. Pharmacogenetic testing for drug metabolizing enzymes: is it happening in practice? Pharmacogenet Genomics 2005;15(5):365–369.

[49] Inglis S, Stewart S. Metabolic therapeutics in angina pectoris: history revisited with perhexilene. Eur J Cardiovasc Nurs 2006;5(2):175–184.

[50] Seierstad M, Agrafiotis DK. A QSAR model of hERG binding using a large, diverse, and internally consistent training set. Chem Biol Drug Des 2006;67(4):284–296.

[51] Sun H, Xia M, Shahane SA, Jadhav A, Austin CP, Huang R. Are hERG channel blockers also phospholipidosis inducers? Bioorg Med Chem Lett 2013;23(16):4587–4590.

[52] Cunningham ML, Lehman-McKeeman L. Applying toxicogenomics in mechanistic and predictive toxicology. Toxicol Sci 2005;83(2):205–206.

[53] van de Water FM, Havinga J, Ravesloot WT, Horbach GJ, Schoonen WG. High content screening analysis of phospholipidosis: validation of a 96-well assay with CHO-K1 and HepG2 cells for the prediction of *in vivo* based phospholipidosis. Toxicol in Vitro 2011;25(8):1870–1882.

[54] Kasahara T, Tomita K, Murano H, Harada T, Tsubakimoto K, Ogihara T, Ohnishi S, Kakinuma C. Establishment of an *in vitro* high-throughput screening assay for detecting phospholipidosis-inducing potential. Toxicol Sci 2006;90(1):133–141.

[55] Bhandari N, Figueroa DJ, Lawrence JW, Gerhold DL. Phospholipidosis assay in HepG2 cells and rat or rhesus hepatocytes using phospholipid probe NBD-PE. Assay Drug Dev Technol 2008;6(3):407–419.

[56] Eskelinen EL. Roles of LAMP-1 and LAMP-2 in lysosome biogenesis and autophagy. Mol Aspects Med 2006;27(5–6):495–502.

[57] Obert LA, Sobocinski GP, Bobrowski WF, Metz AL, Rolsma MD, Altrogge DM, Dunstan RW. An immunohistochemical approach to differentiate hepatic lipidosis from hepatic phospholipidosis in rats. Toxicol Pathol 2007;35(5):728–734.

[58] Asaoka Y, Togashi Y, Imura N, Sai T, Miyoshi T, Miyamoto Y. Immunohistochemistry of LAMP-2 and adipophilin for phospholipidosis in liver and kidney in ketoconazole-treated mice. Exp Toxicol Pathol 2013;65(6):817–823.

[59] Larrain S, Rinella ME. A myriad of pathways to NASH. Clin Liver Dis 2012;16(3): 525–548.

[60] Luo D. Ultracytochemical observation on lysosome of hair cells in kanamycin ototoxic guinea pigs [in Chinese]. Zhonghua Er Bi Yan Hou Ke Za Zhi 1990;25(5):281–282, 318.

[61] Dennis AT, Wang L, Wan H, Nassal D, Deschenes I, Ficker E. Molecular determinants of pentamidine-induced hERG trafficking inhibition. Mol Pharmacol 2012;81(2): 198–209.

[62] Kuryshev YA, Ficker E, Wang L, Hawryluk P, Dennis AT, Wible BA, Brown AM, Kang J, Chen XL, Sawamura K, Reynolds W, Rampe D. Pentamidine-induced long QT syndrome and block of hERG trafficking. J Pharmacol Exp Therap 2005;312(1):316–323.

[63] Bravo-Cuellar A, Balercia G, Levesque JP, Liu XH, Osculati F, Orbach-Arbouys . Enhanced activity of peritoneal cells after aclacinomycin injection: effect of pretreatment with superoxide dismutase on aclacinomycin-induced cytological alterations and antitumoral activity. Cancer Res 1989;49(6):1578–1586.

[64] Lowe R, Mussa HY, Nigsch F, Glen RC, Mitchell JBO. Predicting the mechanism of phospholipidosis. J Chemoinform 2012;4:2. DOI: 10.1186/1758-2946-4-2.

[65] Hiraoka M, Abe A, Lu Y, Yang K, Han X, Gross RW, Shayman JA. Lysosomal phospholipase A2 and phospholipidosis. Mol Cell Biol 2006;26(16):6139–6148.

[66] Joshi UM, Kodavanti PR, Coudert B, Dwyer TM, Mehendale HM. Types of interaction of amphiphilic drugs with phospholipid vesicles. J Pharmacol Experim Therap 1988;246(1):150–157.

[67] Morissette G, Lodge R, Marceau F. Intense pseudotransport of a cationic drug mediated by vacuolar ATPase: procainamide-induced autophagic cell vacuolization. Toxicol Appl Pharmacol 2008;228(3):364–377.

[68] El Mouedden M, Laurent G, Mingeot-Leclercq MP, Taper HS, Cumps J, Tulkens PM. Apoptosis in renal proximal tubules of rats treated with low doses of aminoglycosides. Antimicrob Agents Chemother 2000;44(3):665–675.

[69] Kishore BK, Lambricht P, Laurent G, Maldague P, Wagner R, Tulkens PM. Mechanism of protection afforded by polyaspartic acid against gentamicin-induced phospholipidosis. II. Comparative *in vitro* and *in vivo* studies with poly-L-aspartic, poly-L-glutamic and poly-D-glutamic acids. J Pharmacol Exp Therap 1990;255(2):875–885.

[70] Vejux A, Guyot S, Montange T, Riedinger JM, Kahn E, Lizard G. Phospholipidosis and down-regulation of the PI3-K/PDK-1/Akt signalling pathway are vitamin E inhibitable events associated with 7-ketocholesterol-induced apoptosis. J Nutr Biochem 2009;20(1):45–61.

[71] Card JW, Racz WJ, Brien JF, Massey TE. Attenuation of amiodarone-induced pulmonary fibrosis by vitamin E is associated with suppression of transforming growth factor-beta1 gene expression but not prevention of mitochondrial dysfunction. J Pharmacol Exp Therap 2003;304(1):277–283.

[72] Scuntaro I, Kientsch U, Wiesmann UN, Honegger UE. Inhibition by vitamin E of drug accumulation and of phospholipidosis induced by desipramine and other cationic amphiphilic drugs in human cultured cells. Br J Pharmacol 1996;119(5):829–834.

[73] Lounkine E, Keiser MJ, Whitebread S, Mikhailov D, Hamon J, Jenkins JL, Lavan P, Weber E, Doak AK, Côté S, Shoichet BK, Urban L. Large-scale prediction and testing of drug activity on side-effect targets. Nature 2012;486(7403):361–367.

[74] Hanumegowda UM, Wenke G, Regueiro-Ren A, Yordanova R, Corradi JP, Adams SP. Phospholipidosis as a function of basicity, lipophilicity, and volume of distribution of compounds. Chem Res Toxicol 2010;23(4):749–755.

[75] Price DA, Blagg J, Jones L, Greene N, Wager T. Physicochemical drug properties associated with *in vivo* toxicological outcomes: a review. Expert Opin Drug Metab Toxicol 2009;5(8):921–931.

[76] Ratcliffe AJ. Medicinal chemistry strategies to minimize phospholipidosis. Curr Med Chem 2009;16(22):2816–2823.

[77] Mathew R, Karantza-Wadsworth V, White E. Role of autophagy in cancer. Nature Reviews. Cancer 2007;7(12):961–967.

[78] Zhou L, Geraci G, Hess S, Yang L, Wang J, Argikar U. Predicting phospholipidosis: a fluorescence noncell based *in vitro* assay for the determination of drug-phospholipid complex formation in early drug discovery. Anal Chem 2011;83(18):6980–6987.

[79] Naisbitt DJ, Ruscoe JE, Williams D, O'Neill PM, Pirmohamed M, Park BK. Disposition of amodiaquine and related antimalarial agents in human neutrophils: implications for drug design. Journal of Pharmacology and Experimental Therapeutics 1997;280(2):884–893.

[80] Chatman LA, Morton D, Johnson TO, Anway SD. A strategy for risk management of drug-induced phospholipidosis. Toxicol Pathol 2009;37(7):997–1005.

[81] Kuroda Y, Saito M. Prediction of phospholipidosis-inducing potential of drugs by *in vitro* biochemical and physicochemical assays followed by multivariate analysis. Toxicol in Vitro 2010;24(2):661–668.

[82] Smith DA, van de Waterbeemd H, Walker DK. Pharmocokinetics and metabolism in drug design. In: Mannhold R, Kubinyi H, Folkers G, editors. *Methods And Principles In Medicinal Chemistry*. Berlin: Wiley-VCH Verlag GmbH; 2001.

[83] United States Food and Drug Administration. 2005. Guidance for industry: premarketing risk assessment. Available at http://www.fda.gov/downloads/RegulatoryInformation/Guidances/UCM126958.pdf.

[84] Barnett A, Brewer GJ. Autophagy in aging and Alzheimer's disease: pathologic or protective? J Alzheimers Dis 2011;25(3):385–394.

[85] Mesens N, Desmidt M, Verheyen GR, Starckx S, Damsch S, De Vries R, Verhemeldonck M, Van Gompel J, Lampo A, Lammens L. Phospholipidosis in rats treated with amiodarone: serum biochemistry and whole genome micro-array analysis supporting the lipid traffic jam hypothesis and the subsequent rise of the biomarker BMP. Toxicol Pathol 2012;40(3):491–503.

INDEX

Lysosomes: Biology, Diseases, and Therapeutics, First Edition.
Edited by Frederick R. Maxfield, James M. Willard and Shuyan Lu.

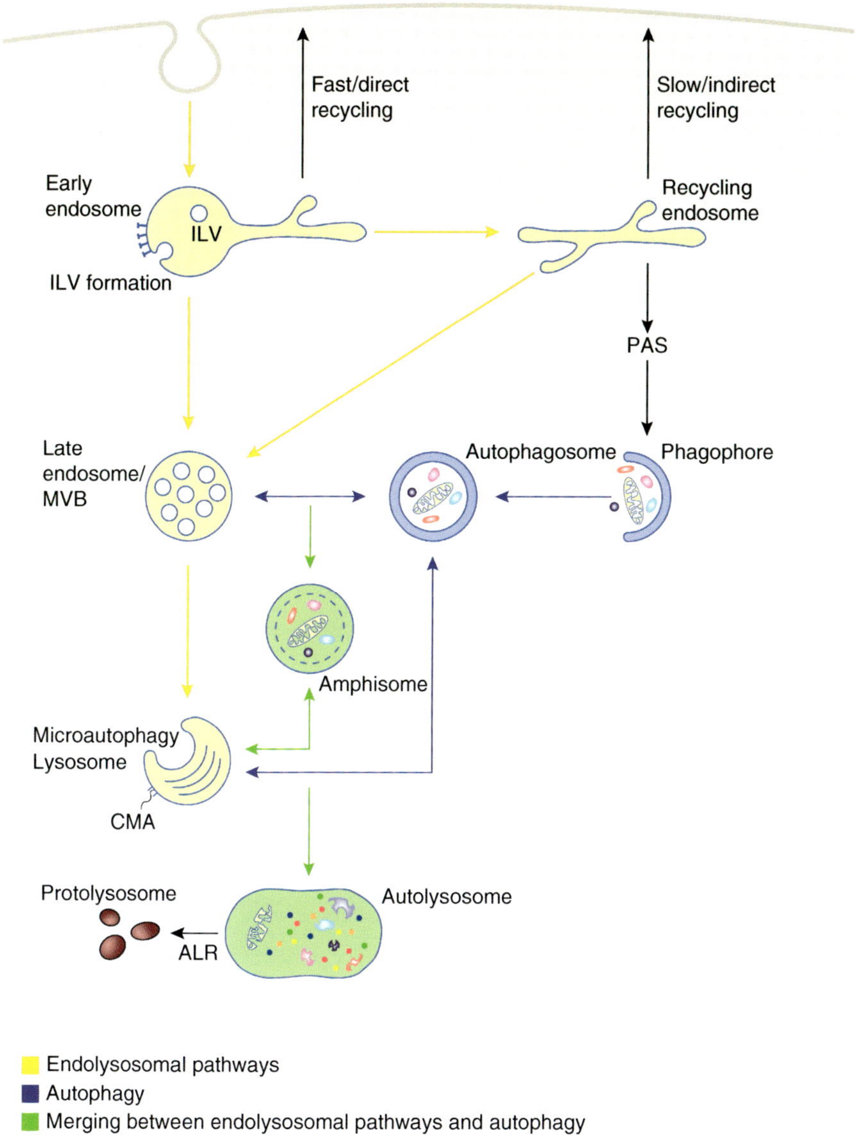

Figure 2.1 Schematic drawing depicting the endocytic and autophagy pathways to the lysosomes. Note: Refer to page 8 for full caption.

Lysosomes: Biology, Diseases, and Therapeutics, First Edition.
Edited by Frederick R. Maxfield, James M. Willard and Shuyan Lu.

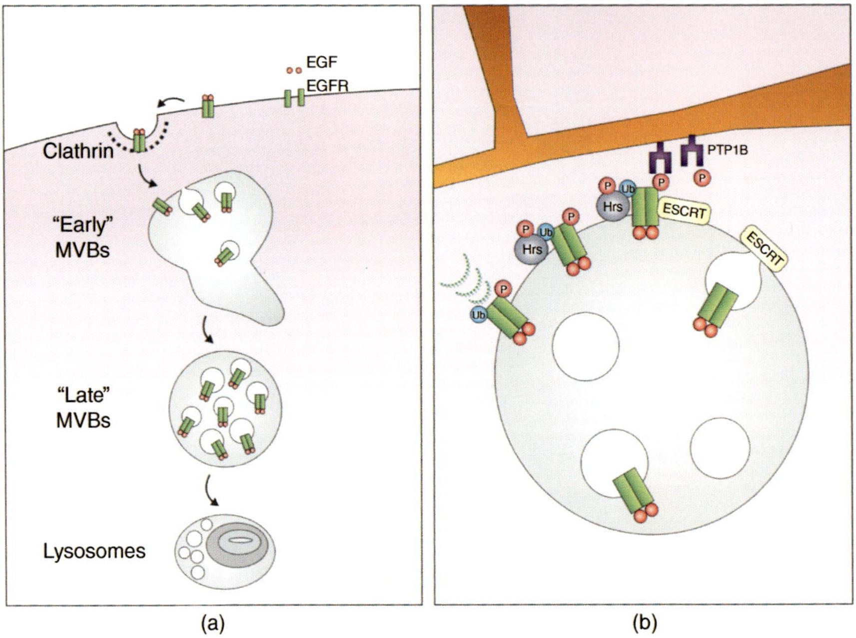

Figure 3.2 Downregulation of EGFR signaling by sorting onto ILVs. Note: Refer to page 37 for full caption.

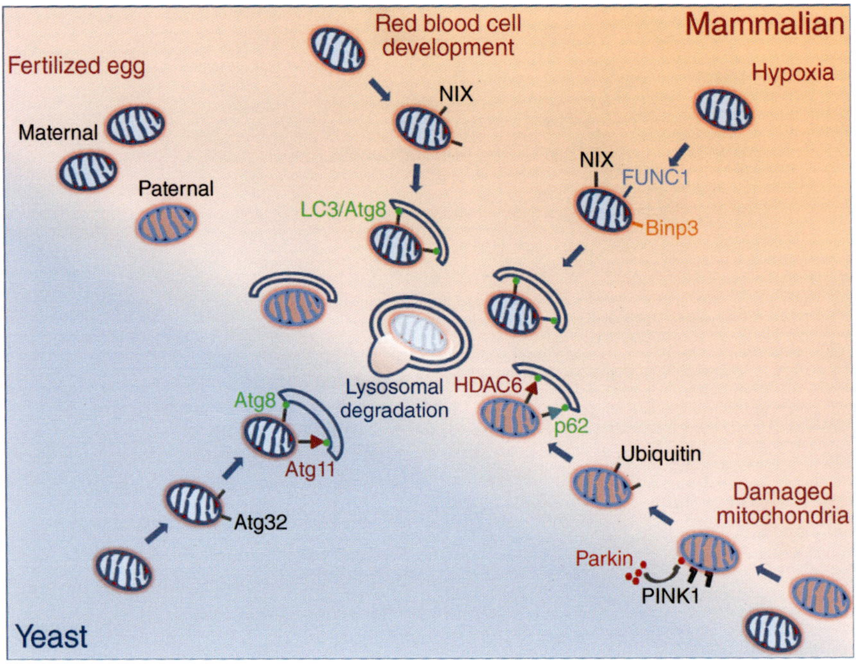

Figure 4.1 Summary of mitophagy in yeast and mammalian cells. Note: Refer to page 55 for full caption.

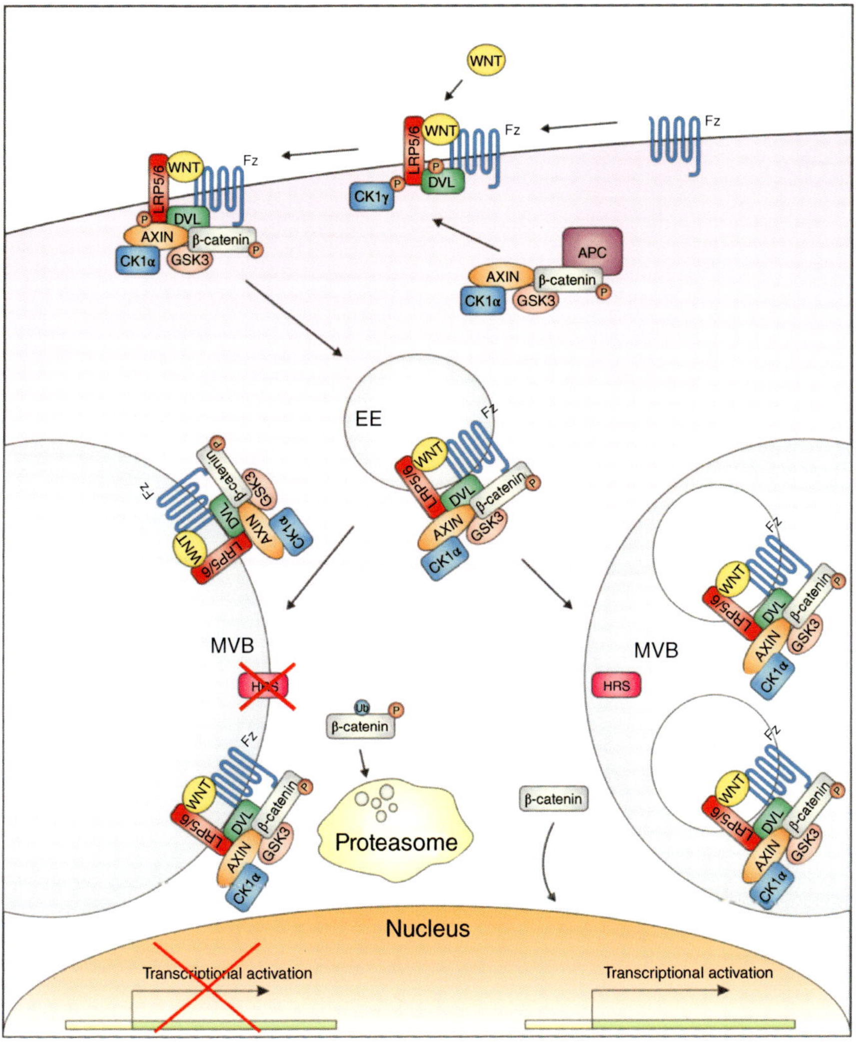

Figure 3.3 Upregulation of Wnt signaling by sorting onto ILVs. Note: Refer to page 40 for full caption.

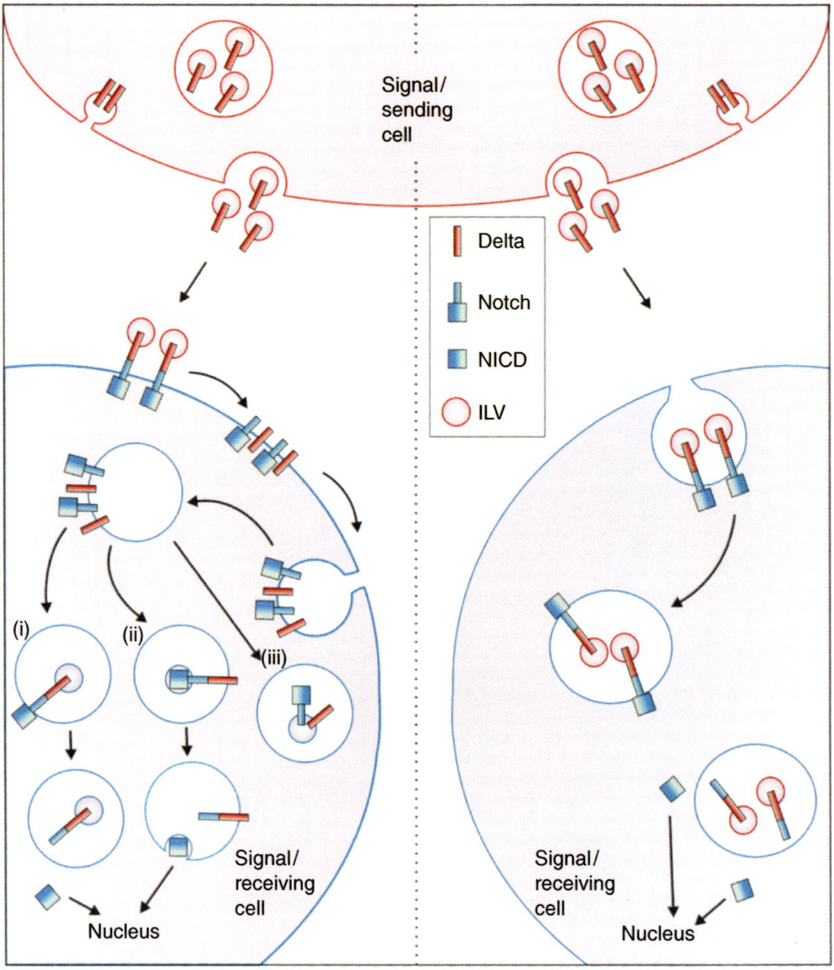

Figure 3.4 Intercellular Notch signaling mediated by ligand-bearing exosomes. Note: Refer to page 43 for full caption.

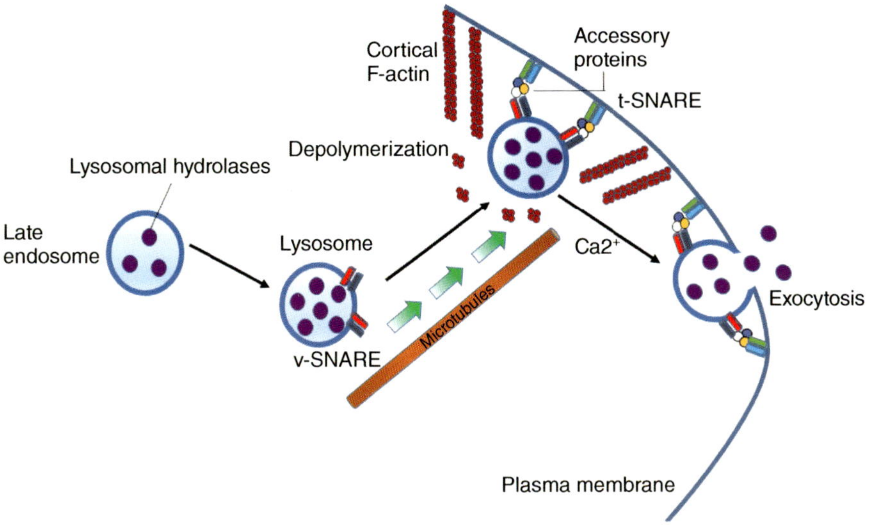

Figure 5.2 Schematic of mechanisms involved in lysosome exocytosis. Note: Refer to page 69 for full caption.

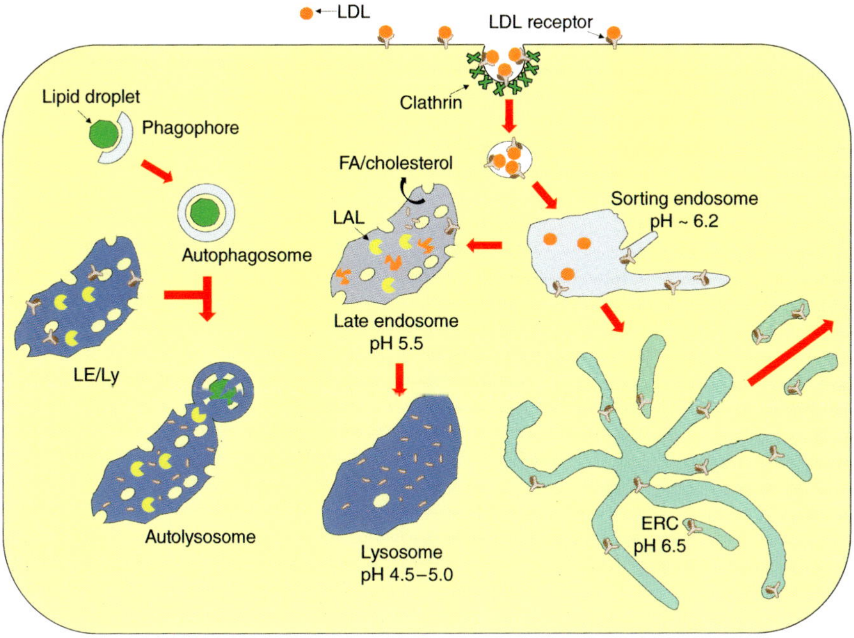

Figure 6.1 Schematic diagram of lipoprotein and lipid droplet degradation by lysosomes. Note: Refer to page 88 for full caption.

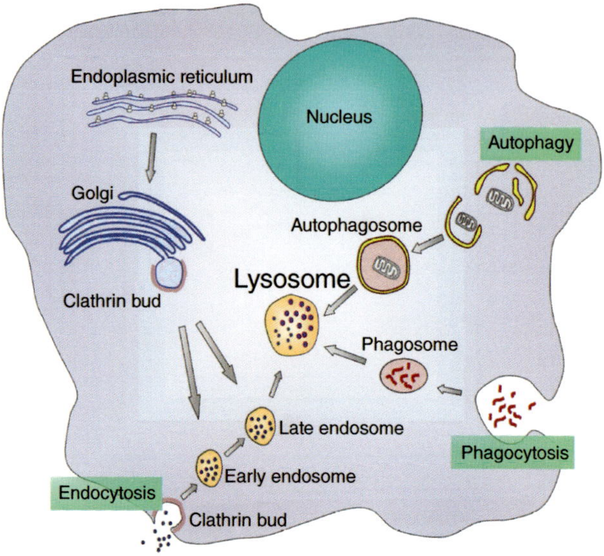

Figure 7.1 Central role of lysosomes in key cellular processes.

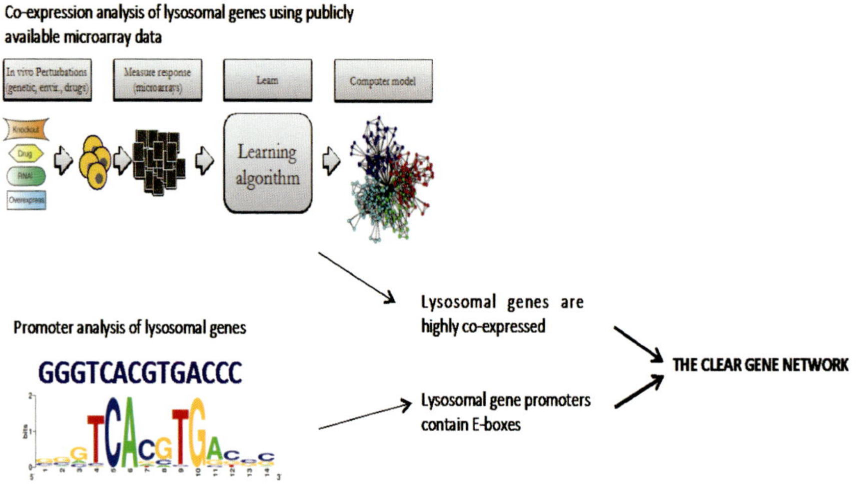

Figure 7.2 Systems biology approach used to discover the CLEAR network. Note: Refer to page 103 for full caption.

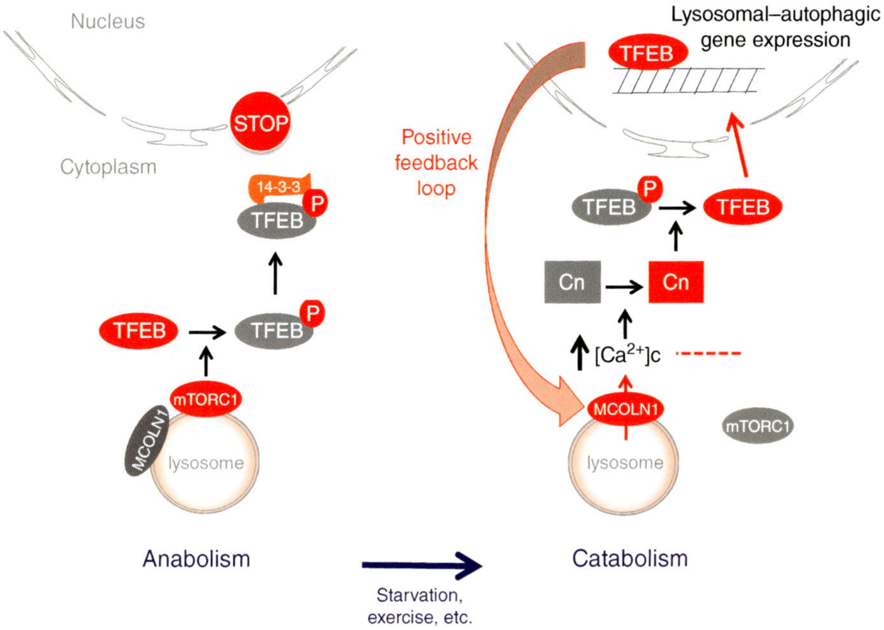

Figure 7.3 Model depicting Ca^{2+}-mediated regulation of TFEB. Note: Refer to page 107 for full caption.

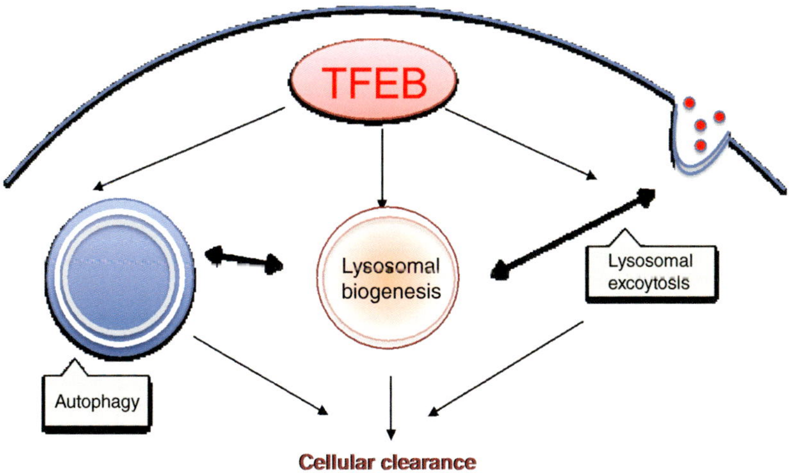

Figure 7.4 TFEB activation on cellular mechanisms that lead to cellular clearance. Note: Refer to page 108 for full caption.

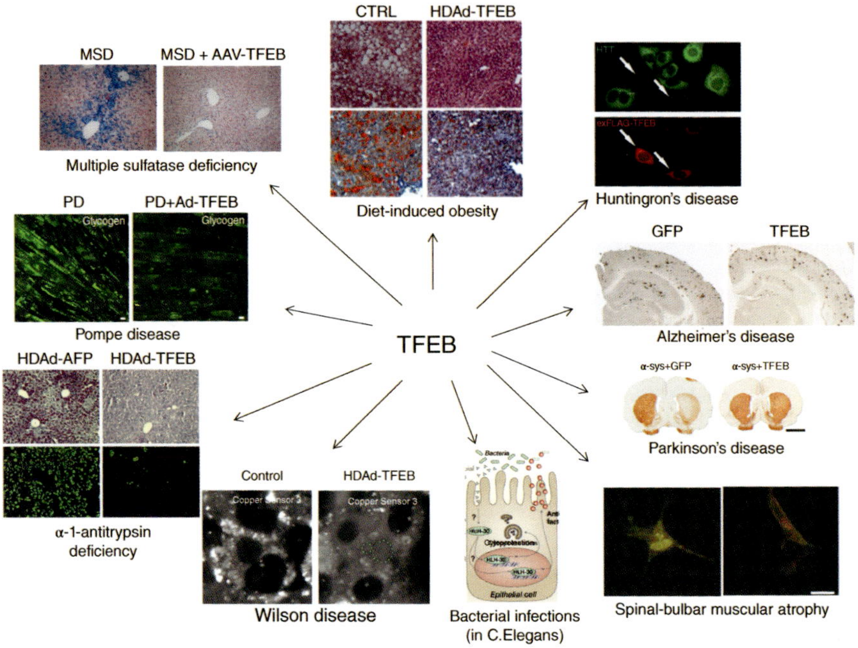

Figure 7.5 Diseases that respond to TFEB-mediated clearance. Note: Refer to page 108 for full caption.

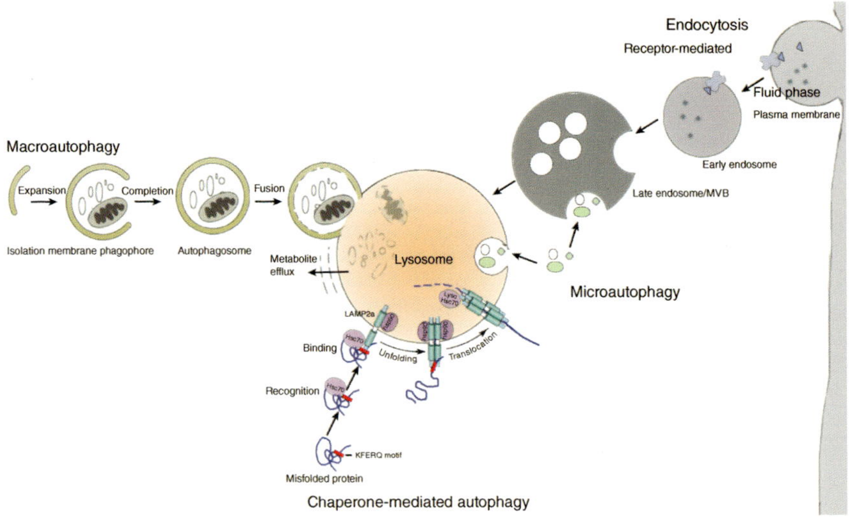

Figure 9.1 Major routes of substrate delivery to lysosomes. Note: Refer to page 138 for full caption.

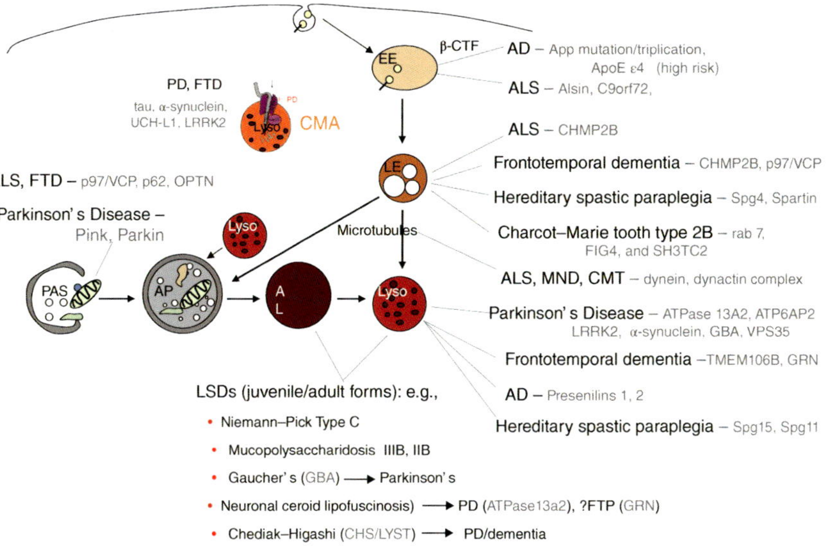

Figure 9.2 Genetic evidence strongly implicates the lysosomal network in the pathogenesis of neurodegenerative disease. Note: Refer to page 152 for full caption.

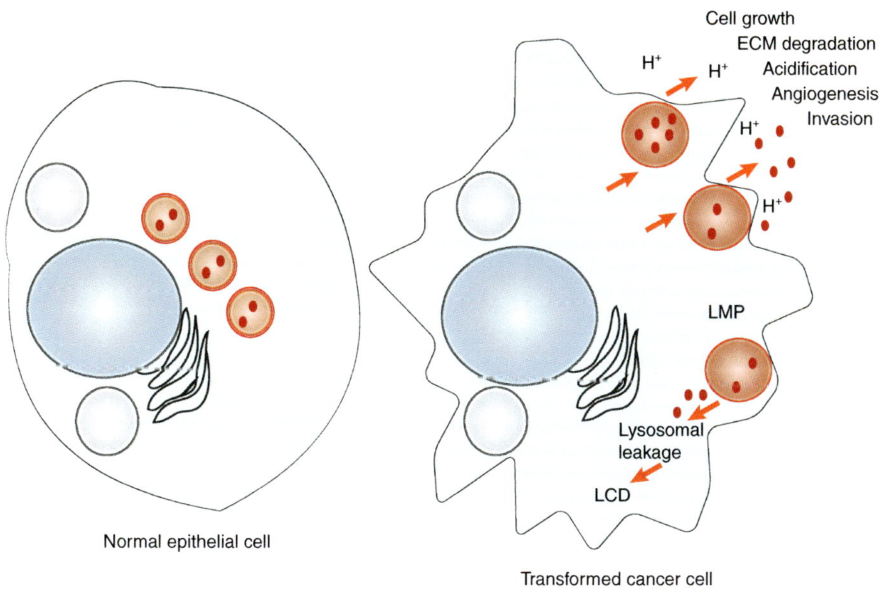

Figure 10.1 Cancer development induces changes in the lysosomal function. Cancer development and progression induces lysosomal biogenesis increasing the expression of various lysosomal hydrolases. It alters the lysosomal membrane integrity, sensitizing them to LMP and lysosomal leakage that can lead to LCD. Note: Refer to page 183 for full caption.

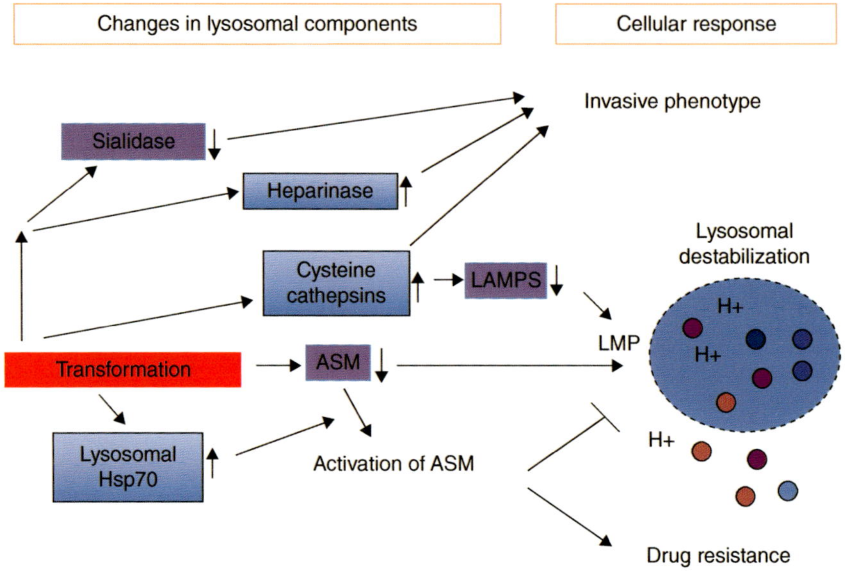

Figure 10.2 Cancer induces changes in lysosomal composition. These include changes in the expression levels and activity of several lysosomal hydrolases. Note: Refer to page 186 for full caption.

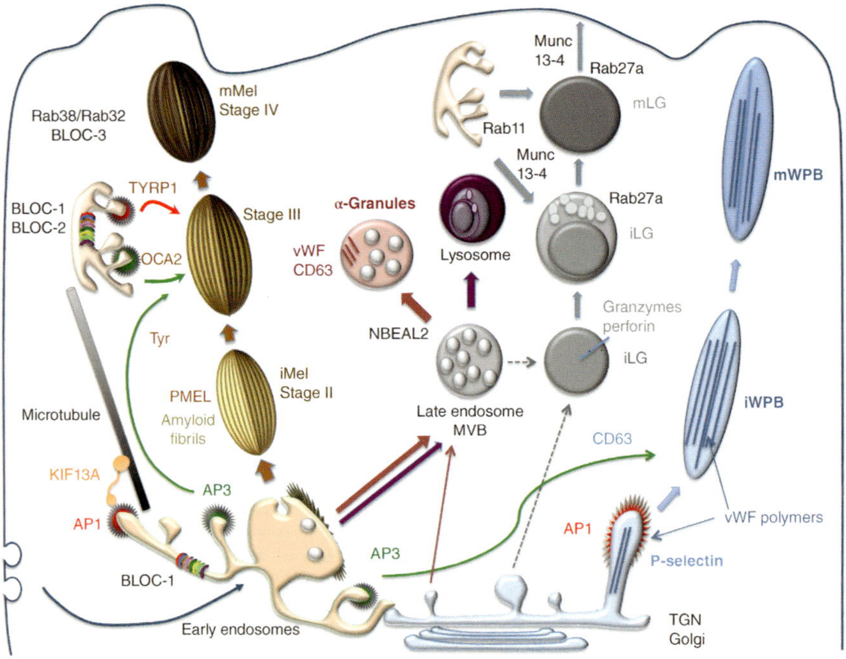

Figure 12.2 Model for biogenesis of four vertebrate LROs. Note: Refer to page 245 for full caption.

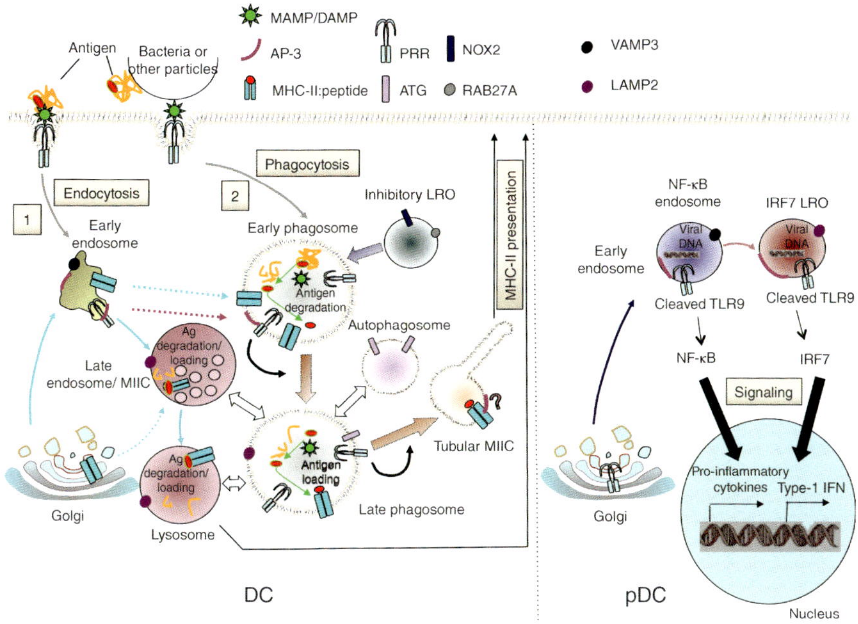

Figure 12.3 Model for LRO function in conventional DCs and pDCs. Left, LROs and phagosome maturation in DCs. 1. Note: Refer to page 257 for full caption.

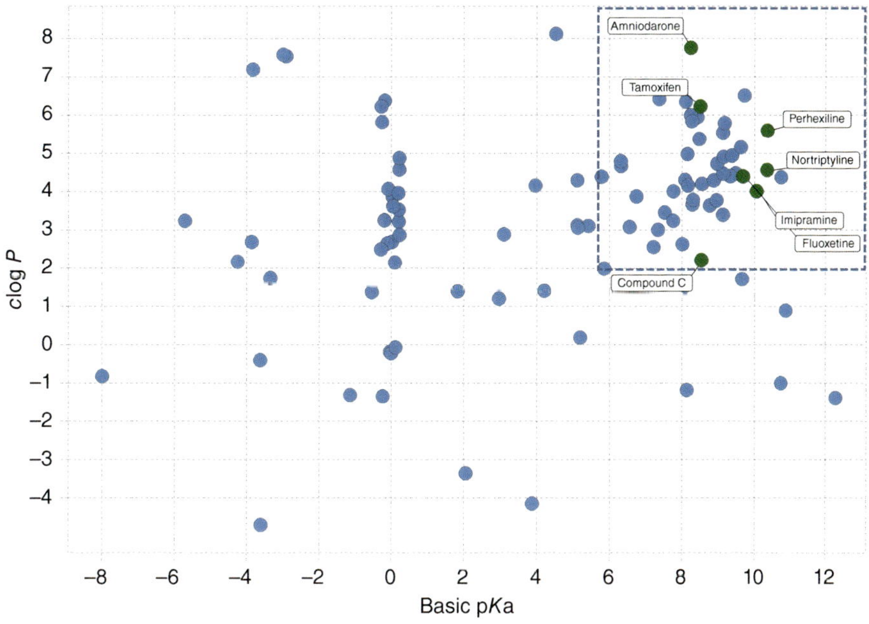

Figure 14.1 Correlation of physicochemical properties with autophagy screening: a scatter plot showing the distribution of autophagy enhancers within the clog P-basic pK_a physicochemical property space. Note: Refer to page 328 for full caption.

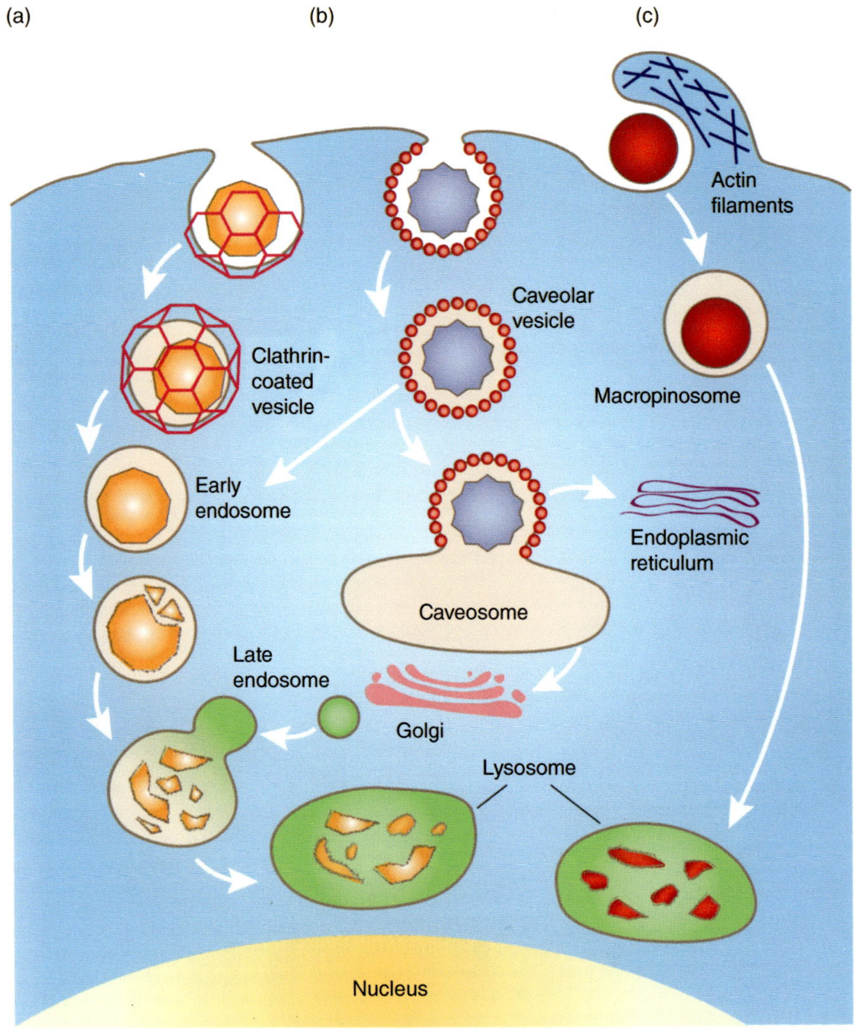

Figure 17.1 Internalization pathways mediating ADC uptake. Note: Refer to page 411 for full caption.

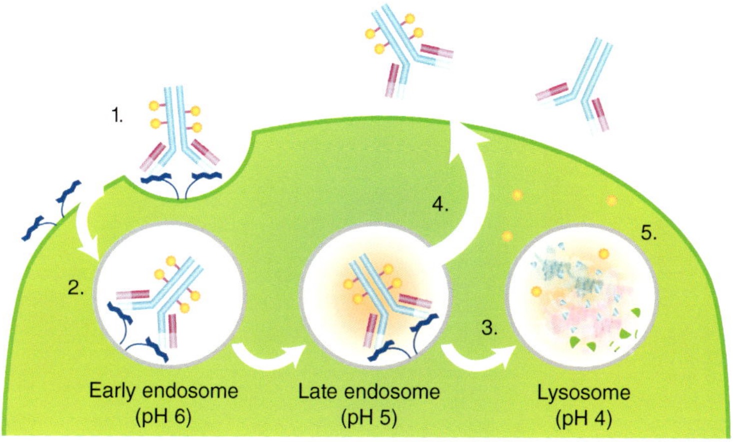

Figure 17.2 ADC receptor-mediated internalization. Note: Refer to page 412 for full caption.

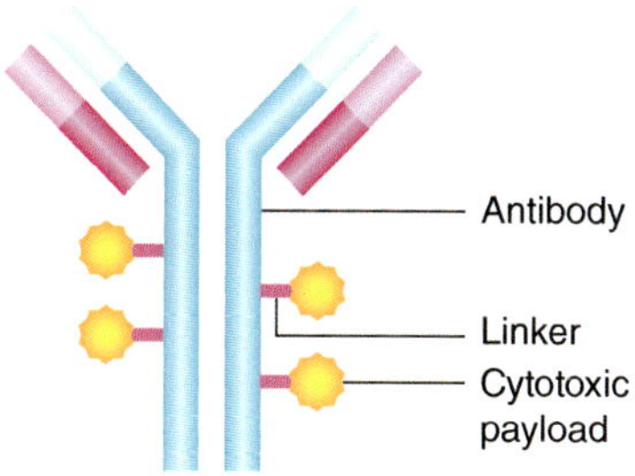

Figure 17.3 ADC structure. Note: Refer to page 413 for full caption.

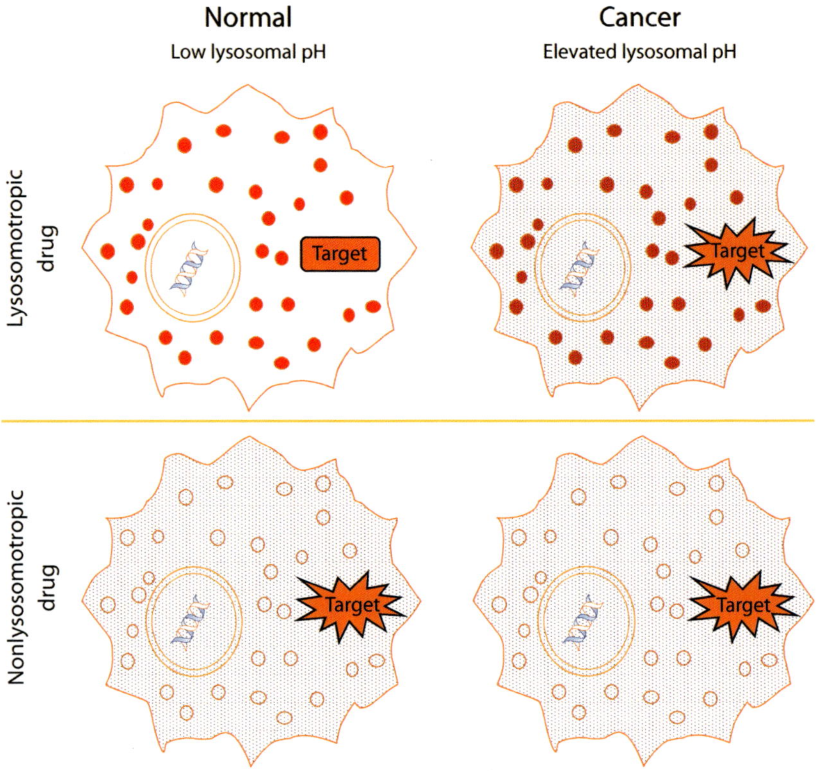

Figure 18.7 Overview of intracellular distribution-based anticancer drug targeting platform.

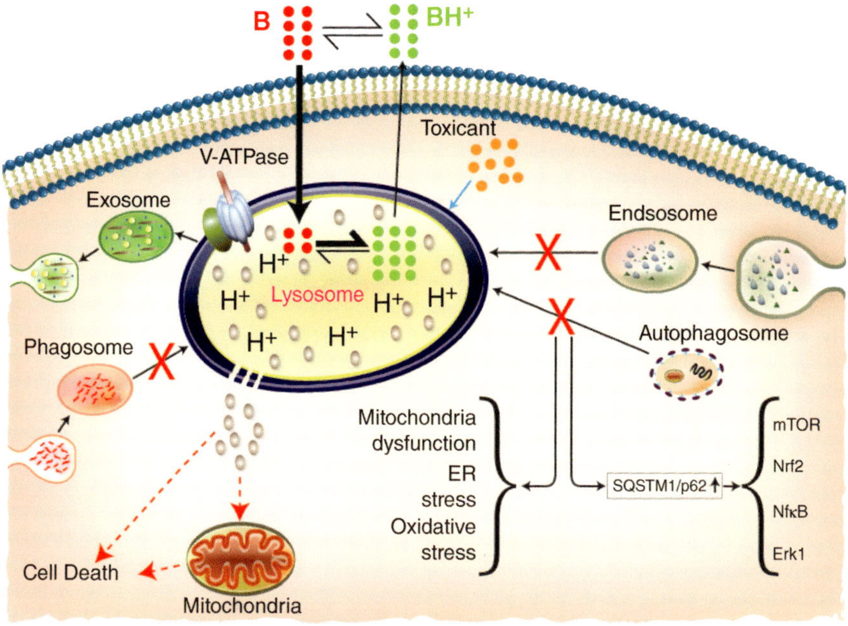

Figure 19.2 Biological impacts of the lysosomotropic compounds. Note: Refer to page 461 for full caption.

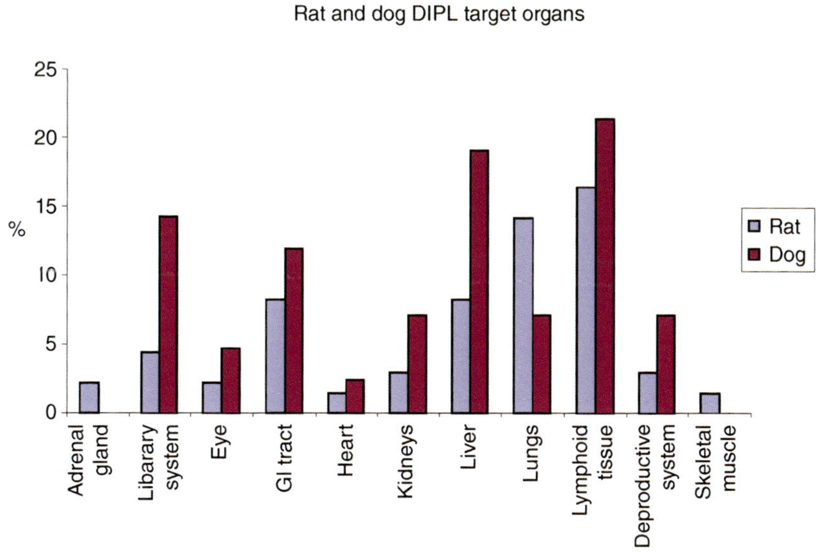

Figure 20.1 Comparison of dog (42 studies) and rat (134 studies) tissue distributions of drug-induced phospholipidosis from investigational new drug (IND) and new drug application (NDA) submissions to the USFDA from pharmaceutical companies [38]. Note: Refer to page 495 for full caption.

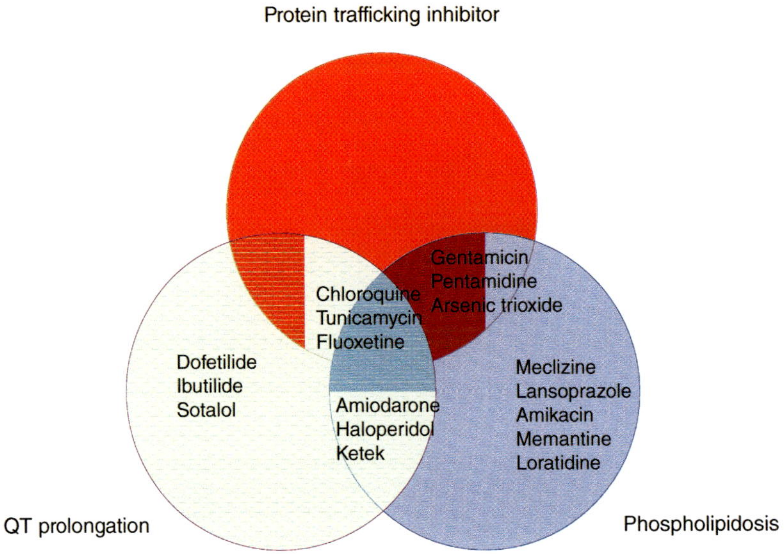

Figure 20.3 Relation between phospholipidosis, QT prolongation, protein trafficking effects, and adverse events. Note: Refer to page 504 for full caption.

Printed in the USA/Agawam, MA
September 16, 2026

910661.030